W9-DFM-873

THE HORSE, THE WHEEL, AND LANGUAGE

THE

HORSE

THE

WHEEL

AND

LANGUAGE

HOW
BRONZE-AGE RIDERS
FROM THE
EURASIAN STEPPES
SHAPED THE
MODERN WORLD

LIBRARY
FRANKLIN PIERCE UNIVERSITY
RINDGE, NH 03461

DAVID W. ANTHONY

Princeton University Press
Princeton and Oxford

Copyright © 2007 by Princeton University Press

Published by Princeton University Press, 41 William Street, Princeton, New Jersey 08540

In the United Kingdom: Princeton University Press, 3 Market Place, Woodstock, Oxfordshire OX20 1SY

All Rights Reserved

ISBN-13: 978-0-691-05887-0

Library of Congress Control Number: 2007932082

British Library Cataloging-in-Publication Data is available

This book has been composed in Adobe Caslon

Printed on acid-free paper ∞

press.princeton.edu

Printed in the United States of America

3   5   7   9   10   8   6   4

CONTENTS

Chapter Four
Language and Time 2:
Wool, Wheels, and Proto-Indo-European 59

Chapter Five
Language and Place:
The Location of the Proto-Indo-European Homeland 83

Chapter Six
The Archaeology of Language 102

PART TWO
The Opening of the Eurasian Steppes 121

Chapter Seven
How to Reconstruct a Dead Culture 123

ACKNOWLEDGMENTS

This book would not have been written without the love and support of my mother and father, David F. and Laura B. Anthony. Laura B. Anthony read and commented on every chapter. Bernard Wailes drew me into the University of Pennsylvania, led me into my first archaeological excavation, and taught me to respect the facts of archaeology. I am blessed with Dorcas Brown as my partner, editor, critic, fellow archaeologist, field excavation co-director, lab director, illustrator, spouse, and best friend through thick and thin. She edited every chapter multiple times. All the maps and figures are by D. Brown. Much of the content in chapters 10 and 16 was the product of our joint research, published over many years. Dorcas's brother, Dr. Ben Brown, also helped to read and edit the ms.

The bit-wear research described in chapter 10 and the field work associated with the Samara Valley Project (chapter 16) was supported by grants from Hartwick College, the Freedman and Fortis Foundations, the American Philosophical Society, the Wenner-Gren Foundation, the National Geographic Society, the Russian Institute of Archaeology (Moscow), the Institute for the History and Archaeology of the Volga (Samara), and the National Science Foundation (United States), with assistance for chapter 10 from the State University of New York at Cobleskill. We are particularly grateful to the National Science Foundation (NSF).

Support to write this book was provided by a fellowship from the National Endowment for the Humanities in 1999–2000 and a membership in the School of Historical Studies at the Institute for Advanced Study (IAS) at Princeton, New Jersey, in 2006, where Nicola DiCosmo and Patricia Crone made us welcome. The term at the IAS was crucial.

People who have helped me in numerous different ways include:

Near East and East Asia: Kathy Linduff, Victor Mair, Oscar Muscarella, Karen Rubinson, Chris Thornton, Lauren Zych, C. C. Lamberg-Karlovsky, Fred Hiebert, Phil Kohl, Greg Possehl, Glenn Schwartz, David Owen, Mitchell Rothman, Emmy Bunker, Nicola DiCosmo, and Peter Golden.

Horses and wheeled vehicles: Dexter Perkins and Pat Daly; Şandor Bökönyi, Sandra Olsen, Mary Littauer and Joost Crouwel (my instructors in ancient transport); and Peter Raulwing, Norbert Benecke, and Mindy Zeder.

Bit wear and the riding experiment: Mindy Zeder, Ron Keiper; the Bureau of Land Management, Winnemucca, Nevada; Cornell University Veterinary School; University of Pennsylvania New Bolton Center; the Assateague Island Wildlife Refuge; and, at the State University of New York at Cobleskill, Steve MacKenzie, Stephanie Skargensky, and Michelle Beyea.

Linguistics: Ward Goodenough, Edgar Polomé, Richard Diebold, Winfrid Lehmann, Alexander Lubotsky, Don Ringe, Stefan Zimmer, and Eric Hamp. A special thanks to Johanna Nichols, who helped edit chapter 5, and J. Bill Darden and Jim Mallory, who reviewed the first draft.

Eastern European archaeology: Petar Glumac (who made me believe I could read Russian-language sources), Peter Bogucki, Douglass Bailey (who reviewed chapter 11), Ruth Tringham (who gave me my first field experience in Eastern Europe), Victor Shnirelman (our first guide in Russia), Dimitri Telegin (my first source on steppe archaeology), Natalya Belan, Oleg Zhuravlev, Yuri Rassamakin, Mikhail Videiko, Igor Vasiliev, Pavel Kuznetsov, Oleg Mochalov, Aleksandr Khokhlov, Pavel Kosintsev, Elena Kuzmina, Sergei Korenevskii, Evgeni Chernykh, R. Munchaev, Nikolai Vinogradov, Victor Zaibert, Stanislav Grigoriev, Andrei Epimakhov, Valentin Dergachev, and Ludmila Koryakova. Of these I owe the deepest debts to Telegin (my first guide) and my colleagues in Samara: Vasiliev, Kuznetsov, Mochalov, Khokhlov, and (honorary Samaran) Kosintsev.

The errors I have made are mine alone; these people tried their best.

least recognized his legal authority, the Indians obeyed an already functioning and ancient system of Hindu law, which was regularly cited in court by Hindu legal scholars, or pandits (the source of our term *pundit*). English judges could not determine if the laws the pandits cited really existed. Sanskrit was the ancient language of the Hindu legal texts, like Latin was for English law. If the two legal systems were to be integrated, one of the new Supreme Court justices had to learn Sanskrit. That was Jones.

He went to the ancient Hindu university at Nadiya, bought a vacation cottage, found a respected and willing pandit (Rāmalocana) on the faculty, and immersed himself in Hindu texts. Among these were the *Vedas*, the ancient religious compositions that lay at the root of Hindu religion. The *Rig Veda*, the oldest of the Vedic texts, had been composed long before the Buddha's lifetime and was more than two thousand years old, but no one knew its age exactly. As Jones pored over Sanskrit texts his mind made comparisons not just with Persian and English but also with Latin and Greek, the mainstays of an eighteenth-century university education; with Gothic, the oldest literary form of German, which he had also learned; and with Welsh, a Celtic tongue and his boyhood language which he had not forgotten. In 1786, three years after his arrival in Calcutta, Jones came to a startling conclusion, announced in his third annual discourse to the Asiatic Society of Bengal, which he had founded when he first arrived. The key sentence is now quoted in every introductory textbook of historical linguistics (punctuation mine):

> The Sanskrit language, whatever be its antiquity, is of a wonderful structure: more perfect than the Greek, more copious than the Latin, and more exquisitely refined than either; yet bearing to both of them a stronger affinity, both in the roots of verbs and in the forms of grammar, than could possibly have been produced by accident; so strong indeed, that no philologer could examine them all three, without believing them to have sprung from some common source, which, perhaps, no longer exists.

Jones had concluded that the Sanskrit language originated from the same source as Greek and Latin, the classical languages of European civilization. He added that Persian, Celtic, and German probably belonged to the same family. European scholars were astounded. The occupants of India, long regarded as the epitome of Asian exotics, turned out to be long-lost cousins. If Greek, Latin, and Sanskrit were relatives, descended from the same ancient parent language, what was that language? Where had it been it spoken? And by whom? By what historical circumstances did it generate daughter tongues that became the dominant languages spoken from Scotland to India?

These questions resonated particularly deeply in Germany, where popular interest in the history of the German language and the roots of German traditions were growing into the Romantic movement. The Romantics wanted to discard the cold, artificial logic of the Enlightenment to return to the roots of a simple and authentic life based in direct experience and community. Thomas Mann once said of a Romantic philosopher (Schlegel) that his thought was contaminated too much by reason, and that he was therefore a poor Romantic. It was ironic that William Jones helped to inspire this movement, because his own philosophy was quite different: "The race of man . . . cannot long be happy without virtue, nor actively virtuous without freedom, nor securely free without rational knowledge."[3] But Jones had energized the study of ancient languages, and ancient language played a central role in Romantic theories of authentic experience. In the 1780s J. G. Herder proposed a theory later developed by von Humboldt and elaborated in the twentieth century by Wittgenstein, that language creates the categories and distinctions through which humans give meaning to the world. Each particular language, therefore, generates and is enmeshed in a closed social community, or "folk," that is at its core meaningless to an outsider. Language was seen by Herder and von Humboldt as a vessel that molded community and national identities. The brothers Grimm went out to collect "authentic" German folk tales while at the same time studying the German language, pursuing the Romantic conviction that language and folk culture were deeply related. In this setting the mysterious mother tongue, Proto-Indo-European, was regarded not just as a language but as a crucible in which Western civilization had its earliest beginnings.

After the 1859 publication of Charles Darwin's *The Origin of Species*, the Romantic conviction that language was a defining factor in national identity was combined with new ideas about evolution and biology. Natural selection provided a scientific theory that was hijacked by nationalists and used to rationalize why some races or "folks" ruled others—some were more "fit" than others. Darwin himself never applied his theories of fitness and natural selection to such vague entities as races or languages, but this did not prevent unscientific opportunists from suggesting that the less "fit" races could be seen as a source of genetic weakness, a reservoir of barbarism that might contaminate and dilute the superior qualities of the races that were more "fit." This toxic mixture of pseudo-science and

Romanticism soon produced its own new ideologies. Language, culture, and a Darwinian interpretation of race were bundled together to explain the superior biological–spiritual–linguistic essence of the northern Europeans who conducted these self-congratulatory studies. Their writings and lectures encouraged people to think of themselves as members of long-established, biological–linguistic nations, and thus were promoted widely in the new national school systems and national newspapers of the emerging nation-states of Europe. The policies that forced the Welsh (including Sir William Jones) to speak English, and the Bretons to speak French, were rooted in politicians' need for an ancient and "pure" national heritage for each new state. The ancient speakers of Proto-Indo-European soon were molded into the distant progenitors of such racial–linguistic–national stereo-types.[4]

Proto-Indo-European, the linguistic problem, became "the Proto-Indo-Europeans," a biological population with its own mentality and personality: "a slim, tall, light-complexioned, blonde race, superior to all other peoples, calm and firm in character, constantly striving, intellectually brilliant, with an almost ideal attitude towards the world and life in general".[5] The name *Aryan* began to be applied to them, because the authors of the oldest religious texts in Sanskrit and Persian, the *Rig Veda* and *Avesta*, called themselves Aryans. These Aryans lived in Iran and eastward into Afghanistan–Pakistan–India. The term *Aryan* should be confined only to this Indo-Iranian branch of the Indo-European family. But the *Vedas* were a newly discovered source of mystical fascination in the nineteenth century, and in Victorian parlors the name Aryan soon spread beyond its proper linguistic and geographic confines. Madison Grant's *The Passing of the Great Race* (1916), a best-seller in the U.S., was a virulent warning against the thinning of superior American "Aryan" blood (by which he meant the British–Scots–Irish–German settlers of the original thirteen colonies) through interbreeding with immigrant "inferior races," which for him included Poles, Czechs, and Italians as well as Jews—all of whom spoke Indo-European languages (Yiddish is a Germanic language in its basic grammar and morphology).[6]

The gap through which the word *Aryan* escaped from Iran and the Indian subcontinent was provided by the *Rig Veda* itself: some scholars found passages in the *Rig Veda* that seemed to describe the Vedic Aryans as invaders who had conquered their way into the Punjab.[7] But from where? A feverish search for the "Aryan homeland" began. Sir William Jones placed it in Iran. The Himalayan Mountains were a popular choice in the early nineteenth century, but other locations soon became the subject of animated debates. Amateurs and experts alike joined the search, many hoping to prove that their own nation had given birth to the Aryans. In the second decade of the twentieth century the German scholar Gustav Kossinna attempted to demonstrate on archaeological grounds that the Aryan homeland lay in northern Europe—in fact, in Germany. Kossinna illustrated the prehistoric migrations of the "Indo-Germanic" Aryans with neat black arrows that swept east, west, and south from his presumed Aryan homeland. Armies followed the pen of the prehistorian less than thirty years later.[8]

The problem of Indo-European origins was politicized almost from the beginning. It became enmeshed in nationalist and chauvinist causes, nurtured the murderous fantasy of Aryan racial superiority, and was actually pursued in archaeological excavations funded by the Nazi SS. Today the Indo-European past continues to be manipulated by causes and cults. In the books of the Goddess movement (Marija Gimbutas's *Civilization of the Goddess*, Riane Eisler's *The Chalice and the Blade*) the ancient "Indo-Europeans" are cast in archaeological dramas not as blonde heroes but as patriarchal, warlike invaders who destroyed a utopian prehistoric world of feminine peace and beauty. In Russia some modern nationalist political groups and neo-Pagan movements claim a direct linkage between themselves, as Slavs, and the ancient "Aryans." In the United States white supremacist groups refer to themselves as Aryans. There actually were Aryans in history—the composers of the *Rig Veda* and the *Avesta*—but they were Bronze Age tribal people who lived in Iran, Afghanistan, and the northern Indian subcontinent. It is highly doubtful that they were blonde or blue-eyed, and they had no connection with the competing racial fantasies of modern bigots.[9]

The mistakes that led an obscure linguistic mystery to erupt into racial genocide were distressingly simple and therefore can be avoided by anyone who cares to avoid them. They were the equation of race with language, and the assignment of superiority to some language-and-race groups. Prominent linguists have always pleaded against both these ideas. While Martin Heidegger argued that some languages—German and Greek—were unique vessels for a superior kind of thought, the linguistic anthropologist Franz Boas protested that no language could be said to be superior to any other on the basis of objective criteria. As early as 1872 the great linguist Max Müller observed that the notion of an Aryan skull was not just unscientific but anti-scientific; languages are not white-skinned or long-headed. But then how can the Sanskrit language be connected with a skull type? And how did the Aryans themselves define

"Aryan"? According to their own texts, they conceived of "Aryan-ness" as a *religious–linguistic* category. Some Sanskrit-speaking chiefs, and even poets in the *Rig Veda*, had names such as Balbūtha and Bṛbu that were foreign to the Sanskrit language. These people were of non-Aryan origin and yet were leaders among the Aryans. So even the Aryans of the *Rig Veda* were not genetically "pure"—whatever that means. The *Rig Veda* was a ritual canon, not a racial manifesto. If you sacrificed in the right way to the right gods, which required performing the great traditional prayers in the traditional language, you were an Aryan; otherwise you were not. The *Rig Veda* made the *ritual* and *linguistic* barrier clear, but it did not require or even contemplate racial purity.[10]

Any attempt to solve the Indo-European problem has to begin with the realization that the term *Proto-Indo-European* refers to a language community, and then work outward. Race really cannot be linked in any predictable way with language, so we cannot work from language to race or from race to language. Race is poorly defined; the boundaries between races are defined differently by different groups of people, and, since these definitions are cultural, scientists cannot describe a "true" boundary between any two races. Also, archaeologists have their own, quite different definitions of race, based on traits of the skull and teeth that often are invisible in a living person. However race is defined, languages are not normally sorted by race—all racial groups speak a variety of different languages. So skull shapes are almost irrelevant to linguistic problems. Languages and genes are correlated only in exceptional circumstances, usually at clear geographic barriers such as significant mountain ranges or seas—and often not even there.[11] A migrating population did not have to be genetically homogeneous even if it did recruit almost exclusively from a single dialect group. Anyone who *assumes* a simple connection between language and genes, without citing geographic isolation or other special circumstances, is wrong at the outset.

THE LURE OF THE MOTHER TONGUE

The only aspect of the Indo-European problem that has been answered to most peoples' satisfaction is how to define the language family, how to determine which languages belong to the Indo-European family and which do not. The discipline of linguistics was created in the nineteenth century by people trying to solve this problem. Their principal interests were comparative grammar, sound systems, and syntax, which provided the basis for classifying languages, grouping them into types, and otherwise

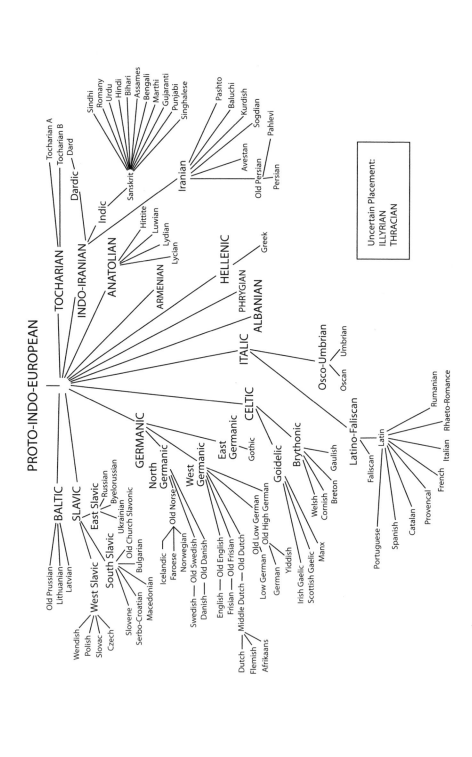

PROTO-INDO-EUROPEAN

Uncertain Placement:
ILLYRIAN
THRACIAN

defining the relationships between the tongues of humanity. No one had done this before. They divided the Indo-European language family into twelve major branches, distinguished by innovations in phonology or pronunciation and in morphology or word form that appeared at the root of each branch and were maintained in all the languages of that branch (figure 1.1). The twelve branches of Indo-European included most of the languages of Europe (but not Basque, Finnish, Estonian, or Magyar); the Persian language of Iran; Sanskrit and its many modern daughters (most important, Hindi and Urdu); and a number of extinct languages including Hittite in Anatolia (modern Turkey) and Tocharian in the deserts of Xinjiang (northwestern China) (figure 1.2). Modern English, like Yiddish and Swedish, is assigned to the Germanic branch. The analytic methods invented by nineteenth-century philologists are today used to describe, classify, and explain language variation worldwide.

Historical linguistics gave us not just static classifications but also the ability to reconstruct at least parts of extinct languages for which no written evidence survives. The methods that made this possible rely on regularities in the way sounds change inside the human mouth. If you collect Indo-European words for *hundred* from different branches of the language family and compare them, you can apply the myriad rules of sound change to see if all of them can be derived by regular changes from a single hypothetical ancestral word at the root of all the branches. The proof that Latin *kentum* (hundred) in the Italic branch and Lithuanian *shimtas* (hundred) in the Baltic branch are genetically related cognates is the construction of the ancestral root *k'mtom-. The daughter forms are compared sound by sound, going through each sound in each word in each branch, to see if they can converge on one unique sequence of sounds that could have evolved into all of them by known rules. (I explain how this is done in the next chapter.) That root sequence of sounds, if it can be found, is the proof that the terms being compared are genetically related cognates. A reconstructed root is the residue of a successful comparison.

Figure 1.1 The twelve branches of the Indo-European language family. Baltic and Slavic are sometimes combined into one branch, like Indo-Iranian, and Phrygian is sometimes set aside because we know so little about it, like Illyrian and Thracian. With those two changes the number of branches would be ten, an acceptable alternative. A tree diagram is meant to be a sketch of broad relationships; it does not represent a complete history.

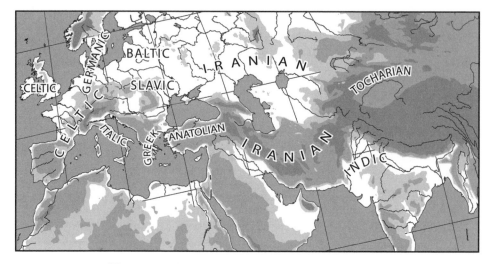

Figure 1.2 The approximate geographic locations of the major Indo-European branches at about 400 BCE.

Linguists have reconstructed the sounds of more than fifteen hundred Proto-Indo-European roots.[12] The reconstructions vary in reliability, because they depend on the surviving linguistic evidence. On the other hand, archeological excavations have revealed inscriptions in Hittite, Mycenaean Greek, and archaic German that contained words, never seen before, displaying precisely the sounds previously reconstructed by comparative linguists. That linguists accurately predicted the sounds and letters later found in ancient inscriptions confirms that their reconstructions are not entirely theoretical. If we cannot regard reconstructed Proto-Indo-European as literally "real," it is at least a close approximation of a prehistoric reality.

The recovery of even fragments of the Proto-Indo-European language is a remarkable accomplishment, considering that it was spoken by nonliterate people many thousands of years ago and never was written down. Although the grammar and morphology of Proto-Indo-European are most important in typological studies, it is the reconstructed vocabulary, or lexicon, that holds out the most promise for archaeologists. The reconstructed lexicon is a window onto the environment, social life, and beliefs of the speakers of Proto-Indo-European.

For example, reasonably solid lexical reconstructions indicate that Proto-Indo-European contained words for otter, beaver, wolf, lynx, elk, red deer, horse, mouse, hare, and hedgehog, among wild animals; goose, crane, duck, and eagle, among birds; bee and honey; and cattle (also cow, ox, and steer), sheep (also wool and weaving), pig (also boar, sow, and piglet), and dog among the domestic animals. The horse was certainly known to the speakers of Proto-Indo-European, but the lexical evidence alone is insufficient to determine if it was domesticated. All this lexical evidence might also be attested in, and compared against, archaeological remains to reconstruct the environment, economy, and ecology of the Proto-Indo-European world.

But the proto-lexicon contains much more, including clusters of words, suggesting that the speakers of PIE inherited their rights and duties through the father's bloodline only (patrilineal descent); probably lived with the husband's family after marriage (patrilocal residence); recognized the authority of chiefs who acted as patrons and givers of hospitality for their clients; likely had formally instituted warrior bands; practiced ritual sacrifices of cattle and horses; drove wagons; recognized a male sky deity; probably avoided speaking the name of the bear for ritual reasons; and recognized two senses of the sacred ("that which is imbued with holiness" and "that which is forbidden"). Many of these practices and beliefs are simply unrecoverable through archaeology. The proto-lexicon offers the hope of recovering some of the details of daily ritual and custom that archaeological evidence alone usually fails to deliver. That is what makes the solution of the Indo-European problem important for archaeologists, and for all of us who are interested in knowing our ancestors a little better.

A New Solution for an Old Problem

Linguists have been working on cultural-lexical reconstructions of Proto-Indo-European for almost two hundred years. Archaeologists have argued about the archaeological identity of the Proto-Indo-European language community for at least a century, probably with less progress than the linguists. The problem of Indo-European origins has been intertwined with European intellectual and political history for considerably more than a century. Why hasn't a broadly acceptable union between archaeological and linguistic evidence been achieved?

Six major problems stand in the way. One is that the recent intellectual climate in Western academia has led many serious people to question the entire idea of proto-languages. The modern world has witnessed increasing cultural fusion in music (Black Ladysmith Mombasa and Paul Simon, Pavarotti and Sting), in art (Post-Modern eclecticism), in information services (News-Gossip), in the mixing of populations (international migration is at an all-time high), and in language (most of the people in the world are now bilingual or trilingual). As interest in the phenomenon of cultural convergence increased during the 1980s, thoughtful academics began to reconsider languages and cultures that had once been interpreted as individual, distinct entities. Even standard languages began to be seen as creoles, mixed tongues with multiple origins. In Indo-European studies this movement sowed doubt about the very concept of language families and the branching tree models that illustrated them, and some declared the search for any proto-language a delusion. Many ascribed the similarities between the Indo-European languages to convergence between neighboring languages that had distinct historical origins, implying that there never was a single proto-language.[13]

Much of this was creative but vague speculation. Linguists have now established that the similarities between the Indo-European languages are not the kinds of similarities produced by creolization and convergence. None of the Indo-European languages looks at all like a creole. The Indo-European languages must have replaced non–Indo-European languages rather than creolizing with them. Of course, there was interlanguage borrowing, but it did not reach the extreme level of mixing and structural simplification seen in all creoles. The similarities that Sir William Jones noted among the Indo-European languages can *only* have been produced by descent from a common proto-language. On that point most linguists agree.

So we should be able to use the reconstructed Proto-Indo-European vocabulary as a source of clues about where it was spoken and when. But then the second problem arises: many archaeologists, apparently, do not believe that it is possible to reliably reconstruct any portion of the Proto-Indo-European lexicon. They do not accept the reconstructed vocabulary as real. This removes the principal reason for pursuing Indo-European origins and one of the most valuable tools in the search. In the next chapter I offer a defense of comparative linguistics, a brief explanation of how it works, and a guide to interpreting the reconstructed vocabulary.

The third problem is that archaeologists cannot agree about the antiquity of Proto-Indo-European. Some say it was spoken in 8000 BCE, others say as late as 2000 BCE, and still others regard it as an abstract idea that exists only in linguists' heads and therefore cannot be assigned to any one time. This makes it impossible, of course, to focus on a specific era. But the principal reason for this state of chronic disagreement is that most archaeologists do not pay much attention to linguistics. Some have proposed solutions that are contradicted by large bodies of linguistic evidence. By solving the second problem, regarding the question of reliability and reality, we will advance significantly toward solving problem number 3—the question of when—which occupies chapters 3 and 4.

The fourth problem is that archaeological methods are underdeveloped in precisely those areas that are most critical for Indo-European origin studies. Most archaeologists believe it is impossible to equate prehistoric language groups with archaeological artifacts, as language is not reflected in any consistent way in material culture. People who speak different languages might use similar houses or pots, and people who speak the same language can make pots or houses in different ways. But it seems to me that language and culture *are* predictably correlated under some circumstances. Where we see a *very clear* material-culture frontier—not just different pots but also different houses, graves, cemeteries, town patterns, icons, diets, and dress designs—that *persists* for centuries or millennia, it tends also to be a linguistic frontier. This does not happen everywhere. In fact, such *ethno-linguistic* frontiers seem to occur rarely. But where a robust material-culture frontier does persist for hundreds, even thousands of years, language tends to be correlated with it. This insight permits us to identify at least *some* linguistic frontiers on a map of purely archaeological cultures, which is a critical step in finding the Proto-Indo-European homeland.

Another weak aspect of contemporary archaeological theory is that archaeologists generally do not understand migration very well, and migration is an important vector of language change—certainly not the only cause but an important one. Migration was used by archaeologists before World War II as a simple explanation for any kind of change observed in prehistoric cultures: if pot type A in level one was replaced by pot type B in level two, then it was a migration of B-people that had caused the change. That simple assumption was proven to be grossly inadequate by a later generation of archaeologists who recognized the myriad *internal* catalysts of change. Shifts in artifact types were shown to be caused by changes in the size and complexity of social gatherings, shifts in economics, reorganization in the way crafts were managed, changes in the social function of crafts, innovations in technology, the introduction of new trade and exchange commodities, and so on. "Pots are not people" is a rule taught to every Western archaeology student since the 1960s. Migration disappeared entirely from the explanatory toolkit of Western archaeologists in the 1970s and 1980s. But migration is a hugely important human behavior, and you cannot understand the Indo-European problem if you ignore migration or pretend it was unimportant in the past. I have tried to use modern migration theory to understand prehistoric migrations and their probable role in language change, problems discussed in chapter 6.

Problem 5 relates to the specific homeland I defend in this book, located in the steppe grasslands of Russia and Ukraine. The recent prehistoric archaeology of the steppes has been published in obscure journals and books, in languages understood by relatively few Western archaeologists, and in a narrative form that often reminds Western archaeologists of the old "pots are people" archaeology of fifty years ago. I have tried to understand this literature for twenty-five years with limited success, but I can say that Soviet and post-Soviet archaeology is not a simple repetition of any phase of Western archaeology; it has its own unique history and guiding assumptions. In the second half of this book I present a selective and unavoidably imperfect synthesis of archaeology from the Neolithic, Copper, and Bronze Ages in the steppe zone of Russia, Ukraine, and Kazakhstan, bearing directly on the nature and identity of early speakers of Indo-European languages.

Horses gallop onstage to introduce the final, sixth problem. Scholars noticed more than a hundred years ago that the oldest well-documented Indo-European languages—Imperial Hittite, Mycenaean Greek, and the most ancient form of Sanskrit, or Old Indic—were spoken by militaristic societies that seemed to erupt into the ancient world driving chariots pulled by swift horses. Maybe Indo-European speakers invented the chariot. Maybe they were the first to domesticate horses. Could this explain the initial spread of the Indo-European languages? For about a thousand years, between 1700 and 700 BCE, chariots were the favored weapons of pharaohs and kings throughout the ancient world, from Greece to China. Large numbers of chariots, in the dozens or even hundreds, are mentioned in palace inventories of military equipment, in descriptions of battles, and in proud boasts of loot taken in warfare. After 800 BCE chariots were gradually abandoned as they became vulnerable to a new kind of warfare conducted by disciplined troops of mounted archers, the earliest cavalry. If Indo-European speakers were the first to have chariots, this could explain their early expansion; if they were the first to domesticate horses, then this could explain the central role horses played as symbols of strength and power in the rituals of the Old Indic Aryans, Greeks, Hittites, and other Indo-European speakers.

But until recently it has been difficult or impossible to determine when and where horses were domesticated. Early horse domestication left very few marks on the equine skeleton, and all we have left of ancient horses is their bones. For more than ten years I have worked on this problem with my research partner, and also my wife, Dorcas Brown, and we believe we now know where and when people began to keep herds of tamed horses. We also think that horseback riding began in the steppes long before chariots were invented, in spite of the fact that chariotry preceded cavalry in the warfare of the organized states and kingdoms of the ancient world.

LANGUAGE EXTINCTION AND THOUGHT

The people who spoke the Proto-Indo-European language lived at a critical time in a strategic place. They were positioned to benefit from innovations in transport, most important of these the beginning of horseback riding and the invention of wheeled vehicles. They were in no way superior to their neighbors; indeed, the surviving evidence suggests that their economy, domestic technology, and social organization were simpler than those of their western and southern neighbors. The expansion of their language was not a single event, nor did it have only one cause.

Nevertheless, that language did expand and diversify, and its daughters—including English—continue to expand today. Many other language families have become extinct as Indo-European languages spread. It is possible that the resultant loss of linguistic diversity has narrowed and channeled habits of perception in the modern world. For example, all Indo-European languages force the speaker to pay attention to tense and number when talking about an action: you *must* specify whether the action is past, present, or future; and you *must* specify whether the actor is singular or plural. It is impossible to use an Indo-European verb without deciding on these categories. Consequently speakers of Indo-European languages habitually frame all events in terms of when they occurred and whether they involved multiple actors. Many other language families do not *require* the speaker to address these categories when speaking of an action, so tense and number can remain unspecified.

On the other hand, other language families require that other aspects of reality be constantly used and recognized. For example, when describing an event or condition in Hopi you *must* use grammatical markers that specify whether you witnessed the event yourself, heard about it from someone else, or consider it to be an unchanging truth. Hopi speakers are forced by Hopi grammar to habitually frame all descriptions of reality in terms of the source and reliability of their information. The constant and automatic use of such categories generates habits in the perception and framing of the world that probably differ between people who use fundamentally different grammars.[14] In that sense, the spread of Indo-European grammars has perhaps reduced the diversity of human perceptual habits. It might also have caused this author, as I write this book, to frame my observations in a way that repeats the perceptual habits and categories of a small group of people who lived in the western Eurasian steppes more than five thousand years ago.

CHAPTER TWO

How to Reconstruct a Dead Language

Proto-Indo-European has been dead as a spoken language for at least forty-five hundred years. The people who spoke it were nonliterate, so there are no inscriptions. Yet, in 1868, August Schleicher was able to tell a story in reconstructed Proto-Indo-European, called "The Sheep and the Horses," or *Avis akvasas ka*. A rewrite in 1939 by Herman Hirt incorporated new interpretations of Proto-Indo-European phonology, and the title became *Owis ek'woses-k^{w}e*. In 1979 Winfred Lehmann and Ladislav Zgusta suggested only minor new changes in their version, *Owis ekwosk^{w}e*. While linguists debate increasingly minute details of pronunciation in exercises like these, most people are amazed that anything can be said about a language that died without written records. Amazement, of course, is a close cousin of suspicion. Might the linguists be arguing over a fantasy? In the absence of corroborative evidence from documents, how can linguists be sure about the accuracy of reconstructed Proto-Indo-European?[1]

Many archaeologists, accustomed to digging up real things, have a low opinion of those who merely reconstruct hypothetical phonemes—what is called "linguistic prehistory." There are reasons for this skepticism. Both linguists and archaeologists have made communication across the disciplines almost impossible by speaking in dense jargons that are virtually impenetrable to anyone but themselves. Neither discipline is at all simple, and both are riddled with factions on many key questions of interpretation. Healthy disagreement can resemble confusion to an outsider, and most archaeologists, including this author, are outsiders in linguistics. Historical linguistics is not taught regularly in graduate archaeology programs, so most archaeologists know very little about the subject. Sometimes we make this quite clear to linguists. Nor is archaeology taught to graduate students in linguistics. Linguists' occasional remarks about archaeology can sound simplistic and naïve to archaeologists, making some of us suspect that the entire field of historical linguistics may be riddled with simplistic and naïve assumptions.

The purpose of these first few chapters is to clear a path across the no-man's land that separates archaeology and historical linguistics. I do this with considerable uncertainty—I have no more formal training in linguistics than most archaeologists. I am fortunate that a partial way has already been charted by Jim Mallory, perhaps the only doubly qualified linguist-archaeologist in Indo-European studies. The questions surrounding Indo-European origins are, at their core, about linguistic evidence. The most basic linguistic problem is to understand how language changes with time.[2]

LANGUAGE CHANGE AND TIME

Imagine that you had a time machine. If you are like me, there would be many times and places that you would like to visit. In most of them, however, no one spoke English. If you could not afford the Six-Month-Immersion Trip to, say, ancient Egypt, you would have to limit yourself to a time and place where you could speak the language. Consider, perhaps, a trip to England. How far back in time could you go and still be understood? Say we go to London in the year 1400 CE.

As you emerge from the time machine, a good first line to speak, something reassuring and recognizable, might be the opening line of the Lord's Prayer. The first line in a conservative, old-fashioned version of Modern Standard English would be, "*Our Father, who is in heaven, blessed be your name.*" In the English of 1400, as spoken by Chaucer, you would say, "*Oure fadir that art in heuenes, halwid be thy name.*" Now turn the dial back another four hundred years to 1000 CE, and in Old English, or Anglo-Saxon, you would say, "*Fæader ure thu the eart on heofonum, si thin nama gehalgod.*" A chat with Alfred the Great would be out of the question.

Most normal spoken languages over the course of a thousand years undergo enough change that speakers at either end of the millennium, attempting a conversation, would have difficulty understanding each other. Languages like Church Latin or Old Indic (the oldest form of Sanskrit), frozen in ritual, would be your only hope for effective communication with people who lived more than a thousand years ago. Icelandic is a frequently cited example of a spoken language that has changed little in a thousand years, but it is spoken on an island isolated in the North Atlantic by people whose attitude to their old sagas and poetry has been one approaching religious reverence. Most languages undergo significantly more changes than Icelandic over far fewer than a thousand years for two reasons: first, no two people speak the same language exactly alike; and, second, most people meet a lot more people who speak differently than do the Icelanders. A language that borrows many words and phrases from another language changes more rapidly than one with a low borrowing rate. Icelandic has one of the lowest borrowing rates in the world.[3] If we are exposed to a number of different ways of speaking, our own way of speaking is likely to change more rapidly. Fortunately, however, although the speed of language change is quite variable, the structure and sequence of language change is not.

Language change is not random; it flows in the direction of accents and phrases admired and emulated by large numbers of people. Once a target accent is selected, the structure of the sound changes that moves the speaker away from his own speech to the target is governed by rules. The same rules apparently exist in all our minds, mouths, and ears. Linguists just noticed them first. If rules define how a given innovation in pronunciation affects the old speech system—if sound shifts are predictable—then we should be able to play them backward, in effect, to hear earlier language states. That is more or less how Proto-Indo-European was reconstructed.

Most surprising about sound change is its regularity, its conformation to rules no one knows consciously. In early Medieval French there probably was a time when *tsent'm* 'hundred' was heard as just a dialectical pronunciation of the Latin word *kentum* 'hundred'. The differences in sound between the two were *allophones*, or different sounds that did not create different meanings. But because of other changes in how Latin was spoken, [ts-] began to be heard as a different sound, a phoneme distinct from [k-] that could change the meaning of a word. At that point people had to decide whether *kentum* was pronounced with a [k-] or a [ts-]. When French speakers decided to use [ts-], they did so not just for the word *kentum* but in every word where Latin had the sound *k-* before a front vowel like *-e-*. And once this happened, *ts-* became confused with initial *s-*, and people had to decide again whether *tsentum* was pronounced with a [ts-] or [s-]. They chose [s-]. This sequence of shifts dropped below the level of consciousness and spread like a virus through all pre-French words with analogous sequences of sounds. Latin *cera* 'wax', pronounced [kera], became French *cire*, pronounced [seer]; and Latin *civitas* 'community', pronounced [kivitas], became French *cité*, pronounced [seetay]. Other sound changes happened, too, but they all followed the same unspoken and unconscious rules—the sound shifts were not idiosyncratic or confined to certain words; rather, they spread systematically to all similar sounds in the language. Peoples' ears were very discriminating in identifying words that fit or did not fit the analogy. In words where the Latin *k-* was followed by a *back* vowel like *-o* it remained a *k-*, as in Latin *costa*>French *côte*.

Sound changes are rule-governed probably because all humans instinctively search for order in language. This must be a hard-wired part of all human brains. We do it without committee meetings, dictionaries, or even literacy, and we are not conscious of what we are doing (unless we are linguists). Human language is defined by its rules. Rules govern sentence construction (syntax), and the relationship between the sounds of words (phonology and morphology) and their meaning. Learning these rules changes our awareness from that of an infant to a functioning member of the human tribe. Because language is central to human evolution, culture, and social identity, each member of the tribe is biologically equipped to cooperate in converting novel changes into regular parts of the language system.[4]

Historical linguistics was created as a discipline in the nineteenth century, when scholars first exposed and analyzed the rules we follow when speaking and listening. I do not pretend to know these rules adequately, and if I did I would not try to explain them all. What I hope to do is indicate, in a general way, how some of them work so that we can use the "reconstructed vocabulary" of Proto-Indo-European with some awareness of its possibilities and limitations.

We begin with phonology. Any language can be separated into several interlocking systems, each with its own set of rules. The vocabulary, or *lexicon*, composes one system; *syntax*, or word order, and sentence construction compose another; *morphology*, or word form, including much of what is called "grammar" is the third; and *phonology*, or the rules about which sounds are acceptable and meaningful, is the fourth. Each system has its own peculiar tendencies, although a change in one (say, phonology) can bring about changes in another (say, morphology).[5] We will look most closely at phonology and the lexicon, as these are the most important in understanding how the Proto-Indo-European vocabulary has been reconstructed.

PHONOLOGY: HOW TO RECONSTRUCT A DEAD SOUND

Phonology, or the study of linguistic sounds, is one of the principal tools of the historical linguist. Phonology is useful as a historical tool, because the sounds people utter tend to change over time in certain directions and not in others.

The direction of phonetic change is governed by two kinds of constraints: those that are generally applicable across most languages, and those specific to a single language or a related group of languages. General constraints are imposed by the mechanical limits of the human vocal anatomy, the need to issue sounds that can be distinguished and understood by listeners, and the tendency to simplify sound combinations that are difficult to pronounce. Constraints within languages are imposed by the limited range of sounds that are acceptable and meaningful for that language. Often these language-specific sounds are very recognizable. Comedians can make us laugh by speaking nonsense if they do it in the characteristic phonology of French or Italian, for example. Armed with a knowledge of both the *general* tendencies in the direction of phonetic change and the *specific* phonetic conventions within a given language group, a linguist can arrive at reliable conclusions about which phonetic variants are early pronunciations and which come later. This is the first step in reconstructing the phonological history of a language.

We know that French developed historically from the dialects of Latin spoken in the Roman province of Gaul (modern France) during the waning centuries of the Roman Empire around 300–400 CE. As late as the 1500s vernacular French suffered from low prestige among scholars, as it was considered nothing more than a corrupt form of Latin. Even if we knew nothing about that history, we could examine the Latin *centum* (pronounced [kentum]), and the French *cent* (pronounced [sohnt]), both meaning "hundred," and we could say that the sound of the Latin word makes it the older form, that the Modern French form could have developed from it according to known rules of sound change, and that an intermediate pronunciation, [tsohnt], probably existed before the modern form appeared—and we would be right.

Some Basic Rules of Language Change: Phonology and Analogy

Two general phonetic rules help us make these decisions. One is that initial hard consonants like *k* and hard *g* tend to change toward soft sounds like *s* and *sh* if they change at all, whereas a change from *s* to *k* would generally be unusual. Another is that a consonant pronounced as a stop in the back of the mouth (*k*) is particularly likely to shift toward the front of the mouth (*t* or *s*) in a word where it is followed by a vowel that is pronounced in the front of the mouth (*e*). Pronounce [ke-] and [se-], and note the position of your tongue. The *k* is pronounced by using the back of the tongue and both *e* and *s* are formed with the middle or the tip of the tongue, which makes it easier to pronounce the segment *se-* than the segment *ke-*. Before a front vowel like *-e* we might expect the *k-* to shift forward to [ts-] and then to [s-] but not the other way around.

This is an example of a general phonetic tendency called *assimilation*: one sound tends to assimilate to a nearby sound in the same word, simplifying the needed movements. The specific type of assimilation seen here is called *palatalization*—a back consonant (*k*) followed by a front vowel (*e*) was assimilated in French toward the front of the palate, changing the [k] to [s]. Between the Latin [k] (pronounced with the back of the tongue at the back of the palate) and the Modern French [s] (tip of the tongue at the front of the palate) there should have been an intermediate pronunciation *ts* (middle of the tongue at the middle of the palate). Such sequences permit historical linguists to reconstruct undocumented intermediate stages in the evolution of a language. Palatalization has been systematic in the development of French from Latin. It is responsible for much of the distinctive phonology of the French language.

Assimilation usually changes the quality of a sound, or sometimes removes sounds from words by slurring two sounds together. The opposite process is the *addition* of new sounds to a word. A good example of an innovation of this kind is provided by the variable pronunciations of the word *athlete* in English. Many English speakers insert [-uh] in the middle of the word, saying [ath-uh-lete], but most are not aware they are doing so. The inserted syllable always is pronounced precisely the same way, as [-uh], because it assimilates to the tongue position required to pronounce the following *-l*. Linguists could have predicted that some speakers would insert a vowel in a difficult cluster of consonants like *-thl* (a phenomenon called *epenthesis*) and that the vowel inserted in *athlete* always would be pronounced [-uh] because of the rule of assimilation.

Another kind of change is *analogical* change, which tends to affect grammar quite directly. For example, the *-s* or *-es* ending for the plural of English nouns was originally limited to one class of Old English nouns: *stān* for *stone* (nominative singular), *stānas* for *stones* (nominative plural). But when a series of sound changes (see note 5) resulted in the loss of the phonemes that had once distinguished nouns of different classes, the *-s* ending began to be reinterpreted as a *general* plural indicator and was attached to all nouns. Plurals formed with *-n* (oxen), with a zero change (sheep), and with a vowel change in the stem (women) remain as relics of Old English, but the shift to *-s* is driving out such "irregular" forms and has been doing so for eight hundred years. Similar analogical changes have affected verbs: *help/helped* has replaced Old English *help/holp* as the *-ed* ending has been reinterpreted as a general ending for the past tense, reducing the once large number of strong verbs that formed their past with a vowel change. Analogical changes can also create new words or forms by analogy with old ones. Words formed with *-able* and *-scape* exist in such great numbers in English because these endings, which were originally bound to specific words (*measurable, landscape*), were reinterpreted as suffixes that could be removed and reattached to any stem (*touchable, moonscape*).

Phonological and analogical change are the internal mechanisms through which novel forms are incorporated into a language. By examining a sequence of documents within one language lineage from several different points in the past—inscriptions in, say, classical Latin, late vulgar Latin, early Medieval French, later Medieval French, and modern French—linguists have defined virtually all the sound changes and analogical shifts in the evolution of French from Latin. Regular, systematic rules, applicable also to other cases of language change in other languages, explain most of these shifts. But how do linguists replay these shifts "backward" to discover the origins of modern languages? How can we reconstruct the sounds of a language like Proto-Indo-European, for which there are *no* documents, a language spoken before writing was invented?

"Hundred": An Example of Phonetic Reconstruction

Proto-Indo-European words were not reconstructed to create a dictionary of Proto-Indo-European vocabulary, although they are extraordinarily useful in this way. The real aim in reconstruction is to prove that a list of daughter terms are cognates, descended from the same mother term. The reconstruction of the mother term is a by-product of the comparison, the proof that every sound in every daughter word can be derived from a sound in the common parent. The first step is to gather up the suspected daughters: you must make a list of all the variants of the word you can find in the Indo-European languages (table 2.1). You have to know the rules of phonological change to do even this successfully, as some variants of the word might have changed radically in sound. Just recognizing the candidates and making up a good list can be a challenge. We will try this with the Proto-Indo-European word for "hundred." The Indo-European roots for numbers, especially 1 to 10, 100, and 1,000, have been retained in almost all the Indo-European daughters.

Our list includes Latin *centum*, Avestan *satəm*, Lithuanian *šimtas*, and Old Gothic *hunda-* (a root much like *hunda-* evolved into the English word *hundred*). Similar-looking words meaning "hundred" in other

TABLE 2.1

Indo-European Cognates for the Root "Hundred"

Branch	Language	Term	Meaning
Celtic	Welsh	cant	hundred
	Old Irish	cēt	hundred
Italic	Latin	centum	hundred
Tocharian	TochA	känt	hundred
	TochB	kante	hundred
Greek	Greek	ἑκατόν	hundred
Germanic	Old English	hund	hundred
	OldHighGerm.	hunt	hundred
	Gothic	hunda	100, 120
	OldSaxon	hunderod	(long) hundred
Baltic	Lithuanian	šimtas	hundred
	Latvian	simts	hundred
Slavic	OldChurchSlav.	sŭto	hundred
	Bulgarian	sto	hundred
Anatolian	Lycian	sñta	unit of 10 or 100
Indo-Iranian	Avestan	satəm	hundred
	OldIndic	śatám	hundred

Indo-European languages should be added, and I have already referred to the French word *cent*, but I will use only four for simplicity's sake. The four words I have chosen come from four Indo-European branches: Italic, Indo-Iranian, Baltic, and Germanic.

The question we must answer is this: Are these words phonetically transformed daughters of a single parent word? If the answer is yes, they are cognates. To prove they are cognates, we must be able to reconstruct an ancestral sequence of phonemes that could have developed into all the documented daughter sounds through known rules. We start with the first sound in the word.

The initial [k] phoneme in Latin *centum* could be explained if the parent term began with a [k] sound as well. The initial soft consonants ([s] [sh])

in Avestan *satəm* and Lithuanian *šimtas* could have developed from a Proto-Indo-European word that began with a hard consonant [k], like Latin *centum*, since hard sounds generally tend to shift toward soft sounds if they change at all. The reverse development ([s] or [sh] to [k]) would be very unlikely. Also, palatalization and sibilation (shifting to a 's' or 'sh' sound) of initial hard consonants is expected in both the Indic branch, of which Vedic Sanskrit is a member; and the Baltic branch, of which Lithuanian is a member. The general direction of sound change and the specific conventions in each branch permit us to say that the Proto-Indo-European word from which all three of these developed could have begun with 'k'.

What about *hunda*? It looks quite different but, in fact, the *h* is expected—it follows a rule that affected all initial [k] sounds in the Germanic branch. This shift involved not just *k* but also eight other consonants in Pre-Germanic.[6] The consonant shift spread throughout the prehistoric Pre-Germanic language community, giving rise to a new Proto-Germanic phonology that would be retained in all the later Germanic languages, including, ultimately, English. This consonant shift was described by and named after Jakob Grimm (the same Grimm who collected fairy tales) and so is called Grimm's Law. One of the changes described in Grimm's Law was that the archaic Indo-European sound [k] shifted in most phonetic environments to Germanic [h]. The Indo-European *k* preserved in Latin *centum* shifted to *h* in Old Gothic *hunda-*; the initial *k* seen in Latin *caput* 'head' shifted to *h* in Old English *hafud* 'head'; and so on throughout the vocabulary. (*Caput > hafud* shows that *p* also changed to *f*, as in *pater > fater*). So, although it looks very different, *hunda-* conforms: its first consonant can be derived from *k* by Grimm's Law.

The first sound in the Proto-Indo-European word for "hundred" probably was *k*. (An initial [k] sound satisfies the other Indo-European cognates for "hundred" as well.)[7] The second sound should have been a vowel, but which vowel?

The second sound was a vowel that does not exist in English. In Proto-Indo-European resonants could act as vowels, similar to the resonant *n* in the colloquial pronunciation of *fish'n'* (as in *Bob's gone fish'n'*). The second sound was a resonant, either **m* or **n*, both of which occur among the daughter terms being compared. (An asterisk is used before a reconstructed form for which there is no direct evidence.) *M* is attested in the Lithuanian cognate *šimtas*. An *m* in the Proto-Indo-European parent could account for the *m* in Lithuanian. It could have changed to *n* in Old Indic, Germanic, and other lineages by assimilating to the following *t* or *d*, as both *n* and *t* are articulated on the teeth. (Old Spanish *semda* 'path' changed to modern Spanish *senda* for the same reason.) A shift from an original *m* to an *n* before a *t* is explicable, but a shift from an original *n* to an *m* is much less likely. Therefore, the original second sound probably was m̥. This consonant could have been lost entirely in Sanskrit *satam* by yet another assimilative tendency called total assimilation: after the *m* changed to *n*, giving **santam*, the *n* was completely assimilated to the following *t*, giving *satam*. The same process was responsible for the loss of the [k] sound in the shift from Latin *octo* to modern Italian *otto* 'eight'.

I will stop here, with an ancestral **k'm̥ -*, in my discussion of the Proto-Indo-European ancestor of *centum*. The analysis should continue through the phonemes that are attested in all the surviving cognates to reconstruct an acceptable ancestral root. By applying such rules to all the cognates, linguists have been able to reconstruct a Proto-Indo-European sequence of phonemes, **k'm̥tom*, that could have developed into all the attested phonemes in all the attested daughter forms. The Proto-Indo-European root **k'm̥tom* is the residue of a successful comparison—it is the proof that the daughter terms being compared are indeed cognates. It is also likely to be a pretty good approximation of the way this word was pronounced in at least some dialects of Proto-Indo-European.

The Limitations and Strengths of Reconstruction

The comparative method will produce the *sound* of the ancestral root and confirm a genetic relationship *only* with a group of cognates that has evolved regularly according to the rules of sound change. The result of a comparative analysis is either a demonstration of a genetic connection, if every phoneme in every cognate can be derived from a mutually acceptable parental phoneme; or no *demonstrable* connection. In many cases sounds may have been borrowed into a language from a neighboring language, and those sounds might replace the predicted shifts. The comparative method cannot force a regular reconstruction on an irregular set of sounds. Much of the Proto-Indo-European vocabulary, perhaps most of it, never will be reconstructed. Regular groups of cognates permit us to reconstruct a Proto-Indo-European root for the word *door* but not for *wall*; for *rain* but not for *river*; for *foot* but not for *leg*. Proto-Indo-European certainly had words for these things, but we cannot safely reconstruct how they sounded.

The comparative method cannot prove that two words are *not* related, but it can fail to produce proof that they *are*. For example, the Greek god Ouranos and the Indic deity Varuna had strikingly similar mythological attributes, and their names sound somewhat alike. Could Ouranos and

Varuna be reflexes of the name of some earlier Proto-Indo-European god? Possibly—but the two names cannot be derived from a common parent by the rules of sound change known to have operated in Greek and Old Indic. Similarly Latin *deus* (god) and Greek *théos* (god) look like obvious cognates, but the comparative method reveals that Latin *deus*, in fact, shares a common origin with Greek *Zéus*.[8] If Greek *théos* were to have a Latin cognate it should begin with an [f] sound (*festus* 'festive' has been suggested, but some of the other sounds in this comparison are problematic). It is still possible that *deus* and *théos* were historically related in some irregular way, but we cannot prove it.

In the end, how can we be sure that the comparative method accurately reconstructs undocumented stages in the phonological history of a language? Linguists themselves are divided on the question of the "reality" of reconstructed terms.[9] A reconstruction based on cognates from eight Indo-European branches, like *k'ṃtom-*, is much more reliable and probably more "true" than one based on cognates in just two branches. Cognates in at least three branches, including an ancient branch (Anatolian, Greek, Avestan Iranian, Old Indic, Latin, some aspects of Celtic) should produce a reliable reconstruction. But how reliable? One test was conceived by Robert A. Hall, who reconstructed the shared parent of the Romance languages using just the rules of sound change, and then compared his reconstruction to Latin. Making allowances for the fact that the actual parents of the Romance languages were several provincial Vulgar Latin dialects, and the Latin used for the test was the classical Latin of Cicero and Caesar, the result was reassuring. Hall was even able to reconstruct a contrast between two sets of vowels although none of the modern daughters had retained it. He was unable to identify the feature that distinguished the two vowel sets as length—Latin had long vowels and short vowels, a distinction lost in all its Romance daughters—but he was able to rebuild a system with two contrasting sets of vowels and many of the other, more obvious aspects of Latin morphology, syntax, and vocabulary. Such clever exercises aside, the best proof of the realism of reconstruction lies in several cases where linguists have suggested a reconstruction and archaeologists have subsequently found inscriptions that proved it correct.[10]

For example, the oldest recorded Germanic cognates for the word *guest* (Gothic *gasts*, Old Norse *gestr*, Old High German *gast*) are thought to be derived from a reconstructed late Proto-Indo-European *ghos-ti-* (which probably meant both "host" and "guest" and thus referred to a relationship of hospitality between strangers rather than to one of its roles) through a Proto-Germanic form reconstructed as *gastiz*. None of the known forms of the word in the later Germanic languages contained the *i* before the final consonant, but rules of sound change predicted that the *i* should theoretically have been there in Proto-Germanic. Then an archaic Germanic inscription was found on a gold horn dug from a grave in Denmark. The inscription *ek hlewagastiz holitijaz* (or *holtingaz*) *horna tawido* is translated "I, Hlewagasti of Holt (or Holting) made the horn." It contained the personal name Hlewagastiz, made up of two stems, *Hlewa-* 'fame' and *gastiz* 'guest'. Linguists were excited not because the horn was a beautiful golden artifact but because the stem contained the predicted *i*, verifying the accuracy of both the reconstructed Proto-Germanic form and its late Proto-Indo-European ancestor. Linguistic reconstruction had passed a real-world test.

Similarly linguists working on the development of the Greek language had proposed a Proto-Indo-European labiovelar $*k^w$ (pronounced [kw-]) as the ancestral phoneme that developed into Greek *t* (before a front vowel) or *p* (before a back vowel). The reconstruction of $*k^w$ was a reasonable but complex solution for the problem of how the Classical Greek consonants were related to their Proto-Indo-European ancestors. It remained entirely theoretical until the discovery and decipherment of the Mycenaean Linear B tablets, which revealed that the earliest form of Greek, Mycenaean, had the predicted k^w where later Greek had *t* or *p* before front and back vowels.[11] Examples like these confirm that the reconstructions of historical linguistics are more than just abstractions.

A reconstructed term is, of course, a phonetic idealization. Reconstructed Proto-Indo-European cannot capture the variety of dialectical pronunciations that must have existed more than perhaps one thousand years when the language was living in the mouths of people. Nevertheless, it is a remarkable victory that we can now pronounce, however stiffly, thousands of words in a language spoken by nonliterate people before 2500 BCE.

THE LEXICON: HOW TO RECONSTRUCT DEAD MEANINGS

Once we have reconstructed the *sound* of a word in Proto-Indo-European, how do we know what it *meant*? Some archaeologists have doubted the reliability of reconstructed Proto-Indo-European, as they felt that the original meanings of reconstructed terms could never be known confidently.[12] But we can assign reliable meanings to many reconstructed Proto-Indo-European terms. And it is in the meanings of their words that we find the best evidence for the material culture, ecological environment, social relations, and spiritual beliefs of the speakers of Proto-Indo-European. Every meaning is worth the struggle.

Three general rules guide the assignment of meaning. First, look for the most ancient meanings that can be found. If the goal is to retrieve the meaning of the original Proto–Indo–European word, modern meanings should be checked against meanings that are recorded for ancient cognates.

Second, if one meaning is consistently attached to a cognate in all language branches, like *hundred* in the example I have used, that is clearly the least problematic meaning we can assign to the original Proto-Indo-European root. It is difficult to imagine how that meaning could have become attached to all the cognates unless it were the meaning attached to the ancestral root.

Third, if the word can be broken down into roots that point to the same meaning as the one proposed, then that meaning is doubly likely. For example, Proto-Indo-European *$k'mtom$ probably was a shortened version of *$dek'mtom$, a word that included the Proto-Indo-European root *$dek'm$ 'ten'. The sequence of sounds in *$dek'm$ was reconstructed independently using the cognates for the word *ten*, so the fact that the reconstructed roots for *ten* and *hundred* are linked in both meaning and sound tends to verify the reliability of both reconstructions. The root *$k'mtom$ turns out to be not just an arbitrary string of Proto-Indo-European phonemes but a meaningful compound: "(a unit) of tens." This also tells us that the speakers of Proto-Indo-European had a decimal numbering system and counted to one hundred by tens, as we do.

In most cases the meaning of a Proto-Indo-European word changed and drifted as the various speech communities using it became separated, centuries passed, and daughter languages evolved. Because the association between word and meaning is arbitrary, there is less regular directionality to change in meaning than there is in sound change (although some semantic shifts are more probable than others). Nevertheless, general meanings can be retrieved. A good example is the word for "wheel."

"Wheel": An Example of Semantic Reconstruction

The word *wheel* is the modern English descendant of a PIE root that had a sound like *$k^w\acute{e}k^wlos$ or *$k^wek^wl\acute{o}s$. But what, exactly, did *$k^w\acute{e}k^wlos$ mean in Proto-Indo-European? The sequence of phonemes in the root *$k^w\acute{e}k^wlos$ was pieced together by comparing cognates from eight old Indo-European languages, representing five branches. Reflexes of this word survived in Old Indic and Avestan (from the Indo-Iranian branch), Old Norse and

Old English (from the Germanic branch), Greek, Phrygian, and Tocharian A and B. The meaning "wheel" is attested for the cognates in Sanskrit, Avestan, Old Norse, and Old English. The meaning of the Greek cognate had shifted to "circle" in the singular but in the plural still meant "wheels." In Tocharian and Phrygian the cognates meant "wagon" or "vehicle." What was the original meaning? (table 2.2).

Five of the eight $*k^w ek^w los$ cognates have "wheel" or "wheels" as an attested meaning, and in those languages (Phrygian, Greek, Tocharian A & B) where the meaning drifted away from "wheel(s)," it had not drifted far ("circle," "wagon," or "vehicle"). Moreover, the cognates that preserve the meaning "wheel" are found in languages that are geographically isolated from one another (Old Indic and Avestan in Iran were neighbors, but neither had any known contact with Old Norse or Old English). The meaning "wheel" is unlikely to have been borrowed into Old Norse from Old Indic, or vice versa.

Some shifts in meaning are unlikely, and others are common. It is common to name a whole ("vehicle," "wagon") after one of its most characteristic parts ("wheels"), as seems to have happened in Phrygian and Tocharian. We do the same in modern English slang when we speak of someone's car as their "wheels," or clothing as their "threads." A shift in meaning in the other direction, using a word that originally referred to the whole to refer to one of its parts (using *wagon* to refer to *wheel*), is much less probable.

The meaning of *wheel* is given additional support by the fact that it has an Indo-European etymology, like the root for $*k'mtom$. It was a word created from another Indo-European root. That root was $*k^w el-$, a verb that meant "to turn." So $*k^w ek^w los$ is not just a random string of phonemes reconstructed from the cognates for *wheel*; it meant "the thing that turns." This not only tends to confirm the meaning "wheel" rather than "circle" or "vehicle" but it also indicates that the speakers of Proto-Indo-European made up their own words for wheels. If they learned about the invention of the wheel from others they did not adopt the foreign name for it, so the social setting in which the transfer took place probably was brief, between people who remained socially distant. The alternative, that wheels were invented within the Proto-Indo-European language community, seems unlikely for archaeological and historical reasons, though it remains possible (see chapter 4).

One more rule helps to confirm the reconstructed meaning. If it fits within a semantic field consisting of other roots with closely related reconstructed meanings, we can at least be relatively confident that such a word

TABLE 2.2

Proto-Indo-European Roots for Words Referring to Parts of a Wagon

PIE Root Word	Wagon Part	Daughter Languages
*kʷekʷlos	(wheel)	*Old Norse* hvēl 'wheel'; *Old English* hweohl 'wheel'; *Middle Dutch* wiel 'wheel'; *Avestan Iranian* čaxtra- 'wheel'; *Old Indic* cakrá 'wheel, Sun disc'; *Greek* kuklos 'circle' and kukla (plural) 'wheels'; *Tocharian A* kukal 'wagon'; *Tocharian B* kokale 'wagon'
*rot-eh₂-	(wheel)	*Old Irish* roth 'wheel'; *Welsh* rhod 'wheel'; *Latin* rota 'wheel'; *Old High German* rad 'wheel'; *Lithuanian* rātas 'wheel'; *Latvian* rats 'wheel' and rati (plural) 'wagon'; *Albanian* rreth 'ring, hoop, carriage tire'; *Avestan Iranian* ratha 'chariot, wagon'; *Old Indic* rátha 'chariot, wagon'
*ak*s-, or	(axle)	*Latin* axis 'axle, axis'; *Old English* eax 'axle'; *Old High German* *hₐek*s- ahsa 'axle'; *Old Prussian* assis 'axle'; *Lithuanian* ašís 'axle'; *Old Church Slavonic* osi 'axle'; *Mycenaean Greek* a-ko-so-ne 'axle'; *Old Indic* áks*a 'axle'
*ei-/*oi-, or	(thill)	*Old English* ār- 'oar'; *Russian* vojë 'shaft'; *Slovenian* oje 'shaft'; *Hittite* h₂ih₃s or hišša- 'pole, harnessing shaft'; *Greek* oisioi* 'tiller, rudderpost'; *Avestan Iranian* aēša 'pair of shafts, plow-pole'; *Old Indic* is*a 'pole, shaft'
*wéĝheti-	(ride)	*Welsh* amwain 'drive about'; *Latin* vehō 'bear, convey'; *Old Norse* vega 'bring, move'; *Old High German* wegan 'move, weigh'; *Lithuanian* vežù 'drive'; *Old Church Slavonic* vezǫ 'drive'; *Avestan Iranian* vazaiti 'transports, leads'; *Old Indic* váhati 'transports, carries, conveys'. Derivative nouns have the meaning "wagon" in *Greek, Old Irish, Welsh, Old High German,* and *Old Norse.*

could have existed in Proto-Indo-European. "Wheel" is part of a semantic field consisting of *words for the parts of a wagon or cart* (table 2.2). Happily, at least four other such words can be reconstructed for Proto-Indo-European. These are:

1. *$rot-eh_2$-*, a second term for "wheel," with cognates in Old Indic and Avestan that meant "chariot," and cognates that meant "wheel" in Latin, Old Irish, Welsh, Old High German, and Lithuanian.

2. *aks-* (or perhaps *h_2eks-*) 'axle' attested by cognates that had not varied in meaning over thousands of years, and still meant "axle" in Old Indic, Greek, Latin, Old Norse, Old English, Old High German, Lithuanian, and Old Church Slavonic.

3. *h_2ih_3s-* 'thill' (the harness pole) attested by cognates that meant "thill" in Hittite and Old Indic.

4. *wégheti*, a verb meaning "to convey or go in a vehicle," attested by cognates carrying this meaning in Old Indic, Avestan, Latin, Old English, and Old Church Slavonic and by cognate-derived nouns ending in *-no-* meaning "wagon" in Old Irish, Old English, Old High German, and Old Norse.

These four additional terms constitute a well-documented semantic field (*wheel, axle, thill,* and *wagon* or *convey in a vehicle*) that increases our confidence in reconstructing the meaning "wheel" for *$k^wék^wlos$*. Of the five terms assigned to this semantic field, all but *thill* have clear Indo-European etymologies in independently reconstructed roots. The speakers of Proto-Indo-European were familiar with wheels and wagons, and used words of their own creation to talk about them.

Fine distinctions, shades of meaning, and the word associations that enriched Proto-Indo-European poetry may be forever lost, but gross meanings are recoverable for at least fifteen hundred Proto-Indo-European roots such as *dekm-* 'ten', and for additional thousands of other words derived from them, such as *kmtom-* 'hundred'. Those meanings provide a window into the lives and thoughts of the speakers of Proto-Indo-European.

Syntax and Morphology: The Shape of a Dead Language

I will not try to describe in any detail the grammatical connections between the Indo-European languages. The reconstructed vocabulary is most important for our purposes. But grammar, the bedrock of language classification, provides the primary evidence for classifying languages and determining relationships between them. Grammar has two aspects: *syntax*, or the rules governing the order of words in sentences; and *morphology*, or the rules governing the forms words must take when used in particular ways.

Proto-Indo-European grammar has left its mark on all the Indo-European languages to one degree or another. In all the Indo-European language branches, nouns are declined; that is, the noun changes form depending on how it is used in a sentence. English lost most of these declinations during its evolution from Anglo-Saxon, but all the other languages in the Germanic branch retain them, and we have kept some use-dependent pronouns (*masculine*: he, his, him/*feminine*: she, hers, her). Moreover, most Indo-European nouns are declined in similar ways, with endings that are genetically cognate, and with the same formal system of cases (nominative, genitive, accusative, etc.) that intersect in the same way with the same three gender classes (masculine, feminine, neuter); and with similar formal classes, or declensions, of nouns that are declined in distinctive ways. Indo-European verbs also share similar conjugation classes (first person, second person or familiar, third person or formal, singular, plural, past tense, present tense, etc.), similar stem alterations (run-ran, give-gave), and similar endings. This particular constellation of formal categories, structures, transformations, and endings is not at all necessary or universal in human language. It is unique, as a system, and is found only in the Indo-European languages. The languages that share this grammatical system certainly are daughters of a single language from which that system was inherited.

One example shows how unlikely it would be for the Indo-European languages to share these grammatical structures by random chance. The verb *to be* has one form in the first-person singular ([I] *am*) and another in the third-person singular ([he/she/it] *is*). Our English verbs are descended from the archaic Germanic forms *im* and *ist*. The Germanic forms have exact, proven cognates in Old Indic *ásmi* and *ásti*; in Greek *eimí* and *estí*; and in Old Church Slavonic *jesmi* and *jestŭ*. All these words are derived from a reconstructable Proto-Indo-European pair, *h_1e'smi* and *h_1e'sti*. That all these languages share the same system of verb classes (first person, second person or familiar, and third person), and that they use the same basic roots and endings to identify those classes, confirms that they are genetically related languages.

Conclusion: Raising a Language from the Dead

It will always be difficult to work with Proto-Indo-European. The version we have is uncertain in many morphological details, phonetically idealized, and fragmentary, and can be difficult to decipher. The meanings of some terms will never be fully understood, and for others only an approximate definition is possible. Yet reconstructed Proto-Indo-European captures key parts of a language that actually existed.

Some dismiss reconstructed Proto-Indo-European as nothing more than a hypothesis. But the limitations of Proto-Indo-European apply equally to the written languages of ancient Egypt and Mesopotamia, which are universally counted among the great treasures of antiquity. No curator of Assyrian records would suggest that we should discard the palace archives of Nineveh because they are incomplete, or because we cannot know the exact sound and meaning of many terms, or because we are uncertain about how the written court language related to the 'real' language spoken by the people in the street. Yet these same problems have convinced many archaeologists that the study of Proto-Indo-European is too speculative to yield any real historical value.

Reconstructed Proto-Indo-European is a long, fragmentary list of words used in daily speech by people who created no other texts. That is why it is important. The list becomes useful, however, only if we can determine where it came from. To do that we must locate the Proto-Indo-European homeland. But we cannot locate the Proto-Indo-European homeland until we first locate Proto-Indo-European in time. We have to know *when* it was spoken. Then it becomes possible to say where.

CHAPTER THREE

Language and Time 1
The Last Speakers of Proto-Indo-European

Time changes everything. Reading to my young children, I found that in mid-sentence I began to edit and replace words that suddenly looked archaic to me, in stories I had loved when I was young. The language of Robert Louis Stevenson and Jules Verne now seems surprisingly stiff and distant, and as for Shakespeare's English—we all need the glossary. What is true for modern languages was true for prehistoric languages. Over time, they changed. So what do we mean by Proto-Indo-European? If it changed over time, is it not a moving target? However we define it, for how long was Proto-Indo-European spoken? Most important, when was it spoken? How do we assign a date to a language that left no inscriptions, that died without ever being written down? It helps to divide any problem into parts, and this one can easily be divided into two: the birth date and the death date.

This chapter concentrates on the death date, the date *after which* Proto-Indo-European must have ceased to exist. But it helps to begin by considering how long a period probably preceded that. Given that the time between the birth and death dates of Proto-Indo-European could not have been infinite, precisely how long a time was it? Do languages, which are living, changing things, have life expectancies?

THE SIZE OF THE CHRONOLOGICAL WINDOW:
HOW LONG DO LANGUAGES LAST?

If we were magically able to converse with an English speaker living a thousand years ago, as proposed in the last chapter, we would not understand each other. Very few natural languages, those that are learned and spoken at home, remain sufficiently unchanged after a thousand years to be considered the "same language." How can the rate of change be measured?

Languages normally have dialects—regional accents—and, within any region, they have innovating social sectors (entertainers, soldiers, traders) and conservative sectors (the very rich, the very poor). Depending on who you are, your language might be changing very rapidly or very slowly. Unstable conditions—invasions, famines, the fall of old prestige groups and the rise of new ones—increase the rate of change. Some parts of language change earlier and faster, whereas other parts are resistant. That last observation led the linguist Morris Swadesh to develop a standard word list chosen from the most resistant vocabulary, a group of words that tend to be retained, not replaced, in most languages around the world, even after invasions and conquests. Over the long term, he hoped, the average rate of replacement in this resistant vocabulary might yield a reliable standardized measurement of the speed of language change, what Swadesh called *glottochronology*.[1]

Between 1950 and 1952 Swadesh published a hundred-word and a two-hundred-word *basic core vocabulary*, a standardized list of resistant terms. All languages, he suggested, tend to retain their own words for certain kinds of meanings, including body parts (blood, foot); lower numerals (one, two, three); some kinship terms (mother, father); basic needs (eat, sleep); basic natural features (sun, moon, rain, river); some flora and fauna (tree, domesticated animals); some pronouns (this, that, he, she); and conjunctions (and, or, if). The content of the list can be and has been modified to suit vocabularies in different languages—in fact, the preferred two-hundred-meaning list in English contains 215 words. The English core vocabulary has proven extremely resistant to change. Although English has borrowed more than 50% of its *general* vocabulary from the Romance languages, mainly from French (reflecting the conquest of Anglo-Saxon England by the French-speaking Normans) and Latin (from centuries of technical and professional vocabulary training in courts, churches, and schools), only 4% of the English *core* vocabulary is borrowed from Romance. In its core vocabulary English remains a Germanic language, true to its origins among the Anglo-Saxons who migrated from northern Europe to Britain after the fall of the Roman Empire.

Comparing core vocabularies between old and new phases in languages with long historical records (Old English/Modern English, Middle Egyptian/Coptic, Ancient Chinese/Modern Mandarin, Late Latin/Modern French, and nine other pairs), Swadesh calculated an average replacement rate of 14% per thousand years for the hundred-word list, and 19% per thousand years for the two-hundred-word list. He suggested that 19% was an acceptable average for all languages (usually rounded to 20%). To illustrate what that number means, Italian and French have distinct, unrelated words for 23% of the terms in the two-hundred-word list, and Spanish and Portuguese show a difference of 15%. As a general rule, if more than 10% of the core vocabulary is different between two dialects, they are either mutually unintelligible or approaching that state, that is, they are distinct languages or emerging languages. On average, then, with a replacement rate of 14–19% per thousand years in the core vocabulary, we should expect that most languages—including this one— would be incomprehensible to our own descendants a thousand years from now.

Swadesh hoped to use the replacement rate in the core vocabulary as a standardized clock to establish the date of splits and branches in unwritten languages. His own research involved the splits between American Indian language families in prehistoric North America, which were undatable by any other means. But the reliability of his standard replacement rate wilted under criticism. Extreme cases like Icelandic (very slow change, with a replacement rate of only 3–4% per thousand years) and English (very rapid, with a 26% replacement rate per thousand years) challenged the utility of the "average" rate.[2] The mathematics was affected if a language had multiple words for one meaning on the list. The dates given by glottochronology for many language splits contradicted known historical dates, generally by giving a date much later than it should have been. This direction in the errors suggested that real language change often was slower than Swadesh's model suggested—less than 19% per thousand years. A devastating critique of Swadesh's mathematics by Chretien, in 1962, seemed to drive a stake through the heart of glottochronology.

But in 1972 Chretien's critique was itself shown to be incorrect, and, since the 1980s, Sankoff and Embleton have introduced equations that include as critical values borrowing rates, the number of geographic borders with other languages, and a similarity index between the compared languages (because similar languages borrow in the core more easily then dissimilar languages). Multiple synonyms can each be given a fractional score. Studies incorporating these improved methods succeeded better in producing dates for splits between known languages that matched historical facts. More important, comparisons between most Indo-European languages still yielded replacement rates in the core vocabulary of about 10–20% per thousand years. Comparing the core vocabularies in ninety-five Indo-European languages, Kruskal and Black found that the most frequent date for the first splitting of Proto–Indo–European was about

3000 BCE. Although this estimate cannot be relied on absolutely, it is probably "in the ballpark" and should not be ignored.[3]

One simple point can be extracted from these debates: if the Proto-Indo-European core vocabulary changed at a rate ≥10% per millennium, or at the lower end of the expected range, Proto-Indo-European did not exist as a single language with a single grammar and vocabulary for as long as a thousand years. Proto-Indo-European grammar and vocabulary should have changed quite substantially over a thousand years. Yet the grammar of Proto-Indo-European, as reconstructed by linguists, is remarkably homogeneous both in morphology and phonology. Proto-Indo-European nouns and pronouns shared a set of cases, genders, and declensions that intersect with dozens of cognate phonological endings. Verbs had a shared system of tenses and aspects, again tagged by a shared set of phonological vowel changes (*run-ran*) and endings. This shared system of grammatical structures and phonological ways of labeling them looks like a single language. It suggests that reconstructed Proto-Indo-European probably refers to less than a thousand years of language change. It took less than a thousand years for late Vulgar Latin to evolve into seven Romance languages, and Proto-Indo-European does not contain nearly enough internal grammatical diversity to represent seven distinct grammars.

But considering that Proto-Indo-European is a fragmentary reconstruction, not an actual language, we should allow it more time to account for the gaps in our knowledge (more on this in chapter 5). Let us assign a nominal lifetime of two thousand years to the phase of language history represented by reconstructed Proto-Indo-European. In the history of English two thousand years would take us all the way back to the origins of the sound shifts that defined Proto-Germanic, and would include all the variation in all the Germanic languages ever spoken, from Hlewagasti of Holt to Puff Daddy of hip-hop fame. Proto-Indo-European does not seem to contain that much variation, so two thousand years probably is too long. But for archaeological purposes it is quite helpful to be able to say that the time period we are trying to identify is no longer than two thousand years.

What is the end date for that two-thousand-year window of time?

THE TERMINAL DATE FOR PROTO-INDO-EUROPEAN: THE MOTHER BECOMES HER DAUGHTERS

The terminal date for reconstructed Proto-Indo-European—the date after which it becomes an anachronism—should be close to the date when its oldest daughters were born. Proto-Indo-European was reconstructed on the basis of systematic comparisons between all the Indo-European daughter languages. The mother tongue cannot be placed later than the daughters. Of course, it would have survived after the detachment and isolation of the oldest daughter, but as time passed, if that daughter dialect remained isolated from the Proto-Indo-European speech community, each would have developed its own peculiar innovations. The image of the mother that is retained through each of the daughters is the form the mother had *before* the detachment of that daughter branch. Each daughter, therefore, preserves a somewhat different image of the mother.

Linguists have exploited this fact and other aspects of internal variation to identify chronological phases within Proto-Indo-European. The number of phases defined by different linguists varies from three (early, middle, late) to six.[4] But if we define Proto-Indo-European as the language that was ancestral to *all* the Indo-European daughters, then it is the *oldest* reconstructable form, the *earliest* phase of Proto-Indo-European, that we are talking about. The later daughters did not evolve directly from this early kind of Proto-Indo-European but from some intermediate, evolved set of late Indo-European languages that preserved aspects of the mother tongue and passed them along.

So when did the oldest daughter separate? The answer to that question depends very much on the accidental survival of written inscriptions. And the oldest daughter preserved in written inscriptions is so peculiar that it is probably safer to rely on the image of the mother preserved within the second set of daughters. What's wrong with the oldest daughter?

The Oldest and Strangest Daughter (or Cousin?): Anatolian

The oldest written Indo-European languages belonged to the Anatolian branch. The Anatolian branch had three early stems: Hittite, Luwian, and Palaic.[5] All three languages are extinct but once were spoken over large parts of ancient Anatolia, modern Turkey (figure 3.1). Hittite is by far the best known of the three, as it was the palace and administrative language of the Hittite Empire.

Inscriptions place Hittite speakers in Anatolia as early as 1900 BCE, but the empire was created only about 1650–1600 BCE, when Hittite warlords conquered and united several independent native Hattic kingdoms in central Anatolia around modern Kayseri. The name *Hittite* was given to them by Egyptian and Syrian scribes who failed to distinguish the Hittite kings from the Hattic kings they had conquered. The Hittites called themselves *Neshites* after the Anatolian city, Kanesh, where they

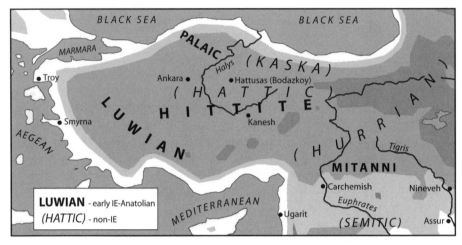

Figure 3.1 The ancient languages of Anatolia at about 1500 BCE.

rose to power. But Kanesh had earlier been a Hattic city; its name was Hattic. Hattic-speakers also named the city that became the capital of the Hittite Empire, Hattušas. Hattic was a non–Indo-European language, probably linked distantly to the Caucasian languages. The Hittites borrowed Hattic words for throne, lord, king, queen, queen mother, heir apparent, priest, and a long list of palace officials and cult leaders—probably in a historical setting where the Hattic languages were the languages of royalty. Palaic, the second Anatolian language, also borrowed vocabulary from Hattic. Palaic was spoken in a city called Pala probably located in north-central Anatolia north of Ankara. Given the geography of Hattic place-names and Hattic→ Palaic/Hittite loans, Hattic seems to have been spoken across all of central Anatolia before Hittite or Palaic was spoken there. The early speakers of Hittite and Palaic were intruders in a non–Indo-European central Anatolian landscape dominated by Hattic speakers who had already founded cities, acquired literate bureaucracies, and established kingdoms and palace cults.[6]

After Hittite speakers usurped the Hattic kingdom they enjoyed a period of prosperity enriched by Assyrian trade, and then endured defeats that later were dimly but bitterly recalled. They remained confined to the center of the Anatolian plateau until about 1650 BCE, when Hittite armies became mighty enough to challenge the great powers of the Near East and the imperial era began. The Hittites looted Babylon, took other cities from the Assyrians, and fought the Egyptian pharaoh Ramses II to a standstill at the greatest chariot battle of ancient times, at Kadesh, on the banks of the Orontes River in Syria, in 1286 BCE. A Hittite monarch married an Egyptian princess. The Hittite kings also knew and negotiated with the princes who ruled Troy, probably the place referred to in the Hittite archives as *steep Wilusa* (*Ilios*).[7] The Hittite capital city, Hattušas, was burned in a general calamity that brought down the Hittite kings, their army, and their cities about 1180 BCE. The Hittite language then quickly disappeared; apparently only the ruling élite ever spoke it.

The third early Anatolian language, Luwian, was spoken by more people over a larger area, and it continued to be spoken after the end of the empire. During the later Hittite empire Luwian was the dominant spoken language even in the Hittite royal court. Luwian did not borrow from Hattic and so might have been spoken originally in western Anatolia, outside the Hattic core region—perhaps even in Troy, where a Luwian inscription was found on a seal in Troy level VI—the Troy of the Trojan War. On the other hand, Luwian did borrow from other, unknown non–Indo-European language(s). Hittite and Luwian texts are abundant from the empire period, 1650–1180 BCE. These are the earliest complete texts in any Indo-European language. But individual Hittite and Luwian words survive from an earlier era, before the empire began.[8]

The oldest Hittite and Luwian names and words appeared in the business records of Assyrian merchants who lived in a commercial district, or *karum*, outside the walls of Kanesh, the city celebrated by the later Hittites as the place where they first became kings. Archaeological excavations here, on the banks of the Halys River in central Anatolia, have shown that the Assyrian *karum*, a foreigners' enclave that covered more than eighty acres outside the Kanesh city walls, operated from about 1920 to 1850 BCE (level II), was burned, rebuilt, and operated again (level Ib) until about 1750 BCE, when it was burned again. After that the Assyrians abandoned the karum system in Anatolia, so the Kanesh karum is a closed archaeological deposit dated between 1920 and 1750 BCE. The Kanesh karum was the central office for a network of literate Assyrian merchants who oversaw trade between the Assyrian state and the warring kingdoms of Late Bronze Age Anatolia. The Assyrian decision to make Kanesh their distribution center greatly increased the power of its Hittite and Luwian occupants.

Most of the local names recorded by the merchants in the Kanesh karum accounts were Hittite or Luwian, beginning with the earliest records of about 1900 BCE. Many still were Hattic. But Hittite speakers seem to have controlled business with the Assyrian karum. The Assyrian merchants were so accustomed to doing business with Hittite speakers that they adopted Hittite words for *contract* and *lodging* even in their private correspondence. Palaic, the third language of the Anatolian branch, is not known from the Kanesh records. Palaic died out as a spoken language probably before 1500 BCE. It presumably was spoken in Anatolia during the karum period but not at Kanesh.

Hittite, Luwian, and Palaic had evolved already by 1900 BCE. This is a critical piece of information in any attempt to date Proto-Indo-European. All three were descended from the same root language, Proto-Anatolian. The linguist Craig Melchert described Luwian and Hittite of the empire period, ca. 1400 BCE, as sisters about as different as twentieth-century Welsh and Irish.[9] Welsh and Irish probably share a common origin of about two thousand years ago. If Luwian and Hittite separated from Proto-Anatolian two thousand years before 1400 BCE, then Proto-Anatolian should be placed at about 3400 BCE. What about *its* ancestor? When did the root of the Anatolian branch separate from the rest of Proto-Indo-European?

Dating Proto-Anatolian: The Definition of Proto- and Pre-Languages

Linguists do not use the term *proto-* in a consistent way, so I should be clear about what I mean by Proto-Anatolian. Proto-Anatolian is the language that was *immediately ancestral* to the three known daughter languages in the Anatolian branch. Proto-Anatolian can be described fairly accurately on the basis of the shared traits of Hittite, Luwian, and Palaic. But Proto-Anatolian occupies just the *later portion* of an undocumented period of linguistic change that must have occurred between it and Proto-Indo-European. The hypothetical language stage in between can be called *Pre-Anatolian*. Proto-Anatolian is a fairly concrete linguistic entity closely related to its known daughters. But Pre-Anatolian represents an *evolutionary period*. Pre-Anatolian is a phase defined by Proto-Anatolian at one end and Proto-Indo-European at the other. How can we determine when Pre-Anatolian separated from Proto-Indo-European?

The ultimate age of the Anatolian branch is based partly on objective external evidence (dated documents at Kanesh), partly on presumed rates of language change over time, and partly on internal evidence within the Anatolian languages. The Anatolian languages are quite different phonologically and grammatically from all the other known Indo-European daughter languages. They are so peculiar that many specialists think they do not really belong with the other daughters.

Many of the peculiar features of Anatolian look like archaisms, characteristics thought to have existed in an extremely early stage of Proto-

Indo-European. For example, Hittite had a kind of consonant that has become famous in Indo-European linguistics (yes, consonants can be famous): h_2, a guttural sound or *laryngeal*. In 1879 a Swiss linguist, Ferdinand de Saussure, realized that several seemingly random differences in vowel pronunciation between the Indo-European languages could be brought under one explanatory rule if he assumed that the pronunciation of these vowels had been affected by a "lost" consonant that no longer existed in any Indo-European language. He proposed that such a lost sound had existed in Proto-Indo-European. It was the first time a linguist had been so bold as to reconstruct a feature for Proto-Indo-European that no longer existed in any Indo-European language. The discovery and decipherment of Hittite forty years later proved Saussure right. In a stunning confirmation of the predictive power of comparative linguistics, the Hittite laryngeal h_2 (and traces of a slightly different laryngeal, h_3) appeared in Hittite inscriptions in just those positions Saussure had predicted for his "lost" consonant. Most Indo-Europeanists now accept that archaic Proto-Indo-European contained laryngeal sounds (probably three different ones, usually transcribed as $*h_1$, $*h_2$, $*h_3$) that were preserved clearly only in the Anatolian branch.[10] The best explanation for why Anatolian has laryngeals is that Pre-Anatolian speakers became separated from the Proto-Indo-European language community at a very early date, when a laryngeal-rich phonology was still characteristic of archaic Proto-Indo-European. But then what does *archaic* mean? What, exactly, did Pre-Anatolian separate from?

The Indo-Hittite Hypothesis

The Anatolian branch either lost or never possessed other features that were present in all other Indo-European branches. In verbs, for example, the Anatolian languages had only two tenses, a present and a past, whereas the other ancient Indo-European languages had as many as six tenses. In nouns, Anatolian had just animate and neuter; it had no feminine case. The other ancient Indo-European languages had feminine, masculine, and neuter cases. The Anatolian languages also lacked the dual, a form that was used in other early Indo-European languages for objects that were doubled like eyes or ears. (Example: Sanskrit dēvas 'one god', but dēvau 'double gods'.) Alexander Lehrman identified ten such traits that probably were innovations in Proto-Indo-European after Pre-Anatolian split away.[11]

For some Indo-Europeanists these traits suggest that the Anatolian branch did not develop from Proto-Indo-European at all but rather evolved from an older Pre-Proto-Indo-European ancestor. This ancestral language was called Indo-Hittite by William Sturtevant. According to the Indo-Hittite hypothesis, Anatolian is an Indo-European language only in the broadest sense, as it did not develop from Proto-Indo-European. But it did preserve, uniquely, features of an earlier language community from which they both evolved. I cannot solve the debate over the categorization of Anatolian here, although it is obviously true that Proto-Indo-European must have evolved from an earlier language community, and we can use *Indo-Hittite* to refer to that hypothetical earlier stage. The Proto-Indo-European language community was a chain of dialects with both geographic and chronological differences. The Anatolian branch seems to have separated from an archaic chronological stage in the evolution of Proto-Indo-European, and it probably separated from a different geographic dialect as well, but I will call it archaic Proto-Indo-European rather than Indo-Hittite.[12]

A substantial period of time is needed for the Pre-Anatolian phase. Craig Melchert and Alexander Lehrman agreed that a separation date of about 4000 BCE between Pre-Anatolian and the archaic Proto-Indo-European language community seems reasonable. The millennium or so around 4000 BCE, say 4500 to 3500 BCE, constitutes the *latest* window within which Pre-Anatolian is likely to have separated.

Unfortunately the oldest daughter of Proto-Indo-European looks so peculiar that we cannot be certain she is a daughter rather than a cousin. Pre-Anatolian could have emerged from Indo-Hittite, not from Proto-Indo-European. So we cannot confidently assign a terminal date to Proto-Indo-European based on the birth of Anatolian.

The Next Oldest Inscriptions: Greek and Old Indic

Luckily we have well-dated inscriptions in two other Indo-European languages from the same era as the Hittite empire. The first was Greek, the language of the palace-centered Bronze Age warrior kings who ruled at Mycenae, Pylos, and other strongholds in Greece beginning about 1650 BCE. The Mycenaean civilization appeared rather suddenly with the construction of the spectacular royal Shaft Graves at Mycenae, dated about 1650 BCE, about the same time as the rise of the Hittite empire in Anatolia. The Shaft Graves, with their golden death masks, swords, spears, and images of men in chariots, signified the elevation of a new Greek-speaking dynasty of unprecedented wealth whose economic power depended on long-distance sea trade. The Mycenaean kingdoms were destroyed during the same period of unrest and pillage that brought down the Hittite Empire about 1150 BCE. Mycenaean Greek, the language of palace administration as recorded in the Linear B tablets, was clearly Greek, not Proto-Greek, by 1450 BCE, the date of the oldest preserved inscriptions. The people who spoke it were the models for Nestor and Agamemnon, whose deeds, dimly remembered and elevated to epic, were celebrated centuries later by Homer in the *Iliad* and the *Odyssey*. We do not know when Greek speakers appeared in Greece, but it happened no later than 1650 BCE. As with Anatolian, there are numerous indications that Mycenaean Greek was an intrusive language in a land where non-Greek languages had been spoken before the Mycenaean age.[13] The Mycenaeans almost certainly were unaware that another Indo-European language was being used in palaces not far away.

Old Indic, the language of the *Rig Veda*, was recorded in inscriptions not long after 1500 BCE but in a puzzling place. Most Vedic specialists agree that the 1,028 hymns of the *Rig Veda* were compiled into what became the sacred form in the Punjab, in northwestern India and Pakistan, probably between about 1500 and 1300 BCE. But the deities, moral concepts, and Old Indic language of the *Rig Veda* first appeared in written documents not in India but in *northern Syria*.[14]

The Mitanni dynasty ruled over what is today northern Syria between 1500 and 1350 BCE. The Mitanni kings regularly spoke a *non*–Indo-European language, Hurrian, then the dominant local language in much of northern Syria and eastern Turkey. Like Hattic, Hurrian was a native language of the Anatolian uplands, related to the Caucasian languages. But all the Mitanni kings, first to last, took Old Indic throne names, even if they had Hurrian names before being crowned. Tus'ratta I was Old Indic *Tvesa-ratha* 'having an attacking chariot', Artatama I was *Rta-dhaaman* 'having the abode of r'ta', Artas's'umara was *Rta-smara* 'remembering r'ta', and S'attuara I was *Satvar* 'warrior'.[15] The name of the Mitanni capital city, Waššukanni, was Old Indic *vasu-khani*, literally "wealth-mine." The Mitanni were famous as charioteers, and, in the oldest surviving horse-training manual in the world, a Mitanni horse trainer named Kikkuli (a Hurrian name) used many Old Indic terms for technical details, including horse colors and numbers of laps. The Mitanni military aristocracy was composed of chariot warriors called *maryanna*, probably from an Indic term *márya* meaning "young man," employed in the *Rig Veda* to refer to the heavenly war-band assembled around Indra. Several royal Mitanni names contained the Old Indic term *r'ta*, which meant "cosmic order and truth," the central moral concept of the *Rig Veda*. The Mitanni king

Kurtiwaza explicitly named four Old Indic gods (Indra, Varuna, Mithra, and the Nāsatyas), among many native Hurrian deities, to witness his treaty with the Hittite monarch around 1380 BCE. And these were not just any Old Indic gods. Three of them—Indra, Varuna, and the Nāsatyas or Divine Twins—were the three most important deities in the *Rig Veda*. So the Mitanni texts prove not only that the Old Indic language existed by 1500 BCE but also that the central religious pantheon and moral beliefs enshrined in the *Rig Veda* existed equally early.

Why did Hurrian-speaking kings in Syria use Old Indic names, words, and religious terms in these ways? A good guess is that the Mitanni kingdom was founded by Old Indic-speaking mercenaries, perhaps charioteers, who regularly recited the kinds of hymns and prayers that were collected at about the same time far to the east by the compilers of the *Rig Veda*. Hired by a Hurrian king about 1500 BCE, they usurped his throne and founded a dynasty, a very common pattern in Near Eastern and Iranian dynastic histories. The dynasty quickly became Hurrian in almost every sense but clung to a tradition of using Old Indic royal names, some Vedic deity names, and Old Indic technical terms related to chariotry long after its founders faded into history. This is, of course, a guess, but something like it seems almost necessary to explain the distribution and usage of Old Indic by the Mitanni.

The Mitanni inscriptions establish that Old Indic was being spoken before 1500 BCE in the Near East. By 1500 BCE Proto-Indo-European had differentiated into at least Old Indic, Mycenaean Greek, and the three known daughters of Proto-Anatolian. What does this suggest about the terminal date for Proto-Indo-European?

Counting the Relatives: How Many in 1500 BCE?

To answer this question we first have to understand where Greek and Old Indic are placed among the known branches of the Indo-European family. Mycenaean Greek is the oldest recorded language in the Greek branch. It is an isolated language; it has no recorded close relatives or sister languages. It probably had unrecorded sisters, but none survived in written records. The appearance of the Shaft-Grave princes about 1650 BCE represents the latest possible arrival of Greek speakers in Greece. The Shaft-Grave princes probably already spoke an early form of Greek, not Proto-Greek, since their descendants' oldest preserved inscriptions at about 1450 BCE were in Greek. Proto-Greek might be dated *at the latest* between about 2000 and 1650 BCE. Pre-Greek, the phase that preceded

Proto-Greek, probably originated as a dialect of late Proto-Indo-European *at least* five hundred to seven hundred years before the appearance of Mycenaean Greek, and very probably earlier—*minimally* about 2400–2200 BCE. The terminal date for Proto-Indo-European can be set at about 2400–2200 BCE—it could not have been later than this—from the perspective of the Greek branch. What about Old Indic?

Unlike Mycenaean Greek, Old Indic *does* have a known sister language, Avestan Iranian, which we must take into account. Avestan is the oldest of the Iranian languages that would later be spoken by Persian emperors and Scythian nomads alike, and today are spoken in Iran and Tajikistan. Avestan Iranian was the language of the *Avesta,* the holiest text of Zorastrianism. The oldest parts of the *Avesta,* the *Gathas,* probably were composed by Zoroaster (the Greek form of the name) or by Zarathustra (the original Iranian form) himself. Zarathustra was a religious reformer who lived in eastern Iran, judging from the places he named, probably between 1200 and 1000 BCE.[16] His theology was partly a reaction against the glorification of war and blood sacrifice by the poets of the *Rig Veda.* One of the oldest Gathas was "the lament of the cow," a protest against cattle stealing from the cow's point of view. But the *Avesta* and the *Rig Veda* were closely related in both language and thought. They used the same deity names (although Old Indic gods were demonized in the *Avesta*), employed the same poetic conventions, and shared specific rituals. For example, they used a cognate term for the ritual of spreading straw for the seat of the attending god before a sacrifice (Vedic *barhis,* Avestan *baresman*); and both traditions termed a pious man "one who spread the straw." In many small details they revealed their kinship in a shared Indo-Iranian past. The two languages, Avestan Iranian and Old Indic, developed from a shared parent language, Indo-Iranian, which is not documented.

The Mitanni inscriptions establish that Old Indic had appeared as a distinct language by 1500 BCE. Common Indo-Iranian must be earlier. It probably dates back *at least* to 1700 BCE. Proto-Indo-Iranian—a dialect that had some of the innovations of Indo-Iranian but not yet all of them— has to be placed earlier still, at or before 2000 BCE. Pre-Indo-Iranian was an eastern dialect of Proto-Indo-European, and must then have existed at the *latest* around 2500–2300 BCE. As with Greek, the period from 2500 to 2300 BCE, give or take a few centuries, is the *minimal* age for the separation of Pre-Indo-Iranian from Proto-Indo-European.

So the terminal date for Proto-Indo-European—the date after which our reconstructed form of the language becomes an anachronism—can be set around 2500 BCE, more or less, from the perspective of Greek and

Old Indic. It might be extended a century or two later, but, as far as *these two languages* are concerned, a terminal date *much* later than 2500 BCE—say, as late as 2000 BCE—is impossible. And, of course, Anatolian must have separated long before 2500 BCE. By about 2500 BCE Proto-Indo-European had changed and fragmented into a variety of late dialects and daughter languages—including at least the Anatolian group, Pre-Greek and Pre-Indo-Iranian. Can other daughters be dated to the same period? How many other daughters existed by 2500 BCE?

More Help from the Other Daughters: Who's the Oldest of Them All?

In fact, some other daughters not only *can* be placed this early—they *must* be. Again, to understand why, we have to understand where Greek and Old Indic stand within the known branches of the Indo-European language family. Neither Greek nor Indo-Iranian can be placed among the very oldest Indo-European daughter branches. They are the oldest daughters to survive in inscriptions (along with Anatolian), but that is an accident of history (table 3.1). From the perspective of historical linguistics, Old Indic and Greek must be classified as *late* Indo-European daughters. Why?

Linguists distinguish older daughter branches from younger ones on the basis of shared innovations and archaisms. Older branches seem to have separated earlier because they lack innovations characteristic of the later branches, and they retain archaic features. Anatolian is a good example; it retains some phonetic traits that definitely are archaic (laryngeals) and lacks other features that probably represent innovations. Indo-Iranian, on the other hand, exhibits three innovations that identify it as a later branch.

Indo-Iranian shared one innovation with a group of languages that linguists labeled the *satəm* group: Indo-Iranian, Slavic, Baltic, Albanian, Armenian, and perhaps Phrygian. Among the *satəm* languages, Proto-Indo-European *k-* before a front vowel (like *k'mtom* 'hundred') was regularly shifted to *š-* or *s-* (like Avestan Iranian *satəm*). This same group of languages exhibited a second shared innovation: Proto-Indo-European *kʷ-* (called a labiovelar, pronounced like the first sound in *queen*) changed to *k-*. The third innovation was shared between just a subgroup within the *satəm* languages: Indo-Iranian, Baltic, and Slavic. It is called the *ruki*-rule: the original sound [*-s] in Proto-Indo-European was shifted to [*-sh] after the consonants *r*, *u*, *k*, and *i*. Language branches that do not share these innovations are assumed to have split away and lost regular contact with the *satəm* and *ruki* groups before they occurred.

TABLE 3.1
The First Appearance in Written Records of the Twelve Branches of Indo-European

Language Branch	Oldest Documents or Inscriptions	Diversity at That Date	Latest Date for Proto-Language for the Branch	Grouped with
Anatolian	1920 BCE	Three closely related languages	2800–2300 BCE	No close sisters
Indo-Iranian	1450 BCE	Two very closely related languages	2000–1500 BCE	Greek, Balto–Slavic
Greek	1450 BCE	One dialect recorded, but others probably existed	2000–1500 BCE	Indo–Iranian, Armenian
Phrygian	750 BCE	Poorly documented	1200–800 BCE	Greek? Italo–Celtic?
Italic	600–400 BCE	Four languages, grouped into two quite distinct sub-branches	1600–1100 BCE	Celtic
Celtic	600–300 BCE	Three broad groups with different SVO syntax	1350–850 BCE	Italic
Germanic	0–200 CE	Low diversity; probably the innovations that defined Germanic were recent and still spreading through the Pre–Germanic speech community	500–0 BCE	Baltic/Slavic

TABLE 3.1 (continued)

Language Branch	Oldest Documents or Inscriptions	Diversity at That Date	Latest Date for Proto-Language for the Branch	Grouped with
Armenian	400 CE	Only one dialect documented, but Armina was a Persian province ca. 500 BCE so other dialects probably existed 400 CE	500 BCE–0 CE?	Greek, Phrygian?
Tocharian	500 CE	Two (perhaps three) quite distinct languages	500 BCE–0 CE	No close sisters
Slavic	865 CE	Only one dialect documented (OCS), but the West, South, and East Slavic branches must have existed already	0–500 CE	Baltic
Baltic	1400 CE	Three languages	0–500 CE	Slavic
Albanian	1480 CE	Two dialects	0–500 CE	Dacian–Thracian? No close sisters

The Celtic and Italic branches do not display the *satəm* innovations or the *ruki* rule; both exhibit a number of archaic features and also share a few innovations. Celtic languages, today limited to the British Isles and nearby coastal France, were spoken over much of central and western Europe, from Austria to Spain, around 600–300 BCE, when the earliest records of Celtic appeared. Italic languages were spoken in the Italian peninsula at about 600–500 BCE, but today, of course, Latin has many daughters—the Romance languages. In most comparative studies of the Indo-European languages, Italic and Celtic would be placed among the earliest branches to separate from the main trunk. The people who spoke Pre-Celtic and Pre-Italic lost contact with the eastern and northern groups of Indo-European speakers before the *satəm* and *ruki* innovations occurred. We cannot yet discuss where the boundaries of these linguistic regions were, but we can say that Pre-Italic and Pre-Celtic departed to form a western regional–chronological block, whereas the ancestors of Indo-Iranian, Baltic, Slavic, and Armenian stayed behind and shared a set of later innovations. Tocharian, the easternmost Indo-European language, spoken in the Silk Road caravan cities of the Tarim Basin in northwestern China, also lacked the *satəm* and *ruki* innovations, so it seems to have departed equally early to form an eastern branch.

Greek shared a series of linguistic features uniquely with the Indo-Iranian languages, but it did not adopt the *satəm* innovation or the *ruki* rule.[17] Pre-Greek and Pre-Indo-Iranian must have developed in neighboring regions, but the speakers of Pre-Greek departed before the *satəm* or the *ruki* innovations appeared. The shared features included morphological innovations, conventions in heroic poetry, and vocabulary. In morphology, Greek and Indo-Iranian shared two important innovations: the augment, a prefix *e-* before past tenses (although, because it is not well attested in the earliest forms of Greek and Indo-Iranian, the augment *might* have developed independently in each branch much later); and a mediopassive verb form with a suffixed *-i*. In weapon vocabulary they shared common terms for *bow* (**taksos*), *arrow* (**eis-*), *bowstring* (**jya-*), and *club* (**uágros*), or *cudgel*, the weapon specifically associated with Indra and his Greek counterpart Herakles. In ritual they shared a unique term for a specific ritual, the *hecatomb*, or sacrifice of a hundred cows; and they referred to the gods with the same shared epithet, *those who give riches*. They retained shared cognate names for at least three deities: (1) *Erinys/Saraṇ, yū*, a horse-goddess in both traditions, born of a primeval creator-god and the mother of a winged horse in Greek or of the Divine Twins in

Indo-Iranian, who are often represented as horses; (2) *Kérberos/Śárvara*, the multiheaded dog that guarded the entrance to the Otherworld; and (3) *Pan/Pūṣán*, a pastoral god that guarded the flocks, symbolically associated in both traditions with the goat. In both traditions, goat entrails were the specific funeral offering made to the hell-hound *Kérberos/Śárvara* during a funeral ceremony. In poetry, ancient Greek, like Indo-Iranian, had two kinds of verse: one with a twelve-syllable line (the Sapphic/Alcaic line) and another with an eight-syllable line. No other Indo-European poetic tradition shared both these forms. They also shared a specific poetic formula, meaning "fame everlasting," applied to heroes, found in this exact form only in the *Rig Veda* and Homer. Both Greek and Indo-Iranian used a specific verb tense, the imperfect, in poetic narratives about past events.[18]

It is unlikely that such a large bundle of common innovations, vocabulary, and poetic forms arose independently in two branches. Therefore, Pre-Greek and Pre-Indo-Iranian almost certainly were neighboring late Indo-European dialects, spoken near enough to each other so that words related to warfare and ritual, names of gods and goddesses, and poetic forms were shared. Greek did not adopt the *ruki* rule or the *satəm* shift, so we can define two strata here: the older links Pre-Greek and Pre-Indo-Iranian, and the later separates Proto-Greek from Proto-Indo-Iranian.

The Birth Order of the Daughters and the Death of the Mother

The *ruki* rule, the *centum/satəm* split, and sixty-three possible variations on seventeen other morphological and phonological traits were analyzed mathematically to generate thousands of possible branching diagrams by Don Ringe, Wendy Tarnow, and colleagues at the University of Pennsylvania.[19] The cladistic method they used was borrowed from evolutionary biology but was adapted to compare linguistic innovations rather than genetic ones. A program selected the trees that emerged most often from among *all possible* evolutionary trees. The evolutionary trees identified by this method agreed well with branching diagrams proposed on more traditional grounds. The oldest branch to split away was, without any doubt, Pre-Anatolian (figure 3.2). Pre-Tocharian probably separated next, although it also showed some later traits. The next branching event separated Pre-Celtic and Pre-Italic from the still evolving core. Germanic has some archaic traits that suggest an initial separation at about the same time as Pre-Celtic and Pre-Italic, but then later it was strongly affected by

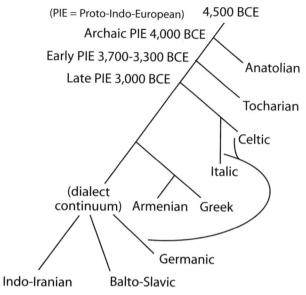

Figure 3.2 The best branching diagram according to the Ringe–Warnow–Taylor (2002) cladistic method, with the minimal separation dates suggested in this chapter. Germanic shows a mixture of archaic and derived traits that make its place uncertain; it could have branched off at about the same time as the root of Italic and Celtic, although here it is shown branching later because it also shared many traits with Pre-Baltic and Pre-Slavic.

borrowing from Celtic, Baltic, and Slavic, so the precise time it split away is uncertain. Pre-Greek separated after Italic and Celtic, followed by Indo-Iranian. The innovations of Indo-Iranian were shared (perhaps later) with several language groups in southeastern Europe (Pre-Armenian, Pre-Albanian, partly in Pre-Phrygian) and in the forests of northeastern Europe (Pre-Baltic and Pre-Slavic). Common Indo-Iranian, we must remember, is dated *at the latest* to about 1700 BCE. The Ringe-Tarnow branching diagram puts the separations of Anatolian, Tocharian, Italic, Celtic, German, and Greek before this. Anatolian probably had split away before 3500 BCE, Italic and Celtic before 2500 BCE, Greek after 2500 BCE, and Proto-Indo-Iranian by 2000 BCE. Those are not meant to be exact dates, but they are in the right sequence, are linked to dated inscriptions in three places (Greek, Anatolian, and Old Indic), and make sense.

By 2500 BCE the language that has been reconstructed as Proto-Indo-European had evolved into something else or, more accurately, into a variety of things,—late dialects such as Pre-Greek and Pre-Indo-Iranian that continued to diverge in different ways in different places. The Indo-European languages that evolved after 2500 BCE did not develop from Proto-Indo-European but from a set of intermediate Indo-European languages that preserved and passed along aspects of the mother tongue. By 2500 BCE Proto-Indo-European was a dead language.

CHAPTER FOUR

Language and Time 2
Wool, Wheels, and Proto-Indo-European

If Proto-Indo-European was dead as a spoken language by 2500 BCE, when was it born? Is there a date *after which* Proto-Indo-European must have been spoken? This question can be answered with surprising precision. Two sets of vocabulary terms identify the date after which Proto-Indo-European must have been spoken: words related to woven wool textiles, and to wheels and wagons. Neither woven wool textiles nor wheeled vehicles existed before about 4000 BCE. It is possible that neither existed before about 3500 BCE. Yet Proto-Indo-European speakers spoke regularly about wheeled vehicles and some sort of wool textile. This vocabulary suggests that Proto-Indo-European was spoken after 4000–3500 BCE. As the Proto-Indo-European vocabulary for wheeled vehicles has already been described in chapter 2, let us begin here with the Proto-Indo-European terms for wool.

THE WOOL VOCABULARY

Woven woolen textiles are made from long wool fibers of a type that did not grow on wild sheep. Sheep with long wooly coats are genetic mutants bred just for that trait. If Proto-Indo-European contained words referring unequivocally to woven woolen textiles, then those words had to have entered Proto-Indo-European after the date when wool sheep were developed. But if we are to use the wool vocabulary as a dating tool, we need to know both the exact meaning of the reconstructed roots and the date when wool sheep first appeared. Both issues are problematic.

Proto-Indo-European contained roots that meant "sheep," "ewe," "ram," and "lamb"—a developed vocabulary that undoubtedly indicates familiarity with domesticated sheep. It also had a term that in most daughter cognates meant "wool". The root *$HwlHn$- is based on cognates in almost all branches from Welsh to Indic and including Hittite, so it goes back to the archaic Proto-Indo-European era before the Anatolian branch split away. The stem is unusually long, however, suggesting to Bill Darden of the University of Chicago that it was either borrowed or derived by the addition of the -*n*- suffix from a shorter, older root. He suggested that the shorter root, and the *earliest* form, was **Hwel-* or **Hwol-* (transcribed as **Hw(e/o)l*). Its cognates in Baltic, Slavic, Greek, Germanic, and Armenian meant "felt," "roll," "beat," and "press." "Felt" seems to be the meaning that unites them, since the verbs describe operations in the manufacture of felt. Felt is made by beating or pressing wool fibers until they are pounded into a loose mat. The mat is then rolled up and pressed tightly, unrolled and wetted, then rolled and pressed again, all this repeated until the mat is tight. Wool fibers are curly, and they interlock during this pressing process. The resulting felt textile is quite warm. The winter tents of Eurasian nomads and the winter boots of Russian farmers (made to fit over regular shoes) were traditionally made from felt. If Darden is right, the most ancient Pre-Proto-Indo-European wool root, **Hw(e/o)l-*), was connected with felt. The derivative stem **HwlHn-*, the root retained in both Anatolian and classic Proto-Indo-European, meant "wool" or something made of wool, but we cannot be certain that it referred to a woven wool textile. It could have referred to the short, natural wool that grew on wild sheep or to some kind of felt textile made of short wool.[1]

Sheep (*Ovis orientalis*) were domesticated in the period from about 8000 to 7500 BCE in eastern Anatolia and western Iran as a captive source of meat, which is all they were used for during the first four thousand years of sheepherding. They were covered not with wool but with long, coarse hair called *kemp*. Wool grew on these sheep as an insulating undercoat of very short curly fibers that, in the words of textile specialist Elizabeth Barber, were "structurally unspinnable." This "wild" short wool was molted at the end of the winter. In fact, the annual shedding of short wild wool might have created the first crude (and smelly) felts, when sheep slept on their own damp sheddings. The next step would have been to intentionally pluck the wool when it loosened, just before it was shed. But *woven* wool textiles required wool *thread*.

Wool thread could only be made from unnaturally long wool fibers, as the fibers had to be long enough to cling to each other when pulled apart. A spinner of wool would pull a clump of fibers from a mass of long-fiber wool and twist them into a thread by handfeeding the strand onto a twirling weighted stick, or hand spindle (the spinning wheel was a much later invention). The spindle was suspended in the air and kept twirling with a motion of the wrist. The spindle weights are called *spindle whorls*, and they are just about the only evidence that survives of ancient thread making, although it is difficult to distinguish spindle whorls used for making woolen thread from those used for making flaxen thread, apparently the oldest kind of thread made by humans. Linen made from flax was the oldest woven textile. Woolen thread was invented only after spinners of flax and other plant fibers began to obtain the longer animal fibers that grew on mutant wool sheep. When did this genetic alteration happen? The conventional wisdom is that wool sheep appeared about 4000–3500 BCE.[2]

In southern Mesopotamia and western Iran, where the first city-based civilizations appeared, woven wool textiles were an important part of the earliest urban economies. Wool absorbed dye much better than linen did, so woolen textiles were much more colorful, and the color could be woven in with differently colored threads rather than stamped on the textile surface (apparently the oldest kind of textile decoration). But almost all the evidence for wool production appears in the Late Uruk period or later, after about 3350 BCE.[3] Because wool itself is rarely preserved, the evidence comes from animal bones. When sheep are raised for their wool, the butchering pattern should show three features: (1) sheep or goats (which differ only in a few bones) or both should make up the majority of the herded animals; (2) sheep, the wool producers, should greatly outnumber goats, the best milk producers; and (3) the sheep should have been butchered at an advanced age, after years of wool production. Susan Pollock's review of the faunal data from eight Uruk-period sites in southern Mesopotamia, northern Mesopotamia, and western Iran showed that the shift to a wool-sheep butchering pattern occurred in this heartland of cities no earlier than the Late Uruk period, after 3350 BCE (figure 4.1). Early and Middle Uruk sheep (4000–3350 BCE) did not show a wool-butchering pattern. This Mesopotamian/western Iranian date for wool sheep was confirmed at Arslantepe on the upper Euphrates in eastern Anatolia. Here, herds were dominated by cattle and goats before 3350 BCE (phase VII), but in the next phase (VIa) Late Uruk pottery appeared, and sheep suddenly rose to first place, with more than half of them living to maturity.[4]

The animal-bone evidence from the Near East suggests that wool sheep appeared after about 3400 BCE. Because sheep were not native to Europe, domesticated Near Eastern sheep were imported to Europe by the first farmers who migrated to Europe from Anatolia about 6500 BCE. But the mutation for longer wool might have appeared as an adaptation to cold winters after domesticated sheep were introduced to northern climates, so

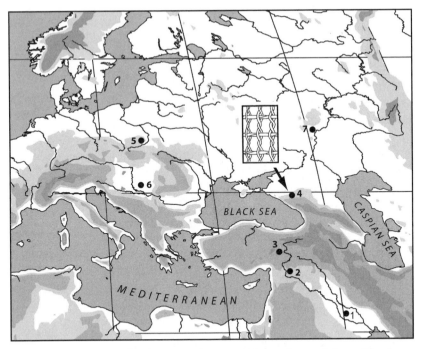

Figure 4.1 Locations of early sites with some evidence for wool sheep. The drawing is from a microscopic image of the oldest known woven wool textile published by N. Shishlina: (1) Uruk; (2) Hacinebi; (3) Arslantepe; (4) Novosvobodnaya; (5) Bronocice; (6) Kétegyháza; (7) Khvalynsk. After Shishlina 1999.

it would not be surprising if the earliest long-wool sheep were bred in Europe. At Khvalynsk, a cemetery dated about 4600–4200 BCE on the middle Volga in Russia, sheep were the principal animal sacrificed in the graves, and most of them were mature, as if being kept alive for wool or milk. But animals chosen for sacrifice might have been kept alive for a ritual reason. At Svobodnoe, a farming settlement in the North Caucasus piedmont in what is now southern Russia, dated between about 4300 and 3700 BCE, sheep were the dominant domesticated animal, and sheep outnumbered goats by 5 to 1. This is a classic wool-sheep harvesting pattern. But at other settlements of the same age in the North Caucasus this pattern is not repeated. A new large breed of sheep appeared in eastern Hungary at Kétegyháza in the Cernavoda III–Boleraz period, dated 3600–3200 BCE, which Sandor Bökönyi suggested was introduced from Anatolia and Mesopotamia; at Bronocice in southern Poland, in levels dated to the same period, sheep greatly outnumbered goats by 20 to 1. But beyond these tantalizing cases there was no broad or widespread shift to sheep keeping or to a wool-butchering pattern in Europe until after about 3300–3100 BCE, about the same time it occurred in the Near East.[5]

No actual woven woolen textiles are firmly dated before about 3000 BCE, but they were very widespread by 2800 BCE. A woven woolen textile fragment that might predate 3000 BCE was found in a grave in the North Caucasus Mountains, probably a grave of the Novosvobod-naya culture (although there is some uncertainty about the provenience). The wool fibers were dyed dark brown and beige, and then a red dye was painted on the finished fabric. The Novosvobodnaya culture is dated between 3400 and 3100 BCE, but this fabric has not been directly dated. At Shar-i Sokhta, a Bronze Age semi-urban trading center in east-central Iran, woven woolens were the only kinds of textiles recovered in levels dated 2800–2500 BCE. A woven wool fragment was found at Clairvaux-les-lacs Station III in France, dated 2900 BCE, so wool sheep and woven wool textiles were known from France to central Iran by 2900–2500 BCE.[6]

The preponderance of the evidence suggests that woven wool textiles appeared in Europe, as in the Near East, after about 3300 BCE, although wool sheep may have appeared earlier than this, about 4000 BCE, in the North Caucasus Mountains and perhaps even in the steppes. But if the root *HwlHn-* referred to the short undercoat wool of "natural" sheep, it could have existed before 4000 BCE. This uncertainty in meaning weakens the reliability of the wool vocabulary for dating Proto-Indo-European. The wheeled vehicle vocabulary is different. It refers to very definite objects (wheels, axles), and the earliest wheeled vehicles are very well dated. Unlike wool textiles, wagons required an elaborate set of metal tools (chisels, axes) that preserve well, the images of wagons are easier to categorize, and the wagons themselves preserve more easily than textiles.

The Wheel Vocabulary

Proto-Indo-European contained a set of words referring to wheeled vehicles—wagons or carts or both. We can say with great confidence that wheeled vehicles were not invented until after 4000 BCE; the surviving evidence suggests a date closer to 3500 BCE. Before 4000 BCE there were no wheels or wagons to talk about.

Proto-Indo-European contained at least five terms related to wheels and wagons, as noted in chapter 2: two words for *wheel* (perhaps for different kinds of wheels), one for *axle*, one for *thill* (the pole to which the animals

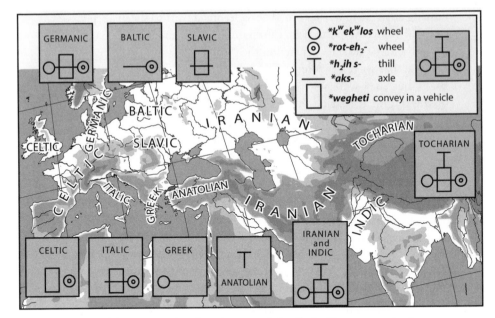

Figure 4.2 The geographic distribution of the Indo-European wheel-wagon vocabulary.

were yoked), and a verb meaning "to go or convey in a vehicle." Cognates for these terms occur in all the major branches of Indo-European, from Celtic in the west to Vedic Sanskrit and Tocharian in the east, and from Baltic in the north to Greek in the south (figure 4.2). Most of the terms have a kind of vowel structure called an o-stem that identifies a late stage in the development of Proto-Indo-European; *axle* was an older n-stem derived from a word that meant "shoulder." The o-stems are important, since they appeared only during the later end of the Proto-Indo-European period. Almost all the terms are derived from Proto-Indo-European roots, so the vocabulary for wagons and wheels was not imported from the outside but was created within the Proto-Indo-European speech community.[7]

The only branch that might *not* contain a convincing wheeled-vehicle vocabulary is Anatolian, as Bill Darden observed. Two possible Proto-Indo-European wheeled-vehicle roots are preserved in Anatolian. One (*hurki-* 'wheel') is thought to be descended from a Proto-Indo-European root, because the same root might have yielded Tocharian A *wärkänt* and Tocharian B *yerkwanto*, both meaning "wheel." Tocharian is an extinct Indo-European branch consisting of two (perhaps three) known languages, called A and B (and perhaps C), recorded in documents written in about 500–700 CE by Buddhist monks in the desert caravan cities of the Tarim Basin in northwestern China. But Tocharian specialist Don Ringe sees serious difficulties in deriving either Tocharian term from the same root that yielded Anatolian *ḫurki-*, suggesting that the Tocharian and Anatolian terms were unrelated and therefore do not require a Proto-Indo-European root.[8] The other Anatolian vehicle term (*ḫišša-* 'thill' or 'harness-pole') has a good Indo-European source, **ei-/*oi-* or perhaps **h₂ih₃s-*, but its original meaning might have referred to plow shafts rather than wagon shafts. So we cannot be certain that archaic Proto-Indo-European, as partially preserved in Anatolian, had a wheeled-vehicle vocabulary. But the rest of Proto-Indo-European did.

When Was the Wheel Invented?

How do we know that wheeled vehicles did not exist before 4000 BCE? First, a wheeled vehicle required not just wheels but also an axle to hold the vehicle. The wheel, axle, and vehicle together made a complicated combination of load-bearing moving parts. The earliest wagons were planed and chiseled entirely from wood, and the moving parts had to fit precisely. In a wagon with a fixed axle and revolving wheels (apparently the earliest type), the axle arms (the ends of the axle that passed through the center of the wheel) had to fit snugly, but not too snugly, in the hole through the nave, or hub. If the fit was too loose, the wheels would wobble as they turned. If it was too tight, there would be excessive drag on the revolving wheel.

Then there was the problem of the draft—the total weight, with drag, pulled by the animal team. Whereas a sledge could be pulled using traces, or flexible straps and ropes, a wagon or cart had to have a rigid draft pole, or thill, and a rigid yoke. The weight of these elements increased the overall draft. One way to reduce the draft was to reduce the diameter of the axle arms to fit a smaller hole in the wheel. A large-diameter axle was strong but created more friction between the axle arms and the revolving wheel. A smaller-diameter axle arm would cause less drag but would break easily unless the wagon was very narrow. The first wagon-wrights had to calculate the relationship between drag, axle diameter/strength, axle length/rigidity, and the width of the wagon bed. In a work vehicle meant to carry heavy loads, a short axle with small-diameter axle arms and a narrow wagon bed made good engineering sense, and, in fact, this is what the earliest wagons looked like, with a bed only about 1 m wide. Another way to reduce the draft was to reduce the number of wheels from four to two—to make a *wagon* into a *cart*. The draft of a modern two-wheeled cart is 40% less than a four-wheeled wagon *of the same weight*, and we can assume that an advantage of approximately the same magnitude applied to ancient carts. Carts were lighter and easier to pull, and on rough ground were less likely to get stuck. Large loads probably still needed wagons, but carts would have been useful for smaller loads.[9]

Archaeological and inscriptional evidence for wheeled vehicles is widespread after about 3400 BCE. One uncertain piece of evidence, a track preserved under a barrow grave at Flintbek in northern Germany, might have been made by wheels, and might be as old as 3600 BCE. But the real explosion of evidence begins about 3400 BCE. Wheeled vehicles appeared in four different media dated between about 3400 and 3000 BCE—a written sign for wagons, two-dimensional images of wagons and carts, three-dimensional models of wagons, and preserved wooden wheels and wagon parts themselves. These four independent kinds of evidence appeared across the ancient world between 3400 and 3000 BCE, about the same time as wool sheep, and clearly indicate when wheeled vehicles became widespread. The next four sections discuss the four kinds of evidence.[10]

Mesopotamian Wagons: The Oldest Written Evidence

Clay tablets with "wagon" signs impressed on them were found in the Eanna temple precinct in Uruk, one of the first cities created by humans. About thirty-nine hundred tablets were recovered from level IVa, the end of Late Uruk. In these texts, among the oldest documents in the world, a pictograph (figure 4.3.f) shows a four-wheeled wagon with some kind of canopy or superstructure. The "wagon" sign occurred just three times in thirty-nine hundred texts, whereas the sign for "sledge"—a similar kind of transport, but dragged on runners not rolled on wheels—occurred thirty-eight times. Wagons were not yet common.

The Eanna precinct tablets were inside Temple C when it burned down. Charcoal from the Temple C roof timbers yielded four radiocarbon dates averaging about 3500–3370 BCE. A radiocarbon date tells us when the dated material, in this case wood, died, not when it was burned. The wood in the center of any tree is actually dead (something few people realize); only the outer ring of bark and the sappy wood just beneath it are alive. If the timbers in Temple C were made from the center of a large tree, the wood might have died a century or two before the building was burned down, so the actual age of the Temple C tablets is later than the radiocarbon date, perhaps 3300–3100 BCE. Sledges still were far more common than wagons in the city of Uruk at that date. Ox-drawn canopied sledges might have preceded canopied wagons as a form of transport (in parades or processions? harvest rituals?) used by city officials.

A circular clay object that *might* be a model wheel, perhaps from a small ceramic model of a wagon, was found at the site of Arslantepe in eastern Turkey, in the ruins of a temple-palace from level VIa at the site, also dated 3400–3100 BCE (figure 4.3.c). Arslantepe was one of a string of native strongholds along the upper Euphrates River in eastern Anatolia that entered into close relations with faraway Uruk during the Late Uruk period. Although the kind of activities that lay behind this "Uruk expansion" northward up the Euphrates valley is not known (see chapter 12), the possible clay wheel model at Arslantepe *could* indicate that wagons were being used in eastern Anatolia during the period of Late Uruk influence.

Wagons and Carts from the Rhine to the Volga: The Oldest Pictorial Evidence

A two-dimensional image that seems to portray a four-wheeled wagon, harness pole, and yoke was incised on the surface of a decorated clay mug of the Trichterbecker (TRB) culture found at the settlement of Bronocice in southern Poland, dated about 3500–3350 BCE (figure 4.3.b). The TRB culture is recognized by its distinctive pottery shapes and tombs, which are found over a broad region in modern Poland, eastern Germany, and southern Denmark. Most TRB people were simple farmers who lived in small agricultural villages, but the Bronocice settlement was unusually large, a TRB town covering fifty-two hectares. The cup or mug with the wagon image incised on its surface was found in a rubbish pit containing animal bones, the broken sherds of five clay vessels, and flint tools. Only this cup had a wagon image. The design is unusual for TRB pottery, not an accidental combination of normal decorative motifs. The cup's date is the subject of some disagreement. A cattle bone found in the same pit yielded an average age of about 3500 BCE, whereas six of the seven other radiocarbon dates for the settlement around the pit average 150 years later, about 3350 BCE. The excavators accept an age range spanning these results, about 3500–3350 BCE. The Bronocice wagon image is the oldest well-dated image of a wheeled vehicle in the world.

Two other images could be about the same age, although they probably are somewhat later. An image of two large-horned cattle pulling what seems to be a two-wheeled cart was scratched on the wall of a Wartberg culture stone tomb at Lohne-Züschen I, Hesse, central Germany (figure 4.3.e). The

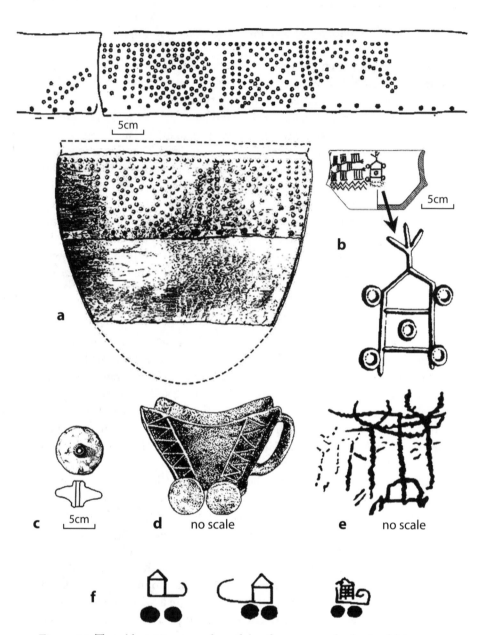

Figure 4.3 The oldest images and models of wagons and wheels: (a) bronze kettle from Evdik kurgan, lower Volga, Russia, with a design that could represent, from the left, a yoke, cart, wheel, X-braced floor, and animal head; (b) image of a four-wheeled wagon on a ceramic vessel from Bronocice, southern Poland; (c) ceramic wheel (from a clay model?) at Arslantepe, eastern Anatolia; (d) ceramic wagon model from Baden grave 177 at Budakalász,

grave was reused over a long period of time between about 3400 and 2800 BCE, so the image could have been carved any time in that span. Far away to the east, a metal cauldron from the Evdik kurgan near the mouth of the Volga River bears a repoussé image that might show a yoke, a wheel, a cart, and a draft animal; it was found in a grave with objects of the Novosvobodnaya culture, dated between 3500 and 3100 BCE (figure 4.3.a). These images of carts and wagons are distributed from central Germany through southern Poland to the Russian steppes.

Hungarian Wagons: The Oldest Clay Models

The Baden culture is recognized by its pottery and to a certain extent by its distinctive copper tools, weapons, and ornaments. It appeared in Hungary about 3500 BCE, and the styles that define it then spread into northern Serbia, western Romania, Slovakia, Moravia, and southern Poland. Baden-style polished and channeled ceramic mugs and small pots were used across southeastern Europe about 3500–3000 BCE. Similarities between Baden ceramics and those of northwestern Anatolia in the centuries before Troy I suggest one route by which wheeled vehicles could have spread between Mesopotamia and Europe. Three-dimensional ceramic models of four-wheeled wagons (figure 4.3.d) were included in sacrificial deposits associated with two graves of the Late Baden (Pécel) culture at Budakalász (Grave 177) and Szigetszentmárton in eastern Hungary, dated about 3300–3100 BCE. Paired oxen, almost certainly a team, were found sacrificed in Grave 3 at Budakalász and in other Late Baden graves in Hungary. Paired oxen also were placed in graves of the partly contemporary Globular Amphorae culture (3200–2700 BCE) in central and southern Poland. The Baden wagon models are the oldest well-dated three-dimensional models of wheeled vehicles.

Steppe and Bog Vehicles: The Oldest Actual Wagons

Remains of about 250 wagons and carts have been discovered under earthen burial mounds, or kurgans, in the steppe grasslands of Russia and Ukraine, dated about 3000–2000 BCE (figures 4.4 and 4.5). The wheels

---

Figure 4.3 (continued) Hungary; (e) cart image with two cattle incised on stone, from a tomb at Lohne-Züschen I, Hesse, central Germany; (f) earliest written symbols for a wagon, on clay tablets from Uruk IVa, southern Iraq. After (a) Shilov and Bagautdinov 1997; (b, d, e) Milisauskas 2002; (c,f) Bakker et al. 1999.

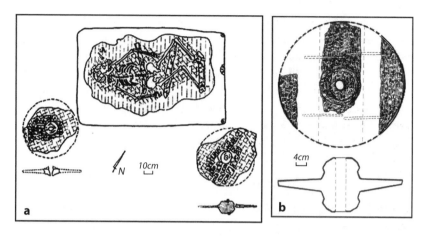

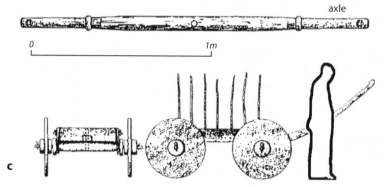

Figure 4.4 Preserved wagon parts and wheels: (a) two solid wooden wheels at the corners of grave 57, Bal'ki kurgan, Ukraine, radiocarbon dated 3330–2900 BCE; (b) Catacomb-culture tripartite wheel with dowels, probably 2600–2200 BCE; (c) preserved axle and reconstructed wagon from various preserved wheel and wagon fragments in bog deposits in northwestern Germany and Denmark dated about 3000–2800 BCE. After (a) Lyashko and Otroshchenko 1988; (b) Korpusova and Lyashko 1990; (c) Hayen 1989.

were 50–80 cm in diameter. Some were made of a single plank cut vertically from the trunk of a tree, with the grain (not like a salami). Most
steppe wheels, however, were made of two or three planks cut into circular
segments and then doweled together with mortice-and-tenon joints. In
the center were long tapered naves (hubs), about 20–30 cm wide at the
base and projecting outward about 10–20 cm on either side of the wheel.
The naves were secured to the axle arms by a lynchpin that pinned the

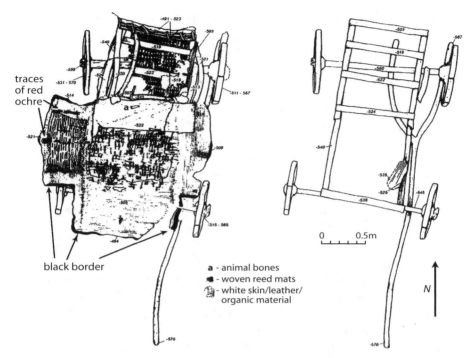

traces
of red
ochre

black border

a - animal bones
- woven reed mats
- white skin/leather/
  organic material

0   0.5m

N

Figure 4.5 The best-preserved wagon graves in the steppes are in the Kuban River region in southern Russia. This wagon was buried under Ostannii kurgan 1. Radiocarbon dated about 3300–2900 BCE, the upper part of the wagon is on the left and the lower part, on the right. After Gei 2000, figure 53.

nave to the axle, and between them they kept the wheel from wobbling. The axles had rounded axle arms for the wheel mounts and were about 2 m long. The wagons themselves were about 1 m wide and about 2 m long. The earliest radiocarbon dates on wood from steppe wagons average around 3300–2800 BCE. A wagon or cart grave at Bal'ki kurgan (grave 57) on the lower Dnieper was dated 4370±120 BP, or 3330–2880 BCE; and wood from a wagon buried in Ostanni kurgan 1 (grave 160) on the Kuban River was dated 4440 ± 40 BP, or 3320–2930 BCE. The probability distributions for both dates lie predominantly before 3000 BCE, so both vehicles probably date before 3000 BCE. But these funeral vehicles can hardly have been the very first wagons used in the steppes.

Other wooden wheels and axles have been discovered preserved in bogs or lakes in central and northern Europe. In the mountains of Switzerland and southwestern Germany wagon-wrights made the axle arms square and mortised them into a square hole in the wheel. The middle of the axle was circular and revolved under the wagon. This revolving-axle design created more drag and was less efficient than the revolving-wheel design, but it did not require carving large wooden naves and so the Alpine wheels were much easier to make. One found near Zurich in a waterlogged settlement of the Horgen culture (the Pressehaus site) was dated about 3200 BCE by associated tree-ring dates. The Pressehaus wheel tells us that separate regional European design traditions for wheel making already existed before 3200 BCE. Wooden wheels and axles also have been found in bogs in the Netherlands and Denmark, providing important evidence on the construction details of early wagons, but dated after 3000 BCE. They had fixed axles and revolving wheels, like those of the steppes and central Europe.

The Significance of the Wheel

It would be difficult to exaggerate the social and economic importance of the first wheeled transport. Before wheeled vehicles were invented, really heavy things could be moved efficiently only on water, using barges or rafts, or by organizing a large hauling group on land. Some of the heavier items that prehistoric, temperate European farmers had to haul across land all the time included harvested grain crops, hay crops, manure for fertilizer, firewood, building lumber, clay for pottery making, hides and leather, and people. In northern and western Europe, some Neolithic communities celebrated their hauling capacities by moving gigantic stones to make megalithic community tombs and stone henges; other communities hauled earth, making massive earthworks. These constructions demonstrated in a visible, permanent way the solidity and strength of the communities that made them, which depended in many ways on human hauling capacities. The importance and significance of the village community as a group transport device changed profoundly with the introduction of wagons, which passed on the burden of hauling to animals and machines, where it has remained ever since.

Although the earliest wagons were slow and clumsy, and probably required teams of specially trained oxen, they permitted single families to carry manure out to the fields and to bring firewood, supplies, crops, and people back home. This reduced the need for cooperative communal labor and made single-family farms viable. Perhaps wagons contributed to the disappearance of large nucleated villages and the dispersal of many farming populations across the European landscape after about 3500 BCE. Wagons were useful in a different way in the open grasslands of the steppes, where the economy depended more on herding than on agriculture. Here wagons made portable things that had never been portable in bulk—shelter, water, and food. Herders who had always lived in the forested river valleys and grazed their herds timidly on the edges of the steppes now could take their tents, water, and food supplies to distant pastures far from the river valleys. The wagon was a mobile home that permitted herders to follow their animals deep into the grasslands and live in the open. Again, this permitted the dispersal of communities, in this case across interior steppes that earlier had been almost useless economically. Significant wealth and power could be extracted from larger herds spread over larger pastures.

Andrew Sherratt bundled the invention of the wheel together with the invention of the plow, wool sheep, dairying, and the beginning of horse transport to explain a sweeping set of changes that occurred among European societies about 3500–3000 BCE. The Secondary Products Revolution (now often shortened to SPR), as Sherratt described it in 1981, was an economic explanation for widespread changes in settlement patterns, economy, rituals, and crafts, many of which had been ascribed by an older generation of archaeologists to Indo-European migrations. ("Secondary products" are items like wool, milk, and muscular power that can be harvested continuously from an animal without killing it, in contrast to "primary products" such as meat, blood, bone, and hides.) Much of the subject matter discussed in arguments over the SPR—the diffusion of wagons, horseback riding, and wool sheep—was also central in discussions of Indo-European expansions, but, in Sherratt's view, all of them were derived by diffusion from the civilizations of the Near East rather than from Indo-Europeans. Indo-European languages were no longer central or even necessary to the argument, to the great relief of many archaeologists. But Sherrat's proposal that all these innovations came from the Near East and entered Europe at about the same time quickly fell apart. Scratch-plows and dairying appeared in Europe long before 3500 BCE, and horse domestication was a local event in the steppes. An important fragment of the SPR survives in the conjoined diffusion of wool sheep and wagons across much of the ancient Near East and Europe between 3500 and 3000 BCE, but we do not know where either of these innovations started.[11]

The clearest proof of the wheel's impact was the speed with which wagon technology spread (figure 4.6), so rapidly, in fact, that we cannot even say where the wheel-and-axle principle was invented. Most specialists assume that the earliest wagons were produced in Mesopotamia, which was urban and therefore more sophisticated than the tribal societies of Europe; indeed, Mesopotamia had sledges that served as prototypes. But we

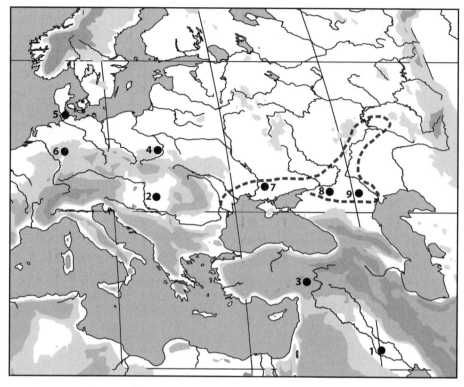

Figure 4.6 Sites with early evidence for wheels or wagons: (1) Uruk; (2) Buda-kalasz; (3) Arslantepe; (4) Bronicice; (5) Flintbek; (6) Lohne-Zuschen I; (7) Bal'ki kurgan; (8) Ostannii kurgan; (9) Evdik kurgan. Dashed line indicates the distribution of about 250 wagon graves in the Pontic-Caspian steppes.

really don't know. Another prototype existed in Europe in the form of Mesolithic and Neolithic bent-wood sleds, doweled together with fine mortice-and-tenon joints; in much of eastern Europe, in fact, right up to the twentieth century, it made sense to park your wagon or carriage in the barn for the winter and resort to sleds, far more effective than wheels in snow and ice. Bent-wood sleds were at least as useful in prehistoric Europe as in Mesopotamia, and they began to appear in northern Europe as early as the Mesolithic; thus the skills needed to make wheels and axles existed in both Europe and the Near East.[12]

Regardless of where the wheel-and-axle principle was invented, the technology spread rapidly over much of Europe and the Near East between 3400 and 3000 BCE. Proto-Indo-European speakers talked about wagons and wheels using their own words, created from Indo-European roots. Most of these words were o-stems, a relatively late development in Proto-Indo-European phonology. The wagon vocabulary shows that late Proto-Indo-European was spoken certainly after 4000 BCE, and probably after 3500 BCE. Anatolian is the only major early Indo-European branch that has a doubtful wheeled-vehicle vocabulary. As Bill Darden suggested, perhaps Pre-Anatolian split away from the archaic Proto-Indo-European dialects before wagons appeared in the Proto-Indo-European homeland. Pre-Anatolian could have been spoken before 4000 BCE. Late Proto-Indo-European, including the full wagon vocabulary, probably was spoken after 3500 BCE.

Wagons and the Anatolian Homeland Hypothesis

The wagon vocabulary is a key to resolving the debate about the place and time of the Proto-Indo-European homeland. The principal alternative to a homeland in the steppes dated 4000–3500 BCE is a homeland in Anatolia and the Aegean dated 7000–6500 BCE. Colin Renfrew proposed that Indo-Hittite (Pre-Proto-Indo-European) was spoken by the first farmers in southern and western Anatolia at sites such as Çatal Höyük dated about 7000 BCE. In his scenario, a dialect of Indo-Hittite was carried to Greece with the first farming economy by pioneer farmers from Anatolia about 6700–6500 BCE. In Greece, the language of the pioneer farmers developed into Proto-Indo-European and spread through Europe and the Mediterranean Basin with the expansion of the earliest agricultural economy. By linking the dispersal of the Indo-European languages with the diffusion of the first farming economy, Renfrew achieved an appealingly elegant solution to the problem of Indo-European origins. Since 1987 he and others have shown convincingly that the migrations of pioneer farmers were one of the principal vectors for the spread of many ancient languages around the world. The "first-farming/language-dispersal" hypothesis, therefore, was embraced by many archaeologists. But it required that the first split between parental Indo-Hittite and Proto-Indo-European began about 6700–6500 BCE, when Anatolian farmers first migrated to Greece. By 3500 BCE, the earliest date for wagons in Europe, the Indo-European language family should have been bushy, multi-branched, and three thousand years old, well past the period of sharing a common vocabulary for anything.[13]

The Anatolian—origin hypothesis raises other problems as well. The first Neolithic farmers of Anatolia are thought to have migrated there from northern Syria, which, according to Renfrew's first-farming/

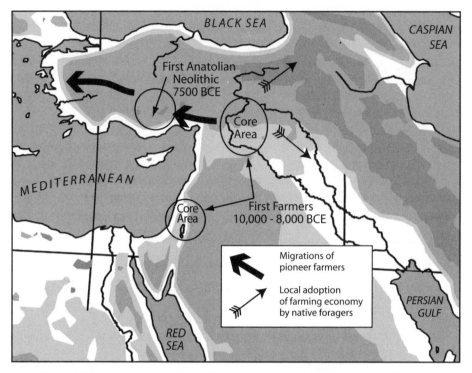

Figure 4.7 The spread of the first farming economy into Anatolia, probably by migration from the Core Area in northern Syria, about 7500 BCE. The first pioneer farmers probably spoke an Afro-Asiatic language. After Bar-Yosef 2002.

language-dispersal hypothesis, should have resulted in the spread of a north Syrian Neolithic language to Anatolia (figure 4.7). The indigenous languages of northern Syria probably belonged to the Afro-Asiatic language phylum, like Semitic and most languages of the lowland Near East. If the first Anatolian farmers spoke an Afro-Asiatic language, it was that language, not Proto-Indo-European, that should have been carried to Greece.[14] The earliest Indo-European languages documented in Anatolia—Hittite, Palaic, and Luwian—showed little diversity, and only Luwian had a significant number of speakers by 1500 BCE. All three borrowed extensively from non–Indo-European languages (Hattic, Hurrian, and perhaps others) that seem to have been older, more prestigious, and more widely spoken. The Indo-European languages of Anatolia did not have the established population base of speakers, and also lacked the kind of diversity that would be expected had they been evolving there since the Neolithic.

Phylogenetic Approaches to Dating Proto-Indo-European

Still, the Anatolian-origin hypothesis has support from new methods in phylogenetic linguistics. Cladistic methods borrowed from biology have been used for two purposes: to arrange the Indo-European languages in a chronological *order* of branching events (discussed in the previous chapter); and to estimate *dates* for the separation between any two branches, or for the root of all branches which is a much riskier proposition. Attaching time estimates to language branches using evolutionary models based on biological change is, at best, an uncertain procedure. People intentionally reshape their speech all the time but cannot intentionally reshape their genes. The way a linguistic innovation is reproduced in a speech community is quite different from the way a mutation is reproduced in a breeding population. The topography of language splits and rejoinings is much more complex and the speed of language branching far more variable. Whereas genes spread as whole units, the spread of language is always a *modular process*, and some modules (grammar and phonology) are more resistant to borrowing and spread than others (words).

Russell Gray and Quentin Atkinson attempted to work around these problems by processing a cocktail of cladistic and linguistic methods through computer programs. They suggested that pre-Anatolian detached from the rest of the Indo-European community about 6700 BCE (plus or minus twelve hundred years). Pre-Tocharian separated next (about 5900 BCE), then pre-Greek/Armenian (about 5300 BCE), and then pre–Indo-Iranian/Albanian (about 4900 BCE). Finally, a super-clade that included the ancestors of pre–Balto-Slavic and pre–Italo-Celto-Germanic separated about 4500 BCE. Archaeology shows that 6700–6500 BCE was about when the first pioneer farmers left Anatolia to colonize Greece. One could hardly ask for a closer match between archaeological and phylogentic dates.[15] But how can the presence of the wagon vocabulary in Proto-Indo-European be synchronized with a first-dispersal date of 6500 BCE?

The Slow Evolution Hypothesis

The wagon vocabulary cannot have been created *after* Proto-Indo-European was dead and the daughter languages differentiated. The wagon/wheel terms do not contain the sounds that would be expected had they been created in a later daughter language and then borrowed into the others, whereas they do contain the sounds predicted if they were inherited into the daughter branches from Proto-Indo-European. The Proto-Indo-European origin of the wagon vocabulary cannot be rejected, as it consists of at least five classic reconstructions. If they are in fact false, then the core methods of comparative linguistics—those that determine "genetic" relatedness—would be so unreliable as to be useless, and the question of Indo-European origins would be moot.

But could the wagon/wheel vocabularies have been created *independently* by the speakers of each branch from the same Proto-Indo-European roots? In the example of *$k^w ek^w los$* 'wheel', Gray suggested (in a comment on his homepage) that the semantic development from the verb *$kwel$*- 'turn' to the noun *wheel* 'the turner' was so natural that it could have been repeated independently in each branch. One difficulty here is that at least four different verbs meaning "turn" or "roll" or "revolve" are reconstructed for Proto-Indo-European, which makes the repeated independent choice of *$kwel$*- problematic.[16] More critical, the Proto-Indo-European pronunciations of *$kwel$*- and the other wagon terms would not have survived unchanged through time. They could not have been available frozen in their Proto-Indo-European phonetic forms to speakers of nine or ten branches that originated at different times across thousands of years. We cannot assume stasis in phonetic development for the wheel vocabulary when all the rest of the vocabulary changed normally with time. But what if all the other vocabulary also changed very slowly?

This is the solution Renfrew offered (figure 4.8). For the wagon/wheel vocabulary to be brought into synchronization with the first-farming/language-dispersal hypothesis, Proto-Indo-European must have been spoken for thirty-five hundred years, requiring a very long period when Proto-Indo-European changed very little. Pre-Proto-Indo-European or Indo-Hittite was spoken in Anatolia before 6500 BCE. Archaic Proto-Indo-European evolved as the language of the pioneer farmers in Greece about 6500–6000 BCE. As their descendants migrated northward and westward, and established widely scattered Neolithic communities from Bulgaria to Hungary and Ukraine, the language they carried remained a single language, Archaic Proto-Indo-European. Their descendants paused for several centuries, and then a second wave of pioneer migration pushed across the Carpathians into the North European plain between about 5500 and 5000 BCE with the Linear Pottery farmers. These farming migrations created Renfrew's Stage 1 of Proto-Indo-European, which was spoken across most of Europe between 6500 and 5000 BCE, from the Rhine to the Dnieper and from Germany to Greece. During Renfrew's Proto-Indo-European Stage 2, between 5000 and 3000 BCE, archaic Proto-Indo-European spread into the steppes

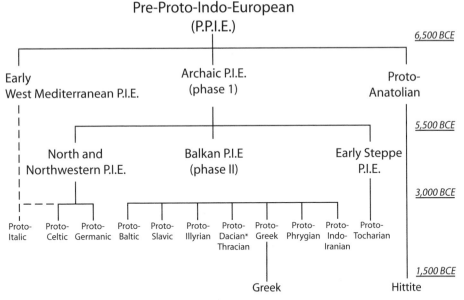

Figure 4.8 If Proto-Indo-European spread across Europe with the first farmers about 6500–5500 BCE, it must have remained almost unchanged until about 3500 BCE, when the wheeled vehicle vocabulary appeared. This diagram illustrates a division into just three dialects in three thousand years. After Renfrew 2001.

and was carried to the Volga with the adoption of herding economies. Late Proto-Indo-European dialectical features developed, including the appearance of "thematic" inflections such as o-stems, which occur in all the wagon/wheel terms. These late features were shared across the Proto-Indo-European–speaking region, which comprised two-thirds of prehistoric Europe. The wagon vocabulary appeared late in Stage 2 and was adopted from the Rhine to the Volga.[17]

It seems to me that this conception of Proto-Indo-European contains three fatal flaws. First, for Proto-Indo-European to have remained a unified dialect chain for more than thirty-five hundred years, from 6500 to 3000 BCE, would require that all its dialects changed at about the same rate and that the rate was extraordinarily slow. A *homogeneous rate of change* across most of Neolithic Europe is very unlikely, as the rate of language change is affected by a host of local factors, as Sheila Embleton showed, and these would have varied from one region to the next. And for Proto-Indo-European only to have evolved from its earlier form to its later form in thirty-five hundred years would require a pan-European condition of near stasis in the speed of language change during the Neolithic/Eneolithic, a truly unrealistic demand. In addition, Neolithic Europe evinces an almost incredible *diversity in material culture.* "This bewildering diversity," as V. Gordon Childe observed, "though embarrassing to the student and confusing on a map, is yet a significant feature in the pattern of European prehistory."[18] Long-established, undisturbed tribal languages tend to be *more* varied than tribal material cultures (see chapter 6). One would therefore expect that the linguistic diversity of Neolithic/Eneolithic Europe should have been even more bewildering than its material-culture diversity, not less so, and certainly not markedly less.

Finally, this enormous area was just too big for the survival of a single language under the conditions of tribal economics and politics, with foot travel the only means of land transport. Mallory and I discussed the likely scale of tribal language territories in Neolithic/Eneolithic Europe, and Nettles described tribal language geographies in West Africa.[19] Most tribal cultivators in West Africa spoke languages distributed over less than 10,000 km². Foragers around the world generally had much larger language territories than farmers had, and shifting farmers in poor environments had larger language territories than intensive farmers had in rich environments. Among most tribal farmers the documented size of language *families*—not languages but language families like Indo-European or Uralic—has usually been significantly less than 200,000 km². Mallory used an average of 250,000–500,000 km² for Neolithic European language families just to make room on the large end for the many uncertainties involved. Still, that resulted in twenty to forty language families for Neolithic Europe.

The actual number of language families in Europe at 3500 BCE probably was less than this, as the farming economy had been introduced into Neolithic Europe through a series of migrations that began about 6500 BCE. The dynamics of long-distance migration, particularly among pioneer farmers, *can* lead to the rapid spread of an unusually homogeneous language over an unusually large area for a few centuries (see chapter 6), but then local differentiation should have set in. In Neolithic Europe several distinct migrations flowed from different demographic recruiting pools and went to different places, where they interacted with different Mesolithic forager language groups. This should have produced incipient language differentiation among the immigrant farmers within five hundred to a thousand years, by 6000–5500 BCE. In comparison, the migrations of Bantu-speaking cattle herders across central and southern Africa occurred about two thousand years ago, and Proto-Bantu has diversified since then into more than five hundred modern Bantu languages assigned to nineteen branches, still interspersed today with enclaves belonging to non-Bantu language families. Europe in 3500 BCE, two thousand to three thousand years after the initial farming migrations, probably had at least the linguistic diversity of modern central and southern Africa— hundreds of languages that were descended from the original Neolithic farmers' speech, interspersed with pre-Neolithic language families of different types. The language of the original migrants to Greece cannot have remained a single language for three thousand years after its speakers were dispersed over many millions of square kilometers and several climate zones. Ethnographic or historic examples of such a large, stable language territory among tribal farmers simply do not exist.

That the speakers of Proto-Indo-European had wagons and a wagon vocabulary cannot be brought into agreement with a dispersal date as early as 6500 BCE. The wagon vocabulary is incompatible with the first-farming/language-dispersal hypothesis. Proto-Indo-European cannot have been spoken in Neolithic Greece and still have existed three thousand years later when wagons were invented. Proto-Indo-European therefore did not spread with the farming economy. Its first dispersal occurred much later, after 4000 BCE, in a European landscape that was already densely occupied by people who probably spoke hundreds of languages.

The Birth and Death of Proto-Indo-European

The historically known early Indo-European languages set one chronological limit on Proto-Indo-European, a *terminus ante quem*, and the reconstructed vocabulary related to wool and wheels sets another limit, a *terminus post quem*. The latest possible date for Proto-Indo-European can be set at about 2500 BCE (chapter 3). The evidence of the wool and wagon/wheel vocabularies establishes that late Proto-Indo-European was spoken after about 4000–3500 BCE, probably after 3500 BCE. If we include in our definition of Proto-Indo-European the end of the archaic Anatolian-like stage, without a securely documented wheeled-vehicle vocabulary, and the dialects spoken at the beginning of the final dispersal about 2500 BCE, the maximum window extends from about 4500 to about 2500 BCE. This two thousand-year target guides us to a well-defined archaeological era.

Within this time frame the archaeology of the Indo-European homeland is probably consistent with the following sequence, which makes sense also in terms of both traditional branching studies and cladistics. Archaic Proto-Indo-European (partly preserved only in Anatolian) probably was spoken before 4000 BCE; early Proto-Indo-European (partly preserved in Tocharian) was spoken between 4000 and 3500 BCE; and late Proto-Indo-European (the source of Italic and Celtic with the wagon/wheel vocabulary) was spoken about 3500–3000 BCE. Pre-Germanic split away from the western edge of late Proto-Indo-European dialects about 3300 BCE, and Pre-Greek split away about 2500 BCE, probably from a different set of dialects. Pre-Baltic split away from Pre-Slavic and other northwestern dialects about 2500 BCE. Pre-Indo-Iranian developed from a northeastern set of dialects between 2500 and 2200 BCE.

Now that the target is fixed in time, we can solve the old and bitter debate about *where* Proto-Indo-European was spoken.

CHAPTER FIVE

Language and Place
The Location of the Proto-Indo-European Homeland

The Indo-European homeland is like the Lost Dutchman's Mine, a legend of the American West, discovered almost everywhere but confirmed nowhere. Anyone who claims to know its *real* location is thought to be just a little odd—or worse. Indo-European homelands have been identified in India, Pakistan, the Himalayas, the Altai Mountains, Kazakhstan, Russia, Ukraine, the Balkans, Turkey, Armenia, the North Caucasus, Syria/Lebanon, Germany, Scandinavia, the North Pole, and (of course) Atlantis. Some homelands seem to have been advanced just to provide a historical precedent for nationalist or racist claims to privileges and territory. Others are enthusiastically zany. The debate, alternately dryly academic, comically absurd, and brutally political, has continued for almost two hundred years.[1]

This chapter lays out the linguistic evidence for the location of the Proto-Indo-European homeland. The evidence will take us down a well-worn path to a familiar destination: the grasslands north of the Black and Caspian Seas in what is today Ukraine and southern Russia, also known as the Pontic-Caspian steppes (figure 5.1). Certain scholars, notably Marija Gimbutas and Jim Mallory, have argued persuasively for this homeland for the last thirty years, each using criteria that differ in some significant details but reaching the same end point for many of the same reasons.[2] Recent discoveries have strengthened the Pontic-Caspian hypothesis so significantly, in my opinion, that we can reasonably go forward on the assumption that this was the homeland.

PROBLEMS WITH THE CONCEPT OF "THE HOMELAND"

At the start I should acknowledge some fundamental problems. Many of my colleagues believe that it is impossible to identify *any* homeland for Proto-Indo-European, and the following are their three most serious concerns.

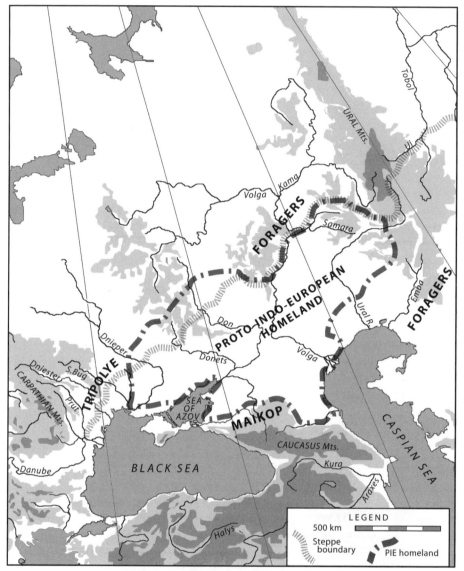

Figure 5.1 The Proto-Indo-European homeland between about 3500–3000 BCE.

Problem #1. Reconstructed Proto-Indo-European is merely a linguistic hypothesis, and hypotheses do not have homelands.

This criticism concerns the "reality" of reconstructed Proto-Indo-European, a subject on which linguists disagree. We should not imagine, some remind us, that reconstructed Proto-Indo-European was ever actually spoken anywhere. R.M.W. Dixon commented that if we cannot have "abso-lute certainty" about the grammatical type of a reconstructed language, it throws doubt over "every detail of the putative reconstruction."[3] But this is an extreme demand. The only field in which we can find absolute certainty is religion. In all other activities we must be content with the best (meaning both the simplest and the most data-inclusive) interpretation we can ad-vance, given the data as they now stand. After we accept that this is true in *all* secular inquiries, the question of whether Proto-Indo-European can be thought of as "real" boils down to three sharper criticisms:

a. Reconstructed Proto-Indo-European is *fragmentary* (most of the language it represents never will be known).
b. The part that is reconstructed is *homogenized,* stripped of many of the peculiar sounds of its individual dialects, by the comparative method (although in reconstructed Proto-Indo-European some evidence of dialect survives).
c. Proto-Indo-European is not a snapshot of a moment in time but rather is "timeless": it *averages together centuries or even millennia* of development. In that sense, it is an accurate picture of no single era in language history.

These seem to be serious criticisms. But if their effect is to make Proto-Indo-European a mere fantasy, then the English language as presented in the Merriam-Webster Dictionary is a fantasy, too. My dictionary con-tains the English word *ombre* (a card game popular in the seventeenth and eighteenth centuries) as well as *hard disk* (a phrase that first appeared in the 1978 edition). So its vocabulary averages together at least three hundred years of the language. And its phonology, the "proper" pronuncia-tion it describes, is quite restricted. Only one pronunciation is given for *hard disk,* and it is not the Bostonian *hard* [haahd]. The English of Merriam-Webster has never been spoken in its entirety by any one per-son. Nevertheless we all find it useful as a guide to real spoken English. Reconstructed Proto-Indo-European is similar, a dictionary version of a language. It is not, in itself, a real language, but it certainly *refers* to one. And we should remember that Sumerian cuneiform documents and Egyptian hieroglyphs present exactly the same problems as reconstructed Proto-Indo-European: the written scripts do not clearly indicate every sound, so their phonology is uncertain; they contain only royal or priestly dialects; and they might preserve archaic linguistic forms, like Church Latin. They are not, in themselves, real languages; they only *refer* to real languages. Reconstructed Proto-Indo-European is not so different from cuneiform Sumerian.

If Proto-Indo-European is like a dictionary, then it cannot be "time-less." A dictionary is easily dated by its most recent entries. A dictionary containing the term *hard disk* is dated after 1978 in just the way that the wagon terminology in Proto-Indo-European dates it to a time after about 4000–3500 BCE. It is more dangerous to use negative information as a dating tool, since many words that really existed in Proto-Indo-European will never be reconstructed, but it is at least interesting that Proto-Indo-European does not contain roots for items like spoke, iron, cotton, chariot, glass, or coffee—things that were invented after the evolution and disper-sal of the daughter languages, or, in the metaphor we are using, after the dictionary was printed.

Of course, the dictionary of reconstructed Proto-Indo-European is much more tattered than my copy of Merriam-Webster's. Many pages have been torn out, and those that survive are obscured by the passage of time. The problem of the missing pages bothers some linguists the most. A recon-structed proto-language can seem a disappointing skeleton with a lot of bones missing and the placement of others debated between experts. The complete language the skeleton once supported certainly is a theoretical construct. So is the flesh-and-blood image of any dinosaur. Nevertheless, like the paleontologist, I am happy to have even a fragmentary skeleton. I think of Proto-Indo-European as a partial grammar and a partial set of pronunciation rules attached to the abundant fragments of a very ancient dictionary. To some linguists, that might not add up to a "real" language. But to an archaeologist it is more valuable than a roomful of potsherds.

Problem #2. The entire concept of "reconstructed Proto-Indo-European" is a fantasy: the similarities between the Indo-European languages could just as well have come about by gradual convergence over thousands of years between languages that had very different origins.

This is a more radical criticism then the first one. It proposes that the comparative method is a rigged game that automatically produces a proto-language as its outcome. The comparative method is said to ignore the linguistic changes that result from inter-language borrowing and conver-gence. Gradual convergence between originally diverse tongues, these scholars claim, might have produced the similarities between the Indo-European languages.[4] If this were true or even probable there would indeed be no reason to pursue a single parent of the Indo-European lan-guages. But the Russian linguist who inspired this line of questioning, Nikolai S. Trubetzkoy, worked in the 1930s before linguists really had the tools to investigate his startling suggestion.

Since then, quite a few linguists have taken up the problem of convergence between languages. They have greatly increased our understanding of how convergence happens and what its linguistic effects are. Although they disagree strongly with one another on some subjects, all recent studies of convergence accept that the Indo-European languages owe their essential similarities to descent from a common ancestral language, and not to convergence.[5] Of course, some convergence has occurred between neighboring Indo-European languages—it is not a question of all or nothing—but specialists agree that the basic structures that define the Indo-European language family can only be explained by common descent from a mother tongue.

There are three reasons for this unanimity. First, the Indo-European languages are the most thoroughly studied languages in the world—simply put, we know a lot about them. Second, linguists know of no language where bundled similarities of the kinds seen among the Indo-European languages have come about through borrowing or convergence between languages that were originally distinct. And, finally, the features known to typify creole languages—languages that *are* the product of convergence between two or more originally distinct languages—are not seen among the Indo-European languages. Creole languages are characterized by greatly reduced noun and pronoun inflections (no case or even single/plural markings); the use of pre-verbal particles to replace verb tenses ("we bin get" for "we got"); the general absence of tense, gender, and person inflections in verbs; a severely reduced set of prepositions; and the use of repeated forms to intensify adverbs and adjectives. In each of these features Proto-Indo-European was the *opposite* of a typical creole. It is not possible to classify Proto-Indo-European as a creole by any of the standards normally applied to creole languages.[6]

Nor do the Indo-European daughter languages display the telltale signs of creoles. This means that the Indo-European vocabularies and grammars replaced competing languages rather than creolizing with them. Of course, some back-and-forth borrowing occurred—it always does in cases of language contact—but superficial borrowing and creolization are very different things. Convergence simply cannot explain the similarities between the Indo-European languages. If we discard the mother tongue, we are left with *no* explanation for the regular correspondences in sound, morphology, and meaning that define the Indo-European language family.

Problem #3. Even if there was *a homeland where Proto-Indo-European was spoken, you cannot use the reconstructed vocabulary to find it because the reconstructed vocabulary is full of anachronisms that never existed in Proto-Indo-European.*

This criticism, like the last one, reflects concerns about recent interlanguage borrowing, focused here on just the vocabulary. Of course, many borrowed words are known to have spread through the Indo-European daughter languages long after the period of the proto-language—recent examples are *coffee* (borrowed from Arabic through Turkish) and *tobacco* (from Carib). The words for these items sound alike and have the same meanings in the different Indo-European languages, but few linguists would mistake them for ancient inherited words. Their phonetics are non–Indo-European, and their forms in the daughter branches do not represent what would be expected from inherited roots.[7] Terms like *coffee* are not a significant source of contamination.

Historical linguists do not ignore borrowing between languages. An understanding of borrowing is essential. For example, subtle inconsistencies embedded within German, Greek, Celtic, and other languages, including such fleeting sounds as the word-initial [kn-] (*knob*) can be identified as phonetically uncharacteristic of Indo-European. These fragments from extinct non–Indo-European languages are preserved only *because* they were borrowed. They can help us create maps of pre–Indo-European place-names, like the places ending with [-ssos] or [-nthos] (Corinthos, Knossos, Parnassos), borrowed into Greek and thought to show the geographic distribution of the pre-Greek language(s) of the Aegean and western Anatolia. Borrowed non–Indo-European sounds also were used to reconstruct some aspects of the long-extinct non–Indo-European languages of northern and eastern Europe. All that is left of these tongues is an occasional word or sound in the Indo-European languages that replaced them. Yet we can still identify their fragments in words borrowed thousands of years ago.[8]

Another regular use of borrowing is the study of "areal" features like *Sprachbund*s. A *Sprachbund* is a region where several different languages are spoken interchangeably in different situations, leading to their extensive borrowing of features. The most famous *Sprachbund* is in southeastern Europe, where Albanian, Bulgarian, Serbo-Croat, and Greek share many features, with Greek as the dominant element, probably because of its association with the Greek Orthodox Church. Finally, borrowing is an ever-present factor in any study of "genetic" relatedness. Whenever a linguist tries to decide whether cognate terms in two daughter languages are inherited from a common source, one alternative that must be excluded is that one language borrowed the term from the other. Many of the methods of comparative linguistics *depend* on the accurate identification of borrowed words, sounds, and morphologies.

When a root of similar sound and similar meaning shows up in widely separated Indo-European languages (including an ancient language), and phonological comparison of its forms yields a single ancestral root, that root term can be assigned with some confidence to the Proto-Indo-European vocabulary. No single reconstructed root should be used as the basis for an elaborate theory about Proto-Indo-European culture, but we do not need to work with single roots; we have clusters of terms with related meanings. At least fifteen hundred unique Proto-Indo-European roots have been reconstructed, and many of these unique roots appear in multiple reconstructed Proto-Indo-European words, so the total count of reconstructed Proto-Indo-European terms is much greater than fifteen hundred. Borrowing is a specific problem that affects specific reconstructed roots, but it does not cancel the usefulness of a reconstructed vocabulary containing thousands of terms.

The Proto-Indo-European homeland is not a racist myth or a purely theoretical fantasy. A real language lies behind reconstructed Proto-Indo-European, just as a real language lies behind any dictionary. And that language is a guide to the thoughts, concerns, and material culture of real people who lived in a definite region between about 4500 and 2500 BCE. But where was that region?

FINDING THE HOMELAND: ECOLOGY AND ENVIRONMENT

Regardless of where they ended up, most investigators of the Indo-European problem all started out the same way. The first step is to identify roots in the reconstructed Proto-Indo-European vocabulary referring to animal and plant species or technologies that existed only in certain places at particular times. The vocabulary itself should point to a homeland, at least within broad limits. For example, imagine that you were asked to identify the home of a group of people based only on the knowledge that a linguist had recorded these words in their normal daily speech:

armadillo	*sagebrush*	*cactus*
stampede	*steer*	*heifer*

calf	branding-iron	chuck-wagon
stockyard	rail-head	six-gun
saddle	lasso	horse

You could identify them fairly confidently as residents of the American southwest, probably during the late nineteenth or early twentieth centuries (*six-gun* and the absence of words for trucks, cars, and highways are the best chronological indicators). They probably were cowboys—or pretending to be. Looking closer, the combination of *armadillo*, *sagebrush*, and *cactus* would place them in west Texas, New Mexico, or Arizona.

Linguists have long tried to find animal or plant names in the reconstructed Proto-Indo-European vocabulary referring to species that lived in just one part of the world. The reconstructed Proto-Indo-European term for *salmon*, **lók*s*, was once famous as definite proof that the "Aryan" homeland lay in northern Europe. But animal and tree names seem to narrow and broaden in meaning easily. They are even reused and recycled when people move to a new environment, as English colonists used *robin* for a bird in the Americas that was a different species from the robin of England. The most specific meaning most linguists would now feel comfortable ascribing to the reconstructed term **lók*s*- is "trout-like fish." There are fish like that in the rivers across much of northern Eurasia, including the rivers flowing into the Black and Caspian Seas. The reconstructed Proto-Indo-European root for *beech* has a similar history. Because the copper beech, *Fagus silvatica*, did not grow east of Poland, the Proto-Indo-European root **bhágo*- was once used to support a northern or western European homeland. But in some Indo-European languages the same root refers to other tree species (oak or elder), and in any case the common beech (*Fagus orientalis*) grows also in the Caucasus, so its original meaning is unclear. Most linguists at least agree that the fauna and flora designated by the reconstructed vocabulary are temperate-zone types (*birch, otter, beaver, lynx, bear, horse*), not Mediterranean (no *cypress, olive*, or *laurel*) and not tropical (no *monkey, elephant, palm*, or *papyrus*). The roots for *horse* and *bee* are most helpful.

Bee and *honey* are very strong reconstructions based on cognates in most Indo-European languages. A derivative of the term for honey, **medhu-*, was also used for an intoxicating drink, mead, that probably played a prominent role in Proto-Indo-European rituals. Honeybees were not native east of the Ural Mountains, in Siberia, because the hardwood trees (lime and oak, particularly) that wild honeybees prefer as nesting sites were rare or absent east of the Urals. If bees and honey did not exist in Siberia, the homeland could not have been there. That removes all of Siberia and much of northeastern Eurasia from contention, including the Central Asian steppes of Kazakhstan. The horse, **ek*wo-*, is solidly reconstructed and seems also to have been a potent symbol of divine power for the speakers of Proto-Indo-European. Although horses lived in small, isolated pockets throughout prehistoric Europe, the Caucasus, and Anatolia between 4500 and 2500 BCE, they were rare or absent in the Near East, Iran, and the Indian subcontinent. They were numerous and economically important only in the Eurasian steppes. The term for horse removes the Near East, Iran, and the Indian subcontinent from serious contention, and encourages us to look closely at the Eurasian steppes. This leaves temperate Europe, including the steppes west of the Urals, and the temperate parts of Anatolia and the Caucasus Mountains.[9]

FINDING THE HOMELAND: THE ECONOMIC AND SOCIAL SETTING

The speakers of Proto-Indo-European were farmers and stockbreeders: we can reconstruct words for *bull, cow, ox, ram, ewe, lamb, pig,* and *piglet.* They had many terms for milk and dairy foods, including *sour milk, whey,* and *curds.* When they led their cattle and sheep out to the *field* they walked with a faithful *dog.* They knew how to *shear wool,* which they used to *weave* textiles (probably on a horizontal band loom). They tilled the earth (or they knew people who did) with a scratch-plow, or *ard,* which was pulled by *oxen* wearing a *yoke.* There are terms for *grain* and *chaff,* and perhaps for *furrow.* They turned their grain into flour by *grind*ing it with a hand *pestle,* and cooked their food in clay *pots* (the root is actually for *cauldron,* but that word in English has been narrowed to refer to a metal cooking vessel). They divided their possessions into two categories: movables and immovables; and the root for *movable wealth* (**peku-*, the ancestor of such English words as *pecuniary*) became the term for *herds* in general.[10] Finally, they were not averse to increasing their herds at their neighbors' expense, as we can reconstruct verbs that meant "to drive cattle," used in Celtic, Italic, and Indo-Iranian with the sense of cattle raiding or "rustling."

What was social life like? The speakers of Proto-Indo-European lived in a world of tribal politics and social groups united through kinship and marriage. They lived in households (**dómh₂*), containing one or more families (**génh₁es-*) organized into clans (**weiḱ-*), which were led by clan leaders, or chiefs (*weik-potis*). They had no word for *city*. Households appear to have been male-centered. Judging from the reconstructed kin terms, the important named kin were predominantly on the father's side, which suggests patrilocal marriages (brides moved into the husband's household). A group identity above the level of the clan was probably *tribe* (*$h_4erós$*), a root that developed into *Aryan* in the Indo-Iranian branch.[11]

The most famous definition of the basic divisions in Proto-Indo-European society was the tripartite scheme of Georges Dumézil, who suggested that there was a fundamental three-part division between the ritual specialist or priest, the warrior, and the ordinary herder/cultivator. Colors might have been associated with these three roles: white for the priest, red for the warrior, and black or blue for the herder/cultivator; and each role might have been assigned a specific type of ritual/legal death: strangulation for the priest, cutting/stabbing for the warrior, and drowning for the herder/cultivator. A variety of other legal and ritual distinctions seem to have applied to these three identities. It is unlikely that Dumézil's three divisions were groups with a limited membership. Probably they were something much less defined, like three age grades through which all males were expected to pass—perhaps herders (young), warriors (older), and lineage elders/ritual leaders (oldest), as among the Maasai in east Africa. The warrior category was regarded with considerable ambivalence, often represented in myth by a figure who alternated between a protector and a berserk murderer who killed his own father (Hercules, Indra, Thor). Poets occupied another respected social category. Spoken words, whether poems or oaths, were thought to have tremendous power. The poet's praise was a mortal's only hope for immortality.

The speakers of Proto-Indo-European were tribal farmers and stockbreeders. Societies like this lived across much of Europe, Anatolia, and the Caucasus Mountains after 6000 BCE. But regions where hunting and gathering economies persisted until after 2500 BCE are eliminated as possible homelands, because Proto-Indo-European was a dead language by 2500 BCE. The northern temperate forests of Europe and Siberia are excluded by this stockbreeders-before-2500 BCE rule, which cuts away one more piece of the map. The Kazakh steppes east of the Ural Mountains are excluded as well. In fact, this rule, combined with the exclusion of tropical regions and the presence of honeybees, makes a homeland anywhere east of the Ural Mountains unlikely.

FINDING THE HOMELAND: URALIC AND CAUCASIAN CONNECTIONS

The possible homeland locations can be narrowed further by identifying the neighbors. The neighbors of the speakers of Proto-Indo-European can be identified through words and morphologies borrowed between Proto-Indo-European and other language families. It is a bit risky to discuss borrowing between reconstructed proto-languages—first, we have to reconstruct a phonological system for each of the proto-languages, then identify roots of similar form and meaning in both proto-languages, and finally see if the root in one proto-language meets all the expectations of a root borrowed from the other. If neighboring proto-languages have the same roots, reconstructed independently, and one root can be explained as a predictable outcome of borrowing from the other, then we have a strong case for borrowing. So who borrowed words from, or loaned words into, Proto-Indo-European? Which language families exhibit evidence of early contact and interchange with Proto-Indo-European?

Uralic Contacts

By far the strongest linkages can be seen with Uralic. The Uralic languages are spoken today in northern Europe and Siberia, with one southern offshoot, Magyar, in Hungary, which was conquered by Magyar-speaking invaders in the tenth century. Uralic, like Indo-European, is a broad language family; its daughter languages are spoken across the northern forests of Eurasia from the Pacific shores of northeastern Siberia (Nganasan, spoken by tundra reindeer herders) to the Atlantic and Baltic coasts (Finnish, Estonian, Saami, Karelian, Vepsian, and Votian). Most linguists divide the family at the root into two super-branches, Finno-Ugric (the western branch) and Samoyedic (the eastern), although Salminen has argued that this binary division is based more on tradition than on solid linguistic evidence. His alternative is a "flat" division of the language family into nine branches, with Samoyedic just one of the nine.[12]

The homeland of Proto-Uralic probably was in the forest zone centered on the southern flanks of the Ural Mountains. Many argue for a homeland west of the Urals and others argue for the east side, but almost all Uralic linguists and Ural-region archaeologists would agree that Proto-Uralic was spoken somewhere in the birch-pine forests between the Oka River on the west (around modern Gorky) and the Irtysh River on the east (around modern Omsk). Today the Uralic languages spoken in this core region include, from west to east, Mordvin, Mari, Udmurt, Komi, and Mansi, of which two (Udmurt and Komi) are stems on the same branch (Permian). Some linguists have proposed homelands located farther east (the Yenisei River) or farther west (the Baltic), but the evidence for these extremes has not convinced many.[13]

The reconstructed Proto-Uralic vocabulary suggests that its speakers lived far from the sea in a forest environment. They were foragers who hunted and fished but possessed no domesticated plants or animals except the dog. This correlates well with the archaeological evidence. In the region between the Oka and the Urals, the Lyalovo culture was a center of cultural influences and interchanges among forest-zone forager cultures, with inter-cultural connections extending from the Baltic to the eastern slopes of the Urals during approximately the right period, 4500–3000 BCE.

The Uralic languages show evidence of very early contact with Indo-European languages. How that contact is interpreted is a subject of debate. There are three basic positions. First, the *Indo-Uralic* hypothesis suggests that the morphological linkages between the two families are so deep (shared pronouns), and the kinds of shared vocabulary so fundamental (words for *water* and *name*), that Proto-Indo-European and Proto-Uralic must have inherited these shared elements from some very ancient common linguistic parent—perhaps we might call it a "grandmother-tongue." The second position, the *early loan* hypothesis, argues that the forms of the shared proto-roots for terms like *name* and *water*, as reconstructed in the vocabularies of both Proto-Uralic and Proto-Indo-European, are much too similar to reflect such an ancient inheritance. Inherited roots should have undergone sound shifts in each developing family over a long period, but these roots are so similar that they can only be explained as loans from one proto-language into the other—and, in all cases, the loans went from Proto-Indo-European into Proto-Uralic.[14] The third position, the *late loan* hypothesis, is the one perhaps encountered most frequently in the general literature. It claims that there is little or no convincing evidence for borrowings even as old as the respective proto-languages; instead, the oldest well-documented loans should be assigned to contacts between Indo-Iranian and late Proto-Uralic, long after the Proto-Indo-European period. Contacts with Indo-Iranian could not be used to locate the Proto-Indo-European homeland.

At a conference dedicated to these subjects held at the University of Helsinki in 1999, not one linguist argued for a strong version of the late-loan hypothesis. Recent research on the earliest loans has reinforced the case for an early period of contact at least as early as the level of the proto-languages. This is well reflected in vocabulary loans. Koivulehto discussed at least thirteen words that are probable loans from Proto-Indo-European (PIE) into Proto-Uralic (P-U):

1. *to give* or *to sell*; P-U *mexe from PIE **h₂mey-gʷ-* 'to change', 'exchange'
2. *to bring, lead,* or *draw*; P-U **wetä-* from PIE **wedʰ-e/o-* 'to lead', 'to marry', 'to wed'
3. *to wash*; P-U **mośke-* from PIE **mozg-eye/o-* 'to wash', 'to submerge'
4. *to fear*; P-U **pele-* from PIE **pelh₁-* 'to shake', 'cause to tremble'
5. *to plait, to spin*; P-U **puna-* from PIE **pn.H-e/o-* 'to plait', 'to spin'
6. *to walk, wander, go*; P-U **kulke-* from PIE **kʷelH-e/o-* 'it/he/she walks around', 'wanders'
7. *to drill, to bore*; P-U **pura-* from PIE **bhṛH-* 'to bore', 'to drill'
8. *shall, must, to have to*; P-U **kelke-* from PIE **skelH-* 'to be guilty', 'shall', 'must'
9. *long thin pole*; P-U **śalka-* from PIE **gʰalgʰo-* 'well-pole', 'gallows', 'long pole'
10. *merchandise, price*; P-U *wosa from PIE *wosā 'merchandise', 'to buy'
11. *water;* P-U *wete from PIE **wed-er/en,* 'water', 'river'
12. *sinew;* P-U *sōne from PIE **sneH(u)-* 'sinew'
13. *name;* P-U **nime-* from PIE **h₃neh₃mn-* 'name'

Another thirty-six words were borrowed from differentiated Indo-European daughter tongues into early forms of Uralic prior to the emergence of differentiated Indic and Iranian—before 1700–1500 BCE at the latest. These later words included such terms as *bread, dough, beer, to winnow,* and *piglet,* which might have been borrowed when the speakers of Uralic languages began to adopt agriculture from neighboring Indo-European–speaking farmers and herders. But the loans between the proto-languages are the important ones bearing on the location of the Proto-Indo-European homeland. And that they are so similar in form does suggest that they were loans rather than inheritances from some very ancient common ancestor.

This does not mean that there is no evidence for an older level of shared ancestry. Inherited similarities, reflected in shared pronoun forms and some noun endings, might have been retained from such a common ancestor. The pronoun and inflection forms shared by Indo-European and Uralic are the following:

Proto-Uralic		**Proto-Indo-European**
*te-nä	(*thou*)	*ti (?)
*te	(*you*)	*ti (clitic dative)
*me-nä	(*I*)	*mi
*tä-/to-	(*this/that*)	*te-/to-
*ke-, ku-	(*who, what*)	*kʷe/o-
*-m	(*accusative sing.*)	*-m
*-n	(*genitive plural*)	*-om

These parallels suggest that Proto-Indo-European and Proto-Uralic shared two kinds of linkages.[15] One kind, revealed in pronouns, noun endings, and shared basic vocabulary, could be ancestral: the two proto-languages shared some quite ancient common ancestor, perhaps a broadly related set of intergrading dialects spoken by hunters roaming between the Carpathians and the Urals at the end of the last Ice Age. The relationship is so remote, however, that it can barely be detected. Johanna Nichols has called this kind of very deep, apparently genetic grouping a "quasi-stock."[16] Joseph Greenberg saw Proto-Indo-European and Proto-Uralic as particularly close cousins within a broader set of such language stocks that he called "Eurasiatic."

The other link between Proto-Indo-European and Proto-Uralic seems cultural: some Proto-Indo-European words were borrowed by the speakers of Proto-Uralic. Although they seem odd words to borrow, the terms *to wash*, *price*, and *to give* or *to sell* might have been borrowed through a trade jargon used between Proto-Uralic and Proto-Indo-European speakers. These two kinds of linguistic relationship—a possible common ancestral origin and inter-language borrowings—suggest that the Proto-Indo-European homeland was situated near the homeland of Proto-Uralic, in the vicinity of the southern Ural Mountains. We also know that the speakers of Proto-Indo-European were farmers and herders whose language had disappeared by 2500 BCE. The people living east of the Urals did not adopt domesticated animals until *after* 2500 BC. Proto-Indo-European must therefore have been spoken somewhere to the *south and west of the Urals*, the only region close to the Urals where farming and herding was regularly practiced before 2500 BCE.

Caucasian Contacts and the Anatolian Homeland

Proto-Indo-European also had contact with the languages of the Caucasus Mountains, primarily those now classified as South Caucasian or Kartvelian, the family that produced modern Georgian. These connections have suggested to some that the Proto-Indo-European homeland should be placed in the Caucasus near Armenia or perhaps in nearby eastern Anatolia. The links between Proto-Indo-European and Kartvelian are said to appear in both phonetics and vocabulary, although the phonetic link is controversial. It depends on a brilliant but still problematic revision of the phonology of Proto-Indo-European proposed by the linguists T. Gamkrelidze and V. Ivanov, known as the glottalic theory.[17] The glottalic theory made Proto-Indo-European phonology sound somewhat similar to that of Kartvelian, and even to the Semitic languages (Assyrian, Hebrew, Arabic) of the ancient Near East. This opened the possibility that Proto-Indo-European, Proto-Kartvelian, and Proto-Semitic might have evolved in a region where they shared certain areal phonological features. But by itself the glottalic phonology cannot prove a homeland in the Caucasus, even if it is accepted. And the glottalic phonology still has failed to convince many Indo-European linguists.[18]

Gamkrelidze and Ivanov have also suggested that Proto-Indo-European contained terms for panther, lion, and elephant, and for southern tree species. These animals and trees could be used to exclude a northern homeland. They also compiled an impressive list of loan words which they said were borrowed from Proto-Kartvelian and the Semitic languages into Proto-Indo-European. These relationships suggested to them that Proto-Indo-European had evolved in a place where it was in close contact with both the Semitic languages and the languages of the Southern Caucasus. They suggested Armenia as the most probable Indo-European homeland. Several archaeologists, prominently Colin Renfrew and Robert Drews, have followed their general lead, borrowing some of their linguistic arguments but placing the Indo-European homeland a little farther west, in central or western Anatolia.

But the evidence for a Caucasian or Anatolian homeland is weak. Many of the terms suggested as loans from Semitic into Proto-Indo-European have been rejected by other linguists. The few Semitic-to-Proto-Indo-European loan words that are widely accepted, words for items like silver and bull, might be words that were carried along trade and migration routes far from the Semites' Near Eastern homeland. Johanna Nichols has shown from the phonology of the loans that the Proto-Indo-European/Proto-Kartvelian/Proto-Semitic contacts were indirect—all the loan words passed through unknown intermediaries between the known three. One intermediary is required by chronology, as Proto-Kartvelian is generally thought to have existed after Proto-Indo-European and Proto-Semitic.[19]

The Semitic and Caucasian vocabulary that was borrowed into Proto-Indo-European through Kartvelian therefore contains roots that belonged to some *Pre-Kartvelian* or *Proto-Kartvelian* language in the Caucasus. This language had relations, through unrecorded intermediaries, with Proto-Indo-European on one side and Proto-Semitic on the other. That is not a particularly close lexical relationship. If Proto-Kartvelian was spoken on the south side of the North Caucasus Mountain range, as seems likely, it might have been spoken by people associated with the Early Transcaucasian Culture (also known as the Kura-Araxes culture), dated about 3500–2200 BCE. They could have had indirect relations with the speakers of Proto-Indo-European through the Maikop culture of the North Caucasus region. Many experts agree that Proto-Indo-European shared some features with a language ancestral to Kartvelian but not necessarily through a direct face-to-face link. Relations with the speakers of Proto-Uralic were closer.

So who were the neighbors? Proto-Indo-European exhibits strong links with Proto-Uralic and weaker links with a language ancestral to Proto-Kartvelian. The speakers of Proto-Indo-European lived somewhere between the Caucasus and Ural Mountains but had deeper linguistic relationships with the people who lived around the Urals.

The Location of the Proto-Indo-European Homeland

The speakers of Proto-Indo-European were tribal farmers who cultivated grain, herded cattle and sheep, collected honey from honeybees, drove wagons, made wool or felt textiles, plowed fields at least occasionally or knew people who did, sacrificed sheep, cattle, and horses to a troublesome array of sky gods, and fully expected the gods to reciprocate the favor. These traits guide us to a specific kind of material culture—one with wagons, domesticated sheep and cattle, cultivated grains, and sacrificial deposits with the bones of sheep, cattle, and horses. We should also look for a specific kind of ideology. In the reciprocal exchange of gifts and favors between their patrons, the gods, and human clients, humans offered a portion of their herds through sacrifice, accompanied by well-crafted verses of praise; and the gods in return provided protection from disease and misfortune, and the blessings of power and prosperity. Patron-client reciprocity of this kind is common among chiefdoms, societies with institutionalized differences in prestige and power, where some clans or lineages claim a right of patronage over others, usually on grounds of holiness or historical priority in a given territory.

Knowing that we are looking for a society with a specific list of material culture items and institutionalized power distinctions is a great help in locating the Proto-Indo-European homeland. We can exclude all regions where hunter-gatherer economies survived up to 2500 BCE. That eliminates the northern forest zone of Eurasia and the Kazakh steppes east of the Ural Mountains. The absence of honeybees east of the Urals eliminates any part of Siberia. The temperate-zone flora and fauna in the reconstructed vocabulary, and the absence of shared roots for Mediterranean or tropical flora and fauna, eliminate the tropics, the Mediterranean, and the Near East. Proto-Indo-European exhibits some very ancient links with the Uralic languages, overlaid by more recent lexical borrowings into Proto-Uralic from Proto-Indo-European; and it exhibits less clear linkages to some Pre- or Proto-Kartvelian language of the Caucasus region. All these requirements would be met by a Proto-Indo-European homeland placed west of the Ural Mountains, between the Urals and the Caucasus, in the steppes of eastern Ukraine and Russia. The internal coherence of reconstructed Proto-Indo-European—the absence of evidence for radical internal variation in grammar and phonology—indicates that the period of language history it reflects was less than two thousand years, probably less than one thousand. The heart of the Proto-Indo-European period probably fell between 4000 and 3000 BCE, with an early phase that might go back to 4500 BCE and a late phase that ended by 2500 BCE.

What does archaeology tell us about the steppe region between the Caucasus and the Urals, north of the Black and Caspian Seas—the Pontic-Caspian region—during this period? First, archaeology reveals a set of cultures that fits all the requirements of the reconstructed vocabulary: they sacrificed domesticated horses, cattle, and sheep, cultivated grain at least occasionally, drove wagons, and expressed institutionalized status distinctions in their funeral rituals. They occupied a part of the world—the steppes—where the sky is by far the most striking and magnificent part of

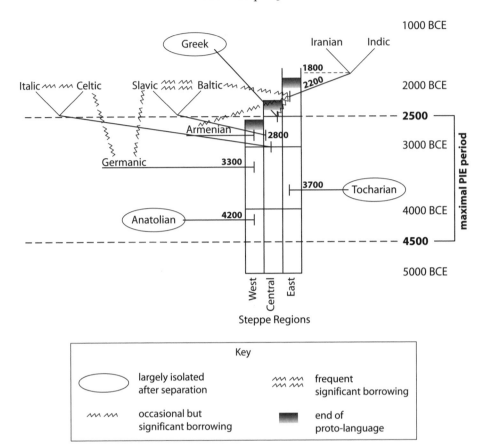

Figure 5.2 A diagram of the sequence and approximate dates of splits in early Indo-European as proposed in this book, with the maximal window for Proto-Indo-European indicated by the dashed lines. The dates of splits are determined by archaeological events described in chapters 11 (Anatolian) through16 (Iranian and Indic).

the landscape, a fitting environment for people who believed that all their most important deities lived in the sky. Archaeological evidence for migrations from this region into neighboring regions, both to the west and to the east, is well established. The sequence and direction of these movements matches the sequence and direction suggested by Indo-European linguistics and geography (figure 5.2). The first identifiable migration out of the Pontic-Caspian steppes was a movement toward the west about 4200–3900 BCE that could represent the detachment of the Pre-Anatolian branch, at a time before wheeled vehicles were introduced to the steppes (see chapter

4). This was followed by a movement toward the east (about 3700–3300 BCE) that could represent the detachment of the Tocharian branch. The next visible migration out of the steppes flowed toward the west. Its earliest phase might have separated the Pre-Germanic branch, and its later, more visible phase detached the Pre-Italic and Pre-Celtic dialects. This was followed by movements to the north and east that probably established the Baltic-Slavic and Indo-Iranian tongues. The remarkable match between the archaeologically documented pattern of movements out of the steppes and that expected from linguistics is fascinating, but it has absorbed, for too long, most of the attention and debate that is directed at the archaeology of Indo-European origins. Archaeology also adds substantially to our cultural and economic understanding of the speakers of Proto-Indo-European. Once the homeland has been located with linguistic evidence, the archaeology of that region provides a wholly new kind of information, a new window onto the lives of the people who spoke Proto-Indo-European and the process by which it became established and began to spread.

Before we step into the archaeology, however, we should pause and think for a moment about the gap we are stepping across, the void between linguistics and archaeology, a chasm most Western archaeologists feel cannot be crossed. Many would say that language and material culture are completely unrelated, or are related in such changeable and complicated ways that it is impossible to use material culture to identify language groups or boundaries. If that is true, then even if we *can* identify the place and time of the Indo-European homeland using the reconstructed vocabulary, the link to archaeology is impossible. We cannot expect any correlation with material culture. But is such pessimism warranted? Is there *no* predictable, regular link between language and material culture?

Chapter Six

The Archaeology of Language

A language homeland implies a bounded space of some kind. How can we define those boundaries? Can ancient linguistic frontiers be identified through archaeology?

Let us first define our terms. It would be helpful if anthropologists used the same vocabulary used in geography. According to geographers, the word *border* is neutral—it has no special or restricted meaning. A *frontier* is a specific kind of border—a transitional zone with some depth, porous to cross-border movement, and very possibly dynamic and moving. A frontier can be cultural, like the Western frontier of European settlement in North America, or ecological. An *ecotone* is an ecological frontier. Some ecotones are very subtle and small-scale—there are dozens of tiny ecotones in any suburban yard—and others are very large-scale, like the border between steppe and forest running east-west across central Eurasia. Finally, a sharply defined border that limits movement in some way is a *boundary*; for example, the political borders of modern nations are boundaries. But nation-like political and linguistic boundaries were unknown in the Pontic-Caspian region between 4500 and 2500 BCE. The cultures we are interested in were tribal societies.[1]

Archaeologists' interpretations of premodern tribal borders have changed in the last forty years. Most pre-state tribal borders are now thought to have been porous and dynamic—frontiers, not boundaries. More important, most are thought to have been ephemeral. The tribes Europeans encountered in their colonial ventures in Africa, South Asia, the Pacific, and the Americas were at first assumed to have existed for a long time. They often claimed antiquity for themselves. But many tribes are now believed to have been transient political communities of the historical moment. Like the Ojibwa, some might have crystallized only after contact with European agents who wanted to deal with bounded groups to facilitate the negotiation of territorial treaties. And the same critical attitude toward bounded tribal territories is applied to European history. Ancient European tribal identities—Celt, Scythian, Cimbri, Teuton, and Pict—are now frequently seen as convenient names for chameleon-like political alliances that had no true ethnic identity, or as brief ethnic phenomena that were *unable* to persist for any length of time, or even as entirely imaginary later inventions.[2]

Pre-state language borders are thought to have been equally fluid, characterized by intergrading local dialects rather than sharp boundaries. Where language and material culture styles (house type, town type, economy, dress, etc.) did coincide geographically to create a tribal ethnolinguistic frontier, we should expect it to have been short-lived. Language and material culture can change at different speeds for different reasons, and so are thought to grow apart easily. Historians and sociologists from Eric Hobsbawm to Anthony Giddens have proposed that there were no really distinct and stable ethnolinguistic borders in Europe until the late eighteenth century, when the French Revolution ushered in the era of nation-states. In this view of the past only the state is accorded both the need and the power to warp ethnolinguistic identity into a stable and persistent phenomenon, like the state itself. So how can we hope to identify ephemeral language frontiers in 3500 BCE? Did they even exist long enough to be visible archaeologically?[3]

Unfortunately this problem is compounded by the shortcomings of archaeological methods. Most archaeologists would agree that we do not really know how to recognize tribal ethnolinguistic frontiers, even if they *were* stable. Pottery styles were often assumed by pre–World War II archaeologists to be an indicator of social identity. But we now know that no simple connection exists between pottery types and ethnicity; as noted in chapter 1, every modern archaeology student knows that "pots are not people." The same problem applies to other kinds of material culture. Arrow-point types did seem to correlate with language families among the San hunter-gatherers of South Africa; however, among the Contact-period Native Americans in the northeastern U.S., the "Madison"-type arrow point was used by both Iroquoian and Algonkian speakers—its distribution had no connection to language. Almost any object could have been used to signal linguistic identity, or not. Archaeologists have therefore rejected the possibility that language and material culture are correlated in any predictable or recognizable way.[4]

But it seems that language and material culture are related in at least two ways. One is that tribal languages are generally more numerous in any long-settled region than tribal material cultures. Silver and Miller noticed, in 1997, that most tribal regions had more languages than material cultures. The Washo and Shoshone in the Great Basin had very different languages, of distinct language families, but similar material cultures; the Pueblo Indians had more languages than material cultures; the California Indians had more languages than stylistic groups; and the Indians of the central Amazon are well known for their amazing linguistic variety and broadly similar material cultures. A Chicago Field Museum study of language and material culture in northern New Guineau, the most detailed of its type, confirmed that regions defined by material culture were crisscrossed with numerous materially invisible language borders.[5] But the opposite pattern seems to be rare: a homogeneous tribal language is rarely separated into two very distinct bundles of material culture. This regularity seems discouraging, as it guarantees that many prehistoric language borders must be archaeologically invisible, but it does help to decide such questions as whether one language could have covered all the varied material culture groups of Copper Age Europe (probably not; see chapter 4).

The second regularity is more important: language is correlated with material culture at very long-lasting, distinct material-culture borders.

Persistent Frontiers

Persistent cultural frontiers have been ignored, because, I believe, they were dismissed on theoretical grounds.[6] They are not supposed to be there, since pre-state tribal borders are interpreted today as ephemeral and unstable. But archaeologists have documented a number of remarkably long-lasting, prehistoric, material-culture frontiers in settings that must have been tribal. A robust, persistent frontier separated Iroquoian and Algonkian speakers along the Hudson Valley, who displayed different styles of smoking pipes, subtle variations in ceramics, quite divergent house and settlement types, diverse economies, and very different languages for at least three centuries prior to European contact. Similarly the Linear Pottery/Lengyel farmers created a robust material-culture frontier between themselves and the indigenous foragers in northern Neolithic Europe, a moving border that persisted for at least a thousand years; the Criş/Tripolye cultures were utterly different from the Dnieper-Donets culture on a moving frontier between the Dniester and Dnieper Rivers in Ukraine for twenty-five hundred years during the Neolithic and Eneolithic; and the Jastorf and Halstatt cultures maintained distinct identities for centuries on either side of the lower Rhine in the Iron Age.[7] In each of these cases cultural norms changed; house designs, decorative aesthetics, and religious rituals were not frozen in a single form on either side. It was the *persistent opposition of bundles of customs* that defined the frontier rather than any one artifact type.

Persistent frontiers need not be stable geographically—they can move, as the Romano-Celt/Anglo-Saxon material-culture frontier moved across Britain between 400 and 700 CE, or the Linear Pottery/forager frontier moved across northern Europe between 5400 and 5000 BCE. Some material-culture frontiers, described in the next chapters, survived for millennia, in a pre-state social world governed just by tribal politics—no border guards, no national press. Particularly clear examples defined the edges of the Pontic-Caspian steppes on the west (Tripolye/Dnieper), on the north (Russian forest forager/steppe herder), and on the east (Volga-Ural steppe herder/Kazakh steppe forager). These were the borders of the region that probably was the homeland of Proto-Indo-European. If ancient ethnicities were ephemeral and the borders between them short-lived, how do we understand premodern tribal material-culture frontiers that persisted for thousands of years? And can language be connected to them?

I think the answer is yes. Language is strongly associated with persistent material-culture frontiers that are defined by bundles of opposed customs, what I will call *robust* frontiers.[8] The migrations and frontier formation processes that followed the collapse of the Roman Empire in western Europe provide the best setting to examine this association, because documents and place-names establish the linguistic identity of the migrants, the locations of newly formed frontiers, and their persistence over many centuries in political contexts where centralized state governments were weak or nonexistent. For example, the cultural frontier between the Welsh (Celtic branch) and the English (Germanic branch) has persisted since the Anglo-Saxon conquest of Romano-Celtic Britain during the sixth century. Additional conquests by Norman-English feudal barons after 1277 pushed the frontier back to the *landsker*, a named and overtly recognized ethnolinguistic frontier between Celtic Welsh-speaking and Germanic English-speaking populations that persisted to the present day. They spoke different languages (Welsh/English), built different kinds of churches (Celtic/Norman English), managed agriculture differently and with different tools, used diverse systems of land measurement, employed dissimilar standards of justice, and maintained a wide variety of distinctions in dress, food, and custom. For many centuries men rarely married across this border, maintaining a genetic difference between modern Welsh and English men (but not women) in traits located on the male Y chromosome.

Other post-Roman ethnolinguistic frontiers followed the same pattern. After the fall of Rome German speakers moved into the northern cantons of Switzerland, and the Gallic kingdom of Burgundy occupied what had been Gallo-Roman western Switzerland. The frontier between them still separates ecologically similar regions within a single modern state that differ in language (German-French), religion (Protestant-Catholic), architecture, the size and organization of landholdings, and the nature of the agricultural economy. Another post-Roman migration created the Breton/French frontier across the base of the peninsula of Brittany, after Romano-Celts migrated to Brittany from western Britain around 400–600 CE, fleeing the Anglo-Saxons. For more than fifteen hundred years the Celtic-speaking Bretons have remained distinct from their French-speaking neighbors in rituals, dress, music, and cuisine. Finally, migrations around 900–1000 CE brought German speakers into what is now northeastern Italy, where the persistent frontier between Germans and Romance speakers inside Italy was studied by Eric Wolf and John Cole in the 1960s. Although in this case both cultures were Catholic Christians, after a thousand years they still maintained different languages, house types, settlement organizations, land tenure and inheritance systems, attitudes toward authority and cooperation, and quite unfavorable stereotypes of each other. In all these cases documents and inscriptions show that the ethnolinguistic oppositions were not recent or invented but deeply historical and persistent.[9]

These examples suggest that most persistent, robust material-culture frontiers were ethnolinguistic. Robust, persistent, material-culture frontiers are not found everywhere, so only exceptional language frontiers can be identified. But that, of course, is better than nothing.

Population Movement across Persistent Frontiers

Unlike the men of Wales and England, most people moved back and forth across persistent frontiers easily. A most interesting fact about stable ethnolinguistic frontiers is that they were not necessarily biological; they persisted for an extraordinarily long time despite people regularly moving across them. As Warren DeBoer described in his study of native pottery styles in the western Amazon basin, "ethnic boundaries in the Ucayali basin are highly permeable with respect to bodies, but almost inviolable with respect to style."[10] The back-and-forth movement of *people* is indeed the principal focus of most contemporary borderland studies. The persistence of the *borders* themselves has remained understudied, probably because modern nation-states insist that all borders are permanent and inviolable, and many nation-states, in an attempt to naturalize their borders, have tried to argue that they have persisted from ancient times. Anthropologists and historians alike dismiss this as a fiction; the borders I have discussed frequently persist *within* modern nation-states rather than corresponding to their modern boundaries. But I think we have failed to recognize that we have internalized the modern nation-state's basic premise by insisting that ethnic borders must be inviolable boundaries or they did not really exist.

If people move across an ethno-linguistic frontier freely, then the frontier is often described in anthropology as, in some sense, a fiction. Is this just because it was not a boundary *like that of a modern nation?* Eric Wolf used this very argument to assert that the North American Iroquois did not exist as a distinct tribe during the Colonial period; he called them a multiethnic trading company. Why? Because their communities were full of captured and adopted non-Iroquois. But if biology is independent of language and culture, then the simple movement of Delaware and Nanticoke *bodies* into Iroquoian towns should not imply a dilution of Iroquoian *culture*. What matters is how the immigrants acted. Iroquoian adoptees were required to behave as Iroquois or they might be killed. The Iroquoian cultural identity remained distinct, and it was long established and persistent. The idea that European nation-states created the Iroquois "nation" in their own European image is particularly ironic in view of the fact that the five nations or tribes of the pre-European Northern Iroquois can be traced back archaeologically *in their traditional five tribal territories* to 1300 CE, more than 250 years before European contact. An Iroquois might argue that the borders of the original five nations of the Northern Iroquois were demonstrably older than those of many European nation-states at the end of the sixteenth century.[11]

Language frontiers in Europe are not generally strongly correlated with genetic frontiers; people mated across them. But persistent ethno-linguistic frontiers probably did originate in places where relatively *few* people moved between neighboring mating and migration networks. Dialect borders usually are correlated with borders between socioeconomic "functional zones," as linguists call a region marked by a strong network of intra-migration and socioeconomic interdependence. (Cities usually are divided into several distinct socioeconomic-linguistic functional zones.) Labov, for example, showed that dialect borders in central Pennsylvania correlated with reduced cross-border traffic flow densities at the borders of functional zones. In some places, like the Welsh/English border, the cross-border flow of people was low enough to appear genetically as a contrast in gene pools, but at other persistent frontiers there was enough cross-border movement to blur genetic differences. What, then, maintained the frontier itself, the persistent sense of difference?[12]

Persistent, robust premodern ethnolinguistic frontiers seem to have survived for long periods under one or both of two conditions: at *large-scale ecotones* (forest/steppe, desert/savannah, mountain/river bottom, mountain/ coast) and at places where long-distance migrants stopped migrating and formed a *cultural frontier* (England/Wales, Britanny/France, German Swiss/ French Swiss). Persistent identity depended partly on the continuous confrontation with Others that was inherent in these kinds of borders, as Frederik Barth observed, but it also relied on a home culture behind the border, a font of imagined tradition that could continuously feed those contrasts, as Eric Wolf recognized in Italy.[13] Let us briefly examine how these factors worked together to create and maintain persistent frontiers. We begin with borders created by long-distance migration.

Migration as a Cause of Persistent Material-Culture Frontiers

During the 1970s and 1980s the very idea of folk migrations was avoided by Western archaeologists. Folk migrations seemed to represent the boiled-down essence of the discredited idea that ethnicity, language, and material culture were packaged into neatly bounded societies that careened across the landscape like self-contained billiard balls, in a famously dismissive simile. Internal causes of social change—shifts in production and the means of production, in climate, in economy, in access to wealth and prestige, in political structure, and in spiritual beliefs—all got a good long look by archaeologists during these decades. While archaeologists were ignoring migration, modern demographers became very good at picking apart the various causes, recruiting patterns, flow dynamics, and targets of modern migration streams. Migration models moved far beyond the billiard ball analogy. The acceptance of modern migration models in the archaeology of the U.S. Southwest and in Iroquoian archaeology in the Northeast during the 1990s added new texture to the interpretation of Anasazi/Pueblo and Iroquoian societies, but in most other parts of the world the archaeological database was simply not detailed enough to test the very specific behavioral predictions of modern migration theories.[14] History, on the other hand, contains a very detailed record of the past, and among modern historians migration is accepted as a cause of persistent cultural frontiers.

The colonization of North America by English speakers is one prominent example of a well-studied, historical connection between migration and ethnolinguistic frontier formation. Decades of historical research have shown, surprisingly, that while the borders separating Europeans and Native Americans were important, those that separated different British cultures were just as significant. Eastern North America was colonized by four distinct migration streams that originated in four different parts of the British Isles. When they touched down in eastern North America, they created four clearly bounded ethnolinguistic regions between about 1620 and 1750. The Yankee dialect was spoken in New England. The same region also had a distinctive form of domestic architecture—the salt-box clapboard house—as well as its own barn and church architecture, a distinctive town type (houses clustered around a common grazing green), a peculiar cuisine (often baked, like Boston baked beans), distinct fashions in clothing, a famous style of gravestones, and a fiercely legalistic approach to politics and power. The geographic boundaries of the New England folk-culture region, drawn by folklorists on the basis of these traits, and the Yankee dialect region, drawn by linguists, coincide almost exactly. The Yankee dialect was a variant of the dialect of East Anglia, the region from which most of the early Pilgrim migrants came; and New England folk culture was a simplified version of East Anglian folk culture. The other three regions also exhibited strongly correlated dialects and folk cultures, as defined by houses, barn types, fence types, the frequency of towns and their organization, food preferences, clothing styles, and religion. One was the mid-Atlantic region (Pennsylvania Quakers from the English Midlands), the third was the Virginia coast (Royalist Anglican tobacco planters from southern England, largely Somerset and Wessex), and the last was the interior Appalachians (borderlanders from the Scotch-Irish borders). Both dialect and folk culture are traceable in each case to a particular region in the British Isles from which the first effective European settlers came.[15]

The four ethnolinguistic regions of Colonial eastern North America were created by four separate migration streams that imported people with distinctive ethnolinguistic identities into four different regions where simplified versions of their original linguistic and material differences were established, elaborated, and persisted for centuries (table 6.1). In some ways, including modern presidential voting patterns, the remnants of these four regions survive even today. But can modern migration patterns be applied to the past, or do modern migrations have purely modern causes?

TABLE 6.1
Migration Streams to Colonial North America

Colonial Region	Source	Religion
New England	East Anglia/Kent	Puritan
Mid–Atlantic	English Midlandss/ Southern Germany	Quaker/German Protestant
Tidewater Virginia–Carolina	Somerset/Wessex	Anglican
Southern Appalachian	Scots–Irish borderlands	Calvinist/Celtic church

The Causes of Migration

Many archaeologists think that modern migrations are fueled principally by overpopulation and the peculiar boundaries of modern nation-states, neither of which affected the prehistoric world, making modern migration studies largely irrelevant to prehistoric societies.[16] But migrations have many causes besides overpopulation within state borders. People do not migrate, even in today's crowded world, simply because there are too many at home. Crowding would be called a "push" factor by modern demographers, a negative condition at home. But there are other kinds of "push" factors—war, disease, crop failure, climate change, institutionalized raiding for loot, high bride-prices, the laws of primogeniture, religious intolerance, banishment, humiliation, or simple annoyance with the neighbors. Many causes of today's migrations and those in the past were social, not demographic. In ancient Rome, feudal Europe, and many parts of modern Africa, *inheritance rules* favored older siblings, condemning the younger ones to find their own lands or clients, a strong motive for them to migrate.[17] Pushes could be even more subtle. The persistent outward migrations and conquests of the pre-Colonial East African Nuer were caused, according to Raymond Kelley, not by overpopulation within Nuerland but rather by a cultural system of *bride-price regulations* that made it very expensive for young Nuer men to obtain a socially desirable bride. A bride-price was a payment made by the groom to the bride's family to compensate for the loss of her labor. Escalation in bride-prices encouraged Nuer men to raid their non-Nuer neighbors for cattle (and pastures to support them) that could be used to pay the elevated bride-price for a high-status marriage. Tribal status rivalries supported by high brideprices in an arid, low-productivity environment led to out-migration and the rapid territorial expansion of the Nuer.[18] Grassland migrations among tribal pastoralists can be "pushed" by many things other than absolute resource shortages.

Regardless of how "pushes" are defined, *no* migration can be adequately explained by "pushes" alone. Every migration is affected as well by "pull" factors (the alleged attractions of the destination, regardless of whether they are true), by communication networks that bring information to potential migrants, and by transport costs. Changes in any of these factors will raise or lower the threshold at which migration becomes an attractive option. Migrants weigh these dynamics, for far from being an instinctive response to overcrowding, migration is often a *conscious social strategy* meant to improve the migrant's position in competition for status and riches. If possible, migrants recruit clients and followers among the people at home, convincing them also to migrate, as Julius Caesar described the recruitment speeches of the chiefs of the Helvetii prior to their migration from Switzerland into Gaul. Recruitment in the homeland by potential and already departed migrants has been a continuous pattern in the expansion and reproduction of West African clans and lineages, as Igor Kopytoff noted. There is every reason to believe that similar social calculations have inspired migrations since humans evolved.

Effects: The Archaeological Identification of Ancient Migrations

Large, sustained migrations, particularly those that moved a long distance from one cultural setting into a very different one, or *folk migrations*, can be identified archaeologically. Emile Haury knew most of what to look for already in his excavations in Arizona in the 1950s: (1) the sudden appearance of a new material culture that has no local antecedents or prototypes; (2) a simultaneous shift in skeletal types (biology); (3) a neighboring territory where the intrusive culture evolved earlier; and (4) (a sign not recognized by Haury) the introduction of new *ways* of making things, new technological styles, which we now know are more "fundamental" (like the core vocabulary in linguistics) than decorative styles.

Smaller-scale migrations by specialists, mercenaries, skilled craft workers, and so on, are more difficult to identify. This is partly because archaeologists have generally stopped with the four simple criteria just described and neglected to analyze the internal workings even of folk migrations. To really understand why and how folk migrations occurred, and to have any hope of identifying small-scale migrations, archaeologists have to study the internal structure of long-distance migration streams, both large and small.

The organization of migrating groups depends on the identity and social connections of the scouts (who select the target destination); the social organization of information sharing (which determines who gets access to the scouts' information); transportation technology (cheaper and more effective transport makes migration easier); the targeting of destinations (whether they are many or few); the identity of the first effective settlers (also called the "charter group"); return migration (most migrations have a counterflow going back home); and changes in the goals and identities of migrants who join the stream later. If we look for all these factors we can better understand why and how migrations happened. Sustained migrations, particularly by pioneers looking to settle in new homes, can create very long-lasting, persistent ethnolinguistic frontiers.

The Simplification of Dialect and Culture among Long-distance Migrants

Access to the scouts' information defines the pool of potential migrants. Studies have found that the first 10% of new migrants into a region is an accurate predictor of the social makeup of the population that will follow them. This restriction on information at the source produces two common behaviors: leapfrogging and chain migration. In leapfrogging, migrants go only to those places about which they have heard good things, skipping over other possible destinations, sometimes moving long distances in one leap. In chain migration, migrants follow kin and co-residents to familiar places with social support, not to the objectively "best" place. They jump to places where they can rely on people they know, from point to targeted point. Recruitment usually is relatively restricted, and this is clearly audible in their speech.

Colonist speech generally is more homogeneous than the language of the homeland they left behind. Dialectical differences were fewer among Colonial-era English speakers in North America than they were in the British Isles. The Spanish dialects of Colonial South America were more homogeneous than the dialects of Southern Spain, the home region of most of the original colonists. Linguistic simplification has three causes. One is chain migration, where colonists tend to recruit family and friends from the same places and social groups that the colonists came from. Simplification also is a normal linguistic outcome of mixing between dialects in a contact situation at the destination.[19] Finally, simplification is encouraged among long-distance migrants by the social influence of the charter group.

The first group to establish a viable social system in a new place is called the *charter group*, or the first effective settlers.[20] They generally get the best land. They might claim rights to perform the highest-status rituals, as among the Maya of Central America or the Pueblo Indians of the American Southwest. In some cases, for example, Puritan New England, their councils choose who is permitted to join them. Among Hispanic migrants in the U.S. Southwest, charter groups were called *apex families* because of their structural position in local prestige hierarchies. Many later migrants were indebted to or dependent on the charter group, whose dialect and material culture provided the cultural capital for a new group identity. Charter groups leave an inordinate cultural imprint on later generations, as the latter copy the charter group's behavior, at least publicly. This explains why the English language, English house forms, and English settlement types were retained in nineteenth-century Ohio, although the overwhelming majority of later immigrants was German. The charter group, already established when the Germans arrived, was English. It also explains why East Anglian English traits, typical of the earliest Puritan immigrants, continued to typify New England dialectical speech and domestic architecture long after the majority of later immigrants arrived from other parts of England or Ireland. As a font of tradition and success in a new land, the charter group exercised a kind of historical cultural hegemony over later generations. Their genes, however, could easily be swamped by later migrants, which is why it is often futile to pursue a genetic fingerprint associated with a particular language.

The combination of chain migration, which restricted the pool of potential migrants at home, and the influence of the charter group, which encouraged conformity at the destination, produced a leveling of differences among many colonists. Simplification (fewer variants than in the home region) and leveling (the tendency toward a standardized form) affected both dialect *and* material culture. In material culture, domestic architecture and settlement organization—the external form and construction of the house and the layout of the settlement—particularly tended toward standardization, as these were the most visible signals of identity in any social landscape.[21] Those who wished to declare their membership in the mainstream culture adopted its external domestic forms, whereas those who retained their old house and barn styles (as did some Germans in Ohio) became political, as well as architectural and linguistic, minorities. Linguistic and cultural homogeneity among long-distance migrants facilitated stereotyping by Others, and strengthened the illusion of shared interests and origins among the migrants.

Ecological Frontiers: Different Ways of Making a Living

Franz Boas, the father of American anthropology, found that the borders of American Indian tribes rarely correlated with geographic borders. Boas decided to study the diffusion of cultural ideas and customs *across* borders. But a certain amount of agreement between ecology and culture is not at all surprising, particularly among people who were farmers and animal herders, which Boas's North American tribes generally were not. The length of the frost-free growing season, precipitation, soil fertility, and topography affect many aspects of daily life and custom among farmers: herding systems, crop cultivation, house types, the size and arrangement of settlements, favorite foods, sacred foods, the size of food surpluses, and the timing and richness of public feasts. At large-scale ecotones these basic differences in economic organization, diet, and social life can blossom into oppositional ethnic identities, which sometimes are complementary and mutually supportive, sometimes are hostile, and often are both. Frederick Barth, after working among the societies of Iran and Afghanistan, was among the first anthropologists to argue that ethnic identity was continuously created, even invented, at frontiers, rather than residing in the genes or being passively inherited from the ancestors. Oppositional politics crystallize who we are *not*, even if we are uncertain who we *are*, and therefore play a large role in the definition of ethnic identities. Ecotones were places where contrasting identities were likely to be reproduced and maintained for long periods because of structural differences in how politics and economics were played.[22]

Ecotones coincide with ethnolinguistic frontiers at many places. In France the Mediterranean provinces of the South and the Atlantic provinces of the North have been divided by an ethnolinguistic border for at least eight hundred years; the earliest written reference to it dates to 1284. The flat, tiled roofs of the South sheltered people who spoke the *langue d'oc*, whereas the steeply pitched roofs of the North were home to people who spoke the *langue d'oil*. They had different cropping systems, and different legal systems as well until they were forced to conform to a national legal standard. In Kenya the Nilotic-speaking pastoralist Maasai maintained a purely cattle-herding economy (or at least that was their ideal) in the dry plains and plateaus, whereas Bantu-speaking farmers occupied moister environments on the forested slopes of the mountains or in low wetlands. Probably the most famous anthropological example of this type was described by Sir Edmund Leach in his classic *Political Systems of High-*

land Burma. The upland Kachin forest farmers, who lived in the hills of Burma (Myanmar), were distinct linguistically, and also in many aspects of ritual and material culture, from the Thai-speaking Shan paddy farmers who occupied the rich bottomlands in the river valleys. Some Kachin leaders adopted Shan identities on certain occasions, moving back and forth between the two systems. But the broader distinction between the two cultures, Kachin and Shan, persisted, a distinction rooted in different ecologies, for example, the contrasting reliability and predictability of crop surpluses, the resulting different potentials for surplus wealth, and the dissimilar social organizations required for upland forest and lowland paddy farming. Cultural frontiers rooted in ecological differences could survive for a long time, even with people regularly moving across them.[23]

Language Distributions and Ecotones

Why do some language frontiers follow ecological borders? Does language just ride on the coattails of economy? Or is there an independent relationship between ecology and the way people speak? The linguists Daniel Nettle at Oxford University and Jane Hill at the University of Arizona proposed, in 1996 (independently, or at least without citing each other), that the geography of language reflects an underlying ecology of social relationships.[24]

Social ties require a lot of effort to establish and maintain, especially across long distances, and people are unlikely to expend all that energy unless they think they *need* to. People who are self-sufficient and fairly sure of their economic future tend to maintain *strong* social ties with a small number of people, usually people very much like themselves. Jane Hill calls this a *localist* strategy. Their own language, the one they grew up with, gets them everything they need, and so they tend to speak only that language—and often only one dialect of that language. (Most college-educated North Americans fit nicely in this category.) Secure people like this tend to live in places with productive natural ecologies or at least secure access to pockets of high productivity. Nettles showed that the average size of language groups in West Africa is inversely correlated with agricultural productivity: the richer and more productive the farmland, the smaller the language territory. This is one reason why a single pan-European Proto-Indo-European language during the Neolithic is so improbable.

But people who are moderately uncertain of their economic future, who live in less-productive territories and have to rely on multiple sources of income (like the Kachin in Burma or most middle-class families with two income earners), maintain numerous *weak* ties with a wider variety of people. They often learn two or more languages or dialects, because they need a wider network to feel secure. They pick up new linguistic habits very rapidly; they are innovators. In Jane Hill's study of the Papago Indians in Arizona, she found that communities living in rich, productive environments adopted a "localist" strategy in both their language and social relations. They spoke just one homogeneous, small-territory Papago dialect. But communities living in more arid environments knew many different dialects, and combined them in a variety of nonstandard ways. They adopted a "distributed" strategy, one that distributed alliances of various kinds, linguistic and economic, across a varied social and ecological terrain. She proposed that arid, uncertain environments were natural "spread zones," where new languages and dialects would spread quickly between communities that relied on diverse social ties and readily picked up new dialects from an assortment of people. The Eurasian steppes had earlier been described by the linguist Johanna Nichols as the prototypical linguistic spread zone; Hill explained why. Thus the association between language and ecological frontiers is not a case of language passively following culture; instead, there are independent socio-linguistic reasons why language frontiers tend to break along ecological frontiers.[25]

Summary: Ecotones and Persistent Ethnolinguistic Frontiers

Language frontiers did not universally coincide with ecological frontiers or natural geographic barriers, even in the tribal world, because migration and all the other forms of language expansion prevented that. But the heterogeneity of languages—the number of languages per $1,000\,km^2$—certainly was affected by ecology. Where an ecological frontier separated a predictable and productive environment from one that was unpredictable and unproductive, societies could not be organized the same way on both sides. Localized languages and small language territories were found among settled farmers in ecologically productive territories. More variable languages, fuzzier dialect boundaries, and larger language territories appeared among mobile hunter-gatherers and pastoralists occupying territories where farming was difficult or impossible. In the Eurasian steppes the ecological frontier between the steppe (unproductive, unpredictable, occupied principally by hunters or herders) and the neighboring agricultural lands (extremely productive and reliable, occupied by rich farmers) was a linguistic frontier through recorded history. Its persistence was one of the guiding factors in the history of China at one end of the steppes and of eastern Europe at the other.[26]

SMALL-SCALE MIGRATIONS, ELITE RECRUITMENT, AND LANGUAGE SHIFT

Persistent ecological and migration-related frontiers surrounded the Proto-Indo-European homeland in the Pontic-Caspian steppes. But the spread of the Indo-European languages *beyond* that homeland probably did not happen principally through chain-type folk migrations. A folk movement is not required to establish a new language in a strange land. Language change flows in the direction of accents that are admired and emulated by large numbers of people. Ritual and political elites often introduce and popularize new ways of speaking. Small elite groups can encourage widespread language shift toward their language, even in tribal contexts, in places where they succeed at introducing a new religion or political ideology or both while taking control of key territories and trade commodities. An ethnohistorical study of such a case in Africa among the Acholi illustrates how the introduction of a new ideology and control over trade can result in language spread even where the initial migrants were few in number.[27]

The Acholi are an ethnolinguistic group in northern Uganda and southern Sudan. They speak Luo, a Western Nilotic language. In about 1675, when Luo-speaking chiefs first migrated into northern Uganda from the south, the overwhelming majority of people living in the area spoke Central Sudanic or Eastern Nilotic languages—Luo was very much a minority language. But the Luo chiefs imported symbols and regalia of royalty (drums, stools) that they had adopted from Bantu kingdoms to the south. They also imported a new ideology of chiefly religious power, accompanied by demands for tribute service. Between about 1675 and 1725 thirteen new chiefdoms were formed, none larger than five villages. In these islands of chiefly authority the Luo-speaking chiefs recruited clients from among the lineage elders of the egalitarian local populations, offering them positions of prestige in the new hierarchy. Their numbers grew through marriage alliances with the locals, displays of wealth and generosity, assistance for local families in difficulty, threats of violence, and, most important, control over the inter-regional trade in iron prestige objects used to pay bride-prices. The Luo language spread slowly through recruitment.[28] Then an external stress, a severe drought beginning in 1790–1800, affected the region. One ecologically favored Lou chiefdom—an old one, founded by one of the first Luo charter groups—rose to paramount status as its wealth was maintained through the crisis. The Luo language then spread rapidly. When European traders arrived from Egypt in the 1850s they designated the local people by the name of this widely spoken language, which they called *Shooli*, which became *Achooli*. The paramount chiefs acquired so much wealth through trade with the Europeans that they quickly became an aristocracy. By 1872 the British recorded a single Luo-speaking tribe called the Acholi, an inter-regional ethnic identity that had not existed two hundred years earlier.

Indo-European languages probably spread in a similar way among the tribal societies of prehistoric Europe. Out-migrating Indo-European chiefs probably carried with them an ideology of political clientage like that of the Acholi chiefs, becoming patrons of their new clients among the local population; and they introduced a new ritual system in which they, in imitation of the gods, provided the animals for public sacrifices and feasts, and were in turn rewarded with the recitation of praise poetry—all solidly reconstructed for Proto-Indo-European culture, and all effective public recruiting activities. Later Proto-Indo-European migrations also introduced a new, mobile kind of pastoral economy made possible by the combination of ox-drawn wagons and horseback riding. Expansion beyond a few islands of authority might have waited until the new chiefdoms successfully responded to external stresses, climatic or political. Then the original chiefly core became the foundation for the development of a new regional ethnic identity. Renfrew has called this mode of language shift *elite dominance* but *elite recruitment* is probably a better term. The Normans conquered England and the Celtic Galatians conquered central Anatolia, but both failed to establish their languages among the local populations they dominated. Immigrant elite languages are adopted only where an elite status system is not only dominant but is also open to recruitment and alliance. For people to change to a new language, the shift must provide a key to integration within the new system, and those who join the system must see an opportunity to rise within it.[29]

A good example of how an open social system can encourage recruitment and language shift, cited long ago by Mallory, was described by Frederik Barth in eastern Afghanistan. Among the Pathans (today usually called Pashtun) on the Kandahar plateau, status depended on agricultural surpluses that came from circumscribed river-bottom fields. Pathan landowners competed for power in local councils (*jirga*) where no man admitted to being subservient and all appeals were phrased as requests among equals. The Baluch, a neighboring ethnic group, lived in the arid mountains and were, of necessity, pastoral herders. Although poor, the Baluch had an openly hierarchical political system, unlike the Pathan. The Pathan had more weapons than the Baluch, more people, more wealth, and generally more power and status. Yet, at the Baluch-Pathan frontier, many dispossessed Pathans crossed over to a new life as clients of Baluchi chiefs. Because Pathan status was tied to land ownership, Pathans who had lost their land in feuds were doomed to menial and peripheral lives. But Baluchi status was linked to herds, which could grow rapidly if the herder was lucky; and to political alliances, not to land. All Baluchi chiefs were the clients of more powerful chiefs, up to the office of *sardar*, the highest Baluchi authority, who himself owed allegiance to the khan of Kalat. Among the Baluch there was no shame in being the client of a powerful chief, and the possibilities for rapid economic and political improvement were great. So, in a situation of chronic low-level warfare at the Pathan-Baluch frontier, former agricultural refugees tended to flow toward the pastoral Baluch, and the Baluchi language thus gained new speakers. Chronic tribal warfare might generally favor pastoral over sedentary economies as herds can be defended by moving them, whereas agricultural fields are an immobile target.

Migration and the Indo-European Languages

Folk migrations by pioneer farmers brought the first herding-and-farming economies to the edge of the Pontic-Caspian steppes about 5800 BCE. In the forest-steppe ecological zone northwest of the Black Sea the incoming pioneer farmers established a cultural frontier between themselves and the native foragers. This frontier was robust, defined by bundles of cultural and economic differences, and it persisted for about twenty-five hundred years. If I am right about persistent frontiers and language, it was a linguistic frontier; if the other arguments in the preceding chapters are correct, the incoming pioneers spoke a non–Indo-European language, and the foragers spoke a Pre-Proto-Indo-European language. Selected aspects of the new farming economy (a little cattle herding, a little grain cultivation) were adopted by the foragers who lived on the frontier, but away from the frontier the local foragers kept hunting and fishing for many centuries. At the frontier both societies could reach back to very different sources of tradition in the lower Danube valley or in the steppes, providing a continuously renewed source of contrast and opposition.

Eventually, around 5200–5000 BCE, the new herding economy was adopted by a few key forager groups on the Dnieper River, and it then diffused very rapidly across most of the Pontic-Caspian steppes as far east as the Volga and Ural rivers. This was a revolutionary event that transformed not just the economy but also the rituals and politics of steppe societies.

A new set of dialects and languages probably spread across the Pontic-Caspian steppes with the new economic and ritual-political system. These dialects were the ancestors of Proto-Indo-European.

With a clearer idea of how language and material culture are connected, and with specific models indicating how migrations work and how they might be connected with language shifts, we can now begin to examine the archaeology of Indo-European origins.

PART TWO

The Opening of the Eurasian Steppes

CHAPTER SEVEN

How to Reconstruct a Dead Culture

The archaeology of Indo-European origins usually is described in terms that seem arcane to most people, and that even archaeologists define differently. So I offer a short explanation of how I approach the archaeological evidence. To begin at the beginning, surprisingly enough, we must start out in Denmark.

In 1807 the kingdom of Denmark was unsure of its prospects for survival. Defeated by Britain, threatened by Sweden, and soon to be abandoned by Norway, it looked to its glorious past to reassure its citizens of their greatness. Plans for a National Museum of Antiquities, the first of its type in Europe, were developed and promoted. The Royal Cabinet of Antiquities quickly acquired vast collections of artifacts that had been plowed or dug from the ground under a newly expanded agricultural policy. Amateur collectors among the country gentry, and quarrymen or ditch diggers among the common folk, brought in glimmering hoards of bronze and boxes of flint tools and bones.

In 1816, with dusty specimens piling up in the back room of the Royal Library, the Royal Commission for the Preservation of Danish Antiquities selected Christian J. Thomsen, a twenty-seven-year-old without a university degree but known for his practicality and industry, to decide how to arrange this overwhelming trove of strange and unknown objects in some kind of order for its first display. After a year of cataloguing and thinking, Thomsen elected to put the artifacts in three great halls. One would be for the stone artifacts, which seemed to come from graves or sediments belonging to a Stone Age, lacking any metals at all; one for the bronze axes, trumpets, and spears of the Bronze Age, which seemed to come from sites that lacked iron; and the last for the iron tools and weapons, made during an Iron Age that continued into the era of the earliest written references to Scandinavian history. The exhibit opened in 1819 and was a triumphant success. It inspired an animated discussion among

European intellectuals about whether these three ages truly existed in this chronological order, how old they were, and whether a science of archaeology, like the new science of historical linguistics, was possible. Jens Worsaae, originally an assistant to Thomsen, proved, through careful excavation, that the Three Ages indeed existed as distinct prehistoric eras, with some qualifications. But to do this he had to dig much more carefully than the ditch diggers, borrowing stratigraphic methods from geology. Thus professional field archaeology was born to solve a problem, not to acquire things.[1]

It was no longer possible, after Thomsen's exhibit, for an educated person to regard the prehistoric past as a single undifferentiated era into which mammoth bones and iron swords could be thrown together. Forever after time was to be divided, a peculiarly satisfying task for mortals, who now had a way to triumph over their most implacable foe. Once chronology was discovered, tinkering with it quickly became addictive. Even today chronological arguments dominate archaeological discussions in Russia and Ukraine. Indeed, a chief problem preventing Western archaeologists from really understanding steppe archaeology is that Thomsen's Three Ages are defined differently in the steppes than in western Europe. The Bronze Age seems like a simple concept, but if it began at different times in places very close to each other, it can be complicated to apply.

The Bronze Age can be said to begin when bronze tools and ornaments began to appear regularly in excavated graves and settlements. But what is bronze? It is an alloy, and the oldest bronze was an alloy of copper and arsenic. Arsenic, recognized by most of us simply as a poison, is in fact a naturally occurring whitish mineral typically in the form of arsenopyrite, which is frequently associated with copper ores in quartzitic copper deposits, and is probably how the alloy was discovered. In nature, arsenic rarely comprises more than about 1% of a copper ore, and usually much less than that. Ancient metalsmiths discovered that, if the arsenic content was boosted to about 2–8% of the mixture, the finished metal was lighter in color than pure copper, harder when cool, and, when molton, less viscous and easier to cast. A bronze alloy even lighter in color, harder, and more workable was copper and about 2–8% tin, but tin was rare in the ancient Old World, so tin-bronzes only appeared later, after tin deposits were discovered. The Bronze Age, therefore, marks that moment when metalsmiths regularly began to mix molten minerals to make alloys that were superior to naturally occurring copper. From that perspective, it immediately becomes clear that the Bronze Age would have started in different places at different times.

THE THREE AGES IN THE PONTIC-CASPIAN STEPPES

The oldest Bronze Age in Europe began about 3700–3500 BCE, when smiths started to make arsenical bronze in the North Caucasus Mountains, the natural frontier between the Near East and the Pontic-Caspian steppes. Arsenical bronzes, and the Bronze Age they signaled, appeared centuries later in the steppes and eastern Europe including the lower Danube valley, beginning about 3300–3200 BCE; and the beginning of the Bronze Age in central and western Europe was delayed a thousand years after that, starting only about 2400–2200 BCE. Yet, an archaeologist trained in western Europe may commonly ask why a Caucasian culture dated 3700 BCE is called a Bronze Age culture, when this would be the Stone Age (or Neolithic) in Britain or France. The answer is that bronze metallurgy appeared first in eastern Europe and then spread to the west, where it was adopted only after a surprisingly long delay. The Bronze Age began in the Pontic-Caspian steppes, the probable Indo-European homeland, much earlier than in Denmark.

The age preceding the Bronze Age in the steppes is called the Eneolithic; Christian Thomsen did not recognize that period in Denmark. The Eneolithic was a Copper Age, when metal tools and ornaments were used widely but were made of unalloyed copper. This was the first age of metal, and it lasted a long time in southeastern Europe, where European copper metallurgy was invented. The Eneolithic did not appear in northern or western Europe, which skipped directly from the Neolithic to the Bronze Age. Experts in southeastern Europe disagree on how to divide the Eneolithic internally; the chronological boundaries of the Early, Middle, and Late Eneolithic are set at different times by different archaeologists in different regions. I have tried to follow what I see as an emerging interregional consensus among Russian and Ukrainian archaeologists, and between them and the archaeologists of eastern Poland, Bulgaria, Romania, Hungary, and the former Yugoslavia.[2]

Before the Eneolithic was the Neolithic, the later end of Thomsen's Stone Age. Eventually the Stone Age was divided into the Old, Middle, and New Stone Ages, or the Paleolithic, Mesolithic, and Neolithic. In Soviet archaeology and in current Slavic or post-Soviet terminology the word *Neolithic* is applied to prehistoric societies that made pottery but had not yet discovered how to make metal. The invention of ceramics defined the beginning of the Neolithic. Pottery, of course, was an important discovery. Fire-resistant clay pots made it possible to cook stews and soups all day over a low fire, breaking down complex starches and proteins so that they were easier to digest for people with delicate stomachs—babies and elders. Soups that simmered in clay pots helped infants survive and kept old people alive longer. Pottery also is a convenient "type fossil" for archaeologists, easily recognized in archaeological sites. But Western archaeologists defined the Neolithic differently. In Western archaeology, societies can only be called *Neolithic* if they had economies based on food production—herding or farming or both. Hunters and gatherers who had pottery are called *Mesolithic*. It is oddly ironic that capitalist archaeologists made the mode of production central to their definition of the Neolithic, and Marxist archaeologists ignored it. I'm not sure what this might say about archaeologists and their politics, but here I must use the Eastern European definition of the Neolithic—which includes both foragers and early farmers who made pottery but used no metal tools or ornaments—because this is what *Neolithic* means in Russian and Ukrainian archaeology.

Dating and the Radiocarbon Revolution

Radiocarbon dating created a revolution in prehistoric archaeology. From Christian Thomsen's museum exhibit until the mid-twentieth century archaeologists had no clear idea how old their artifacts were, even if they knew how to place them in a sequence of types. The only way even to guess their age was to attempt to relate dagger or ornament styles in Europe to similar styles of known age in the Near East, where inscriptions provided dates going back to 3000 BCE. These long-distance stylistic comparisons, risky at best, were useless for dating artifacts older than the earliest Near Eastern inscriptions. Then, in 1949, Willard Libby demonstrated that the absolute age (literally the number of years since death) of any organic material (wood, bone, straw, shell, skin, hair, etc.) could be determined by counting its ^{14}C content, and thus radiocarbon dating was born. A radiocarbon date reveals when the dated sample died. Of course, the sample had to have been alive at some point, which disqualified Libby's discovery for dating rocks or minerals, but archaeologists often found charred wood from ancient fireplaces or discarded animal bones in places where humans had lived. Libby was awarded a Nobel Prize, and Europe acquired its own prehistory independent of the civilizations of the Near East. Some important events such as the invention of copper metallurgy were shown to have happened so early in Europe that influence from the Near East was almost ruled out.[3]

Chronological schemes based on radiocarbon dates have struggled through several significant changes in methods since 1949 (see the appendix in this volume). The most significant changes were the introduction of a new method (Accelerator Mass Spectrometry, or AMS) for counting how much ^{14}C remained in a sample, which made all dates much more accurate; and the realization that all radiocarbon dates, regardless of counting method, had to be corrected using calibration tables, which revealed large errors in old, uncalibrated dates. These periodic changes in methods and results slowed the scientific reception of radiocarbon dates in the former Soviet Union. Many Soviet archaeologists resisted radiocarbon dating, partly because it sometimes contradicted their theories and chronologies; partly because the first radiocarbon dates were later proved wrong by changes in methods, making it possible that all radiocarbon dates might soon be proved wrong by a newer refinement; and partly because the dates themselves, even when corrected and calibrated, sometimes made no sense—the rate of error in radiocarbon dating in Soviet times seemed high.

A new problem affecting radiocarbon dates in the steppes is that old carbon in solution in river water is absorbed by fish and then enters the bones of people who eat a lot of fish. Many steppe archaeological sites are cemeteries, and many radiocarbon dates in steppe archaeology are from human bones. Analysis of ^{15}N isotopes in human bone can tell us how much fish a person ate. Measurements of ^{15}N in skeletons from early steppe cemeteries show that fish was very important in the diet of most steppe societies, including cattle herders, often accounting for about 50% of the food consumed. Radiocarbon dates measured on the bones of these humans might come out too old, contaminated by old carbon in the fish they ate. This is a newly realized problem, one still without a solution widely agreed on. The errors should be in the range of 100–500 radiocarbon years too old, meaning that the person actually died 100–500 years *after* the date given by the count of ^{14}C. I note in the text places where old carbon contamination might be a problem making the dates measured on human bones too old, and, in the appendix, I explain my own interim approach to fixing the problem.[4]

Attitudes toward radiocarbon dating in the CIS have changed since 1991. The major universities and institutes have thrown themselves into new radiocarbon dating programs. The field collection of samples for dating has become more careful and more widespread, laboratories continuously improve their methods, and the error rate has fallen. It is difficult now to keep up with the flow of new radiocarbon dates. They have overthrown many old ideas and chronologies, including my own. Some of the chronological relationships outlined in my 1985 Ph.D. dissertation have now been proved wrong, and entire cultures I barely knew about in 1985 have become central to any understanding of steppe archaeology.[5]

But to understand people we need to know more than just *when* they lived; we also need to know something about their economy and culture. And in the specific case of the people of the Pontic-Caspian region, some of the most important questions are about *how* they lived—whether they were wandering nomads or lived in one place all year, whether they had chiefs or lived in egalitarian groups without formal full-time leaders, and how they went about getting their daily bread, if indeed they ate bread at all. But to talk about these matters I first need to introduce some additional methods archaeologists use.

What Did They Eat?

One of the most salient signals of cultural identity is food. Long after immigrants give up their native clothing styles and languages, they retain and even celebrate their traditional food. How the members of a society get food is, of course, a central organizing fact of life for all humans. The supermarkets we use so casually today are microcosms of modern Western life: they would not exist without a highly specialized, capital-financed, market-based economic structure; a consumer-oriented culture of profligate consumption (Do we really need fifteen kinds of mushrooms?); interstate highways; suburbs; private automobiles; and dispersed nuclear families lacking a grandma at home who could wash, chop, process, and prepare meat and produce. Long ago, before all these modern conveniences appeared, getting food determined how people spent much of their day, every day: what time they woke in the morning, where they went to work, what skills and knowledge they needed there, whether they could live in independent family homes or needed the much larger communal labor resources of a village, how long they were away from home, what kind of ecological resources they needed, what cooking and food-preparation skills they had to know, and even what foods they offered to the gods. In a world dominated by the rhythms and values of raising crops and caring for animals, clans with productive fields or large herds of cattle were the envy of everyone. Wealth and the political power it conveyed were equated with cultivated land and pasture.

To understand ancient agricultural and herding economies, archaeologists have to collect the animal bones from ancient garbage dumps with the same care they devote to broken pottery, and they must also make special efforts to recover carbonized plant remains. Luckily ancient people often buried their food trash in dumps or pits, restricting it to one place where archaeologists can find it more easily. Although cow bones and charred seeds cannot easily be displayed in the national museum, archaeology is not about collecting pretty things but about solving problems, so in the following pages much attention is devoted to animal bones and charred seeds.

Archaeologists count animal bones in two principal ways. Many bones in garbage dumps had been broken into such small pieces for cooking that they cannot be assigned to a specific animal species. Those that are big enough or distinctive enough to assign to a definite species constitute the NISP, or the "number of identified specimens," where *identified* means assignable to a species. Thus, the NISP count, which describes the number of bones found for each species, is the first way to count bones: three hundred cattle, one hundred sheep, five horse. The second counting method is to calculate the MNI, or the "minimum number of individuals" those bones represent. If the five horse bones were each from a different animal, they would represent five horses, whereas the hundred sheep bones might all be from a single skeleton. The MNI is used to convert bones into minimum meat weights—how much beef, for example, would be represented, minimally, by a certain number of cattle bones. Meat weight, comprised of fat and muscle, in most adult mammals averages about half the live body weight, so by identifying the minimum number, age, and species of animals butchered at the site, the minimum meat weight, with some qualifications, can be estimated.

Seeds, like wheat and barley, were often parched by charring them lightly over a fire to help preserve them for storage. Although many charred seeds are accidentally lost in this process, without charring they would soon rot into dust. The seeds preserved in archaeological sites have been charred just enough to carbonize the seed hull. Seeds tell us which plant foods were eaten, and can reveal the nature of the area's gardens, fields, forests, groves, and vineyards. The recovery of charred seeds from excavated sediments requires a flotation tank and a pump to force water through the tank. Excavated dirt is dumped into the tank and the moving water helps the seeds to float to the surface. They are then collected in screens as the water flows out the top of the tank through an exit spout. In the laboratory the species of plants are identified and counted, and domesticated varieties of wheat, barley, millet, and oats are distinguished from wild plant seeds. Flotation was rarely used in Western archaeology before the late 1970s and was almost never used in Soviet archaeology. Soviet paleobotanical experts relied on chance finds of seeds charred in burned pots or on seed impressions preserved in the damp clay of a pot before it had been fired. These lucky finds occur rarely. A true understanding of the importance of plant foods in the steppes will come only after flotation methods are widely used in excavations.

ARCHAEOLOGICAL CULTURES AND LIVING CULTURES

The story that follows is populated rarely by individuals and more often by cultures, which, although created and reproduced by people, act quite differently than people do. Because "living cultures" contain so many subgroups and variants, anthropologists have difficulty describing them in the abstract, leading many anthropologists to discard the concept of a "unitary culture" entirely. However, when cultural identities are contrasted with other bordering cultures, they are much easier to describe.

Frederik Barth's investigations of border identities in Afghanistan suggested that the reproduction and perhaps even the invention of cultural identities often was generated by the continuous confrontation with Others inherent in border situations. Today many anthropologists find this a productive way to understand cultural identities, that is, as responses to particular historical situations rather than as long-term phenomena, as noted in the previous chapter. But cultural identities also carry emotional and historical weight in the hearts of those who believe in them, and the source of this shared emotional attachment is more complicated. It must be derived from a shared set of customs and historical experiences, a font of tradition that, even if largely imagined or invented, provides the fuel that feeds border confrontations. If that font of tradition is given a geographic location or a homeland it is often away from the border, dispersed, for example, across shrines, burial grounds, coronation sites, battlefields, and landscape features like mountains and forests, all thought to be imbued with culture-specific spiritual forces.[6]

Archaeological cultures are defined on the basis of potsherds, grave types, architecture, and other material remains, so the relationship between archaeological cultures and living cultures might seem tenuous. When Christian Thomsen and Jens Worsaae first began to divide artifacts into types, they were trying to arrange them in a chronological sequence; they soon realized, however, that a lot of regional variation also cut across the chronological types. Archaeological cultures are meant to capture and define that regional variation. An archaeological culture is a recurring set of artifact types that co-occur in a particular region during a set time period.

In practice, pottery types are often used as the key identifiers of archaeological cultures, as they are easy to find and recognize even in small excavations, whereas the recognition of distinct house types, for example, requires much larger exposures. But archaeological cultures should never be defined on the basis of pottery alone. What makes an archaeological culture interesting, and meaningful, is the co-occurrence of many similar customs, crafts, and dwelling styles across a region, including, in addition to ceramics, grave types, house types, settlement types (the arrangement of houses in the typical settlement), tool types, and ritual symbols (figurines, shrines, and deities.) Archaeologists worry about individual types changing through time and shifting their areas of distribution, and we *should* worry about these things, but we should not let problems with defining individual tree species and ranges convince us that the forest is not there. Archaeological cultures (like forests) are particularly recognizable and definable at their borders, whereas regional variation in the back country, away from the borders, might often present a more confusing picture. It is at robust borders, defined by bundles of material-culture contrasts, where archaeological cultures and living cultures or societies might actually correspond. As I argued in the previous chapter, robust borders that persist for centuries probably were not just archaeological or cultural but also linguistic.

Within archaeological cultures a few traits, archaeologists have learned, are particularly important as keys to cultural identity. Most Western archaeologists accept that technological style, or the way an object is made, is a more fundamental indicator of craft tradition than the way it is decorated, its decorative style. The technology of production is more culture-bound and resistant to change, rather like the core vocabulary in linguistics. So clay tempering materials and firing methods usually are better indicators of a potter's cultural origin than the decorative styles the potter produced, and the same probably was true for metallurgy, weaving, and other crafts.[7]

One important alternative to archaeological cultures is the archaeological *horizon*. A *horizon*, more like a popular fashion than a culture, can be defined by a single artifact type or cluster of artifact types that spreads suddenly over a very wide geographic area. In the modern world the blue jeans and T-shirt complex is a horizon style, superimposed on diverse populations and cultures around the planet but still representing an important diffusion of cultural influence, particularly youth culture, from an area of origin in the United States. It is important, as it tells us something about the place the United States occupied in world youth culture at the moment of initial diffusion (the 1960s and 1970s), but it is not a migration or cultural replacement. Similarly the Beaker horizon in Late Neolithic Europe is defined primarily by a widespread style of decorated drinking cups (beakers) and in many places by a few weapon types (copper daggers, polished stone wrist-guards) that diffused with a new fashion in social drinking. In most places these styles were superimposed on preexisting archaeological cultures. A horizon is different from an archaeological culture because it is less robust—it is defined on the basis of just a few traits—and is often superimposed on local archaeological cultures. Horizons were highly significant in the prehistoric Eurasian steppes.

The Big Questions Ahead

We will proceed on the assumption that Proto-Indo-European probably was spoken in the steppes north of the Black and Caspian Seas, the Pontic-Caspian steppes, broadly between 4500 and 2500 BCE. But we have to start somewhat earlier to understand the evolution of Indo-European-speaking societies. The speakers of Proto-Indo-European were a cattle-keeping people. Where did the cattle come from? Both cattle and sheep were introduced from outside, probably from the Danube valley (although we also have to consider the possibility of a diffusion route through the Caucasus Mountains). The Neolithic pioneers who imported domesticated cattle and sheep into the Danube valley probably spoke non–Indo-European languages ultimately derived from western Anatolia. Their arrival in the eastern Carpathians, northwest of the Black Sea, around 5800 BCE, created a cultural frontier between the native foragers and the immigrant farmers that persisted for more than two thousand years.

The arrival of the first pioneer farmers and the creation of this cultural frontier is described in chapter 8. A recurring theme will be the development of the relationship between the farming cultures of the Danube valley and the steppe cultures north of the Black Sea. Marija Gimbutas called the Danubian farming cultures "Old Europe." The agricultural towns of Old Europe were the most technologically advanced and aesthetically sophisticated in all of Europe between about 6000 and 4000 BCE.

Chapter 9 describes the diffusion of the earliest cattle-and-sheep-herding economy across the Pontic-Caspian steppes after about 5200–5000 BCE. This event laid the foundation for the kinds of power politics and rituals that defined early Proto-Indo-European culture. Cattle herding was not just a new way to get food; it also supported a new division of society between high-status and ordinary people, a social hierarchy that had not existed when daily sustenance was based on fishing and hunting. Cattle and the cleavage of society into distinct statuses appeared together. Right away, cattle, sheep—and horses—were offered together in sacrifices at the funerals of a select group of people, who also carried unusual weapons and ornamented their bodies in unique and ostentatious ways. They were the new leaders of a new kind of steppe society.

Chapter 10 describes the discovery of horseback riding—a subject of intense controversy—by these archaic steppe herding societies, probably before 4200 BCE. The intrusion into Old Europe of steppe herders, probably mounted on horses, who either caused or took advantage of the collapse of Old Europe, is the topic of chapter 11. Their spread into the lower Danube valley about 4200–4000 BCE likely represented the initial expansion of archaic Proto-Indo-European speakers into southeastern Europe, speaking dialects that were ancestral to the later Anatolian languages.

Chapter 12 considers the influence of the earliest Mesopotamian urban civilizations on steppe societies—and vice versa—at a very early age, about 3700–3100 BCE. The chiefs who lived in the North Caucasus Mountains overlooking the steppes grew incredibly rich from long-distance trade with the southern civilizations. The earliest wheeled vehicles, the first wagons, probably rolled into the steppes through these mountains.

The societies that probably spoke classic Proto-Indo-European—the herders of the Yamnaya horizon—are introduced in chapter 13. They were the first people in the Eurasian steppes to create a herding economy that required regular seasonal movements to new pastures throughout the year. Wagons pulled by cattle allowed them to carry tents, water, and food into the deep steppes, far from the river valleys, and horseback riding enabled them to scout rapidly and over long distances and to herd on a large scale, necessities in such an economy. Herds were spread out across the enormous grasslands between the river valleys, making those grasslands useful, which led to larger herds and the accumulation of greater wealth.

Chapters 14 through 16 describe the initial expansions of societies speaking Proto-Indo-European dialects, to the east, the west, and finally to the south, to Iran and the Indian subcontinent. I do not attempt to follow what happened after the initial migrations of these groups; my effort is just to understand the development and the first dispersal of speakers of Proto-Indo-European and, along the way, to investigate the influence of technological innovations in transportation—horseback riding, wheeled vehicles, and chariots—in the opening of the Eurasian steppes.

Chapter Eight

First Farmers and Herders
The Pontic-Caspian Neolithic

At the beginning of time there were two brothers, twins, one named Man (*Manu*, in Proto-Indo-European) and the other Twin (*Yemo*). They traveled through the cosmos accompanied by a great cow. Eventually Man and Twin decided to create the world we now inhabit. To do this, Man had to sacrifice Twin (or, in some versions, the cow). From the parts of this sacrificed body, with the help of the sky gods (*Sky Father, Storm God of War, Divine Twins*), Man made the wind, the sun, the moon, the sea, earth, fire, and finally all the various kinds of people. Man became the first priest, the creator of the ritual of sacrifice that was the root of world order.

After the world was made, the sky-gods gave cattle to "Third man" (*Trito*). But the cattle were treacherously stolen by a three-headed, six-eyed serpent (*Ng^{w}hi*, the Proto-Indo-European root for *negation*). Third man entreated the storm god to help get the cattle back. Together they went to the cave (or mountain) of the monster, killed it (or the storm god killed it alone), and freed the cattle. *Trito* became the first warrior. He recovered the wealth of the people, and his gift of cattle to the priests insured that the sky gods received their share in the rising smoke of sacrificial fires. This insured that the cycle of giving between gods and humans continued.[1]

These two myths were fundamental to the Proto-Indo-European system of religious belief. *Manu* and *Yemo* are reflected in creation myths preserved in many Indo-European branches, where *Yemo* appears as Indic *Yama*, Avestan *Yima*, Norse *Ymir*, and perhaps Roman *Remus* (from *iemus*, the archaic Italic form of *yemo*, meaning "twin"); and Man appears as Old Indic *Manu* or Germanic *Mannus*, paired with his twin to create the world. The deeds of *Trito* have been analyzed at length by Bruce Lincoln, who found the same basic story of the hero who recovered primordial lost cattle from a three-headed monster in Indic, Iranian, Hittite, Norse, Roman and Greek myths. The myth of Man and Twin established the importance of the sacrifice and the priest who regulated it. The myth of the "Third one" defined the role of the warrior, who obtained animals for the people and the gods. Many other themes are also reflected in these two stories: the Indo-European fascination with binary doublings combined with triplets, two's and three's, which reappeared again and again, even in the metric structure of Indo-European poetry; the theme of pairs who represented magical and legal power (Twin and Man, Varuna-Mitra, Odin-Tyr); and the partition of society and the cosmos between three great functions or roles: the priest (in both his magical and legal aspects), the warrior (the Third Man), and the herder/cultivator (the cow or cattle).[2]

For the speakers of Proto-Indo-European, domesticated cattle were basic symbols of the generosity of the gods and the productivity of the earth. Humans were created from a piece of the primordial cow. The ritual duties that defined "proper" behavior revolved around the value, both moral and economic, of cattle. Proto-Indo-European mythology was, at its core, the worldview of a male-centered, cattle-raising people—not necessarily cattle nomads but certainly people who held sons and cattle in the highest esteem. Why were cattle (and sons) so important?

Domesticated Animals and Pontic-Caspian Ecology

Until about 5200–5000 BCE most of the people who lived in the steppes north of the Black and Caspian Seas possessed no domesticated animals at all. They depended instead on gathering nuts and wild plants, fishing, and hunting wild animals; in other words, they were foragers. But the environment they were able to exploit profitably was only a small fraction of the total steppe environment. The archaeological remains of their camps are found almost entirely in river valleys. Riverine gallery forests provided shelter, shade, firewood, building materials, deer, aurochs (European wild cattle), and wild boar. Fish supplied an important part of the diet. Wider river valleys like the Dnieper or Don had substantial gallery forests, kilometers wide; smaller rivers had only scattered groves. The wide grassy plateaus between the river valleys, the great majority of the steppe environment, were forbidding places occupied only by wild equids and saiga antelope. The foragers were able to hunt the wild equids, including horses. The wild horses of the steppes were stout-legged, barrel-chested, stiff-maned animals that probably looked very much like modern Przewalski horses, the only truly wild horses left in the world.[3] The most efficient hunting method would have been to ambush horse bands in a ravine, and the easiest opportunity would have been when they came into the river valleys to drink or to find shelter. In the steppe regions, where wild horses were most numerous, wild equid hunting was common. Often it supplied most of the foragers' terrestrial meat diet.

The Pontic-Caspian steppes are at the western end of a continuous steppe belt which rolls east all the way to Mongolia. It is possible, if one is so inclined, to walk, 5,000 km from the Danube delta across the center of the Eurasian continent to Mongolia without ever leaving the steppes. But a person on foot in the Eurasian steppes feels very small. Every footfall raises the scent of crushed sage, and a puff of tiny white grasshoppers skips ahead of your boot. Although the flowers that grow among the fescue and feathergrass (*Festuca* and *Stipa*) make a wonderful boiled tea, the grass is inedible, and outside the forested river valleys there is not much else to eat. The summer temperature frequently rises to 110–120°F (43–49°C), although it is a dry heat and usually there is a breeze, so it is surprisingly tolerable. Winter, however, kills quickly. The howling, snowy winds drive temperatures below −35°F (−37°C). The bitter cold of steppe winters (think North Dakota) is the most serious limiting factor for humans and animals, more restricting even than water, since there are shallow lakes in most parts of the Eurasian steppes.

The dominant mammal of the interior steppes at the time our account begins was the wild horse, *Equus caballus*. In the moister, lusher western steppes of Ukraine, north of the Black Sea (the North Pontic steppes), there was another, smaller equid that ranged into the lower Danube valley and down to central Anatolia, *Equus hydruntinus*, the last one hunted to extinction between 4000 and 3000 BCE. In the drier, more arid steppes of the Caspian Depression was a third ass-like, long-eared equid, the onager, *Equus hemionus*, now endangered in the wild. Onagers then lived in Mesopotamia, Anatolia, Iran, and in the Caspian Depression. Pontic-Caspian foragers hunted all three.

The Caspian Depression was itself a sign of another important aspect of the Pontic-Caspian environment: its instability. The Black and Caspian Seas were not placid and unchanging. Between about 14,000 and 12,000 BCE the warming climate that ended the last Ice Age melted the northern glaciers and the permafrost, releasing their combined meltwater in a torrential surge that flowed south into the Caspian basin. The late Ice-Age Caspian ballooned into a vast interior sea designated the Khvalynian Sea. For two thousand years the northern shoreline stood near Saratov on the middle Volga and Orenburg on the Ural River, restricting east-west movement south of the Ural Mountains. The Khvalynian Sea separated the already noticeably different late-glacial forager cultures that prospered east and west of the Ural Mountains.[4] Around 11,000–9,000 BCE the water finally rose high enough to overflow catastrophically through a southwestern outlet, the Manych Depression north of the North Caucasus Mountains, and a violent flood poured into the Black Sea, which was then well below the world ocean level. The Black Sea basin filled up until it overflowed, also through a southwestern outlet, the narrow Bosporus valley, and finally poured into the Aegean. By 8000 BCE the Black Sea, now about the size of California and seven thousand feet deep, was in equilibrium with the Aegean and the world ocean. The Caspian had fallen back into its own basin and remained isolated thereafter. The Black Sea became the Pontus Euxeinos of the Greeks, from which we derive the term *Pontic* for the Black Sea region in general. The North Caspian Depression, once the bottom of the northern end of the Khvalynian Sea, was left an enormous flat plain of salty clays, incongruous beds of sea shells, and sands, dotted with brackish lakes and covered with dry steppes that graded into red sand deserts (the Ryn Peski) just north of the Caspian Sea. Herds of saiga antelopes, onagers, and horses were hunted across these saline plains by small bands of post-glacial Mesolithic and Neolithic hunters. But, by the time the sea receded, they had become very different culturally and probably linguistically on the eastern and western sides of the Ural-Caspian frontier. When domesticated cattle were accepted by societies west of the Urals, they were rejected by those east of the Urals, who remained foragers for thousands of years.[5]

Domesticated cattle and sheep started a revolutionary change in how humans exploited the Pontic-Caspian steppe environment. Because cattle and sheep were cultured, like humans, they were part of everyday work and worry in a way never approached by wild animals. Humans identified with their cattle and sheep, wrote poetry about them, and used them as a currency in marriage gifts, debt payments, and the calculation of social status. And they were grass processors. They converted plains of grass, useless and even hostile to humans, into wool, felt, clothing, tents, milk, yogurt, cheese, meat, marrow, and bone—the foundation of both life and wealth. Cattle and sheep herds can grow rapidly with a little luck. Vulnerable to bad weather and theft, they can also decline rapidly. Herding was a volatile, boom-bust economy, and required a flexible, opportunistic social organization.

Because cattle and sheep are easily stolen, unlike grain crops, cattle-raising people tend to have problems with thieves, leading to conflict and warfare. Under these circumstances brothers tend to stay close together. In Africa, among Bantu-speaking tribes, the spread of cattle raising seems to have led to the loss of matrilineal social organizations and the spread of male-centered patrilineal kinship systems.[6] Stockbreeding also created entirely new kinds of political power and prestige by making possible elaborate public sacrifices and gifts of animals. The connection between animals, brothers, and power was the foundation on which new forms of male-centered ritual and politics developed among Indo-European-speaking societies. That is why the cow (and brothers) occupied such a central place in Indo-European myths relating to how the world began.

So where did the cattle come from? When did the people living in the Pontic-Caspian steppes begin to keep and care for herds of dappled cows?

The First Farmer-Forager Frontier in the Pontic-Caspian Region

The first cattle herders in the Pontic-Caspian region arrived about 5800–5700 BCE from the Danube valley, and they probably spoke languages unrelated to Proto-Indo-European. They were the leading edge of a broad movement of farming people that began around 6200 BCE when pioneers from Greece and Macedonia plunged north into the temperate forests of the Balkans and the Carpathian Basin (figure 8.1). Domesticated sheep and cattle had been imported from Anatolia to Greece by their ancestors centuries before, and now were herded northward into forested southeastern Europe. Genetic research has shown that the cattle did interbreed with the native European aurochs, the huge wild cattle of Europe, but only the male calves (traced on the Y chromosome) of aurochs were kept, perhaps because they could improve the herd's size or resistance to disease without affecting milk yields. The cows, probably already kept for their milk, all were descended from mothers that had come from Anatolia (traced through MtDNA). Wild aurochs cows probably were relatively poor milk producers and might have been temperamentally difficult to milk, so Neolithic European farmers made sure that all their cows were born of long-domesticated mothers, but they did not mind a little crossbreeding with native wild bulls to obtain larger domestic bulls.[7]

Comparative studies of chain migration among recent and historical pioneer farmers suggest that, in the beginning, the farming-and-herding groups that first moved into temperate southeastern Europe probably spoke similar dialects and recognized one another as cultural cousins. The thin native population of foragers was certainly seen as culturally and linguistically Other, regardless of how the two cultures interacted.[8] After an initial rapid burst of exploration (sites at Anzabegovo, Karanovo

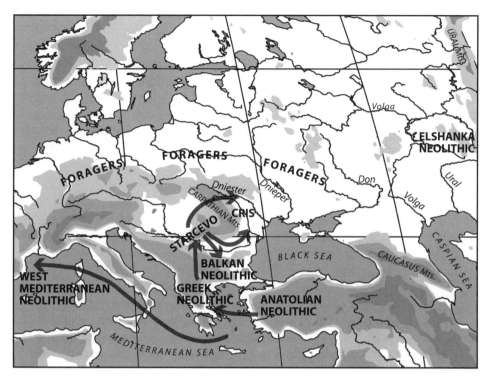

Figure 8.1 The migrations of pioneer farmers into Greece and across Europe between 6500 and 5500 BCE, including the colonization of the eastern Carpathian piedmont by the Criş culture.

I, Gura Baciului, Cirçea) pioneer groups became established in the Middle Danube plains north of Belgrade, where the type site of Starčevo and other similar Neolithic settlements are located. This central Danubian lowland produced two streams of migrants that leapfrogged in one direction down the Danube, into Romania and Bulgaria, and in the other up the Mureş and Körös Rivers into Transylvania. Both migration streams created similar pottery and tool types, assigned today to the Criş culture (figure 8.2).[9]

First Farmers in the Pontic Region: The Criş Culture

The names Criş in Romania and Körös in eastern Hungary are two variants of the same river name and the same prehistoric culture. The northern Criş people moved up the Hungarian rivers into the mountains of Transylvania and then pushed over the top of the Carpathian ridges into

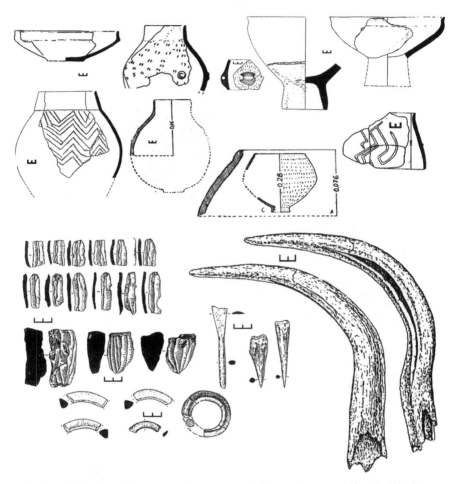

Figure 8.2 Criş-culture ceramic shapes and decorative motifs (*top half*), flint blades and cores (*left*), antler and bone tools (*right*), and ceramic rings (*bottom*) dated 5700–5300 BCE. After Dergachev 1999; and Ursulescu 1984.

an ecologically rich and productive piedmont region east of the Carpathians. They herded their cattle and sheep down the eastern slopes into the upper valleys of the Seret and Prut rivers about 5800–5700 BCE. (Criş radiocarbon dates are unaffected by reservoir effects because they were not measured on human bone; see table 8.1.) The other migration stream in the lower Danube valley moved into the same eastern Carpathian piedmont from the south. These two groups created a northern and a southern variant of the East Carpathian Criş culture, which survived from about 5800 to about 5300 BCE. Criş farms in the East Carpathian piedmont

TABLE 8.1

Radiocarbon Dates for the Late Mesolithic and Early Neolithic of the Pontic-Caspian Region.

Lab Number	BP Date	Sample	Calibrated Date
1. Criş Culture Farming Settlements			
Trestiana (Romania), phase III of the Criş culture			
GrN-17003	6665±45	Charcoal	5640–5530 BCE
Cârcea-Viaduct (Romania), phase IV of the Criş culture			
Bln-1981	6540±60	?	5610–5390 BCE
Bln-1982	6530±60	?	5610–5380 BCE
Bln-1983	6395±60	?	5470–5310 BCE
2. Linear Pottery (LBK) Farming Settlements			
Tirpeşti, Siret River, (Romania)			
Bln-800	6170±100	?	5260–4960 BCE
Bln-801	6245±100	?	5320–5060 BCE
3. Bug-Dniester Mesolithic-Neolithic Settlements			
Soroki II, level 1 early Bug–Dniester, Dniester valley			
Bln-586	6825±150	?	5870–5560 BCE
Soroki II, level 2 pre-ceramic Bug–Dniester, Dniester valley			
Bln-587	7420±80	?	6400–6210 BCE
Savran settlement, late Bug–Dniester, Dniester valley			
Ki-6654	6985±60	?	5980–5790 BCE
Bazkov Ostrov settlement, with early ceramics, South Bug valley			
Ki-6651	7235±60	?	6210–6010 BCE
Ki-6696	7215±55	?	6200–6000 BCE
Ki-6652	7160±55	?	6160–5920 BCE
Sokolets II settlement, with early ceramics, South Bug valley			
Ki-6697	7470±60	?	6400–6250 BCE
Ki-6698	7405±55	?	6390–6210 BCE
4. Early Neolithic Elshanka-type Settlements, Middle Volga Region*			
Chekalino 4, Sok River, Samara oblast			
Le-4781	8990±100	shell	8290–7960 BCE

TABLE 8.1 (*continued*)

Lab Number	BP Date	Sample	Calibrated Date
GrN–7085	8680±120	shell	7940–7580 BCE
Le–4783	8050±120	shell	7300–6700 BCE
Le–4782	8000±120	shell	7080–6690 BCE
GrN–7086	7950±130	shell	7050–6680 BCE
Le–4784	7940±140	shell	7050–6680 BCE

Chekalino 6, Sok River, Samara oblast

Le–4883	7940±140	shell	7050–6650 BCE

Ivanovka, upper Samara River, Orenburg oblast

Le–2343	8020±90	bone	7080–6770 BCE

5. Steppe Early Neolithic Settlements*

Matveev Kurgan I, very primitive ceramics, Azov steppes

GrN–7199	7505±210	charcoal	6570–6080 BCE
Le–1217	7180±70	charcoal	6160–5920 BCE

Matveev Kurgan II, same material culture, Azov steppes

Le–882	5400±200	charcoal	4450–3980 BCE

Varfolomievka, Layer 3 (bottom ceramic layer), North Caspian steppes

GIN–6546	6980±200	charcoal	6030–5660 BCE

Kair-Shak III, North Caspian steppes

GIN–5905	6950±190	?	6000–5660 BCE
GIN 5927	6720±80	?	5720–5550 BCE

Rakushechni Yar, lower Don shell midden, layers 14–15

Ki–6479	6925±110	?	5970–5710 BCE
Ki–6478	6930±100	?	5970–5610 BCE
Ki–6480	7040±100	?	6010–5800 BCE

Surskii Island, Dnieper Rapids forager settlement

Ki–6688	6980±65	?	5980–5780 BCE
Ki–6989	7125±60	?	6160–5910 BCE
Ki–6690	7195±55	?	6160–5990 BCE
Ki–6691	7245±60	?	6210–6020 BCE

were the source of the first domesticated cattle in the North Pontic region. The Criş pioneers moved eastward through the forest-steppe zone in the piedmont northwest of the Black Sea, where rainfall agriculture was possible, avoiding the lowland steppes on the coast and the lower courses of the rivers that ran through them into the sea.

Archaeologists have identified at least thirty Criş settlement sites in the East Carpathian piedmont, a region of forests interspersed with natural meadows cut by deep, twisting river valleys (figure 8.3). Most Criş farming hamlets were built on the second terraces of rivers, overlooking the floodplain; some were located on steep-sided promontories above the floodplain (Suceava); and a few farms were located on the high forested ridges between the rivers (Sakarovka I). Houses were one room, built with timber posts and beams, plaster-on-wattle walls, and probably reed-thatched roofs. Larger homes, sometimes oval in outline, were built over dug-out floors and contained a kitchen with a domed clay oven; lighter, smaller structures were built on the surface with an open fire in the center. Most villages consisted of just a few families living in perhaps three to ten smoky thatched pit-dwellings, surrounded by agricultural fields, gardens, plum orchards, and pastures for the animals. No Criş cemeteries are known. We do not know what they did with their dead. We do know, however, that they still prized and wore white shell bracelets made from imported *Spondylus*, an Aegean species that was first made into bracelets by the original pioneers in Early Neolithic Greece.[10]

Criş families cultivated barley, millet, peas, and four varieties of wheat (emmer, einkorn, spelt, and bread wheats). Wheat and peas were not native to southeastern Europe; they were exotics, domesticated in the Near East, carried into Greece by sea-borne immigrant farmers, and propagated through Europe from Greece. Residues inside pots suggest that grains were often eaten in the form of a soup thickened with flour. Charred fragments of Neolithic bread from Germany and Switzerland suggest that wheat flour was also made into a batter that was fried or baked, or the grains were moistened and pressed into small whole-grain baked loaves. Criş harvesting sickles used a curved red deer antler inset with flint blades 5–10 cm long, angled so that their corners formed teeth. Their working corners show "sickle gloss" from cutting grain. The same type of sickle and flint blade is found in all the Early Neolithic farming settlements of the Danube-Balkans-Carpathians. Most of the meat in the East Carpathian Criş diet was from cattle and pigs, with red deer a close third, followed by sheep—a distribution of species reflecting their largely forested environment. Their small-breed cows and pigs were slightly different from the

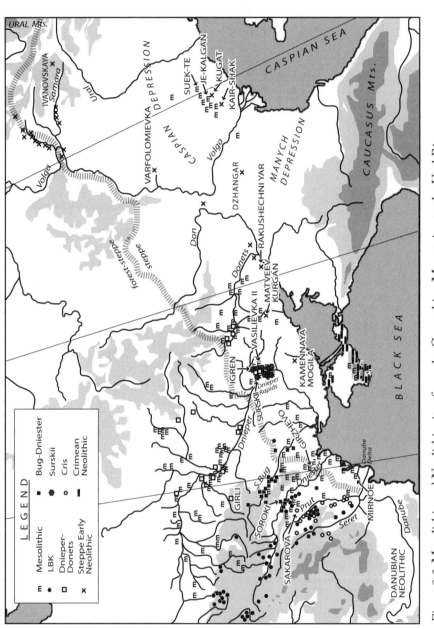

Figure 8.3 Mesolithic and Neolithic sites from the Carpathian Mountains to the Ural River.

local wild aurochs or wild boar but not markedly so. The sheep, however, were exotic newcomers, an invasive species like wheat and peas, brought into the steep Carpathian valleys by strange people whose voices made a new kind of sound.[11]

Criş ceramic vessels were hand-made by the coiling method, and included plain pots for cooking and storage, and a variety of fine wares with polished reddish-brown surfaces—tureens, bowls, and cups on pedestals (figure 8.2). Decorative designs were incised with a stick on the clay surface before firing or were impressed with a fingernail. Very rarely they were painted in broad brown stripes. The shapes and designs made by Criş settlers in the East Carpathians were characteristic of periods III and IV of the Criş culture; older sites of stages I and II are found only in eastern Hungary, the Danube valley, and Transylvania.

Criş farmers never penetrated east of the Prut-Dniester watershed. In the Dniester valley they came face-to-face with a dense population of local foragers, known today as the Bug-Dniester culture, named after the two river valleys (Dniester and South Bug) where most of their sites are found. The Bug-Dniester culture was the filter through which farming and stock-breeding economies were introduced to Pontic-Caspian societies farther east (figure 8.3).

The Criş people were different from their Bug-Dniester neighbors in many ways: Criş flint tool kits featured large blades and few scrapers, whereas the foragers used microlithic blades and many scrapers; most Criş villages were on the better-drained soils of the second terrace, convenient for farming, and most foragers lived on the floodplain, convenient for fishing; whereas Criş woodworkers used polished stone axes, the foragers used chipped flint axes; Criş pottery was distinct both in the way it was made and its style of decoration; and Criş farmers raised and ate various exotic foods, including mutton, which has a distinctive taste. Four forged cylindrical copper beads were found at the Criş site of Selishte, dated 5800–5600 BCE (6830±100 BP).[12] They show an early awareness of the metallic minerals in the mountains of Transylvania (copper, silver, gold) and the Balkans (copper), something the foragers of southeastern Europe had never noticed.

Some archaeologists have speculated that the East Carpathian Criş culture could have been an acculturated population of local foragers who had adopted a farming economy, rather than immigrant pioneers.[13] This is unlikely given the numerous similarities between the material culture and economy of Criş sites in the Danube valley and the East Carpathians, and the sharp differences between the East Carpathian Criş culture and the local foragers. But it really is of no consequence—no one seriously believes that the East Carpathian Criş people were *genetically* "pure" anyway. The important point is that the people who lived in Criş villages in the East Carpathians were *culturally* Criş in almost all the material signs of their identity, and given how they got there, almost certainly in nonmaterial signs like language as well. The Criş *culture* came, without any doubt, from the Danube valley.

The Language of the Criş Culture

If the Starcevo–Criş–Karanovo migrants were at all similar to pioneer farmers in North America, Brazil, southeast Asia, and other parts of the world, it is very likely that they retained the language spoken in their parent villages in northern Greece. Forager languages were more apt to decline in the face of agricultural immigration. Farmers had a higher birth rate; their settlements were larger, and were occupied permanently. They produced food surpluses that were easier to store over the winter. Owning and feeding "cultured" animals has always been seen as an utterly different ethos from hunting wild ones, as Ian Hodder emphasized. The material and ritual culture and economy of the immigrant farmers were imposed on the landscapes of Greece and southeastern Europe and persisted there, whereas the external signs of forager identity disappeared. The language of the foragers might have had substrate effects on that of the farmers, but it is difficult to imagine a plausible scenario under which it could have competed with the farmers' language.[14]

What languages were spoken by Starčevo, Criş, and Karanovo I pioneers? The parent language for all of them was spoken in the Thessalian plain of Greece, where the first Neolithic settlements were founded about 6700–6500 BCE probably by seafarers who island-hopped from western Anatolia in open boats. Katherine Perlés has convincingly demonstrated that the material culture and economy of the first farmers in Greece was transplanted from the Near East or Anatolia. An origin somewhere in western Anatolia is suggested by similarities in pottery, flint tools, ornaments, female figurines, pintadera stamps, lip labrets, and other traits. The migrants leapfrogged to the Thessalian plain, the richest agricultural land in Greece, almost certainly on the basis of information from scouts (probably Aegean fishermen) who told their relatives in Anatolia about the destination. The population of farmers in Thessaly grew rapidly. At least 120 Early Neolithic settlements stood on the Thessalian plain by 6200–6000 BCE, when pioneers began to move north into the temperate forests of southeastern Europe. The Neolithic villages of Thessaly provided the original breeds of domesticated sheep, cattle, wheat, and barley, as well as red-on-white pottery, female-centered domestic rituals, bracelets and beads made of Aegean *Spondylus* shell, flint tool types, and other traditions that were carried into the Balkans. The language of Neolithic Thessaly probably was a dialect of a language spoken in western Anatolia about 6500 BCE. Simplification and leveling should have occurred among the first colonist dialects in Thessaly, so the 120 villages occupied five hundred years later spoke a language that had passed through a bottleneck and probably was just beginning to separate again into strongly differentiated dialects.[15]

The tongue spoken by the first Criş farmers in the East Carpathian foothills about 5800–5600 BCE was removed from the parent tongue spoken by the first settlers in Thessaly by less than a thousand years—the same interval that separates Modern American English from Anglo-Saxon. That was long enough for several new Old European Neolithic languages to have emerged from the Thessalian parent, but they would have belonged to a single language family. That language family was not Indo-European. It came from the wrong place (Anatolia and Greece) at the wrong time (before 6500 BCE). Curiously a fragment of that lost language might be preserved in the Proto-Indo-European term for bull, *tawro-s*, which many linguists think was borrowed from an Afro-Asiatic term. The Afro-Asiatic super-family generated both Egyptian and Semitic in the Near East, and one of its early languages might have been spoken in Anatolia by the earliest farmers. Perhaps the Criş people spoke a language of Afro-Asiatic type, and as they drove their cattle into the East Carpathian valleys they called them something like *tawr-*.[16]

FARMER MEETS FORAGER: THE BUG-DNIESTER CULTURE

The first indigenous North Pontic people to adopt Criş cattle breeding and perhaps also the Criş word for bull were the people of the Bug-Dniester culture, introduced a few pages ago. They occupied the frontier where the expansion of the Criş farmers came to a halt, apparently blocked by the Bug-Dniester culture itself. The initial contact between farmers and foragers must have been a fascinating event. The Criş immigrants brought herds of cultured animals that wandered up the hillsides among the deer. They introduced sheep, plum orchards, and hot wheat-cakes. Their families lived in the same place all year, year after year; they cut down the trees to make houses and orchards and gardens; and they spoke a foreign language.

The foragers' language might have been part of the broad language family from which Proto-Indo-European later emerged, although, since the ultimate fate of the Bug-Dniester culture was extinction and assimilation, their dialect probably died with their culture.[17]

The Bug-Dniester culture grew out of Mesolithic forager cultures that dwelt in the region since the end of the last Ice Age. Eleven Late Mesolithic technological-typological groups have been defined by differences in flint tool kits just in Ukraine; other Late Mesolithic flint tool-based groups have been identified in the Russian steppes east of the Don River, in the North Caspian Depression, and in coastal Romania. Mesolithic camps have been found in the lower Danube valley and the coastal steppes northwest of the Black Sea, not far from the Criş settlement area. In the Dobruja, the peninsula of rocky hills skirted by the Danube delta at its mouth, eighteen to twenty Mesolithic surface sites were found just in one small area northwest of Tulcea on the southern terraces of the Danube River. Late Mesolithic groups also occupied the northern side of the estuary. Mirnoe is the best-studied site here. The Late Mesolithic hunters at Mirnoe hunted wild aurochs (83% of bones), wild horse (14%), and the extinct *Equus hydruntinus* (1.1%). Farther up the coast, away from the Danube delta, the steppes were drier, and at Late Mesolithic Girzhevo, on the lower Dniester, 62% of the bones were of wild horses, with fewer aurochs and *Equus hydruntinus*. There is no archaeological trace of contact between these coastal steppe foragers and the Criş farmers who were advancing into the upland forest-steppe.[18]

The story is different in the forest-steppe. At least twenty-five Bug-Dniester sites have been excavated in the forest-steppe zone in the middle and upper parts of the South Bug and Dniester River valleys, in the transitional ecological zone where rainfall was sufficient for the growth of forests but there were still open meadows and some pockets of steppe. This environment was favored by the Criş immigrants. In it the native foragers had for generations hunted red deer, roe deer, and wild boar, and caught riverine fish (especially the huge river catfish, *Siluris glanis*). Early Bug-Dniester flint tools showed similarities both to coastal steppe groups (Grebenikov and Kukrekskaya types of tool kits) and northern forest groups (Donets types).

Pottery and the Beginning of the Neolithic

The Bug-Dniester culture was a Neolithic culture; Bug-Dniester people knew how to make fired clay pottery vessels. The first pottery in the Pontic-Caspian region, and the beginning of the Early Neolithic, is asso-

ciated with the Elshanka culture in the Samara region in the middle
Volga River valley. It is dated by radiocarbon (on shell) about 7000–6500
BCE, which makes it, surprisingly, the oldest pottery in all of Europe.
The pots were made of a clay-rich mud collected from the bottoms of stag-
nant ponds. They were formed by the coiling method and were baked in
open fires at 450–600°C (figure 8.4).[19] From this northeastern source ceramic technology diffused south and westward. It was adopted widely by
most foraging and fishing bands across the Pontic-Caspian region about
6200–6000 BCE, before any clear contact with southern farmers. Early
Neolithic pottery tempered with vegetal material and crushed shells appeared at Surskii Island in the Dnieper Rapids in levels dated about
6200–5800 BCE. In the lower Don River valley a crude vegetal-tempered
pottery decorated with incised geometric motifs appeared at Rakushechni
Yar and other sites such as Samsonovka in levels dated 6000–5600 BCE.[20]
Similar designs and vessel shapes, but made with a shell-tempered clay
fabric, appeared on the lower Volga, at Kair Shak III dated about 5700–
5600 BCE (6720±80 BP). Older pottery was made in the North Caspian
at Kugat, where a different kind of pottery was stratified beneath Kair
Shak-type pottery, possibly the same age as the pottery at Surskii Island.
Primitive, experimental ceramic fragments appeared about 6200 BCE
also at Matveev Kurgan in the steppes north of the Sea of Azov. The oldest pottery south of the middle Volga appeared at the Dnieper Rapids
(Surskii), on the lower Don (Rakushechni Yar), and on the lower Volga
(Kair Shak III, Kugat) at about the same time, around 6200–6000 BCE
(figure 8.4).

The earliest pottery in the South Bug valley was excavated by Danilenko
at Bas'kov Ostrov and Sokolets II, dated by five radiocarbon dates about
6200–6000 BCE, about the same age as Surskii on the Dnieper.[21] In the
Dniester River valley, just west of the South Bug, at Soroki II, archaeologists excavated two stratified Late Mesolithic occupations (levels 2 and 3)
dated by radiocarbon to about 6500–6200 BCE. They contained no pottery. Pottery making was adopted by the early Bug-Dniester culture about
6200 BCE, probably the same general time it appeared in the Dnieper
valley and the Caspian Depression.

Farmer-Forager Exchanges in the Dniester Valley

After about 5800–5700 BCE, when Criș farmers moved into the East
Carpathian foothills from the west, the Dniester valley became a frontier between two very different ways of life. At Soroki II the uppermost

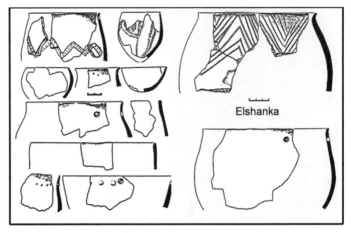

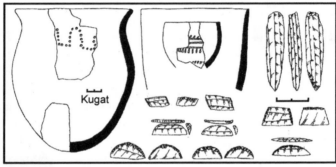

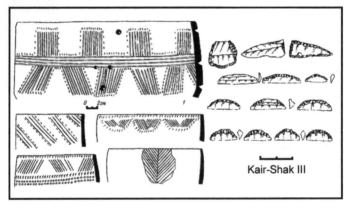

Figure 8.4 Top: Early Neolithic ceramics of Elshanka type on the middle Volga (7000–6500 BCE); *middle*: ceramics and flint tools from Kugat (perhaps 6000 BCE), North Caspian; *bottom*: ceramics and flint tools from Kair-Shak III (5700–5600 BCE) North Caspian. After (*top*) Mamonov 1995; and (*middle and bottom*) Barynkin and Kozin 1998.

occupation level (1) was left by Bug-Dniester people who clearly had made contact with the incoming Criş farmers, dated by good radiocarbon dates at about 5700–5500 BCE. Some of the ceramic vessels in level 1 were obvious copies of Criş vessels—round-bodied, narrow-mouthed jars on a ring base and bowls with carinated sides. But they were made locally, using clay tempered with sand and plant fibers. The rest of the pottery in level 1 looked more like indigenous bag-shaped South Bug ceramics (figure 8.5). Continuity in the flint tools between level 1 and the older levels 2 and 3 suggests that it was the same basic culture, and all three levels are traditionally assigned to the Bug-Dniester culture.

The Bug-Dniester people who lived at Soroki II in the level 1 camp copied more than just Criş pottery. Botanists found seed impressions in the clay vessels of three kinds of wheat. Level 1 also yielded a few bones from small domesticated cattle and pigs. This was the beginning of a significant shift—the adoption of an imported food-production economy by the native foragers. It is perhaps noteworthy that the exotic ceramic types copied by Soroki II potters were small Criş pedestaled jars and bowls, probably used to serve drink and food rather than to store or cook it. Perhaps Criş foods were served to visiting foragers in jars and bowls like these inside Criş houses, inspiring some Bug-Dniester families to re-create both the new foods and the vessels in which they were served. But the original decorative motifs on Bug-Dniester pottery, the shapes of the largest pots, the vegetal and occasional shell temper in the clay, and the low-temperature firing indicate that early Bug-Dniester potters knew their own techniques, clays, and tempering formulas. The largest pots they made (for cooking? storage?) were shaped like narrow-mouthed baskets, unlike any shape made by Criş potters.

Three kinds of wheat impressions appeared in the clay of early Bug-Dniester pots at two sites in the Dniester valley: Soroki II/level 1 and Soroki III. Both sites had impressions of emmer, einkorn, and spelt.[22] Was the grain actually grown locally? Both sites had a variety of wheats, with impressions of chaff and spikelets, parts removed during threshing. The presence of threshing debris suggests that at least some grain was grown and threshed locally. The foragers of the Dniester valley seem to have cultivated at least small plots of grain very soon after their initial contact with Criş farmers. What about the cattle?

In three Early Bug-Dniester Neolithic sites in the Dniester valley occupied about 5800–5500 BCE, domesticated cattle and swine averaged 24% of the 329 bones recovered from garbage pits, if each bone is counted for the NISP; or 20% of the animals, if the bones are converted into a

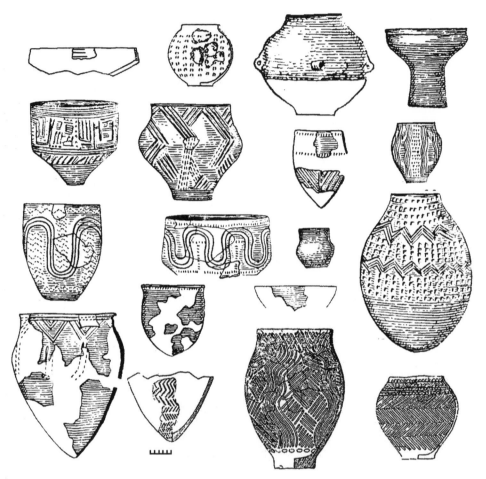

Figure 8.5 Pottery types of the Bug-Dniester culture. The four vessels in the top row appear to have been copied after Criş types seen in Figure 8.2. After Markevich 1974; and Dergachev 1999.

minimum number of individuals, or MNI. Red deer and roe deer remained more important than domesticated animals in the meat diet. Middle Bug-Dniester sites (Samchin phase), dated about 5600–5400 BCE, contained more domesticated pigs and cattle: at Soroki I/level 1a, a Middle-phase site, cattle and swine made up 49% of the 213 bones recovered (32% MNI). By the Late (Savran) phase, about 5400–5000 BCE, domesticated pigs and cattle totaled 55% of the animal bones (36% MNI) in two sites.[23] In contrast, the Bug-Dniester settlement sites in the South

Bug valley, farther away from the source of the domesticated animals, never showed more than 10% domesticated animal bones. But even in the South Bug valley a few domesticated cattle and pigs appeared at Bas'kov Ostrov and Mit'kov Ostrov very soon after the Criş farmers entered the Eastern Carpathian foothills. The "availability" phase, in Zvelebil's three-phase description of farmer-forager interactions, was very brief.[24] Why? What was so attractive about Criş foods and even the pottery vessels in which they were served?

There are three possibilities: intermarriage, population pressure, and status competition. Intermarriage is an often-repeated but not very convincing explanation for incremental changes in material culture. In this case, imported Criş-culture wives would be the vehicle through which Criş-culture pottery styles and foods should have appeared in Bug-Dniester settlements. But Warren DeBoer has shown that wives who marry into a foreign tribe among tribal societies often feel so exposed and insecure that they become hyper-correct imitators of their new cultural mores rather than a source of innovation. And the technology of Bug-Dniester ceramics, the method of manufacture, was local. Technological styles are often better indicators of ethnic origin than decorative styles. So, although there may have been intermarriage, it is not a persuasive explanation for the innovations in pottery or economy on the Dniester frontier.[25]

Was it population pressure? Were the pre-Neolithic Bug-Dniester foragers running out of good hunting and fishing grounds, and looking for ways to increase the amount of food that could be harvested within their hunting territories? Probably not. The forest-steppe was an ideal hunting territory, with maximal amounts of the forest-edge environment preferred by deer. The abundant tree pollen in Criş-period soils indicates that the Criş pioneers had little impact on the forest around them, so their arrival did not greatly reduce deer populations. A major component of the Bug-Dniester diet was riverine fish, some of which supplied as much meat as a small adult pig, and there is no evidence that fish stocks were falling. Cattle and pigs might have been acquired by cautious foragers as a hedge against a bad year, but the immediate motive probably was not hunger.

The third possibility is that the foragers were impressed by the abundance of food available for feasting and seasonal festivals among Criş farmers. Perhaps some Bug-Dniester locals were invited to such festivals by the Criş farmers in an attempt to encourage peaceful coexistence. Socially ambitious foragers might have begun to cultivate gardens and raise cattle to sponsor feasts among their own people, even making serving bowls and cups like those used in Criş villages—a political explanation, and one that also explains why Criş pots were copied. Unfortunately neither culture had cemeteries, and so we cannot examine graves to look for evidence of a growing social hierarchy. Status objects seem to have been few, with the possible exception of food itself. Probably both economic insurance and social status played roles in the slow but steady adoption of food production in the Dniester valley.

The importance of herding and cultivation in the Bug-Dniester diet grew very gradually. In Criş settlements domesticated animals contributed 70–80% of the bones in kitchen middens. In Bug-Dniester settlements domesticated animals exceeded hunted wild game only in the latest phase, and only in the Dniester valley, immediately adjacent to Criş settlements. Bug-Dniester people never ate mutton—not one single sheep bone has been found in a Bug-Dniester site. Early Bug-Dniester bakers did not use Criş-style saddle querns to grind their grain; instead, they initially used small, rhomboidal stone mortars of a local style, switching to Criş-style saddle querns only in the middle Bug-Dniester phase. They preferred their own chipped flint axe types to the smaller polished stone Criş axes. Their pottery was quite distinctive. And their historical trajectory led directly back to the local Mesolithic populations, unlike the Criş culture.

Even after 5500–5200 BCE, when a new farming culture, the Linear Pottery culture, moved into the East Carpathian piedmont from southern Poland and replaced the Criş culture, the Dniester valley frontier survived. No Linear Pottery sites are known east of the Dniester valley.[26] The Dniester was a cultural frontier, not a natural one. It persisted despite the passage of people and trade goods across it, and through significant cultural changes on each side. Persistent cultural frontiers, particularly at the edges of ancient migration streams, usually are ethnic and linguistic frontiers. The Bug-Dniester people may well have spoken a language belonging to the language family that produced Pre-Proto-Indo-European, while their Criş neighbors spoke a language distantly related to those of Neolithic Greece and Anatolia.

Beyond the Frontier: Pontic-Caspian Foragers before Cattle Arrived

The North Pontic societies east of the Dniester frontier continued to live as they always had, by hunting, gathering wild plants, and fishing until about 5200 BCE. Domesticated cattle and hot wheatcakes might have seemed irresistibly attractive to the foragers who were in direct contact with the farmers who presented and legitimized them, but, away from that active frontier, North Pontic forager-fishers were in no rush to become animal tenders. Domesticated animals can only be raised by people who are committed morally and ethically to watching their families go hungry rather than letting them eat the breeding stock. Seed grain and breeding stock must be saved, not eaten, or there will be no crop and no calves the next year. Foragers generally value immediate sharing and generosity over miserly saving for the future, so the shift to keeping breeding stock was a moral as well as an economic one. It probably offended the old morals. It is not surprising that it was resisted, or that when it did begin it was surrounded by new rituals and a new kind of leadership, or that the new leaders threw big feasts and shared food when the deferred investment paid off. These new rituals and leadership roles were the foundation of Indo-European religion and society.[27]

The most heavily populated part of the Pontic-Caspian steppes was the place where the shift to cattle keeping happened next after the Bug-Dniester region. This was around the Dnieper Rapids. The Dnieper Rapids started at modern Dnepropetrovsk, where the Dnieper River began to cut down to the coastal lowlands through a shelf of granite bedrock, dropping 50 m in elevation over 66 km. The Rapids contained ten major cascades, and in early historical accounts each one had its own name, guardian spirits, and folklore. Fish migrating upstream, like the sudak (*Lucioperca*), could be taken in vast quantities at the Rapids, and the swift water between the cascades was home to wels (*Silurus glanis*), a type of catfish that grows to 16 feet. The bones of both types of fish are found in Mesolithic and Neolithic camps near the Rapids. At the southern end of the Rapids there was a ford near Kichkas where the wide Dnieper could be crossed relatively easily on foot, a strategic place in a world without bridges.

The Rapids and many of the archaeological sites associated with them were inundated by dams and reservoirs built between 1927 and 1958. Among the many sites discovered in connection with reservoir construction was Igren 8 on the east bank of the Dnieper. Here the deepest level F contained Late Mesolithic Kukrekskaya flint tools; levels E and E1 above contained Surskii Early Neolithic pottery (radiocarbon dated 6200–5800 BCE); and stratum D1 above that contained Middle Neolithic Dnieper-Donets I pottery tempered with plant fibers and decorated with incised chevrons and small comb stamps (probably about 5800–5200 BCE but not directly dated by radiocarbon). The animal bones in the Dnieper-Donets I garbage were from red deer and fish. The shift to cattle keeping had not yet begun. Dnieper-Donets I was contemporary with the Bug-Dniester culture.[28]

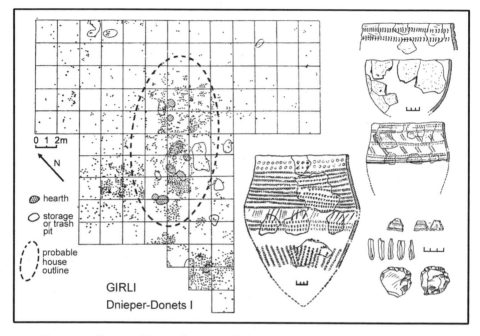

Figure 8.6 Dnieper-Donets I camp at Girli, Ukraine, probably about 5600–5200 BCE. After Neprina 1970, Figures 3, 4, and 8.

Campsites of foragers who made Dnieper-Donets I (DDI) pottery have been excavated on the southern borders of the Pripet Marshes in the northwest and in the middle Donets valley in the east, or over much of the forest-steppe and northern steppe zone of Ukraine. At Girli (figure 8.6) on the upper Teterev River near Zhitomir, west of Kiev, a DDI settlement contained eight hearths arranged in a northeast-southwest line of four pairs, each pair about 2–3 m apart, perhaps representing a shelter some 14 m long for four families. Around the hearths were thirty-six hundred flint tools including microlithic blades, and sherds of point-based pots decorated with comb-stamped and pricked impressions. The food economy depended on hunting and gathering. Girli was located on a trail between the Dnieper and South Bug rivers, and the pottery was similar in shape and decoration to some Bug-Dniester ceramics of the middle or Samchin phase. But DDI sites did not contain domesticated animals or plants, or even polished stone axes like those of the Criş and late Bug-Dniester cultures; DDI axes were still chipped from large pieces of flint.[29]

Forager Cemeteries around the Dnieper Rapids

Across most of Ukraine and European Russia post-glacial foragers did not create cemeteries. The Bug-Dniester culture was typical: they buried their dead by ones and twos, often using an old campsite, perhaps the one where the death occurred. Graveside rituals took place but not in places set aside just for them. Cemeteries were different: they were formal plots of ground reserved just for funerals, funeral monuments, and public remembrance of the dead. Cemeteries were visible statements connecting a piece of land with the ancestors. During reservoir construction around the Dnieper Rapids archaeologists found eight Mesolithic and forager Neolithic cemeteries, among them Vasilievka I (twenty-four graves), Vasilievka II (thirty-two graves), Vasilievka III (forty-five graves), Vasilievka V (thirty-seven graves), Marievka (fifteen graves), and Volos'ke (nineteen graves). No comparable cluster of forager cemeteries exists anywhere else in the Pontic-Caspian region.

Several different forager populations seem to have competed with one another around the Dnieper Rapids at the end of the Ice Age. Already by about 8000 BCE, as soon as the glaciers melted, at least three skull-and-face types, a narrow-faced gracile type (Volos'ke), a broad-faced medium-weight type (Vasilievka I), and a broad-faced robust type (Vasilievka III) occupied different cemeteries and were buried in different poses (contracted and extended). Two of the nineteen individuals buried at Volos'ke and two (perhaps three) of the forty-five at Vasilevka III were wounded by weapons tipped with Kukrekskaya-type microlithic blades. The Vasilievka III skeletal type and burial posture ultimately spread over the whole Rapids during the Late Mesolithic, 7000–6200 BCE. Two cemeteries that were assumed to be Early Neolithic (Vasilievka II and Marievka) because of the style of the grave now are dated by radiocarbon to 6500–6000 BCE, or the Late Mesolithic.

Only one of the Dnieper Rapids cemeteries, Vasilievka V, is dated to the Middle Neolithic DDI period by radiocarbon dates (5700–5300 BCE). At Vasilevka V thirty-seven skeletons were buried in supine positions (on their backs) with their hands near the pelvis, with their heads to the northeast. Some were buried singly in individual pits, and others apparently were layered in reused graves. Sixteen graves in the center of the cemetery seem to represent two or three superimposed layers of burials, the first hint of a collective burial ritual that would be elaborated greatly in the following centuries. Eighteen graves out of thirty-seven were sprinkled with red ochre, again a hint of things to come. The grave gifts at Vasilievka V, however, were very simple, limited to microlithic flint blades and flint scrapers. These were the last people on the Dnieper Rapids who clung to the old morality and rejected cattle keeping.[30]

Foragers on the Lower Volga and Lower Don

Different styles of pottery were made among the Early Neolithic foragers who lived even farther east, a longer distance away from the forager/farmer frontier on the Dniester. Forager camps on the lower Volga River dated between 6000 and 5300 BCE contained flat-based open bowls made of clay tempered with crushed shell and vegetal material, and were decorated by stabbing rows of impressions with a triangular-ended stick or drawing incised diamond and lozenge shapes. These decorative techniques were different from the comb-stamps used to decorate DDI pottery in the Dnieper valley. Flint tool kits on the Volga contained many geometric microliths, 60–70% of the tools, like the flint tools of the earlier Late Mesolithic foragers. Important Early Neolithic sites included Varfolomievka level 3 (radiocarbon dated about 5900–5700 BCE) and Kair-Shak III (also dated about 5900–5700 BCE) in the lower Volga region; and the lower levels at Rakushechni Yar, a dune on the lower Don (dated 6000–5600 BCE).[31] At Kair Shak III, located in an environment that was then semi-desert, the economy was based almost entirely on hunting onagers (*Equus hemionus*). The animal bones at Varfolomievka, located in a small river valley in the dry steppe, have not been reported separately by level, so it is impossible to say what the level 3 Early Neolithic economy was, but half of all the animal bones at Varfolomievka were of horses (*Equus caballus*), with some bones of aurochs (*Bos primigenius*). Fish scales (unidentified) were found on the floors of the dwellings. At Rakushechni Yar, then surrounded by broad lower-Don valley gallery forests, hunters pursued red deer, wild horses, and wild pigs. As I noted in several endnotes in this chapter, some archaeologists have claimed that the herding of cattle and sheep began earlier in the lower Don-Azov steppes, but this is unlikely. Before 5200 BCE the forager-farmer frontier remained confined to the Dniester valley.[32]

THE GODS GIVE CATTLE

The Criş colonization of the Eastern Carpathians about 5800 BCE created a robust and persistent cultural frontier in the forest-steppe zone at the Dniester valley. Although the Bug-Dniester culture quickly acquired at least some domesticated cereals, pigs, and cattle, it retained an economy based primarily on hunting and gathering, and remained culturally and economically distinct in most ways. Beyond it, both in the forest-steppe zone and the steppe river valleys to the east, no other indigenous societies seem to have adopted cereal cultivation or domesticated animals until after about 5200 BCE.

In the Dniester valley, native North Pontic cultures had direct, face-to-face contact with farmers who spoke a different language, had a different religion, and introduced an array of invasive new plants and animals as if they were something wonderful. The foragers on the frontier itself rapidly accepted some cultivated plants and animals but rejected others, particularly sheep. Hunting and fishing continued to supply most of the diet. They did not display obvious signs of a shift to new rituals or social structures. Cattle keeping and wheat cultivation seem to have been pursued part-time, and were employed as an insurance policy against bad years and perhaps as a way of keeping up with the neighbors, not as a replacement of the foraging economy and morality. For centuries even this halfway shift to partial food production was limited to the Dniester valley, which became a narrow and well-defined frontier. But after 5200 BCE a new threshold in population density and social organization seems to have been crossed among European Neolithic farmers. Villages in the East Carpathian piedmont adopted new customs from the larger towns in the lower Danube valley, and a new, more complex culture appeared, the Cucuteni-Tripolye culture. Cucuteni-Tripolye villages spread eastward. The Dniester frontier was breached, and large western farming communities pushed into the Dniester and South Bug valleys. The Bug-Dniester culture, the original frontier society, disappeared into the wave of Cucuteni-Tripolye immigrants.

But away to the east, around the Dnieper Rapids, the bones of domesticated cattle, pigs, and, remarkably, even sheep began to appear regularly in garbage dumps. The Dnieper Rapids was a strategic territory, and the clans that controlled it already had more elaborate rituals than clans elsewhere in the steppes. When they accepted cattle keeping it had rapid economic and social consequences across the steppe zone.

CHAPTER NINE

Cows, Copper, and Chiefs

The Proto-Indo-European vocabulary contained a compound word (*weik-potis*) that referred to a village chief, an individual who held power within a residential group; another root (*reg̑-*) referred to another kind of powerful officer. This second root was later used for *king* in Italic (*rēx*), Celtic (*rīx*), and Old Indic (*raj-*), but it might originally have referred to an official more like a priest, literally a "*regulator*" (from the same root) or "one who makes things *right*" (again the same root), possibly connected with drawing "*correct*" (same root) boundaries. The speakers of Proto-Indo-European had institutionalized offices of power and social ranks, and presumably showed deference to the people who held them, and these powerful people, in return, sponsored feasts at which food and gifts were distributed.[1] When did a hierarchy of social power first appear in the Pontic-Caspian region? How was it expressed? And who were these powerful people?

Chiefs first appeared in the archaeological record of the Pontic-Caspian steppes when domesticated cattle, sheep, and goats first became widespread, after about 5200–5000 BCE.[2] An interesting aspect of the spread of animal keeping in the steppes was the concurrent rapid rise of chiefs who wore multiple belts and strings of polished shell beads, bone beads, beaver-tooth and horse-tooth beads, boars tusk pendants, boars-tusk caps, boars-tusk plates sewed to their clothing, pendants of crystal and porphyry, polished stone bracelets, and gleaming copper rings. Their ornaments must have clacked and rustled when they walked. Older chiefs carried maces with polished stone mace-heads. Their funerals were accompanied by the sacrifice of sheep, goats, cattle, and horses, with most of the meat and bones distributed to the celebrants so only a few symbolic lower leg pieces and an occasional skull, perhaps attached to a hide, remained in the grave. No such ostentatious leaders had existed in the old hunting and gathering bands of the Neolithic. What made their sudden rise even more intriguing is that the nitrogen levels in their bones suggest that more than 50% of their meat diet continued to come from fish. In the Volga region the bones of horses, the preferred wild prey of the earlier hunters, still outnumbered cattle and sheep in kitchen trash. The domesticated cattle and sheep that played such a large ritual role were eaten only infrequently, particularly in the east.

What seems at first to be the spread of a new food economy on second look appears to be deeply interwined in new rituals, new values associated with them, and new institutions of social power. People who did not accept the new animal currency, who remained foragers, did not even use formal cemeteries, much less sponsor such aggrandizing public funeral feasts. Their dead still were buried simply, in plain clothing, in their old camping places. The cultural gap widened between those who tended domesticated animals, including foreign sheep and goats, and those who hunted native wild animals.

The northern frontier of the new economy coincided with the ecological divide between the forests in the north and the steppes in the south. The northern hunters and fishers refused to be shackled to domesticated animals for another two thousand years. Even in the intervening zone of forest-steppe the percentage of domesticated animal bones declined and the importance of hunted game increased. In contrast, the eastern frontier of the new economy did not coincide with an ecotone but instead ran along the Ural River, which drained the southern flanks of the Ural Mountains and flowed south through the Caspian Depression into the Caspian Sea. East of the Ural River, in the steppes of northern Kazakhstan, steppe foragers of the Atbasar type continued to live by hunting wild horses, deer, and aurochs. They lived in camps sheltered by grassy bluffs on low river terraces or on the marshy margins of lakes in the steppes. Their rejection of the new western economy possibly was rooted in ethnic and linguistic differences that had sharpened during the millennia between 14,000 and 9,000 BCE, when the Khvalynian Sea had divided the societies of the Kazakh and the Russian steppes. Regardless of its cause, the Ural valley became a persistent frontier dividing western steppe societies that accepted domesticated animals from eastern steppe societies that rejected them.

Copper ornaments were among the gifts and baubles traded eastward across the steppes from the Danube valley to the Volga-Ural region with the first domesticated animals. The regular, widespread appearance of copper in the Pontic-Caspian steppes signals the beginning of the Eneolithic. The copper was Balkan in origin and probably was obtained with the animals through the same trade networks. From this time forward Pontic-Caspian steppe cultures were drawn into increasingly complicated social, political, and economic relations with the cultures of the Balkans and the lower Danube valley. The gulf between them, however, only intensified. By 4400–4200 BCE, when the Old European cultures were at their peak of economic productivity, population size, and stability, their frontier with the Pontic-Caspian herding cultures was the most pronounced cultural divide in prehistoric Europe, an even starker contrast than that between the northern forest hunters and the steppe herders. The Neolithic and Eneolithic cultures of the Balkans, Carpathians, and middle and lower Danube valley had more productive farming economies in an age when that really mattered, their towns and houses were much more substantial, and their craft techniques, decorative aesthetics, and metallurgy were more sophisticated than those of the steppes. The Early Eneolithic herding cultures of the steppes certainly were aware of the richly ornamented and colorfully decorated people of Old Europe, but steppe societies developed in a different direction.[3]

The Early Copper Age in Old Europe

There is an overall rhythm to the Eneolithic over most of southeastern Europe: a rise to a new level of social and technological complexity, its flourishing, and its subsequent disintegration into smaller-scale, more mobile, and technologically simpler communities at the opening of the Bronze Age. But it began, developed, and ended differently in different places. Its beginning is set at about 5200–5000 BCE in Bulgaria, which was in many ways the heart and center of Old Europe. Pontic-Caspian steppe societies were pulled into the Old European copper-trade network at least as early as 4600 BCE, more than six hundred years before copper was regularly used in Germany, Austria, or Poland.[4]

The scattered farming hamlets of Bulgaria and southern Romania, about 5200–5000 BCE, blossomed into increasingly large and solidly built agricultural villages of large multiroomed timber and mud-plaster houses, often two-storied, set in cleared and cultivated landscapes surrounded by herds of cattle, pigs, and sheep. Cattle pulled ards, primitive scratch-plows, across the fields.[5] In the Balkans and the fertile plains of the lower Danube valley, villages were rebuilt on the same spot generation after generation, creating stratified tells that grew to heights of 30–50 feet, lifting the village above its surrounding fields. Marija Gimbutas has made Old Europe famous for the ubiquity and variety of its goddesses. Household cults symbolized by broad-hipped female figurines were practiced everywhere. Marks incised on figurines and pots suggest the appearance of a notation system.[6] Fragments of colored plaster suggest that house walls were painted with the same swirling, curvilinear designs that appeared on decorated pottery. Potters invented kilns that reached temperatures of 800–1100°C. They used a low-oxygen reducing atmosphere to create a black ceramic surface that was painted with graphite to make silver designs; or a bellows-aided high-oxygen atmosphere to create a red or orange surface, intricately painted in white ribbons bordered with black and red.

Pottery kilns led to metallurgy. Copper was extracted from stone by mixing powdered green-blue azurite or malachite minerals (possibly used for pigments) with powdered charcoal and baking the mixture in a bellows-aided kiln, perhaps accidentally at first. At 800°C the copper separated from the powdered ore in tiny shining beads. It could then be tapped out, reheated, forged, welded, annealed, and hammered into a wide variety of tools (hooks, awls, blades) and ornaments (beads, rings, and other pendants). Ornaments of gold (probably mined in Transylvania and coastal Thrace) began to circulate in the same trade networks. The early phase of copper working began before 5000 BCE.

Balkan smiths, about 4800–4600 BCE, learned to fashion molds that withstood the heat of molten copper, and began to make cast copper tools and weapons, a complicated process requiring a temperature of 1,083°C to liquefy copper metal. Molten copper must be stirred, skimmed, and poured correctly or it cools into a brittle object full of imperfections. Well-made cast copper tools were used and exchanged across southeastern Europe by about 4600–4500 BCE in eastern Hungary with the Tiszapolgar culture; in Serbia with the Vinča D culture; in Bulgaria at Varna and in the Karanovo VI tell settlements; in Romania with the Gumelnitsa culture; and in Moldova and eastern Romania with the Cucuteni-Tripolye culture. Metallurgy was a new and different kind of craft. It was obvious to anyone that pots were made of clay, but even after being told that a shiny copper ring was made from a green-stained rock, it was difficult to see how. The magical aspect of copperworking set metalworkers apart, and the demand for copper objects increased trade. Prospecting, mining, and long-distance trade for ore and finished products introduced a new era in inter-regional politics and interdependence that quickly reached deep into the steppes as far as the Volga.[7]

Kilns and smelters for pottery and copper consumed the forests, as did two-storied timber houses and the bristling palisade walls that protected many Old European settlements, particularly in northeastern Bulgaria. At Durankulak and Sabla Ezerec in northeastern Bulgaria and at Tîrpeşti in Romania, pollen cores taken near settlements show significant reductions in local forest cover.[8] The earth's climate reached its post-glacial thermal maximum, the Atlantic period, about 6000–4000 BCE, and was at its warmest during the late Atlantic (paleoclimatic zone A3), beginning about 5200 BCE. Riverine forests in the steppe river valleys contracted because of increased warmth and dryness, and grasslands expanded. In the forest-steppe uplands majestic forests of elm, oak, and lime trees spread from the Carpathians to the Urals by 5000 BCE. Wild honeybees, which preferred lime and oak trees for nests, spread with them.[9]

The Cucuteni-Tripolye Culture

The Cucuteni-Tripolye culture occupied the frontier between Old Europe and the Pontic-Caspian cultures. More than twenty-seven hundred Cucuteni-Tripolye sites have now been discovered and examined with small excavations, and a few have been entirely excavated (figure 9.1). The Cucuteni-Tripolye culture first appeared around 5200–5000 BCE and survived a thousand years longer than any other part of the Old European world. Tripolye people were still creating large houses and villages, advanced pottery and metals, and female figurines as late as 3000 BCE. They were the sophisticated western neighbors of the steppe people who probably spoke Proto-Indo-European.

Cucuteni-Tripolye is named after two archaeological sites: Cucuteni, discovered in eastern Romania in 1909, and Tripolye, discovered in central Ukraine in 1899. Romanian archaeologists use the name Cucuteni and Ukrainians use Tripolye, each with its own system of internal chronological divisions, so we must use cumbersome labels like Pre-Cucuteni III/Tripolye A to refer to a single prehistoric culture. There is a Borges-like dreaminess to the Cucuteni pottery sequence: one phase (Cucuteni C) is not a phase at all but rather a type of pottery probably made outside the Cucuteni-Tripolye culture; another phase (Cucuteni A1) was defined before it was found, and never was found; still another (Cucteni A5) was created in 1963 as a challenge for future scholars, and is now largely forgotten; and the whole sequence was first defined on the assumption, later proved wrong, that the Cucuteni A phase was the oldest, so later archaeologists had to invent the Pre-Cucuteni phases I, II, and III, one of which (Pre-Cucuteni I) might not exist. The positive side of this obsession with pottery types and phases is that the pottery is known and studied in minute detail.[10]

The Cucuteni-Tripolye culture is defined most clearly by its decorated pottery, female figurines, and houses. They first appeared about 5200–5000 BCE in the East Carpathian piedmont. The late Linear Pottery

Figure 9.1 Early Eneolithic sites in the Pontic-Caspian region.

people of the East Carpathians acquired these new traditions from the late Boian-Giuleşti and late Hamangia cultures of the lower Danube valley. They adopted Boian and Hamangia design motifs in pottery, Boian-style female figurines, and some aspects of Boian house architecture (a clay floor fired before the walls were raised, called a *ploshchadka* floor in Russian). They acquired objects made of Balkan copper and Dobrujan flint, again from the Danube valley. The borrowed customs were core aspects of any tribal farming culture—domestic pottery production, domestic architecture, and domestic female-centered rituals—and so it seems likely that at least some Boian people migrated up into the steep, thickly forested valleys at the peakline of the East Carpathians. Their appearance defined the beginning of the Cucuteni-Tripolye culture—phases Pre-Cucuteni I (?) and II (about 5200–4900 BCE).

The first places that showed the new styles were clustered near high Carpathian passes, and perhaps attracted migrants partly because they controlled passage through the mountains. From these high Carpathian valleys the new styles and domestic rituals spread quickly northeastward to Pre-Cucuteni II settlements located as far east as the Dniester valley. As the culture developed (during pre-Cucteni III/Tripolye A) it was carried across the Dniester, erasing a cultural frontier that had existed for six hundred to eight hundred years, and into the South Bug River valley in Ukraine. Bug-Dniester sites disappeared. Tripolye A villages occupied the South Bug valley from about 4900–4800 BCE to about 4300–4200 BCE.

The Cucuteni-Tripolye culture made a visible mark on the forest-steppe environment, reducing the forest and creating pastures and cultivated fields over wider areas. At Floreşti, on a tributary of the Seret River, the remains of a late Linear Pottery homestead, radiocarbon dated about 5200–5100 BCE, consisted of a single house with associated garbage pits, set in a clearing in an oak-elm forest—tree pollen was 43% of all pollen. Stratified above it was a late Pre-Cucuteni III village, dated about 4300 BCE, with at least ten houses set in a much more open landscape—tree pollen was only 23%.[11]

Very few Bug-Dniester traits can be detected in early Cucuteni-Tripolye artifacts. The late Bug-Dniester culture was absorbed or driven away, removing the buffer culture that had mediated interchanges on the frontier.[12] The frontier shifted eastward to the uplands between the Southern Bug and Dnieper rivers. This soon became the most clearly defined, high-contrast cultural frontier in all of Europe.

The Early Cucuteni-Tripolye Village at Bernashevka

A good example of an early Cucuteni-Tripolye farming village on that moving frontier is the site of Bernashevka, wholly excavated by V. G. Zbenovich between 1972 and 1975.[13] On a terrace overlooking the Dniester River floodplain six houses were built in a circle around one large structure (figure 9.2). The central building, 12 by 8 m, had a foundation of horizontal wooden beams, or sleeper beams, probably with vertical wall posts morticed into them. The walls were wattle-and-daub, the roof thatched, and the floor made of smooth fired clay 8–17 cm thick on a subfloor of timber beams (a *ploshchadka*). The door had a flat stone threshold, and inside was the only domed clay oven in the settlement—perhaps a

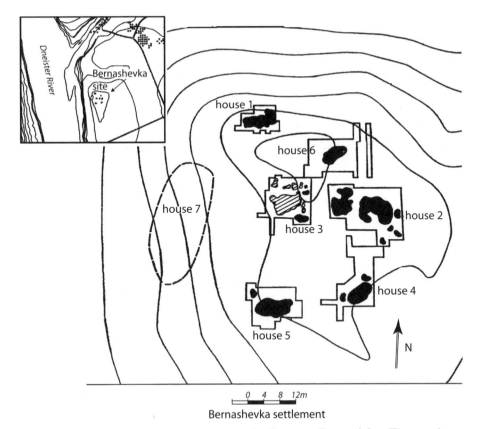

Figure 9.2 Bernashevka settlement on the Dniester River. After Zbenovich 1980, figure 3.

central bakery and work building for the village. The houses ranged from 30 m² to 150 m² in floor area. The population of the village probably was forty to sixty people. Two radiocarbon dates (5500–5300 BCE) seem two hundred years too old (table 9.1), perhaps because the dated wood fragments were from burned heartwood that had died centuries before the village was occupied.

No cemetery was found at Bernashevka or at any other Cucuteni-Tripolye village. Like the Criş people, the Cucuteni-Tripolye people did not ordinarily bury their dead. Parts of human skeletons are occasionally found in ritual deposits beneath house floors, human teeth were used occasionally as beads, and at Drăguşeni (Cucuteni A4, about 4300–4000 BCE) loose human bones were found in the litter between houses. Perhaps

TABLE 9.1
Early Eneolithic Radiocarbon Dates

Lab Number	BP Date	Sample	Calibrated Date
1. Pre-Cucuteni II Settlements			
Bernashevka			
Ki-6670	6440±60	?	5490–5300 BCE
Ki-6681	6510±55	?	5620–5360 BCE
Okopi			
Ki-6671	6330±65	?	5470–5210 BCE
2. Tripolye A Settlements			
Sabatinovka 2			
Ki-6680	6075±60	?	5060–4850 BCE
Ki-6737	6100±55	?	5210–4850 BCE
Luka Vrublevetskaya			
Ki-6684	5905±60	?	4850–4710 BCE
Ki-6685	5845±50	?	4780–4610 BCE
Grenovka			
Ki-6683	5860±45	?	4790–4620 BCE
Ki-6682	5800±50	?	4720–4550 BCE

3. Dnieper-Donets II Cemeteries (average $^{15}N = 11.8$, average offset 228±30 too old)

Osipovka cemetery		*Skeleton #*	
OxA6168	7675±70	skeleton 20, bone (invalid?)*	6590–6440 BCE
Ki 517	6075±125	skeleton 53	5210–4800 BCE
Ki 519	5940±420	skeleton 53	5350–4350 BCE
Nikol'skoe cemetery		*Grave Pit, Skeleton #*	
OxA 5029	6300±80	E, skeleton 125	5370–5080 BCE
OxA 6155	6225±75	Z, skeleton 94	5300–5060 BCE
Ki 6603	6160±70	E, skeleton 125	5230–4990 BCE
OxA 5052	6145±70	Z, skeleton 137	5210–4950 BCE
Ki 523	5640±400	skeleton ?	4950–4000 BCE
Ki 3125	5560±30	Z, bone	4460–4350 BCE

TABLE 9.1 (*continued*)

Lab Number	BP Date	Sample	Calibrated Date
Ki 3575	5560±30	B, skeleton 1	4460–4350 BCE
Ki 3283	5460±40	E, skeleton 125 (invalid?)	4450–4355 BCE
Ki 5159	5340±50	Z, skeleton 105 (invalid?)	4250–4040 BCE
Ki 3158	5230±40	Z, bone (invalid?)	4220–3970 BCE
Ki 3284	5200±30	E, skeleton 115 (invalid?)	4040–3970 BCE
Ki 3410	5200±30	D, skeleton 79a (invalid?)	4040–3970 BCE
Yasinovatka cemetery			
OxA 6163	6465±60	skeleton 5	5480–5360 BCE
OxA 6165	6370±70	skeleton 19	5470–5290 BCE
Ki-6788	6310±85	skeleton 19	5470–5080 BCE
OxA 6164	6360±60	skeleton 45	5470–5290 BCE
Ki-6791	6305±80	skeleton 45	5370–5080 BCE
Ki-6789	6295±70	skeleton 21	5370–5080 BCE
OxA 5057	6260±180	skeleton 36	5470–4990 BCE
Ki-1171	5800±70	skeleton 36	4770–4550 BCE
OxA 6167	6255±55	skeleton 18	5310–5080 BCE
Ki-3032	5900±90	skeleton 18	4910–4620 BCE
Ki-6790	5860±75	skeleton 39	4840–4610 BCE
Ki-3160	5730±40	skeleton 15	4670–4490 BCE
Dereivka 1 cemetery			
OxA 6159	6200±60	skeleton 42	5260–5050 BCE
OxA 6162	6175±60	skeleton 33	5260–5000 BCE
Ki-6728	6145±55	skeleton 11	5210–4960 BCE

4. Rakushechni Yar Settlement, Lower Don River

Bln 704	6070±100	level 8, charcoal	5210–4900 BCE
Ki-955	5790±100	level 5, shell	4790–4530 BCE
Ki-3545	5150±70	level 4, ?	4040–3800 BCE
Bln 1177	4360±100	level 3, ?	3310–2880 BCE

5. Khvalynsk Cemetery (average ^{15}N = 14.8, average offset 408±52 too old)

AA12571	6200±85	cemetery II, grave 30	5250–5050 BCE
AA12572	5985±85	cemetery II, grave 18	5040–4780 BCE
OxA 4310	6040±80	cemetery II, ?	5040–4800 BCE

Table 9.1 (*continued*)

Lab Number	BP Date	Sample	Calibrated Date
OxA 4314	6015±85	cemetery II, grave 18	5060–4790 BCE
OxA 4313	5920±80	cemetery II, grave 34	4940–4720 BCE
OxA 4312	5830±80	cemetery II, grave 24	4840–4580 BCE
OxA 4311	5790±80	cemetery II, grave 10	4780–4570 BCE
UPI119	5903±72	cemetery I, grave 4	4900–4720 BCE
UPI120	5808±79	cemetery I, grave 26	4790–4580 BCE
UPI132	6085±193	cemetery I, grave 13	5242–4780 BCE

6. Lower Volga Cultures

Varfolomievka settlement, North Caspian

Lu2642	6400±230	level 2B, unknown material	5570–5070 BCE
Lu2620	6090±160	level 2B, "	5220–4840 BCE
Ki-3589	5430±60	level 2A, "	4350–4170 BCE
Ki-3595	5390±60	level 2A, "	4340–4050 BCE

Kombak-Te, Khvalynsk hunting camp in the North Caspian

GIN 6226	6000±150	?	5210–4710 BCE

Kara-Khuduk, Khvalynsk hunting camp in the North Caspian

UPI 431	5110±45	?	3800–3970 BCE

*"Invalid" means the date was contradicted by stratigraphy or by another date.

bodies were exposed and permitted to return to the birds somewhere near the village. As Gimbutas noted, some Tripolye female figurines seem to be wearing bird masks.

Half the pottery at Bernashevka was coarse ware: thick-walled, relatively crude vessels tempered with sand, quartz, and grog (crushed ceramic sherds) decorated with rows of stabbed impressions or shallow channels impressed with a spatula in swirling patterns (figure 9.3). Some of these were perforated strainers, perhaps used for making cheese or yogurt. Another 30% were thin-walled, fine-tempered jugs, lidded bowls, and ladles. The last 20% were very fine, thin-walled, quite beautiful lidded jugs and bowls (probably for individual servings of food), ladles (for serving), and hollow-pedestaled "fruit-stands" (perhaps for food presentation), elaborately decorated over the entire surface with stamped, incised, and channeled motifs, some enhanced with white paint against the orange clay.

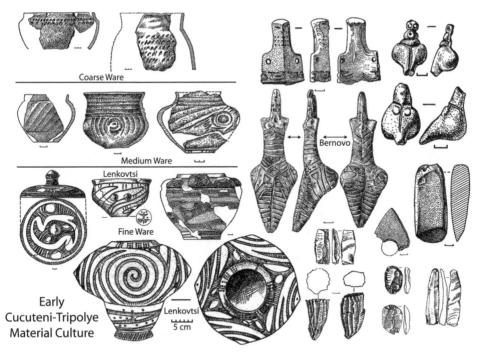

Coarse Ware

Medium Ware

Lenkovtsi

Bernovo

Fine Ware

Early
Cucuteni-Tripolye
Material Culture

Lenkovtsi

5 cm

Figure 9.3 Artifacts of the Pre-Cucuteni II/III-Tripolye A period from the sites of Bernashevka (most), Bernovo (labeled), and Lenkovtsi (labeled). After Zbenovich 1980, figures 55, 57, 61, 69, 71, 75, 79; and Zbenovich 1989, figure 65, 74.

Lidded bowls and jugs imply that food was served in individual containers at some distance from the hearth where it was cooked, and their careful decoration implies that the presentation of food involved an element of social theater, an unveiling.

Every house at Bernashevka contained fragmented ceramic female figurines with joined legs, exaggerated hips and buttocks, and schematic rod-like heads, about 10 cm long (figure 9.3). Simple incisions indicated the pubis and a girdle or waistband. Figurines were found at various places on the house floors; there was no obvious domestic shrine or altar. The number of figurines per house ranged from one to twenty-one, but four houses had nine or more. Almost two thousand similar figurines have been found in other Pre-Cucuteni II-III/Tripolye A sites, occasionally arranged in groups seated in chairs. At the Tripolye A site of Luka-Vrublevetskaya on the Dniester, they were made of clay tempered with a mixture of wheat, barley, and millet grains—all the grains cultivated in the village—and with finely ground flour. These, at least, seem to have symbolized the generative fertility of cultivated grain. But they were only one aspect of domestic cults. Under every house at Bernashevka was the skull of a domesticated cow or bull. One house also had wild animal symbols: the skull of a wild aurochs and the antlers of a red deer. Preconstruction foundation deposits of cattle horns and skulls, and occasionally of human skulls, are found in many Tripolye A villages. Bovine and female spirit powers were central to domestic household cults.

The Bernashevka farmers cultivated emmer and spelt wheats, with some barley and millet. Fields were prepared with mattocks made of antler (nineteen examples were found) and polished slate (twenty examples); some of these might have been attached to ards, which were primitive plows. The grain was harvested with flint blades of the Karanovo type (figure 9.3).

The animal bones from Bernashevka are the largest sample from any early Cucuteni-Tripolye site: 12,657 identifiable bones from a minimum of 804 animals. About 50% of the bones (60% of the individuals) were from wild animals, principally red deer (*Cervus elaphus*) and wild pig. Roe deer (*Capreolus capreolus*) and the wild aurochs (*Bos primigenius*) were hunted occasionally. Many early Cucuteni-Tripolye sites have about 50% wild animal bones. Like Bernashevka, most were frontier settlements established in places not previously cleared or farmed. In contrast, at the long-settled locale of Tirpeşti the Pre-Cucuteni III settlement produced 95% domesticated animal bones. And even in frontier settlements like Bernashevka, about 50% of all animal bones were from cattle, sheep/goat, and pigs. Cattle and pigs were more important in heavily forested areas like Bernashevka, where cattle constituted 75% of the domesticated animal bones, whereas sheep and goats were more important in villages closer to the steppe border.

Pre-Cucuteni II Bernashevka was abandoned before copper tools and ornaments became common enough to lose casually; no copper artifacts were left in the settlement. But only a few centuries later small copper artifacts became common. At Tripolye A Luka-Vrublevetskaya, probably occupied about 4800–4600 BCE, 12 copper objects (awls, fishhooks, a bead, a ring) were found among seven houses in piles of discarded shellfish, animal bones, and broken crockery. At Karbuna, near the steppe boundary, probably occupied about 4500–4400 BCE, a spectacular hoard of 444 copper objects was buried in a fine late Tripolye A pot closed with a Tripolye A bowl (figure 9.4). The hoard contained two cast copper hammer-axes 13–14 cm long, hundreds of copper beads, and dozens of flat

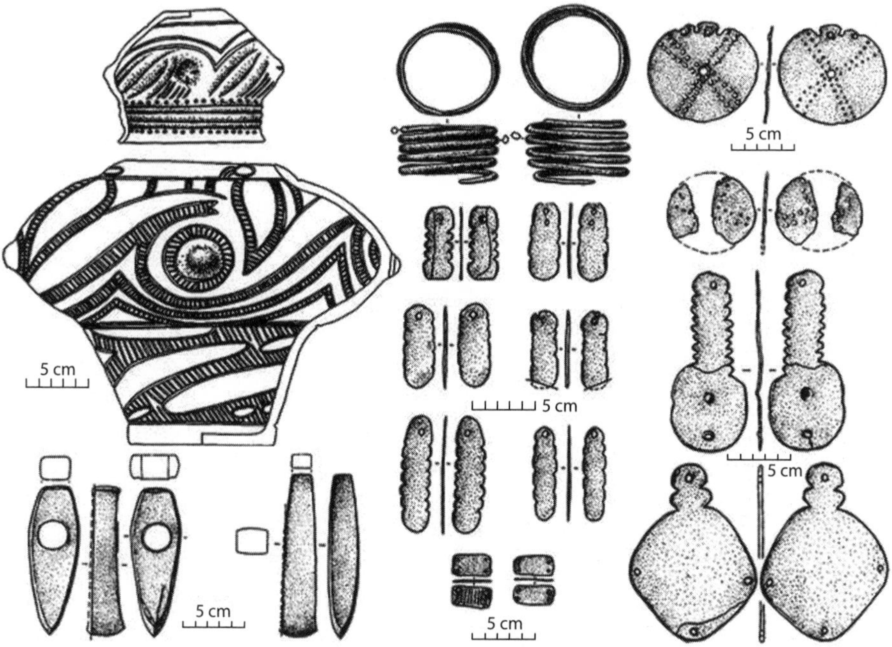

Figure 9.4 Part of the Karbuna hoard with the Tripolye A pot and bowl-lid in which it was found. All illustrated objects except the pot and lid are copper, and all are the same scale. After Dergachev 1998.

"idols," or wide-bottomed pendants made of flat sheet copper; two hammer-axes of marble and slate with drilled shaft-holes for the handle; 127 drilled beads made of red deer teeth; 1 drilled human tooth; and 254 beads, plaques, or bracelets made of *Spondylus* shell, an Aegean shell used for ornaments continuously from the first Greek Neolithic through the Old European Eneolithic. The Karbuna copper came from Balkan ores, and the Aegean shell was traded from the same direction, probably through the tell towns of the lower Danube valley. By about 4500 BCE social prestige had become closely linked to the accumulation of exotic commodities, including copper.[14]

As Cucuteni-Tripolye farmers moved eastward out of the East Carpathian piedmont they began to enter a more open, gently rolling, drier landscape. East of the Dniester River annual precipitation declined and the forests thinned. The already-old cultural frontier moved to the Southern Bug river valley. The Tripolye A town of Mogil'noe IV, among the first established in the South Bug valley, had more than a hundred buildings and covered 15–20 hectares, with a population of perhaps between four hundred and seven hundred. East of the Southern Bug, in the Dnieper valley, were people of a very different cultural tradition: the Dnieper-Donets II culture.

The Dnieper-Donets II Culture

Dimitri Telegin defined the Dnieper-Donets II culture based on a series of excavated cemeteries and settlement sites in the Dnieper valley, in the steppes north of the Sea of Azov, and in the Donets valley. Dnieper-Donets II societies created large, elaborate cemeteries, made no female figurines, had open fires rather than kilns or ovens in their homes, lived in bark-covered huts rather than in large houses with fired clay floors, had no towns, cultivated little or no grain, and their pottery was very different in appearance and technology from Tripolye ceramics. The trajectory of the Cucuteni-Tripolye culture led back to the Neolithic societies of Old Europe, and that of Dnieper-Donets II led to the local Mesolithic foragers. They were fundamentally different people and almost certainly spoke different languages. But around 5200 BCE, the foragers living around the Dnieper Rapids began to keep cattle and sheep.

The bands of fishers and hunters whose cemeteries had overlooked the Rapids since the Early Mesolithic might have been feeling the pinch of growing populations. Living by the rich resources of the Rapids they might have become relatively sedentary, and woman, when they live a settled life, generally have more children. They controlled a well-known, strategic area in a productive territory. Their decision to adopt cattle and sheep herding could have opened the way for many others in the Pontic-Caspian steppes. In the following two or three centuries domesticated cattle, sheep, and goats were walked and traded from the Dnieper valley eastward to the Volga-Ural steppes, where they had arrived by about 4700–4600 BCE. The evidence for any cereal cultivation east of the Dnieper before about 4200 BCE is thin to absent, so the initial innovation seems to have involved animals and animal herding.

Dating the Shift to Herding

The traditional Neolithic/Eneolithic chronology of the Dnieper valley is based on several sites near the Dnieper Rapids; the important ones are Igren 8, Pokhili, and Vovchok, where a repeated stratigraphic sequence was found. At the bottom were Surskii-type Neolithic pots and microlithic flint tools associated with the bones of hunted wild animals, principally red deer, wild pigs, and fish. These assemblages defined the Early Neolithic (dated about 6200–5700 BCE). Above them were Dnieper-Donets phase I occupations with comb-impressed and vegetal-tempered pottery, still associated with wild fauna; they defined the Middle Neolithic (probably about 5700–5400 BCE, contemporary with the Bug-Dniester culture). Stratified above these deposits were layers with Dnieper-Donets II pottery, sand-tempered with "pricked" or comb-stamped designs, and large flint blade tools, associated with the bones of domesticated cattle and sheep. These DDII assemblages represented the beginning of the Early Eneolithic and the beginning of herding economies east of the Dnieper River.[15]

Unlike the dates from DDI and Surskii, most DDII radiocarbon dates were measured on human bone from cemeteries. The average level of ^{15}N in DDII human bones from the Dnieper valley is 11.8%, suggesting a meat diet of about 50% fish. Correcting the radiocarbon dates for this level of ^{15}N, I obtained an age range of 5200–5000 BCE for the oldest DDII graves at the Yasinovatka and Dereivka cemeteries near the Dnieper Rapids. This is probably about when the DDII culture began. Imported pots of the late Tripolye A^2 Borisovka type have been found in DDII settlements at Grini, Piliava, and Stril'cha Skelia in the Dnieper valley, and sherds from three Tripolye A pots were found at the DDII Nikol'skoe cemetery. Tripolye A^2 is dated about 4500–4200 BCE by good dates (not on human bone) in the Tripolye heartland, and late DDII radiocarbon dates (when corrected for ^{15}N) agree with this range. The DDII period began about 5200–5000 BCE and lasted until about 4400–4200 BCE. Contact with Tripolye A people seems to have intensified after about 4500 BCE.[16]

The Evidence for Stockbreeding and Grain Cultivation

Four Dnieper-Donets II settlement sites in the Dnieper valley have been studied by zoologists—Surskii, Sredni Stog 1, and Sobachki in the steppe zone near the Rapids; and Buz'ki in the moister forest-steppe to the north (table 9.2). Domesticated cattle, sheep/goat, and pig accounted for 30–75% of the animal bones in these settlements. Sheep/goat contributed more than 50% of the bones at Sredni Stog 1 and 26% at Sobachki. Sheep finally were accepted into the meat diet in the steppes. Perhaps they were already being plucked for felt making; the vocabulary for wool might have first appeared among Pre-Proto-Indo-European speakers at about this time. Wild horses were the most important game (?) animal at Sredni Stog 1 and Sobachki, whereas red deer, roe deer, wild pig, and beaver were hunted in the more forested parts of the river at Buz'ki and Surskii 2–4.

TABLE 9.2

Dnieper–Donets II Animal Bones from Settlements

	Sobachki	*Sredni Stog 1*	*Buz'ki*
Mammal Bones		*(Bones / MNI)**	
Cattle	56/5	23/2	42/3
Sheep/goat	54/8	35/4	3/1
Pig	10/3	1/1	4/1
Dog	9/3	12/1	8/2
Horse	48/4	8/1	—
Onager	1/1	—	—
Aurochs	2/1	—	—
Red deer	16/3	12/1	16/3
Roe deer	—	—	28/4
Wild pig	3/1	—	27/4
Beaver	—	—	34/5
Other mammal	8/4	—	7/4
Domestic	129 bones / 62%	74 bones / 78%	57 bones / 31%
Wild	78 bones / 38%	20 bones / 22%	126 bones / 69%

*MNI = minimum number of individuals

Fishing net weights and hooks suggest that fish remained important. This is confirmed by levels of ^{15}N in the bones of people who lived on the Dnieper Rapids, which indicate a meat diet containing more than 50% fish. Domesticated cattle, pig, and sheep bones occurred in all DDII settlements and in several cemeteries, and constituted more than half the bones at two settlement sites (Sredni Stog I and Sobachki) in the steppe zone. Domesticated animals seem indeed to have been an important addition to the diet around the Dnieper Rapids.[17]

Flint blades with sickle gloss attest to the harvesting of cereals at DDII settlements. But they could have been wild seed plants like *Chenopodium* or *Amaranthus*. If cultivated cereals were harvested there was very little evidence found. Two impressions of barley (*Hordeum vulgare*) were recovered on a potsherd from a DDII settlement site at Vita Litovskaya, near Kiev, west of the Dnieper. In the forests northwest of Kiev, near the Pripet marshes, there were sites with pottery that somewhat resembled DDII pottery but there were no elaborate cemeteries or other traits of the DDII culture. Some of these settlements (Krushniki, Novosilki, Obolon') had pottery with a few seed impressions of wheat (*T. monococcum and T. dicoccum*) and millet (*Panicum sativum*). These sites probably should be dated before 4500 BCE, since Lengyel-related cultures replaced them in Volhynia and the Polish borderlands after about that date. Some forest-zone farming seems to have been practiced in the southern Pripet forests west of the Dnieper. But in steppe-zone DDII cemeteries east of the Dnieper, Malcolm Lillie recorded almost no dental caries, suggesting that the DDII people ate a low-carbohydrate diet similar to that of the Mesolithic. No cultivated cereal imprints have been found east of the Dnieper River in pots dated before about 4000 BCE.[18]

Pottery and Settlement Types

Pottery was more abundant in DDII living sites than it had been in DDI, and appeared for the first time in cemeteries (figure 9.5). The growing importance of pottery perhaps implies a more sedentary lifestyle, but shelters were still lightly built and settlements left only faint footprints. A typical DDII settlement on the Dnieper River was Buz'ki. It consisted of five hearths and two large heaps of discarded shellfish and animal bones. No structures were detected, although some kind of shelter probably did exist.[19] Pots here and in other DDII sites were made in larger sizes (30–40 cm in diameter) with flat bottoms (pots seen in DDI sites had mainly pointed or rounded bottoms) and an applied collar around the rim. Decoration usually covered the entire outside of the vessel, made by pricking the surface with a stick, stamping designs with a small comb-stamp, or incising thin lines in horizontal-linear and zig-zag motifs—quite different from the spirals and swirls of Tripolye A potters. The application of a "collar" to thicken the rim was a popular innovation, widely adopted across the Pontic-Caspian steppes about 4800 BCE.

Polished (not chipped) stone axes now became common tools, perhaps for felling forests, and long unifacial flint blades (5–15 cm long) also became increasingly common, perhaps as a standardized part of a trade or gift package, since they appeared in graves and in small hoards in settlements.

Dnieper-Donets II Funeral Rituals

DDII funerals were quite different from those of the Mesolithic or Neolithic. The dead usually were exposed, their bones were collected, and they were finally buried in layers in communal pits. Some individuals were

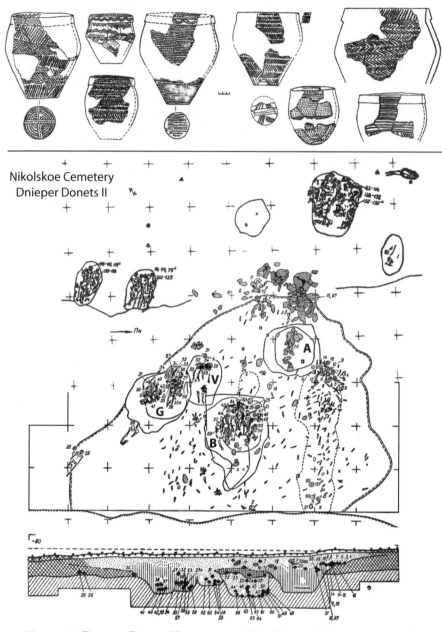

Nikolskoe Cemetery
Dnieper Donets II

Figure 9.5 Dnieper-Donets II cemetery at Nikol'skoe with funerary ceramics. Pits A,B,G, and V were in an area deeply stained with red ochre. The other five burial pits were on a slightly higher elevation. Broken pots and animal bones were found near the cluster of rocks in the center. After Telegin 1991, figures 10, 20; and Telegin 1968, figure 27.

buried in the flesh, without exposure. This communal pit type of cemetery, with several treatments of the body in one pit, spread to other steppe regions. The thirty known DDII communal cemeteries were concentrated around the Dnieper Rapids but occurred also in other parts of the Dnieper valley and in the steppes north of the Sea of Azov. The largest cemeteries were three times larger than those of any earlier era, with 173 bodies at Dereivka, 137 at Nikol'skoe, 130 at Vovigny II, 124 at Mariupol, 68 at Yasinovatka, 50 at Vilnyanka, and so on. Pits contained up to four layers of burials, some whole and in an extended supine position, others consisting of only skulls. Cemeteries contained up to nine communal burial pits. Traces of burned structures, perhaps charnel houses built to expose dead bodies, were detected near the pits at Mariupol and Nikol'skoe. At some cemeteries, including Nikol'skoe (figure 9.5), loose human bones were widely scattered around the burial pits.

At Nikol'skoe and Dereivka some layers in the pits contained only skulls, without mandibles, indicating that some bodies were cleaned to the bone long before final burial. Other individuals were buried in the flesh, but the pose suggests that they were tightly wrapped in some kind of shroud. The first and last graves in the Nikol'skoe pits were whole skeletons. The standard burial posture for a body buried in the flesh was extended and supine, with the hands by the sides. Red ochre was densely strewn over the entire ritual area, inside and outside the grave pits, and pots and animal bones were broken and discarded near the graves.[20]

The funerals at DDII cemeteries were complex events that had several phases. Some bodies were exposed, and sometimes just their skulls were buried. In other cases whole bodies were buried. Both variants were placed together in the same multilayered pits, strewn with powdered red ochre. The remains of graveside feasts—cattle and horse bones—were thrown in the red-stained soil at Nikol'skoe, and cattle bones were found in grave 38, pit A, at Vilnyanka.[21] At Nikol'skoe almost three thousand sherds of pottery, including three Tripolye A cups, were found among the animal bones and red ochre deposited over the graves.

Power and Politics

The people of the DDII culture looked different than people of earlier periods in two significant respects: the profusion of new decorations for the human body and the clear inequality in their distribution. The old fisher-gatherers of the Dnieper Rapids were buried wearing, at most, a few beads of deer or fish teeth. But in DDII cemeteries a few individuals were

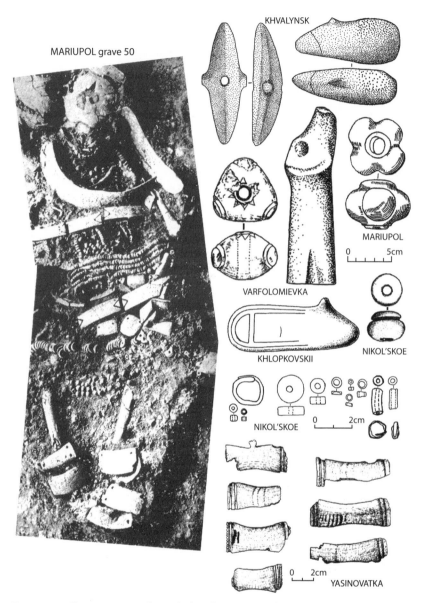

Figure 9.6 Ornaments and symbols of power in the Early Eneolithic, from Dnieper-Donets II graves, Khvalynsk, and Varfolomievka. The photo of grave 50 at Mariupol, skull at the top, is adapted from Gimbutas 1956, plate 8. The beads from Nikol'skoe include two copper beads and a copper ring on the left, and a gold ring on the lower right. The other beads are polished and drilled stone. The maces from Mariupol and Nikol'skoe, and beads from Nikol'skoe are after Telegin 1991, figures 29, 38; and Telegin and Potekhina 1987, figure 39.

buried with thousands of shell beads, copper and gold ornaments, imported crystal and porphyry ornaments, polished stone maces, bird-bone tubes, and ornamental plaques made of boar's tusk (figure 9.6). Boar's-tusk plaques were restricted to very few individuals. The tusks were cut into rectangular flat pieces (not an easy thing to do), polished smooth, and pierced or incised for attachment to clothing. They may have been meant to emulate Tripolye A copper and *Spondylus*-shell plaques, but DDII chiefs found their own symbols of power in the tusks of wild boars.

At the Mariupol cemetery 310 (70%) of the 429 boar's-tusk plaques accompanied just 10 (8%) of the 124 individuals. The richest individual (gr. 8) was buried wearing forty boars-tusk plaques sewn to his thighs and shirt, and numerous belts made of hundreds of shell and mother-of-pearl beads. He also had a polished porphyry four-knobbed mace head (figure 9.6), a bull figurine carved from bone, and seven bird-bone tubes. At Yasinovatka, only one of sixty-eight graves had boars-tusk plaques: an adult male wore nine plaques in grave 45. At Nikol'skoe, a pair of adults (gr. 25 and 26) was laid atop a grave pit (B) equipped with a single boar's-tusk plaque, a polished serpentine mace head, four copper beads, a copper wire ring, a gold ring, polished slate and jet beads, several flint tools, and an imported Tripolye A pot. The copper contained trace elements that identify it as Balkan in origin. Surprisingly few children were buried at Mariupol (11 of 124 individuals), suggesting that a selection was made—not all children who died were buried here. But one was among the richest of all the graves: he or she (sex is indeterminate in immature skeletons) wore forty-one boar's-tusk plaques, as well as a cap armored with eleven whole boar's tusks, and was profusely ornamented with strings of shell and bone beads. The selection of only a few children, including some who were very richly ornamented, implies the inheritance of status and wealth. Power was becoming institutionalized in families that publicly advertised their elevated status at funerals.

The valuables that signaled status were copper, shell, and imported stone beads and ornaments; boars-tusk plaques; polished stone maceheads; and bird-bone tubes (function unknown). Status also might have been expressed through the treatment of the body after death (exposed, burial of the skull/not exposed, burial of the whole body); and by the

---

Figure 9.6 (continued) The Varfolomievka mace (or pestle?) is after Yudin 1988, figure 2; Khvalynsk maces are after Agapov, Vasliev, and Pestrikova 1990, figure 24. Boars-tusk plaques, at the bottom, are after Telegin 1991, figure 38.

public sacrifice of domesticated animals, particularly cattle. Similar markers of status were adopted across the Pontic-Caspian steppes, from the Dnieper to the Volga. Boars-tusk plaques with exactly the same flower-like projection on the upper edge (figure 9.6, top plaque from Yasinovatka) were found at Yasinovatka in the Dnieper valley and in a grave at S'yezzhe in the Samara valley, 400 km to the east. Ornaments made of Balkan copper were traded across the Dnieper and appeared on the Volga. Polished stone mace-heads had different forms in the Dnieper valley (Nikol'skoe), the middle Volga (Khvalynsk), and the North Caspian region (Varfolomievka), but a mace is a weapon, and its wide adoption as a symbol of status suggests a change in the politics of power.

The Khvalynsk Culture on the Volga

The initial spread of stockbreeding in the Pontic-Caspian steppes was notable for the various responses it provoked. The DDII culture, where the shift began, incorporated domesticated animals not just as a ritual currency but also as an important part of the daily diet. Other people reacted in quite different ways, but they were all clearly interacting, perhaps even competing, with one another. A key regional variant was the Khvalynsk culture.

A prehistoric cemetery was discovered at Khvalynsk in 1977 on the west bank of the middle Volga. Threatened by the water impounded behind a Volga dam, it was excavated by teams led by Igor Vasiliev of Samara (figure 9.7). Its location has since been completely destroyed by bank erosion. Sites of the Khvalynsk type are now known from the Samara region southward along the banks of the Volga into the Caspian Depression and the Ryn Peski desert in the south. The characteristic pottery included open bowls and bag-like, round-bottomed pots, thick-walled and shell-tempered, with very distinctive sharply everted thick "collars" around the rims. They were densely embellished with bands of pricked and comb-stamped decoration that often covered the entire exterior surface. Early Khvalynsk, well documented at the Khvalynsk cemetery, began around 4700–4600 BCE in the middle Volga region (after adjusting the dates downward for the ^{15}N content of the humnan bones on which the dates were measured). Late Khvalynsk on the lower Volga is dated 3900–3800 BCE at the site of Kara-Khuduk but probably survived even longer than this on the lower Volga.[22]

The first excavation at the Khvalynsk cemetery, in 1977–79 (excavation I), uncovered 158 graves; the second excavation in 1980–85 (excavation II) recovered, I have been told, 43 additional graves.[23] Only Khvalynsk I has

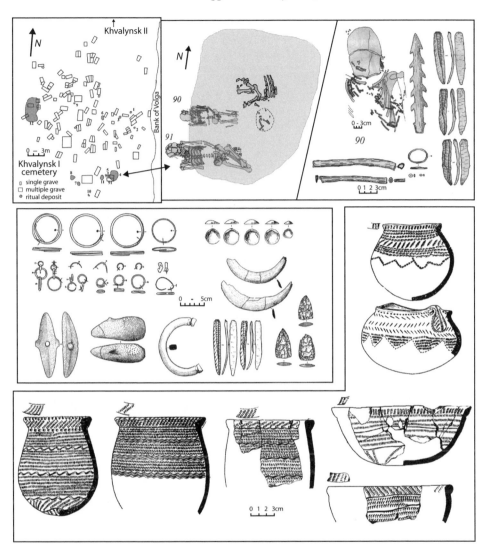

Figure 9.7 Khvalynsk cemetery and grave gifts. Grave 90 contained copper beads and rings, a harpoon, flint blades, and a bird-bone tube. Both graves (90 and 91) were partly covered by Sacrificial Deposit 4 with the bones from a horse, a sheep, and a cow.

Center: grave goods from the Khvalynsk cemetery—copper rings and bracelets, polished stone mace heads, polished stone bracelet, Cardium shell ornaments, boars tusk chest ornaments, flint blades, and bifacial projectile points.

Bottom: shell-tempered pottery from the Khvalynsk cemetery. After Agapov, Vasiliev, and Pestrikova 1990; and Ryndina 1998, Figure 31.

been published, so all statistics here are based on the first 158 graves (figure 9.7). Khvalynsk was by far the largest excavated Khvalynsk-type cemetery; most others had fewer than 10 graves. At Khvalynsk most of the deceased were layered in group pits, somewhat like DDII graves, but the groups were much smaller, containing only two to six individuals (perhaps families) buried on top of one another. One-third of the graves were single graves, a move away from the communal DDII custom. Only mature males, aged thirty to fifty, were exposed and disarticulated prior to burial, probably an expression of enhanced male status, associated with the introduction of herding economies elsewhere in the world.[24] Few children were buried in the cemetery (13 of 158), but those who were included some of the most profusely ornamented individuals, again possibly indicating that status was inherited. The standard burial posture was on the back with the knees raised, a distinctive pose. Most had their heads to the north and east, a consistent orientation that was absent in DDII cemeteries. Both the peculiar posture and the standard orientation later became widespread in steppe funeral customs.

Khvalynsk had many more animal sacrifices than any DDII cemetery: 52 (or 70) sheep/goat, 23 cattle, and 11 horses, to accompany the burials of 158 humans. (The published reports are inconsistent on the number of sheep/goat.) The head-and-hoof form of sacrifice appeared for the first time: at least 17 sheep/goat and 9 cattle were slaughtered and only the skull and lower leg bones were buried, probably still attached to the animal's hide. In later steppe funerals the custom of hanging a hide containing the head and hooves over the grave or burying it in the grave was very common. The head and hide symbolized a gift to the gods, and the flesh was doled out to guests at the funeral feast. Parts of domesticated animals were offered in all phases of the funerals at Khvalynsk: on the grave floor, in the grave fill, at the edge of the grave, and in twelve special sacrificial deposits stained with red ochre, found above the graves (figure 9.7). The distribution of animal sacrifices was unequal: 22 graves of 158 (14 percent) had animal sacrifices in the grave or above it, and enough animals were sacrificed to supply about half of the graves were they distributed equally. Only 4 graves (100, 127, 139, and 55–57) contained multiple species (cattle and sheep, sheep and horse, etc.) and all four of those also were covered by ochre-stained ritual deposits above the grave, with additional sacrifices. About one in five people had sacrificed domestic animals, and one in forty had multiple domestic animals.

The role of the horse in the Khvalynsk sacrifices is intriguing. The only animals sacrificed at Khvalysnk I were domesticated sheep/goat, domesticated cattle, and horses. Horse leg parts occurred by themselves, without other animal bones, in eight graves. They were included with a sheep/goat head-and-hoof offering in grave 127, and were included with sheep/goat and cattle remains in sacrificial deposit 4 (figure 9.7). It is not possible to measure the bones—they were discarded long ago—but horses certainly were treated symbolically like domesticated animals at Khvalynsk: they were grouped with cattle and sheep/goat in human funeral rituals that excluded obviously wild animals. Carved images of horses were found at other cemeteries dated to this same period (see below). Horses certainly had a new ritual and symbolic importance at Khvalynsk. If they were domesticated, they would represent the oldest domesticated horses.[25]

There is much more copper at Khvalynsk than is known from the entire DDII culture, and the copper objects there are truly remarkable (figure 9.7). Unfortunately most of it, an astonishing 286 objects, came from the 43 (?) graves of the Khvalynsk II excavation, still unpublished though analyses of some of the objects have been published by Natalya Ryndina. The Khvalynsk I excavation yielded 34 copper objects found in 11 of the 158 published graves. The copper from excavations I and II showed the same trace elements and technology, the former characteristic of Balkan copper. Ryndina's study of 30 objects revealed three technological groups: 14 objects made at 300–500°C, 11 made at 600–800°C, and 5 made at 900–1,000°C. The quality of welding and forging was uniformly low in the first two groups, indicating local manufacture, but was strongly influenced by the methods of the Tripolye A culture. The third group, which included two thin rings and three massive spiral rings, was technically identical to Old European status objects from the cemeteries of Varna and Durankulak in Bulgaria. These objects were made in Old Europe and were traded in finished form to the Volga. In the 158 graves of Khvalynsk I, adult males had the most copper objects, but the number of graves with *some* copper was about equal between the sexes, five adult male graves and four adult female graves. An adolescent (gr. 90 in figure 9.7) and a child were also buried with copper rings and beads.[26]

Polished stone mace-heads and polished serpentine and steatite stone bracelets appeared with copper as status symbols. Two polished stone maces occurred in one adult male grave (gr. 108) and one in another (gr. 57) at Khvalynsk. Grave 108 also contained a polished steatite bracelet. Similar bracelets and mace-heads were found in other Khvalynsk-culture cemeteries on the Volga, for example, at Krivoluchie (Samara oblast) and Khlopkovskii (Saratov oblast). Some mace heads were given "ears" that made them seem vaguely zoomorphic, and some observers have seen horse heads in them. A clearly zoomorphic polished stone mace head appeared at Varfolomievka, part of a different culture group on the lower Volga. Maces, copper, and elaborate decoration of the body appeared with domesticated animals, not before.[27]

Khvalynsk settlements have been found at Gundurovka and Lebyazhinka I on the Sok River, north of the Samara. But the Khvalynsk artifacts and pottery are mixed with artifacts of other cultures and ages, making it difficult to isolate features or animal bones that can be ascribed to the Khvalynsk period alone. We do know from the bones of the Khvalynsk people themselves that they ate a lot of fish; with an average ^{15}N measurement of 14.8%, fish probably represented 70% of their meat diet. Pure Khvalynsk camps have been found on the lower Volga in the Ryn Peski desert, but these were specialized hunters' camps where onagers and saiga antelope were the quarry, comprising 80–90 percent of the animal bones. Even here, at Kara Khuduk I, we find a few sheep/goat and cattle bones (6–9 %), perhaps provisions carried by Khvalynsk hunters.

In garbage dumps found at sites of other steppe cultures of the same period east of the Don (see below), horse bones usually made up more than half the bones found, and the percentage of cattle and sheep was usually under 40%. In the east, cattle and sheep were more important in ritual sacrifices than in the diet, as if they were initially regarded as a kind of ritual currency used for occasional (seasonal?) sanctified meals and funeral feasts. They certainly were associated with new rituals at funerals, and probably with other new religious beliefs and myths as well. The set of cults that spread with the first domesticated animals was at the root of the Proto-Indo-European conception of the universe as described at the beginning of chapter 8.

Nalchik and North Caucasian Cultures

Many archaeologists have wondered if domesticated cattle and sheep might have entered the steppes through the Eneolithic farmers of the Caucasus as well as from Old Europe.[28] Farming cultures had spread from the Near East into the southern Caucasus Mountains (Shulaveri, Arukhlo, and Shengavit) by 5800–5600 BCE. But these earliest farming communities in the Caucasus were not widespread; they remained concentrated in a few riverbottom locations in the upper Kura and Araxes River valleys. No bridging sites linked them to the distant European steppes, more than 500 km to the north and west. The permanently glaciated North Caucasus Mountains, the highest and most impassable mountain range in Europe, stood between them and the steppes. The bread wheats (*Triticum aestivum*) preferred in the

Caucasus were less tolerant of drought conditions than the hulled wheats (emmer, einkorn) preferred by Criş, Linear Pottery, and Bug-Dniester cultivators. The botanist Zoya Yanushevich observed that the cultivated cereals that appeared in Bug-Dniester sites and later in the Pontic-Caspian steppe river valleys were a Balkan/Danubian crop suite, not a Caucasian crop suite.[29] Nor is there an obvious stylistic connection between the pottery or artifacts of the earliest Caucasian farmers at Shulaveri and those of the earliest herders in the steppes off to the north. If I had to guess at the linguistic identity of the first Eneolithic farmers at Shulaveri, I would link them with the ancestors of the Kartvelian language family.

The Northwest Caucasian languages, however, are quite unlike Kartvelian. Northwest Caucasian seems to be an isolate, a survival of some unique language stock native to the northern slopes of the North Caucasus Mountains. In the western part of the North Caucasian piedmont, overlooking the steppes, the few documented Eneolithic communities had stone tools and pottery somewhat like those of their northern steppe neighbors; these communities were southern participants in the steppe world, not northern extensions of Shulaveri-type Caucasian farmers. I would guess they spoke languages ancestral to Northwest Caucasian, but only a few early sites are published. The most important is the cemetery at Nalchik.

Near Nalchik, in the center of the North Caucasus piedmont, was a cemetery containing 147 graves with contracted skeletons lying on their sides in red ochre–stained pits in groups of two or three under stone cairns. Females lay in a contracted pose on the left side and males on their right.[30] A few copper ornaments, beads made of deer and cattle teeth, and polished stone bracelets (like those found in grave 108 at Khvalynsk and at Krivoluchie) accompanied them. One grave yielded a date on human bone of 5000–4800 BCE (possibly too old by a hundred to five hundred years, if the dated sample was contaminated by old carbon in fish). Five graves in the same region at Staronizhesteblievsk were provided with boars-tusk plaques of the DDII Mariupol type, animal-tooth beads, and flint blades that seem at home in the Early Eneolithic.[31] An undated cave occupation in the Kuban valley at Kamennomost Cave, level 2, which could be of the same date, has yielded sheep/goat and cattle bones stratified beneath a later level with Maikop-culture materials. Carved stone bracelets and ornamental stones from the Caucasus—black jet, rock crystal, and porphyry—were traded into Khvalynsk and Dnieper-Donets II sites, perhaps from people like those at Nal'chik and Kamennomost Cave 2. The Nalchik-era sites clearly represent a community that had at least a few domesticated cattle and sheep/goats, and was in contact with Khvalynsk.

They probably got their domesticated animals from the Dnieper, as the Khvalynsk people did.

The Lower Don and North Caspian Steppes

In the steppes between Nalchik and Khvalynsk many more sites, of different kinds, are dated to this period. Rakushechni Yar on the lower Don, near the Sea of Azov, is a deeply stratified settlement site with a cluster of six graves at the edge of the settlement area. The lowest cultural levels, with shell-tempered pottery lightly decorated with incised linear motifs and impressions made with a triangular-ended stick, probably dated about 5200–4800 BCE, contained the bones of sheep/goat and cattle. But in the interior steppes, away from the major river valleys, equid hunting was still the focus of the economy. In the North Caspian Depression the forager camp of Dzhangar, also dated 5200 BCE (on animal bone) and with pottery similar to Rakushechni Yar, yielded only the bones of wild horses and onagers.[32]

On the eastern side of the lower Volga, sites such as Varfolomievka were interspersed with Khvalynsk hunters' camps such as Kara Khuduk I.[33] The settlement at Varfolomievka is stratified and well dated by radiocarbon, and clearly shows the transition from foraging to herding in the North Caspian Depression. Varfolomievka was first occupied around 5800–5600 BCE by pottery-making foragers who hunted onagers and horses (level 3). The site was reoccupied twice more (levels 2B and 2A). In level 2B, dated about 5200–4800 BCE, people constructed three pit-houses. They used copper (one copper awl and some amorphous lumps of copper were found) and kept domesticated sheep/goats, though "almost half" the animal bones at Varfolomievka were of horses. Bone plaques were carved in the shape of horses, and horse metacarpals were incised with geometric decorations. Three polished stone mace-head fragments were found here. One was carved into an animal head at one end, perhaps a horse (figure 9.6). Four graves were dug rather casually into abandoned house depressions at Varfolomievka, like the similar group of graves at the edge of Rakushechni Yar. Hundreds of beads made of drilled and polished horse teeth were deposited in ochre-stained sacrificial deposits near the human graves. There were also a few deer teeth, several kinds of shell beads, and whole boars' tusk ornaments.

These sites in the southern steppes, from the lower Don to the lower Volga, are dated 5200–4600 BCE and exhibit the bones of sheep/goat and occasionally cattle, small objects of copper, and casual disposal of the dead. Small settlements provide most of the data, unlike the cemetery-based archaeological record for Khvalynsk. Pots were shell-tempered and decorated with designs incised or pricked with a triangular-ended stick. Motifs included diamond-like lozenges and, rarely, incised meanders filled with pricked ornament. Most rims were simple but some were thickened on the inside. A. Yudin has grouped these sites together under the name of the Orlovka culture, after the settlement of Orlovka, excavated in 1974, on the Volga. Nalchik seems to have existed at the southern fringe of this network.[34]

THE FOREST FRONTIER: THE SAMARA CULTURE

One other culture interacted with northern Khvalynsk in the middle Volga region, along the forest-steppe boundary (see figure 9.1). The Samara Neolithic culture, distinguished by its own variety of "collared" pots covered with pricked, incised, and rocker-stamped motifs, developed at the northern edge of the steppe zone along the Samara River. The pottery, tempered with sand and crushed plants, was similar to that made on the middle Don River. Dwellings at Gundurovka near Samara had dug-out floors, 20 m by 8 m, with multiple hearths and storage pits in the floors (this settlement also contained Khvalynsk pottery). Domesticated sheep/goat (13% of 3,602 bones) and cattle (21%) were identified at Ivanovskaya on the upper Samara River, although 66% of the bones were of horses. The settlement of Vilovatoe on the Samara River yielded 552 identifiable bones, of which 28.3% were horse, 19.4% were sheep/goat, and 6.3% were cattle, in addition to beaver (31.8%) and red deer (12.9%). The Samara culture showed some forest-culture traits: it had large polished stone adzes like those of forest foragers to the north.

Samara people created formal cemeteries (figure 9.8). The cemetery at S'yezzhe (see-YOZH-yay) contained nine burials in an extended position on their backs, different from the Khvalynsk position and more like that of DDII. Above the graves at the level of the original ground surface was a ritual deposit of red ochre, broken pottery, shell beads, a bone harpoon, and the skulls and lower leg bones (astragali and phalanges) of two horses—funeral-feast deposits like the above-grave deposits at Khvalynsk. S'yezzhe had the oldest horse head-and-hoof deposit in the steppes. Near the horse head-and-hoof deposit, but outside the area of ochre-stained soil, were two figurines of horses carved on flat pieces of bone, similar to others found at Varfolomievka, and one bone figurine of a bull. The S'yezzhe people wore boar's-tusk plaques like those of the Dnieper-Donets II culture, one of which was shaped exactly like one found at the DDII cemetery of Yasinovatka in the Dnieper valley.[35]

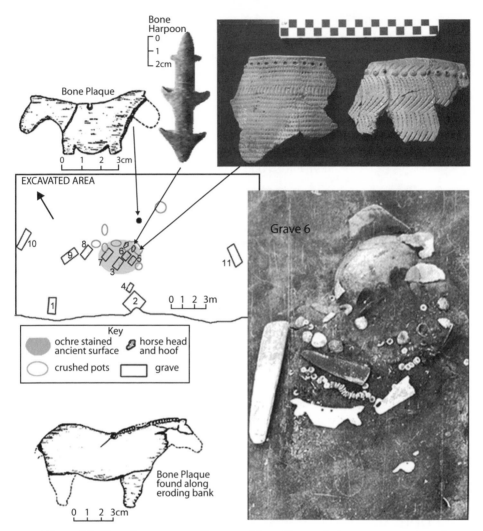

Figure 9.8 S'yezzhe cemetery, Samara oblast. Graves 1–9 were a cemetery of the Samara culture, Early Eneolithic. Graves 10 and 11 were later. After Vasiliev and Matveeva 1979.

Cows, Social Power, and the Emergence of Tribes

It is impossible to say how much the people buried at Khvalynsk really *knew* of the societies of Old Europe, but they certainly were connected by a trade network of impressive reach. Cemeteries across the Pontic-Caspian steppes (DDII, Khvalynsk, S'yezzhe, Nalchik) became larger or appeared for the first time, suggesting the growth of larger, more stable communities.

Cattle and sheep were important in the diet at some DDII settlements on the Dnieper River, but farther east they seem initially to have been more important in funeral rituals than in the daily diet, which was still dominated by horse meat. In the east, domesticated cattle and sheep seem to have served as a kind of currency in a new set of rituals and religious beliefs.

Participation in long-distance trade, gift exchange, and a new set of cults requiring public sacrifices and feasting became the foundation for a new kind of social power. Stockbreeding is by nature a volatile economy. Herders who lose animals always borrow from those who still have them. The social obligations associated with these loans are institutionalized among the world's pastoralists as the basis for a fluid system of status distinctions. Those who loaned animals acquired power over those who borrowed them, and those who sponsored feasts obligated their guests. Early Proto-Indo-European included a vocabulary about verbal contracts bound by oaths (*$h_1óitos$-), used in later religious rituals to specify the obligations between the weak (humans) and the strong (gods). Reflexes of this root were preserved in Celtic, Germanic, Greek, and Tocharian. The model of political relations it references probably began in the Eneolithic. Only a few Eneolithic steppe people wore the elaborate costumes of tusks, plaques, beads, and rings or carried the stone maces that symbolized power, but children were included in this exceptional group, suggesting that the rich animal loaners at least tried to see that their children inherited their status. Status competition between regional leaders *weikpotis or *reg̑- in later Proto-Indo-European resulted in a surprisingly widespread set of shared status symbols. As leaders acquired followers, political networks emerged around them—and this was the basis for tribes.

Societies that did not accept the new herding economy became increasingly different from those that did. The people of the northern forest zone remained foragers, as did those who lived in the steppes east of the Ural Mountains. These frontiers probably were linguistic as well as economic, given their persistence and clarity. The Pre-Proto-Indo-European language family probably expanded with the new economy during the Early Eneolithic in the western steppes. Its sister-to-sister linguistic links may well have facilitated the spread of stockbreeding and the beliefs that went with it.

One notable aspect of the Pontic-Caspian Early Eneolithic is the importance of horses, in both diet and funeral symbolism. Horse meat was a major part of the meat diet. Images of horses were carved on bone plaques at Varfolomievka and S'yezzhe. At Khvalynsk, horses were included with cattle and sheep in funeral rituals that excluded obviously wild animals. But, zoologically, we cannot say whether they looked very different from wild horses—the bones no longer exist. The domestication of the horse, an enormously important event in human history, is not at all well under- stood. Recently, however, a new kind of evidence has been obtained straight from the horse's mouth.

CHAPTER TEN

The Domestication of the Horse
and the Origins of Riding
The Tale of the Teeth

The importance of the horse in human history is matched only by the difficulties
inherent in its study; there is hardly an incident in the story which is not the
subject of controversy, often of a violent nature.
—Grahame Clark, 1941

In the summer of 1985 I went with my wife Dorcas Brown, a fellow archaeologist, to the Veterinary School at the University of Pennsylvania to
ask a veterinary surgeon a few questions. Do bits create pathologies on
horse teeth? If they do, then shouldn't we be able to see the signs of
bitting—scratches or small patches of wear—on ancient horse teeth?
Wouldn't that be a good way to identify early bitted horses? Could he point
us toward the medical literature on the dental pathologies associated with
horse bits? He replied that there really was no literature on the subject. A
properly bitted horse wearing a well-adjusted bridle, he said, really *can't*
take the bit in its teeth very easily, so contact between the bit and the teeth
would have been too infrequent to show up with any regularity. Nice idea,
but it wouldn't work. We decided to get a second opinion.

At the Veterinary School's New Bolton Center for large mammals, outside Philadelphia, the trainers, who worked every day with horses, responded very differently. Horses chewed their bits all the time, they said.
Some rolled the bit around in their mouths like candy. You could hear it
clacking against their teeth. Of course, it was a vice—properly trained and
harnessed horses were not supposed to do it, but they did. And we should
talk to Hillary Clayton, formerly at New Bolton, who had gone to a university job somewhere in Canada. She had been studying the mechanics of
bits in horses' mouths.

We located Hillary Clayton at the University of Saskatchewan and found that she had made X-ray fluoroscopic videos of horses chewing bits (figure 10.1). She bitted horses and manipulated the reins from a standing position behind. An X-ray fluoroscope mounted beside the horses' heads took pictures of what was happening inside their mouths. No one had done this before. She sent us two articles co-authored with colleagues in Canada.[1] Their images showed just how horses manipulated a bit inside their mouths and precisely where it sat between their teeth. A well-positioned bit is supposed to sit on the tongue and gums in the space between the front and back teeth, called the "bars" of the mouth. When the rider pulls the reins, the bit presses the tongue and the gums into the lower jaw, squeezing the sensitive gum tissue between the bit and the underlying bone. That hurts. The horse will dip its head toward a one-sided

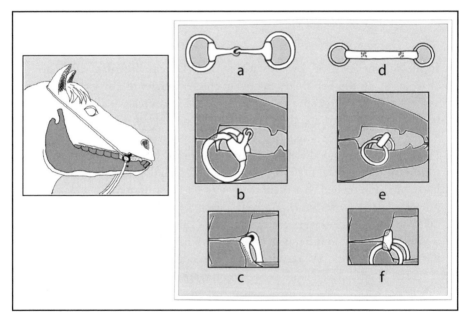

Figure 10.1 A modern metal bit in a horse's mouth. Mandible bone tinted gray. (a) jointed snaffle bit; (b) X-ray of jointed snaffle sitting on the tongue in proper position; (c) X-ray of snaffle being grasped in the teeth; (d) bar bit showing chewing wear; (e) X-ray of bar bit sitting on the tongue in proper position; (f) X-ray of bar bit being grasped in the teeth. After Clayton and Lee 1984; and Clayton 1985.

pull (a turn) or lower its chin into a two-sided pull (a brake) to avoid the bit's pressure on its tongue and gums.

Clayton's X-rays showed how horses use their tongues to elevate the bit and then retract it, pushing it back into the grip of their premolars, where it can no longer cause pressure on soft tissue no matter how hard the rider pulls on the reins. The soft corners of the mouth are positioned in front of the molars, so in order to get a bit into its teeth the horse has to force it back against the corners of its mouth. These stretched tissues act like a spring. If the bit is not held *very* firmly between the tips of the teeth it will pop forward again onto the bars of the mouth. It seemed likely to us that this repeated back-and-forth movement over the tips of the front premolars should affect the lower teeth more than the uppers just because of gravity—the bit sat on the lower jaw. The wear from bit chewing should be concentrated on one small part of two teeth (the lower second premolars, or P_2s), unlike the wear from chewing anything else. Clayton's X-rays made it possible, for the first time, to say positively that a specific part of a single tooth was the place to look for bit wear. We found several published photographs of archaeological horse P_2s with wear facets or bevels on precisely that spot. Two well-known archaeological zoologists, Juliet Clutton-Brock in London and Antonio Azzaroli in Rome, had described this kind of wear as "possibly" made by a bit. Other zoologists thought it was impossible for horses to get a bit that far back into their mouth with any frequency, like our first veterinary surgeon. No one knew for sure. But they had not seen Clayton's X-rays.[2]

Encouraged and excited, we visited the anthropology department at the Smithsonian Museum of Natural History in Washington, and asked Melinda Zeder, then a staff archaeozoologist, if we could study some never-bitted ancient wild horse teeth—a control sample—and if she could offer us some technical advice about how to proceed. We were not trained as zoologists, and we did not know much about horse teeth. Zeder and a colleague who knew a lot about dental microwear, Kate Gordon, sat us down in the staff cafeteria. How would we distinguish bit wear from tooth irregularities caused by malocclusion? Or from dietary wear, created by normal chewing on food? Would the wear caused by a bit survive very long, or would it be worn away by dietary wear? How long would that take? How fast do horse teeth grow? Aren't they the kind of teeth that grow out of the jaw and are worn away at the crown until they become little stubs? Would that change bit wear facets with increasing age? What about rope or leather bits—probably the oldest kind? Do they cause wear? What kind? Is the action of the bit different when a horse is ridden from when it pulls a chariot? And what, exactly, causes wear—if it exists? Is it the rider pulling the bit into the *front* of the tooth, or is it the horse chewing on the bit, which would cause wear on the *occlusal* (chewing) surface of the tooth? Or is it both? And if we did find wear under the microscope, how would we describe it so that the difference between a tooth with and without wear could be quantified?

Mindy Zeder took us through her collections. We made our first molds of ancient equid P_2s, from the Bronze Age city of Malyan in Iran, dated about 2000 BCE. They had wear facets on their mesial corners; later we would be able to say that the facets were created by a hard bit of bone or metal. But we didn't know that yet, and, as turned out, there really was not a large collection of never-bitted wild horse teeth at the Smithsonian. We had to find our own, and we left thinking that we could do it if we took one problem at a time. Twenty years later we still feel that way.[3]

Where Were Horses First Domesticated?

Bit wear is important, because other kinds of evidence have proven uncertain guides to early horse domestication. Genetic evidence, which we might hope would solve the problem, does not help much. Modern horses are genetically schizophrenic, like cattle (chapter 8) but with the genders reversed. The *female* bloodline of modern domesticated horses shows extreme diversity. Traits inherited through the mitochondrial DNA, which passes unchanged from mother to daughter, show that this part of the bloodline is so diverse that *at least* seventy-seven ancestral mares, grouped into seventeen phylogenetic branches, are required to account for the genetic variety in modern populations around the globe. Wild mares must have been taken into domesticated horse herds in many different places at different times. Meanwhile, the *male* aspect of modern horse DNA, which is passed unchanged on the Y chromosome from sire to colt, shows remarkable homogeneity. It is possible that just a single wild stallion was domesticated. So horse keepers apparently have felt free to capture and breed a variety of wild mares, but, according to these data, they universally rejected wild males and even the male progeny of any wild stallions that mated with domesticated mares. Modern horses are descended from very few original wild males, and many, varied wild females.[4]

Why the Difference?

Wildlife biologists have observed the behavior of feral horse bands in several places around the world, notably at Askania Nova, Ukraine, on the barrier islands of Maryland and Virginia (the horses described in the childrens' classic *Misty of Chincoteague*), and in northwestern Nevada. The standard feral horse band consists of a stallion with a harem of two to seven mares and their immature offspring. Adolescents leave the band at about two years of age. Stallion-and-harem bands occupy a home range, and stallions fight one another, fiercely, for control of mares and territory. After the young males are expelled they form loose associations called "bachelor bands," which lurk at the edges of the home range of an established stallion. Most bachelors are unable to challenge mature stallions or keep mares successfully until they are more than five years old. Within established bands, the mares are arranged in a social hierarchy led by the lead mare, who chooses where the band will go during most of the day and leads it in flight if there is a threat, while the stallion guards the flanks or the rear. Mares are therefore instinctively disposed to accept the dominance of others, whether dominant mares, stallions— or humans. Stallions are headstrong and violent, and are instinctively disposed to challenge authority by biting and kicking. A relatively docile and controllable mare could be found at the bottom of the pecking order in many wild horse bands, but a relatively docile and controllable stallion was an unusual individual—and one that had little hope of reproducing in the wild. Horse domestication might have depended on a lucky coincidence: the appearance of a relatively manageable and docile male in a place where humans could use him as the breeder of a domesticated bloodline. From the horse's perspective, humans were the only way he could get a girl. From the human perspective, he was the only sire they wanted.

Where Did He Live? And When?

Animal domestication, like marriage, is the culmination of a long prior relationship. People would not invest the time and energy to attempt to care for an animal they were unfamiliar with. The first people to think seriously about the benefits of keeping, feeding, and raising tame horses must have been familiar with wild horses. They must have lived in a place where humans spent a lot of time hunting wild horses and learning their behavior. The part of the world where this was possible contracted significantly about ten thousand to fourteen thousand years ago, when the Ice Age steppe—a favorable environment for horses—was replaced by dense forest over much of the Northern Hemisphere. The horses of North America became extinct as the climate shifted, for reasons still poorly understood. In Europe and Asia large herds of wild horses survived only in the

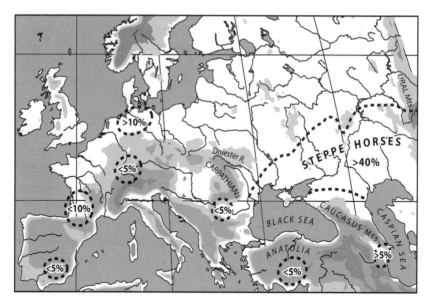

Figure 10.2 Map of the distribution of wild horses (*Equus caballus*) in the mid-Holocene, about 5000 BCE. The numbers show the approximate frequencies of horse bones in human kitchen garbage in each region, derived from charts in Benecke 1994 and from various Russian sources.

steppes in the center of the Eurasian continent, leaving smaller populations isolated in pockets of naturally open pasture (marsh-grass meadows, alpine meadows, arid *mesetas*) in Europe, central Anatolia (modern Turkey), and the Caucasus Mountains. Horses disappeared from Iran, lowland Mesopotamia, and the Fertile Crescent, leaving these warm regions to other equids (onagers and asses) (figure 10.2).

In western and central Europe, central Anatolia, and the Caucasus the isolated pockets of horses that survived into the Holocene never became important in the human food quest—there just weren't enough of them. In Anatolia, for example, a few wild horses probably were hunted occasionally by the Neolithic occupants of Catal Hüyök, Pinarbaşi, and other farming villages in the central plateau region between about 7400 and 6200 BCE. But most of the equids hunted at these sites were *Equus hydruntinus* (now extinct) or *Equus hemionus* (onagers), both ass-like equids smaller than horses. Only a few bones are large enough to qualify as possible horses. Horses were not present in Neolithic sites in western Anatolia, or in Greece or Bulgaria, or in the

Mesolithic and Early Neolithic of Austria, Hungary, or southern Poland. In western and northern Europe, Mesolithic foragers hunted horses occasionally. But horse bones accounted for more than 5% of the animals in only a few post-Glacial sites in the coastal plain of Germany/ Poland and in the uplands of southern France. In the Eurasian steppes, on the other hand, wild horses and related wild equids (onagers, *E. hydruntinus*) were the most common wild grazing animals. In early Holocene steppe archaeological sites (Mesolithic and early Neolithic) wild horses regularly account for more than 40% of the animal bones, and probably more than 40% of the meat diet because horses are so big and meaty. For this reason alone we should look first to the Eurasian steppes for the earliest episode of domestication, the one that probably gave us our modern male bloodline.[5]

Early and middle Holocene archaeological sites in the Pontic-Caspian steppes contain the bones of three species of equids. In the Caspian Depression, at Mesolithic sites such as Burovaya 53, Je-Kalgan, and Istai IV, garbage dumps dated before 5500 BCE contain almost exclusively the bones of horses and onagers (see site map, figure 8.3). The onager, *Equus hemionus*, also called a "hemione" or "half-ass," was a fleet-footed, long-eared animal smaller than a horse and larger than an ass. The natural range of the onager extended from the Caspian steppes across Central Asia and Iran and into the Near East. A second equid, *Equus hydruntinus*, was hunted in the slightly moister North Pontic steppes in Ukraine, where its bones occur in small percentages in Mesolithic and Early Neolithic components at Girzhevo and Matveev Kurgan, dated to the late seventh millennium BCE. This small, gracile animal, which then lived from the Black Sea steppes westward into Bulgaria and Romania and south into Anatolia, became extinct before 3000 BCE. The true horse, *Equus caballus*, ranged across both the Caspian Depression and the Black Sea steppes, and it survived in both environments long after both *E. hemionus* and *E. hydruntinus* were hunted out. Horse bones contributed more than 50% of the identified animal bones at Late Mesolithic Girzhevo in the Dniester steppes and Meso/Neolithic Matveev Kurgan and Kammenaya Mogila in the Azov steppes; also at Neo/Eneolithic Varfolomievka and Dzhangar in the Caspian Depression, Ivanovskaya on the Samara River, and Mullino in the southern foothills of the Ural Mountains. The long history of human dependence on wild equids in the steppes created a familiarity with their habits that would later make the domestication of the horse possible.[6]

WHY WERE HORSES DOMESTICATED?

The earliest evidence for possible horse domestication in the Pontic-Caspian steppes appeared after 4800 BCE, long after sheep, goats, pigs, and cattle were domesticated in other parts of the world. What was the incentive to tame wild horses if people already had cattle and sheep? Was it for transportation? Almost certainly not. Horses were large, powerful, aggressive animals, more inclined to flee or fight than to carry a human. Riding probably developed only after horses were already familiar as domesticated animals that could be controlled. The initial incentive probably was the desire for a cheap source of winter meat.

Horses are easier to feed through the winter than cattle or sheep, as cattle and sheep push snow aside with their noses and horses use their hard hooves. Sheep can graze on winter grass through soft snow, but if the snow becomes crusted with ice than their noses will get raw and bloody, and they will stand and starve in a field where there is ample winter forage just beneath their feet. Cattle do not forage through even soft snow if they cannot see the grass, so a snow deep enough to hide the winter grass will kill range cattle if they are not given fodder. Neither cattle nor sheep will break the ice on frozen water to drink. Horses have the instinct to break through ice and crusted snow with their hooves, not their noses, even in deep snows where the grass cannot be seen. They paw frozen snow away and feed themselves and so do not need water or fodder. In 1245 the Franciscan John of Plano Carpini journeyed to Mongolia to meet Güyük Khan (the successor to Genghis) and observed the steppe horses of the Tartars, as he called them, digging for grass from under the snow, "since the Tartars have neither straw nor hay nor fodder." During the historic blizzard of 1886 in the North American Plains hundreds of thousands of cattle were lost on the open range. Those that survived followed herds of mustangs and grazed in the areas they opened up.[7] Horses are supremely well adapted to the cold grasslands where they evolved. People who lived in cold grasslands with domesticated cattle and sheep would soon have seen the advantage in keeping horses for meat, just because the horses did not need fodder or water. A shift to colder climatic conditions or even a particularly cold series of winters could have made cattle herders think seriously about domesticating horses. Just such a shift to colder winters occurred between about 4200 and 3800 BCE (see chapter 11).

Cattle herders would have been particularly well suited to manage horses because cattle and horse bands both follow the lead of a dominant female. Cowherds already knew they needed only to control the lead cow to control the whole herd, and would easily have transferred that knowledge to controlling lead mares. Males presented a similar management problem in both species, and they had the same iconic status as symbols of virility and strength. When people who depended on equid-hunting began to keep domesticated cattle, someone would soon have noticed these similarities and applied cattle-management techniques to wild horses. And that would quickly have produced the earliest domesticated horses.

This earliest phase of horse keeping, when horses were primarily a recalcitrant but convenient source of winter meat, may have begun as early as 4800 BCE in the Pontic-Caspian steppes. This was when, at Khvalynsk and S'yezzhe in the middle Volga region, and Nikol'skoe on the Dnieper Rapids, horse heads and/or lower legs were first joined with the heads and/or lower legs of cattle and sheep in human funeral rituals; and when bone carvings of horses appeared with carvings of cattle in a few sites like S'yezzhe and Varfolomievka. Certainly horses were linked symbolically with humans and the cultured world of domesticated animals by 4800 BCE. Horse keeping would have added yet another element to the burst of economic, ritual, decorative, and political innovations that swept across the western steppes with the initial spread of stockbreeding about 5200–4800 BCE.

What Is a Domesticated Horse?

We decided to investigate bit wear on horse teeth, because it is difficult to distinguish the bones of early domesticated horses from those of their wild cousins. The Russian zoologist V. Bibikova tried to define a domesticated skull type in 1967, but her small sample of horse skulls did not define a reliable type for most zoologists.

The bones of wild animals usually are distinguished from those of domesticated animals by two quantifiable measurements: measurements of variability in size, and counts of the ages and sexes of butchered animals. Other criteria include finding animals far outside their natural range and detecting domestication-related pathologies, of which bit wear is an example. Crib biting, a stall-chewing vice of bored horses, might cause another domestication-related pathology on the incisor teeth of horses kept in stalls, but it has not been studied systematically. Marsha Levine of the McDonald Institute at Cambridge University has examined riding-related pathologies in vertebrae, but vertebrae are difficult to study. They break and rot easily, their frequency is low in most archaeological samples, and only eight caudal thoracic vertebrae (T11–18) are known to exhibit pathologies from riding. Discussions of horse domestication still tend to focus on the first two methods.[8]

The Size-Variability Method

The size-variability method depends on two assumptions: (1) domesticated populations, because they are protected, should contain a wider variety of sizes and statures that survive to adulthood, or *more variability*; and (2) the average size of the domesticated population as a whole should decline, because penning, control of movement, and a restricted diet should *reduce average stature*. Measurements of leg bones (principally the width of the condyle and shaft) are used to look for these patterns. This method seems to work quite well with the leg bones of cattle and sheep: an increase in variability and reduction in average size does apparently identify domesticated cattle and sheep.

But the underlying assumptions are not known to apply to the earliest domesticated horses. American Indians controlled their horses not in a corral but with a "hobble" (a short rope tied between the two front legs, permitting a walk but not a run). The principal advantage of early horse keeping—its low cost in labor—could be realized only if horses were permitted to forage for themselves. Pens and corrals would defeat this purpose. Domesticated horses living and grazing in the same environment with their wild cousins probably would not show a reduction in size, and might not show an increase in variability. These changes could be expected if and when horses were restricted to shelters and fed fodder over the winter, like cattle and sheep were, or when they were separated into different herds that were managed and trained differently, for example, for riding, chariot teams, or meat and milk production.

During the earliest phase of horse domestication, when horses were free-ranging and kept for their meat, any size reductions caused by human control probably would have been obscured by natural variations in size between different regional wild populations. The scattered wild horses living in central and western Europe were smaller than the horses that lived in the steppes. In figure 10.3, the three bars on the left of the graph represent wild horses from Ice Age and Early Neolithic Germany. They were quite small. Bars 4 and 5 represent wild horses from forest-steppe and steppe-edge regions, which were significantly bigger. The horses from Dereivka, in the central steppes of Ukraine, were bigger still; 75% stood between 133 and 137 cm at the withers, or between 13 and 14 hands. The

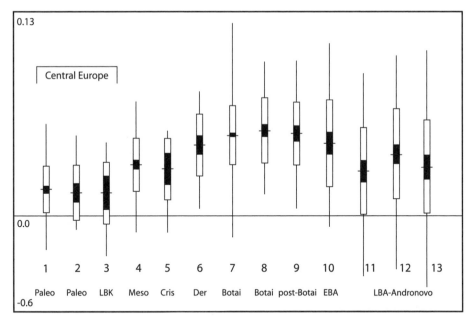

Figure 10.3 The size-variability method for identifying the bones of domesticated horses. The box-and-whisker graphs show the thickness of the leg bones for thirteen archaeological horse populations, with the oldest sites (Paleolithic) on the left and the youngest (Late Bronze Age) on the right. The whiskers, showing the extreme measurements, are most affected by sample size and so are unreliable indicators of population variability. The white boxes, showing two standard deviations from the mean, are reliable indicators of variability, and it is these that are usually compared. The increase in this measurement of variability in bar 10 is taken as evidence for the beginning of horse domestication. After Benecke and von den Dreisch 2003, figures 6.7 and 6.8 combined.

horses of Botai in northern Kazakhstan were even bigger, often over 14 hands. West-east movements of horse populations could cause changes in their average sizes, without any human interference. This leaves an increase in variability as the only indicator of domestication during the earliest phase. And variability is very sensitive to sample size—the larger the sample of bones, the better the chance of finding very small and very large individuals—so changes in variability alone are difficult to separate from sample-size effects.

The domestication of the horse is dated about 2500 BCE by the size-variability method. The earliest site that shows both a significant decrease in average size and an increase in variability is the Bell Beaker settlement of Csepel-Háros in Hungary, represented by bar 10 in figure 10.3, and dated about 2500 BCE. Subsequently many sites in Europe and the steppes show a similar pattern. The absence of these statistical indicators at Dereivka in Ukraine, dated about 4200–3700 BCE (see chapter 11), and at Botai-culture sites in northern Kazakhstan, dated about 3700–3000 BCE, are widely accepted as evidence that horses were not domesticated before about 2500 BCE. But marked regional size differences among early wild horses, the sensitivity of variability measurements to sample size effects, and the basic question of the applicability of these methods to the earliest domesticated horses are three reasons to look at other kinds of evidence. The appearance of significant new variability in horse herds after 2500 BCE could reflect the later development of specialized breeds and functions, not the earliest domestication.[9]

Age-at-Death Statistics

The second quantifiable method is the study of the ages and sexes of butchered animals. The animals selected for slaughter from a domesticated herd should be different ages and sexes from those obtained by hunting. Herders would probably cull young males as soon as they reached adult meat weight, at about two to three years of age. A site occupied by horse herders might contain very few obviously male horses, since the eruption of the canine teeth in males, the principal marker of gender in horse bones, happens at about age four or five, after the age when the males should have been slaughtered for food. Females should have been kept alive as breeders, up to ten years old or more. In contrast, hunters prey on the most predictable elements of a wild herd, so they would concentrate their efforts on the standard wild horse social group, the stallion-with-harem bands, which move along well-worn paths and trails within a defined territory. Regular hunting of stallion-with-harem bands would yield a small number of prime stallions (six to nine years old) and a large number of breeding-age females (three to ten years old) and their immature young.[10]

But many other hunting and culling patterns are possible, and might be superimposed on one another in a long-used settlement site. Also, only a few bones in a horse's body indicate sex—a mature male (more than five years old) has canine teeth whereas females usually do not, and the pelvis of a mature female is distinctive. Horse jaws with the canines still embedded are not often preserved, so data on gender are spotty. Age is estimated based on molar teeth, which preserve well, so the sample for age estimation usually is bigger. But assigning a precise age to a loose horse molar, not found in the jaw, is difficult, and teeth are often found loose in archaeological sites. We had to invent a way to narrow down the very broad range of ages that could be assigned to each tooth. Further, teeth are part of the head, and heads may receive special treatment. If the goal of the analysis is to determine which horses were culled for food, heads are not necessarily the most direct indicators of the human diet. If the occupants of the site kept and used the heads of prime-age stallions for rituals, the teeth found in the site would reflect that, and not culling for food.[11]

Marsha Levine studied age and sex data at Dereivka in Ukraine (4200–3700 BCE) and Botai in northern Kazakhstan (3700–3000 BCE), two critical sites for the study of horse domestication in the steppes. She concluded that the horses at both sites were wild. At Dereivka the majority of the teeth were from animals whose ages clustered between five and seven years old, and fourteen of the sixteen mandibles were from mature males.[12] This suggested that most of the horse heads at Dereivka came from prime-age stallions, not the butchering pattern expected for a managed population. But, in fact, it is an odd pattern for a hunted population as well. Why would hunters kill only prime stallions? Levine suggested that the Dereivka hunters had *stalked* wild horse bands, drawing the attention of the stallions, which were killed when they advanced to protect their harems. But stalking in the open steppe is probably the least productive way for a pedestrian hunter to attack a wild horse band, as stallions are more likely to alarm their band and run away than to approach a predator. Pedestrian hunters should have used ambush methods, shooting at short range on a habitually used horse trail. Moreover, the odd stallion-centered slaughter pattern of Dereivka closely matches the slaughter pattern at the Roman military cemetery at Kestren, the Netherlands (figure 10.4), where the horses certainly were domesticated. At Botai, in contrast, the age-and-sex profile matched what would be expected if whole wild herds were slaughtered en masse, with no selection for age or sex. The two profiles were dissimilar, yet Levine concluded that horses were wild at both places. Age and sex profiles are open to many different interpretations.

If it is difficult to distinguish wild from domesticated horses, it is doubly problematic to distinguish the bones of a mount from those of a horse merely eaten for dinner. Riding leaves few traces on horse bones. But a bit leaves marks on the teeth, and teeth usually survive very well. Bits are used only to guide horses from behind, to drive or to ride. They are not used if the horse is pulled from the front, as a packhorse is, as this would just pull the bit out of the mouth. Thus bit wear on the teeth indicates

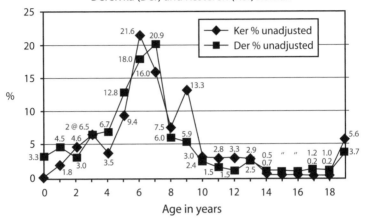

Figure 10.4 The age-at-death method for identifying the bones of domesticated horses. This graph compares the age-at-death statistics for Late Eneolithic horses from Dereivka, Ukraine, to domesticated horses from the Roman site of Kesteren, Netherlands. The two graphs are strikingly similar, but one is interpreted as a "wild" profile and the other is "domesticated." After Levine 1999, figure 2.21.

riding or driving. The *absence* of bit wear means nothing, since other forms of control (nosebands, hackamores) might leave no evidence. But its *presence* is an unmistakable sign of riding or driving. That is why we pursued it. Bit wear could be the smoking gun in the long argument over the origins of horseback riding and, by extension, in debates over the domestication of the horse.

BIT WEAR AND HORSEBACK RIDING

After Brown and I left the Smithsonian in 1985 we spent several years gathering a collection of horse lower second premolars (P_2s), the teeth most affected by bit chewing. Eventually we collected 139 P_2s from 72 modern horses. Forty were domesticated horses processed through veterinary autopsy labs at the University of Pennsylvania and Cornell University. All had been bitted with modern metal bits. We obtained information on their age, sex, and usage—hunting, leisure, driving, racing, or draft—and for some horses we even knew how often they had been bitted, and with what kind of bit. Thirteen additional horses came from the Horse

Training and Behavior program at the State University of New York at Cobleskill. Some had never been bitted. We made casts of their teeth in their mouths, much as a dentist makes an impression to fit a crown—we think that we were the first people to do this to a living horse. A few feral horses, never bitted, were obtained from the Atlantic barrier island of Assateague, MD. Their bleached bones and teeth were found by Ron Keiper of Penn State, who regularly followed and studied the Assateague horses and generously gave us what he had found. Sixteen Nevada mustangs, killed in 1988 by ranchers, supplied most of our never-bitted P_2s. I read about the event, made several telephone calls, and was able to get their mandibles from the Bureau of Land Management after the kill sites were documented. Many years later, in a separate study, Christian George at the University of Florida applied our methods to 113 more never-bitted P_2s from a minimum of 58 fossil equids 1.5 million years old. These animals, of the species *Equus "leidyi,"* were excavated from a Pleistocene deposit near Leisey, Florida. George's Leisey equids (the same size, diet, and dentition as modern horses) had never seen a human, much less a bit.[13]

We studied high-resolution casts or replicas of all the P_2s under a Scanning Electron Microscope (SEM). The SEM revealed that the vice of bit chewing was amazingly widely practiced (figure 10.5). More than 90% of the bitted horses showed some wear on their P_2s from chewing the bit, often just on one side. Their bits also showed wear from being chewed. Riding creates the same wear as driving, because it is not the rider or driver who creates bit wear—it is the horse grasping and releasing the bit between its teeth. A metal bit or even a bone bit creates distinctive microscopic abrasions on the occlusal enamel of the tooth, usually confined to the first or metaconid cusp, but extending back to the second cusp in many cases. These abrasions (type "a" wear, in our terminology) are easily identified under a microscope. All bits, whether hard (metal or bone) or soft (rope or leather) also create a second kind of wear: a wear facet or bevel on the front (mesial) corner of the tooth. The facet is caused both by direct pressure (particularly with a hard bit of bone or metal), which weakens and cracks the enamel when the bit is squeezed repeatedly between the teeth; and by the bit slipping back and forth over the front or mesial corner of the P_2. Metal bits create both kinds of wear: abrasions on the occlusal enamel and wear facets on the mesial corner of the tooth. But rope bits probably were the earliest kind. Can a rope bit alone create visible wear on the enamel of horse teeth?

With a grant from the National Science Foundation and the cooperation of the State University of New York (SUNY) at Cobleskill we acquired

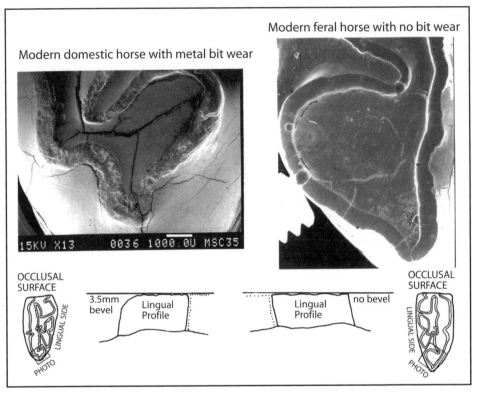

Figure 10.5 Bit wear and no wear on the lower second premolars (P$_2$s) of modern horses.

Left: a Scanning Electron Micrograph (SEM) taken at 13x of "a-wear" abrasions on the first cusp of a domesticated horse that was bitted with a metal bit. The profile shows a 3.5 mm bevel or facet on the same cusp.

Right: An SEM taken at 15x of the smooth surface of the first cusp of a feral horse from Nevada, never bitted. The profile shows a 90° angle with no bevel.

four horses that had never been bitted. They were kept and ridden at SUNY Cobleskill, which has a Horse Training and Behavior Program and a thirty-five-horse stable. They ate only hay and pasture, no soft feeds, to mimic the natural dental wear of free-range horses. Each horse was ridden with a different organic bit—leather, horsehair rope, hemp rope, or bone—for 150 hours, or 600 hours of riding for all four horses. The horse with the horsehair rope bit was bitted by tying the rope around its lower jaw in the classic "war bridle" of the Plains Indians, yet

Figure 10.6 Brown and Anthony removing a high-resolution mold of the P$_2$ of a horse bitted with an organic bit at State University of New York, Cobleskill, in 1992.

it was still able to loosen the loop with its tongue and chew the rope. The other horses' bits were kept in place by antler cheek-pieces made with flint tools. At four intervals each horse was anaesthetized by a bemused veterinarian, and we propped open its mouth, brushed its teeth, dried them, pulled its tongue to the side, and made molds of its P$_2$s (figure 10.6). We tracked the progress of bit wear over time, and noted the differences between the wear made by the bone bit (hard) and the leather and rope bits (soft).[14]

The riding experiment demonstrated that soft bits do create bit wear. The actual cause of wear might have been microscopic grit trapped in and under the bit, since all the soft bits were made of materials softer than enamel. After 150 hours of riding, bits made of leather and rope wore away about 1 mm of enamel on the first cusp of the P$_2$ (figure 10.7). The mean bevel measurement for the three horses with rope or leather bits at the end of the experiment was more than 2 standard deviations greater than the pre-experiment mean.[15] The rope and leather mouthpieces stood up well to chewing, although the horse with the hemp rope bit chewed through it several times. The horses bitted with soft bits showed the same

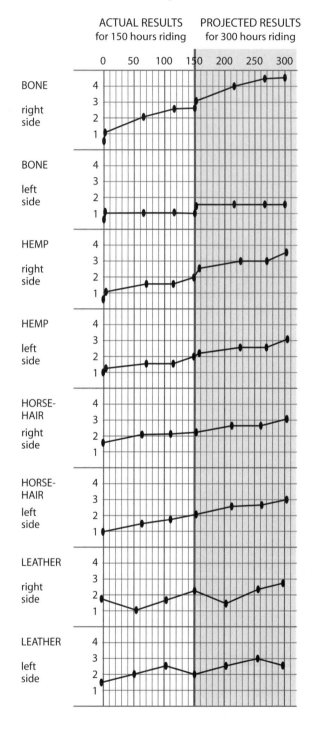

TABLE 10.1

Bevel Measurements on the P$_2$s of Bitted and Never–Bitted Mature (>3yr) Horses

	Never–Bitted, Feral and Domestic (16 horses / 31 teeth)	Pleistocene Leisey equids (44 h. / 74t.)	Domestic Bitted (39 h. / 73 t.)	Domestic Bitted Daily (13 h. / 24t.)
Median	0.5 mm	1.1 mm	2.5 mm	4.0 mm
Mean	0.79 mm	1.1 mm	3.11 mm	3.6 mm
Standard Deviation	0.63 mm	0.71 mm	1.93 mm	1.61 mm
Range	0–2 mm	0–2.9 mm	0–10 mm	1–7 mm

wear facet on the same part of the P$_2$ as horses bitted with metal and bone bits, but the surface of the facet was microscopically smooth and polished, not abraded. Hard bits, including our experimental bone bit, create distinctive "a" wear on the occlusal enamel of the facet, but soft bits do not. Soft bit wear is best identified by measuring the depth of the wear facet or bevel on the P$_2$, not by looking for abrasions on its surface.

Table 10.1 shows bevel measurements for modern horses that never were bitted (left column); Pleistocene North American equids that never were bitted (center left column); domestic horses that were bitted, including some that were bitted infrequently (center right column); and a smaller sub-group of domestic horses that were bitted at least five times a week up to the day we made molds of their teeth (right column). Measurements of the depth of the wear facet easily distinguished the 73 teeth of bitted horses from the 105 teeth of never-bitted horses. The never-bitted/bitted means are different at better than the .001 level of significance. The never-bitted/daily-bitted means are more than 4 standard deviations apart. Bevel measurements segregate mature bitted from mature never-bitted horses, *as populations.*[16]

We set a bevel measurement of 3.0 mm as the minimum threshold for recognizing bit wear on archaeological horse teeth (figure 10.8). More than half of our occasionally bitted teeth did not exhibit a bevel measuring as much as 3 mm . But all horses in our sample with a bevel of 3 mm or more

Figure 10.7 Graph showing the increase in bevel measurements in millimeters caused by organic bits over 150 hours of riding, with projections of measurements if riding had continued for 300 hours.

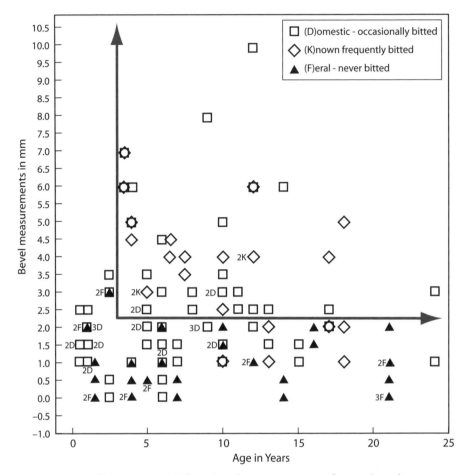

Figure 10.8 From our 1998 data: bevel measurements of never bitted, occasionally bitted, and frequently bitted horse teeth plotted against age. All domesticated horses had precisely known ages; all feral horses were aged by examining entire mandibles with intact incisor teeth. The line excludes feral horses and horses aged ≤3yr. and includes only bitted horses. After Brown and Anthony 1998.

had been bitted. So the last question was, how adequate was our sample? Could a 3 mm -wear facet occur naturally on a wild horse P_2, caused by malocclusion? Criticisms of bit wear have centered on this problem.[17]

Very young horses with newly erupted permanent premolars do display natural dips and rises on their teeth. New permanent premolars are uneven because they have not yet been worn flat by occlusion with the opposing tooth. We had to exclude the teeth of horses two to three years old for that reason. But among the 105 measurable P_2s from *mature* equids that had never been bitted, Pleistocene to modern, we found that a "natural" bevel measurement of more than 2.0 mm is unusual (less than 3% of teeth), and a bevel of 2.5 mm is exceedingly rare (less than 1%). Only one of the 105 never-bitted teeth had a bevel measurement greater than 2.5 mm—a single tooth from the Leisey equids with a mesial bevel of 2.9 mm (the next-nearest bevel was 2.34 mm). In contrast, bevels of 2.5 mm and more occurred in 58% of the teeth of mature horses that were bitted.[18]

A bevel of 3 mm or more on the P_2 of a mature horse is evidence for either an exceedingly rare malocclusion or a very common effect of bitting. If even one mature horse from an archaeological site shows a bevel ≥3 mm bit wear is suggested, but is not a closed case. If multiple mature horses from a single site show mesial bevel measurements of 3 mm or more, they probably were bitted. I should stress that our method depends on the accurate measurement of a very small feature—a bevel or facet just a few millimeters deep. According to our measurements on 178 P_2 teeth of mature equids the difference between a 2 mm and a 3 mm bevel is extremely important. In any discussion of bit wear, precise measurements are required and young animals must be eliminated. But until someone finds a population of mature wild horses that displays many P_2 teeth with bevels ≥3 mm , bit wear as we have defined it indicates that a horse has been ridden or driven.[19]

INDO-EUROPEAN MIGRATIONS AND BIT WEAR AT DEREIVKA

Many archaeologists and historians in the first half of the twentieth century thought that horses were first domesticated by Indo-European–speaking peoples, often specifically characterized as Aryans, who also were credited with inventing the horse-drawn chariot. This fascination with the Aryans, or *Ariomania*, to use Peter Raulwing's term, dominated the study of horseback riding and chariots before World War II.[20]

In 1964 Dimitri Telegin discovered the head-and-hoof bones of a seven- to eight-year-old stallion buried together with the remains of two dogs at Dereivka in Ukraine, apparently a cultic deposit of some kind (see figure 11.9). The Dereivka settlement contained three excavated structures of the Sredni Stog culture and the bones of a great many horses, 63% of the bones found. Ten radiocarbon dates placed the Sredni Stog settlement about 4200–3700 BCE, after the Dnieper-Donets II and Early Khvalynsk era. V. I. Bibikova, the chief paleozoologist at the Kiev

Institute of Archaeology, declared the stallion a domesticated horse in 1967. The respected Hungarian zoologist and head of the Hungarian Institute of Archaeology, Sandor Bökönyi, agreed, noting the great variabity in the leg dimensions of the Dereivka horses. The German zoologist G. Nobis also agreed. During the late 1960s and 1970s horse domestication at Dereivka was widely accepted.[21]

For Marija Gimbutas of UCLA, the domesticated horses at Dereivka were part of the evidence which proved that horse-riding, Indo-European–speaking "Kurgan-culture" pastoralists had migrated in several waves out of the steppes between 4200 and 3200 BCE, destroying the world of egalitarian peace and beauty that she imagined for the Eneolithic cultures of Old Europe. But the idea of Indo-European migrations sweeping westward out of the steppes was not accepted by most Western archaeologists, who were increasingly suspicious of any migration-based explanation for culture change. During the 1980s Gimbutas's scenario of massive "Kurgan-culture" invasions into eastern and central Europe was largely discredited, notably by the German archaeologist A. Häusler. Jim Mallory's 1989 masterful review of Indo-European archaeology retained Gimbutas's steppe homeland and her three waves as periods of increased movement in and around the steppes, but he was much less optimistic about linking specific archaeological cultures with specific migrations by specific Indo-European branches. Others, myself included, criticized both Gimbutas's archaeology and Bibikova's interpretation of the Dereivka horses. In 1990 Marsha Levine seemed to nail the coffin shut on the horse-riding, Kurgan-culture invasion hypothesis when she declared the horse age and sex ratios at Dereivka to be consistent with a wild, hunted population.[22]

Brown and I visited the Institute of Zoology in Kiev in 1989, the year after Levine, learning of her trip only after we arrived. With the cheerful help of Natalya Belan, a senior zoologist, we made molds of dozens of horse P_2s from many archaeological sites in Ukraine. We examined one P_2 from Early Eneolithic Varfolomievka in the Caspian Depression (no wear), one from the Tripolye A settlement of Luka Vrublevetskaya (no wear), several from Mesolithic and Paleolithic sites in Ukraine (no wear), many from Scythian and Roman-era graves (a lot of bit wear, some of it extreme), and those of the cult stallion and four other horse P_2s from Dereivka. As soon as we saw the Dereivka cult stallion we knew it had bit wear. Its P_2s had bevels of 3.5 mm and 4 mm , and the enamel on the first cusp was deeply abraded. Given its stratigraphic position at the base of a Late Eneolithic cultural level almost 1 m deep, dated by ten radiocarbon dates to 4200–3700 BCE, the cult stallion should have been about two thousand years older than the previously known oldest evidence for horseback riding. Only four other P_2s still survived in the Dereivka collection: two deciduous teeth from horses less than 2.5 years old (not measurable), and two others from adult horses but with no bit wear. So our case rested on a single horse. But it was very clear wear—surprisingly similar to modern metal bit wear. In 1991 we published articles in *Scientific American* and in the British journal *Antiquity* announcing the discovery of bit wear at Dereivka. Levine's conclusion that the Dereivka horses were wild had been published just the year before. Briefly we were too elated to worry about the argument that would follow.[23]

It began when A. Häusler challenged us at a conference in Berlin in 1992. He did not think the Dereivka stallion was Eneolithic or cultic; he deemed it a Medieval garbage deposit, denying there was evidence for a horse cult anywhere in the steppes during the Eneolithic. That the wear looked like metal bit wear was part of the problem, since a metal bit was improbable in the Eneolithic. Häusler's target was bigger than bit wear or even horse domestication: he had dedicated much of his career to refuting Gimbutas's "Kurgan-culture" migrations and the entire notion of a steppe Indo-European homeland.[24] The horses at Dereivka were just a small piece in a larger controversy. But criticisms like his forced us to obtain a direct date on the skull itself.

Telegin first sent us a bone sample from the same excavation square and level as the stallion. It yielded a date between 90 BCE and 70 BCE (OxA 6577), our first indication of a problem. He obtained another anomalous radiocarbon date, ca. 3000 BCE, on a piece of bone that, like our first sample, seems not to have been from the stallion itself (Ki 5488). Finally, he sent us one of the bit-worn P_2s from the cult stallion. The Oxford radiocarbon laboratory obtained a date of 410–200 BCE from this tooth (OxA 7185). Simultaneously the Kiev radiocarbon laboratory obtained a date of 790–520 BCE on a piece of bone from the skull (Ki 6962). Together these two samples suggest a date between 800 and 200 BCE.

The stallion-and-dog deposit at Dereivka was of the Scythian era. No wonder it had metal bit wear—so did many other Scythian horse teeth. It had been placed in a pit dug into the Eneolithic settlement between 800 and 200 BCE. The archaeologists who excavated this part of the site in 1964 did not see the intrusive pit. In 2000, nine years after our initial publication in *Antiquity*, we published another *Antiquity* article retracting the early date for bit wear at Dereivka. We were disappointed, but by then Dereivka was no longer the only prehistoric site in the steppes with bit wear.[25]

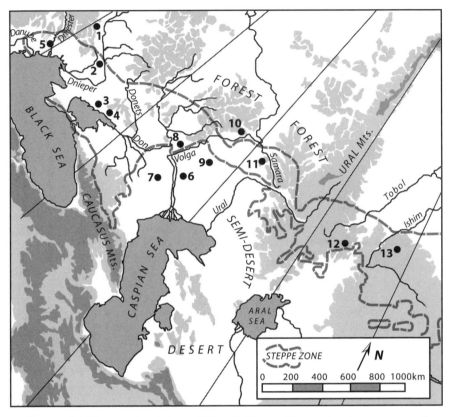

Figure 10.9 Horse-related sites of Eneolithic or older age in the western and central Eurasian steppes. The steppe ecological zone is enclosed in dashed lines.
(1) Moliukhor Bugor; (2) Dereivka; (3) Mariupol; (4) Matveev Kurgan; (5) Girzhevo; (6) Kair Shak; (7) Dzhangar; (8) Orlovka; (9) Varfolomievka; (10) Khvalynsk; (11) S'yezzhe; (12) Tersek; (13) Botai

BOTAI AND ENEOLITHIC HORSEBACK RIDING

The oldest horse P_2s showing wear facets of 3 mm and more are from the Botai and Tersek cultures of northern Kazakhstan (figure 10.9). Excavated through the 1980s by Victor Zaibert, Botai was a settlement of specialized hunters who rode horses to hunt horses, a peculiar kind of economy that existed only between 3700 and 3000 BCE, and only in the steppes of northern Kazakhstan. Sites of the Botai type, east of the Ishim River, and of the related Tersek type, west of the Ishim, contain

Figure 10.10 A concentration of horse bones in an excavated house pit at the Botai settlement in north-central Kazakhstan, dated about 3700–3000 BCE. Archaeozoologist Lubomir Peske takes measurements during an international conference held in Kazakhstan in 1995 "Early Horsekeepers of the Eurasian Steppe 4500–1500 BC." Photo by Asko Parpola.

65–99.9%/horse bones. Botai had more than 150 house-pits (figure 10.10) and 300,000 animal bones, 99.9% of them horse. A partial list of the other species represented at Botai (primarily by isolated teeth and phalanges) includes a very large bovid, probably bison but perhaps aurochs, as well as elk, red deer, roe deer, boar, bear, beaver, saiga antelope, and gazelle. Horses, not the easiest prey for people on foot, were overwhelmingly preferred over these animals.[26]

We visited Zaibert's lab in Petropavlovsk, Kazakhstan, in 1992, again unaware that Marsha Levine had arrived the year before. Among the forty-two P_2s we examined from Botai, nineteen were acceptable for study (many had heavily damaged surfaces, and others were from horses younger than three years old). Five of these nineteen teeth, representing at least three different horses, had significant bevel measurements: two 3 mm, one 3.5 mm, one 4 mm, and one 6 mm . Wear facets on undamaged portions of the Botai P_2s were polished smooth, the same kind of polish created by "soft" bits in our experiment. The five teeth were found in different places across the settlement—they did not come from a single intrusive pit. The

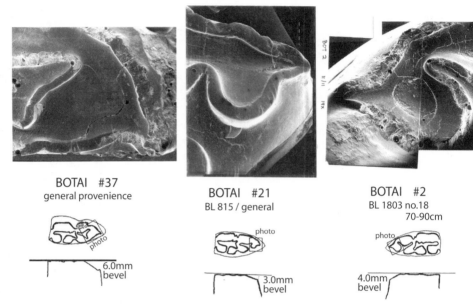

BOTAI #37
general provenience

BOTAI #21
BL 815 / general

BOTAI #2
BL 1803 no.18
70-90cm

6.0mm
bevel

3.0mm
bevel

4.0mm
bevel

Figure 10.11 Three horse P_2s with bit wear from the Botai settlement. The photos show extensive postmortem damage to the occlusal surfaces. The undamaged middle tooth showed smooth enamel surfaces but had a significant wear facet, like a horse ridden with a "soft" bit of rope or leather.

proportion of P_2s exhibiting bit wear at Botai was 12% of the entire sample of P_2s provided, or 26% of the nineteen measurable P_2s. Either number was just too high to explain by appealing to a rare natural malocclusion (figure 10.11). We also examined the horse P_2s from a Tersek site, Kozhai 1, dated to the same period, 3700–3000 BCE. At Kozhai 1 horses accounted for 66.1% of seventy thousand identified animal bones (others were saiga antelope at 21.8%, onager at 9.4%, and bison, perhaps including some very large domesticated cattle, at 2.1%). We found a 3 mm wear facet on two P_2s of the twelve we examined from Kozhai 1. Most of the P_2s at Botai and Kozhai 1 did not exhibit bit wear, but a small percentage (12–26%) did, consistent with the interpretation that the Botai-Tersek people were mounted horse hunters.[27]

Botai attracted the attention of everyone interested in early horse domestication. Two field excavations by Western archaeologists (Marsha Levine and Sandra Olsen) have occurred at Botai or Botai-culture sites. The original excavator, Victor Zaibert, the Kazakh zoologist L.A.

Makarova, and the American archaeozoologist Sandra Olsen of the Carnegie Museum of Natural History in Pittsburgh all concluded that at least some of the Botai horses were domesticated. In opposition, the archaeozoologists N. M. Ermolova, Marsha Levine, and the German team Norbert Benecke and Angela von den Dreisch concluded that all the Botai horses were wild.[28] Levine found some pathologies in the Botai vertebrae but attributed them to age. Benecke and von den Dreisch showed that the Botai horses exhibited a narrow range of variability in size, like Paleolithic wild populations. The ages and sexes of the Botai horses were typical of a wild population, with a 1:1 ratio between the sexes, including all age groups, even colts and pregnant mares with gestating fetuses. Everyone agrees that whole herds of wild horses were killed by the Botai people, using herd-driving hunting techniques that had never been used before in the Kazakh steppes, certainly not on this scale. Were the hunters riding or on foot? Native American hunters on foot drove bison herds over cliffs before the introduction of horses to the Americas by Europeans, so herd driving was possible without riding.

Sandra Olsen of the Carnegie Museum concluded that at least some Botai horses were used for transport, because whole horse carcasses were butchered regularly over the course of several centuries *in the settlement* at Botai.[29] How would pedestrian hunters drag eight-hundred-pound carcasses to the settlement, not just once or twice but as a regular practice that continued for centuries? Pedestrian hunters who used herd-driving hunting methods in the European Paleolithic at Solutré (where Olsen had worked earlier) and in the North American Plains butchered large animals where they died at the kill site. But the Botai settlement is located on the open, south-facing slope of a broad ridge top in a steppe environment—wild horses could not have been trapped in the settlement. Either some horses were tamed and could be led into the settlement or horses were used to drag whole carcasses of killed animals into the settlement, perhaps on sleds. Olsen's interpretation was supported by soil analysis from a house pit at Botai (Olsen's excavation 32) that revealed a distinct layer of soil filled with horse dung. This "must have been the result of redeposition of material from stabling layers," according to the soil scientists who examined it.[30] This dung-rich soil was removed from a horse stable or corral. The stabling of horses at Botai obviously suggests domestication.

One more argument for horseback riding is that the slaughter of wild populations with a 1:1 sex ratio could only be achieved by sweeping up both stallion-with-harem bands *and* bachelor bands, and these two kinds of social groups normally live far apart in the wild. If stallion-with-harem bands were driven into traps, the female:male ratio would be more than 2:1. The only way to capture both bachelor bands and harem bands in herd drives is to actively search and sweep up all the wild horses in a very large region. That would be impossible on foot.

Finally, the beginning of horseback riding provides a good explanation for the economic and cultural changes that appeared with the Botai-Tersek cultures. Before 3700 BCE foragers in the northern Kazkah steppes lived in small groups at temporary lakeside camps such as Vinogradovka XIV in Kokchetav district and Tel'manskie in Tselinograd district. Their remains are assigned to the Atbasar Neolithic.[31] They hunted horses but also a variety of other game: short-horned bison, saiga antelope, gazelle, and red deer. The details of their foraging economy are unclear, as their camp sites were small and ephemeral and have yielded relatively few animal bones. Around 3700–3500 BCE they shifted to specialized horse hunting, started to use herd-driving hunting methods, and began to aggregate in large settlements—a new hunting strategy and a new settlement pattern. The number of animal bones deposited at each settlement rose to tens or even hundreds of thousands. Their stone tools changed from microlithic tool kits to large bifacial blades. They began to make large polished stone weights with central perforations, probably for manufacturing multi-stranded rawhide ropes (weights are hung from each strand as the strands are twisted together). Rawhide thong manufacture was one of the principal activities Olsen identified at Botai based on bone tool microwear. For the first time the foragers of the northern Kazakh steppes demonstrated the ability to drive and trap whole herds of horses and transport their carcasses into new, large communal settlements. No explanation other than the adoption of horseback riding has been offered for these changes.

The case for horse management and riding at Botai and Kozhai 1 is based on the presence of bit wear on seven Botai-Tersek horse P_2s from two different sites, carcass transport and butchering practices, the discovery of horse-dung–filled stable soils, a 1:1 sex ratio, and changes in economy and settlement pattern consistent with the beginning of riding. The case against riding is based on the low variability in leg thickness and the absence of riding-related pathologies in a small sample of horse vertebrae, possibly from wild hunted horses, which probably made up 75–90% of the horse bones at Botai. We are reasonably certain that horses were bitted and ridden in northern Kazakhstan beginning about 3700–3500 BCE.

The Origin of Horseback Riding

Horseback riding probably did not begin in northern Kazakhstan. The Botai-Tersek people were mounted foragers. A few domesticated cattle (?) bones might be found in some Tersek sites, but there were none in Botai sites, farther east; and neither had sheep.[32] It is likely that Botai-Tersek people acquired the idea of domesticated animal management from their western neighbors, who had been managing domesticated cattle and sheep, and probably horses, for a thousand years before 3700–3500 BCE.

The evidence for riding at Botai is not isolated. Perhaps the most interesting parallel from beyond the steppes is a case of severe wear on a mesial horse P_2 with a bevel much deeper than 3 mm , on a five-year-old stallion jaw excavated from Late Chalcolithic levels at Mokhrablur in Armenia, dated 4000–3500 BCE. This looks like another case of early bit wear perhaps even older than Botai, but we have not examined it for confirmation.[33] Also, after about 3500 BCE horses began to appear in greater numbers or appeared regularly for the first time outside the Pontic-Caspian steppes. Between 3500 and 3000 BCE horses began to show up regularly in settlements of the Maikop and Early Transcaucasian Culture (ETC) in the Caucasus, and also for the first time in the lower and middle Danube valley in settlements of the Cernavoda III and Baden-Boleraz cultures as at Cernavoda and Kétegyháza. Around 3000 BCE horse bones rose to about 10–20% of the bones in Bernberg sites in central Germany and to more than 20% of the bones at the Cham site of Galgenberg in Bavaria. The Galgenburg horses included a native small type and a larger type probably imported from the steppes. This general increase in the importance of horses from Kazakhstan to the Caucasus, the Danube valley, and Germany after 3500 BCE suggests a significant change in the relationship between humans and horses. Botai and Tersek show what that change was: people had started to ride.[34]

Over the long term it would have been very difficult to manage horse herds without riding them. Anywhere that we see a sustained, long-term dependence on domesticated horses, riding is implied for herd management alone. Riding began in the Pontic-Caspian steppes before 3700 BCE, or before the Botai-Tersek culture appeared in the Kazakh steppes. It may well have started before 4200 BCE. It spread outside the Pontic-Caspian steppes between 3700 and 3000 BCE, as shown by increases in horse bones in southeastern Europe, central Europe, the Caucasus, and northern Kazakhstan.

The Economic and Military Effects
of Horseback Riding

A person on foot can herd about two hundred sheep with a good herding dog. On horseback, with the same dog, that single person can herd about five hundred.[35] Riding greatly increased the efficiency and therefore the scale and productivity of herding in the Eurasian grasslands. More cattle and sheep could be owned and controlled by riders than by pedestrian herders, which permitted a greater accumulation of animal wealth. Larger herds, of course, required larger pastures, and the desire for larger pastures would have caused a general renegotiation of tribal frontiers, a series of boundary conflicts. Victory in tribal warfare depended largely on forging alliances and mobilizing larger forces than your enemy, and so intensified warfare stimulated efforts to build alliances through feasts and the redistribution of wealth. Gifts were effective both in building alliances before conflicts and in sealing agreements after them. An increase in boundary conflicts would thus have encouraged more long-distance trade to acquire prestigious goods, as well as elaborate feasts and public ceremonies to forge alliances. This early phase of conflict, caused partly by herding on horseback, might be visible archaeologically in the horizon of polished stone mace-heads and body decorations (copper, gold, boars-tusk, and shell ornaments) that spread across the western steppes with the earliest herding economies about 5000–4200 BCE.[36]

Horses were valuable and easily stolen, and riding increased the efficiency of stealing cattle. When American Indians in the North American Plains first began to ride, chronic horse-stealing raids soured relationships even between tribes that had been friendly. Riding also was an excellent way to retreat quickly; often the most dangerous part of tribal raiding on foot was the running retreat after a raid. Eneolithic war parties might have left their horses under guard and attacked on foot, as many American Indians did in the early decades of horse warfare in the Plains. But even if horses were used for nothing more than transportation to and from the raid, the rapidity and reach of mounted raiders would have changed raiding tactics, status-seeking behaviors, alliance-building, displays of wealth, and settlement patterns. Thus riding cannot be cleanly separated from warfare.[37]

Many experts have suggested that horses were not ridden in warfare until after about 1500–1000 BCE, but they failed to differentiate between *mounted raiding*, which probably is very old, and *cavalry*, which was invented in the Iron Age after about 1000 BCE.[38] Eneolithic tribal herders probably rode horses in inter-clan raids before 4000 BCE, but they were not like the Huns sweeping out of the steppes on armies of shaggy horses. What is intriguing about the Huns and their more ancient cousins, the Scythians, was that they formed armies. During the Iron Age the Scythians, essentially tribal in most other aspects of their political organization, became organized in their military operations like the formal armies of urban states. That required a change in ideology—how a warrior thought about himself, his role, and his responsibilities—as well as in the technology of mounted warfare—how weapons were used from horseback. Probably the change in weapons came first.

Mounted archery probably was not yet very effective before the Iron Age, for three reasons. The bows reconstructed from their traces in steppe Bronze Age graves were more than 1 m long and up to 1.5 m, or almost five feet, in length, which would clearly have made them clumsy to use from horseback; the arrowheads were chipped from flint or made from bone in widely varying sizes and weights, implying a nonstandardized, individualized array of arrow lengths and weights; and, finally, the bases of most arrowheads were made to fit into a hollow or split shaft, which weakened the arrow or required a separate hollow foreshaft for the attachment of the point. The more powerful the bow, and the higher the impact on striking a target, the more likely the arrow was to split, if the shaft had already been split to secure the point. Stemmed and triangular flint points, common before the Iron Age, were made to be inserted into a separate foreshaft with a hollow socket made of reed or wood (for stemmed points), or were set into a split shaft (for triangular points). The long bows, irregular arrow sizes, and less-than-optimal attachments between points and arrows together reduced the military effectiveness of early mounted archery. Before the Iron Age mounted raiders could harass tribal war bands, disrupt harvests in farming villages, or steal cattle, but that is not the same as defeating a disciplined army. Tribal raiding by small groups of riders in eastern Europe did not pose a threat to walled cities in Mesopotamia, and so was ignored by the kings and generals of the Near East and the eastern Mediterranean.[39]

The invention of the short, recurved, compound bow (the "cupid" bow) around 1000 BCE made it possible for riders to carry a powerful bow short enough to swing over the horse's rear. For the first time arrows could be fired behind the rider with penetrating power. This maneuver, later known as the "Parthian shot," was immortalized as the iconic image of the steppe archer. Cast bronze socketed arrowheads of standard weights and sizes also appeared in the Early Iron Age. A socketed arrowhead did not require a split-shaft mount, so arrows with socketed arrowheads did not split despite the power of the bow; they also did not need a separate fore-shaft, and so arrows could be simpler and more streamlined. Reusable moulds were invented so that smiths could produce hundreds of socketed arrowheads of standard weight and size. Archers now had a much wider field of fire—to the rear, the front, and the left—and could carry dozens of standardized arrows. An army of mounted archers could now fill the sky with arrows that struck with killing power.[40]

But organizing an army of mounted archers was not a simple matter. The technical advances in bows, arrows, and casting were meaningless without a matching change in mentality, in the identity of the fighter, from a heroic single warrior to a nameless soldier. An ideological model of fighting appropriate for a state had to be grafted onto the mentality of tribal horseback riders. Pre-Iron-Age warfare in the Eurasian steppes, from what we can glean from sources like the *Iliad* and the *Rig Veda*, probably emphasized personal glory and heroism. Tribal warfare generally was conducted by forces that never drilled as a unit, often could choose to ignore their leaders, and valued personal bravery above following orders.[41] In contrast, the tactics and ideology of state warfare depended on large disciplined units of anonymous soldiers who obeyed a general. These tactics, and the soldier mentality that went with them, were not applied to riders before 1000 BCE, partly because the short bows and standardized arrows that would make mounted archery truly threatening had not yet been invented. As mounted archers gained in firepower, someone on the edge of the civilized world began to organize them into armies. That seems to have occurred about 1000–900 BCE. Cavalry soon swept chariotry from the battlefield, and a new era in warfare began. But it would be grossly inappropriate to apply that later model of mounted warfare to the Eneolithic.

Riding began in the region identified as the Proto-Indo-European homeland. To understand how riding affected the spread of Indo-European languages we have to pick up the thread of the archaeological narrative that ended in chapter 9.

CHAPTER ELEVEN

The End of Old Europe and the Rise of the Steppe

By 4300–4200 BCE Old Europe was at its peak. The Varna cemetery in eastern Bulgaria had the most ostentatious funerals in the world, richer than anything of the same age in the Near East. Among the 281 graves at Varna, 61 (22%) contained more than three thousand golden objects together weighing 6 kg (13.2 lb). Two thousand of these were found in just four graves (1, 4, 36, and 43). Grave 43, an adult male, had golden beads, armrings, and rings totaling 1,516 grams (3.37 lb), including a copper axe-adze with a gold-sheathed handle.[1] Golden ornaments have also been found in tell settlements in the lower Danube valley, at Gumelniţa, Vidra, and at Hotnitsa (a 310-gm cache of golden ornaments). A few men in these communities played prominent social roles as chiefs or clan leaders, symbolized by the public display of shining gold ornaments and cast copper weapons.

Thousands of settlements with broadly similar ceramics, houses, and female figurines were occupied between about 4500 and 4100 BCE in eastern Bulgaria (Varna), the upland plains of Balkan Thrace (KaranovoVI), the upper part of the Lower Danube valley in western Bulgaria and Romania (Krivodol-Sălcuţa), and the broad riverine plains of the lower Danube valley (Gumelniţa) (figure 11.1). Beautifully painted ceramic vessels, some almost 1 m tall and fired at temperatures of over 800°C, lined the walls of their two-storied houses. Conventions in ceramic design and ritual were shared over large regions. The crafts of metallurgy, ceramics, and even flint working became so refined that they must have required master craft specialists who were patronized and supported by chiefs. In spite of this, power was not obviously centralized in any one village. Perhaps, as John Chapman observed, it was a time when the restricted resources (gold, copper, *Spondylus* shell) were not critical, and the critical resources (land, timber, labor, marriage partners) were not seriously restricted. This could have prevented any one region or town from dominating others.[2]

225

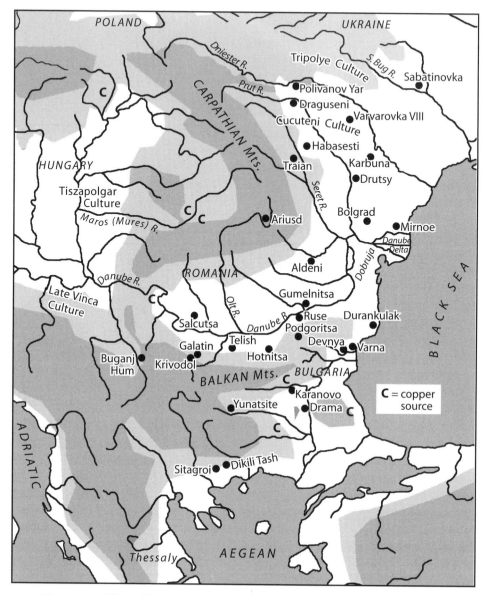

Figure 11.1 Map of Old Europe at 4500–4000 BCE.

Towns in the high plains atop the Balkans and in the fertile lower Danube valley formed high tells. Settlements fixed in one place for so long imply fixed agricultural fields and a rigid system of land tenure around each tell. The settlement on level VI at Karanovo in the Balkans was the type site for the period. About fifty houses crowded together in orderly rows inside a protective wooden palisade wall atop a massive 12-m (40-ft) tell. Many tells were surrounded by substantial towns. At Bereket, not far from Karanovo, the central part of the tell was 250 m in diameter and had cultural deposits 17.5 m (57 ft) thick, but even 300–600 m away from this central eminence the occupation deposits were 1–3 m thick. Surveys at Podgoritsa in northeastern Bulgaria also found substantial off-tell settlement.[3]

Around 4200–4100 BCE the climate began to shift, an event called the Piora Oscillation in studies of Swiss alpine glaciers. Solar insolation decreased, glaciers advanced in the Alps (which gave this episode its name), and winters became much colder.[4] Variations in temperature in the northern hemisphere are recorded in the annual growth rings in oaks preserved in bogs in Germany and in annual ice layers in the GISP2 glacial ice core from Greenland. According to these sources, extremely cold years happened first in 4120 and 4040 BCE. They were harbingers of a 140-year-long, bitterly cold period lasting from 3960 to 3821 BCE, with temperatures colder than at any time in the previous two thousand years. Investigations led by Douglass Bailey in the lower Danube valley showed that floods occurred more frequently and erosion degraded the riverine floodplains where crops were grown. Agriculture in the lower Danube valley shifted to more cold-tolerant rye in some settlements.[5] Quickly these and perhaps other stresses accumulated to create an enormous crisis.

Between about 4200 and 3900 BCE more than six hundred tell settlements of the Gumelnița, Karanovo VI, and Varna cultures were burned and abandoned in the lower Danube valley and eastern Bulgaria. Some of their residents dispersed temporarily into smaller villages like the Gumelnița B1 hamlet of Jilava, southwest of Bucharest, with just five to six houses and a single-level cultural deposit. But Jilava was burned, apparently suddenly, leaving behind whole pots and many other artifacts.[6] People scattered and became much more mobile, depending for their food on herds of sheep and cattle rather than fixed fields of grain. The forests did not regenerate; in fact, pollen cores show that the countryside became even more open and deforested.[7] Relatively mild climatic conditions returned after 3760 BCE according to the German oaks, but by then the cultures of the lower Danube valley and the Balkans had changed dramatically. The cultures that appeared after about 3800 BCE did not regularly use female figurines in domestic rituals, no longer wore copper spiral bracelets or *Spondylus*-shell ornaments, made relatively plain pottery in a limited number of shapes, did not live on tells, and depended more on stockbreeding. Metallurgy, mining, and ceramic technology declined sharply in both volume and technical skill, and ceramics and metal objects changed markedly in style. The copper mines in the Balkans abruptly ceased production; copper-using cultures in central Europe and the Carpathians switched to Transylvanian and Hungarian ores about 4000 BCE, at the beginning of the Bodrogkeresztur culture in Hungary (see ore sources in figure 11.1). Oddly this was when metallurgy really began in western Hungary and nearby in Austria and central Europe.[8] Metal objects now were made using new arsenical bronze alloys, and were of new types, including new weapons, daggers being the most important. "We are faced with the complete replacement of a culture," the foremost expert on Eneolithic metallurgy E. N. Chernykh said. It was "a catastrophe of colossal scope . . . a complete cultural caesura," according to the Bulgarian archaeologist H. Todorova.[9]

The end of Old Europe truncated a tradition that began with the Starcevo-Criş pioneers in 6200 BCE. Exactly what happened to Old Europe is the subject of a long, vigorous debate. Graves of the Suvorovo type, ascribed to immigrants from the steppes, appeared in the lower Danube valley just before the destruction of the tells. Settlements of the Cernavoda I type appeared just after. They regularly contain horse bones and ceramics exhibiting a mixture of steppe technology and indigenous Danubian shapes, and are ascribed to a mixed population of steppe immigrants and people from the tells. The number of abandoned sites and the rapid termination of many long-standing traditions in crafts, domestic rituals, decorative customs, body ornaments, housing styles, living arrangements, and economy suggest not a gradual evolution but an abrupt and probably violent end. At Hotnitsa on the Danube in north-central Bulgaria the burned houses of the final Eneolithic occupation contained human skeletons, interpreted as massacred inhabitants. The final Eneolithic destruction level at Yunatsite on the Balkan upland plain contained forty-six human skeletons. It looks like the tell towns of Old Europe fell to warfare, and, somehow, immigrants from the steppes were involved. But the primary causes of the crisis could have included climate change and related agricultural failures, or soil erosion and environmental degradation accumulated from centuries of intensive farming, or internecine warfare over declining timber and copper resources, or a combination of all these.[10]

The crisis did not immediately affect all of southeastern Europe. The most widespread settlement abandonments occurred in the lower Danube valley

(Gumelniţa, northeastern Bulgaria, and the Bolgrad group), in eastern Bulgaria (Varna and related cultures), and in the mountain valleys of the Balkans (Karanovo VI), east of the Yantra River in Bulgaria and the Olt in Romania. This was where tell settlements, and the stable field systems they imply, were most common. In the Balkans, a well-cultivated, densely populated landscape occupied since the earliest Neolithic, no permanent settlements can be dated between 3800 and 3300 BCE. People probably still lived there, but herds of sheep grazed on the abandoned tells.

The traditions of Old Europe survived longer in western Bulgaria and western Romania (Krivodol-Sălcuţa IV–Bubanj Hum Ib). Here the settlement system had always been somewhat more flexible and less rooted; the sites of western Bulgaria usually did not form high tells. Old European ceramic types, house types, and figurine types were abandoned gradually during Sălcuţa IV, 4000–3500 BCE. Settlements that were occupied during the crisis, places like Telish-Redutite III and Galatin, moved to high, steep-sided promontories, but they retained mud-brick architecture, two-story houses, and cult and temple buildings.[11] Many caves in the region were newly occupied, and since herders often use upland caves for shelter, this might suggest an increase in upland-lowland seasonal migrations by herders. The Krivodol–Salcutsa–Bubanj Hum Ib people reoriented their external trade and exchange connections to the north and west, where their influence can be seen on the Lasinja-Balaton culture in western Hungary.

The Old European traditions of the Cucuteni-Tripolye culture also survived and, in fact, seemed curiously reinvigorated. After 4000 BCE, in its Tripolye B2 phase, the Tripolye culture expanded eastward toward the Dnieper valley, creating ever larger agricultural towns, although none was rebuilt in one place long enough to form a tell. Domestic cults still used female figurines, and potters still made brightly painted fine lidded pots and storage jars 1 m high. Painted fine ceramics were mass-produced in the largest towns (Varvarovka VIII), and flint tools were mass-produced at flint-mining villages like Polivanov Yar on the Dniester.[12] Cucuteni AB/Tripolye B2 settlements such as Veseli Kut (150 ha) contained hundreds of houses and apparently were preeminent places in a new settlement hierarchy. The Cucuteni-Tripolye culture forged new relationships with the copper-using cultures of eastern Hungary (Borogkeresztur) in the west and with the tribes of the steppes in the east.

The languages spoken by those steppe tribes, around 4000 BCE, probably included archaic Proto-Indo-European dialects of the kind partly preserved later in Anatolian. The steppe people who spoke in that way probably already rode horses. Were the Suvorovo sites in the lower Danube valley created by Indo-European invaders on horseback? Did they play a role in the destruction of the tell settlements of the lower Danube valley, as Gimbutas suggested? Or did they just slip into an opening created by climate change and agricultural failures? In either case, why did the Cucuteni-Tripolye culture survive and even prosper? To address these questions we first have to examine the Cucuteni-Tripolye culture and its relations with steppe cultures.

Warfare and Alliance: The Cucuteni-Tripolye Culture and the Steppes

The crisis in the lower Danube valley corresponded to late Cucuteni A3/Tripolye B1, around 4300–4000 BCE. Tripolye B1 was marked by a steep increase in the construction of fortifications—ditches and earthen banks—to protect settlements (figure 11.2). Fortifications might have appeared just about when the climate began to deteriorate and the collapse of Old Europe occurred, but Cucuteni-Tripolye fortifications then *decreased* during the coldest years of the Piora Oscillation, during Tripolye B2, 4000–3700 BCE. If climate change destabilized Old Europe and caused the initial construction of Cucuteni-Tripolye fortifications, the first phase of change was sufficient by itself to tip the system into crisis. Probably there was more to it than just climate.

Only 10% of Tripolye B1 settlements were fortified even in the worst of times. But those that *were* fortified required substantial labor, implying a serious, chronic threat. Fortified Cucteni-Tripolye villages usually were built at the end of a steep-sided promontory, protected by a ditch dug across the promontory neck. The ditches were 2–5 m wide and 1.5–3 m deep, made by removing 500–1,500 m³ of earth. They were relocated and deepened as settlements grew in size, as at Traian and Habaşeşti I. In a database of 2,017 Cucuteni/Tripolye settlements compiled by the Moldovan archaeologist V. Dergachev, half of *all* fortified Cucuteni/Tripolye sites are dated just to the Tripolye B1 period. About 60% of all the flint projectile points from all the Cucuteni/Tripolye culture also belonged just to the Tripolye B1 period. There was no corresponding increase in hunting during Tripolye B1 (no increase in wild animal bones in settlements), and so the high frequency of projectile points was not connected with hunting. Probably it was associated with increased warfare.

The number of Cucuteni-Tripolye settlements increased from about 35 settlements per century during Tripolye A to about 340 (!) during Tripolye

Figure 11.2 Habaşesti I, a fortified Tripolye B1 village. After Chernysh 1982.

B1, a tenfold rise in the number of settlements without a significant expansion of the area settled (figure 11.3b).[13] Part of this increase in settlement density during Tripolye B1 might be ascribed to refugees fleeing from the towns of the Gumelniţa culture. At least one Tripolye B1 settlement in the Prut drainage, Drutsy 1, appears to have been attacked. More than one hundred flint points (made of local Carpathian flint) were found around the walls of the three excavated houses as if they had been peppered with arrows.[14] Compared to its past and its future, the Tripolye B1 period was a time of sharply increased conflict in the Eastern Carpathians.

Contact with Steppe Cultures during Tripolye B: Cucuteni C Ware

Simultaneously with the increase in fortifications and weapons, Tripolye B1 towns showed widespread evidence of contact with steppe cultures. A new pottery type, Cucuteni C ware,[15] shell-tempered and similar to steppe pottery, appeared in Tripolye B1 settlements of the South Bug valley (Sabatinovka I) and in Romania (Draguşeni and Fedeleşeni, where Cucuteni C ware amounted to 10% of the ceramics). Cucuteni C ware is usually thought to indicate contact with and influence from steppe pottery traditions (figure 11.4).[16] Cucuteni C ware might have been used in ordinary homes with standard Cucuteni-Tripolye fine wares as a new kind of coarse or kitchen pottery, but it did not replace traditional coarse kitchen wares tempered with grog (ground-up ceramic sherds). Some Cucuteni C pots look very much like steppe pottery, whereas others had shell-temper,

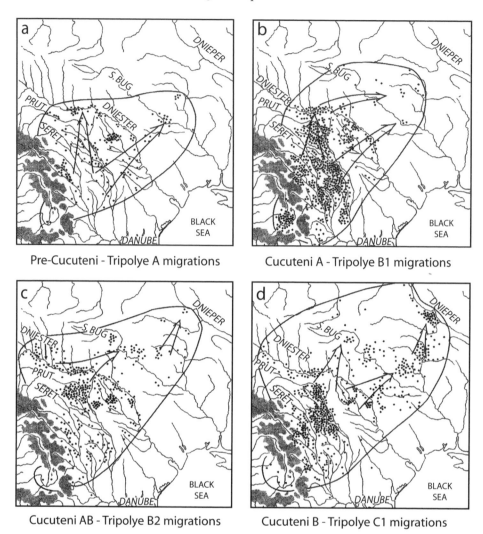

Figure 11.3. Tripolye B1-B2 migrations. After Dergachev 2002, figure 6.2.

gray-to-brown surface color and some typical steppe decorative techniques (like "caterpillar" impressions, made with a cord-wrapped, curved pressing tool) but were made in typical Cucuteni-Tripolye shapes with other decorative elements typical of Cucuteni-Tripolye wares.

The origin of Cucuteni C ware is disputed. There were good utilitarian reasons for Tripolye potters to adopt shell-tempering. Shell-temper in the clay can increase resistance to heat shock, and shell-tempered pots

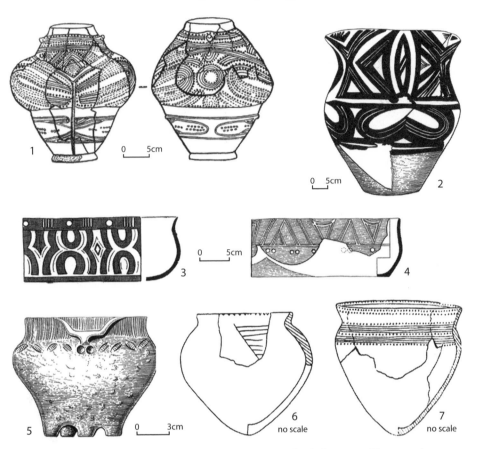

Figure 11.4 Cucuteni C (bottom row) and standard Cucuteni B wares (top two rows): (1) fine ware, Novye Ruseshti I$_{1a}$ (Tripolye B1); (2) fine ware, Geleshti (Tripolye B2); (3–4) fine ware, Frumushika I (Tripolye B1); (5) Cucuteni C ware, Frumushika II (Tripolye B2); (6–7) Cucuteni C ware, Berezovskaya GES. After Danilenko and Shmagli 1972, Figure 7; Chernysh 1982, Figure LXV.

can harden at lower firing temperatures, which could save fuel.[17] Changes in the organization of pottery making could also have encouraged the spread of Cucuteni C wares. Ceramic production was beginning to be taken over by specialized ceramic-making towns during Tripolye B1 and B2, although local household production also continued in most places. Rows of reusable two-chambered kilns appeared at the edges of a few settlements, with 11 kilns at Ariuşd in southeastern Transylvania. If fine painted wares were beginning to be produced in villages that specialized in making pottery and the coarse wares remained locally produced, the change in coarse wares could have reflected the changing organization of production.

On the other hand, these particular coarse wares obviously resembled the pottery of steppe tribes. Many Cucuteni C pots look like they were made by Sredni Stog potters. This suggests familiarity with steppe cultures and even the presence of steppe people in some Tripolye B villages, perhaps as hired herders or during seasonal trade fairs. Although it is unlikely that *all* Cucuteni C pottery was made by steppe potters—there is just too much of it—the appearance of Cucuteni C ware suggests intensified interactions with steppe communities.

Steppe Symbols of Power: Polished Stone Maces

Polished stone maces were another steppe artifact type that appeared in Tripolye B1 villages. A mace, unlike an axe, cannot really be used for anything except cracking heads. It was a new weapon type and symbol of power in Old Europe, but maces had appeared across the steppes centuries earlier in DDII, Khvalynsk, and Varfolomievka contexts. There were two kinds—zoomorphic and eared types—and both had steppe prototypes that were older (figure 11.5; also see figure 9.6). Mace heads carved and polished in the shape of horse heads were found in two Cucuteni A3/A4-Tripolye B1 settlements, Fitioneşti and Fedeleşeni, both of which also had significant amounts of Cucuteni C ware. The eared type appeared at the Cucuteni-Tripolye settlements of Obarşeni and Berezovskaya GES, also with Cucuteni C ware that at Berezovskaya looked like it was imported from steppe communities. Were steppe people present in these Tripolye B1 towns? It seems likely. The integration of steppe pottery and symbols of power into Cucuteni-Tripolye material culture suggests some kind of social integration, but the maintenance of differences in economy, house form, fine pottery, metallurgy, mortuary rituals, and domestic rituals indicates that it was limited to a narrow social sector.[18]

Other Signs of Contact

Most settlements of the Tripolye B period, even large ones, continued to dispose of their dead in unknown ways. But inhumation graves appeared in or at the edge of a few Tripolye B1 settlement sites. A grave in the settle-

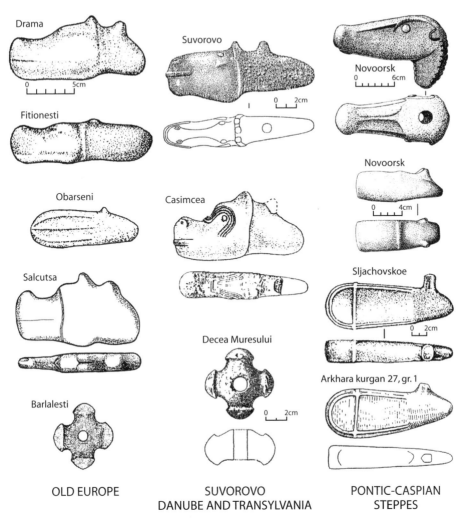

Drama

Fitionesti

Obarseni

Salcutsa

Barlalesti

Suvorovo

Casimcea

Decea Muresului

Novoorsk

Novoorsk

Sljachovskoe

Arkhara kurgan 27, gr. 1

OLD EUROPE   SUVOROVO   PONTIC-CASPIAN
                 DANUBE AND TRANSYLVANIA   STEPPES

Figure 11.5 Eared and horse-head maces of Old Europe, the Suvorovo migrants, and the Pontic-Caspian steppes. Stone mace heads appeared first and were more common in the steppes. After Telegin et al. 2001; Dergachev 1999; Gheorgiu 1994; Kuzmina 2003.

ment of Nezvisko contained a man with a low skull and broad, thick-boned face like those of steppe people—a type of skull-and-face configuration called "Proto-Europoid" by Eastern European physical anthropologists. Tripolye, Varna, and Gumelniţa people generally had taller heads, narrower faces, and more gracile facial bones, a configuration called "Mediterranean."[19]

Another indicator of movement across the steppe border was the little settlement near Mirnoe in the steppes north of the Danube delta. This is the only known classic-period Tripolye settlement in the coastal steppe lowlands. It had just a few pits and the remains of a light structure containing sherds of Tripolye B1 and Cucuteni C pots, a few bones of cattle and sheep, and more than a hundred grape seeds, identified as wild grapes. Mirnoe seems to have been a temporary Tripolye B1 camp in the steppes, perhaps for grape pickers.[20] Some people, though not many, were moving across the cultural-ecological frontier in both directions.

During Tripolye B2, around 4000–3700 BCE, there was a significant migration out of the Prut-Seret forest-steppe uplands, the most densely settled part of the Tripolye B1 landscape, eastward into the South Bug and Dnieper valleys (figure 11.3c). Settlement density in the Prut-Seret region declined by half.[21] Tripolye, the type site first explored in 1901, was an eastern frontier village of the Tripolye B2 period, situated on a high terrace overlooking the broad, fertile valley of the Dnieper River. The population consolidated into fewer, larger settlements (only about 180 settlements per century during Tripolye B2). The number of fortified settlements decreased sharply.

These signs of demographic expansion and reduced conflict appeared after the tell settlements of the Danube valley were burned and abandoned. It appears that any external threat from the steppes, if there was one, turned away from Cucuteni-Tripolye towns. Why?

Steppe Riders at the Frontiers of Old Europe

Frontiers can be envisioned as peaceful trade zones where valuables are exchanged for the mutual benefit of both sides, with economic need preventing overt hostilities, or as places where distrust is magnified by cultural misunderstandings, negative stereotypes, and the absence of bridging institutions. The frontier between agricultural Europe and the steppes has been seen as a border between two ways of life, farming and herding, that were implacably opposed. Plundering nomads like the Huns and Mongols are old archetypes of savagery. But this is a misleading stereotype, and one derived from a specialized form of militarized pastoral nomadism that did not exist before about 800 BCE. As we saw in the previous chapter, Bronze Age riders in the steppes used bows that were too long for effective mounted archery. Their arrows were of varied weights and sizes. And Bronze Age war bands were not organized like armies. The Hunnic invasion analogy is anachronistic, yet that does not mean that mounted raiding never occurred in the Eneolithic.[22]

There is persuasive evidence that steppe people rode horses to hunt horses in Kazakhstan by about 3700–3500 BCE. Almost certainly they were not the first to ride. Given the symbolic linkage between horses, cattle, and sheep in Pontic-Caspian steppe funerals as early as the Khvalynsk period, horseback riding might have begun in a limited way before 4500 BCE. But western steppe people began to *act* like they were riding only about 4300–4000 BCE, when a pattern consistent with long-distance raiding began, seen most clearly in the Suvorovo-Novodanilovka horizon described at the end of this chapter. Once people began to ride, there was nothing to prevent them from riding into tribal conflicts—not the supposed shortcomings of rope and leather bits (an organic bit worked perfectly well, as our students showed in the organic-bit riding experiment, and as the American Indian "war bridle" demonstrated on the battlefield); not the size of Eneolithic steppe horses (most were about the size of Roman cavalry horses, big enough); and certainly not the use of the wrong "seat" (an argument that early riders sat on the rump of the horse, perhaps for millennia, before they discovered the more natural forward seat—based entirely on Near Eastern images of riders probably made by artists who were unfamiliar with horses).[23]

Although I *do* see evidence for mounted raiding in the Eneolithic, I do *not* believe that any Eneolithic army of pitiless nomads ever lined up on the horizon mounted on shaggy ponies, waiting for the command of their bloodthirsty general. Eneolithic warfare was tribal warfare, so there were no armies, just the young men of this clan fighting the young men of that clan. And early Indo-European warfare seems from the earliest myths and poetic traditions to have been conducted principally to gain glory—*imperishable fame*, a poetic phrase shared between Pre-Greek and Pre-Indo-Iranian. If we are going to indict steppe raiders in the destruction of Old Europe, we first have to accept that they did not fight like later cavalry. Eneolithic warfare probably was a strictly seasonal activity conducted by groups organized more like modern neighborhood gangs than modern armies. They would have been able to disrupt harvests and frighten a sedentary population, but they were not nomads. Steppe Eneolithic settlements like Dereivka cannot be interpreted as pastoral nomadic camps. After nomadic cavalry is removed from the picture, how do we understand social and political relations across the steppe/Old European frontier?

A mutualist interpretation of steppe/farming-zone relations is one alternative. Conflict is not denied, but it is downplayed, and mutually beneficial trade and exchange are emphasized.[24] Mutualism might well explain the relationship between the Cucuteni-Tripolye and Sredni Stog cultures during the Tripolye B period. Among historically known pastoralists in close contact with farming populations there has been a tendency for wealthy herd owners to form alliances with farmers to acquire land as insurance against the loss of their more volatile wealth in herds. In modern economies, where land is a market commodity, the accumulation of property could lead the wealthiest herders to move permanently into towns. In a pre-state tribal world this was not possible because agricultural land was not for sale, but the strategy of securing durable alliances and assets in agricultural communities as insurance against future herd losses could still work. Steppe herders might have taken over the management of some Tripolye herds in exchange for metal goods, linen textiles, or grain; or steppe clans might have attended regular trading fairs at agricultural towns. Annual trading fairs between mounted hunters and river-valley corn farmers were a regular feature of life in the northern Plains of the U.S.[25] Alliances and trade agreements sealed by marriages could account for the increased steppe involvement in Tripolye communities during Tripolye B1, about 4400–4000 BCE. The institutions that normalized these cross-cultural relations probably included gift partnerships. In archaic Proto-Indo-European as partly preserved in Hittite, the verb root that in all other Indo-European languages meant "give" (*dō-) meant "take" and another root (*pai*) meant "give." From this give-and-take equivalence and a series of other linguistic clues Emile Benveniste concluded that, during the archaic phase of Proto-Indo-European, "exchange appears as a round of gifts rather than a genuine commercial operation."[26]

On the other hand, mutualism cannot explain everything, and the end of the Varna-Karanovo VI–Gumelniţa culture is one of those events it does not explain. Lawrence Keeley sparked a heated debate among archaeologists by insisting that warfare was common, deadly, and endemic among prehistoric tribal societies. Tribal frontiers might be creative places, as Frederik Barth realized, but they often witnessed pretty nasty behavior. Tribal borders commonly were venues for insults: the Sioux called the Bannock the "Filthy-Lodge People"; the Eskimo called the Ingalik "Nit-heads"; the Hopi called the Navaho "Bastards"; the Algonkian called the Mohawk "Maneaters"; the Shuar called the Huarani "Savages"; and the simple but eloquent "Enemies" is a very common meaning for names given by neighboring tribes. Because tribal frontiers displayed things people needed just beyond the limits of their own society, the temptation to take them by force was strong. It was doubly strong when those things had legs, like cattle.[27]

Cattle raiding was encouraged by Indo-European beliefs and rituals. The myth of Trito, the warrior, rationalized cattle theft as the recovery of cattle that the gods had *intended* for the people who sacrificed properly. Proto-Indo-European initiation rituals included a requirement that boys initiated into manhood *had* to go out and become like a band of dogs or wolves—to raid their enemies.[28] Proto-Indo-European also had a word for bride-price, **üedmo-*.[29] Cattle, sheep, and probably horses would have been used to pay bride-prices, since they generally are valued higher than other currencies for bride-price payments in pastoral societies without formal money.[30] Already in the preceding centuries domesticated animals had become the proper gifts for gods at funerals (e.g., at Khvalynsk). A relatively small elite already competed across very large regions, adopting the same symbols of status—maces with polished stone heads, boar's tusk plaques, copper rings and pendants, shell disc beads, and bird-bone tubes. When bride-prices escalated as one aspect of this competition, the result would be increased cattle raiding by unmarried men. Combined with the justification provided by the Trito myth and the institution of male-initiation-group raiding, rising bride-prices calculated in animals would have made cross-border raiding almost inevitable.

If they were on foot, Eneolithic steppe cattle raiders might have attacked one another or attacked neighboring Tripolye settlements. But, if they were mounted, they could pick a distant target that did not threaten valued gift partnerships. Raiding parties of a dozen riders could move fifty to seventy-five head of cattle or horses fairly quickly over hundreds of kilometers.[31] Thieving raids would have led to deaths, and then to more serious killing and revenge raids. A cycle of warfare evolving from thieving to revenge raids probably contributed to the collapse of the tell towns of the Danube valley.

What kinds of societies lived on the steppe side of the frontier? Is there good archaeological evidence that they were indeed deeply engaged with Old Europe and the Cucuteni-Tripolye culture in quite different ways?

THE SREDNI STOG CULTURE: HORSES AND RITUALS FROM THE EAST

The Sredni Stog culture is the best-defined Late Eneolithic archaeological culture in steppe Ukraine. Sredni Stog, or "middle stack," was the name of a small haystack-shaped island in the Dnieper at the southern end of the Dnieper Rapids, the central one of three. All were inundated by a dam, but before that happened, archaeologists found and excavated a site there in 1927. It contained a stratified sequence of settlements with Early

Eneolithic (DDII) pottery in level I and Late Eneolithic pottery in level II.[32] Sredni Stog II became the type site for this Late Eneolithic kind of pottery. Sredni Stog–style pottery was found stratified above older DDII settlements at several other sites, including Strilcha Skelya and Aleksandriya. Dimitri Telegin, who had earlier defined the Dnieper-Donets culture, in 1973 first pulled together and mapped all the sites with Sredni Stog material culture, about 150 in all (figure 11.6). He found Sredni Stog sites across the Ukrainian steppes from the Ingul valley, west of the Dnieper, on the west to the lower Don on the east.

The Sredni Stog culture became the archaeological foundation for the Indo-European steppe pastoralists of Marija Gimbutas. The horse bones from the Sredni Stog settlement of Dereivka, excavated by Telegin, played a central role in the ensuing debates between pro-Kurgan-culture and anti-Kurgan-culture archaeologists. I described in the last chapter how Gimbutas's interpretation of the horses of Dereivka was challenged by Levine. Simultaneously Yuri Rassamakin challenged Telegin's concept of the Sredni Stog culture.[33]

Rassamakin separated Telegin's Sredni Stog culture into at least three separate cultures, reordered and redated some of the resulting pieces, and refocused the central cause of social and political change away from the development of horse riding and agro-pastoralism in the steppes (Telegin's themes) to the integration of steppe societies into the cultural sphere of Old Europe, which was Rassamakin's new mutualist theme. But Rassamakin assigned well-dated sites like Dereivka and Khvalynsk to periods inconsistent with their radiocarbon dates.[34] Telegin's groupings seem to me to be better documented and explained, so I retain the Sredni Stog culture as a framework for ordering Eneolithic sites in Ukraine, while disagreeing with Telegin in some details.

This was the critical era when innovative early Proto-Indo-European dialects began to spread across the steppes. The principal causes of change in the steppes included both the internal maturation of new economic systems and new social networks (Telegin's theme) and the inauguration of new interactions with Old Europe (Rassamakin's theme).

The Origins and Development of the Sredni Stog Culture

We should not imagine that Sredni Stog, or any other archaeological culture, appeared or disappeared everywhere at the same time. Telegin defined four broad phases (Ia, Ib, IIa, IIb) in its evolution, but a phase might last longer in some regions than others. In his scheme, the settlements at

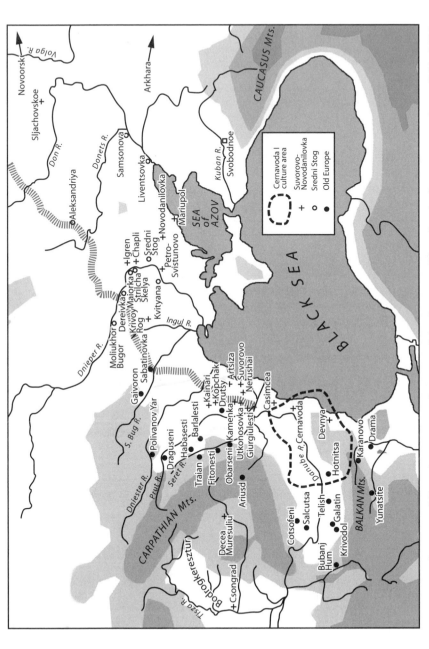

Figure 11.6 Steppe and Danubian sites at the time of the Suvorovo-Novodanilovka intrusion, about 4200–3900 BCE.

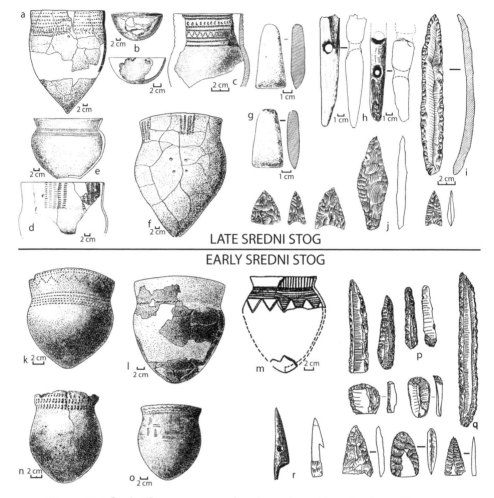

Figure 11.7 Sredni Stog pottery and tools, early and late. Perforated bone or antler artifacts like (h) were identified as cheekpieces for horse bits, but this identification is speculative. After Telegin 2002, figure 3.1.

Sredni Stog and Strilcha Skelya on the Dnieper represented an early phase (Ib), which Rassamakin called the Skelya culture. The pottery of this phase lacked cord-impressed decoration. The settlements at Dereivka (IIa) and Moliukhor Bugor (IIb) on the Dnieper represented the late phases, with braided cord impressions on the pottery (figure 11.7). Early Sredni Stog (phase I) was contemporary with the violent era of Tripolye B1 and the crisis in the Danube valley. Tripolye B1 painted pottery was found at

TABLE 11.2

Radiocarbon Dates for Late Eneolithic Cultures from the Lower Danube to the North Caucasus

Lab Number	BP Date	Sample	Calibrated Date
1. Sredni Stog culture			
Dereivka, Dnieper Valley			
Ki 2195	6240±100	settlement, shell	5270–5058 BCE
UCLA 1466a	5515±90	settlement, bone	4470–4240 BCE
Ki 2193	5400±100	settlement, shell	4360–4040 BCE
OxA 5030	5380±90	cemetery, grave 2	4350–4040 BCE
KI 6966	5370±70	settlement, bone	4340–4040 BCE
Ki 6960	5330±60	settlement, bone	4250–4040 BCE
KI 6964	5260±75	settlement, bone	4230–3990 BCE
Ki 2197	5230±95	settlement, bone	4230–3970 BCE
Ki 6965	5210±70	settlement, bone	4230–3960 BCE
UCLA 1671a	4900±100	settlement, bone	3900–3530 BCE
Ki 5488	4330±120	cult horse skull??	3300–2700 BCE
Ki 6962	2490±95	cult horse skull	790–520 BCE
OxA 7185	2295±60	cult horse tooth with bit wear	410–200 BCE
OxA 6577	1995±60	bone near cult horse	90 BCE–70CE
Aleksandriya, Donets Valley			
Ki-104	5470±300	?	4750–3900 BCE
2. North Caucasian Eneolithic			
Svobodnoe settlement			
Le-4531	5400±250	?	4500–3950 BCE
Le-4532	5475±100	?	4460–4160 BCE
3. Varna Culture, Bulgaria, lower Danube			
Durankulak tell settlement			
Bln-2122	5700±50	settlement, level 5	4600–4450 BCE
Bln-2111	5495±60	settlement, house 7	4450–4250 BCE
Bln-2121	5475±50	settlement, level 4	4360–4240 BCE
Pavelyanovo 1 tell settlement			
Bln-1141	5591±100	settlement	4540–4330 BCE

TABLE 11.2 (*continued*)

Lab Number	BP Date	Sample	Calibrated Date
4. Gumelnitsa culture, Romania, lower Danube			
Vulcanesti II, Bolgrad group			
MO-417	5110±150	settlement	4050–3700 BCE
Le-640	5300±60	settlement	4230–4000 BCE
Gumelnitsa, tell settlement			
GrN-3025	5715±70	settlement, charcoal	4680–4450 BCE
Bln-605	5675±80	settlement, charcoal	4620–4360 BCE
Bln-604	5580±100	settlement, charcoal	4540–4330 BCE
Bln-343	5485±120	settlement, charcoal	4460–4110 BCE
GrN-3028	5400±90	settlement, charred grain	4340–4050 BCE
5. Suvorovo Group, lower Danube			
Giurgiuleşti, cemetery, lower Prut/Danube			
Ki-7037	5398±69*	?	4340–4050 BCE

*This date was printed in Telegin et al. 2001 as 4398±69 BP, but I was told that this was a misprint and that the actual reported date was 5398+69 BP.

Strilcha Skelya. The stylistic changes that identified late Sredni Stog (phase II) probably began while the crisis in the Danube valley was going on, but then most of the late Sredni Stog period occurred after the collapse of Old Europe. Imported Tripolye B2 bowls were found in graves in the phase IIa cemeteries at Dereivka and Igren, and a Tripolye C1 vessel was found at the phase IIb Moliukhor Bugor settlement. The Dereivka settlement (phase IIa) is dated between 4200 and 3700 BCE by ten radiocarbon dates (table 11.2). The latest Sredni Stog period (IIb) is dated as late as 3600–3300 BCE by four radiocarbon dates at Petrovskaya Balka on the Dnieper. Early Sredni Stog probably began around 4400 BCE; late Sredni Stog probably lasted until 3400 BCE in some places on the Dnieper.

The origin of the Sredni Stog culture is poorly understood, but people from the east, perhaps from the Volga steppes, apparently played a role. Round-bottomed Sredni Stog shell-tempered pots were quite different from DDII pots of the Early Eneolithic, which were sand-tempered and flat-based (see figure 9.5). Almost all early Sredni Stog vessels had round or pointed bases and flaring, everted rims. Flat-based pots appeared only in the late period. Simple open bowls, probably food bowls, were the other common shape, usually undecorated. Sredni Stog pots were decorated just on the upper third of the vessel with rows of comb-stamped impressions, incised triangles, and cord impressions. Rows of U-shaped "caterpillar" impressions made with a U-shaped, cord-wrapped tool were typical (figure 11.7d). One pot shape, with a rounded body and a short vertical neck decorated with vertically combed lines (figure 11.7m) was copied directly from a common Tripolye B1 type. The round-based pots and shell temper seem to reflect influence from the east, from the Azov-Caspian or Volga regions, where there was a long tradition of shell-tempered, round-bottomed, everted-rim, impressed pottery beginning in the Neolithic and continuing through Eneolithic Khvalynsk.

Sredni Stog funeral rituals also were new. The new Sredni Stog burial posture (on the back with the knees raised) and standard orientation (head to the east-northeast) copied that of the Khvalynsk culture on the Volga (figure 11.8). The communal collective grave pits of DDII were abandoned. Individual single graves took their place. Cemeteries also became much smaller. The DDII cemetery near Dereivka had contained 173 individuals, most of them in large communal grave pits. The Sredni Stog cemetery near Dereivka contained only 12 graves, all single burials. Sredni Stog communities probably were smaller and more mobile. Graves had no surface marker, as at Dereivka, or exhibited a new surface treatment: some were surrounded by a small circle of stones and covered by a low stone or earth mound—a very modest kurgan—as at Kvityana or Maiorka. These probably were the earliest kurgans in the steppes. Stone circles and mounds were features that isolated and emphasized individuals. The shift from a communal funeral ritual to an individual ritual probably was a symptom of broader changes toward more openly self-aggrandizing social values, which were also reflected in a series of rich graves of the Suvorovo-Novodanilovka type discussed separately below.

Sredni Stog skull types also exhibited new traits. The DDII population had been a single homogeneous type, with a very broad, thick-boned face of the Proto-Europoid configuration. Sredni Stog populations included people with a more gracile bone structure and medium-width faces that showed the strongest statistical similarity to the Khvalynsk population. Immigrants from the Volga seem to have arrived in the Dnieper-Azov steppes at the beginning of the shift from DDII to Sredni Stog, instigating

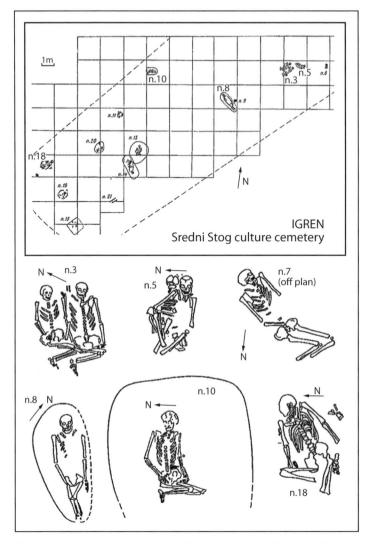

Figure 11.8 Sredni Stog graves, Igren cemetery, Dnieper Rapids. Graves were quite scattered. After Telegin et al. 2001.

changes in both funeral customs and pottery making. Perhaps they arrived on horseback.[35]

The places where people lived and put their cemeteries did not change markedly when Sredni Stog began. Sredni Stog settlements were stratified above DDII settlements at several sites near the Dnieper Rapids and on the Donets. Sredni Stog graves were located in or near DDII cemeteries at Mariupol, Igren, and Dereivka. Stone tools also showed continuity; lamellar flint blades, triangular flint points, and large almond-shaped flint points were made in both periods. Long unifacial flint blades were occasionally found in hoards in DDII sites but were found in much larger hoards in Sredni Stog sites, where some single hoards (Goncharovka) contained more than a hundred flint blades up to 20 cm long. These blades were typical grave gifts in Sredni Stog graves. Similar long flint blades became popular trade items across eastern Europe, appearing also in Funnel Beaker (TRB) sites in Poland and in Bodrogkeresztur sites in Hungary.

The Sredni Stog Economy: Horses and Agro-Pastoralism

Sredni Stog settlements had, on average, more than twice as many horse bones as DDII settlements in the Dnieper valley, where most of the studied sites are located. This increase in the use of horses for food could have been connected with the colder climate of the period 4200–3800 BCE, since domesticated horses are easier to maintain than cattle and sheep in snowy conditions (chapter 10). The maintenance advantage would, of course, have been gained only with domesticated horses. Horses were by far the most important source of meat at the Sredni Stog settlement of Dereivka. The 2,408 horse bones counted by Bibikova represented at least fifty-one animals (MNI)—more than half the mammals butchered at the site—and 9,000 kg of meat.[36]

Domesticated cattle, sheep, and pigs accounted for between 12% and 84% of the bones (NISP) from the settlements of Sredni Stog II, Dereivka, Aleksandriya, and Moliukhor Bugor (table 11.1). If horses are counted as domesticated animals, the percentage of domesticated animals at these settlements rises to 30–93%. The percentage of horse bones ranged from 7–63% of all bones found (average 54% NISP but with much variation). The highest percentage (63 percent of the mammal bones NISP, 28% of the individual mammals MNI) was at Dereivka, which was also the site with the largest sample of animal bones.[37] Sheep or goats were by far the most common animals (61% of mammals) in the southernmost site, Sredni Stog, in the driest steppe environment; and hunted game was most important (70% of mammals) at Moliukhor Bugor, the northenmost site, in the most forested environment. In the north, where forest resources were richer, deer hunting remained important, and in the steppe river valleys, where gallery forests were confined to the valley bottoms, sheep herding necessarily supplied a larger proportion of the diet.

TABLE 11.1
Mammal Bones from Sredni Stog Culture

	% horse	% cattle	% caprine	% pig	% dog	% horse
	(% of all bones, NISP/ % of individuals, MNI)*					
Sredni Stog II	7/12	21/12	61/47	2/6	3/11	7/22
Dereivka	63/52	16/8	2/7	3/4	1/2	17/45
Aleksandriya	29/24	37/20	7/12	—	—	27/44
Moliukhor BugorII	18/9	10/9	—	2/6	—	70/76

*NISP=number of identified species; MNI=minimum number of individuals.

Dereivka is the Sredni Stog settlement with the largest archaeological exposure, about 2000 m². It was located west of the Dnieper in the northern steppes. A scattered cemetery of twelve Sredni Stog graves was found half a kilometer upstream from the settlement.[38] Three shallow ovoid house pits, measuring about 12 m by 5 m, surrounded an open area used for ceramic manufacture, flint working, and other tasks (figure 11.9). A thick midden of river shellfish shells (*Unio* and *Paludinae*) enclosed one side. Only a part of the settlement was excavated, so we do not know how large it was. The mammal bones would have provided 1 kilo of meat per house, for the three houses, every day for more than eight years, indicating that Dereivka was occupied many times or for many years. On the other hand, the ephemeral nature of the Dereivka architectural remains and the small size of the nearby cemetery suggest that it was not a permanent settlement. Probably it was a favored living site that was revisited over many years by people who had large herds of horses (62% NISP) and cattle (16% NISP), hunted red deer (10% NISP), trapped or shot ducks (mallard and pintail), fished for wels catfish (*Silurus glanis*) and perch (*Lucioperca lucioperca*), and cultivated a little grain.

The ceramics from the Dereivka settlement have not been examined systematically for seed imprints, but Dereivka had flint blades with sickle gloss; three flat, ovoid grinding stones; and six polished schist mortars. Cultivated wheat, barley, and millet (*T. dicoccum, T. monococcum, H. vulgare, P. miliaceum*) have been identified in ceramic imprints at the phase IIb settlement of Moliukhor Bugor. Probably some grain cultivation occurred at Dereivka also, perhaps the first grain cultivation practiced east of the Dnieper.

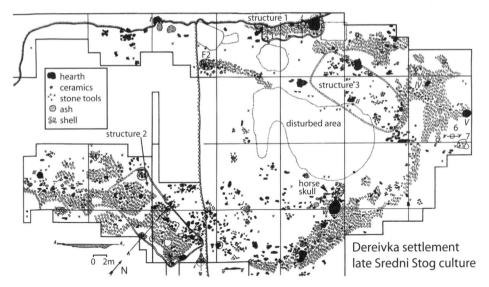

Figure 11.9 Dereivka settlement, Sredni Stog culture, 4200–3700 BCE. The location of the intrusive horse skull with bit wear is noted. The top edge is an eroded riverbank. After Telegin 1986.

Were the people of the Sredni Stog culture horse riders? Without bit wear or some other pathology associated with riding we cannot be certain. Objects from Dereivka tentatively identified as antler cheekpieces for bits (figure 11.7h) could have had other functions.[39] One way to approach this question is to ask if the steppe societies of the Late Eneolithic *behaved* like horseback riders. It looks to me like they did. Increased mobility (implied by smaller cemeteries), more long-distance trade, increased prestige and power for prominent individuals, status weapons appearing in graves, and heightened warfare against settled agricultural communities are all things we would expect to occur after horseback riding started, and we see them most clearly in cemeteries of the Suvorovo-Novodanilovka type.

MIGRATIONS INTO THE DANUBE VALLEY: THE SUVOROVO-NOVODANILOVKA COMPLEX

About 4200 BCE herders who probably came from the Dnieper valley appeared on the northern edge of the Danube delta. The lake country north of the delta was then occupied by Old European farmers of the Bolgrad culture. They left quickly after the steppe people showed up. The immigrants

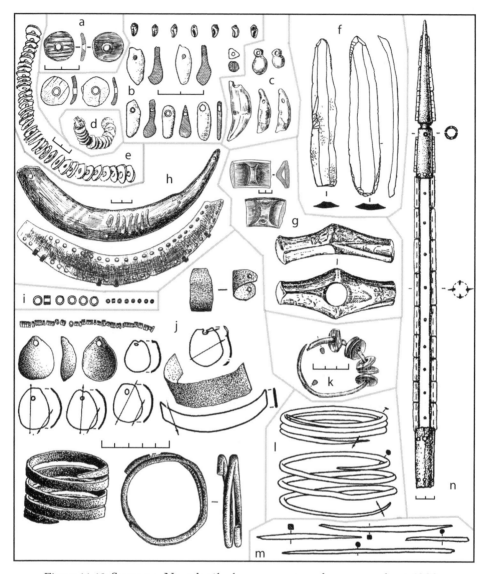

Figure 11.10 Suvorovo-Novodanilovka ornaments and weapons, about 4200–3900 BCE. (a, c) Vinogradni shell and canine tooth beads; (b) Suvorovo shell and deer tooth beads; (d) Decea Muresului shell beads; (e) Krivoy Rog shell beads; (f) Chapli lamellar flint blades; (g) Petro-Svistunovo, bone button and cast copper axe; (h) Petro-Svistunovo boar's tusk (*top*), Giurgiulesti copper-sheathed boar's tusk (*bottom*); (j) Chapli copper ornaments, including copper imitations of *Cardium* shells; (i) Utkonosovka bone beads; (k) Kainari copper "torque" with shell beads; (l) Petro-Svistunovo copper bracelet; (m) Suvorovo

built kurgan graves and carried maces with stone heads shaped like horse heads, objects that quickly appeared in a number of Old European towns. They acquired, either by trade or as loot, copper from the tell towns of the lower Danube valley, much of which they directed back into the steppes around the lower Dnieper. Their move into the lower Danube valley probably was the historical event that separated the Pre-Anatolian dialects, spoken by the migrants, from the archaic Proto-Indo-European language community back in the steppes.

The archaeology that documents this event emerged into the literature in small bits and pieces over the last fifty years, and it is still is not widely known. The steppe culture involved in the migration has been called variously the Skelya culture, the Suvorovo culture, the Utkonsonovka group, and the Novodanilovka culture. I will call it the Suvorovo-Novodanilovka complex (see figure 11.6). One cluster of graves, created by the migrants, is concentrated near the Danube delta. This was the Suvorovo group. Their relatives back home in the North Pontic steppes were the Novodanilovka group. Only graves are known for either group. About thirty-five to forty cemeteries are assigned to the complex, most containing fewer than ten graves and many, like Novodanilovka itself, represented by just a single rich burial. They first appeared during early Sredni Stog, around 4300–4200 BCE, and probably ceased before 3900 BCE.

In his earliest discussions Telegin interpreted the Novodanilovka graves (his term) as a wealthy elite element within the Sredni Stog culture. Later he changed his mind and made them a separate culture. I agree with his original position: the Suvorovo-Novodanilovka complex represents the chiefly elite within the Sredni Stog culture. Novodanilovka graves are distributed across the same territory as graves and settlements designated Sredni Stog, and many aspects of grave ritual and lithics are identical. The Suvorovo-Novodanilovka elite was involved in raiding and trading with the lower Danube valley during the Tripolye B1 period, just before the collapse of Old Europe.[40]

The people buried in these graves wore long belts and necklaces of shell disc beads, copper beads, and horse or deer tooth beads; copper rings; copper shell-shaped pendants; and copper spiral bracelets (figure 11.10). They bent thick pieces of copper wire into neckrings ("torques") decorated with shell beads, used copper awls, occasionally carried solid cast copper shaft-hole axes

---

Figure 11.10 (continued) and Aleksandriya copper awls; (n) Giurgiuleşti composite spear-head, bone with flint microblade edges and tubular copper fittings. After Ryndina 1998, figure 76; and Telegin et al. 2001.

(cast in a two-part mold), and put copper and gold fittings around the dark wood of their spears and javelins. In 1998 N. Ryndina counted 362 objects of copper and 1 of gold from thirty Suvorovo-Novodanilovka graves. They also carried polished stone mace heads made in several shapes, including horse heads (see figure 11.5). They used large triangular flint points, probably for spears/javelins; small round-butted flint axes with the cutting edge ground sharp; and long lamellar flint blades, often made of gray flint quarried from outcrops on the Donets River.

Most Suvorovo-Novodanilovka graves contained no pottery, and so they are difficult to link to a ceramic type. Imported ceramics were found in several graves: a Tripolye B1 pot in the Kainari kurgan, between the Prut and Dniester; a late Gumelniţa vessel in the Kopchak kurgan, not far from Kainari; another late Gumelniţa vessel in grave 2 at Giurgiuleşti, on the lower Prut; and a long-traveled pot of North Caucasian Svobodnoe type in the Novodanilovka grave in the Dnieper-Azov steppes. These imported pots were all the same age, dated roughly 4400–4000 BCE, and so are useful chronologically, but they throw no light on the cultural affiliation of the individuals in the graves. Only a few potsherds actually seem to have been made by the people who built the graves. One of the principal graves (gr. 1) at Suvorovo had two small sherds of a pot made of gray, shell-tempered clay, decorated with a small-toothed stamp and incised diagonal lines (figure 11.11). An analogous pot was found in Utkonosovka, kurgan 3, grave 2, near Suvorovo. These sherds resembled Cucuteni C ceramics: round body, round base, everted rim, shell-tempered, with diagonal incised and comb-stamped surface decoration.[41]

The Suvorovo graves around the Danube delta always were marked by the erection of a mound or kurgan, probably to increase their visibility on a disputed frontier, but possibly also as a visual response to the tells of the lower Danube valley (figure 11.11). Suvorovo kurgans were among the first erected in the steppes. Back in the Dnieper-Azov steppes, most Novodanilovka graves also had a surface marker of some kind, but earthen kurgans were less common than small stone cairns piled above the grave (Chapli, Yama). Kurgans in the Danube steppes were rarely were more than 10 m in diameter, and often were surrounded by a ring of small stones or a cromlech (retaining wall) of large stones. The grave pit was usually rectangular but sometimes oval. The Sredni Stog burial posture (on the back with knees raised) appeared in most (Csongrad, Chapli, Novodanilovka, Giurgiuleşti, Suvorovo grave 7) but not all graves. In some the body was laid out extended (Suvorovo grave 1) or contracted on the side (Utkonosovka). Animal sacrifices occurred in some graves (cattle at

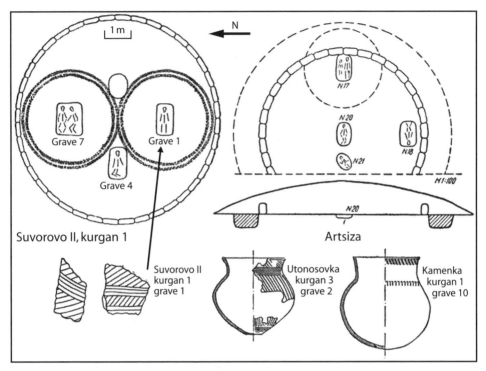

Figure 11.11 Suvorovo-type kurgan graves and pots. Most Suvorovo graves contained no pottery or contained pots made by other cultures, and so these few apparently self-made pots are important: *left*, Suvorovo cemetery II kurgan 1; *right*, Artsiza kurgan; *bottom*, sherds and pots from graves. After Alekseeva 1976, figure 1.

Giurgiuleşti, cattle and sheep at Chapli, and cattle at Krivoy Rog). The people buried in Novodanilovka graves in the Pontic steppes were wide-faced Proto-Europoid types, like the dominant element in Sredni Stog graves, whereas at least some of those buried in Suvorovo graves such as Giurgiuleşti had narrow faces and gracile skulls, suggesting intermarriage with local Old European people.[42]

The copper from Suvorovo-Novodanilovka graves helps to date them. Trace elements in the copper from Giurgiuleşti and Suvorovo in the lower Danube, and from Chapli and Novodanilovka in the Pontic steppes, are typical of the mines in the Bulgarian Balkans (Ai Bunar and/or Medni Rud) that abruptly ceased production when Old Europe collapsed. The eastern European copper trade shifted to chemically distinctive Hungarian and

Transylvanian ores during Tripolye B2, after 4000 BCE.[43] So Suvorovo-
Novodanilovka is dated before 4000 BCE by its copper. On the other hand,
Suvorovo kurgans replaced the settlements of the Bolgrad group north of
the Danube delta, which were still occupied during early Tripolye B1, or
after about 4400–4300 BCE. These two bookends (after the abandonment
of Bolgrad, before the wider Old European collapse) restrict Suvorovo-
Novodanilovka to a period between about 4300 and 4000 BCE.

Polished stone mace-heads shaped like horse heads were found in the
main grave at Suvorovo and at Casimcea in the Danube delta region (figure 11.5). Similar mace-heads occurred at two Tripolye B1 settlements, at
two late Karanovo VI settlements, and up the Danube valley at the settlement of Sălcuţa IV—all of them in Old European towns contemporary
with the Suvorovo intrusion. Similar horse-head mace-heads were found
in the Volga-Ural steppes and in the Kalmyk steppes north of the Terek
River at Terekli-Mekteb.[44] "Eared" stone mace heads appeared first in
several cemeteries of the Khvalynsk culture (Khvalynsk, Krivoluchie) and
then somewhat later at several eastern steppe sites contemporary with
Suvorovo-Novodanilovka (Novorsk, Arkhara, and Sliachovsko) and in two
Tripolye B1 towns. Cruciform mace heads appeared first in the grave of a
DDII chief at Nikol'skoe on the Dnieper (see figure 9.6), and then reap-
peared centuries later with the Suvorovo migration into Transylvania at
Decea Mureşului and Ocna Sibiului; one example also appeared at a Tri-
polye settlement on the Prut (Bârlăleşti).

Polished stone maces were typical steppe prestige objects going back to
Khvalynsk, Varfolomievka, and DDII, beginning ca. 5000–4800 BCE.
They were not typical prestige objects for earlier Tripolye or Gumelniţa
societies.[45] Maces shaped into horse-heads probably were made by people
for whom the horse was a powerful symbol. Horse bones averaged only
3–6% of mammal bones in Tripolye B1 settlements and even less in
Gumelniţa, and so horses were not important in Old European diets. The
horse-head maces signaled a new iconic status for the horse just when the
Suvorovo people appeared. If horses were *not* being ridden into the Danube valley, it is difficult to explain their sudden symbolic importance in
Old European settlements.[46]

The Causes and Targets of the Migrations

Winters began to get colder in the interior steppes after about 4200 BCE.
The marshlands of the Danube delta are the largest in Europe west of the
Volga. Marshes were the preferred winter refuge for nomadic pastoralists

in the Black Sea steppes during recorded history, because they offered good winter forage and cover for cattle. The Danube delta was richer in this resource than any other place on the Black Sea. The first Suvorovo herders who appeared on the northern edge of the Danube delta about 4200–4100 BCE might have brought some of their cattle south from the Dnieper steppes during a period of particularly cold winters.

Another attraction was the abundant copper that came from Old European towns. The archaeologist Susan Vehik argued that increased levels of conflict associated with climatic deterioration in the southwestern U.S. Plains around 1250 BCE created an increased demand for gift-wealth (to attract and retain allies in tribal warfare) and therefore stimulated long-distance trade for prestige goods.[47] But the Suvorovo immigrants did not establish gift exchanges like those I have hypothesized for their relations with Cucuteni-Tripolye people. Instead, they seem to have chased the locals away.

The thirty settlements of the Bolgrad culture north of the Danube delta were abandoned and burned soon after the Suvorovo immigrants arrived. These small agricultural villages were composed of eight to ten semi-subterranean houses with fired clay hearths, benches, and large storage pots set in pits in the floor. Graphite-painted fine pottery and numerous female figurines show a mixture of Gumelniţa (Aldeni II type) and Tripolye A traits.[48] They were occupied mainly during Tripolye A, then were abandoned and burned during early Tripolye B1, probably around 4200–4100 BCE. Most of the abandonments apparently were planned, since almost everything was picked up. But at Vulcaneşti II, radiocarbon dated 4200–4100 BCE (5300 ± 60 BP), abandonment was quick, with many whole pots left to burn. This might date the arrival of the Suvorovo migrants.[49]

A second and seemingly smaller migration stream branched off from the first and ran westward to the Transylvanian plateau and then down the copper-rich Mureş River valley into eastern Hungary. These migrants left cemeteries at Decea Mureşului in the Mureş valley and at Csongrad in the plains of eastern Hungary. At Decea Mureşului, near important copper deposits, there were fifteen to twenty graves, posed on the back with the knees probably originally raised but fallen to the left or right, colored with red ochre, with Unio shell beads, long flint blades (up to 22 cm long), copper awls, a copper rod "torque," and two four-knobbed mace heads made of black polished stone (see figure 11.10). The migrants arrived at the end of the Tiszapolgar and the beginning of the Bodrog-keresztur periods, about 4000–3900 BCE, but seemed not to disrupt the local cultural traditions. Hoards of large golden and copper ornaments of

Old European types were hidden at Hencida and Mojgrad in eastern Hungary, probably indicating unsettled conditions, but otherwise there was a lot of cultural continuity between Tiszapolgar and Bodrogkeresztur.[50] This was no massive folk migration but a series of long-distance movements by small groups, exactly the kind of movement expected among horseback riders.

The Suvorovo Graves

The Suvorovo kurgan (Suvorovo II k.1) was 13 m in diameter and covered four Eneolithic graves (see figure 11.11).[51] Stones a meter tall formed a cromlech around the base of the mound. Within the cromlech two smaller stone circles were built on a north-south axis, each surrounding a central grave (gr. 7 and 1). Grave 7 was the double grave of an adult male and female buried supine with raised legs, heads to the east. The floor of the grave was covered with red ochre, white chalk, and black fragments of charcoal. A magnificent polished stone mace shaped like the head of a horse lay on the pelvis of the male (see figure 11.5). Belts of shell disk beads draped the female's hips. The grave also contained two copper awls made of Balkan copper, three lamellar flint blades, and a flint end scraper. Grave 1, in the other stone circle, contained an adult male in an extended position and two sherds of a shell-tempered pot.

The Suvorovo cemetery at Giurgiuleşti, near the mouth of the Prut, contained five graves grouped around a hearth full of burned animal bones.[52] Above grave 4, that of the adult male, was another deposit of cattle skulls and bones. Graves 4 and 5 were those of an adult male and female; graves 1, 2, and 3, contained three children, apparently a family group. The graves were covered by a mound, but the excavators were uncertain if the mound was built for these graves or was made later. The pose in four of the five graves was on the back with raised knees (grave 2 contained disarticulated bones), and the grave floors were painted with red ochre. Two children (gr. 1 and 3) and the adult woman (gr. 5) together wore nineteen copper spiral bracelets and five boars-tusk pendants, one of which was covered in sheet copper (see figure 11.10:h). Grave 2 contained a late Gumelniţa pot. The children and adult female also had great numbers (exact count not published) of copper beads, shell disc beads, beads of red deer teeth, two beads made of Aegean coral, flint blades, and a flint core. Six of eight metal objects analyzed by N. Ryndina were made from typical Varna-Gumelniţa Balkan ores. One bracelet and one ring were made of an intentional arsenic-copper alloy (respectively, 1.9% and 1.2% arsenic) that had never occurred in Varna or Gumelniţa metals. The adult male buried in grave 4 had two gold rings and two composite projectile points, each more than 40 cm long, made with microlithic flint blades slotted along the edges of a bone point decorated with copper and gold tubular fittings (see figure 11.10:n). They probably were for two javelins, perhaps the preferred weapons of Suvorovo riders.

Kurgans also appeared south of the Danube River in the Dobruja at Casimcea, where an adult male was buried in an ochre-stained grave on his back with raised knees, accompanied by a polished stone horse-head mace (see figure 11.5), five triangular flint axes, fifteen triangular flint points, and three lamellar flint blades. Another Suvorovo grave was placed in an older Varna-culture cemetery at Devnya, near Varna. This single grave contained an adult male in an ochre-stained grave on his back with raised knees, accompanied by thirty-two golden rings, a copper axe, a copper decorative pin, a copper square-sectioned chisel 27 cm long, a bent copper wire 1.64 m long, thirty-six flint lamellar blades, and five triangular flint points.

A separate (about 80–90 km distant) but contemporary cluster of kurgans was located between the Prut and Dniester valleys near the Tripolye frontier (Kainari, Artsiza, and Kopchak). At Kainari, only a dozen kilometers from the Tripolye B1 settlement of Novi Ruşeşti, a kurgan was erected over a grave with a copper "torque" strung with *Unio* shell disc beads (see figure 11.10:k); long lamellar flint blades, red ochre, and a Tripolye B1 pot.

The Novodanilovka Group

Back in the steppes north of the Black Sea the elite were buried with copper spiral bracelets, rings, and bangles; copper beads of several types; copper shell-shaped pendants; and copper awls, all containing Balkan trace elements and made technologically just like the objects at Giurgiuleşti and Suvorovo.[53] Copper shell-shaped pendants, a very distinctive steppe ornament type, occurred in both Novodanilovka (Chapli) and Suvorovo (Giurgiuleşti) graves (see figure 11.10:j): The grave floors were strewn with red ochre or with a chunk of red ochre. The body was positioned on the back with raised knees and the head oriented toward the east or northeast. Surface markers were a small kurgan or stone cairn, often surrounded by a stone circle or cromlech. The following were among the richest:

Novodanilovka, a single stone-lined cist grave containing two adults
   at Novodanilovka in the dry hills between the Dnieper and the Sea
   of Azov with two copper spiral bracelets, more than a hundred

Unio shell beads, fifteen lamellar flint blades, and a pot imported from the North Caucasian Svobodnoe culture;

Krivoy Rog, in the Ingulets valley, west of the Dnieper, a kurgan covering two graves (1 and 2) with flint axes, flint lamellar blades, a copper spiral bracelet, two copper spiral rings, hundreds of copper beads, a gold tubular shaft fitting, *Unio* disc beads, and other objects;

Chapli (see figure 11.10) at the north end of the Dnieper Rapids, with five rich graves. The richest of these (1a and 3a) were children's graves with two copper spiral bracelets, thirteen shell-shaped copper pendants, more than three hundred copper beads, a copper foil headband, more than two hundred *Unio* shell beads, one lamellar flint blade, and one boars-tusk pendant like those at Giurgiuleşti; and

Petro-Svistunovo (see figure 11.10), a cemetery of twelve cromlechs at the south end of the Dnieper Rapids largely destroyed by erosion, with Grave 1 alone yielding two copper spiral bracelets, more than a hundred copper beads, three flint axes, and a flint lamellar blade, and the other graves yielding three more spiral bracelets, a massive cast copper axe comparable to some from Varna, and boars-tusk pendants like those at Chapli and Giurgiuleşti.

About eighty Sredni Stog cemeteries looked very similar in ritual and occurred in the same region but did not contain the prestige goods that appeared in the Novodanilovka graves, which probably were the graves of clan chiefs. The chiefs redistributed some of their imported Balkan wealth. For example, in the small Sredni Stog cemetery at Dereivka, grave 1 contained three small copper beads and grave 4 contained an imported Tripolye B1 bowl. The other graves contained no grave gifts at all.

Warfare, Climate Change, and Language Shift in the Lower Danube Valley

The colder climate of 4200–3800 BCE probably weakened the agricultural economies of Old Europe at the same time that steppe herders pushed into the marshes and plains around the mouth of the Danube. Climate change probably played a significant role in the ensuing crisis, because virtually all the cultures that occupied tell settlements in southeastern Europe abandoned them about 4000 BCE—in the lower Danube valley, the Balkans, on the Aegean coast (the end of Sitagroi III), and even in Greece (the end of Late Neolithic II in Thessaly).[54]

But even if climatic cooling and crop failures must have been significant causes of these widespread tell abandonments, they were not the only cause. The massacres at Yunatsite and Hotnitsa testify to conflict. Polished stone mace heads were status weapons that glorified the cracking of heads. Many Suvorovo-Novodanilovka graves contained sets of lanceolate flint projectile points, flint axes, and, in the Giurgiuleşti chief's grave, two fearsome 40-cm javelin heads decorated with copper and gold. Persistent raiding and warfare would have made fixed settlements a strategic liability. Raids by Slavic tribes caused the abandonment of all the Greek-Byzantine cities in this same region over the course of less than a hundred years in the sixth century CE. Crop failures exacerbated by warfare would have encouraged a shift to a more mobile economy.[55] As that shift happened, the pastoral tribes of the steppes were transformed from scruffy immigrants or despised raiders to chiefs and patrons who were rich in the animal resources that the new economy required, and who knew how to manage larger herds in new ways, most important among these that herders were mounted on horseback.

The Suvorovo chiefs displayed many of the behaviors that fostered language shift among the Acholi in East Africa: they imported a new funeral cult with an associated new mortuary ideology; they sponsored funeral feasts, always events to build alliances and recruit allies; they displayed icons of power (stone maces); they seem to have glorified war (they were buried with status weapons); and it was probably their economic example that prompted the shift to pastoral economies in the Danube valley. Proto-Indo-European religion and social structure were both based on oath-bound promises that obligated patrons (or the gods) to provide protection and gifts of cattle and horses to their clients (or humans). The oath (*$h_1óitos$*) that secured these obligations could, in principle, be extended to clients from the Old European tells.

An archaic Proto-Indo-European language, probably ancestral to Anatolian, spread into southeastern Europe during this era of warfare, dislocation, migration, and economic change, around 4200–3900 BCE. In a similar situation, in a context of chronic warfare on the Pathan/Baluch border in western Pakistan, Frederik Barth described a steady stream of agricultural Pathans who had lost their land and then crossed over and joined the pastoral Baluch. Landless Pathan could not regain their status in other Pathan villages, where land was necessary for respectable status. Tells and their fixed field systems might have played a similar limiting role in Old European status hierarchies. Becoming the client of a pastoral patron who offered protection and rewards in exchange for service was an alternative that held the promise of vertical social mobility for the children. The speakers of Proto-Indo-European talked about gifts and honors awarded for great deeds and loot/booty acquired unexpectedly, suggesting that achievement-based honor and wealth could be acquired.[56] Under conditions of chronic warfare, displaced tell dwellers may well have adopted an Indo-European patron and language as they adopted a pastoral economy.

AFTER THE COLLAPSE

In the centuries after 4000 BCE, sites of the Cernavoda I type spread through the lower Danube valley (figure 11.12). Cernavoda I was a settlement on a promontory overlooking the lower Danube. Cernavoda I material culture probably represented the assimilation of migrants from the steppes with local people who had abandoned their tells. Cernavoda I ceramics appeared at Pevec and Hotnitsa-Vodopada in north-central Bulgaria, and at Renie II in the lower Prut region. These settlements were small, with five to ten pit-houses, and were fortified. Cernavoda I pottery also occurred in settlements of other cultural types, as at Telish IV in northwestern Bulgaria. Cernavoda I pottery included simplified versions of late Gumelniţa shapes, usually dark-surfaced and undecorated but made in shell-tempered fabrics. The U-shaped "caterpillar" cord impressions (figure 11.12i), dark surfaces, and shell tempering were typical of Sredni Stog or Cucuteni C.[57]

Prominent among these new dark-surfaced, shell-tempered pottery assemblages were loop-handled drinking cups and tankards called "Scheibenhenkel," a new style of liquid containers and servers that appeared throughout the middle and lower Danube valley. Andrew Sherratt interpreted the Scheibenhenkel horizon as the first clear indicator of a new custom of drinking intoxicating beverages.[58] The replacement of highly decorated storage and serving vessels by plain drinking cups could indicate that new elite drinking rituals had replaced or nudged aside older household feasts.

The Cernavoda I economy was based primarily on the herding of sheep and goats. Many horse bones were found at Cernavoda I, and, for the first time, domesticated horses became a regular element in the animal herds of the middle and lower Danube valley.[59] Greenfield's zoological studies in the middle Danube showed that, also for the first time, animals were

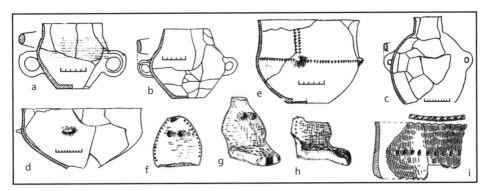

Figure 11.12 Black- or grey-surfaced ceramics from the Cernavoda I settlement, lower Danube valley, about 3900–3600 BCE, including two-handled tankards. After Morintz and Roman 1968.

butchered at different ages in upland and lowland sites. This suggested that herders moved animals seasonally between upland and lowland pastures, a form of herding called "transhumant pastoralism." The new pastoral economy might have been practiced in a new, more mobile way, perhaps aided by horseback riding.[60]

Kurgan graves were created only during the initial Suvorovo penetration. Afterward the immigrants' descendants stopped making kurgans. The flat-grave cemetery of Ostrovul Corbului probably dates to this settling-in period, with sixty-three graves, some displaying a posture on the back with raised knees, others contracted on the side, on the ruins of an abandoned tell. Cernavoda I flat graves also appeared at the Brailiţa cemetery, where the males had wide Proto-Europoid skulls and faces like the steppe Novodanilovka population, and the females had gracile Mediterranean faces, like the Old European Gumelnitsa population.

By about 3600 BCE the Cernavoda I culture developed into Cernavoda III. Cernavoda III was, in turn, connected with one of the largest and most influential cultural horizons of eastern Europe, the Baden-Boleraz horizon, centered in the middle Danube (Hungary) and dated about 3600–3200 BCE. Drinking cups of this culture featured very high strap handles and were made in burnished grey-black fabrics with channeled flutes decorating their shoulders. Somewhat similar drinking sets were made from eastern Austria and Moravia to the mouth of the Danube and south to the Aegean coast (Dikili Tash IIIA–Sitagroi IV). Horse bones appeared almost everywhere, with larger sheep interpreted as wool sheep. At lowland sites in the middle Danube region, 60–91% of the sheep-goat lived to adult ages, suggesting management for secondary products, probably wool. Similarly 40–50% of the caprids were adults in two late TRB sites of this same era (Schalkenburg and Bronocice) in upland southern Poland. After 3600 BCE horses and wool sheep were increasingly common in eastern Europe.

Pre-Anatolian languages probably were introduced to the lower Danube valley and perhaps to the Balkans about 4200–4000 BCE by the Suvorovo migrants. We do not know when their descendants moved into Anatolia. Perhaps pre-Anatolian speakers founded Troy I in northwestern Anatolia around 3000 BCE. In prayers recited by the later Hittites, the sun god of heaven, *Sius* (cognate with Greek *Zeus*), was described as rising from the sea. This has always been taken as a fossilized ritual phrase retained from some earlier pre-Hittite homeland located west of a large sea.[61] The graves of Suvorovo were located west of the Black Sea. Did the Suvorovo people ride their horses down to the shore and pray to the rising sun?

CHAPTER TWELVE

Seeds of Change on the Steppe Borders
Maikop Chiefs and Tripolye Towns

After Old Europe collapsed, the dedication of copper objects in North Pontic graves declined by almost 80%.[1] Beginning in about 3800 BCE and until about 3300 BCE the varied tribes and regional cultures of the Pontic-Caspian steppes seem to have turned their attention away from the Danube valley and toward their other borders, where significant social and economic changes were now occurring.

On the southeast, in the North Caucasus Mountains, spectacularly ostentatious chiefs suddenly appeared among what had been very ordinary small-scale farmers. They displayed gold-covered clothing, gold and silver staffs, and great quantities of bronze weapons obtained from what must have seemed beyond the rim of the earth—in fact, from the newly formed cities of Middle Uruk Mesopotamia, through Anatolian middlemen. The first contact between southern urban civilizations and the people of the steppe margins occurred in about 3700–3500 BCE. It caused a social and political transformation that was expressed archaeologically as the Maikop culture of the North Caucasus piedmont. Maikop was the filter through which southern innovations—including possibly wagons—first entered the steppes. Sheep bred to grow long wool might have passed from north to south in return, a little considered possibility. The Maikop chiefs used a tomb type that looked like an elaborated copy of the Suvorovo-Novodanilovka kurgan graves of the steppes, and some of them seem to have moved north into the steppes. A few Maikop traders might have lived inside steppe settlements on the lower Don River. But, oddly, very little southern wealth was shared with the steppe clans. The gold, turquoise, and carnelian stayed in the North Caucasus. Maikop people might have driven the first wagons into the Eurasian steppes, and they certainly introduced new metal alloys that made a more sophisticated metallurgy possible. We do not know what they took in return—possibly wool, possibly horses, possibly even *Cannabis* or saiga antelope hides, though there is only circumstantial evidence for any of these. But in most parts of the Pontic-Caspian steppes the evidence for contact with Mai-kop is slight—a pot here, an arsenical bronze axe-head there.

On the west, Tripolye (C1) agricultural towns on the middle Dnieper began to bury their dead in cemeteries—the first Tripolye communities to accept the ritual of cemetery burial—and their coarse pottery began to look more and more like late Sredni Stog pottery. This was the first stage in the breakdown of the Dnieper frontier, a cultural border that had existed for two thousand years, and it seems to have signaled a gradual process of cross-border assimilation in the middle Dnieper forest-steppe zone. But while assimilation and incremental change characterized Tripolye towns on the middle Dnieper frontier, Tripolye towns closer to the steppe border on the South Bug River ballooned to enormous sizes, more than 350 ha, and, between about 3600 and 3400 BCE, briefly became the largest human settlements in the world. The super towns of Tripolye C1 were more than 1 km across but had no palaces, temples, town walls, cemeteries, or irrigation systems. They were not cities, as they lacked the centralized political authority and specialized economy associated with cities, but they were actually bigger than the earliest cities in Uruk Mesopotamia. Most Ukrainian archaeologists agree that warfare and defense probably were the underlying reasons why the Tripolye population aggregated in this way, and so the super towns are seen as a defensive strategy in a situation of confrontation and conflict, either between the Tripolye towns or between those towns and the people of the steppes, or both. But the strategy failed. By 3300 BCE all the big towns were gone, and the entire South Bug valley was abandoned by Tripolye farmers.

Finally, on the east, on the Ural River, a section of the Volga-Ural steppe population decided, about 3500 BCE, to migrate eastward across Kazakhstan more than 2000 km to the Altai Mountains. We do not know why they did this, but their incredible trek across the Kazakh steppes led to the appearance of the Afanasievo culture in the western Gorny Altai. The Afanasievo culture was intrusive in the Altai, and it introduced a suite of domesticated animals, metal types, pottery types, and funeral customs that were derived from the Volga-Ural steppes. This long-distance migration almost certainly separated the dialect group that later developed into the Indo-European languages of the Tocharian branch, spoken in Xinjiang in the caravan cities of the Silk Road around 500 CE but divided at that time into two or three quite different languages, all exhibiting archaic Indo-European traits. Most studies of Indo-European sequencing put the separation of Tocharian after that of Anatolian and before any other branch. The Afanasievo migration meets that expectation. The migrants might also have been responsible for introducing horseback riding to the pedestrian foragers of the northern Kazakh steppes, who were quickly transformed into the horse-riding, wild-horse–hunting Botai culture just when the Afanasievo migration began.

By this time, early Proto-Indo-European dialects must have been spoken in the Pontic-Caspian steppes, tongues revealing the innovations that separated all later Indo-European languages from the archaic Proto-Indo-European of the Anatolian type. The archaeological evidence indicates that a variety of different regional cultures still existed in the steppes, as they had throughout the Eneolithic. This regional variability in material culture, though not very robust, suggests that early Proto-Indo-European probably still was a regional language spoken in one part of the Pontic-Caspian steppes—possibly in the eastern part, since this was where the migration that led to the Tocharian branch began. Groups that distinguished themselves by using eastern innovations in their speech probably were engaging in a political act—allying themselves with specific clans, their political institutions, and their prestige—and in a religious act—accepting rituals, songs, and prayers uttered in that eastern dialect. Songs, prayers, and poetry were central aspects of life in all early Indo-European societies; they were the vehicle through which the right way of speaking reproduced itself publicly.

The Five Cultures of the Final Eneolithic in the Steppes

Much regional diversity and relatively little wealth existed in the Pontic-Caspian steppes between about 3800 and 3300 BCE (table 12.1). Regional variants as defined by grave and pot types, which is how archaeologists define them, had no clearly defined borders; on the contrary, there was a lot of border shifting and inter-penetration. At least five Final Eneolithic archaeological cultures have been identified in the Pontic-Caspian steppes (figure 12.1). Sites of these five groups are sometimes found in the same regions, occasionally in the same cemeteries; overlapped in time; shared a number of similarities; and were, in any case, fairly variable. In these circumstances, we cannot be sure that they all deserve recognition as

TABLE 12.1

Selected Radiocarbon Dates for Final Eneolithic Sites in the Steppes and Early Bronze Age Sites in the North Caucasus Piedmont

Lab Number	BP Date	Sample		Calibrated Date

1. Maikop culture

Klady kurgan cemetery, Farsa River valley near Maikop

Le 4529	4960±120	Klady k29/1 late	bone	3940–3640 BCE
OxA 5059	4835±60	Klady k11/50 early	bone	3700–3520 BCE
OxA 5061	4765±65	Klady k11/55 early	bone	3640–3380 BCE
OxA 5058	4675±70	Klady k11/43 early	bone	3620–3360 BCE
OxA 5060	4665±60	Klady k11/48 early		3520–3360 BCE
Le 4528	4620±40	Klady k30/1 late	bone	3500–3350 BCE

Galugai settlement, upper Tersek River

OxA 3779	4930±120	Galugai I		3940–3540 BCE
OxA 3778	4650±80	Galugai I	bone	3630–3340 BCE
OxA 3777	4480±70	Galugai I		3340–3030 BCE

2. Tripolye C1 settlements

BM-495	4940±105	Soroki-Ozero		3940–3630 BCE
UCLA-1642F	4904±300	Novorozanovka 2		4100–3300 BCE
Bln-2087	4890±50	Maidanets'ke	charcoal	3710–3635 BCE
UCLA-1671B	4890±60	Evminka		3760–3630 BCE
BM-494	4792±105	Soroki-Ozero		3690–3370 BCE
UCLA-1466B	4790±100	Evminka		3670–3370 BCE
Bln-631	4870±100	Chapaevka		3780–3520 BCE
Ki-880	4810±140	Chapaevka	charcoal	3760–3370 BCE
Ki-1212	4600±80	Maidanets'ke		3520–3100 BCE

3. Repin culture

Kyzyl-Khak II settlement, North Caspian desert, lower Volga

?	4900±40	house 2	charcoal	3705–3645 BCE

Mikhailovka II settlement, lower part of level II

Ki-8010	4710±80	square 14, 2.06m depth	bone	3630–3370 BCE

Podgorovka settlement, Aidar River, Donets River tributary

Ki-7843	4560±50	?		3490–3100 BCE
Ki-7841	4370±55	?		3090–2900 BCE
Ki-7842	4330±50	?		3020–2880 BCE

4. Late Khvalynsk culture

Kara-Khuduk settlement, North Caspian desert, lower Volga

UPI-431	5100±45	pit-house	charcoal	3970–3800 BCE

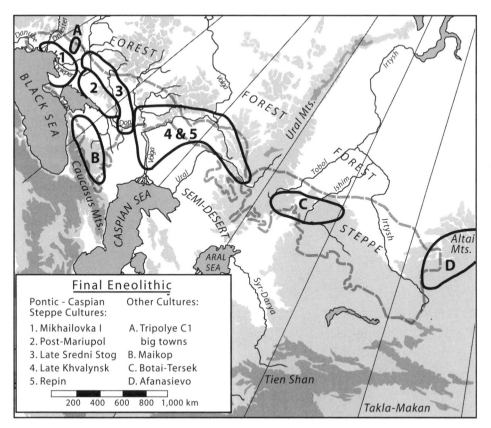

Figure 12.1 Final Eneolithic culture areas from the Carpathians to the Altai, 3800–3300 BCE.

different archaeological cultures. But we cannot understand the archaeological descriptions of this period without them, and together they provide a good picture of what was happening in the Pontic-Caspian steppes between 3800 and 3300 BCE. The western groups were engaged in a sort of two-pronged death dance, as it turned out, with the Cucuteni-Tripolye culture. The southern groups interacted with Maikop traders. And the eastern groups cast off a set of migrants who rode across Kazakhstan to a new home in the Altai, a subject reserved for the next chapter. Horseback riding is documented archaeologically in Botai-Tersek sites in Kazakhstan during this period (chapter 10) and probably appeared earlier, and so we proceed on the assumption that most steppe tribes were now equestrian.

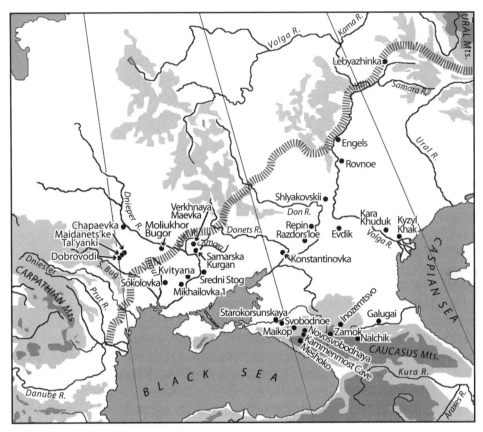

Figure 12.2 Final Eneolithic sites in the steppes and Early Bronze Age sites in the North Caucasus piedmont.

The Mikhailovka I Culture

The westernmost of the five Final Eneolithic cultures of the Pontic-Caspian steppes was the Mikhailovka I culture, also called the Lower Mikhailovka or Nizhnimikhailovkskii culture, named after a stratified settlement on the Dnieper located below the Dnieper Rapids (figure 12.2).[2] Below the last cascade, the river spread out over a broad basin in the steppes. Braided channels crisscrossed a sandy, marshy, forested lowland 10–20 km wide and 100 km long, a rich place for hunting and fishing and a good winter refuge for cattle, now inundated by hydroelectric dams. Mikhailovka overlooked this protected depression at a strategic river crossing. Its initial establishment probably was an outgrowth of increased east-west traffic across the river. It was the most important settlement on the lower Dnieper from the Late Eneolithic through the Early Bronze Age, about 3700–2500 BCE. Mikhailovka I, the original settlement, was occupied about 3700–3400 BCE, contemporary with late Tripolye B2 and early C1, late Sredni Stog, and early Maikop. A few late Sredni Stog and Maikop pottery sherds occurred in the occupation layer at Mikhailovka I. A whole Maikop pot was found in a grave with Mikhailovka I sherds at Sokolovka on the Ingul River, in kurgan 1, grave 6a. Tripolye B2 and C1 pots also are found in Mikhailovka I graves. These exchanges of pottery show that the Mikhailovka I culture had at least sporadic contacts with Tripolye B2/C1 towns, the Maikop culture, and late Sredni Stog communities.[3]

The people of Mikhailovka I cultivated cereal crops. At Mikhailovka I, imprints of cultivated seeds were found on 9 pottery sherds of 2,461 examined, or 1 imprint in 273 sherds.[4] The grain included emmer wheat, barley, millet, and 1 imprint of a bitter vetch seed (*Vicia ervilia*), a crop grown today for animal fodder. Zoologists identified 1,166 animal bones (NISP) from Mikhailovka I, of which 65% were sheep-goat, 19% cattle, 9% horse, and less than 2% pig. Wild boar, aurochs, and saiga antelope were hunted occasionally, accounting for less than 5 percent of the animal bones.

The high number of sheep-goat at Mikhailovka I might suggest that long-wool sheep were present. Wool sheep probably were present in the North Caucasus at Svobodnoe (see below) by 4000 BCE, and almost certainly were in the Danube valley during the Cernavoda III–Boleraz period around 3600–3200 BCE, so wool sheep could have been kept at Mikhailovka I. But even if long-wool sheep *were* bred in the steppes during this period, they clearly were not yet the basis for a widespread new wool economy, because cattle or even deer bones still outnumbered sheep in other steppe settlements.[5]

Mikhailovka I pottery was shell-tempered and had dark burnished surfaces, usually unornamented (figure 12.3). Common shapes were egg-shaped pots or flat-based, wide-shouldered tankards with everted rims. A few silver ornaments and one gold ring, quite rare in the Pontic steppes of this era, were found in Mikhailovka I graves.

Mikhailovka I kurgans were distributed from the lower Dnieper westward to the Danube delta and south to the Crimean peninsula, north and northwest of the Black Sea. Near the Danube they were interspersed with cemeteries that contained Danubian Cernavoda I–III ceramics.[6] Most

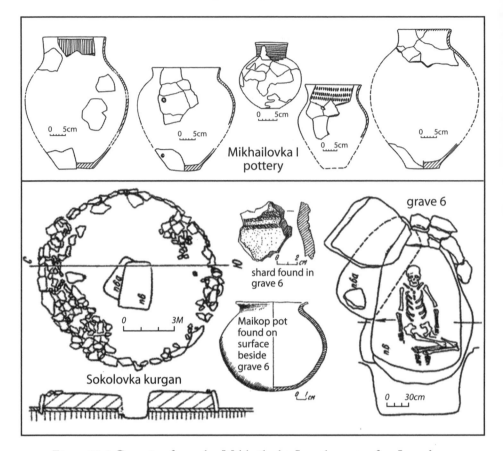

Figure 12.3 Ceramics from the Mikhailovka I settlement, after Lagodovskaya, Shaposhnikova, and Makarevich 1959; and a Mikhailovka I grave (gr. 6) stratified above an older Eneolithic grave (gr. 6a) at Sokolovka kurgan on the Ingul River west of the Dnieper, after Sharafutdinova 1980.

Mikhailovka I kurgans were low mounds of black earth covered by a layer of clay, surrounded by a ditch and a stone cromlech, often with an opening on the southwest side. The graves frequently were in cists lined with stone slabs. The body could be in an extended supine position or contracted on the side or supine with raised knees, although the most common pose was contracted on the side. Occasionally (e.g., Olaneshti, k. 2, gr. 1, on the lower Dniester) the grave was covered by a stone anthropomorphic stela—a large stone slab carved at the top into the shape of a head projecting above rounded shoulders (see figure 13.11). This was the beginning of a long and important North Pontic tradition of decorating some graves with carved stone stelae.[7]

The skulls and faces of some Mikhailovka I people were delicate and narrow. The skeletal anthropologist Ina Potekhina established that another North Pontic culture, the Post-Mariupol culture, looked most like the old wide-faced Suvorovo-Novodanilovka population. The Mikhailovka I people, who lived in the westernmost steppes closest to the Tripolye culture and to the lower Danube valley, seem to have intermarried more with people from Tripolye towns or people whose ancestors had lived in Danubian tells.[8]

The Mikhailovka I culture was replaced by the Usatovo culture in the steppes northwest of the Black Sea after about 3300 BCE. Usatovo retained some Mikhailovka I customs, such as making a kurgan with a surrounding stone cromlech that was open to the southwest. The Usatovo culture was led by a warrior aristocracy centered on the lower Dniester estuary that probably regarded Tripolye agricultural townspeople as tribute-paying clients, and that might have begun to engage in sea trade along the coast. People in the Crimean peninsula retained many Mikhailovka I customs and developed into the Kemi-Oba culture of the Early Bronze Age after about 3300 BCE. These EBA cultures will be described in a later chapter.

The Post-Mariupol Culture

The clumsiest culture name of the Final Eneolithic is the "Post-Mariupol" or "Extended-Position-Grave" culture, both names conveying a hint of definitional uncertainty. Rassamakin called it the "Kvityana" culture. I will use the name "Post-Mariupol." All these names refer to a grave type recognized in the steppes just above the Dnieper Rapids in the 1970s but defined in various ways since then. N. Ryndina counted about three hundred graves of the Post-Mariupol type in the steppes from the Dnieper valley eastward to the Donets. They were covered by low kurgans, occasionally surrounded by a stone cromlech. Burial was in an extended supine position in a narrow oblong or rectangular pit, often lined with stone and covered with wooden beams or stone slabs. Usually there were no ceramics in the grave (although this rule was fortunately broken in a few graves), but a fire was built above the grave; red ochre was strewn heavily on the grave floor; and lamellar flint blades, bone beads, or a few small copper

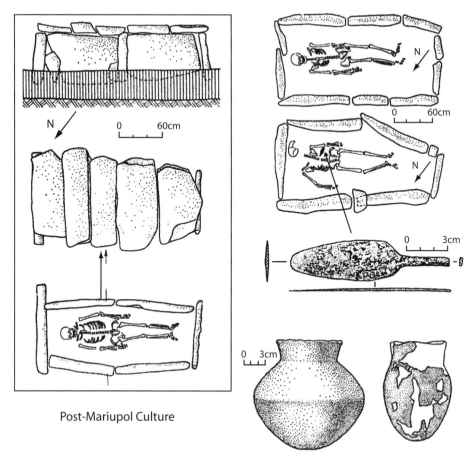

Post-Mariupol Culture

Figure 12.4 Post-Mariupol ceramics and graves: *left*, Marievka kurgan 14, grave 7; *upper right*, Bogdanovskogo Karera Kurgan 2, graves 2 and 17; *lower right*, pots from Chkalovskaya kurgan 3. After Nikolova and Rassamakin 1985, figure 7.

beads or twists were included (figure 12.4). Three cattle skulls, presumably sacrificed at the funeral, were placed at the edge of one grave at Chkalovska kurgan 3. The largest cluster is just north of the Dnieper Rapids on the east side of the Dnieper, between two tributary rivers, the Samara (smaller than the Volga-region Samara River) and the Orel. Two chronological phases are identified: an early (Final Eneolithic) phase contemporary with Tripolye B2/C1, about 3800–3300 BCE; and a later (Early Bronze Age) phase contemporary with Tripolye C2 and the Early Yamnaya horizon, about 3300–2800 BCE.[9]

About 40 percent of the Post-Mariupol graves in the core Orel-Samara region contained copper ornaments, usually just one or two. All forty-six of the copper objects examined by Ryndina from early-phase graves were made from "clean" Transylvanian ores, the same ores used in Tripolye B2 and C1 sites. The copper in the second phase, however, was from two sources: ten objects still were made of "clean" Transylvanian copper but twenty-three were made of arsenical bronze. They were most similar to the arsenical bronzes of the Ustatovo settlement or the late Maikop culture. Only one Post-Mariupol object (a small willow-leaf pendant from Bulakhovka kurgan cemetery I, k. 3, gr. 9) looked metallurgically like a direct import from late Maikop.[10]

Two Post-Mariupol graves were metalsmiths' graves. They contained three bivalve molds for making sleeved axes. (A sleeved axe had a single blade with a cast sleeve hole for the handle on one side.) The molds copied a late Maikop axe type but were locally made.[11] They probably were late Post-Mariupol, after 3300 BCE. They are the oldest known two-sided ceramic molds in the steppes, and they were buried with stone hammers, clay tubes or *tulieres* for bellows attachments, and abrading stones. These kits suggest a new level of technological skill among steppe metalsmiths and the graves began a long tradition of the smith being buried with his tools.

The Late Sredni Stog Culture

The third and final culture group in the *western* part of the Pontic-Caspian steppes was the late Sredni Stog culture. Late Sredni Stog pottery was shell-tempered and often decorated with cord-impressed geometric designs (see figure 11.7), quite unlike the plain, dark-surfaced pots of Mikhailovka I and the Post-Mariupol culture. The late Sredni Stog settlement of Moliukhor Bugor was located on the Dnieper in the forest-steppe zone. A Tripolye C1 vessel was found there. The people of Moliukhor Bugor lived in a house 15 m by 12 m with three internal hearths, hunted red deer and wild boar, fished, kept a lot of horses and a few domesticated cattle and sheep, and grew grain. Eight grain impressions were found among 372 sherds (one imprint in 47 sherds), a higher frequency than at Mikhailovka I. They included emmer wheat, einkorn wheat, millet, and barley. The well-known Sredni Stog settlement at Dereivka was occupied somewhat earlier, about 4000 BCE, but also produced many flint blades with sickle gloss and six stone querns for grinding grain, and so also probably included some grain cultivation.

Horses represented 63% of the animal bones at Dereivka (see chapter 10). The Sredni Stog societies on the Dnieper, like the other western steppe groups, had a mixed economy that combined grain cultivation, stock-breeding, horseback riding, and hunting and fishing.

Late Sredni Stog sites were located in the northern steppe and southern forest-steppe zones on the middle Dnieper, north of the Post-Mariupol and Mikhailovka I groups. Sredni Stog sites also extended from the Dnieper eastward across the middle Donets to the lower Don. The most important stratified settlement on the lower Don was Razdorskoe [raz-DOR-sko-ye]. Level 4 at Razdorskoe contained an early Khvalynsk component, level 5 above it had an early Sredni Stog (Novodanilovka period) occupation, and, after that, levels 6 and 7 had pottery that resembled late Sredni Stog mixed with imported Maikop pottery. A radiocarbon date said to be associated with level 6, on organic material in a core removed for pollen studies, produced a date of 3500–2900 BCE (4490±180 BP). Near Razdorskoe was the fortified settlement at Konstantinovka. Here, in a place occupied by people who made similar lower-Don varieties of late Sredni Stog pottery, there might actually have been a small Maikop colony.[12]

Bodies buried in Sredni Stog graves usually were in the supine-with-raised knees position that was such a distinctive aspect of steppe burials beginning with Khvalynsk. The grave floor was strewn with red ochre, and the body often was accompanied by a unifacial flint blade or a broken pot. Small mounds sometimes were raised over late Sredni Stog graves, but in many cases they were flat.

Repin and Late Khvalynsk in the Lower Don-Volga Steppes

The two eastern groups can be discussed together. They are identified with two quite different kinds of pottery. One type clearly resembled a late variety of Khvalynsk pottery. The other type, called Repin, probably began on the middle Don, and is identified by round-based pots with cord-impressed decoration and decorated rims.

Repin, excavated in the 1950s, was located 250 km upstream from Razdorskoe, on the middle Don at the edge of the feather-grass steppe. At Repin 55% of the animal bones were horse bones. Horse meat was much more important in the diet than the meat of cattle (18%), sheep-goat (9%), pigs (9%), or red deer (9%).[13] Perhaps Repin specialized in raising horses for export to North Caucasian traders (?). The pottery from

Repin defined a type that has been found at many sites in the Don-Volga region. Repin pottery sometimes is found stratified beneath Yamnaya pottery, as at the Cherkasskaya settlement on the middle Don in the Voronezh oblast.[14] Repin components occur as far north as the Samara oblast in the middle Volga region, at sites such as Lebyazhinka I on the Sok River, in contexts also thought to predate early Yamnaya. The Afanasievo migration to the Altai was carried out by people with a Repin-type material culture, probably from the middle Volga-Ural region. On the lower Volga, a Repin antelope hunters' camp was excavated at Kyzyl Khak, where 62% of the bones were saiga antelope (figure 12.5). Cattle were 13%, sheep 9%, and horses and onagers each about 7%. A radiocarbon date (4900±40 BP) put the Repin occupation at Kyzyl-Khak at about 3700–3600 BCE.

Kara Khuduk was another antelope hunters' camp on the lower Volga but was occupied by people who made late Khvalynsk-type pottery (figure 12.5). A radiocarbon date (5100±45 BP, UPI 430) indicated that it was occupied in about 3950–3800 BCE, earlier than the Repin occupation at Kyzyl-Khak nearby. Many large scrapers, possibly for hide processing, were found among the flint tools. Saiga antelope hides seem to have been highly desired, perhaps for trade. The animal bones were 70% saiga antelope, 13% cattle, and 6% sheep. The ceramics (670 sherds from 30–35 vessels) were typical Khvalynsk ceramics: shell-tempered, round-bottomed vessels with thick, everted lips, covered with comb stamps and corded-impressed U-shaped "caterpillar" impressions.

Late Khvalynsk graves without kurgans were found in the 1990s at three sites on the lower Volga: Shlyakovskii, Engels, and Rovnoe. The bodies were positioned on the back with knees raised, strewn with red ochre, and accompanied by lamellar flint blades, flint axes with polished edges, polished stone mace heads of Khvalynsk type, and bone beads. Late Khvalynsk populations lived in scattered enclaves on the lower Volga. Some of them crossed the northern Caspian, perhaps by boat, and established a group of camps on its eastern side, in the Mangyshlak peninsula.

The Volga-Don late Khvalynsk and Repin societies played a central role in the evolution of the Early Bronze Age Yamnaya horizon beginning around 3300 BCE (discussed in the next chapter). One kind of early Yamnaya pottery was really a Repin type, and the other kind was actually a late Khvalynsk type; so, if no other clues are present, it can be difficult to separate Repin or late Khvalynsk pottery from early Yamnaya pottery. The Yamnaya horizon probably was the medium through which late

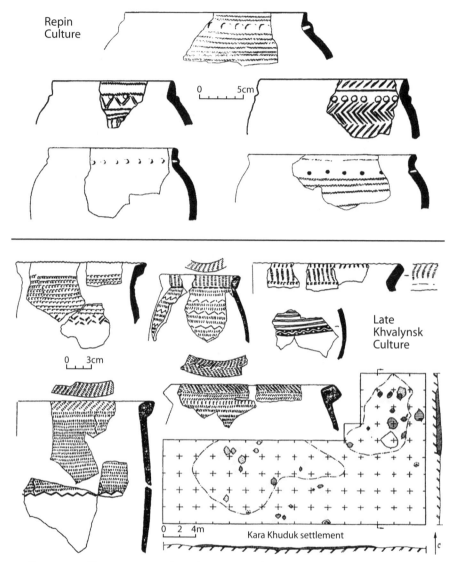

Figure 12.5 Repin pottery from Kyzl-Khak (*top*) and late Khvalynsk pottery and settlement plan from Kara-Khuduk (*bottom*) on the lower Volga. After Barynkin, Vasiliev, and Vybornov 1998, figures 5 and 6.

Proto-Indo-European languages spread across the steppes. This implies that classic Proto-Indo-European dialects were spoken among the Repin and late Khvalynsk groups.[15]

CRISIS AND CHANGE ON THE TRIPOLYE FRONTIER: TOWNS BIGGER THAN CITIES

Two notable and quite different kinds of changes affected the Tripolye culture between about 3700 and 3400 BCE. First, the Tripolye settlements in the forest-steppe zone on the middle Dnieper began to make pottery that looked like Pontic-Caspian ceramics (dark, occasionally shell-tempered wares) and adopted Pontic-Caspian–style inhumation funerals. The Dnieper frontier became more porous, probably through gradual assimilation. But Tripolye settlements on the South Bug River, near the steppe border, changed in very different ways. They mushroomed to enormous sizes, more than 400 ha, twice the size of the biggest cities in Mesopotamia. Simply put, they were the biggest human settlements in the world. And yet, instead of evolving into cities, they were abruptly abandoned.

Contact with Sredni Stog on the Dnieper Frontier

Chapaevka was a Tripolye B2/C1 settlement of eleven dwellings located on a promontory west of the Dnieper valley in the northern forest-steppe zone. It was occupied about 3700–3400 BCE.[16] Chapaevka is the earliest known Tripolye community to adopt cemetery burial (figure 12.6). A cemetery of thirty-two graves appeared on the edge of settlement. The form of burial, in an extended supine position, usually with a pot, sometimes with a piece of red ochre under the head or chest, was not exactly like any of the steppe grave types, but just the acceptance of the burial of the body was a notable change from the Old European funeral customs of the Tripolye culture. Chapaevka also had lightly built houses with dug-out floors rather than houses with plastered log floors (*ploshchadka*). Tripolye C1 pottery was found at Moliukhor Bugor, about 150 km to the south, perhaps the source of some of these new customs.

Most of the ceramics in the Chapaevka houses were well-fired fine wares with fine sand temper or very fine clay fabrics (50–70%), of which a small percentage (1–10%) were painted with standard Tripolye designs; but generally they were black to grey in color, with burnished surfaces, and

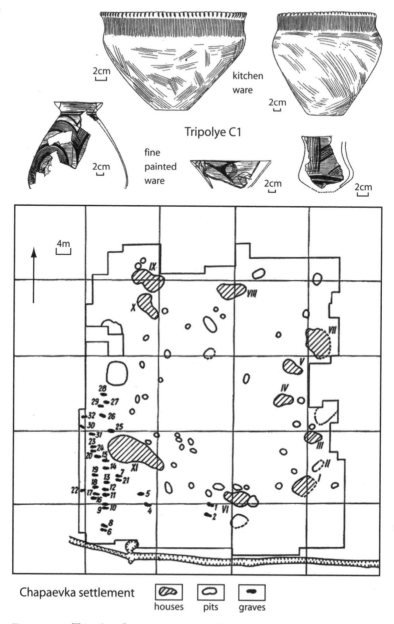

Figure 12.6 Tripolye C1 settlement at Chapaevka on the Dnieper with eleven houses (features I–XI) and cemetery (gr. 1–32) and ceramics. After Kruts 1977, figures 5 and 16.

were often undecorated. They were quite different from the orange wares that had typified earlier Tripolye ceramics. Undecorated grey-to-black ware also was typical of the Mikhailovka I and Post-Mariupol cultures, although their shapes and clay fabrics differed from most of those of the Tripolye C1 culture. One class of Chapaevka kitchen-ware pots with vertical combed decoration on the collars looked so much like late Sredni Stog pots that it is unclear whether this kind of ware was borrowed from Tripolye by late Sredni Stog potters or by Tripolye C1 potters from late Sredni Stog.[17] Around 3700–3500 BCE the Dnieper frontier was becoming a zone of gradual, probably peaceful assimilation between Tripolye villagers and indigenous Sredni Stog societies east of the Dnieper.

Towns Bigger Than Cities: The Tripolye C1 Super Towns

Closer to the steppe border things were quite different. All the Tripolye settlements located between the Dnieper and South Bug rivers, including Chapaevka, were oval, with houses arranged around an open central plaza. Some villages occupied less than 1 ha, many were towns of 8–15 ha, some were more than 100 ha, and a group of three Tripolye C1 sites located within 20 km of one another reached sizes of 250–450 ha between about 3700 and 3400 BCE. These super sites were located in the hills east of the South Bug River, near the edge of the steppe in the southern forest-steppe zone. They were the largest communities not just in Europe but in the world.[18]

The three known super-sites—Dobrovodi (250 ha), Maidanets'ke (250 ha), and Tal'yanki (450 ha)—perhaps were occupied sequentially in that order. None of these sites contained an obvious administrative center, palace, storehouse, or temple. They had no surrounding fortification wall or moat, although the excavators Videiko and Shmagli described the houses in the outer ring as joined in a way that presented an unbroken two-story-high wall pierced only by easily defended radial streets. The most thoroughly investigated of the three, Maidanets'ke, covered 250 ha. Magnetometer testing revealed 1,575 structures (figure 12.7). Most were inhabited simultaneously (there was almost no overbuilding of newer houses over older ones) by a population estimated at fifty-five hundred to seventy-seven hundred people. Using Bibikov's estimate of 0.6 ha of cultivated wheat per person per year, a population of that magnitude would have required 3,300–4,620 ha of cultivated fields each year, which would have necessitated cultivating fields more than 3 km from the town.[18] The houses were built close to one another in concentric oval rings, on a common plan, oriented toward a cen-

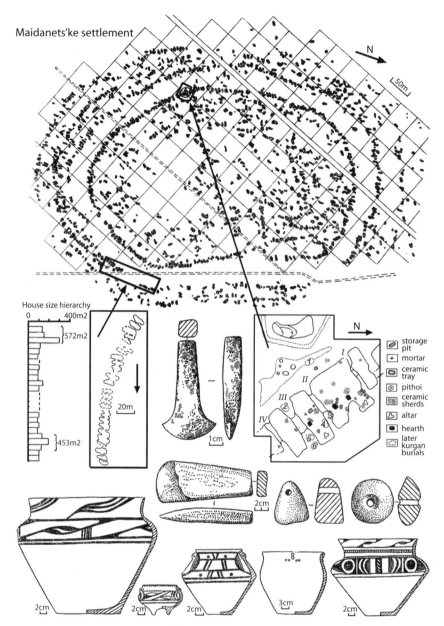

Figure 12.7 The Tripolye C1 Maidanets'ke settlement, with 1,575 structures mapped by magnetometers: *left*: smaller houses cluster around larger houses, thought to be clan or sub-clan centers; *right*: a house group very well preserved by the Yamnaya kurgan built on top of it, showing six inserted late

tral plaza. The excavated houses were large, 5–8 m wide and 20–30 m long, and many were two-storied. Videiko and Shmagli suggested a political organization based on clan segments. They documented the presence of one larger house for each five to ten smaller houses. The larger houses usually contained more female figurines (rare in most houses), more fine painted pots, and sometimes facilities such as warp-weighted looms. Each large house could have been a community center for a segment of five to ten houses, perhaps an extended family (or a "super-family collective," in Videiko's words). If the super towns were organized in this way, a council of 150–300 segment leaders would have made decisions for the entire town. Such an unwieldy system of political management could have contributed to its own collapse. After Maidanests'ke and Tal'yanki were abandoned, the largest town in the South Bug hills was Kasenovka (120 ha, with seven to nine concentric rings of houses), dated to the Tripolye C1/C2 transition, perhaps 3400–3300 BCE. When Kasenovka was abandoned, Tripolye people evacuated most of the South Bug valley.

Specialized craft centers appeared in Tripolye C1 communities for making flint tools, weaving, and manufacturing ceramics. These crafts became spatially segregated both within and between towns.[20] A hierarchy appeared in settlement sizes, comprised of two and perhaps three tiers. These kinds of changes usually are interpreted as signs of an emerging political hierarchy and increasing centralization of political power. But, as noted, instead of developing into cities, the towns were abandoned.

Population concentration is a standard response to increased warfare among tribal agriculturalists, and the subsequent abandonment of these places suggests that warfare and raiding was at the root of the crisis. The aggressors could have been steppe people of Mikhailovka I or late Sredni Stog type. A settlement at Novorozanovka on the Ingul, west of the Dnieper, produced a lot of late Sredni Stog cord-impressed pottery, some Mikhailovka I pottery, and a few imported Tripolye C1 painted fine pots. Mounted raiding might have made it impossible to cultivate fields more than 3 km from the town. Raiding for cattle or captives could have caused the fragmentation and dispersal of the Tripolye population and the abandonment of town-based craft traditions just as it had in the Danube valley

---

Figure 12.7(continued) Yamnaya graves. Artifacts from the settlement: *top center*, a cast copper axe; *central row*, a polished stone axe and two clay loom weights; *bottom row*, selected painted ceramics. After Shmagli and Videiko 1987; and Videiko 1990.

some five hundred years earlier. Farther north, in the forest-steppe zone on the middle Dnieper, assimilation and exchange led ultimately in the same direction but more gradually.

THE FIRST CITIES AND THEIR CONNECTION TO THE STEPPES

Steppe contact with the civilizations of Mesopotamia was, of course, much less direct than contact with Tripolye societies, but the southern door might have been the avenue through which wheeled vehicles first appeared in the steppes, so it was important. Our understanding of these contacts with the south has been completely rewritten in recent years.

Between 3700 and 3500 BCE the first cities in the world appeared among the irrigated lowlands of Mesopotamia. Old temple centers like Uruk and Ur had always been able to attract thousands of laborers from the farms of southern Iraq for building projects, but we are not certain why they began to live around the temples permanently (figure 12.8). This shift in population from the rural villages to the major temples created the first cities. During the Middle and Late Uruk periods (3700–3100 BCE) trade into and out of the new cities increased tremendously in the form of tribute, gift exchange, treaty making, and the glorification of the city temple and its earthly authorities. Precious stones, metals, timber, and raw wool (see chapter 4) were among the imports. Woven textiles and manufactured metal objects probably were among the exports. During the Late Uruk period, wheeled vehicles pulled by oxen appeared as a new technology for land transport. New accounting methods were developed to keep track of imports, exports, and tax payments—cylinder seals for marking sealed packages and the sealed doors of storerooms, clay tokens indicating package contents, and, ultimately, writing.

The new cities had enormous appetites for copper, gold, and silver. Their agents began an extraordinary campaign, or perhaps competing campaigns by different cities, to obtain metals and semiprecious stones. The native chiefdoms of Eastern Anatolia already had access to rich deposits of copper ore, and had long been producing metal tools and weapons. Emissaries from Uruk and other Sumerian cities began to appear in northern cities like Tell Brak and Tepe Gawra. South Mesopotamian garrisons built and occupied caravan forts on the Euphrates in Syria at Habubu Kabira. The "Uruk expansion" began during the Middle Uruk period about 3700 BCE and greatly intensified during Late Uruk, about 3350–3100 BCE. The city of Susa in southwestern Iran might have become an Uruk colony. East of Susa on the Iranian plateau a series of large mudbrick edifices

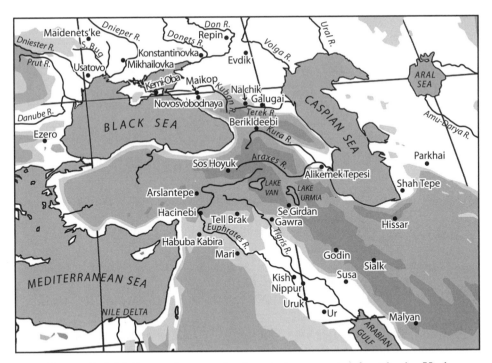

Figure 12.8 Maikop culture and selected sites associated with the Uruk expansion.

rose above the plains, protecting specialized copper production facilities that operated partly for the Uruk trade, regulated by local chiefs who used the urban tools of trade management: seals, sealed packages, sealed storerooms, and, finally, writing. Copper, lapis lazuli, turquoise, chlorite, and carnelian moved under their seals to Mesopotamia. Uruk-related trade centers on the Iranian plateau included Sialk IV_1, Tal-i-Iblis V–VI, and Hissar II in central Iran. The tentacles of trade reached as far northeast as the settlement of Sarazm in the Zerafshan Valley of modern Tajikistan, probably established to control turquoise deposits in the deserts nearby.

The Uruk expansion to the northwest, toward the gold, silver, and copper sources in the Caucasus Mountains, is documented at two important local strongholds on the upper Euphrates. Hacinebi was a fortified center with a large-scale copper production industry. Its chiefs began to deal with Middle Uruk traders during its phase B2, dated about 3700–3300 BCE. More than 250 km farther up the Euphrates, high in the mountains of Eastern Anatolia, the stronghold at Arslantepe expanded in wealth and size at about the same time (Phase VII), although it retained its own native system of seals, architecture, and administration. It also had its own large-scale copper production facilities based on local ores. Phase VIA, beginning about 3350 BCE, was dominated by two new pillared buildings similar to Late Uruk temples. In them officials regulated trade using some Uruk-style seals (among many local-style seals) and gave out stored food in Uruk-type, mass-produced ration bowls. The herds of Arslantepe VII had been dominated by cattle and goats, but in phase VIA sheep rose suddenly to become the most numerous and important animal, probably for the new industry of wool production. Horses also appeared, in very small numbers, at Arslantepe VII and VIA and Hacinebi phase B, but they seem not to have been traded southward into Mesopotamia. The Uruk expansion ended abruptly about 3100 BCE for reasons that remain obscure. Arslantepe and Hacinebi were burned and destroyed, and in the mountains of eastern Anatolia local Early Trans-Caucasian (ETC) cultures built their humble homes over the ruins of the grand temple buildings.[21]

Societies in the mountains to the north of Arslantepe responded in various ways to the general increase in regional trade that began about 3700–3500 BCE. Novel kinds of public architecture appeared. At Berikldeebi, northwest of modern Tbilisi in Georgia, a settlement that had earlier consisted of a few flimsy dwellings and pits was transformed about 3700–3500 BCE by the construction of a massive mudbrick wall that enclosed a public building, perhaps a temple, measuring 14.5×7.5 m (50×25 ft). At Sos level Va near Erzerum in northeastern Turkey there were similar architectural hints of increasing scale and power.[22] But neither prepares us for the funerary splendor of the Maikop culture.

The Maikop culture appeared about 3700–3500 BCE in the piedmont north of the North Caucasus Mountains, overlooking the Pontic-Caspian steppes. The semi-royal figure buried under the giant Maikop chieftan's kurgan acquired and wore Mesopotamian ornaments in an ostentatious funeral display that had no parallel that has been preserved even in Mesopotamia. Into the grave went a tunic covered with golden lions and bulls, silver-sheathed staffs mounted with solid gold and silver bulls, and silver sheet-metal cups. Wheel-made pottery was imported from the south, and the new technique was used to make Maikop ceramics similar to some of the vessels found at Berikldeebi and at Arslantepe VII/VIA.[23] New high-nickel arsenical bronzes and new kinds of bronze weapons (sleeved axes, tanged daggers) also spread into the North Caucasus from the south, and a cylinder seal from the south was worn as a bead in another Maikop grave. What kinds of societies lived in the North Caucasus when this contact began?

THE NORTH CAUCASUS PIEDMONT:
ENEOLITHIC FARMERS BEFORE MAIKOP

The North Caucasian piedmont separates naturally into three geographic parts. The western part is drained by the Kuban River, which flows into the Sea of Azov. The central part is a plateau famous for its bubbling hot springs, with resort towns like Mineralnyi Vody (Mineral Water) and Kislovodsk (Sweet Water). The eastern part is drained by the Terek River, which flows into the Caspian Sea. The southern skyline is dominated by the permanently glaciated North Caucasus Mountains, which rise to icy peaks more than 5,600 m (18,000 ft) high; and off to the north are the rolling brown plains of the steppes.

Herding, copper-using cultures lived here by 5000 BCE. The Early Eneolithic cemetery at Nalchik and the cave occupation at Kammenomost Cave (chapter 9) date to this period. Beginning about 4400–4300 BCE the people of the North Caucasus began to settle in fortified agricultural villages such as Svobodnoe and Meshoko (level 1) in the west, Zamok on the central plateau, and Ginchi in Dagestan in the east, near the Caspian. About ten settlements of the Svobodnoe type, of thirty to forty houses each, are known in the Kuban River drainage, apparently the most densely settled region. Their earthen or stone walls enclosed central plazas surrounded by solid wattle-and-daub houses. Svobodnoe, excavated by A. Nekhaev, is the best-reported site (figure 12.9). Half the animal bones from Svobodnoe were from wild red deer and boar, so hunting was important. Sheep were the most important domesticated animal, and the proportion of sheep to goats was 5:1, which suggests that sheep were kept for wool. But pig keeping also was important, and pigs were the most important meat animals at the settlement of Meshoko.

Svobodnoe pots were brown to orange in color and globular with everted rims, but decorative styles varied greatly between sites (e.g., Zamok, Svobodnoe, and Meshoko are said to have had quite different domestic pottery types). Female ceramic figurines suggest female-centered domestic rituals. Bracelets carved and polished of local serpentine were manufactured in the hundreds at some sites. Cemeteries are almost unknown, but a few individual graves found among later graves under kurgans in the Kuban region have been ascribed to the Late Eneolithic. The Svobodnoe culture differed from Repin or late Khvalynsk steppe cultures in its house

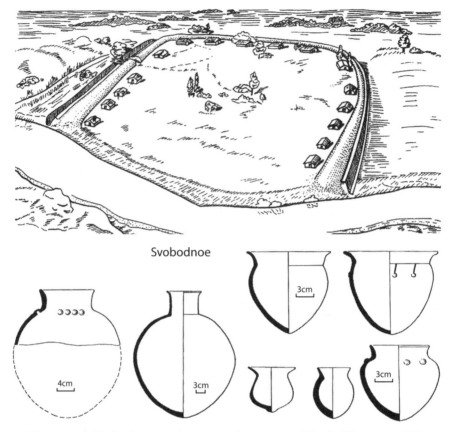

Figure 12.9 Svobodnoe settlement and ceramics, North Caucasus. After Nekhaev 1992.

forms, settlement types, pottery, stone tools, and ceramic female figurines. Probably it was distinct ethnically and linguistically.[24]

Nevertheless, the Svobodnoe culture was in contact with the steppes. A Svobodnoe pot was deposited in the rich grave at Novodanilovka in the Azov steppes, and a copper ring made of Balkan copper, traded through the Novodanilovka network, was found at Svobodnoe. Potsherds that look like early Sredni Stog types were noted at Svobodnoe and Meshoko 1. Green serpentine axes from the Caucasus appeared in several steppe graves and in settlements of the early Sredni Stog culture (Strilcha Skelya, Aleksandriya, Yama). The Svobodnoe-era settlements in the Kuban River valley participated in the eastern fringe of the steppe Suvorovo-Novodanilovka activities around 4000 BCE.

THE MAIKOP CULTURE

The shift from Svobodnoe to Maikop was accompanied by a sudden change in funeral customs—the clear and widespread adoption of kurgan graves—but there was continuity in settlement locations and settlement types, lithics, and some aspects of ceramics. Early Maikop ceramics showed some similarities with Svobodnoe pot shapes and clay fabrics, and some similarities with the ceramics of the Early Trans-Caucasian (ETC) culture south of the North Caucasus Mountains. These analogies indicate that Maikop developed from local Caucasian origins. But some Maikop pots were wheel-made, a new technology introduced from the south, and this new method of manufacture probably encouraged new vessel shapes.

The Maikop chieftain's grave, discovered on the Belaya River, a tributary of the Kuban River, was the first Maikop-culture tomb to be excavated, and it remains the most important early Maikop site. When excavated in 1897 by N. I. Veselovskii, the kurgan was almost 11 m high and more than 100 m in diameter. The earthen center was surrounded by a cromlech of large undressed stones. Externally it looked like the smaller Mikhailovka I and Post-Mariupol kurgans (and, before them, the Suvorovo kurgans), which also had earthen mounds surrounded by stone cromlechs. Internally, however, the Maikop chieftan's grave was quite different. The grave chamber was more than 5 m long and 4 m wide, 1.5 m deep, and was lined with large timbers. It was divided by timber partitions into two northern chambers and one southern chamber. The two northern chambers each held an adult female, presumably sacrificed, each lying in a contracted position on her right side, oriented southwest, stained with red ochre, with one to four pottery vessels and wearing twisted silver foil ornaments.[25]

The southern chamber contained an adult male. He also probably was positioned on his right side, contracted, with his head oriented southwest, the pose of most Maikop burials. He also lay on ground deeply stained with red ochre. With him were eight red-burnished, globular pottery vessels, the type collection for Early Maikop; a polished stone cup with a sheet-gold cover; two arsenical bronze, sheet-metal cauldrons; two small cups of sheet gold; and fourteen sheet-silver cups, two of which were decorated with impressed scenes of animal processions including a Caucasian spotted panther, a southern lion, bulls, a horse, birds, and a shaggy animal (bear? goat?) mounting a tree (figure 12.10). The engraved horse is the oldest clear image of a post-glacial horse, and it looked like a modern Przewalski: thick neck, big head, erect mane, and thick, strong legs. The chieftan also had arsenical

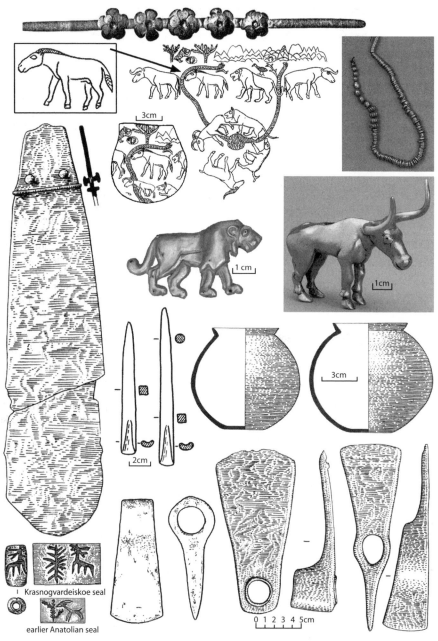

3cm

1 cm

1cm

3cm

2cm

Krasnogvardeiskoe seal

earlier Anatolian seal

0 1 2 3 4 5cm

Figure 12.10 Early Maikop objects from the chieftain's grave at Maikop, the State Hermitage Museum, St. Petersburg; and a seal at lower left from the early Maikop Krasnogvardeiskoe kurgan, with a comparative seal from Chalcolithic Degirmentepe in eastern Anatolia. The lion, bull, necklace, and diadem are gold; the cup with engraved design is silver; the two pots are ceramic; and

bronze tools and weapons. They included a sleeved axe, a hoe-like adze, an axe-adze, a broad spatula-shaped metal blade 47 cm long with rivets for the attachment of a handle, and two square-sectioned bronze chisels with round-sectioned butts. Beside him was a bundle of six (or possibly eight) hollow silver tubes about 1 m long. They might have been silver casings for a set of six (or eight) wooden staffs, perhaps for holding up a tent that shaded the chief. Long-horned bulls, two of solid silver and two of solid gold, were slipped over four of the silver casings through holes in the middle of the bulls, so that when the staffs were erect the bulls looked out at the visitor. Each bull figure was sculpted first in wax; very fine clay was then pressed around the wax figure; this clay was next wrapped in a heavier clay envelope; and, finally, the clay was fired and the wax burned off—the lost wax method for making a complicated metal-casting mold. The Maikop chieftain's grave contained the first objects made this way in the North Caucasus. Like the potter's wheel, the arsenical bronze, and the animal procession motifs engraved on two silver cups, these innovations came from the south.[26]

The Maikop chieftan was buried wearing Mesopotamian symbols of power—the lion paired with the bull—although he probably never saw a lion. Lion bones are not found in the North Caucasus. His tunic had sixty-eight golden lions and nineteen golden bulls applied to its surface. Lion and bull figures were prominent in the iconography of Uruk Mesopotamia, Hacinebi, and Arslantepe. Around his neck and shoulders were 60 beads of turquoise, 1,272 beads of carnelian, and 122 golden beads. Under his skull was a diadem with five golden rosettes of five petals each on a band of gold pierced at the ends. The rosettes on the Maikop diadem had no local prototypes or parallels but closely resemble the eight-petaled rosette seen in Uruk art. The turquoise almost certainly came from northeastern Iran near Nishapur or from the Amu Darya near the trade settlement of Sarazm in modern Tajikistan, two regions famous in antiquity for their turquoise. The red carnelian came from western Pakistan and the lapis lazuli from eastern Afghanistan. Because of the absence of cemeteries in Uruk Mesopotamia, we do not know much about the decorations worn there. The abundant personal ornaments at Maikop, many of them traded up the Euphrates through eastern Anatolia, probably were not made just for the barbarians. They provide an eye-opening glimpse of the kinds of styles that must have been seen in the streets and temples of Uruk.

Figure 12.10 (continued) the other objects are arsenical bronze. The bronze blade with silver rivets is 47 cm long and had sharp edges. After Munchaev 1994 and the Metropolitan Museum of Art, New York.

The Age and Development of the Maikop Culture

The relationship between Maikop and Mesopotamia was misunderstood until just recently. The extraordinary wealth of the Maikop culture seemed to fit comfortably in an age of ostentation that peaked around 2500 BCE, typified by the gold treasures of Troy II and the royal "death-pits" of Ur in Mesopotamia. But since the 1980s it has slowly become clear that the Maikop chieftain's grave probably was constructed about 3700–3400 BCE, during the Middle Uruk period in Mesopotamia—a thousand years *before* Troy II. The archaic style of the Maikop artifacts was recognized in the 1920s by Rostovtseff, but it took radiocarbon dates to prove him right. Rezepkin's excavations at Klady in 1979–80 yielded six radiocarbon dates averaging between 3700 and 3200 BCE (on human bone, so possibly a couple of centuries too old because of old carbon contamination from fish in the diet). These dates were confirmed by three radiocarbon dates also averaging between 3700 and 3200 BCE at the early Maikop-culture settlement of Galugai, excavated by S. Korenevskii between 1985 and 1991 (on animal bone and charcoal, so probably accurate). Galugai's pot types and metal types were exactly like those in the Maikop chieftain's grave, the type site for early Maikop. Graves in kurgan 32 at Ust-Dzhegutinskaya that were stylistically post-Maikop were radiocarbon dated about 3000–2800 BCE. These dates showed that Maikop was contemporary with the first cities of Middle and Late Uruk-period Mesopotamia, 3700–3100 BCE, an extremely surprising discovery.[27]

The radiocarbon dates were confirmed by an archaic cylinder seal found in an early Maikop grave excavated in 1984 at Krasnogvardeiskoe, about 60 km north of the Maikop chieftain's grave. This grave contained an east Anatolian agate cylinder seal engraved with a deer and a tree of life. Similar images appeared on stamp seals at Degirmentepe in eastern Anatolia before 4000 BCE, but cylinder seals were a later invention, appearing first in Middle Uruk Mesopotamia. The one from the kurgan at Kransogvardeiskoe (Red Guards), perhaps worn as a bead, is among the oldest of the type (see figure 12:10).[28]

The Maikop chieftain's grave is the type site for the early Maikop period, dated between 3700 and 3400 BCE. All the richest graves and hoards of the early period were in the Kuban River region, but the innovations in funeral ceremonies, arsenical bronze metallurgy, and ceramics that defined the Maikop culture were shared across the North Caucasus piedmont to the central plateau and as far as the middle Terek River valley. Galugai on the middle Terek River was an early Maikop settlement, with round houses

6–8 m in diameter scattered 10–20 m apart along the top of a linear ridge. The estimated population was less than 100 people. Clay, bell-shaped loom weights indicated vertical looms; four were found in House 2. The ceramic inventory consisted largely of open bowls (probably food bowls) and globular or elongated, round-bodied pots with everted rims, fired to a reddish color; some of these were made on a slow wheel. Cattle were 49% of the animal bones, sheep-goats were 44%, pigs were 3%, and horses (presumably horses that looked like the one engraved on the Maikop silver cup) were 3%. Wild boar and onagers were hunted only occasionally. Horse bones appeared in other Maikop settlements, in Maikop graves (Inozemstvo kurgan contained a horse jaw), and in Maikop art, including a frieze of nineteen horses painted in black and red colors on a stone wall slab inside a late Maikop grave at Klady kurgan 28 (figure 12.11). The widespread appearance of horse bones and images in Maikop sites suggested to Chernykh that horseback riding began in the Maikop period.[29]

The late phase of the Maikop culture probably should be dated about 3400–3000 BCE, and the radiocarbon dates from Klady might support this if they were corrected for reservoir effects. Having no ^{15}N measurements from Klady, I don't know if this correction is justified. The type sites for the late Maikop phase are Novosvobodnaya kurgan 2, located southeast of Maikop in the Farsa River valley, excavated by N. I. Veselovskii in 1898; and Klady (figure 12.11), another kurgan cemetery near Novosvobodnaya, excavated by A. D. Rezepkin in 1979–80. Rich graves containing metals, pottery, and beads like Novosvobodnaya and Klady occurred across the North Caucasus piedmont, including the central plateau (Inozemtsvo kurgan, near Mineralnyi Vody) and in the Terek drainage (Nalchik kurgan). Unlike the sunken grave chamber at Maikop, most of these graves were built on the ground surface (although Nalchik had a sunken grave chamber); and, unlike the timber-roofed Maikop grave, their chambers were constructed entirely of huge stones. In Novosvobodnaya-type graves the central and attendant/gift grave compartments were divided, as at Maikop, but the stone dividing wall was pierced by a round hole. The stone walls of the Nalchik grave chamber incorporated carved stone stelae like those of the Mikhailovka I and Kemi-Oba cultures (see figure 13.11).

Arsenical bronze tools and weapons were much more abundant in the richest late Maikop graves of the Klady-Novosvobodnaya type than they were in the Maikop chieftain's grave. Grave 5 in Klady kurgan 31 alone contained fifteen heavy bronze daggers, a sword 61 cm long (the oldest sword in the world), three sleeved axes and two cast bronze hammer-axes, among many other objects, for one adult male and a seven-year-old child

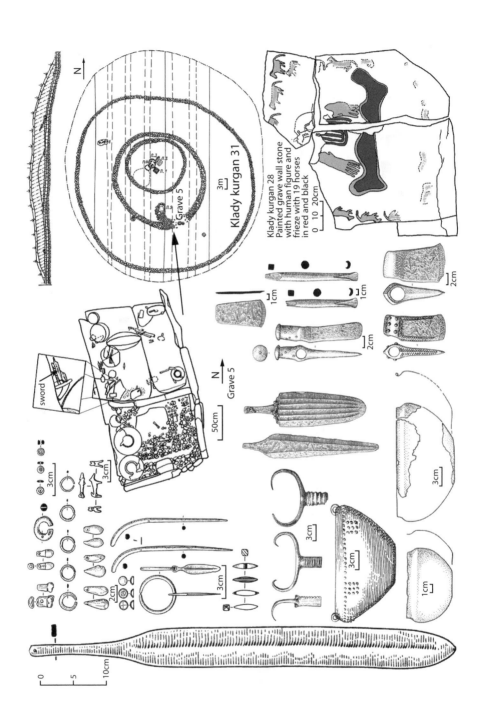

N

Klady kurgan 31

3m

Grave 5

Grave 5

N

50cm

sword

Klady kurgan 28
Painted grave wall stone
with human figure and
frieze with 19 horses
in red and black

0 10 20cm

1cm

1cm

2cm

2cm

3cm

3cm

3cm

3cm

3cm

3cm

3cm

2cm

2cm

1cm

0

5

10cm

(see figure 12.11). The bronze tools and weapons in other Novosvobodnaya-phase graves included cast flat axes, sleeved axes, hammer-axes, heavy tanged daggers with multiple midribs, chisels, and spearheads. The chisels and spearheads were mounted to their handles the same way, with round shafts hammered into four-sided contracting bases that fit into a V-shaped rectangular hole on the handle or spear. Ceremonial objects included bronze cauldrons, long-handled bronze dippers, and two-pronged bidents (perhaps forks for retrieving cooked meats from the cauldrons). Ornaments included beads of carnelian from western Pakistan, lapis lazuli from Afghanistan, gold, rock crystal, and even a bead from Klady made of a human molar sheathed in gold (the first gold cap!). Late Maikop graves contained several late metal types—bidents, tanged daggers, metal hammer-axes, and a spearhead with a tetrahedral tang—that did not appear at Maikop or in other early sites. Flint arrowheads with deep concave bases also were a late type, and black burnished pots had not been in earlier Maikop graves.[30]

Textile fragments preserved in Novosvobodnaya-type graves included linen with dyed brown and red stripes (at Klady), a cotton-like textile, and a wool textile (both at Novosvobodnaya kurgan 2). Cotton cloth was invented in the Indian subcontinent by 5000 BCE; the piece tentatively identified in the Novosvobodnaya royal grave might have been imported from the south.[31]

The Road to the Southern Civilizations

The southern wealth that defined the Maikop culture appeared suddenly in the North Caucasus, and in large amounts. How did this happen, and why?

Figure 12.11 Late Maikop-Novosvobodnaya objects and graves at Klady, Kuban River drainage, North Caucasus: (*Right*) plan and section of Klady kurgan 31 and painted grave wall from Klady kurgan 28 with frieze of red-and-black horses surrounding a red-and-black humanlike figure; (*left and bottom*): objects from grave 5, kurgan 31. These included (*left*) arsenical bronze sword; (*top row, center*) two beads of human teeth sheathed in gold, a gold ring, and three carnelian beads; (*second row*) five gold rings; (*third row*) three rock crystal beads and a cast silver dog; (*fourth row*) three gold button caps on wooden cores; (*fifth row*) gold ring-pendant and two bent silver pins; (*sixth row*) carved bone dice; (*seventh row*) two bronze bidents, two bronze daggers, a bronze hammer-axe, a flat bronze axe, and two bronze chisels; (*eighth row*) a bronze cauldron with repoussé decoration; (*ninth row*) two bronze cauldrons and two sleeved axes. After Rezepkin 1991, figures 1, 2, 4, 5, 6.

The valuables that seemed the most interesting to Mesopotamian urban traders were metals and precious stones. The upper Kuban River is a metal-rich zone. The Elbrusskyi mine on the headwaters of the Kuban, 35 km northwest of Elbruz Mountain (the highest peak in the North Caucasus) produces copper, silver, and lead. The Urup copper mine, on the upper Urup River, a Kuban tributary, had ancient workings that were visible in the early twentieth century. Granitic gold ores came from the upper Chegem River near Nalchik. As the metal prospectors who profited from the Uruk metal trade explored northward, they somehow learned of the copper, silver, and gold ores on the other side of the North Caucasus Mountains. Possibly they also pursued the source of textiles made of long-woolen thread.

It is possible that the initial contacts were made on the Black Sea coast, since the mountains are easy to cross between Maikop and Sochi on the coast, but much higher and more difficult in the central part of the North Caucasus farther east. Maikop ceramics have been found north of Sochi in the Vorontsovskaya and Akhshtyrskaya caves, just where the trail over the mountains meets the coast. This would also explain why the region around Maikop initially had the richest graves—if it was the terminal point for a trade route that passed through eastern Anatolia to western Georgia, up the coast to Sochi, and then to Maikop. The metal ores came from deposits located east of Maikop, so if the main trade route passed through the high passes in the center of the Caucasus ridge we would expect to see more southern wealth near the mines, not off to the west.

By the late Maikop (Novosvobodnaya) period, contemporary with Late Uruk, an eastern route was operating as well. Turquoise and carnelian beads were found at the walled town of Alikemek Tepesi in the Mil'sk steppe in Azerbaijan, near the mouth of the Kura River on the Caspian shore.[32] Alikemek Tepesi possibly was a transit station on a trade route that passed around the eastern end of the North Caucasian ridge. An eastern route through the Lake Urmia basin would explain the discovery in Iran, southwest of Lake Urmia, of a curious group of eleven conical, gravel-covered kurgans known collectively as Sé Girdan. Six of them, up to 8.2 m high and 60 m in diameter, were excavated by Oscar Muscarella in 1968 and 1970. Then thought to date to the Iron Age, they recently have been redated on the basis of their strong similarities to Novosvobodnaya-Klady graves in the North Caucasus.[33] The kurgans and grave chambers were made the same way as those of the Novosvobodnaya-Klady culture; the burial pose was the same; the arsenical bronze flat axes and short-nosed shaft-hole axes were similar in shape and manufacture to Novosvobodnaya-Klady types; and carnelian and gold beads were the same shapes, both containing silver vessels and fragments of silver tubes. The Sé Girdan kurgans could represent the migration southward of a Klady-type chief, perhaps to eliminate troublesome local middlemen. But the Lake Urmia chiefdom did not last. Moscarella counted almost ninety sites of the succeeding Early Trans-Caucasian Culture (ETC) around the southern Urmia Basin, but none of them had even small kurgans.

The power of the Maikop chiefs probably grew partly from the aura of the extraordinary that clung to the exotic objects they accumulated, which were palpable symbols of their personal connection with powers previously unknown.[34] Perhaps the extraordinary nature of these objects was one of the reasons why they were buried with their owners rather than inherited. Limited use and circulation were common characteristics of objects regarded as "primitive valuables." But the supply of new valuables dried up when the Late Uruk long-distance exchange system collapsed about 3100 BCE. Mesopotamian cities began to struggle with internal problems that we can perceive only dimly, their foreign agents retreated, and in the mountains the people of the ETC attacked and burned Arslantepe and Hacinebi on the upper Euphrates. Sé Girdan stood abandoned. This was also the end of the Maikop culture.

Maikop-Novosvobodnaya in the Steppes: Contacts with the North

Valuables of gold, silver, lapis, turquoise, and carnelian were retained exclusively by the North Caucasian individuals in direct contact with the south and perhaps by those who lived near the silver and copper mines that fed the southern trade. But a revolutionary new technology for land transport—wagons—might have been given to the steppes by the Maikop culture. Traces of at least two solid wooden disc wheels were found in a late Maikop kurgan on the Kuban River at Starokorsunskaya kurgan 2, with Novosvobodnaya black-burnished pots. Although not dated directly, the wooden wheels in this kurgan might be among the oldest in Europe.[35] Another Novosvobodnaya grave contained a bronze cauldron with a schematic image that seems to portray a cart. It was found at Evdik.

Evdik kurgan 4 was raised by the shore of the Tsagan-Nur lake in the North Caspian Depression, 350 km north of the North Caucasus piedmont, in modern Kalmykia.[36] Many shallow lakes dotted the Sarpa Depression, an ancient channel of the Volga. At Evdik, grave 20 contained an adult male in a contracted position oriented southwest, the standard Maikop pose, stained with red ochre, with an early Maikop pot by his feet. This was

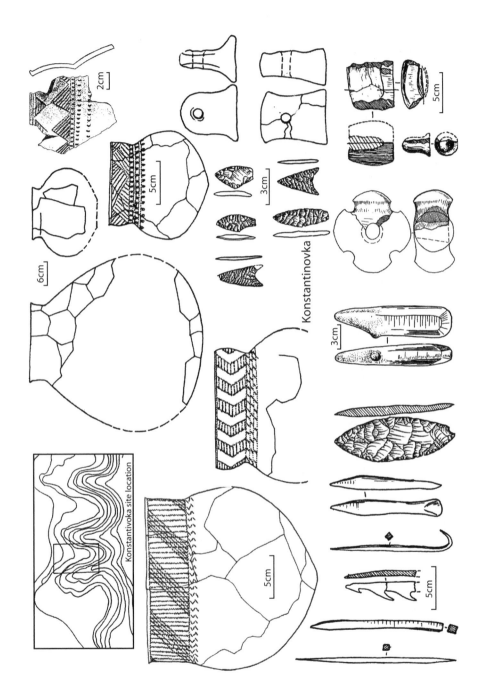

Konstantinovka

Konstantivoka site location

the original grave over which the kurgan was raised. Two other graves followed it, without diagnostic grave goods, after which grave 23 was dug into the kurgan. This was a late Maikop grave. It contained an adult male and a child buried together in sitting positions, an unusual pose, on a layer of white chalk and red ochre. In the grave was a bronze cauldron decorated with an image made in repoussé dots. The image seems to portray a yoke, a wheel, a vehicle body, and the head of an animal (see figure 4.3a). Grave 23 also contained a typical Novosvobodnaya bronze socketed bident, probably used with the cauldron. And it also had a bronze tanged dagger, a flat axe, a gold ring with 2.5 twists, a polished black stone pestle, a whetstone, and several flint tools, all typical Novosvobodnaya artifacts. Evdik kurgan 4 shows a deep penetration of the Novosvobodnaya culture into the lower Volga steppes. The image on the cauldron suggests that the people who raised the kurgan at Evdik also drove carts.

Evdik was the richest of the Maikop-Novosvobodnaya kurgans that appeared in the steppes north of the North Caucasus between 3700 and 3100 BCE. In such places, late Novosvobodnaya people whose speech would probably be assigned to a Caucasian language family met and spoke with individuals of the Repin and Late Khvalynsk cultures who probably spoke Proto-Indo-European dialects. The loans discussed in chapter 5 between archaic Caucasian and Proto-Indo-European languages probably were words spoken during these exchanges. The contact was most obvious, and therefore perhaps most direct, on the lower Don.

Trade across a Persistent Cultural Frontier

Konstantinovka, a settlement on the lower Don River, might have contained a resident group of Maikop people, and there were kurgan graves with Maikop artifacts around the settlement (figure 12.12). About 90% of the settlement ceramics were a local Don-steppe shell-tempered, cord-impressed type connected with the cultures of the Dnieper-Donets steppes to the west (late Sredni Stog, according to Telegin). The other 10% were red-burnished early Maikop wares. Konstantinovka was located on a steep-sided promontory overlooking the strategic lower Don valley, and was protected by a ditch and bank. The gallery forests below it were full of deer (31% of the bones) and

Figure 12.12 Konstantinovka settlement on the lower Don, with topographic location and artifacts. Plain pots are Maikop-like; cord-impressed pots are local. Loom-weights and asymmetrical flint points also are Maikop-like. *Lower right*: crucible and bellows fragments. After Kiashko 1994.

the plateau behind it was the edge of a vast grassland rich in horses (10%), onagers (2%), and herds of sheep/goats (25%). Maikop vistors probably imported the perforated clay loom weights similar to those at Galugai (unique in the steppes), copper chisels like those at Novosvobodnaya (again, unique except for two at Usatovo; see chapter 14), and asymmetrical shouldered flint projectile points very much like those of the Maikop-Novosvobodnaya graves. But polished stone axes and gouges, a drilled cruciform polished stone mace head, and boars-tusk pendants were steppe artifact types. Crucibles and slag show that copper working occurred at the site.

A. P. Nechitailo identified dozens of kurgans in the North Pontic steppes that contained single pots or tools or both that look like imports from Maikop-Novosvobodnaya, distributed from the Dniester River valley on the west to the lower Volga on the east. These widespread northern contacts seem to have been most numerous during the Novosvobodnaya/Late Uruk phase, 3350–3100 BCE. But most of the Caucasian imports appeared singly in local graves and settlements. The region that imported the largest number of Caucasian arsenical bronze tools and weapons was the Crimean Peninsula (the Kemi-Oba culture). The steppe cultures of the Volga-Ural region imported little or no Caucasian arsenical bronze; their metal tools and weapons were made from local "clean" copper. Sleeved, one-bladed metal axes and tanged daggers were made across the Pontic-Caspian steppes in emulation of Maikop-Novosvobodnaya types, but most were made locally by steppe metalsmiths.[37]

What did the Maikop chiefs want from the steppes? One possibility is drugs. Sherratt has suggested that narcotics in the form of *Cannabis* were one of the important exports of the steppes.[38] Another more conventional trade item could have been wool. We still do not know where wool sheep were first bred, although it makes sense that northern sheep from the coldest places would initially have had the thickest wool. Perhaps the Maikop-trained weavers at Konstantinovka were there with their looms to make some of the raw wool into large textiles for payment to the herders. Steppe people had felts or textiles made from narrow strips of cloth, produced on small, horizontal looms, then stitched together. Large textiles made in one piece on vertical looms were novelties.

Another possibility is horses. In most Neolithic and earlier Eneolithic sites across Transcaucasia there were no horse bones. After the evolution of the ETC culture beginning about 3300 BCE horses became widespread, appearing in many sites across Transcaucasia. S. Mezhlumian reported horse bones at ten of twelve examined sites in Armenia dated to the later fourth millennium BCE. At Mokhrablur one horse had severe wear on a P_2 consistent with bit wear. Horses were bitted at Botai and Kozhai 1 in

Kazakhstan during the same period, so bit wear at Mokhrablur would not be unique. At Alikemek Tepesi the horses of the ETC period were thought by Russian zoologists to be domesticated. Horses the same size as those of Dereivka appeared as far south as the Malatya-Elazig region in southeastern Turkey, as at Norşuntepe; and in northwestern Turkey at Demirci Höyük. Although horses were not traded into the lowlands of Mesopotamia this early, they might have been valuable in the steppe-Caucasian trade.[39]

PROTO-INDO-EUROPEAN AS A REGIONAL LANGUAGE IN A CHANGING WORLD

During the middle centuries of the fourth millennium BCE the equestrian tribes of the Pontic-Caspian steppes exhibited a lot of material and probably linguistic variability. They absorbed into their conversations two quite different but equally surprising developments among their neighbors to the south, in the North Caucasus piedmont, and to their west, in the Cucuteni-Tripolye region. From the North Caucasus probably came wagons, and with them ostentatious displays of incredible wealth. In the west, some Tripolye populations retreated into huge planned towns larger than any settlements in the world, probably in response to raiding from the steppes. Other Tripolye towns farther north on the Dnieper began to change their customs in ceramics, funerals, and domestic architecture toward steppe styles in a slow process of assimilation.

Although regionally varied, steppe cultural habits and customs remained distinct from those of the Maikop culture. An imported Maikop or Novosvobodnaya potsherd is immediately obvious in a steppe grave. Lithics and weaving methods were different (no loom weights in the steppes), as were bead and other ornament types, economies and settlement forms, and metal types and sources. These distinctions persisted in spite of significant cross-frontier interaction. When Maikop traders came to Konstantinovka, they probably needed a translator.

The Yamnaya horizon, the material expression of the late Proto-Indo-European community, grew from an eastern origin in the Don-Volga steppes and spread across the Pontic-Caspian steppes after about 3300 BCE. Archaeology shows that this was a period of profound and rapid change along all the old ethnolinguistic frontiers surrounding the Pontic-Caspian steppes. Linguistically based reconstructions of Proto-Indo-European society often suggest a static, homogeneous ideal, but archaeology shows that Proto-Indo-European dialects and institutions spread through steppe societies that exhibited significant regional diversity, during a period of far-reaching social and economic change.

CHAPTER THIRTEEN

Wagon Dwellers of the Steppe
The Speakers of Proto-Indo-European

The sight of wagons creaking and swaying across the grasslands amid herds of wooly sheep changed from a weirdly fascinating vision to a normal part of steppe life between about 3300 and 3100 BCE. At about the same time the climate in the steppes became significantly drier and generally cooler than it had been during the Eneolithic. The shift to drier conditions is dated between 3500 and 3000 BCE in pollen cores in the lower Don, the middle Volga, and across the northern Kazakh steppes (table 13.1). As the steppes dried and expanded, people tried to keep their animal herds fed by moving them more frequently. They discovered that with a wagon you could keep moving indefinitely. Wagons and horseback riding made possible a new, more mobile form of pastoralism. With a wagon full of tents and supplies, herders could take their herds out of the river valleys and live for weeks or months out in the open steppes between the major rivers—the great majority of the Eurasian steppes. Land that had been open and wild became pasture that belonged to someone. Soon these more mobile herding clans realized that bigger pastures and a mobile home base permitted them to keep bigger herds. Amid the ensuing disputes over borders, pastures, and seasonal movements, new rules were needed to define what counted as an acceptable move—people began to manage local migratory behavior. Those who did not participate in these agreements or recognize the new rules became cultural Others, stimulating an awareness of a distinctive Yamnaya identity. That awareness probably elevated a few key behaviors into social signals. Those behaviors crystallized into a fairly stable set of variants in the steppes around the lower Don and Volga rivers. A set of dialects went with them, the speech patterns of late Proto-Indo-European. This is the sequence of changes that I believe created the new way of life expressed archaeologically in the Yamnaya horizon, dated about 3300–2500 BCE (figure 13.1). The spread

TABLE 13.1

Vegetation shifts in steppe pollen cores from the Don to the Irtysh

Site	Razdorskoe, *Lower Don* (Kremenetski 1997)	Buzuluk Forest Pobochnoye peat bog *Middle Volga* (Kremenetski et al. 1999)	Northern Kazakhstan Upper Tobol to *Upper Irtysh* (Kremenetski et al. 1997)
Type	Stratified settlement Pollen core	forest peat bog core	two lake cores and two peat bog cores
Dates Flora	*6500–3800 BCE* Birch-pine forest on sandy river terraces. On floodplain, elm and linden forest with hazelnut & black alder. Oak and hornbeam present after 4300 BCE.	*6000–3800 BCE* Oak trees appear, join elm, hazel, black alder forests around Pobochnoye lake. 4800–3800 BCE lake gets shallower, Typha reeds increase, forest expands.	*6500–3800 BCE* Birch-pine forest evolving to open pine forest in forest-steppe, with willow near waterways. In steppe, Artemesia and Chenopodia.
	3800–3300 BCE Slight reduction in deciduous trees, increase in Ephedra, hazel, lime, and pine on floodplain.	*3800–3300 BCE* Lake slowly converts to sedge-moss swamp. Typha reeds peak. Pine and lime trees peak. Probably warmer.	*3800–3300 BCE* Moist period, forests expand. Lime trees with oak, elm, and black alder also expand. Soils show increased moisture.
	Sub-Boreal 3300–2000 BCE Very dry. Sharp forest decline. Ceralia appears. Chenopodia sharp rise. Maximum aridity 2800–2000 BCE.	*3300–2000 BCE* Reduction in overall forest. In forest, pine down, birch up. Artemesia, an arid herb indicator, increases sharply. Lake is covered by alder shrubs by 2000 BCE.	*3300–2000 BCE* Forest retreats, broadleaf declines. Mokhove bog on the Tobol dries up about 2800 BCE. Steppe grows.

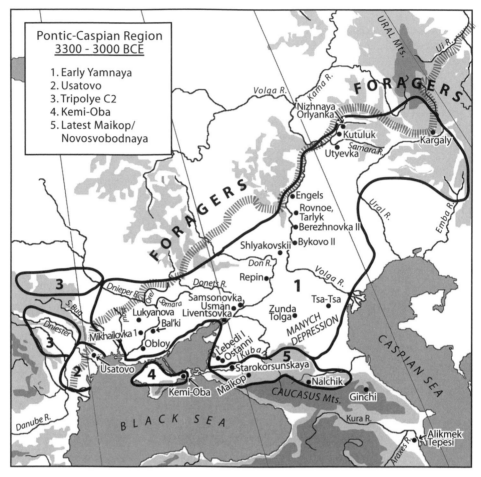

Figure 13.1 Culture areas in the Pontic-Caspian region about 3300–3000 BCE.

of the Yamnaya horizon was the material expression of the spread of late Proto-Indo-European across the Pontic-Caspian steppes.[1]

The behavior that really set the Yamnaya people apart was living on wheels. Their new economy took advantage of two kinds of mobility: wagons for slow bulk transport (water, shelter, and food) and horseback riding for rapid light transport (scouting for pastures, herding, trading and raiding expeditions). Together they greatly increased the potential scale of herding economies. Herders operating out of a wagon could stay with their herds out in the deep steppes, protected by mobile homes that carried tents, water, and food. A diet of meat, milk, yogurt, cheese, and soups made of wild *Chenopodium* seeds and wild greens can be deduced, with a little imagination, from the archaeological evidence. The reconstructed Proto-Indo-European vocabulary tells us that honey and honey-based mead also were consumed, probably on special occasions. Larger herds meant greater disparities in herd wealth, which is reflected in disparities in the wealth of Yamanaya graves. Mobile wagon camps are almost impossible to find archaeologically, so settlements became archaeologically invisible where the new economy took hold.

The Yamnaya horizon is the visible archaeological expression of a social adjustment to high mobility—the invention of the political infrastructure to manage larger herds from mobile homes based in the steppes. A linguistic echo of the same event might be preserved in the similarity between English *guest* and *host*. They are cognates, derived from one Proto-Indo-European root (*ghos-ti-*). (A "ghost" in English was originally a visitor or guest.) The two social roles opposed in English *guest* and *host* were originally two reciprocal aspects of the same relationship. The late Proto-Indo-European guest-host relationship required that "hospitality" (from the same root through Latin *hospes* 'foreigner, guest') and "friend-ship" (*keiwos-*) should be extended by hosts to guests (both *ghos-ti-*), in the knowledge that the receiver and giver of "hospitality" could later reverse roles. The social meaning of these words was then more demanding than modern customs would suggest. The guest-host relationship was bound by oaths and sacrifices so serious that Homer's warriors, Glaukos and Diomedes, stopped fighting and presented gifts to each other when they learned that their *grandfathers* had shared a guest-host relationship. This mutual obligation to provide "hospitality" functioned as a bridge between social units (tribes, clans) that had ordinarily restricted these obligations to their kin or co-residents (*$h_4erós$-*). Guest-host relationships would have been very useful in a mobile herding economy, as a way of separating people who were moving through your territory with your assent from those who were unwelcome, unregulated, and therefore unprotected. The guest-host institution might have been among the critical identity-defining innovations that spread with the Yamnaya horizon.[2]

It is difficult to document a shift to a more mobile residence pattern five thousand years after the fact, but a few clues survive. Increased mobility can be detected in a pattern of brief, episodic use, abandonment, and, much later, re-use at many Yamnaya kurgan cemeteries; the absence of degraded or overgrazed soils under early Yamnaya kurgans; and the first appearance of kurgan cemeteries in the deep steppe, on the dry plateaus between major river valleys. The principal indicator of increased mobility is a negative piece of evidence: the archaeological disappearance of long-term settlements east of the Don River. Yamnaya settlements are known west of the Don in Ukraine, but east of the Don in Russia there are no significant Yamnaya settlements in a huge territory extending to the Ural River containing many hundreds of excavated Yamnaya kurgan cemeteries and probably thousands of excavated Yamnaya graves (I have never seen a full count). The best explanation for the complete absence of settlements is that the eastern Yamnaya people spent much of their lives in wagons.

The Yamnaya horizon was the first more or less unified ritual, economic, and material culture to spread across the entire Pontic-Caspian steppe region, but it was never completely homogeneous even materially. At the beginning it already contained two major variants, on the lower Don and lower Volga, and, as it expanded, it developed other regional variants, which is why most archaeologists are reluctant to call it the Yamnaya "culture." But many broadly similar customs were shared. In addition to kurgan graves, wagons, and an increased emphasis on pastoralism, archaeological traits that defined the early Yamnaya horizon included shell-tempered, egg-shaped pots with everted rims, decorated with comb stamps and cord impressions; tanged bronze daggers; cast flat axes; bone pins of various types; the supine-with-raised-knees burial posture; ochre staining on grave floors near the feet, hips, and head; northeastern to eastern body orientation (usually); and the sacrifice at funerals of wagons, carts, sheep, cattle, and horses. The funeral ritual probably was connected with a cult of ancestors requiring specific rituals and prayers, a connection between language and cult that introduced late Proto-Indo-European to new speakers.

The most obvious material division within the early Yamnaya horizon was between east and west. The eastern (Volga–Ural–North Caucasian steppe) Yamnaya pastoral economy was more mobile than the western one (South Bug–lower Don). This contrast corresponds in an intriguing way to economic and cultural differences between eastern and western Indo-European language branches. For example, impressions of cultivated grain have been found in western Yamnaya pottery, in both settlements and graves, and Proto-Indo-European cognates related to cereal agriculture were well preserved in western Indo-European vocabularies. But grain imprints are absent in eastern Yamnaya pots, just as many of the cognates related to agriculture are missing from the eastern Indo-European languages.[3] Western Indo-European vocabularies contained a few roots that were borrowed from Afro-Asiatic languages, such as the word for the domesticated bull, *tawr-*, and the western Yamnaya groups lived next to the Tripolye culture, which might have spoken a language distantly derived from an Afro-Asiatic language of Anatolia. Eastern Indo-European generally lacked these borrowed Afro-Asiatic roots. Western Indo-European religious and ritual practices were female-inclusive, and western Yamnaya people shared a border with the female-figurine–making Tripolye culture: eastern Indo-European rituals and gods, however, were more male-centered, and eastern Yamnaya people shared borders with northern and eastern foragers who did not make female figurines. In western Indo-European branches the spirit of the domestic hearth was female (Hestia, the Vestal Virgins), and in Indo-Iranian it was male (Agni). Western Indo-European mythologies included strong female deities such as Queen Magb and the Valkyries, whereas in Indo-Iranian the furies of war were male Maruts. Eastern Yamnaya graves on the Volga contained a higher percentage (80%) of males than any other Yamnaya region. Perhaps this east-west tension in attitudes toward gender contributed to the separation of the feminine gender as a newly marked grammatical category in the dialects of the Volga-Ural region, one of the innovations that defined Proto-Indo-European grammar.[4]

Did the Yamnaya horizon spread into neighboring regions in a way that matches the known relationships and sequencing between the Indo-European branches? This also is a difficult subject to follow archaeologically, but the movements of the Yamnaya people match what we would expect surprisingly well. First, just before the Yamnaya horizon appeared, the Repin culture of the Volga-Ural region threw off a subgroup that migrated across the Kazakh steppes about 3700–3500 BCE and established itself in the western Altai, where it became the Afanasievo culture. The separation of the Afanasievo culture from Repin probably represented the separation of Pre-Tocharian from classic Proto-Indo-European. Second, some three to five centuries later, about 3300 BCE, the rapid diffusion of the early Yamnaya horizon across the Pontic-Caspian steppes scattered the speakers of late Proto-Indo-European dialects and sowed the seeds of regional differentiation. After a pause of only a century or two, about 3100–3000 BCE, a large migration stream erupted from within the western Yamnaya region and flowed up the Danube valley and into the Carpathian Basin during the Early Bronze Age. Literally thousands of kurgans can be assigned to this event, which could reasonably have incubated the ancestral dialects for several western Indo-European language branches, including Pre-Italic and Pre-Celtic. After this movement slowed or stopped, about 2800–2600 BCE, late Yamnaya people came face to face with people who made Corded Ware tumulus cemeteries in the east Carpathian foothills, a historic meeting through which dialects ancestral to the northern Indo-European languages (Germanic, Slavic, Baltic) began to spread among eastern Corded Ware groups. Finally, at the end of the Middle Bronze Age, about 2200–2000 BCE, a migration stream flowed from the late Yamnaya/Poltavka cultures of the Middle Volga–Ural region eastward around the southern Urals, creating the Sintashta culture, which almost certainly represented the ancestral Indo-Iranian–speaking community. These migrations are described in chapters 14 and 15.

The Yamnaya horizon meets the expectations for late Proto-Indo-European in many ways: chronologically (the right time), geographically (the right place), materially (wagons, horses, animal sacrifices, tribal pastoralism), and linguistically (bounded by persistent frontiers); and it generated migrations in the expected directions and in the expected sequence. Early Proto-Indo-European probably developed between 4000 and 3500 BCE in the Don–Volga–Ural region. Late Proto-Indo-European, with o-stems and the full wagon vocabulary, expanded rapidly across the Pontic-Caspian steppes with the appearance of the Yamnaya horizon beginning about 3300 BCE. By 2500 BCE the Yamnaya horizon had fragmented into daughter groups, beginning with the appearance of the Catacomb culture in the Don-Kuban region and the Poltavka culture in the Volga-Ural region about 2800 BCE. Late Proto-Indo-European also was so diversified by 2500 BCE that it probably no longer existed (chapter 3). Again, the linkage with the steppe archaeological evidence is compelling.

Why Not a Kurgan Culture?

Marija Gimbutas first articulated her concept of a "Kurgan culture" as the archaeological expression of the Proto-Indo-European language community in 1956.[5] The Kurgan culture combined two cultures first defined by V. A. Gorodtsov, who, in 1901, excavated 107 kurgans in the Don River valley. He divided his discoveries into three chronological groups. The oldest graves, stratified deepest in the oldest kurgans, were the Pit-graves (Yamnaya). They were followed by the Catacomb-graves (Katakombnaya), and above them were the timber-graves (Srubnaya). Gorodtsov's sequence still defines the Early (EBA), Middle (MBA), and Late Bronze Age (LBA) grave types of the western steppes.[6] Gimbutas combined the first two (EBA Pit-graves and MBA Catacomb-graves) into the Kurgan culture. But later she also began to include many other Late Neolithic and

Bronze Age cultures of Europe, including the Maikop culture and many of the Late Neolithic cultures of eastern Europe, as outgrowths or creations of Kurgan culture migrations. The Kurgan culture was so broadly defined that almost any culture with burial mounds, or even (like the Baden culture) without them could be included. Here we are discussing the steppe cultures of the Russian and Ukrainian EBA, just one part of the original core of Gimbutas's Kurgan culture concept. Russian and Ukrainian archaeologists do not generally use the term "Kurgan culture"; rather than lumping EBA Yamnaya and MBA Catacomb-graves together they tend to divide both groups and their associated time periods into ever finer slices. I will seek a middle ground.

The Yamnaya horizon is usually described by Slavic archaeologists not as a "culture" but as a "cultural-historical community." This phrase carries the implication that there was a thread of cultural identity or shared ethnic origin running through the Yamnaya social world, although one that diversified and evolved with the passage of time.[7] Although I agree that this probably was true in this case, I will use the Western term "horizon," which is neutral about cultural identity, in order to avoid using a term loaded toward that interpretation. As I explained in chapter 7, a horizon in archaeology is a style or fashion in material culture that is rapidly accepted by and superimposed on local cultures across a wide area. In this case, the five Pontic-Caspian cultures of the Final Eneolithic (chapter 12) were the local cultures that rapidly accepted, in varying degrees, the Yamnaya lifestyle.

Beyond the Eastern Frontier: The Afanasievo Migration to the Altai

In the last chapter I introduced the subject of the trans-continental, Repin-culture migration that created the Afanasievo culture in the western Altai Mountains and probably detached the Tocharian branch from common Proto-Indo-European. I describe it here because the process of migration and return migration that installed the early Afanasievo culture continued across the north Kazakh steppes during the Yamnaya period. In fact, it is usually discussed as an event connected with the Yamnaya horizon; it is only recently that early Afanasievo radiocarbon dates, and the broadening understanding of the age and geographic extent of the Repin culture, have pushed the beginning of the movement back into the pre-Yamnaya Repin period.

Two or three centuries before the Yamnaya horizon first appeared, the Repin-type communities of the middle Volga-Ural steppes experienced a

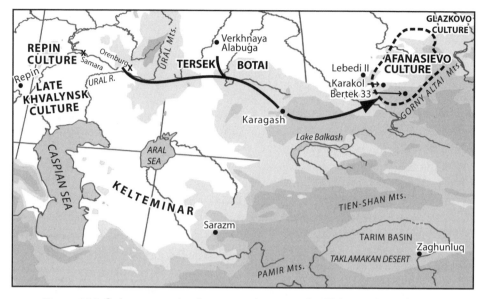

Figure 13.2 Culture areas in the steppes between the Volga and the Altai at the time of the Afanasievo migration, 3700–3300 BCE.

conflict that prompted some groups to move across the Ural River eastward into the Kazakh steppes (figure 13.2). I say a conflict because of the extraordinary distance the migrants eventually put between themselves and their relatives at home, implying a strongly negative push. On the other hand, connections with the Volga-Ural Repin-Yamnaya world were maintained by a continuing round of migrations moving in both directions, so some aspect of the destination must also have exerted a positive pull. It is remarkable that the intervening north Kazakh steppe was not settled, or at least that almost no kurgan cemeteries were constructed there. Instead, the indigenous horse-riding Botai-Tersek culture emerged in the north Kazakh steppe at just the time when the Repin-Afanasievo migration began.

The specific ecological target in this series of movements might have been the islands of pine forest that occur sporadically in the northern Kazakh steppes from the Tobol River in the west to the Altai Mountains in the east. I am not sure why these pine islands would have been targeted other than for the fuel and shelter they offered, but they do seem to correspond with the few site locations linked to Afanasievo in the steppes, and the same peculiar steppe-pine-forest islands occur also in the high mountain valleys of the western Altai where early Afanasievo sites appeared.[8] In the western Altai Mountains broad meadows and mountain steppes dip both westward toward the Irtysh River of western Siberia (probably the route of the first approach) and northward toward the Ob and Yenisei rivers (the later spread). The Afanasievo culture appeared in this beautiful setting, ideal for upland pastoralism, probably around 3700–3400 BCE, during the Repin–late Khvalynsk period.[9] It flourished there until about 2400 BCE, through the Yamnaya period in the Pontic-Caspian steppes.

The Altai Mountains were about 2000 km east of the Ural River frontier that defined the eastern edge of the early Proto-Indo-European world. Only three kurgan cemeteries old enough to be connected with the Afanasievo migrations have been found in the intervening 2000 km of steppes. All three are classified as Yamnaya kurgan cemeteries, although the pottery in some of the graves has Repin traits. Two were on the Tobol, not far east of the Ural River, at Ubagan I and Verkhnaya Alabuga, possibly an initial stopping place. The other, the Karagash kurgan cemetery, was found 1000 km east of the Tobol, southeast of Karaganda in central Kazakhstan. Karagash was on the elevated green slopes of an isolated mountain spur that rose prominently above the horizon, a very visible landmark near Karkaralinsk. The earthen mound of kurgan 2 at Karagash was 27 m in diameter. It covered a stone cromlech circle 23 m in diameter, made of oblong stones 1 m in length, projecting about 60–70 cm above the ground. Some stones had traces of paint on them. A pot was broken inside the southwestern edge of the cromlech on the original ground surface, before the mound was built. The kurgan contained three graves in stone-lined cists; the central grave and another under the southeastern part of the kurgan were later robbed. The lone intact grave was found under the northeastern part of the kurgan. In it were sherds from a shell-tempered pot, a fragment of a wooden bowl with a copper-covered lip, a tanged copper dagger, a copper four-sided awl, and a stone pestle. The skeleton was of a male forty to fifty years old laid on his back with his knees raised, oriented southwest, with pieces of black charcoal and red ochre on the grave floor. The metal artifacts were typical for the Yamnaya horizon; the stone cromlech, stone-lined cist, and pot were similar to Afansievo types. Directly east of Karagash and 900 km away, up the Bukhtarta River valley east of the Irtysh, were the peaks of the western Altai and the Ukok plateau, where the first Afanasievo graves appeared. The Karagash kurgan is unlikely to be a grave of the first migrants—it looks like a Yamnaya-Afanasievo kurgan built by later people still participating in a cross-Kazakhstan circulation of movements—but it probably does mark the initial route, since routes in long-distance migrations tend to be targeted and re-used.[10]

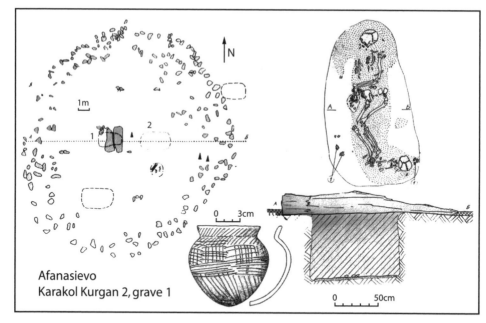

Figure 13.3 Karakol kurgan 2, grave 1, an early Afanasievo grave in the western Gorny Altai. After Kubarev 1988.

The early Afanasievo culture in the Altai introduced fully developed kurgan funeral rituals and Repin-Yamnaya material culture. At Karakol, kurgan 2 in the Gorny Altai, an early Afanasievo grave (gr. 1) contained a small pot similar to pots from the Ural River that are assigned to the Repin variant of early Yamnaya (figure 13.3).[11] Grave 1 was placed under a low kurgan in the center of a stone cromlech 20 m in diameter. Afanasievo kurgans always were marked by a ring of stones, and large stone slabs were used to cover grave pits (early) or to make stone-lined grave cists (late). Early Afanasievo skull types resembled those of Yamnaya and western populations. On the Ukok plateau, where the early Afanasievo cemetery at Bertek 33 was found, the Afanasievo immigrants occupied a virgin landscape—there were no earlier Mesolithic or Neolithic sites. Afanasievo sites also contained the earliest bones of domesticated cattle, sheep, and horses in the Altai. At the Afanasievo settlement of Balyktyul, domesticated sheep-goat were 61% of the bones, cattle were 12%, and horses 8%.[12]

Cemeteries of the local Kuznetsk-Altai foragers like Lebedi II were located in the forest and forest-meadow zone higher up on the slopes of the Altai, and contained a distinct set of ornaments (bear-teeth necklaces and bone carvings of elk and bear), lithics (asymmetrical curved flint knives), antler tools (harpoons), pottery (related to the Serovo-Glazkovo pottery tradition of the Baikal forager tradition), and funeral rituals (no kurgans, no stone slab over the grave). As time passed, Glazkovo forager sites located to the northeast began to show the influence of Afanasievo motifs on their ceramics, and metal objects began to appear in Glazkovo sites.[13]

It is clear that populations continued to circulate between the Ural frontier and the Altai well into the Yamnaya period in the Ural steppes, or after 3300 BCE, bringing many Yamnaya traits and practices to the Altai. About a hundred metal objects have been found in Afanasievo cemeteries in the Altai and Western Sayan Mountains, including three sleeved copper axes of a classic Volga-Ural Yamnaya type, a cast shaft-hole copper hammer-axe, and two tanged copper daggers of typical Yamnaya type. These artifacts are recognized by Chernykh as western types typical of Volga-Ural Yamnaya, with no native local precedents in the Altai region.[14]

Mallory and Mair have argued at book length that the Afanasievo migration detached the Tocharian branch from Proto-Indo-European. A material bridge between the Afanasievo culture and the Tarim Basin Tocharians could be represented by the long-known but recently famous Late Bronze Age Europoid "mummies" (not intentionally mummified but naturally freeze-dried) found in the northern Taklamakan Desert, the oldest of which are dated 1800–1200 BCE. In addition to the funeral ritual (on the back with raised knees, in ledged and roofed grave pits), there was a symbolic connection. On the stone walls of Late Afanasievo graves in the Altai (perhaps dated about 2500 BC) archaeologist V. D. Kubarev found paintings with "solar signs" and headdresses like the one painted on the cheek of one of the Tarim "mummies" found at Zaghunluq, dated about 1200 BCE. If Mallory and Mair were right, as seems likely, late Afanasievo pastoralists were among the first to take their herds from the Altai southward into the Tien Shan; and after 2000 BCE their descendants crossed the Tien Shan into the northern oases of the Tarim Basin.[15]

WAGON GRAVES IN THE STEPPES

We cannot say exactly when wagons first rolled into the Eurasian steppes. But an image of a wagon on a clay cup is securely dated to 3500–3300 BCE at Bronocice in southern Poland (chapter 4). The ceramic wagon models of the Baden culture in Hungary and the Novosvobodnaya wagon grave at Starokorsunskaya kurgan 2 on the Kuban River in the North Caucasus probably are about the same age. The oldest excavated wagon graves in the steppes are radiocarbon dated about 3100–3000 BCE, but it is unlikely that they actually were the first. Wagons probably appeared in the Pontic-Caspian steppes a couple of centuries before the Yamnaya horizon began. It would have taken some time for a new, wagon-dependent herding system to get organized and begin to succeed. The spread of the Yamnaya horizon was the signature of that success.

In a book published in 2000 Aleksandr Gei counted 257 Yamnaya and Catacomb-culture wagon and cart burials in the Pontic-Caspian steppes, dated by radiocarbon between about 3100 and 2200 BCE (see figures 4.4, 4.5, 4.6). Parts of wagons and carts were deposited in less than 5% of excavated Yamnaya-Catacomb graves, and the few graves that had them were concentrated in particular regions. The largest cluster of wagon-graves (120) was in the Kuban steppes north of the North Caucasus, not far from Maikop. Most of the Kuban wagons (115) were in graves of the Novotitorovskaya type, a local Kuban-region EBA culture that developed from early Yamnaya.[16]

Usually the vehicles used in funeral rituals were disassembled and the wheels were placed near the corners of the grave pit, as if the grave itself represented the wagon. But a whole wagon was buried west of the Dnieper in the Yamnaya grave at Lukyanova kurgan, grave 1; and whole wagons were found under nine Novotitorovskaya kurgans in the Kuban steppes. Many construction details can be reconstructed from these ten cases. All ten wagons had a fixed axle and revolving wheels. The wheels were made of two or three planks doweled together and cut in a circular shape about 50–80 cm in diameter. The wagon bed was about 1 m wide and 2–2.5 m long, and the gauge or track width between the wheels was 1.5–1.65 m. The Novotitorovskaya wagon at Lebedi kurgan 2, grave 116, is reconstructed by Gei with a box seat for the driver, supported on a cage of vertical struts doweled into a rectangular frame. Behind the driver was the interior of the wagon, the floor of which was braced with X-crossed planks (like the repoussé image on the Novosvobodnaya bronze cauldron from the Evdik kurgan) (see figure 4.3a). The Lukyanovka wagon frame also was braced with X-crossed planks. The passengers and cargo were protected under a "tilt," a wagon cover made of reed mats painted with red, white, and black stripes and curved designs, possibly sewn to a backing of felt. Similar painted reed mats with some kind of organic backing were placed on the floors of Yamnaya graves (figure 13.4).[17]

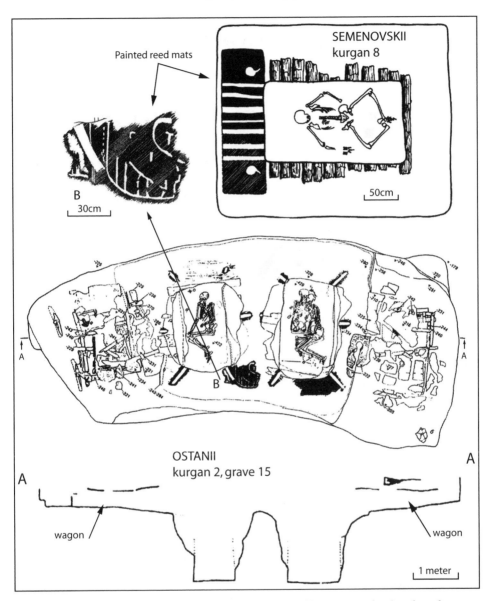

Figure 13.4 Painted reed mats in graves of the Yamnaya and related traditions. Top: Semenovskii kurgan 8, grave 9, late Yamnaya, lower Dniester steppes; bottom, Ostanni kurgan 2, double grave 15 with two wagons, Novotitorovskaya culture, Kuban River steppes. After Subbotin 1985, figure 7.7; and Gei 2000.

Table 13.3

Selected Radiocarbon Dates associated with the Afanasievo Migration and the Yamnaya Horizon

Lab nnumber	BP date	Sample	Calibrated date

1. Afanasievo culture, Altai Mountains (from Parzinger 2002, Figure 10)

Unidentified sites

Bln4764	4409±70	?	3310–2910 BCE
Bln4765	4259±36	?	2920–2780 BCE
Bln4767	4253±36	?	2920–3780 BCE
Bln4766	4205±44	?	2890–2690 BCE
Bln4769	4022+40	?	2580–2470 BCE
Bln4919	3936±35	?	2490–2340 BCE

Kara-Koba I enclosure 3

?	5100±50	?	3970–3800 BCE

Elo-bashi enclosure 5

?	4920±50	?	3760–3640 BCE

2. Yamnaya horizon kurgan cemeteries with multiple kurgans built together and long gaps between construction phases

A. Yamnaya horizon cemeteries in Ukraine (from Telegin et al. 2003)

Avgustnivka cemetery

Phase 1	Ki2118	4800±55	k 1/gr2	3650–3520 BCE
Phase 2	Ki7110	4130±55	k 5/gr2	2870–2590 BCE
	Ki7111	4190±60	k 4/gr2	2890–2670 BCE
	Ki7116	4120±60	k 4/gr1	2870–2570 BCE

Verkhnetarasovka cemetery

Phase 1	Ki602	4070±120	k 9/18	2870–2460 BCE
	Ki957	4090±95	k 70/13	2870–2490 BCE
Phase 2	Ki581	3820±190	k 17/3	2600–1950 BCE
	Ki582	3740±150	k 21/11	2400–1940 BCE

Vinogradnoe cemetery

Phase 1	Ki9414	4340±70	k 3/10	3090–2880 BCE
Phase 2	Ki9402	3970±70	k 3/25	2580–2340 BCE
	Ki987	3950±80	k 2/11	2580–2300 BCE
	Ki9413	3930±70	k 24/37	2560–2300 BCE

TABLE 13.3 (*continued*)

Lab number		BP date	Sample	Calibrated date
Golovkovka cemetery				
Phase 1	Ki6722	3980±60	k 7/4	2580–2350 BCE
	Ki6719	3970±55	k 6/8	2580–2350 BCE
	Ki6730	3960±60	k 5/3	2570–2350 BCE
	Ki6724	3950±50	k 12/3	2560–2340 BCE
	Ki6729	3920±50	k 14/9	2560–2340 BCE
	Ki6727	3910±15	k 14/2	2460–2350 BCE
	Ki6728	3905±55	k 14/7	2470–2300 BCE
	Ki6721	3850±55	k 6/11	2460–2200 BCE
	Ki2726	3840±50	k 4/4	2400–2200 BCE
Dobrovody cemetery				
Phase 1	Ki2129	4160±55	k 2/4	2880–2630 BCE
Phase 2	Ki2107	3980±45	k 2/6	2580–2450 BCE
	Ki7090	3960±60	k 1/6	2570–2350 BCE
Minovka cemetery				
Phase 1	Ki8296	4030±70	k 2/5	2840–2460 BCE
	Ki 421	3970±80	k 1/3	2620–2340 BCE
Novoseltsy cemetery				
Phase 1	Ki1219	4520±70	k 19/7	3360–3100 BCE
Phase 2	Ki1712	4350±70	k 19/15	3090–2880 BCE
Phase 3	Ki7127	4055±65	k 19/19	2840–2470 BCE
	Ki7128	4005±50	k 20/8	2580–2460 BCE
Otradnoe cemetery				
Phase 1	Ki478	3990±100	k 26/9	2850–2300 BCE
Phase 2	Ki 431	3890±105	k 1/17	2550–2200 BCE
	Ki 470	3860±105	k 24/1	2470–2140 BCE
	Ki452	3830±120	k 1/21	2470–2070 BCE
Pereshchepyno cemetery				
Phase 1	Ki9980	4150±70	k 4/13	2880–2620 BCE
	Ki9982	4105±70	k 1/7	2870–2500 BCE
	Ki9981	4080±70	k 1/6	2860–2490 BCE
Svatove cemetery				
Phase 1	Ki585	4000±190	k 1/1	2900–2200 BCE
	Ki586	4010±180	k 2/1	2900–2250 BCE

TABLE 13.3 (*continued*)

Lab number		BP date	Sample	Calibrated date
Talyanki cemetery				
Phase 1	Ki6714	3990±50	k 1/1	2580–2460 BCE
	Ki6716	3950±50	k 1/3	2560–2340 BCE
Phase 2	Ki2612	3760±70	k 2/3	2290–2030 BCE

B. Yamnaya horizon cemeteries in the middle Volga region (Samara Valley Project)

Nizhnaya Orlyanka 1				
Phase 1	AA1257	4520±75	k 4/2	3360–3090 BCE
	OxA**	4510±75	k 1/15	3360–3090 BCE
Grachevka II				
Phase 1	AA53805	4342±56	k 5/2	3020–2890 BCE
	AA53807	4361±65	k 7/1	3090–2890 BCE

C. Poltavka cemetery in the middle Volga region, three kurgans built in a single phase.

Krasnosamarskoe IV cemetery			
AA37034	4306±53	kurgan 1, grave 4	2929–2877 BCE
AA37031	4284±79	kurgan 1, grave 1	3027–2700 BCE
AA37033	4241±70	kurgan 1, grave 3 central	2913–2697 BCE
AA37036	4327±59	kurgan 2, grave 2 central	3031–2883 BCE
AA37041	4236±47	kurgan 3, grave 9 central	2906–2700 BCE
AA37040	4239±49	kurgan 3, grave 8	2910–2701 BCE

The Yamnaya-Poltavka dates show that multiple kurgans were constructed almost simultaneously with long gaps of time between episodes, perhaps indicating episodic use of the associated pastures.

The oldest radiocarbon dates from steppe vehicle graves bracket a century or two around 3000 BCE (table 13.3). One came from Ostannii kurgan 1, grave 160 in the Kuban, a grave of the third phase of the Novotitorovskaya culture dated 4440±40 BP, or 3320–2930 BCE. The other is from Bal'ki kurgan, grave 57, on the lower Dnieper, an early Yamnaya grave dated 4370±120 BP, or 3330–2880 BCE (see figures 4.4, 4.5). The probability distributions for both dates lie predominantly before 3000 BCE, which is why I use the figure 3100 BCE. But almost certainly these were not the first wagons in the steppes.[18]

Wagons probably appeared in the steppes between about 3500 and 3300 BCE, possibly from the west through Europe, or possibly through the late Maikop-Novosvobodnaya culture, from Mesopotamia. Since we cannot really say where the wheel-and-axle principle was invented, we do not know from which direction it first entered the steppes. But it had the greatest effect in the Don-Volga-Ural steppes, the eastern part of the early Proto-Indo-European world, and the Yamnaya horizon had its oldest roots there.

The subsequent spread of the Yamnaya horizon across the Pontic-Caspian steppes probably did not happen primarily through warfare, for which there is only minimal evidence. Rather, it spread because those who shared the agreements and institutions that made high mobility possible became potential allies, and those who did not share these institutions were separated as Others. Larger herds also probably brought increased prestige and economic power, because large herd-owners had more animals to loan or offer as sacrifices at public feasts. Larger herds translated into richer bride-prices for the daughters of big herd owners, which would have intensified social competition between them. A similar competitive dynamic was partly responsible for the Nuer expansion in east Africa (chapter 6). The Don-Volga dialect associated with the biggest and therefore most mobile herd owners probably was late Proto-Indo-European.

Where Did the Yamnaya Horizon Begin?

Why, as I just stated, did the Yamnaya horizon have its oldest roots in the eastern part of the Proto-Indo-European world? The artifact styles and funeral rituals that defined the early Yamnaya horizon appeared earliest in the east. Most archaeologists accept Nikolai Merpert's judgment that the oldest Yamnaya variants appeared in the Volga-Don steppes, the driest and easternmost part of the Pontic-Caspian steppe zone.

The Yamnaya horizon was divided into nine regional groups in Merpert's classic 1974 study. His regions have been chopped into finer and finer pieces by younger scholars.[19] These regional groups, however defined, did not pass through the same chronological stages at the same time. The pottery of the earliest Yamnaya phase (A) is divided by Telegin into two variants, A1 and A2 (figure 13.5).[20] Type A1 pots had a longer collar, decoration was mainly in horizontal panels on the upper third of the vessel, and "pearl" protrusions often appeared on and beneath the collar. Type A1 was like Repin pottery from the Don. Type A2 pots had decorations all over the vessel body, often in vertical panels, and had shorter, thicker, more everted

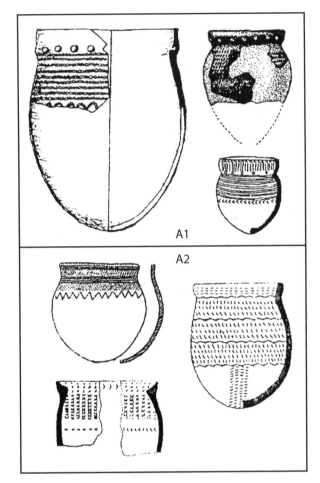

Figure 13.5 Early Yamnaya ceramic types A1 (Repin-related) and A2 (Khvalynsk-related). After Telegin et al. 2003.

rims. Type A2 was like late Khvalynsk pottery from the lower Volga. Repin vessels were made by coiling strips of clay; Type A2 Yamnaya vessels were usually made by pounding strips of clay into bag-shaped depressions or moulds to build up the walls, a very specific technological style. Pots of both subtypes were made of clays mixed with shell. Some of the shell temper seems to have been intentionally added, and some, particularly in Type A2 vessels, came from lake-bottom clays that naturally contained bits of shell and lake snails. Both the A1 and A2 types appeared across the Pontic-Caspian steppes in the earliest Yamnaya graves.

Early Yamnaya on the Lower Volga and Lower Don

Archaeological surveys led by I. V. Sinitsyn on the lower Volga between 1951 and 1953 revealed a regular series of Bronze Age kurgan cemeteries spaced 15–20 km apart along the level plains on the eastern bank between Saratov and Volgograd (then Stalingrad). Some of these kurgans contained stratified sequences of graves, and this stratigraphic evidence was employed to identify the earliest Yamnaya monuments. Important stratified kurgans included Bykovo cemetery II, kurgan 2, grave 1 (with a pot of Telegin's Type A1 stratified beneath later Yamnaya graves) and Berezhnovka cemetery I, kurgans 5 and 32, graves 22 and 2, respectively (with pots of Telegin's Type A2 stratified beneath later graves). In 1956 Gimbutas suggested that the "Kurgan Culture" began on the lower Volga. Merpert's synthesis of the Yamnaya horizon in 1974 supported Gimbutas. Recent excavations have reconfirmed the antiquity of Yamnaya traditions on the lower Volga. Archaic antecedents of both the A1 and A2 types of early Yamnaya pottery have been found in settlements on the lower Volga at Kyzyl Khak and Kara Khuduk (see figure 12.5), dated by radiocarbon between 4000 and 3500 BCE. Graves that seem intermediate between late Khvalynsk and Yamnaya in style and ritual have also been found at Shlyakovskii kurgan, Engels and Tarlyk between Saratov and Volgograd on the lower Volga.

The A1 or Repin style was made earliest in the middle Don–middle Volga region. Repin pottery is stratified beneath Yamnaya pottery at Cherkassky on the middle Don and is dated between 3950 and 3600 BCE at an antelope hunters' camp on the lower Volga at Kyzyl-Khak. The earliest Repin pottery was somewhat similar in form and decoration to the late Sredni Stog–Konstantinovka types on the lower Don, and it is now thought that contact with the late Maikop-Novosvobodnaya culture on the lower Don at places like Konstantinovka stimulated the emergence and spread of the early Repin culture and, through Repin, early Yamnaya. The metal-tanged daggers and sleeved axes of the early Yamnaya horizon certainly were copied after Maikop-Novosvobodnaya types.

The A2 or Khvalynsk style began on the lower Volga among late Khvalynsk populations. This bag-shaped kind of pottery remained the most common type in lower Volga Yamnaya graves, and later spread up the Volga into the middle Volga-Ural steppes, where the A2 style gradually replaced Repin-style Yamnaya pottery. Again, contact with people from the late Maikop-Novosvobodnaya culture, such as the makers of the kurgan at Evdik on the lower Volga, might have stimulated the change from late Khvalynsk to early Yamnaya. One of the stimuli introduced from the North Caucasus might have been wagons and wagon-making skills.[21]

Early Yamnaya on the Dnieper

The type site for early Yamnaya in Ukraine is a settlement, Mikhailovka. That Mikhailovka is a settlement, not a kurgan cemetery, immediately identifies the western Yamnaya way of life as more residentially stable than that of eastern Yamnaya. The strategic hill fort at Mikhailovka (level I) on the lower Dnieper was occupied before 3400 BCE by people who had connections in the coastal steppes to the west (the Mikhailovka I culture). After 3400–3300 BCE Mikhailovka (level II) was occupied by people who made pottery of the Repin-A1 type, and therefore had connections to the east. While Repin-style pottery had deep roots on the middle Don, it was intrusive on the Dnieper, and quite different from the pottery of Mikhailovka I. Mikhailovka II is itself divided into a lower level and an upper level. Lower II was contemporary with late Tripolye C1 and probably should be dated 3400–3300 BCE, whereas upper II was contemporary with early Tripolye C2 and should be dated 3300–3000 BCE. Repin-style pottery was found in both levels. The Mikhailovka II archaeological layer was about 60–70 cm thick. Houses included both dug-outs and surface houses with one or two hearths, tamped clay floors, partial stone wall foundations, and roofs of reed thatch, judging by thick deposits of reed ashes on the floors. This settlement was occupied by people who were newly allied to or intermarried with the Repin-style early Yamnaya communities of the Volga-Don region.

The people of Mikhailovka II farmed much less than those of Mikhailovka I. The frequency of cultivated grain imprints was 1 imprint per 273 sherds at Mikhailovka I but declined to 1 in 604 sherds for early Yamnaya Mikhailovka II, and 1 in 4,065 sherds for late Yamnaya Mikhailovka III, fifteen times fewer than in Mikhailovka I. At the same time food remains in the form of animal bones were forty-five times greater in the Yamnaya levels than in Mikhailovka I.[22] So although the total amount of food debris increased greatly during the Yamnaya period, the contribution of grain to the diet decreased. Grain imprints did occur in late Yamnaya funeral pottery from western Ukraine, as at Belyaevka kurgan 1, grave 20 and Glubokoe kurgan 2, grave 8, kurgans on the lower Dniester. These imprints included einkorn wheat, bread wheat (*Triticum aestivum*), millet (*Panicum miliaceum*), and barley (*Hordeum vulgare*). Some

Yamnaya groups in the Dnieper-Dniester steppes occasionally cultivated small plots of grain, as pastoralists have always done in the steppes. But cultivation declined in importance at Mikhailovka even as the Yamnaya settlement grew larger.[23]

WHEN DID THE YAMNAYA HORIZON BEGIN?

Dimitri Telegin and his colleagues used 210 radiocarbon dates from Yamnaya graves to establish the outlines of a general Yamnaya chronology. The earliest time interval with a substantial number of Yamnaya graves is about 3400–3200 BCE. Almost all the early dates are on wood taken from graves, so they do not need to be corrected for old carbon reservoir effects that can affect human bone. Graves dated in this interval can be found across the Pontic-Caspian steppes: in the northwestern Pontic steppes (Novoseltsy k. 19 gr. 7, Odessa region), the lower Dnieper steppes (Obloy k. 1, gr. 7, Kherson region), the Donets steppes (Volonterivka k. 1, gr. 4, Donetsk region), the lower Don steppes (Usman k. 1, gr. 13, Rostov region), the middle Volga steppes (Nizhnaya Orlyanka I, k. 1, gr. 5 and k. 4, gr. 1), and the Kalmyk steppes south of the lower Volga (Zunda Tolga, k. 1, gr. 15). Early Yamnaya must have spread rapidly across all the Pontic-Caspian steppes between about 3400 and 3200 BCE. The rapidity of the spread is interesting, suggesting both a competitive advantage and an aggressive exploitation of it. Other local cultures survived in pockets for centuries, since radiocarbon dates from Usatovo sites on the Dniester, late Post-Mariupol sites on the Dnieper and Kemi-Oba on the Crimean peninsula overlap with early Yamnaya radiocarbon dates between about 3300 and 2800 BCE. All three groups were replaced by late Yamnaya variants after 2800 BCE.[24]

WERE THE YAMNAYA PEOPLE NOMADS?

Steppe nomads have fascinated and horrified agricultural civilizations since the Scythians looted their way through Assyria in 627 BCE. We still tend to stereotype all steppe nomads as people without towns, living in tents or wagons hung with brilliant carpets, riding shaggy horses among their cattle and sheep, and able to combine their fractious clans into vast pitiless armies that poured out of the steppes at unpredictable intervals for no apparent reason other than pillage. Their peculiar kind of mobile pastoral economy, nomadic pastoralism, is often interpreted by historians as a parasitic adaptation that depended on agriculturally based states. Nomads needed states, according to this *dependency hypothesis*, for grain, metals, and loot. They needed enormous amounts of food and weapons to feed and arm their armies, and huge quantities of loot to maintain their loyalty, and that volume of food and wealth could only be acquired from agricultural states. Eurasian nomadic pastoralism has been interpreted as an opportunistic response to the evolution of centralized states like China and Persia on the borders of the steppe zone. Yamnaya pastoralism, whatever it was, could not have been nomadic pastoralism, because it appeared before there were any states for the Yamnaya people to depend on.[25]

But the dependency model of Eurasian nomadic pastoralism really explains only the *political* and *military* organization of Iron Age and Medieval nomads. The historian Nicola DiCosmo has shown that political and military organizations among nomads were transformed by the evolution of large standing armies that protected the leader—essentially a permanent royal bodyguard that ballooned into an army, with all the costs that implied. As for the *economic* basis of nomadic pastoralism, Sergei Vainshtein, the Soviet ethnographer, and DiCosmo both recognized that many nomads raised a little barley or millet, leaving a few people to tend small valley-bottom fields during the summer migrations. Nomads also mined their own metal ores, abundant in the Eurasian steppes, and made their own metal tools and weapons in their own styles. The metal crafts and subsistence economy that made Eurasian nomadic pastoralism possible did not depend on imported metal or agricultural subsidies from neighboring farmers. Centralized agricultural states like those of Uruk-period Mesopotamia were very good at concentrating wealth, and if steppe pastoralists could siphon off part of that wealth it could radically transform tribal steppe military and political structures, but the everyday subsistence economics of nomadic pastoralism did not require outside support from states.[26]

If nomadic pastoralism is an economic term, referring not to political organization and military confederacies but simply to a form of pastoral economy dependent on high residential mobility, it appeared during the Yamnaya horizon. After the EBA Yamnaya period an increasingly bifurcated economy appeared, with both mobile and settled elements, in the MBA Catacomb culture. This sedentarizing trend then intensified with the appearance of permanent, year-round settlements across the northern Eurasian steppes during the Late Bronze Age (LBA) with the Srubnaya culture. Finally mobile pastoral nomadism of a new militaristic type appeared in the Iron Age with the Scythians. But the Scythians did not invent the first pastoral economy based on mobility. That seems to have been the great innovation of the Yamnaya horizon.

Yamnaya Herding Patterns

An important clue to how the Yamnaya herding system worked is the location of Yamnaya kurgan cemeteries. Most Yamnaya kurgan cemeteries across the Pontic-Caspian region were located in the major river valleys, often on the lowest river terrace overlooking riverine forests and marshes. But at the beginning of the Yamnaya period kurgan cemeteries also began to appear for the first time in the deep steppes, on the plateaus between the major river valleys. If a cemetery can be interpreted as an ancestral claim to property ("here are the graves of my ancestors"), then the appearance of kurgan cemeteries in the deep steppes signaled that deep-steppe pastures had shifted from wild and free to cultured and owned resources. In 1985 V. Shilov made a count of the excavated kurgans located in the deep steppes, on inter-valley plateaus, in the steppe region between the lower Don, the lower Volga, and the North Caucasus. He counted 799 excavated graves in 316 kurgans located in the deep steppes, outside major river valleys. The earliest graves, the first ones to appear in these locations, were Yamnaya graves. Yamnaya accounted for 10% (78) of the graves, and 45% (359) were from MBA cultures related to the Catacomb culture, 7% (58) were from the LBA Srubnaya culture, 29% (230) were of Scytho-Sarmatian origin, and 9% (71) were historical-Medieval. The exploitation of pastures on the plateaus between the river valleys began during the EBA and rapidly reached its all-time peak during the MBA.[27]

N. Shishlina collected seasonal botanical data from kurgan graves in the Kalmyk steppes, north of the North Caucasus, part of the same region that Shilov had studied. Shishlina found that Yamnaya people moved seasonally between valley-bottom pastures (occupied during all seasons) and deep-steppe plateau pastures (probably in the spring and summer) located within 15–50 km of the river valleys. Shishlina emphasized the localized nature of these migratory cycles. Repetitive movements between the valleys and plateau steppes created overgrazed areas with degraded soils (preserved today under MBA kurgan mounds) by the end of the Yamnaya period.

What was the composition of Bronze Age herds in the Don-Volga steppes? Because there are no Yamnaya settlements east of the Don, faunal information has to be extracted from human graves. Of 2,096 kurgan graves reviewed by Shilov in both the river valleys and the inter-valley plateaus—a much bigger sample than just the graves on the plateaus—just 15.2% of Yamnaya graves contained sacrifices of domesticated

TABLE 13.2

Domesticated Animals in Early Bronze Age Graves and Settlements in the Pontic–Caspian Steppes

Culture	Cattle	Sheep/gt	Horse	Pig	Dog
Don–Volga steppe, Yamnaya graves	15%	65%	8%	—	5%
Mikhailovka II/III, Yamnaya settlement	59%	29%	11%	9%	0.7%
Repin (lower Don), settlement	18%	9%	55%	9%	—

Note: Missing % were unidentifiable as to species.

animals. Most of these contained the bones of sheep or goats (65%), with cattle a distant second (15%), horses third (8%) and dogs fourth (5%) (table 13.2).[28]

Yamnaya herding patterns were different in the west, between the Dnieper and Don valleys. One difference was the presence of Yamnaya settlements, implying a less mobile, more settled herding pattern. At Mikhailovka levels II and III, which define early and late Yamnaya in the Dnieper valley, cattle (60%) were more numerous than sheep (29%), unlike the sheep-dominant herds of the east. Kurgan cemeteries penetrated only a few kilometers into the plateaus; most cemeteries were located in the Dnieper valley or its larger tributaries. This riverine cattle-herding economy was tethered to fortified strongholds like Mikhailovka, supported by occasional small grain fields. About a dozen small Yamnaya settlements have been excavated in the Dnieper-Don steppes at places such as Liventsovka and Samsonovka on the lower Don. Most occupy less than 1 ha and were relatively low-intensity occupations, although fortification ditches protected Samsonovka and Mikhailovka, and a stone fortification wall was excavated at Skelya-Kamenolomnya. Cattle are said to predominate in the animal bones from all these places.[29]

East of Repin no Yamnaya settlements have been found. Occasional wind-eroded scatters of microliths and Yamnaya pottery sherds have been observed in valley bottoms and near lakes in the Manych and North Caspian desert-steppes and deserts, but without intact cultural layers. In the lusher grasslands where it is more difficult to see small surface sites, even Yamnaya surface scatters are almost unknown. For example, the Samara oblast on the middle Volga was dotted with known settlements of the Mesolithic, Neolithic, Eneolithic, and Late Bronze Ages, but it had no EBA Yamnaya settlements. In 1996, during the Samara Valley Project, we attempted to find ephemeral Bronze Age camps by digging test pits at twelve favorable-looking places along the bottom of a stream valley, Peschanyi Dol, that had four Yamnaya kurgan cemeteries clustered near its mouth around the village of Utyevka (see figure 16.11 for a map). The Peschanyi Dol valley is today used as a summer pasturing place for cattle herds from three nearby Russian rural villages. We discovered seven ephemeral LBA Srubnaya ceramic scatters in this pleasant valley and a larger Srubnaya settlement, Barinovka, at its mouth. The LBA settlement and one camp also had been occupied during the MBA; each yielded a small handful of MBA ceramic sherds. But we found no EBA sherds—no Yamnaya settlements.

If we cannot find the camps that Yamnaya herders occupied through the winter, when they had to retreat with their herds to the protection of riverine forests and marshes (where most Yamnaya cemeteries were located), then their herds were so large that they had to keep moving even in winter. In a similar northern grassland environment with very cold winters, the fifty bands of the Blackfoot Indians of Canada and Montana had to move a few miles several times each winter just to provide fresh forage for their horses. And the Blackfeet did not have to worry about feeding cattle or sheep. Mongolian herders move their tents and animal herds about once a month throughout the winter. The Yamnaya herding system probably was equally mobile.[30]

Yamnaya herders watched over their herds on horseback. At Repin on the Don, 55% of the animal bones were horse bones. A horse skull was placed in a Yamnaya grave in a kurgan cemetery overlooking the Caspian Depression near Tsa-Tsa, south of the Volga, in kurgan 7, grave 12. Forty horses were sacrificed in a Catacomb-period grave in the same cemetery in kurgan 1, grave 5.[31] The grave probably was dug around 2500 BCE. An adult male was buried in a contracted position on his left side, oriented northeast. Fragments of red ochre and white chalk were placed by his hip. A bronze dagger blade was found under his skull. Above his grave were forty horse skulls arranged in two neat rows. Three ram skulls lay on the floor of the grave. The amount of meat forty horses would have yielded— assuming they were slightly bigger than Przewalskis, or about 400 kg live weight—would be roughly 8,000 k, enough for four thousand portions of 2 k each. This suggests a funeral feast of amazing size. Horses were suitable animals for extraordinary ritual sacrifices.

Wild Seeds and Dairy Foods in the Don–Volga Steppes

A ceramics lab in Samara has microscopically examined many Yamnaya pot-sherds from graves, but no cultivated grain imprints appeared on Yamnaya pottery here or anywhere else east of the Don. Yamnaya people from the middle Volga region had teeth that were entirely free of caries (no caries in 428 adult Yamnaya-Poltavka teeth from Samara oblast [see figure 16.12]), which indicates a diet very low in starchy carbohydrates, like the teeth of foragers.[32] Eastern Yamnaya people might have eaten wild *Chenopodium* and *Amaranthus* seeds and even *Phragmites* reed tubers and rhizomes. Analysis of pollen grains and phytoliths (silica bodies that form inside plant cells) by N. Shishlina from Yamnaya grave floors in the eastern Manych depression, in the steppes north of the North Caucasus, found pollen and phytoliths of *Chenopodium* (goosefoot) and amaranths, which can produce seed yields greater in weight per hectare than einkorn wheat, and without cultivation.[33] Cultivated grain played a small role, if any, in the eastern Yamnaya diet.

Although they were very tall and robust and showed few signs of systemic infections, the Yamnaya people of the middle Volga region exhibited significantly more childhood iron-deficiency anemia (bone lesions called *cribra orbitalia*) than did the skeletons from any earlier or later period (figure 13.6). A childhood diet *too* rich in dairy foods can lead to anemia, since the high phosphorus content of milk can block the absorption of iron.[34] Health often declines in the early phases of a significant dietary change, before the optimal mix of new foods has been established. The anomalous Yamnaya peak in *cribra orbitalia* could also have resulted from an increased parasite load among children, which again would be consistent with a living pattern involving closer contact between animals and people. Recent genetic research on the worldwide distribution of the mutation that created lactose tolerance, which made a dairy-based diet possible, indicates that it probably emerged first in the steppes west of the Ural Mountains between about 4600 and 2800 BCE—the Late Eneolithic (Mikhailovka I) and the EBA Yamnaya periods.[35] Selection for this mutation, now carried by all adults who can tolerate dairy foods, would have been strong in a population that had recently shifted to a mobile herding economy.

The importance of dairy foods might explain the importance of the cow in Proto-Indo-European myth and ritual, even among people who depended largely on sheep. Cattle were sacred because cows gave more milk

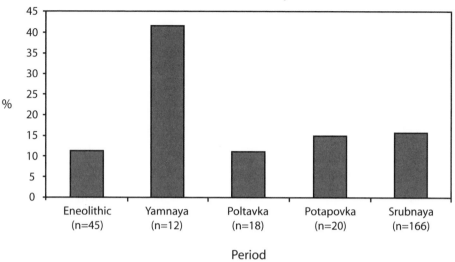

Figure 13.6 Frequencies of cribra orbitalia, associated with anemia, in cultures of the Samara oblast, middle Volga region. After Murphy and Khokhlov 2004.

than any other herd animal in the Eurasian steppe—twice as much as mares and five times more than goats, according to the Soviet ethnographer Vainshtein. He noted that, even among the sheep herders of Tuva in Siberia, an impoverished family of nomads that had lost all its sheep would try to keep at least one cow because that meant they could eat. The cow was the ultimate milk producer, even where herders counted their wealth in sheep.[36]

The Yamnaya wagon-based herding economy seems to have evolved in the steppes east of the Don, like the earliest Yamnaya pottery styles. Unlike the pottery and grave styles, the high-mobility, sheep-herding strategy of eastern Yamnaya pastoralism did not spread westward into the Dnieper steppes or northward into the middle Volga-Ural steppes, where cattle breeding remained the dominant aspect of the herding economies. Instead, it seems that social, religious, and political institutions (guest-host agreements, patron-client contracts, and ancestor cults) spread with the Yamnaya horizon. Some new chiefs from the east probably migrated into the Dnieper steppes, but in the west they added cattle to their herds and lived in fortified home bases.

Yamnaya Social Organization

The speakers of late Proto-Indo-European expressed thanks for sons, fat cattle, and swift horses to Sky Father, **dyew pœter*, a male god whose prominence probably reflected the importance of fathers and brothers in the herding units that composed the core of earthly social organization. The vocabulary for kin relations in Proto-Indo-European was that of a people who lived in a patrilineal, patrilocal social world, meaning that rights, possessions, and responsibilities were inherited only from the father (not the mother), and residence after marriage was with or near the husband's family. Kinship terms referring to grandfather, father, brother, and husband's brother survive in clearly corresponding roots in nearly all Indo-European languages, whereas those relating to wife and wife's family are few, uncertain, and variable. Kinship structure is only one aspect of social organization, but in tribal societies it was the glue that held social units together. We will see, however, that where the linguistic evidence suggests a homogeneous patri-centered Proto-Indo-European kinship system, the archaeological evidence of actual behavior is more variable.

As Jim Mallory admitted years ago, we know very little about the social meanings of kurgan cemeteries, and kurgan cemeteries are all the archaeological evidence left to us over much of the Yamnaya world.[37] We can presume that they were visible claims to territory, but we do not know the rules by which they were first established or who had the right to be buried there or how long they were used before they were abandoned. Archaeologists tend to write about them as static finished objects, but when they were first made they were dynamic, evolving monuments to specific people, clans, and events.

Gender and the Meaning of Kurgan Burial

We can be confident that kurgans were not used as family cemeteries. Mallory's review of 2,216 Yamnaya graves showed that the median Yamnaya kurgan contained fewer than 3 Yamnaya graves. About 25% contained just 1 grave. Children never were buried alone in the central or principal grave—that status was limited to adults. A count of kurgans per century in the well-studied and well-dated Samara River valley, in the middle Volga region, indicated that Yamnaya kurgans were built rarely, only one every five years or so even in regions with many Yamnaya cemeteries. So kurgans commemorated the deaths of special adults, not of everyone in the social group or even of everyone in the distinguished person's family. In the lower Volga, 80% of the Yamnaya graves contained males. E. Murphy and A. Khokhlov have confirmed that 80% of the sexable Yamnaya-Poltavka graves in the middle Volga region also contained males. In Ukraine, males predominated but not as strongly. In the steppes north of the North Caucasus, both in the eastern Manych steppes and in the western Kuban-Azov steppes, females and males appeared about equally in central graves and in kurgan graves generally. Mallory described the near-equal gender distribution in 165 Yamnaya graves in the eastern Manych region, and Gei gave similar gender statistics for 400 Novotitorovskaya graves in the Kuban-Azov steppes. Even in the middle Volga region some kurgans have central graves containing adult females, as at Krasnosamarskoe IV. Males were not always given the central place under kurgans even in regions where they strongly tended to occupy the central grave, and in the steppes north of the North Caucasus (where Maikop influence was strongest before the Yamnaya period) males and females were buried equally.[38]

The male-centered funerals of the Volga-Ural region suggest a more male-centered eastern social variant within the Yamnaya horizon, an archaeological parallel to the male-centered deities reconstructed for eastern Indo-European mythological traditions. But even on the Volga the people buried in central graves were not *exclusively* males. In the patrilocal, patrilineal society reconstructed by linguists for Proto-Indo-European speakers, *all* lineage heads would have been males. The appearance of adult females in one out of five kurgan graves, including central graves, suggests that gender was not the only factor that determined who was buried under a kurgan. Why were adult females buried in central graves under kurgans even on the Volga? Among later steppe societies women could occupy social positions normally assigned to men. About 20% of Scythian-Sarmatian "warrior graves" on the lower Don and lower Volga contained females dressed for battle as if they were men, a phenomenon that probably inspired the Greek tales about the Amazons. It is at least interesting that the frequency of adult females in central graves under Yamnaya kurgans in the same region, but two thousand years earlier, was about the same. Perhaps the people of this region customarily assigned some women leadership roles that were traditionally male.[39]

Kurgan Cemeteries and Mobility

Were the kurgans in a cemetery built together in a rapid sequence and then abandoned, or did people stay around them and use them regularly for longer periods of time? For interval dating *between* kurgans it would be ideal to obtain radiocarbon dates from all the kurgans in a cemetery. In a Yamnaya cemetery, that would usually be from three to as many as forty or fifty kurgans. Very few kurgan cemeteries have been subjected to this intensity of radiocarbon dating.

We can try to approximate the time interval between kurgans from the 210 radiocarbon dates on Yamnaya graves published in 2003 by Telegin and his colleagues. In his list we find nineteen Yamnaya kurgan cemeteries for which there are radiocarbon dates from at least two kurgans in the same cemetery. In eleven of these nineteen, more than half, at least two kurgans yielded radiocarbon dates that are statistically indistinguishable (see table 13.3 for radiocarbon dates). This suggests that kurgans were built rapidly in clusters. In many cases, the cemetery was then abandoned for a period of centuries before it was reused. For example, at the Poltavka cemetery of Krasnosamarskoe IV in the middle Volga region we can show this pattern, because we excavated all three kurgans in a small kurgan group and obtained multiple radiocarbon dates from each (figure 13.7). Like many kurgan groups in Ukraine, all three kurgans here were built within an indistinguishably brief time. The central graves all dated about 2700–2600 BCE (dates reduced by 200 radiocarbon years to account for the measured ^{15}N in the human bone used for the date), and then the cemetery was abandoned. Cemeteries like Krasnosamarskoe IV were used intensively for very short periods.

If pastures were like the cemeteries that marked them, then they were used briefly and abandoned. This episodic pasturing pattern, similar to swidden horticulture, possibly was encouraged by similar conditions—a low-productivity environment demanding frequent relocation. But herding, unlike swidden horticulture, required large pastures for each animal, and it could produce trade commodities (wool, felt, leather) if the herds were sufficiently large. To "rest" pastures under these circumstances would have been attractive only at low population densities.[40] It could have happened when the new Yamnaya economy was expanding into the previously unexploited pastures between the river valleys. But as the population of wagon-driving herders grew during the Early Bronze Age, some pastures began to show signs of overuse. A. A. Golyeva established that EBA Yamnaya kurgans in the Manych steppes were built on pristine soils and grasses, but many MBA Catacomb-culture kurgans were built on soils that had already been overgrazed.[41] Yamnaya kurgan cemeteries were dynamic aspects of a new herding system during its initial expansionary phase.

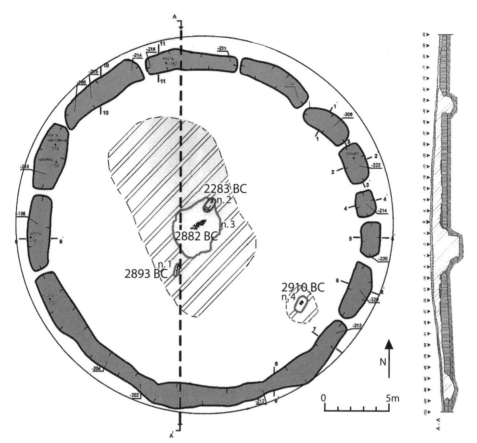

Figure 13.7 Krasnosamarskoe cemetery IV, kurgan 1, early Poltavka culture on the middle Volga. Three graves were created simultaneously when the kurgan was raised, about 2800 BCE: the central grave, covered by a layer of clay, a peripheral grave to its southeast, and an overlying grave in the kurgan. Author's excavation.

Proto-Indo-European Chiefs

The speakers of Proto-Indo-European followed chiefs (**weik-potis*) who sponsored feasts and ceremonies and were immortalized in praise poetry. The richer Yamnaya graves probably commemorated such individuals. The dim outlines of a social hierarchy can be extracted from the amount of labor required to build kurgans. A larger kurgan probably meant that a larger number of people felt obligated to respond to the death of the person buried in the central grave. Most graves contained nothing but the body, or in some cases just the head, with clothing, perhaps a bead or two, reed mats, and wooden beams. The skin of a domestic animal with a few leg or head bones attached was an unusual gift, appearing in about 15% of graves, and a copper dagger or axe was very rare, appearing in less than 5%. Sometimes a few sherds of pottery were thrown into the grave. It is difficult to define social roles on the basis of such slight evidence.

Do big kurgans contain the richest graves? Kurgan size and grave wealth have been compared in at least two regions, in the Ingul River valley west of the Dnieper in Ukraine (a sample of 37 excavated Yamnaya kurgans), and in the Volga-Ural region (a sample of more than 90 kurgans).[42] In both regions kurgans were easily divided into widely disparate size classes—three classes in Ukraine and four on the Volga. In both regions the class 1 kurgans were 50 m or more in diameter, about the width of a standard American football field (or two-thirds the width of a European soccer field), and their construction required more than five hundred man-days, meaning that five hundred people might have worked for one day to build them, or one hundred people for five days, or some other combination totaling five hundred.

The biggest kurgans were not built over the richest central graves in either region. Although the largest class 1 kurgans did contain rich graves, so did smaller kurgans. In both regions wealthy graves occurred both in the central position under a kurgan and in peripheral graves. In the Ingul valley, where there were no metal-rich graves in the study sample, more objects were found in peripheral graves than in central graves. In some cases, where we have radiocarbon dates for many graves under a single kurgan, we can establish through overlapping radiocarbon dates that the central grave and a *richer* peripheral grave were dug simultaneously in a single funeral ceremony, as at Krasnosamarskoe IV. The richest graves in some Novosvobodnaya kurgans, including the Klady cemetery, were peripheral graves, located off-center under the mound. It could be misleading to count the objects in peripheral graves, including some wheeled vehicle sacrifices, as separate from the central grave. In at least some cases, a richer peripheral grave accompanied the central grave in the same funeral ceremony.

Elite status was marked by artifacts as well as architecture, and the most widespread indication of status was the presence of metal grave goods. The largest metal artifact found in any Yamnaya grave was laid on the left arm of a male buried in Kutuluk cemetery I, kurgan 4, overlooking the Kinel River, a tributary of the Samara River in the Samara oblast east of the

Figure 13.8 Kutuluk cemetery I, kurgan 4, grave 1, middle Volga region. An Early Yamnaya male with a large copper mace or club, the heaviest metal object of the Yamnaya horizon. Photograph and excavation by P. Kuznetsov; see Kuznetsov 2005.

Volga (figure 13.8). A solid copper club or mace weighing 750 gm, it was 48.7 cm long and more than 1 cm thick, with a diamond cross-section. The kurgan was medium-sized, 21 m in diameter and less than 1 m high, but the central grave pit (gr. 1) was large. The male was oriented east, positioned supine with raised knees, with ochre at his head, hips, and feet—a classic early Yamnaya grave type. Two samples of bone taken from his

(4200±60, BM-3157). A Yamnaya mining pit has been found at Kargaly with radiocarbon dates of the same era. Almost all the copper objects from the Volga-Ural region were made of "clean" copper from these local sources. Although the cast sleeved single-bladed axes and tanged daggers of the early Yamnaya period imitated Novosvobodnaya originals, they were made locally from local copper ores. North Caucasian arsenical bronze was imported by people buried in graves in the Kalmyk steppe south of the lower Volga and in Kemi-Oba sites on the Crimean peninsula, but not in the Volga-Ural steppes.[45]

The grave at Pershin was not the only smith's grave of the period. Metalworkers were clearly identified in several Yamnaya-period graves, perhaps because metalworking was still a form of shamanic magic, and the tools remained dangerously polluted by the spirit of the dead smith. Two Post-Mariupol smith's graves on the Dnieper (chapter 12) probably were contemporary with early Yamnaya, as was a smith's grave with axe molds, crucibles, and *tulieres* in a Novotitorovskaya-culture grave in the Kuban steppes at Lebedi I (figure 13.10). Copper slag, the residue of metalworking, was included in other graves, as at Utyevka I kurgan 2.[46]

One unappreciated aspect of EBA and MBA steppe metallurgy was its experimentation with iron. The copper pin in Utyevka kurgan 1 with a forged iron head was not unique. A Catacomb-period grave at Gerasimovka on the Donets, probably dated around 2500 BCE, contained a knife with a handle made of arsenical bronze and a blade made of iron. The iron did not contain magnetite or nickel, as would be expected in meteoric iron, so it is thought to have been forged. Iron objects were rare, but they were part of the experiments conducted by steppe metalsmiths during the Early and Middle Bronze Ages, long before iron began to be used in Hittite Anatolia or the Near East.[47]

THE STONE STELAE OF THE NORTH PONTIC STEPPES

The Yamnaya horizon developed in the Pontic-Caspian steppes largely because an innovation in land transport, wagons, was added to horseback riding to make a new kind of herding economy possible. At the same time an innovation in sea transport, the introduction of the multi-oared longboat, probably was responsible for the permanent occupation of the Cycladic Islands by Grotta-Pelos mariners about 3300–3200 BCE, and for the initial development of the northwest Anatolian trading communities such as Kum Tepe that preceded the founding of Troy.[48] These two horizons, one on the sea and the other on a sea of grass, came into contact around the shores of the Black Sea.

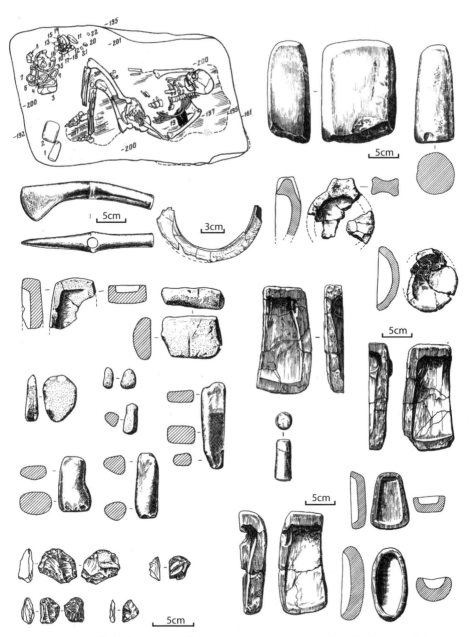

Figure 13.10 Lebedi cemetery I, kurgan 3, grave 10, a metal worker's gave of the late Novotitorovskaya culture, perhaps 2800–2500 BCE, Kuban River steppes. He wore a boars-tusk pendant. Under his arm was a serpentine hammer-axe (*upper left*). By his feet was a complete smithing kit: heavy stone hammers and abraders, sharp-edged flint tools, a round clay crucible (*upper right*), and axe molds for both flat and sleeved axes. After Gei 1986, figures 1, 4, 6, 7, and 9.

Chapter Fourteen

The Western Indo-European Languages

"A wild river full of possibilities flowed from my new tongue."
—Andrew Lam, *Learning a Language, Inventing a Future* 2006

We will not understand the early expansion of the Proto-Indo-European dialects by trying to equate language simply with artifact types. Material culture often has little relationship to language. I have proposed an exception to that rule in the case of robust and persistent frontiers, but that does seem to be an exception. The essence of language expansion is psychological. The initial expansion of the Indo-European languages was the result of widespread cultural shifts in group self-perception. Language replacement always is accompanied by revised self-perceptions, a restructuring of the cultural classifications within which the self is defined and reproduced. Negative evaluations associated with the dying language lead to a descending series of reclassifications by succeeding generations, until no one wants to speak like Grandpa any more. Language shift and the stigmatization of old identities go hand in hand.

The pre–Indo-European languages of Europe were abandoned because they were linked to membership in social groups that became stigmatized. How that process of stigmatization happened is a fascinating question, and the possibilities are much more varied than just invasion and conquest. Increased out-marriage, for example, can lead to language shift. The Gaelic spoken by Scottish "fisher" folk was abandoned after World War II, when increased mobility and new economic opportunities led to out-marriage between Gaelic "fishers" and the surrounding English-speaking population, and the formerly tightly closed and egalitarian "fisher" community became intensely aware both of its low ranking in a larger world and of alternative economic opportunities. Gaelic rapidly disappeared, although only a few people—soldiers, professionals, teachers—moved very far. Similarly, the general situation in Europe after 3300 BCE was one of increased mobility, new pastoral economies, explicitly status-ranked political systems, and inter-regional connectivity—exactly the kind of context that might have led to the stigmatization of the tightly closed identities associated with languages spoken by localized groups of village farmers.[1]

The other side of understanding language shift is to ask why the identities associated with Indo-European languages were emulated and admired. It cannot have been because of some essential quality or inner potential in Indo-European languages or people. Usually language shift flows in the direction of paramount prestige and power. Paramount status can attach to one ethnic group (Celt, Roman, Scythian, Turk, American) for centuries, but eventually it flows away. So we want to know what in this particular era attached prestige and power to the identities associated with Proto-Indo-European speech—Yamnaya identities, principally. At the beginning of this period, Indo-European languages still were spoken principally by pastoral societies from the Pontic-Caspian steppes. Five factors probably were important in enhancing their status:

1. Pontic-Caspian steppe societies were more familiar with horse breeding and riding than anyone outside the steppes. They had many more horses than anywhere else, and measurements show that their steppe horses were larger than the native marsh and mountain ponies of central and western Europe. Larger horses appeared in Baden, Cernavoda III, and Cham sites in central Europe and the Danube valley about 3300–3000 BCE, probably imported from the steppes.[2] Horses began to appear commonly in most sites of the ETC culture in Transcaucasia at the same time, and larger horses appeared among them, as in southeastern Anatolia at Norşuntepe. Steppe horse-breeders might also have had the most manageable male bloodline—the genetic lineage of the original domesticated male founder was preserved even in places with native wild populations (see chapter 10). If they had the largest, strongest, *and* most manageable horses, and they had more than anyone else, steppe societies could have grown rich by trading horses. In the sixteenth century the Bukhara khanate in Central Asia, drawing on horse-breeding grounds in the Ferghana valley, exported one hundred thousand horses *annually* just to one group of customers: the Mughal rulers of India and Pakistan. Although I am not suggesting anything near that scale, the annual demand for steppe horses in Late Eneolithic/Early Bronze Age Europe could easily have totaled thousands of animals during the initial expansion of horseback riding beyond the steppes. That would have made some steppe horse dealers wealthy.[3]

from a homeland in the Pontic-Caspian steppes; and (3) that the separations of Pre-Italic, Pre-Celtic, and Pre-Germanic, at least, from late Proto-Indo-European probably happened at about this time, between 3300 and 2500 BCE (see the conclusions of chapters 3 and 4).

The Roots of the Oldest Western Indo-European Branches

These constraints oblige us to turn our attention to the region just to the west of the early Yamnaya territory, or west of the South Bug River valley, beginning about 3300 BCE. On this frontier we can identify three archaeological cases of cross-cultural contact in which people from the western Pontic steppes established long-term relationships with people outside the steppe zone to their west during the steppe Early Bronze Age, 3300–2800 BCE. Each of these new intercultural meetings provided a context in which language expansion might have occurred, and, given the constraints just described, probably did. But each case happened differently.

The first occurrence involved close integration, noted particularly in pottery but evident in other customs as well, between the steppe Usatovo culture and the late Tripolye villages of the upper Dniester and Prut valleys (figure 14.1). It is fairly clear from the archaeological evidence that the steppe aspect of the integrated culture had separate origins and stood in a position of military dominance over the upland farmers, a situation that would have encouraged the spread of the steppe language into the uplands. In the second case, people of the Yamnaya horizon moved in significant numbers into the lower Danube valley and the Carpathian Basin. This was a true "folk migration," a massive and sustained flow of outsiders into a previously settled landscape. Again there are archaeological signs, in pottery particularly, of integration with the local Cotsofeni culture. Integration with the locals would have provided a medium for language shift. In the third case, the Yamnaya horizon expanded toward the border with the Corded Ware horizon on the headwaters of the Dniester in far northwestern Ukraine. In some places it appears there was no integration at all, but on the east flank of this contact zone, near the middle Dnieper, a hybrid border culture emerged. It is probably safe to assume that the separations of several western Indo-European branches were associated somehow with these events. The linguistic evidence suggests that Italic, Celtic, and Germanic, at least, separated next after Tocharian (discussed in the previous chapter). The probable timing of separations suggests that they happened around this time, and these are the visible events that seem like good candidates.

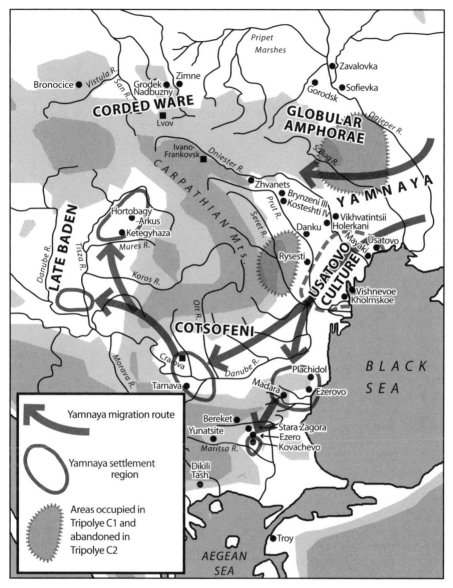

Figure 14.1 Yamnaya migrations into the Danube valley and the east Carpathian piedmont, 3100–2600 BCE. The older western IE branches probably evolved from dialects scattered by these migrations.

TABLE 14.1 (*continued*)

Lab Number	BP Date	Sample	Calibrated Date
Plachidol kurgan cemetery 1, northeast Bulgaria			
Bln-2504	4269±60	charcoal, grave 2 with stela	3010–2700 BCE
Bln-2501	4170±50	charcoal, grave 1 with wagon	2880–2670 BCE
Baia Hamangia, Danube delta, Romania			
GrN-1995	4280±65	charcoal from grave	3020–2700 BCE
Bln-29	4090±160	charcoal from grave	2880–2460 BCE
Ketegyhaza kurgan 3, grave 4 (latest grave in kurgan 3), eastern Hungary			
Bln-609	4265±80	charcoal from grave	3020–2690 BCE

resolved into two geographic groups north and south of the South Bug (see figure 13.1).

The northern Tripolye C2 group was located on the middle Dnieper and its tributaries around Kiev, where the forest-steppe graded into the closed northern forest. Cross-border assimilation with steppe cultures had begun on the middle Dnieper during Tripolye C1, as at Chapaevka (see figures 12.2, 12.6), and this process continued during Tripolye C2. At towns like Gorodsk, west of the Dnieper, and cemeteries like So-fievka, east of the Dnieper, the mix of cultural elements included late Sredni Stog, early Yamnaya, late Tripolye, and various influences from southern Poland (late Baden, late TRB). The hybrid that emerged from all these intercultural meetings slowly became its own distinct culture.

The southern Tripolye C2 group, centered in the Dniester valley, was closely integrated with a steppe culture, the Usatovo culture, described in detail below. The two surviving late Tripolye settlement centers on the Dnieper and Dniester continued to interact—Dniester flint continued to appear in Dnieper sites—but they also slowly grew apart. For reasons that will be clear in the next chapter, I believe that the emerging hybrid culture on the middle Dnieper played an important role in the evolution of both the Pre-Baltic and Pre-Slavic language communities after 2800–2600 BCE. Pre-Germanic is usually assigned an earlier position in branching diagrams. If early Pre-Germanic speakers moved away from the Proto-Indo-European homeland toward the northwest, as seems likely, they moved through one of these Tripolye settlement centers before 2800 BCE. Perhaps it was the other one in the Dniester valley. Its steppe partner was the Usatovo culture.

STEPPE OVERLORDS AND TRIPOLYE CLIENTS: THE USATOVO CULTURE

The Usatovo culture appeared about 3300–3200 BCE in the steppes around the mouth of the Dniester River, a strategic corridor that reached northwest into southern Poland. The rainfall-farming zone in the Dniester valley had been densely occupied by Cucuteni-Tripolye communities for millennia, but they never established settlements in the steppes. Kurgans had overlooked the Dniester estuary in the steppes since the Suvorovo migration about 4000 BCE; these are assigned to various groups including Mikhailovka I and the Cernavoda I–III cultures. Usatovo represented the rapid evolution of a new level of social and political integration between lowland steppe and upland farming communities. The steppe element used Tripolye material culture but clearly declared its greater prestige, wealth, and military power. The upland farmers who lived on the border itself adopted the steppe custom of inhumation burial in a cemetery, but they did not erect kurgans or take weapons to their graves. This integrated culture appeared in the Dniester valley just after the abandonment of all the Tripolye C1 towns in the South Bug valley on one side and the final Cucuteni B2 towns in southern Romania on the other. The chaos caused by the dissolution of hundreds of Cucuteni-Tripolye farming communities probably convinced the Tripolye townspeople of the middle Dniester valley to accept the status of clients. Explicit patronage defined the Usatovo culture.[9]

Cultural Integration between Usatovo and Upland Tripolye Towns

The stone-walled houses of the Usatovo settlement occupied the brow of a grassy ridge overlooking a bay near modern Odessa, the best seaport on the northwest coast of the Black Sea. Usatovo covered about 4–5 ha. A stone defensive wall probably defended the town on its seaward side. The settlement was largely destroyed by modern village construction and limestone quarrying prior to the first excavation by M. F. Boltenko in 1921, but parts of it survived (figure 14.2). Behind the ancient town four separate cemeteries crowned the hillcrest, all of them broadly contemporary. Two were kurgan cemeteries and two were flat-grave cemeteries. In one of the kurgan cemeteries, the one closest to the town, half the central graves contained men buried with bronze daggers and axes. These bronze weapons occurred in no other graves, not even in the second kurgan cemetery. Female figurines were limited to the flat-grave cemeteries and the settlement, never occurring in the kurgan

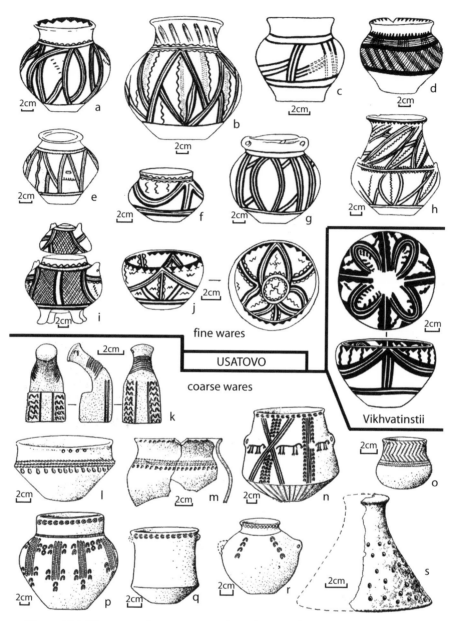

fine wares

USATOVO

coarse wares

Vikhvatinstii

Figure 14.3 Usatovo-culture ceramics (a, e, h, p, q, r) Usatovo kurgan cemetery I; (b) Tudorovo flat grave; (c) Sarata kurgan; (d) Shabablat kurgan; (f)
Parkany kurgan 182; (g, j, l) Usatovo kurgan cemetery II; (i) Parkany kurgan
91; (k) abstract figurine from Usatovo flat grave cemetery II; (m) Mayaki
settlement; (n) Tudorovo kurgan; (o) Usatovo flat grave cemetery II; (s) Mayaki settlement, probably a cheese strainer. Also shown: a painted fine bowl
from the Tripolye C2 cemetery at Vikhvatintsii. After Zbenovich 1968.

Impressions in pottery at the Usatovo settlement showed cultivated wheat (mostly emmer and bread wheats), barley, millet (frequent), oats (frequent), and peas.[12] The settlement also contained grinding stones and flint sickle teeth with characteristic edge gloss from cereal harvesting. This was the first evidence for cereal cultivation in the Dniester steppes, and, in fact, it is surprising, since rainfall agriculture is risky where precipitation is less than 350 mm per year. The grain would have been grown more easily in the upland settlements, perhaps cultivated by Tripolye people who resided part-time at Usatovo.

Tripolye C2 fine pots were particularly valued as grave gifts for the chiefs who died at Usatovo. Tripolye pots with an orange clay fabric, fired at almost 900°C, constituted 18% of the ceramics at the Usatovo settlement but 30% in the kurgan graves (figure 14.3, top). About 80% of the pottery at Usatovo and at other Usatovo-culture settlements was shell-tempered gray or brown ware, undecorated or decorated with cord impressions, and fired at only 700°C. This ware was made like steppe pottery. Though the shapes were like those made in the uplands by late Tripolye potters, some decorative motifs resembled those seen on Yamnaya Mikhailovka II–style pottery. A few of these shell-tempered gray pots at Usatovo were coated with a thick orange slip to make them *look* like fine Tripolye pots, indicating that the two kinds of pottery really were regarded as different.[13]

The painted Tripolye pots in Usatovo kurgan graves were most similar to those of the Tripolye C2 settlements at Brynzeny III on the Prut and Vikhvatintsii on the Dniester. Vikhvatinskii was 175 km up the Dniester from Usatovo near the steppe border, and Brynzeny III was about 350 km distant, hidden in the steep forested valleys of the East Carpathian piedmont. A fine painted pot of Brynzeny type was buried in the central grave of kurgan cemetery I, kurgan 12, at Usatovo, with an imported Maikop pot and a riveted bronze dagger. At this time Brynzeny III still had thirty-seven two-story *ploshchadka* houses, clay ovens, loom weights for large vertical looms, and female figurines. These traditional Tripolye customs survived in towns that showed ceramic connections with Usatovo, perhaps because patron-client agreements protected them. As the identities associated with the dying Tripolye culture were stigmatized and those associated with the Usatovo chiefs were emulated, people who lived at places like Brynzeny III and Vikhvatintsii might well have become bilingual. Their children then shifted to the Usatovo language.

Although fine Tripolye pots were preferred grave gifts for the Usatovo elite, the Tripolye culture itself occupied a secondary position of power

shape, like many contemporary Yamnaya stelae in the South Bug–Dnieper steppes (see figure 13.11). Kurgan 3 (31 m in diameter) had two stelae standing side by side. The larger one (1.1 m tall) was inscribed with the images of a man, a deer, and three horses; the smaller one had just one horse. Kurgan 11 (40 m in diameter, the largest at Usatovo) covered a cromlech circle and inner mound 26 m in diameter surfaced with eighty-five hundred stones. On its southwest border were three stelae, one 2.7 m tall (!) with inscribed images of either dogs or horses. The central grave was robbed.

Only adult men were buried in the central graves of kurgan cemetery I, in a contracted position on the left side oriented east-northeast. Only the central graves and the peripheral graves on the southwestern sector con- tained red ochre. Seven of the fifteen central graves (k. 1, 3, 4, 6, 9, 12, and 14) had arsenical bronze dagger blades with two to four rivet holes for the handle. No other graves at Usatovo contained daggers (figure 14.4). Bronze daggers emerged as new symbols of status here and in the graves of the Yamnaya horizon at this time, but Yamnaya daggers had long tangs for the handle, like Novosvobodnaya daggers and unlike the Usatovo and Sofievka daggers with rivet holes for the handle. The central graves at Usatovo also contained fine Tripolye pots, arsenical bronze awls, flat axes, two Novosvobodnaya-style chisels, adzes, silver rings and spiral twists, flint microlithic blades, and flint hollow-based arrowheads. Bronze weap- ons and tools appeared only in the central graves.

Kurgan cemetery II was about 400 m away from kurgan cemetery I. It originally contained probably ten kurgans, most of them smaller than those in kurgan cemetery I; three were excavated. They yielded no dag- gers, no weapons, only small metal objects (awls, rings), and only a few fine painted Tripolye ceramic vessels. Six individuals had designs painted on their skulls with red ochre (figure 14.5). Three of these were men who had been killed by hammer blows to the head. Hammer wounds did not appear in kurgan cemetery I. Kurgan cemetery II was used for a distinct social group or status, perhaps warriors. But similar red designs were painted on the head of one male in kurgan cemetery I, in a peripheral grave under kurgan 12, grave 2, in the southwestern sector; similar de- signs were painted on the skulls of some Yamnaya graves at the Popilnaya kurgan cemetery on the South Bug.[16]

The flat graves at Usatovo were shallow pits covered by large flat stones, usually containing a body in a contracted position on the left side, oriented east or northeast. The peripheral graves under the kurgans had the same form as flat graves, and two cemeteries contained just flat graves, without

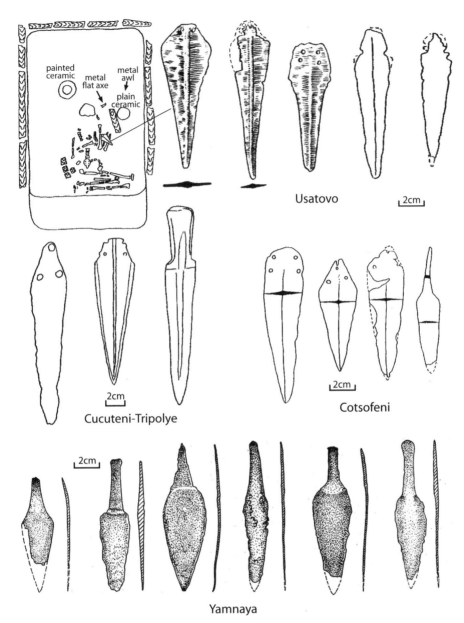

Figure 14.4 Daggers of the EBA, 3300–2800 BCE. *Top row*: Usatovo kurgan cemetery I, kurgan 3, central grave, with midrib dagger; kurgan 1, midrib dagger; Sukleya kurgan, midrib dagger; kurgan 9, lenticular-sectioned dagger; kurgan 6, lenticular-sectioned dagger. *Middle row left*: Werteba Cave, upper Dniester, riveted dagger; Cucuteni B, Moldova, midrib dagger; Werteba Cave, bone dagger carved in the shape of a metal dagger. *Middle row right*, Cotsofeni daggers from the lower Danube valley. *Bottom row*, Yamnaya tanged daggers from the North Pontic steppes. After Anthony 1996; and Nechitailo 1991.

Yamnaya horizon who were able to impose a patron-client relationship on Tripolye farming villages because of the protection that client status offered in a time of great insecurity. The pastoral patrons quickly became closely integrated with the farmers.

Tripolye clients of the Usatovo chiefs could have been the agents through which the Usatovo language spread northward into central Europe. After a few generations of clientage, the people of the upper Dniester might have wanted to acquire their own clients. Nested hierarchies in which clients are themselves patrons of other clients are characteristic of the growth of patron-client systems. The archaeological evidence for some kind of northward spread of people or political relationships consists of pottery exchanges between Tripolye sites on the upper Dniester and late TRB (Trichterbecker or Funnel-Beaker culture) sites in southeastern Poland. Substantial quantities of fine painted Tripolye C2 pottery of the Brynzeny III type occurred in southern Polish settlements of the late TRB culture dated 3000–2800 BCE, importantly at Gródek Nadbużny and Zimne, and late TRB pots were imported into the Tripolye C2 sites of Zhvanets and Brynzeny III.[17] Zhvanets was a production center for fine Tripolye pottery, with seven large two-chambered kilns, a possible source of local economic and political prestige. Conflict accompanied or alternated with exchange, since both the Polish sites and the Tripolye C2 sites closest to southeastern Poland were heavily fortified. The Tripolye C2 settlement of Kosteshti IV had a stone wall 6 m wide and a fortification ditch 5 m wide, and Zhvanets had three lines of fortification walls faced with stone, and both were located on high promontories.[18] Tripolye C2 community leaders whose parents had already adopted the Usatovo language could have attempted to extend to the late TRB communities of southern Poland the same kind of patron-client relationships that the Usatovo chiefs had offered them, an extension that might well have been encouraged or even backed up by paramount Usatovo chiefs.

If I had to hazard a guess I would say that this was how the Proto-Indo-European dialects that would ultimately form the root of Pre-Germanic first became established in central Europe: they spread up the Dniester from the Usatovo culture through a nested series of patrons and clients, and eventually were spoken in some of the late TRB communities between the Dniester and the Vistula. These late TRB communities later evolved into early Corded Ware communities, and it was the Corded Ware horizon (see below) that provided the medium through which the Pre-Germanic dialects spread over a wider area.

THE YAMNAYA MIGRATION UP THE DANUBE VALLEY

About 3100 BCE, during the initial rapid spread of the Yamnaya horizon across the Pontic-Caspian steppes, and while the Usatovo culture was still in its early phase, Yamnaya herders began to move through the steppes past Usatovo and into the lower Danube valley. The initial groups were followed by a regular stream of people that continued for perhaps three hundred years, between 3100 and 2800 BCE.[19] The passage through the Usatovo chiefdoms probably was managed through guest-host relationships. The migrants did not claim any Usatovo territory—at least they did not create their own cemeteries there. Instead, they kept going into the Danube valley, a minimum distance of 600–800 km from where they began in the steppes east of Usatovo—in the South Bug valley and farther east. The largest number of Yamnaya migrants ended up in eastern Hungary, an amazing distance (800–1,300 km depending on the route taken). This was a major, sustained population movement, and, like all such movements, it must have been preceded by scouts who collected information while on some other kind of business, possibly horse trading. The scouts knew just a few areas, and these became the targets of the migrants.[20]

The Yamnaya migrations into the Danube valley were targeted toward at least five specific destinations (see figure 14.1). One cluster of Yamnaya kurgan cemeteries, probably the earliest, appeared on the elevated plain northwest of Varna bay in Bulgaria (kurgan cemeteries at Plachidol, Madara, and other nearby places). This cluster overlooked the fortified coastal settlement at Ezerovo, an important local Early Bronze Age center. The second cluster of kurgan cemeteries appeared in the Balkan uplands 200 km to the southwest (the Kovachevo and Troyanovo cemeteries). They overlooked a fertile plain between the Balkan peaks and the Maritsa River, where many old tells such as Ezero and Mihailich had just been reoccupied and fortified. The third target was 300 km farther up the Danube valley in northwestern Bulgaria (Tarnava), on low ridges overlooking the broad plain of the Danube. These three widely separated clusters in Bulgaria contained at least seventeen Yamnaya cemeteries, each with five to twenty kurgans. Across the Danube and just 100 km west of the northwestern Bulgarian cluster, a larger group of kurgan cemeteries appeared in southwestern Romania, where at least a hundred Yamnaya kurgans dotted the low plains overlooking the Danube around Rast in southern Oltenia, south of Craiova. The Tarnava and Rast kurgans were in the same terrain and can be counted as one group, separated by the Danube River (and a modern international border).

produce a flowback of return migration. The Yamnaya wagon graves (Kholmskoe, Vishnevoe, and others) located in the steppes just north of the Danube delta are stratified above Usatovo graves, so probably were made later than the Yamnaya wagon grave in Bulgaria at Plachidol. The Danube valley migration probably originated east of the Usatovo area, in the steppes around the South Bug, Ingul, and Dnieper valleys. Western-oriented Yamnaya graves are found as a minor variant in Yamnaya cemeteries in the Dnieper–South Bug region. The oldest dated Yamnaya wagon grave (ca. 3000 BCE) at Bal'ki (k. 1 gr. 57) on the lower Dnieper was oriented to the west.[22]

What started this movement? A popular candidate has been a shortage of pasture in the steppes, but I find it hard to believe that there was any absolute shortage of pasture during the initial expansion of a new wagon-based economy. If the migration into the Danube valley began with raiding that then developed into a migration, we have to ask what caused the raiding. In the discussion of the causes of steppe warfare, in chapter 11, I mentioned the Proto-Indo-European *Trito* myth, which legitimized the cattle raid; the likelihood that competition between high-status families would lead to escalating bride-prices calculated in livestock, which might *create* a consumer shortage of animals and pastures in places where no absolute shortage existed; and the Proto-Indo-European initiation ritual that sent all young men out raiding.

The institution of the *Männerbünde* or *korios*, the warrior brotherhood of young men bound by oath to one another and to their ancestors during a ritually mandated raid, has been reconstructed as a central part of Proto-Indo-European initiation rituals.[23] One material trait linked to these ceremonies was the dog or wolf; the young initiates were symbolized by the dog or wolf and in some Indo-European traditions wore dog or wolf skins during their initiation. The canine teeth of dogs were frequently worn as pendants in Yamnaya graves in the western Pontic steppes, particularly in the Ingul valley, one probable region of origin for the Yamnaya migration.[24] A second material trait linked to the *korios* was the belt. The *korios* raiders wore a belt and little else (like the warrior figures in some later Germanic and Celtic art, e.g., the Anglo-Saxon Finglesham belt buckle). The initiates on a raid wore two belts, their leader one, symbolizing that the leader was bound by a single oath to the god of war/ancestors, and the initiates were double-bound to the god/ancestors and to the leader. Stone anthropomorphic stelae were erected over hundreds of Yamnaya graves between the Ingul and the South Bug valleys, in the same region where dog-canine pendants were common. The most common clothing element carved or painted on the stelae was a belt, often with an axe or a pair of sandals attached to it. Usually it was a single belt, perhaps symbolizing the leader of a raid. That stone stelae with belts were erected also by the Yamnaya migrants in Bulgaria near Plachidol provides another link between the migrants and the symbolism of the *korios* raid.[25]

There must also have been other pulls, positive rumors about opportunities in the Danube valley, because the migrants did not just raid but decided to live in the target region. These attractions are difficult to identify now, although the opportunity to acquire clients might have been a powerful pull.

Language Shift and the Yamnaya Migration

The Yamnaya migration occurred at a time of great fluidity and change throughout southeastern Europe. In Bulgaria, the tells in the upland plains of the Balkans at Ezero, Yunatsite, and Dubene-Sarovka were reoccupied about 3300–3200 BCE at the beginning of the Early Bronze Age (EBI) after almost a millennium of abandonment. The reoccupied tell settlements were fortified with substantial stone walls or ditches and palisades. One target of the Yamnaya migration was precisely this region. Yamnaya kurgan cemeteries could be seen for many miles; visually, they dominated the landscapes around them. In contrast, local cemeteries in the lower Danube valley and the Balkans, like the EBI cemetery at the Bereket tell settlement near Stara Zagora, usually had no visible surface monuments.[26]

A series of new artifact types diffused very widely across the lower and middle Danube valleys in connection with the Yamnaya migration. Concave-based arrowheads similar to steppe arrowheads appeared in the newly occupied tell sites in Bulgaria (Ezero) and in Aegean Macedonia (Dikili–Tash IIIB). These possibly were a sign of warfare with intrusive Yamnaya raiding groups. A new ceramic style spread across the entire middle and lower Danube, including the Morava and Struma valleys leading to Greece and the Aegean, and in Aegean Macedonia. The defining trait of this style was cord-impressed pottery encrusted with white paint.[27] White-encrusted, cord-impressed pottery appeared also in the Yamnaya graves. The Yamnaya immigrants could, perhaps, have played a role in joining one region to another and helping to spread this new style. But the pottery styles they spread were not their own. The Yamnaya immigrants usually deposited no pottery in their graves, and, when they did, they borrowed local ceramic styles, so their ceramic footprint is almost invisible.

Many Yamnaya kurgans in the lower Danube valley contained Cotsofeni ceramic vessels. The Cotsofeni culture evolved in mountain refuges in western Romania and Transylvania beginning about 3500 BCE, probably from Old European roots. Cotsofeni settlements were small agricultural hamlets of a few houses. Their owners cremated their dead and buried the ashes in flat graves, some of which contained riveted daggers like Usatovo daggers.[28] When Yamnaya herders reached the plains around Craiova, they probably realized that control over this region was the key to movement up and down the Danube valley through the mountain passes around the Iron Gates. They established alliances or patron-client contracts with the leaders of the Cotsofeni communities, through which they obtained Cotsofeni pottery (and probably other less visible Cotsofeni products), as Usatovo patrons obtained Tripolye pottery. Cotsofeni pottery then was carried into other regions by Yamnaya people. A Cotsofeni vessel was found in a Yamnaya kurgan as far afield as Tarakliya, Moldova, probably in the grave of a returned migrant. In northwestern Bulgaria, kurgan 1 at Tarnava (figure 14.6) contained an unusual concentration of six Cotsofeni pots in six Yamnaya graves.[29] Most of the Yamnaya kurgans in Bulgaria contained no ceramics, but, when they did, they were often Cotsofeni ceramics.

The situation of the Yamnaya chiefs might have been similar to that described by Barth in his account of the Yusufai Pathan invasion of the Swat valley in Pakistan in the sixteenth century. The invader, "faced with the sea of politically undifferentiated villagers proceeds to organize a central island of authority, and from this island he attempts to exercise authority over the surrounding sea. Other landowners establish similar islands, some with overlapping spheres of influence, others having unadministered gaps between them."[30] The mechanism through which the immigrant chief made himself indispensable to the villagers and tied them to him was the creation of a contract in which he guaranteed protection, hospitality, and the recognition of the villagers' rights to agricultural production in exchange for their loyalty, service, and best land. Yamnaya herding groups needed more land for pastures than did farming groups of equal population, and this could have provided a rationale for the Yamnaya people to claim use-rights over most of the available pasture lands and the migration routes that linked them, eventually creating a web of landownership that covered much of southeastern Europe. The reestablishment of tell settlements in the Balkans might have been part of a newly bifurcated economy in which farmers settled on fortified tells and increased grain production in response to reductions in their pastures, taken by their Yamnaya patrons.

The widely separated pockets of Yamnaya settlement in the lower Danube valley and the Balkans established speakers of late Proto-Indo-European dialects in scattered islands where, if they remained isolated from one another, they could have differentiated over centuries into various Indo-European languages. The many thousands of Yamnaya kurgans in eastern Hungary suggest a more continuous occupation of the landscape by a larger population of immigrants, one that could have acquired power and prestige partly just through its numerical weight. This regional group could have spawned both pre-Italic and pre-Celtic. Bell Beaker sites of the Csepel type around Budapest, west of the Yamnaya settlement region, are dated about 2800–2600 BCE. They could have been a bridge between Yamnaya on their east and Austria/Southern Germany to their west, through which Yamnaya dialects spread from Hungary into Austria and Bavaria, where they later developed into Proto-Celtic.[31] Pre-Italic could have developed among the dialects that remained in Hungary, ultimately spreading into Italy through the Urnfield and Villanovan cultures. Eric Hamp and others have revived the argument that Italic and Celtic shared a common parent, so a single migration stream could have contained dialects that later were ancestral to both.[32] Archaeologically, however, the Yamnaya immigrants here, as elsewhere, left no lasting material impression except their kurgans.

YAMNAYA CONTACTS WITH THE CORDED WARE HORIZON

The Corded Ware horizon is often invoked as the archaeological manifestation of the cultures that introduced the northern Indo-European languages to Europe: Germanic, Baltic, and Slavic. The Corded Ware horizon spread across most of northern Europe, from Ukraine to Belgium, after 3000 BCE, with the initial rapid spread happening mainly between 2900 and 2700 BCE. The defining traits of the Corded Ware horizon were a pastoral, mobile economy that resulted in the near disappearance of settlement sites (much like Yamnaya in the steppes), the almost universal adoption of funeral rituals involving single graves under mounds (like Yamnaya), the diffusion of stone hammer-axes probably derived from Polish TRB styles, and the spread of a drinking culture linked to particular kinds of cord-decorated cups and beakers, many of which had local stylistic prototypes in variants of TRB ceramics. The material culture of the Corded Ware horizon was mostly native to northern Europe, but the underlying behaviors were very similar to those of the Yamnaya horizon— the broad adoption of a herding economy based on mobility (using ox-drawn wagons and horses), and a corresponding rise in the ritual prestige and value of livestock.[33] The economy and political structure of the Corded Ware horizon certainly was influenced by what had emerged earlier in the steppes, and, as I just argued, some Corded Ware groups in south-eastern Poland might have evolved from Indo-European–speaking late TRB societies through connections with Usatovo and late Tripolye. The Corded Ware horizon established the material foundation for the evolution of most of the Bronze Age cultures of the northern European plain, so most discussions of Germanic, Baltic, or Slavic origins look back to the Corded Ware horizon.

The Yamnaya and Corded Ware horizons bordered each other in the hills between Lvov and Ivano-Frankovsk, Ukraine, in the upper Dniester piedmont around 2800–2600 BCE (see figure 14.1). At that time early Corded Ware cemeteries were confined to the uppermost headwaters of the Dniester west of Lvov, the same territory that had earlier been occupied by the late TRB communities infiltrated by late Tripolye groups. If Corded Ware societies in this region evolved from local late TRB origins, as many believe, they might already have spoken an Indo-European language. Between 2700 and 2600 BCE Corded Ware and late Yamnaya herders met each other on the upper Dniester over cups of mead or beer.[34] This meeting was another opportunity for language shift, and it is possible that Pre-Germanic dialects either originated here or were enriched by this additional contact.

The wide-ranging pattern of interaction that the Corded Ware horizon inaugurated across northern Europe provided an optimal medium for language spread. Late Proto-Indo-European languages penetrated the eastern end of this medium, either through the incorporation of Indo-European dialects in the TRB base population before the Corded Ware horizon evolved, or through Corded Ware–Yamnaya contacts later, or both. Indo-European speech probably was emulated because the chiefs who spoke it had larger herds of cattle and sheep and more horses than could be raised in northern Europe, and they had a politico-religious culture already adapted to territorial expansion. The dialects that were ancestral to Germanic probably were initially adopted in a small territory between the Dniester and the Vistula and then spread slowly. As we will see in the next chapter, Slavic and Baltic probably evolved from dialects spoken on the middle Dnieper.[35]

THE ORIGINS OF GREEK

The only major post-Anatolian branch that is difficult to derive from the steppes is Greek. One reason for this is chronological: Pre-Greek probably split away from a later set of developing Indo-European dialects and languages, not from Proto-Indo-European itself. Greek shared traits with Armenian and Phrygian, both of which probably descended from languages spoken in southeastern Europe before 1200 BCE, so Greek shared a common background with some southeastern European languages that might have evolved from the speech of the Yamnaya immigrants in Bulgaria. As noted in chapter 3, Pre-Greek also shared many traits with pre–Indo-Iranian. This linguistic evidence suggests that Pre-Greek should have been spoken on the eastern border of southeastern Europe, where it could have shared some traits with Pre-Armenian and Pre-Phrygian on the west and pre–Indo-Iranian on the east. The early western Catacomb culture would fit these requirements (see figure 15.5), as it was in touch with southeastern Europe on one side and with the developing Indo-Iranian world of the east on the other. But it is impossible, as far as I know, to identify a Catacomb-culture migration that moved directly from the western steppes into Greece.

A number of artifact types and customs connect the Mycenaean Shaft Grave princes, the first definite Greek speakers at about 1650 BCE, with steppe or southeastern European cultures. These parallels included specific types of cheekpieces for chariot horses, specific types of socketed spearheads, and even the custom of making masks for the dead, which was common on the Ingul River during the late Catacomb culture, between about 2500 and 2000 BCE. It is very difficult, however, to define the specific source of the migration stream that brought the Shaft Grave princes into Greece. The people who imported Greek or Proto-Greek to Greece might have moved several times, perhaps by sea, from the western Pontic steppes to southeastern Europe to western Anatolia to Greece, making their trail hard to find. The EHII/III transition about 2400–2200 BCE has long been seen as a time of radical change in Greece when new people might have arrived, but the resolution of this problem is outside the scope of this book.[36]

CONCLUSION: THE EARLY WESTERN INDO-EUROPEAN LANGUAGES DISPERSE

There was no Indo-European invasion of Europe. The spread of the Usatovo dialect up the Dniester valley, if it happened as I have suggested, was quite different from the Yamnaya migration into the Danube valley. But even that migration was not a coordinated military invasion. Instead, a succession of Pontic steppe tribal segments fissioned from their home clans and moved toward what they perceived as places with good pastures and opportunities for acquiring clients. The migrating Yamnaya chiefs then organized islands of authority and used their ritual and political institutions to establish control over the lands they appropriated for their herds, which required granting legal status to the local populations nearby, under patron-client contracts. Western Indo-European languages might well have remained confined to scattered islands across eastern and central Europe until after 2000 BCE, as Mallory has suggested.[37] Nevertheless, the movements into the East Carpathians and up the Danube valley occurred in the right sequence, at the right time, and in the right directions to be connected with the detachment of Pre-Italic, Pre-Celtic, and Pre-Germanic—the branch that ultimately gave birth to English.

CHAPTER FIFTEEN

Chariot Warriors of the Northern Steppes

The publication of the book *Sintashta* in 1992 (in Russian) opened a new era in steppe archaeology.[1] Sintashta was a settlement east of the Ural Mountains in the northern steppes. The settlement and the cemeteries around it had been excavated by various archaeologists between 1972 and 1987. But only after 1992 did the significance of the site begin to become clear. Sintashta was a fortified circular town 140 m in diameter, surrounded by a timber-reinforced earthen wall with timber gate towers (figure 15.1). Outside the wall was a V-shaped ditch as deep as a man's shoulders. The Sintashta River, a western tributary of the upper Tobol, had washed away half of it, but the ruins of thirty-one houses remained. The original town probably contained fifty or sixty. Fortified strongholds like this were unprecedented in the steppes. A few smaller fortified settlements had appeared west of the Don (Mikhailovka, for example) during the Yamnaya period. But the walls, gates, and houses of Sintashta were much more substantial than at any earlier fortified site in the steppes. And inside each and every house were the remains of metallurgical activity: slag, ovens, hearths, and copper. Sintashta was a fortified metallurgical industrial center.

Outside the settlement were five funerary complexes that produced spectacular finds (figure 15.2). The most surprising discoveries were the remains of chariots, which radiocarbon dates showed were the oldest chariots known anywhere. They came from a cemetery of forty rectangular grave pits without an obvious kurgan labeled SM for *Sintashta mogila*, or *Sintashta cemetery*. The other four mortuary complexes were a mid-size kurgan (SI, for *Sintashta I*), 32 m in diameter and only 1 m high, that covered sixteen graves; a second flat or non-kurgan cemetery (SII) with ten graves; a second small kurgan (SIII), 16 m in diameter, that covered a single grave containing the partial remains of five individuals; and finally a huge kurgan, 85 m in diameter and 4.5 m high (SB, for *Sintashta bolshoi kurgan*), built over a central grave (robbed in antiquity) constructed of logs and sod on the

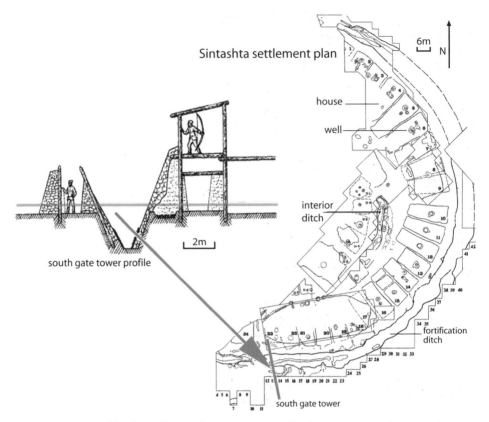

Figure 15.1 The Sintashta settlement: rectangular houses arranged in a circle within a timber-reinforced earthen wall, with excavators' reconstruction of south gate tower and outer defense wall. After Gening, Zdanovich, and Gening 1992, figures 7 and 12.

original ground surface. The southern skirt of the SB kurgan covered, and so was later than, the northern edge of the SM cemetery, although the radiocarbon dates suggest that SM was only slightly older than SB. The forty SM graves contained astounding sacrifices that included whole horses, up to eight in and on a single grave (gr. 5), with bone disc-shaped cheekpieces, chariots with spoked wheels, copper and arsenical bronze axes and daggers, flint and bone projectile points, arsenical bronze socketed spearheads, polished stone mace heads, many ceramic pots, and a few small silver and gold ornaments (figure 15.3). What was impressive in these graves was weaponry, vehicles, and animal sacrifices, not crowns or jewelry.

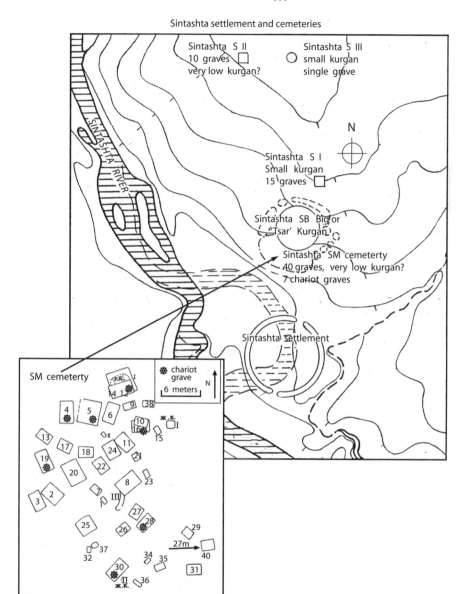

Figure 15.2 The Sintashta settlement landscape, with associated cemeteries, and detail of the SM cemetery. After Gening, Zdanovich, and Gening 1992, figures 2 and 42.

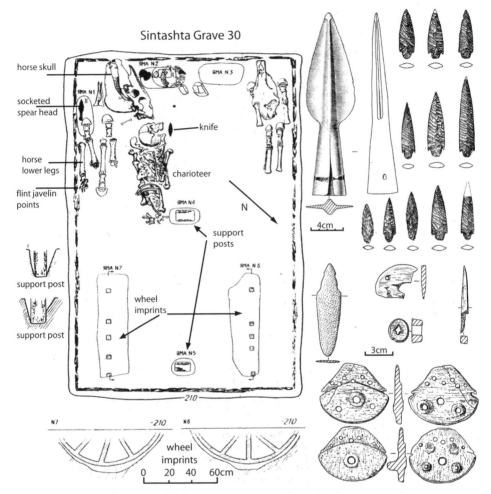

Figure 15.3 Sintashta SM cemetery, grave 30, with chariot wheel impressions, skulls and lower leg bones of horse team, cheekpieces for bits, and weapons. After Gening, Zdanovich, and Gening, figures 111, 113, and 114.

The radiocarbon dates for both the cemeteries and the settlement at Sintashta were worryingly diverse, from about 2800–2700 BCE (4200+100 BP), for wood from grave 11 in the SM cemetery, to about 1800–1600 BCE (3340+60BP), for wood from grave 5 in the SII cemetery. Probably there was an older Poltavka component at Sintashta, as later was found at many other sites of the Sintashta type, accounting for the older dates. Wood from the central grave of the large kurgan (SB) yielded consistent dates (3520+65, 3570+60, and 3720+120), or about 2100–1800 BCE. The same age range was produced by radiocarbon dates from the similar settlement at Arkaim, from several Sintashta cemeteries (Krivoe Ozero, Kammeny Ambar), and from the closely related graves of the Potapovka type in the middle Volga region (table 15.1).

The details of the funeral sacrifices at Sintashta showed startling parallels with the sacrificial funeral rituals of the *Rig Veda*. The industrial scale of metallurgical production suggested a new organization of steppe mining and metallurgy and a greatly heightened demand for copper and bronze. The substantial fortifications implied surprisingly large and determined attacking forces. And the appearance of Pontic-Caspian kurgan rituals, vehicle burials, and weapon types in the steppes east of the Ural River indicated that the Ural frontier had finally been erased.

After 1992 the flow of information about the Sintashta culture grew to a torrent, almost all of it in Russian and much of it still undigested or actively debated as I write.[2] Sintashta was just one of more than twenty related fortified settlements located in a compact region of rolling steppes between the upper Ural River on the west and the upper Tobol River on the east, southeast of the Ural Mountains. The settlement at Arkaim, excavated by G. B. Zdanovich, was not damaged by erosion, and twenty-seven of its fifty to sixty structures were exposed (figure 15.4). All the houses at Arkaim contained metallurgical production facilities. It has become a conference center and national historic monument. Sintashta and Arkaim raised many intriguing questions. Why did these fortified metal-producing towns appear in that place at that time? Why the heavy fortifications—who were they afraid of? Was there an increased demand for copper or just a new organization of copper working and mining or both? Did the people who built these strongholds invent chariots? And were they the original Aryans, the ancestors of the people who later composed the *Rig Veda* and the *Avesta*?[3]

THE END OF THE FOREST FRONTIER: CORDED WARE HERDERS IN THE FOREST

To understand the origins of the Sintashta culture we have to begin far to the west. In what had been the Tripolye region between the Dniester and Dnieper rivers, the interaction between Corded Ware, Globular Amphorae, and Yamnaya populations between 2800 and 2600 BCE produced a complicated checkerboard of regional cultures covering the rolling hills and valleys of the forest-steppe zone (figure 15.5). To the south, in the

TABLE 15.1

Selected radiocarbon dates for the Sintashta–Arkaim (S) and Potapovka (P) cultures in the south Ural steppes and middle Volga steppes.

Lab Number	BP Date	Sample Source	C, K	Calibrated Date
Sintashta SB Big Kurgan (S)				
GIN-6186	3670±40	birch log		2140–1970 BCE
GIN-6187	3510±40	"		1890–1740 BCE
GIN-6188	3510±40	"		1890–1740 BCE
GIN-6189	3260±40	"		1610–1450 BCE
Sintashta SM cemetery (S)				
Ki-653	4200±100	grave 11, wood	K	2900–2620 BC
Ki-658	4100±170	grave 39, wood	K	2900–2450 BC
Ki-657	3760±120	grave 28, wood	C	2400–1970 BC
Ki-864	3560±180	grave 19, wood	C	2200–1650 BCE
Ki-862	3360±70	grave 5, wood	C, K	1740–1520 BC
Krivoe Ozero cemetery, kurgan 9, grave 1 (S)				
AA-9874b	3740±50	horse 1 bone	C, K	2270–2030 BC
AA-9875a	3700±60	horse 2 bone		2200–1970 BC
AA-9874a	3580±50	horse 1 bone		2030–1780 BC
AA-9875b	3525±50	horse 2 bone		1920–1750 BC
Kammeny Ambar 5 (S)				
OxA-12532	3604±31	k2: grave 12, human bone		2020–1890 BCE
OxA-12530	3572±29	k2: grave 6, "	K	1950–1830 BCE
OxA-12533	3555±31	k2: grave 15, "		1950–1780 BCE
OxA-12531	3549±49	k2: grave 8, "	C, K	1950–1770 BCE
OxA-12534	3529±31	k4: grave 3, "		1920–1770 BCE
OxA-12560	3521±28	k4: grave 1, "		1890–1770 BCE
OxA-12535	3498±35	k4: grave 15, "		1880–1740 BCE
Utyevka cemetery VI (P)				
AA-12568	3760±100	k6: grave 4, human bone	K	2340–1980 BC
OxA-4264	3585±80	k6: grave 6, human bone		2110–1770 BC
OxA-4306	3510±80	k6: grave 4, human bone	K	1940–1690 BC
OxA-4263	3470±80	k6: grave 6, human bone	K	1890–1680 BC
Potapovka cemetery I (P)				
AA-12569	4180±85	k5: grave 6, dog bone*		2890–2620 BC

TABLE 15.1 (*continued*)

Lab Number	BP Date	Sample Source	C, K	Calibrated Date
AA–47803	4153±59	k.3: grave 1, human bone*		2880–2620 BC
OxA–4265	3710±80	k5: grave 13, human bone		2270–1960 BC
OxA–4266	3510±80	k5: grave 3, human bone		1940–1690 BC
AA–47802	3536±57	k.3: grave 1, horse skull*		1950–1770 BC
Other Potapovka cemeteries (P)				
AA–53803	4081±54	Kutuluk I, k1:1, human bone		2860–2490 BC
AA–53806	3752±52	Grachevka II k5:3, human bone		2280–2030 BC

*See note 17

Graves that contained chariots are marked C; graves that contained studded disc cheekpieces are marked K.

steppes, late Yamnaya and a few late Usatovo groups continued to erect kurgan cemeteries. Some late Yamnaya groups penetrated northward into the forest-steppe, up the Dniester, South Bug, and Dnieper valleys. Eastern Carpathian groups making Globular Amphorae pottery moved from the upper Dniester region around Lvov eastward into the forest-steppe around Kiev, and then retreated back to the Dniester. Corded Ware groups from southern Poland replaced them around Kiev. Under the influence of this combined Globular Amphorae and Corded Ware expansion to the east, the already complex mixture of Yamnaya-influenced Late Tripolye people in the Middle Dnieper valley created the Middle Dnieper culture in the forest-steppe region around Kiev. This was the first food-producing, herding culture to push into the Russian forests north of Kiev.[4]

The Middle Dnieper and Fatyanovo Cultures

The people of the Middle Dnieper culture carried stockbreeding economies (cattle, sheep, and pigs, depending on the region) north into the forest zone, up the Dnieper and Desna into what is now Belarus (figure 15.5). They followed marshes, open lakes, and riverine floodplains where there were natural openings in the forest. These open places had grass and reeds for the animals, and the rivers supplied plentiful fish. The earliest Middle Dnieper sites are dated about 2800–2600 BCE; the latest ones continued to about 1900–1800 BCE.[5] Early Middle Dnieper pottery showed clear similarities with Carpathian and eastern Polish Corded

Arkaim settlement and finds

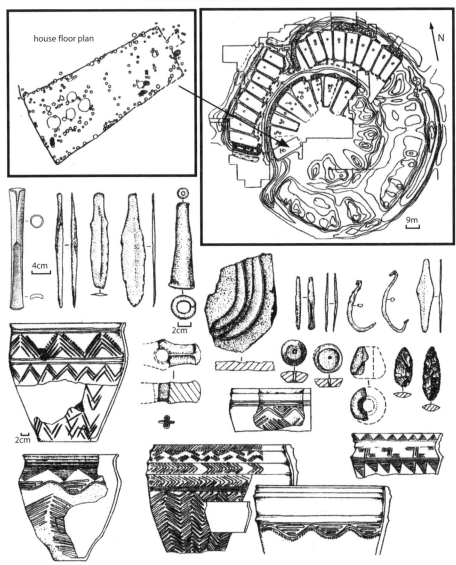

Figure 15.4 Arkaim settlement, house plan, and artifacts, including a mold for casting curved sickle or knife blades. After Zdanovich 1995, figure 6.

Figure 15.5 Culture groups of the Middle Bronze Age, 2800–2200 BCE.

Ware pottery, and Middle Dnieper pots have been found in Corded Ware graves near Grzeda Sokalska between the upper Dniester and the upper Vistula.[6] Some late Sredni Stog or Yamnaya elements also appeared in Middle Dnieper ceramics (figure 15.6). Middle Dnieper cemeteries contained both kurgans and flat-graves, both inhumation burials and cremations, with hollow-based flint arrowheads like those of the Yamnaya and Catacomb cultures, large trapezoidal flint axes like Globular Amphorae, and drilled stone "battle-axes" like those of the Corded Ware cultures. The Middle Dnieper culture clearly emerged from a series of encounters and exchanges between steppe and forest-steppe groups around Kiev, near the strategic fords over the Dnieper.[7]

A second culture, Fatyanovo, emerged at the northeastern edge of the Middle Dnieper culture. After the cattle herders moved out of the south-flowing Dnieper drainage and into the north-flowing rivers such as the Oka that coursed through the pine-oak-birch forests to the Upper Volga, they began to make pottery in distinctive Fatyanovo forms. But Fatyanovo pottery still showed mixed Corded Ware/Globular Amphorae traits, and the Fatyanovo culture probably was derived from an early variant of the Middle Dnieper culture. Ultimately Fatyanovo-type pottery, graves, and the cattle-raising economy spread over almost the entire Upper Volga basin. In the enormous western part of the Fatyanovo territory, from the Dvina to the Oka, very few Fatyanovo settlements are known, but more than three hundred large Fatyanovo flat-grave cemeteries, without kurgans, have been found on hills overlooking rivers or marshes. The Late Eneolithic Volosovo culture of the indigenous forest foragers was quite different in its pottery, economy, and mortuary customs. It disappeared when the Fatyanovo pioneers pushed into the Upper and Middle Volga basin.

The Middle Dnieper and Fatyanovo migrations overlapped the region where river and lake names in Baltic dialects, related to Latvian and Lithuanian, have been mapped by linguists: through the upper and middle Dnieper basin and the upper Volga as far east as the Oka. These names indicate the former extent of Baltic-speaking populations, which once occupied an area much larger than the area they occupy today. The Middle Dnieper and Fatyanovo migrations probably established the populations that spoke pre-Baltic dialects in the Upper Volga basin. Pre-Slavic probably developed between the middle Dnieper and upper Dniester among the populations that stayed behind.[8]

As Fatyanovo groups spread eastward down the Volga they discovered the copper ores of the western Ural foothills, and in this region, around the lower Kama River, they created long-term settlements. The Volga-Kama region,

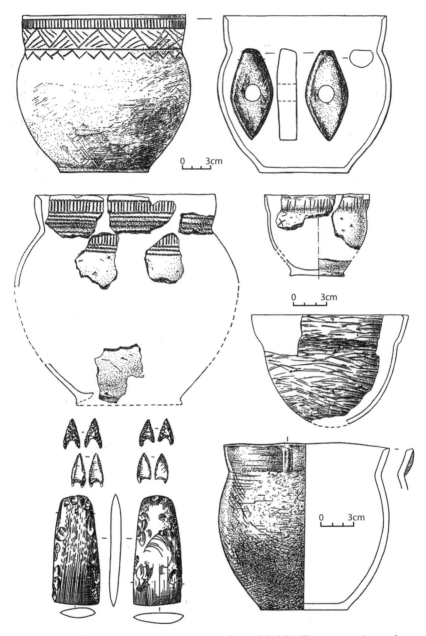

Figure 15.6 Ceramics and stone tools of the Middle Dnieper culture from sites in Belarus. After Kryvaltsevich and Kovalyukh 1999, figures 2 and 3.

which became the metallurgical heartland for almost all Fatyanovo metallurgy, has been separated from the rest of Fatyanovo and designated the Balanovo culture. Balanovo seems to be the settled, metal-working aspect of eastern Fatyanovo. At the southern fringe of Balanovo territory, in the forest-steppe zone of the middle Volga and upper Don where the rivers again flowed south, a fourth group emerged (after Middle Dnieper, Fatyanovo, and Balanovo). This was Abashevo, the easternmost of the Russian forest-zone cultures that were descended from Corded Ware ceramic traditions. The Abashevo culture played an important role in the origin of Sintashta.

The Abashevo Culture

Abashevo probably began about 2500 BCE or a little later. A late Abashevo kurgan at Pepkino on the middle Volga is dated 2400–2200 BCE (3850±95, Ki-7665); I would guess that the grave actually was created closer to 2200 BCE. Late Abashevo traditions persisted west of the Urals probably as late as 1900 BCE, definitely into the Sintashta period, since late Abashevo vessels are found in Sintashta and Potapovka graves. Early Abashevo ceramic styles strongly influenced Sintashta ceramics.

Abashevo sites are found predominantly in the forest-steppe zone, although a few extended into the northern steppes of the middle Volga. Within the forest-steppe, they are distributed between the upper Don on the west, a region with many Abashevo settlements (e.g., Kondrashovka); the middle Volga region in the center, represented largely by kurgan cemeteries (including the type-site, the Abashevo kurgan cemetery); and up the Belaya River into the copper-rich southwestern foothills of the Urals on the east, again with many settlements (like Balanbash, with plentiful evidence of copper smelting). More than two hundred Abashevo settlements are recorded; only two were clearly fortified, and many seem to have been occupied briefly. The easternmost Abashevo sites wrapped around the southern slopes of the Urals and extended into the Upper Ural basin, and it is these sites in particular that played a role in the origins of Sintashta.[9]

Some of the Volosovo foragers who had occupied these regions before 2500 BCE were absorbed into the Abashevo population, and others moved north. At the northern border of Abashevo territory, cord-impressed Abashevo and comb-stamped Volosovo ceramics are occasionally found inside the same structures at sites such as Bolshaya Gora.[10] Contact between late Volosovo and Abashevo populations west of the Urals probably helped to spread cattle-breeding economies and metallurgy into transitional northern forest cultures such as Chirkovska.

Whereas early Abashevo pottery looked somewhat like Fatyanovo/Bala-novo Corded Ware, early Abashevo graves were covered by kurgans, unlike Fatyanovo flat cemeteries. Abashevo kurgans were surrounded by a circular ditch, the grave pit had ledges at the edges, and the body position was either contracted on the side or supine with raised knees—funeral customs derived from the Poltavka culture on the Volga. Abashevo ceramics also showed increasing decorative influences from steppe Catacomb-culture ceramic traditions, in both motifs (horizontal line-and-dot, horizontal flut-ing) and technology (shell tempering). Some Abashevo metal types such as waisted knives copied Catacomb and Poltavka types. A. D. Pryakhin, the preeminent expert on the Abashevo culture, concluded that it origi-nated from contacts between Fatyanovo/Balanovo and Catacomb/Poltavka populations in the southern forest-steppe. In many ways, the Abashevo culture was a conduit through which steppe customs spread northward into the forest-steppe. Most Russian archaeologists interpret the Aba-shevo culture as a border culture associated with Indo-Iranian speakers, unlike Fatyanovo.[11]

Abashevo settlements in the Belaya River valley such as Balanbash contained crucibles, slag, and casting waste. Cast shaft-hole axes, knives, socketed spears, and socketed chisels were made by Abashevo metalsmiths. About half of all analyzed Abashevo metal objects were made of pure cop-per from southwestern Ural sandstone ores (particularly ornaments), and about half were arsenical bronze thought to have been made from south-eastern Ural quartzitic ores (particularly tools and weapons), the same ores later exploited by Sintashta miners. High-status Abashevo graves con-tained copper and silver ornaments, semicircular solid copper and silver bracelets, cast shaft-hole axes, and waisted knives (figure 15.7). High-status Abashevo women wore distinctive headbands decorated with rows of flat and tubular beads interspersed with suspended double-spiral and cast rosette pendants, made of copper and silver. These headbands were unique to the Abashevo culture and probably were signals of ethnic as well as political status.[12]

The clear signaling of identity seen in Abashevo womens' headbands occurred in a context of intense warfare—not just raiding but actual war-fare. At the cemetery of Pepkino, near the northern limit of Abashevo territory on the lower Sura River, a single grave pit 11 m long contained the bodies of twenty-eight young men, eighteen of them decapitated, oth-ers with axe wounds to the head, axe wounds on the arms, and dismem-bered extremities. This mass grave, probably dated about 2200 BCE, also contained Abashevo pottery, a two-part mold for making a shaft-hole axe

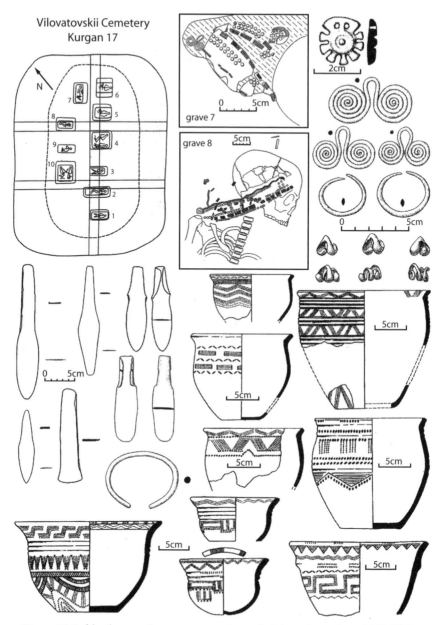

Figure 15.7 Abashevo culture graves and metal objects from the middle Volga forest-steppe (*upper left*), including distinctive cast copper rosettes; and ceramics from the south Ural region (*lower right*). After O. V. Kuzmina 1999, figures 23 and 24 (ceramics); and Bol'shov 1995, figure 13 (grave goods).

of Chernykh's Type V, and a crucible. It was covered by a single kurgan and so probably reflected a single event, clearly a serious battle or massacre. The absence of women or children in the grave indicates that it was not a settlement massacre. If it was the result of a battle, it implies a force of 280 to 560 on the Abashevo side alone, because deaths in tribal battles rarely reached 10% of the fighting force and usually were more like 5%.[13] Forces this size would require a considerable degree of inter-regional political integration. Intense warfare, perhaps on a surprising scale, was part of the political landscape during the late Abashevo era. In this context, the fortifications around Sintashta settlements and the invention of new fighting technologies—including the chariot—begin to make sense.

Linguists have identified loans that were adopted into the early Finno-Ugric (F-U) languages from Pre-Indo-Iranian and Proto-Indo-Iranian (Proto-I-I). Archaeological evidence for Volosovo-Abashevo contacts around the southern Urals probably were the medium through which these loans occurred. Early Proto-Indo-Iranian words that were borrowed into common Finno-Ugric included Proto-I-I *asura-* 'lord, god' > F-U *asera*; Proto-I-I *medʰu-* 'honey' > F-U *mete*; Proto-I-I *čekro-* 'wheel' > F-U *kekrä*; and Proto-I-I *arya-* 'Aryan' > F-U *orya*. Proto-Indo-Iranian *arya-*, the self designation "Aryan," was borrowed into Pre-Saami as *orja-*, the root of *oarji*, meaning "southwest," and of *ārjel*, meaning "southerner," confirming that the Proto-Aryan world lay south of the early Uralic region. The same borrowed *arya-* root developed into words with the meaning "slave" in the Finnish and Permic branches (Finnish, Komi, and Udmurt), a hint of ancient hostility between the speakers of Proto-Indo-Iranian and Finno-Ugric.[14]

PRE-SINTASHTA CULTURES OF THE EASTERN STEPPES

Who lived in the Ural-Tobol steppes during the late Abashevo era, before the Sintashta strongholds appeared there? There are two local antecedents and several unrelated neighbors.

Sintashta Antecedents

Just to the north of the steppe zone later occupied by Sintashta settlements, the southern forest-steppe zone contained scattered settlements of the late Abashevo culture. Abashevo miners regularly worked the quartzitic arsenic-rich copper ores of the Ural-Tobol region. Small settlements of the Ural variant of late Abashevo appeared in the upper Ural River valley and perhaps as far east as the upper Tobol. Geometric meanders first became a significant new decorative motif on Abashevo pottery made in the Ural region [see figure 15.7], and the geometric meander remained popular in Sintashta motifs. Some early Sintashta graves contained late Abashevo pots, and some late Abashevo sites west of the Urals contained Sintashta-type metal weapons and chariot gear such as disc-shaped cheekpieces that might have originated in the Sintashta culture. But Ural Abashevo people did not conduct mortuary animal sacrifices on a large scale, many of their metal types and ornaments were different, and, even though a few of their settlements were surrounded by small ditches, this was unusual. They were not fortified like the Sintashta settlements in the steppes.

Poltavka-culture herders had earlier occupied the northern steppe zone just where Sintashta appeared. The Poltavka culture was essentially a Volga-Ural continuation of the early Yamnaya horizon. Poltavka herding groups moved east into the Ural-Tobol steppes probably between 2800 and 2600 BCE. Poltavka decorative motifs on ceramics (vertical columns of chevrons) were very common on Sintashta pottery. A Poltavka kurgan cemetery (undated) stood on a low ridge 400 m south of the future site of Arkaim before that fortified settlement was built near the marshy bottom of the valley.[15] The cemetery, Aleksandrovska IV, contained twenty-one small (10–20 m in diameter) kurgans, a relatively large Poltavka cemetery (figure 15.8). Six were excavated. All conformed to the typical Poltavka rite: a kurgan surrounded by a circular ditch, with a single grave with ledges, the body tightly contracted on the left or right side, lying on an organic mat, red ochre or white chalk by the head and occasionally around the whole body, with a pot or a flint tool or nothing. A few animal bones occasionally were dropped in the perimeter ditch. A Poltavka settlement was stratified beneath the Sintashta settlement of Kuisak, which is intriguing because Poltavka settlements, like Yamnaya settlements, are generally unknown. Unfortunately this one was badly disturbed by the Sintashta settlement that was built on top of it.[16]

In the middle Volga region, the Potapovka culture was a contemporary sister of Sintashta, with similar graves, metal types, weapons, horse sacrifices, and chariot-driving gear (bone cheekpieces and whip handles), dated by radiocarbon to the same period, 2100–1800 BCE. Potapovka pottery, like Sintashta, retained many Poltavka decorative traits, and Potapovka graves were occasionally situated directly on top of older Poltavka monuments. Some Potapovka graves were dug right through preexisting Poltavka graves, destroying them, as some Sintashta strongholds were built on top of and incorporated older Poltavka settlements.[17] It is difficult to

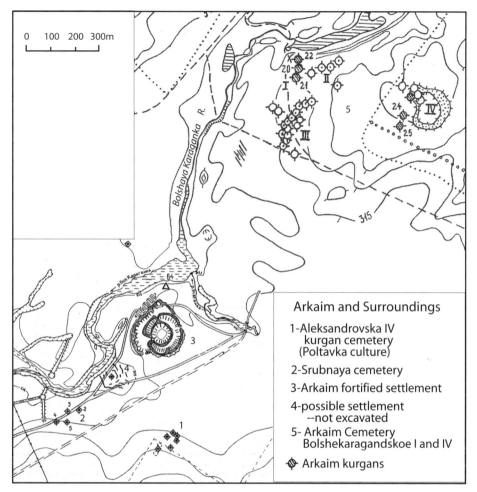

Figure 15.8 Arkaim settlement landscape with the kurgan cemeteries of Aleksandrovka IV (1), an older Poltavka cemetery of six kurgans; and Bolshekaragandskoe I and IV (5), with two excavated Sintashta-culture kurgans (24 and 25). Composite of Zdanovich 2002, Figure 3; and Batanina and Ivanova 1995, figure 2.

imagine that this was accidental. A symbolic connection with old Poltavka clans must have guided these choices.

Poltavka herders might have begun to explore across the vast Kazakh plains toward Sarazm, an outpost of Central Asian urban civilization established before 3000 BCE near modern Samarkand in the Zeravshan valley (see figure 16.1). Its northern location placed it just beyond the range of steppe herders who pushed east of the Urals around 2500 BCE.[18]

Hunters and Traders in Central Asia and the Forest Zone

Between the Poltavka territory in the upper Tobol steppes and Sarazm in the Zeravshan Valley lived at least two distinct groups of foragers. In the south, around the southern, western, and eastern margins of the Aral Sea, was the Kelteminar culture, a culture of relatively sedentary hunters and gatherers who built large reed-covered houses near the marshes and lakes in the steppes and in the riverbank thickets (called *tugai* forest) of the Amu Darya (Oxus) and lower Zeravshan rivers, where huge Siberian tigers still prowled. Kelteminar hunters pursued bison and wild pigs in the *tugai*, and gazelle, onagers, and Bactrian camels in the steppes and deserts. No wild horses ranged south of the Kyzl Kum desert, so Kelteminar hunters never saw horses, but they caught lots of fish, and collected wild pomegranates and apricots. They made a distinctive incised and stamped pottery. Early Kelteminar sites such as Dingil'dzhe 6 had microlithic flint industries much like those of Dzhebel Cave layer IV, dated about 5000 BCE. Kelteminar foragers probably began making pottery about this time, toward the end of the sixth millennium BCE. Late Kelteminar lasted until around 2000 BCE. Kelteminar pottery was found at Sarazm (level II), but the Kyzl Kum desert, north of the Amu Darya River, seems to have been an effective barrier to north-south communication with the northern steppes. Turquoise, which outcropped on the lower Zeravshan and in the desert southeast of the Aral Sea, was traded southward across Iran but not into the northern steppes. Turquoise ornaments appeared at Sarazm, at many early cities on the Iranian plateau, and even in the Maikop chieftain's grave (chapter 12), but not among the residents of the northern steppes.[19]

A second and quite different network of foragers lived in the northern steppes, north of the Aral Sea and the Syr Darya river (the ancient Jaxartes). Here the desert faded into the steppes of central and northern Kazakhstan, where the biggest predators were wolves and the largest grazing mammals were wild horses and saiga antelope (both absent in the Kelteminar region). In the lusher northern steppes, the descendants of the late Botai-Tersek culture still rode horses, hunted, and fished, but some of them now kept a few domesticated cattle and sheep and also worked metal. The post-Botai settlement of Sergeivka on the middle Ishim River is dated by radiocarbon about 2800–2600 BCE (4160±80

BP, OxA-4439). It contained pottery similar to late Botai-Tersek pottery, stone tools typical for late Botai-Tersek, and about 390 bones of horses (87%) but also 60 bones of cattle and sheep (13%), a new element in the economy of this region. Fireplaces, slag, and copper ore also were found. Very few sites like Sergeivka have been recognized in northern Kazakhstan. But Sergeivka shows that by 2800–2600 BCE an indigenous metallurgy and a little herding had begun in northern Kazakhstan. The impetus for these innovations probably was the arrival of Poltavka herders in the Tobol steppes. Pottery similar to that at Sergeivka was found in the Poltavka graves at Aleksandrovska IV, confirming contact between the two.[20]

North of the Ural-Tobol steppes, the foragers who occupied the forested eastern slopes of the Ural Mountains had little effect on the early Sintashta culture. Their natural environment was rich enough to permit them to live in relatively long-term settlements on river banks while still depending just on hunting and fishing. They had no formal cemeteries. Their pottery had complex comb-stamped geometric motifs all over the exterior surface. Ceramic decorations and shapes were somewhat similar between the forest-zone Ayatskii and Lipchinskii cultures on one side and the steppe zone Botai-Tersek cultures on the other. But in most material ways the forest-zone cultures remained distinct from Poltavka and Abashevo, until the appearance of the Sintashta culture, when this relationship changed. Forest-zone cultures adopted many Sintashta customs after about 2200–2100 BCE. Crucibles, slag, and copper rods interpreted as ingots appeared at Tashkovo II and Iska III, forager settlements located on the Tobol River north of Sintashta. The animal bones from these settlements were still from wild game—elk, bear, and fish. Some Tashkovo II ceramics displayed geometric meander designs borrowed from late Abashevo or Sintashta. And the houses at Tashkovo II and Andreevskoe Ozero XIII were built in a circle around an open central plaza, as at Sintashta or Arkaim, a settlement plan atypical of the forest zone.

THE ORIGIN OF THE SINTASHTA CULTURE

A cooler, more arid climate affected the Eurasian steppes after about 2500 BCE, reaching a peak of aridity around 2000 BCE. Ancient pollen grains cored from bogs and lake floors across the Eurasian continent show the effects this event had on wetland plant communities.[21] Forests retreated, open grassland expanded, and marshes dwindled. The steppes southeast of the Ural Mountains, already drier and colder than the Middle Volga grasslands southwest of the Urals, became drier still. Around 2100 BCE a

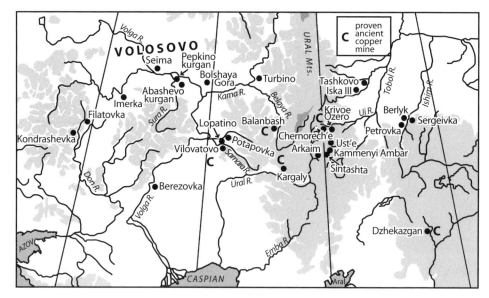

Figure 15.9 Sites of the period 2100–1800 BCE in the northern steppe and southern forest-steppe between the Don and the Ishim, with the locations of proven Bronze Age copper mines. The Sintashta-Potapovka-Filatovka complex probably is the archaeological manifestation of the Indo-Iranian language group.

mixed population of Poltavka and Abashevo herders began to settle in fortified strongholds between the upper Tobol and Ural River valleys, near the shrinking marshes that were vital for wintering their herds (see figure 15.9). Eurasian steppe pastoralists have generally favored marshy regions as winter refuges because of the winter forage and protection offered by stands of *Phragmites* reeds up to three meters tall. In a study of mobility among Late Mesolithic foragers in the Near East, Michael Rosenberg found that mobile populations tended to settle near critical resources when threatened with increased competition and declining productivity. He compared the process to a game of musical chairs,[22] in which the risk of losing a critical resource, in this case, winter marshlands for the cattle, was the impetus for settling down. Most Sintashta settlements were built on the first terrace overlooking the floodplain of a marshy, meandering stream. Although heavily fortified, these settlements were put in marshy, low places rather than on more easily defended hills nearby (see figures 15.2 and 15.8).

More than twenty Sintashta-type walled settlements were erected in the Ural-Tobol steppes between about 2100 and 1800 BCE. Their impressive fortifications indicate that concentrating people and herds near a critical wintering place was not sufficient in itself to protect it. Walls and towers also were required. Raiding must have been endemic. Intensified fighting encouraged tactical innovations, most important the invention of the light war chariot. This escalation of conflict and competition between rival tribal groups in the northern steppes was accompanied by elaborate ceremonies and feasts at funerals conducted within sight of the walls. Competition between rival hosts led to potlatch-type excesses such as the sacrifice of chariots and whole horses.

The geographic position of Sintashta societies at the eastern border of the Pontic-Caspian steppe world exposed them to many new cultures, from foragers to urban civilizations. Contact with the latter probably was most responsible for the escalation in metal production, funeral sacrifices, and warfare that characterized the Sintashta culture. The brick-walled towns of the Bactria-Margiana Archaeological Complex (BMAC) in Central Asia connected the metal miners of the northern steppes with an almost bottomless market for copper. One text from the city of Ur in present-day Iraq, dated to the reign of Rim-Sin of Larsa (1822–1763 BCE), recorded the receipt of 18,333 km (40,417 lb, or 20 tons) of copper in a single shipment, most of it earmarked for only one merchant.[23] This old and well-oiled Asian trade network was connected to the northern Eurasian steppes for the first time around 2100–2000 BCE (see chapter 16 for the contact between Sintashta and BMAC sites).

The unprecedented increase in demand for metal is documented most clearly on the floors of Sintashta houses. Sintashta settlements were industrial centers that specialized in metal production. Every excavated structure at Sintashta, Arkaim, and Ust'e contained the remains of smelting ovens and slag from processing copper ore. The metal in the majority of finished objects was arsenical bronze, usually in alloys of 1–2.5% arsenic; tin-bronzes comprised only 2% or less of metal objects. At Sintashta, 36% of tested objects were made of copper with elevated arsenic (from 0.1–1% arsenic), and 48% were classified as arsenical bronze (over 1% arsenic). Unalloyed copper objects were more frequent at Arkaim, where they constituted almost half the tested objects, than at Sintashta, where they made up only 10% of tested objects. Clay tubular pipes probably for the mouths of the bellows, or *tulieres*, occurred in graves and settlements (see figure 15.4). Pieces of crucibles were found in graves at Krivoe Ozero. Closed two-piece molds were required to cast bronze shaft-hole axes and spear blades (see figure 15.10). Open single-piece molds for casting curved sickles and rod-like copper ingots were found in the Arkaim settlement.

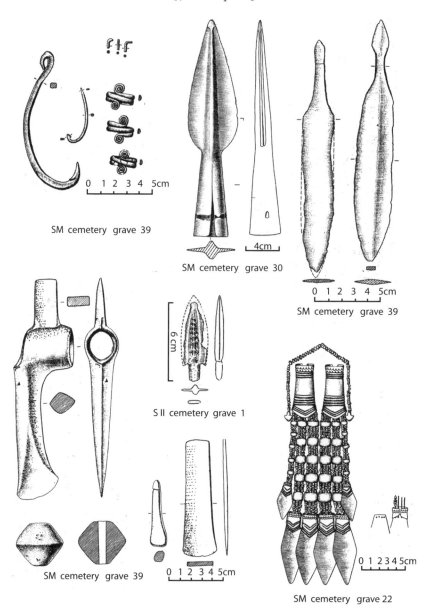

Figure 15.10 Weapons, tools, and ornaments from graves at Sintashta. After Gening, Zdanovich, and Gening 1992, figures 99, 113, 126, and 127.

Ingots or rods of metal weighing 50–130 g might have been produced for export. An estimated six thousand tons of quartzitic rock bearing 2–3% copper was mined from the single excavated mining site of Vorovskaya Yama east of the upper Ural River.[24]

Warfare, a powerful stimulus to social and political change, also shaped the Sintashta culture, for a heightened threat of conflict dissolves the old social order and creates new opportunities for the acquisition of power. Nicola DiCosmo has recently argued that complex political structures arose among steppe nomads in the Iron Age largely because intensified warfare led to the establishment of permanent bodyguards around rival chiefs, and these grew in size until they became armies, which engendered state-like institutions designed to organize, feed, reward, and control them. Susan Vehik studied political change in the deserts and grasslands of the North American Southwest after 1200 CE, during a period of increased aridity and climatic volatility comparable to the early Sintashta era in the steppes. Warfare increased sharply during this climatic downturn in the Southwest. Vehik found that long-distance trade increased greatly at the same time; trade after 1350 CE was more than forty times greater than it had been before then. To succeed in war, chiefs needed wealth to fund alliance-building ceremonies before the conflict and to reward allies afterward. Similarly, during the climatic crisis of the late MBA in the steppes, competing steppe chiefs searching for new sources of prestige valuables probably discovered the merchants of Sarazm in the Zeravshan valley, the northernmost outpost of Central Asian civilization. Although the connection with Central Asia began as an extension of old competitions between tribal chiefs, it created a relationship that fundamentally altered warfare, metal production, and ritual competition among the steppe cultures.[25]

WARFARE IN THE SINTASHTA CULTURE: FORTIFICATIONS AND WEAPONS

A significant increase in the intensity of warfare in the southern Ural steppes is apparent from three factors: the regular appearance of large fortified towns; increased deposits of weapons in graves; and the development of new weapons and tactics. All the Sintashta settlements excavated to date, even relatively small ones like Chernorech'ye III, with perhaps six structures (see figure 15.11), and Ust'e, with fourteen to eighteen structures, were fortified with V-shaped ditches and timber-reinforced earthen walls.[26] Wooden palisade posts were preserved inside the earthen walls at

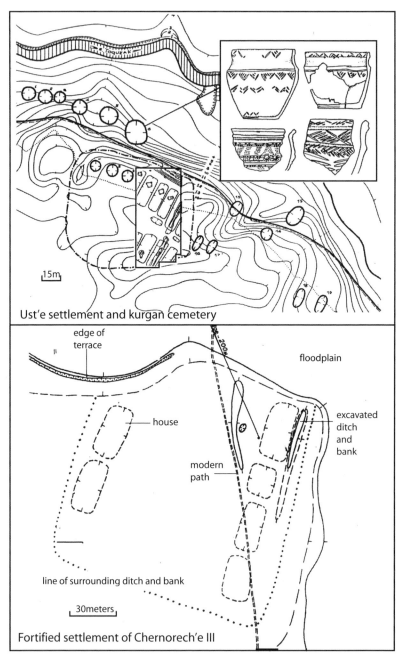

Figure 15.11 Smaller walled settlements of the Sintashta type at Ust'e and Chernorech'e III. After Vinogradov 2003, figure 3.

Ust'ye, Arkaim, and Sintashta. Communities build high walls and gates when they have reason to fear that their homes will come under attack.

The graves outside the walls now also contained many more weapons than in earlier times. The Russian archaeologist A. Epimakhov published a catalogue of excavated graves from five cemeteries of the Sintashta culture: Bol'shekaragandskoe (the cemetery for the Arkaim citadel), Kammeny Ambar 5, Krivoe Ozero, Sintashta, and Solntse II.[27] The catalogue listed 242 individuals in 181 graves. Of these, 65 graves contained weapons. Only 79 of the 242 individuals were adults, but 43 of these, or 54% of all adults, were buried with weapons. Most of the adults in the weapon graves were not assigned a gender, but of the 13 that were, 11 were males. Most adult males of the Sintashta culture probably were buried with weapons. In graves of the Poltavka, Catacomb, or Abashevo cultures, weapons had been unusual. They were more frequent in Abashevo than in the steppe graves, but the great majority of Abashevo graves did not contain weapons of any kind, and, when they did, usually it was a single axe or a projectile point. My reading of reports on kurgan graves of the earlier EBA and MBA suggests to me that less than 10% contained weapons. The frequency of weapons in adult graves of the Sintashta culture (54%) was much higher.

New types of weapons also appeared. Most of the weapon types in Sintashta graves had appeared earlier—bronze or copper daggers, flat axes, shaft-hole axes, socketed spears, polished stone mace heads, and flint or bone projectile points. In Sintashta-culture graves, however, longer, heavier projectile point types appeared, and they were deposited in greater numbers. One new projectile was a spearhead made of heavy bronze or copper with a socketed base for a thick wooden spear handle. Smaller, lighter-socketed spearheads had been used occasionally in the Fatyanovo culture, but the Sintashta spear was larger (see figure 15.3). Sintashta graves also contained two varieties of chipped flint projectile points: lanceolate and stemmed (see figure 15.12). Short lanceolate points with flat or slightly hollow bases became longer in the Sintashta period, and these were deposited in groups for the first time. They might have been for arrows, since prehistoric arrow points were light in weight and usually had flat or hollow bases. Lanceolate flint points with a hollow or flat base occurred in seven graves at Sintashta, with up to ten points in one grave (SM gr. 39). A set of five lanceolate points was deposited in the chariot grave of Berlyk II, kurgan 10.

More interesting were flint points of an entirely new type, with a contracting stem, defined shoulders, and a long, narrow blade with a thick medial ridge, 4–10 cm long. These new stemmed points might have been for javelins. Their narrow, thick blades were ideal for javelin points because the

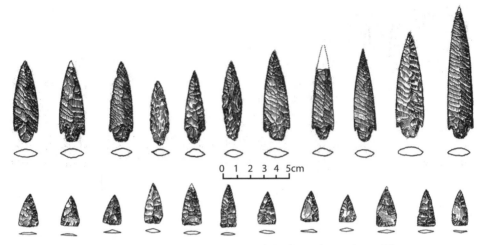

Figure 15.12 Flint projectile point types of the Sintashta culture. The top row was a new type for steppe cultures, possibly related to the introduction of the javelin. The bottom row was an old type in the steppes, possibly used for arrows, although in older EBA and MBA graves it was more triangular. After Gening, Zdanovich, and Gening 1992.

heavier shaft of a javelin (compared to an arrow) causes greater torque stress on the embedded point at the moment of impact; moreover, a narrow, thick point could penetrate deeper before breaking than a thin point could.[28] A stemmed point, by definition, is mounted in a socketed foreshaft, a complex type of attachment usually found on spears or javelins rather than arrows. Smaller stemmed points had existed earlier in Fatyanovo and Balanovo tool kits and were included in occasional graves, as at the Fatyanovo cemetery of Volosovo-Danilovskii, where 1 grave out of 107 contained a stemmed point, but it was shorter than the Sintashta type (only 3–4 cm long). Sintashta stemmed points appeared in sets of up to twenty in a single grave (chariot gr. 20 at the Sintashta SM cemetery), as well as in a few Potapovka graves on the middle Volga. Stemmed points made of cast bronze, perhaps imitations of the flint stemmed ones, occurred in one chariot grave (SM gr. 16) and in two other graves at Sintashta (see figure 15.11).

Weapons were deposited more frequently in Sintashta graves. New kinds of weapons appeared, among them long points probably intended for javelins, and they were deposited in sets that appear to represent warriors' equipment for battle. Another signal of increased conflict is the most hotly debated artifact of this period in the steppes—the light, horse-drawn chariot.

Sintashta Chariots: Engines of War

A chariot is a two-wheeled vehicle with spoked wheels and a standing driver, pulled by bitted horses, and usually driven at a gallop. A two-wheeler with solid wheels or a seated driver is a cart, not a chariot. Carts, like wagons, were work vehicles. Chariots were the first wheeled vehicles designed for speed, an innovation that changed land transport forever. The spoked wheel was the central element that made speed possible. The earliest spoked wheels were wonders of bent-wood joinery and fine carpentry. The rim had to be a perfect circle of joined wood, firmly attached to individually carved spokes inserted into mortices in the outer wheel and a multi-socketed central nave, all carved and planed out of wood with hand tools. The cars also were stripped down to just a few wooden struts. Later Egyptian chariots had wicker walls and a floor of leather straps for shock absorption, with only the frame made of wood. Perhaps originally designed for racing at funerals, the chariot quickly became a weapon and, in that capacity, changed history.

Today most authorities credit the invention of the chariot to Near Eastern societies around 1900–1800 BCE. Until recently, scholars believed that the chariots of the steppes post-dated those of the Near East. Carvings or petroglyphs showing chariots on rock outcrops in the mountains of eastern Kazakhstan and the Russian Altai were ascribed to the Late Bronze Age Andronovo horizon, thought to date after 1650 BCE. Disk-shaped cheekpieces made of antler or bone found in steppe graves were considered copies of older Mycenaean Greek cheekpieces designed for the bridles of chariot teams. Because the Mycenaean civilization began about 1650 BCE, the steppe cheekpieces also were assumed to date after 1650 BCE.[29]

The increasing amount of information about chariot graves in the steppes since about 1992 has challenged this orthodox view. The archaeological evidence of steppe chariots survives only in graves where the wheels were placed in slots that had been dug into the grave floors. The lower parts of the wheels left stains in the earth as they rotted (see figure 15.13). These stains show an outer circle of bent wood 1–1.2 m in diameter with ten to twelve square-sectioned spokes. There is disagreement as to the number of clearly identified chariot graves because the spoke imprints are faint, but even the conservative estimate yields sixteen chariot graves in nine cemeteries. All belonged to either the Sintashta culture in the Ural-Tobol steppes or the Petrovka culture east of Sintashta in northern Kazakhstan. Petrovka was contemporary with late Sintashta, perhaps 1900–1750 BC, and developed directly from it.[30]

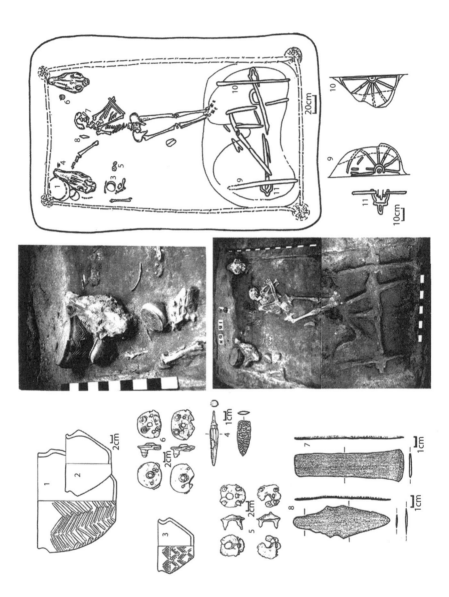

Scholars disagree as to whether steppe chariots were effective instruments of war or merely symbolic vehicles designed only for parade or ritual use, made in barbaric imitation of superior Near Eastern originals.[31] This debate has focused, surprisingly, on the distance between the chariots' wheels. Near Eastern war chariots had crews of two or even three—a driver and an archer, and occasionally a shield-bearer to protect the other two from incoming missiles. The gauge or track width of Egyptian chariots of ca. 1400–1300 BCE, the oldest Near Eastern chariots preserved well enough to measure, was 1.54–1.80 m. The hub or nave of the wheel, a necessary part that stabilized the chariot, projected at least 20 cm along the axle on each side. A gauge around 1.4–1.5 m would seem the minimum to provide enough room between the wheels for the two inner hubs or naves (20 + 20 cm) and a car at least 1 m wide to carry two men. Sintashta and Petrovka-culture chariots with less than 1.4–1.5 m between their wheels were interpreted as parade or ritual vehicles unfit for war.

This dismissal of the functional utility of steppe chariots is unconvincing for six reasons. First, steppe chariots were made in many sizes, including two at Kammeny Ambar 5, two at Sintashta (SM gr. 4, 28) and two at Berlyk (Petrovka culture) with a gauge between 1.4 and 1.6 m, big enough for a crew of two. The first examples published in English, which were from Sintashta (SM gr. 19) and Krivoe Ozero (k. 9, gr. 1), had gauges of only about 1.2–1.3 m, as did three other Sintashta chariots (SM gr. 5, 12, 30) and one other Krivoe Ozero chariot. The argument against the utility of steppe chariots focused on these six vehicles, most of which, in spite of their narrow gauges, were buried with weapons. However, six other steppe vehicles were as wide as some Egyptian war chariots. One (Sintashta SM gr. 28) with a gauge of about 1.5 m was placed in a grave that also contained the partial remains of two adults, possibly its crew. Even if we accept the doubtful assumption that war chariots needed a crew of two, many steppe chariots were big enough.[32]

Second, steppe chariots were not necessarily used as platforms for archers. The preferred weapon in the steppes might have been the javelin. A single

---

Figure 15.13 Chariot grave at Krivoe Ozero, kurgan 9, grave 1, dated about 2000 BCE: (1–3) three typical Sintashta pots; (5–6) two pairs of studded disk cheekpieces made of antler; (4) a bone and a flint projectile point; (7–8) a waisted bronze dagger and a flat bronze axe; (9–10) spoked wheel impressions from wheels set into slots in the floor of the grave; (11) detail of artist's reconstruction of the remains of the nave or hub on the left wheel. After Anthony and Vinogradov 1995, photos by Vinogradov.

warrior-driver could hold the reins in one hand and hurl a javelin with the other. From a standing position in a chariot, a driver-warrior could use his entire body to throw, whereas a man on horseback without stirrups (invented after 300 CE) could use only his arm and shoulder. A javelin-hurling charioteer could strike a man on horseback before the rider could strike him. Unlike a charioteer, a man on horseback could not carry a large sheath full of javelins and so would be at a double disadvantage if his first cast missed. A rider armed with a bow would fare only slightly better. Archers of the steppe Bronze Age seem to have used bows 1.2–1.5 m long, judging by bow remains found at Berezovka (k. 3, gr. 2) and Svatove (k. 12, gr. 12).[33] Bows this long could be fired from horseback only to the side (the left side, for a right-handed archer), which made riders with long bows vulnerable. A charioteer armed with javelins could therefore intimidate a Bronze Age rider on horseback. Many long-stemmed points, suitable for javelins, were found in some chariot graves (Sintashta SM gr. 4, 5, 30). If steppe charioteers used javelins, a single man could use narrower cars in warfare.

Third, if a single driver-warrior needed to switch to a bow in battle, he could fire arrows while guiding the horses with the reins around his hips. Tomb paintings depicted the Egyptian pharaoh driving and shooting a bow in this way. Although it may have been a convention to include only the pharaoh in these illustrations, Littauer noted that a royal Egyptian scribe was also shown driving and shooting in this way, and in paintings of Ramses III fighting the Libyans the archers in the Egyptian two-man chariots had the reins around their hips. Their car-mates helped to drive with one hand and used a shield with the other. Etruscan and Roman charioteers also frequently drove with the reins wrapped around their hips.[34] A single driver-warrior might have used a bow in this manner, although it would have been safer to shift the reins to one hand and cast a javelin.

The fourth reason not to dismiss the functionality of steppe chariots is that most of these chariots, including the narrow-gauge ones, were buried with weapons. I have seen complete inventories for twelve Sintashta and Petrovka chariot graves, and ten contained weapons. The most frequent weapons were projectile points, but chariot graves also contained metal-waisted daggers, flat metal axes, metal shaft-hole axes, polished stone mace heads, and one metal-socketed spearhead 20 cm long (from Sintashta SM gr. 30; see figure 15.3). According to Epimakhov's catalogue of Sintashta graves, cited earlier, all chariot graves where the skeleton could be assigned a gender contained an adult male. If steppe chariots were not designed for war, why were most of them buried with a male driver and weapons?

Fifth, a new kind of bridle cheekpiece appeared in the steppes at the very time that chariots did (see figure 15.14). It was made of antler or bone

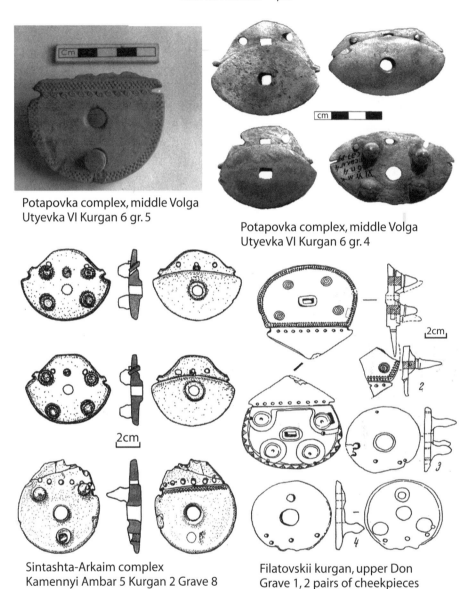

Potapovka complex, middle Volga
Utyevka VI Kurgan 6 gr. 5

Potapovka complex, middle Volga
Utyevka VI Kurgan 6 gr. 4

Sintashta-Arkaim complex
Kamennyi Ambar 5 Kurgan 2 Grave 8

Filatovskii kurgan, upper Don
Grave 1, 2 pairs of cheekpieces

Figure 15.14 Studded disk cheekpieces from graves of the Sintashta, Pota-povka, and Filatovka types. The band of running spirals beneath the checker-board panel on the upper left specimen from Utyevka VI was once thought to be derived from Mycenae. But the steppe examples like this one were older than Mycenae. Photos by the author; drawings after Epimakhov 2002; and Siniuk and Kosmirchuk 1995.

and shaped like an oblong disk or a shield, perforated in the center so that cords could pass through to connect the bit to the bridle and in various other places to allow for attachments to the noseband and cheek-strap. Pointed studs or prongs on its inner face pressed into the soft flesh at the corners of the horse's mouth when the driver pulled the reins on the opposite side, prompting an immediate response from the horse. The development of a new, more severe form of driving control suggests that rapid, precise maneuvers by the driving team were necessary. When disk cheekpieces are found in pairs, different shapes with different kinds of wear are often found together, as if the right and left sides of the horse, or the right and left horses, needed slightly different kinds of control. For example, at Krivoe Ozero (k. 9, gr. 1), the cheekpieces with the left horse had a slot located above the central hole, angled upward, toward the noseband (see figure 15.13). The cheekpieces with the right horse had no such upward-angled slot. A similar unmatched pair, with and without an upward-angled slot, were buried with a chariot team at Kamennyi Ambar 5 (see figure 15.14). The angled slot may have been for a noseband attached to the reins that would pull down on the inside (left) horse's nose, acting as a brake, when the reins were pulled, while the outside (right) horse was allowed to run free—just what a left-turning racing team would need. The chariot race, as described in the *Rig Veda,* was a frequent metaphor for life's challenges, and Vedic races turned to the left. Chariot cheekpieces of the same general design, a bone disk with sharp prongs on its inner face, appeared later in Shaft Grave IV at Mycenae and in the Levant at Tel Haror, made of metal. The oldest examples appeared in the steppes.[35]

Finally, the sixth flaw in the argument that steppe chariots were poorly designed imitations of superior Near Eastern originals is that the oldest examples of the former predate any of the dated chariot images in the Near East. Eight radiocarbon dates have been obtained from five Sintashta-culture graves containing the impressions of spoked wheels, including three at Sintashta (SM cemetery, gr. 5, 19, 28), one at Krivoe Ozero (k. 9, gr. 1), and one at Kammeny Ambar 5 (k. 2, gr. 8). Three of these (3760 ± 120 BP, 3740 ± 50 BP, and 3700 ± 60 BP), with probability distributions that fall predominantly before 2000 BCE, suggest that the earliest chariots *probably* appeared in the steppes before 2000 BCE (table 15.1). Disk-shaped cheekpieces, usually interpreted as specialized chariot gear, also occur in steppe graves of the Sintashta and Potapovka types dated by radiocarbon before 2000 BCE. In contrast, in the Near East the oldest images of true chariots—vehicles with *two spoked* wheels, pulled by *horses* rather than asses or onagers, controlled with *bits* rather than lip- or nose-rings, and guided by a *standing warrior*, not a seated driver—first appeared about 1800 BCE, on Old Syrian seals. The oldest images in Near Eastern art of vehicles with two spoked wheels appeared on seals from Karum Kanesh II, dated about 1900 BCE, but the equids were of an uncertain type (possibly native asses or onagers) and they were controlled by nose-rings (see figure 15.15). Excavations at Tell Brak in northern Syria recovered 102 cart models and 191 equid figurines from the parts of this ancient walled caravan city dated to the late Akkadian and Ur III periods, 2350–2000 BCE by the standard or "middle" chronology. None of the equid figurines was clearly a horse. Two-wheeled carts were common among the vehicle models, but they had built-in seats and solid wheels. No chariot models were found. Chariots were unknown here as they were elsewhere in the Near East before about 1800 BCE.[36]

Chariots were invented earliest in the steppes, where they were used in warfare. They were introduced to the Near East through Central Asia, with steppe horses and studded disk cheekpieces (see chapter 16). The horse-drawn chariot was faster and more maneuverable than the old solid-wheeled battle-cart or battle-wagon that had been pulled into inter-urban battles by ass-onager hybrids in the armies of Early Dynastic, Akkadian, and Ur III kings between 2900 and 2000 BCE. These heavy, clumsy vehicles, mistakenly described as chariots in many books and catalogues, were similar to steppe chariots in one way: they were consistently depicted carrying javelin-hurling warriors, not archers. When horse-drawn chariots appeared in the Near East they quickly came to dominate inter-urban battles as swift platforms for archers, perhaps a Near Eastern innovation. Their wheels also were made differently, with just four or six spokes, apparently another improvement on the steppe design.

Among the Mitanni of northern Syria, in 1500–1350 BC, whose chariot tactics might have been imported with their Old Indic chariot terminology from a source somewhere in the steppes, chariots were organized into squadrons of five or six; six such units (thirty to thirty-six chariots) were combined with infantry under a brigade commander. A similar organization appeared in Chou China a millennium later: five chariots in a squadron, five squadrons in a brigade (twenty-five), with ten to twenty-five support infantry for each chariot.[37] Steppe chariots might also have operated in squadrons supported by individuals on foot or even on horseback, who could have run forward to pursue the enemy with hand weapons or to rescue the charioteer if he were thrown.

Chariots were effective in tribal wars in the steppes: they were noisy, fast, and intimidating, and provided an elevated platform from which a skilled

Near Eastern two-wheelers

Figure 15.15 Two-wheeled, high-speed vehicles of the ancient Near East prior to the appearance of the chariot: (a) cast copper model of a straddle-car with solid wheels pulled by a team of ass-onager–type equids from Tell Agrab, 2700–2500 BCE; (b and c) engraved seal images of vehicles with four-spoked wheels, pulled by equids (?) controlled with lip- or nose-rings from *karum* Kanesh II, 1900 BCE. After Raulwing 2000, figures 7.2 and 10.1.

driver could hurl a sheath full of javelins. As the car hit uneven ground at high speed, the driver's legs had to absorb each bounce, and the driver's weight had to shift to the bouncing side. To drive through a turn, the inside horse had to be pulled in while the outside horse was given rein. Doing this well and hurling a javelin at the same time required a lot of practice. Chariots were supreme advertisements of wealth; difficult to make and requiring great athletic skill *and* a team of specially trained horses to drive, they were available only to those who could delegate much of their daily labor to hired herders. A chariot was material proof that the driver was able to fund a substantial alliance or was supported by someone who had the means. Taken together, the evidence from fortifications, weapon types, and numbers, and the tactical innovation of chariot warfare, all indicate that conflict increased in both scale and intensity in the northern steppes during the early Sintashta period, after about 2100 BCE. It is also apparent that chariots played an important role in this new kind of conflict.

Tournaments of Value

Parallels between the funerals of the Sintashta chiefs and the funeral hymns of the *Rig Veda* (see below) suggest that poetry surrounded chariot burials. Archaeology reveals that feasts on a surprising scale also accompanied chiefly funerals. Poetry and feasting were central to a mortuary performance that emphasized exclusivity, hierarchy, and power—what the anthropologist A. Appadurai called "tournaments of value," ceremonies meant to define membership in the elite and to channel political competition within clear boundaries that excluded most people. In order to understand the nature of these sacrificial dramas, we first have to understand the everyday secular diet.[38]

Flotation of seeds and charcoal from the soils excavated at Arkaim recovered only a few charred grains of barley, too few, in fact, to be certain that they came from the Sintashta-culture site rather than a later occupation. The people buried at Arkaim had no dental caries, indicating that they ate a very low-starch diet, not starchy cereals.[39] Their teeth were like those of hunter-gatherers. Charred millet was found in test excavations at the walled Alands'koe stronghold, indicating that some millet cultivation probably occurred at some sites, and dental decay *was* found in the Krivoe Ozero cemetery population, so some communities might have consumed cultivated grain. Gathering wild seeds from *Chenopodium* and *Amaranthus*, plants that still played an important role in the LBA steppe diet centuries later (see chapter 16 for LBA wild plants), could have supplemented occasional cereal cultivation. Cultivated cereals seem to have played a minor role in the Sintashta diet.[40]

The scale of animal sacrifices in Sintashta cemeteries implies very large funerals. One example was Sacrificial Complex 1 at the northern edge of the Sintashta SM cemetery (see figure 15.16). In a pit 50 cm deep, the heads and hooves of six horses, four cattle, and two rams lay in two rows facing one another around an overturned pot. This single sacrifice provided about six thousand pounds (2,700 kg) of meat, enough to supply each of three thousand participants with two pounds (.9 kg). The Bolshoi Kurgan, built just a few meters to the north, required, by one estimate, three thousand man-days.[41] The workforce required to build the kurgan matched the amount of food provided by Sacrificial Complex 1. However, the Bolshoi Kurgan was unique; the other burial mounds at Sintashta were small and low. If the sacrifices that accompanied the other burials at Sintashta were meant to feed work parties, what they built is not obvious. It seems more likely that most sacrifices were intended to provide food for the funeral guests. With up to eight horses sacrificed for a single funeral, Sintashta feasts would have fed hundreds, even thousands of guests. Feast-hosting behavior is the most common and consistently used avenue to prestige and power in tribal societies.[42]

The central role of horses in Sintashta funeral sacrifices was unprecedented in the steppes. Horse bones had appeared in EBA and earlier MBA graves but not in great numbers, and not as frequently as those of sheep or cattle. The animal bones from the Sintashta and Arkaim settlement refuse middens were 60% cattle, 26% sheep-goat, and 13% horse. Although beef supplied the preponderance of the meat diet, the funeral sacrifices in the cemeteries contained just 23% cattle, 37% sheep-goat, and 39% horse. Horses were sacrificed more than any other animal, and horse bones were three times more frequent in funeral sacrifices than in settlement middens. The zoologist L. Gaiduchenko suggested that the Arkaim citadel specialized in horse breeding for export because the high level of ^{15}N isotopes in human bone suggested that horses, very low in ^{15}N, were not eaten frequently. Foods derived from cattle and sheep, significantly higher in ^{15}N than the horses from these sites, probably composed most of the diet.[43] According to Epimakhov's catalogue of five Sintashta cemeteries, the most frequent animal sacrifices were horses but they were sacrificed in no more than 48 of the 181 graves catalogued, or 27%; multiple horses were sacrificed in just 13% of graves. About one-third of the graves contained weapons, but, among these, two-thirds of graves with horse sacrifices contained weapons, and 83% of graves with multiple horse sacrifices contained weapons. Only a minority of Sintashta graves contained

Sintashta cemetery SM sacrificial complex 1

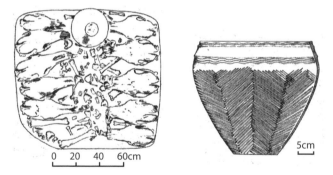

Figure 15.16 Sacrificial complex number 1 at the northern edge of the Sintashta SM cemetery. After Gening, Zdanovich, and Gening 1992, figure 130.

horse sacrifices, but those that did usually also contained weapons, a symbolic association between the ownership of large horse herds, the hosting of feasts, and the warrior's identity.

There is little jewelry or ornaments in Sintashta graves, and no large houses or storage facilities in the settlements. The signs of craft specialization, a signal of social hierarchy, are weak in all crafts except metallurgy, but even in that craft, every household in every settlement seems to have worked metal. The absence of large houses, storage facilities, or craft specialists has led some experts to doubt whether the Sintashta culture had a strong social hierarchy.[44] Sintashta cemeteries contained the graves of a cross-section of the entire age and sex spectrum, including many children, apparently a more inclusive funeral ritual than had been normal in EBA and earlier MBA mortuary ceremonies in the steppes. On the other hand, most Sintashta cemeteries did not contain enough graves to account for more than a small segment of the population of the associated walled settlements. The Sintashta citadel included about fifty to sixty structures, and its associated cemeteries had just sixty-six graves, most of them the graves of children. If the settlement contained 250 people for six generations (150 years), it should have generated more than fifteen hundred graves. Only a few exceptional families were given funerals in Sintashta cemeteries, but the entire family, including children, was honored in this way. This privilege, like the sacrifice of horses and chariots, was not one that everyone could claim. Horses, chariots, weapons, and multiple animal sacrifices identified the graves of the Sintashta chiefs.

The funeral sacrifices of the Simtashta culture are a critical link between archaeology and history. They closely resembled the rituals described in the *Rig Veda*, the oldest text preserved in an Indo-Iranian language.

Sintashta and the Origins of the Aryans

The oldest texts in Old Indic are the "family books," books 2 through 7, of the *Rig Veda* (RV). These hymns and prayers were compiled into "books" or mandalas about 1500–1300 BCE, but many had been composed earlier. The oldest parts of the *Avesta* (AV), the Gathas, the oldest texts in Iranian, were composed by Zarathustra probably about 1200–1000 BCE. The undocumented language that was the parent of both, common Indo-Iranian, must be dated well before 1500 BCE, because, by this date, Old Indic had already appeared in the documents of the Mitanni in North Syria (see chapter 3). Common Indo-Iranian probably was spoken during the Sintashta period, 2100–1800 BCE. Archaic Old Indic probably emerged as a separate tongue from archaic Iranian about 1800–1600 BCE (see chapter 16). The RV and AV agreed that the essence of their shared parental Indo-Iranian identity was linguistic and ritual, not racial. If a person sacrificed to the right gods in the right way using the correct forms of the traditional hymns and poems, that person was an Aryan.[45] Otherwise the individual was a *Dasyu*, again not a racial or ethnic label but a ritual and linguistic one—a person who interrupted the cycle of giving between gods and humans, and therefore a person who threatened cosmic order, *r'ta* (RV) or *aša* (AV). Rituals performed *in the right words* were the core of being an Aryan.

Similarities between the rituals excavated at Sintashta and Arkaim and those described later in the RV have solved, for many, the problem of Indo-Iranian origins.[46] The parallels include a reference in RV 10.18 to a kurgan ("let them . . . bury death in this hill"), a roofed burial chamber supported with posts ("let the fathers hold up this pillar for you"), and with shored walls ("I shore up the earth all around you; let me not injure you as I lay down this clod of earth"). This is a precise description of Sintashta and Potapovka-Filatovka grave pits, which had wooden plank roofs supported by timber posts and plank shoring walls. The horse sacrifice at a royal funeral is described in RV 1.162: "Keep the limbs undamaged and place them in the proper pattern. Cut them apart, calling out piece by piece." The horse sacrifices in Sintashta, Potapovka, and Filatovka graves match this description, with the lower legs of horses carefully cut apart at the joints and placed in and over the grave. The preference for horses as sacrificial animals in Sintashta funeral rituals, a species choice setting Sintashta apart from earlier steppe cultures, was again paralleled in the RV. Another verse in the same hymn read: "Those who see that the racehorse is cooked, who say, 'It smells good! Take it away!' and who wait for the doling out of the flesh of the charger—let their approval encourage us." These lines describe the public feasting that surrounded the funeral of an important person, exactly like the feasting implied by head-and-hoof deposits of horses, cattle, goats, and sheep in Sintashta graves that would have yielded hundreds or even thousands of kilos of meat. In RV 5.85, Varuna released the rain by overturning a pot: "Varuna has poured out the cask, turning its mouth downward. With it the king of the whole universe waters the soil." In Sacrificial Deposit 1 at Sintashta an overturned pot was placed between two rows of sacrificed animals—in a ritual possibly associated with the construction of the enormous Bolshoi Kurgan.[47] Finally, the RV eloquently documents the importance of the poetry and speech making that accompanied all these events. "Let us speak great words as men of power in the sacrificial gathering" was the standard closing attached repeatedly to several different hymns (RV 2.12, 2.23, 2.28) in one of the "family books." These public performances played an important role in attracting and converting celebrants to the Indo-Iranian ritual system and language.

The explosion of Sintashta innovations in rituals, politics, and warfare had a long-lasting impact on the later cultures of the Eurasian steppes. This is another reason why the Sintashta culture is the best and clearest candidate for the crucible of Indo-Iranian identity and language. Both the Srubnaya and the Andronovo horizons, the principal cultural groups of the Late Bronze Age in the Eurasian steppes (see chapter 16), grew from origins in the Potapovka-Sintashta complex.

A Srubnaya site excavated by this author contained surprising evidence for one more parallel between Indo-Iranian (and perhaps even Proto-Indo-European) ritual and archaeological evidence in the steppes: the midwinter New Year's sacrifice and initiation ceremony, held on the winter solstice. Many Indo-European myths and rituals contained references to this event. One of its functions was to initiate young men into the warrior category (*Männerbünde, korios*), and its principal symbol was the dog or wolf. Dogs represented death; multiple dogs or a multi-headed dog (*Cerberus, Saranyu*) guarded the entrance to the Afterworld. At initiation, death came to both the old year and boyhood identities, and as boys became warriors they would feed the dogs of death. In the RV the oath brotherhood of warriors that performed sacrifices at midwinter were called the Vrâtyas, who also were called dog-priests. The ceremonies associated with them featured many contests, including poetry recitation and chariot races.[48]

At the Srubnaya settlement of Krasnosamarskoe (Krasno-sa-MAR-sko-yeh) in the Samara River valley, we found the remains of an LBA midwinter dog sacrifice, a remarkable parallel to the reconstructed midwinter New Year ritual, dated about 1750 BCE. The dogs were butchered only at midwinter, many of them near the winter solstice, whereas the cattle and sheep at this site were butchered throughout the year. Dogs accounted for 40% of all the animal bones from the site. At least eighteen dogs were butchered, probably more. Nerissa Russell's studies showed that each dog head was burned and then carefully chopped into ten to twelve small, neat, almost identical segments with axe blows. The postcranial remains were not chopped into ritually standardized little pieces, and none of the cattle or sheep was butchered like this. The excavated structure at Krasnosamarskoe probably was the place where the dog remains from a midwinter sacrifice were discarded after the event. They were found in an archaeological context assigned to the early Srubnaya culture, but early Srubnaya was a direct outgrowth from Potapovka and Abashevo, the same circle as Sintashta, and nearly the same date. Krasnosamarskoe shows that midwinter dog sacrifices were practiced in the middle Volga steppes, as in the dog-priest initiation rituals described in the RV. Although such direct evidence for midwinter dog rituals has not yet been recognized in Sintashta settlements, many individuals buried in Sintashta graves wore necklaces of dog canine teeth. Nineteen dog canine pendants were found in a single collective grave with eight youths—probably of initiation age—under a Sintashta kurgan at Kammenyi Ambar 5, kurgan 4, grave 2.[49]

In many small ways the cultures between the upper Don and Tobol rivers in the northern steppes showed a common kinship with the Aryans of the *Rig Veda* and *Avesta*. Between 2100 and 1800 BCE they invented the chariot, organized themselves into stronghold-based chiefdoms, armed themselves with new kinds of weapons, created a new style of funeral rituals that involved spectacular public displays of wealth and generosity, and began to mine and produce metals on a scale previously unimagined in the steppes. Their actions reverberated across the Eurasian continent. The northern forest frontier began to dissolve east of the Urals as it had earlier west of the Urals; metallurgy and some aspects of Sintashta settlement designs spread north into the Siberian forests. Chariotry spread west through the Ukrainian steppe MVK culture into southeastern Europe's Monteoru (phase Ic1-Ib), Vatin, and Otomani cultures, perhaps with the *satəm* dialects that later popped up in Armenian, Albanian, and Phrygian, all of which are thought to have evolved in southeastern Europe. (Pre-Greek must have departed before this, as it did not share in the *satəm* innovations.) And the Ural frontier was finally broken—herding economies spread eastward across the steppes. With them went the eastern daughters of Sintashta, the offspring who would later emerge into history as the Iranian and Vedic Aryans. These eastern and southern connections finally brought northern steppe cultures into face-to-face contact with the old civilizations of Asia.

CHAPTER SIXTEEN

The Opening of the Eurasian Steppes

Between about 2300 and 2000 BCE the sinews of trade and conquest began to pull the far-flung pieces of the ancient world together into a single interacting system. The mainspring that drove inter-regional trade was the voracious demand of the Asiatic cities for metal, gems, ornamental stones, exotic woods, leather goods, animals, slaves, and power. Participants gained access to and control over knowledge of the urban centers and their power-attracting abilities—a source of social prestige in most societies.[1] Ultimately, whether through cultural means of emulation and resistance or political means of treaty and alliance, a variety of regional centers linked their fortunes to those of the paramount cities of the Near East, Iran, and South Asia. Regional centers in turn extended their influence outward, partly in a search for raw materials for trade, and partly to feed their own internal appetites for power. On the edges of this expanding, uncoordinated system of consumption and competition were tribal cultures that probably had little awareness of its urban core, at least initially (figures 16.1 and 16.2). But eventually they were drawn in. By 1500 BCE chariot-driving mercenaries not too far removed from the Eurasian steppes, speaking an Old Indic language, created the Mitanni dynasty in northern Syria in the heart of the urban Near East.[2]

How did tribal chiefs from the steppes intrude into the dynastic politics of the Near East? Where else did they go? To understand the crucial role that Eurasian steppe cultures played in the knitting together of the ancient world during the Bronze Age, we should begin in the heartland of cities, where the demand for raw materials was greatest.

BRONZE AGE EMPIRES AND THE HORSE TRADE

About 2350 BCE Sargon of Akkad conquered and united the feuding kingdoms of Mesopotamia and northern Syria into a single super-state—

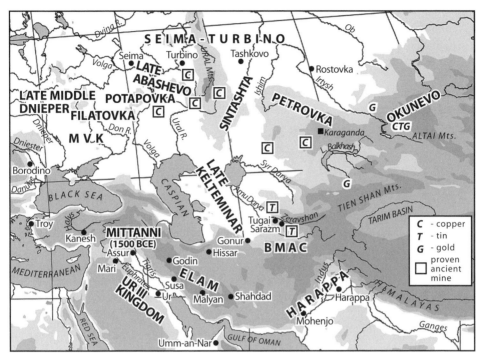

Figure 16.1 Cultures of the steppes and the Asian civilizations between about 2200 and 1800 BCE, with the locations of proven Bronze Age mines in the steppes and the Zeravshan valley.

the first time the world's oldest cities were ruled by one king. The Akkadian state lasted about 170 years. It had economic and political interests in western and central Iran, leading to increased trade, occasionally backed up by military expeditions. Images of horses, distinguished from asses and onagers by their hanging manes, short ears, and bushy tails, began to appear in Near Eastern art during the Akkadian period, although they still were rare and exotic animals. Some Akkadian seals had images of men riding equids in violent scenes of conflict (figure 16.3). Perhaps a few Akkadian horses were acquired from the chiefs and princes of western Iran known to the Akkadians as the Elamites.

Elamite was a non–Indo-European language, now extinct, then spoken across western Iran. A string of walled cities and trade centers stood on the Iranian plateau, revealed by excavations at Godin, Malyan, Konar Sandal, Hissar, Shar-i-Sokhta, Shahdad, and other places. Malyan, the ancient city of Anshan, the largest city on the plateau, certainly was an

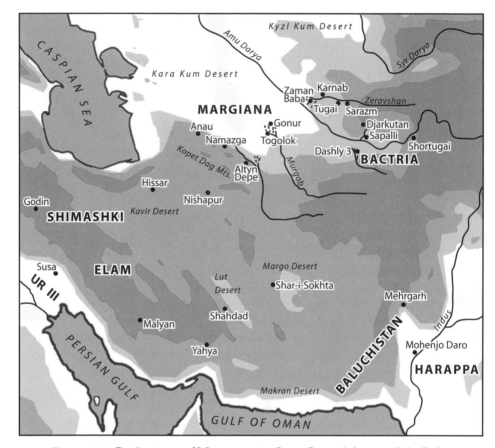

Figure 16.2 Civilizations of Mesopotamia, Iran, Central Asia, and the Indus valley about 2200–1800 BCE.

Elamite city allied to the Elamite king in Susa. Some of the other brick-built towns, almost all of them smaller than Malyan, were part of an alliance called Shimashki, located north of Malyan and south of the Caspian Sea. Among the fifty-nine personal names recorded in the Shimashki alliance, only twelve can be classified as Elamite; the others are from unknown non–Indo-European languages. East of the Iranian plateau, the Harappan civilization of Indo-Pakistan, centered in huge mudbrick cities on the Indus River, used its own script to record a language that has not been definitively deciphered but might have been related to modern Dravidian. The Harappan cities exported precious stones, tropical woods, and metals westward on ships that sailed up the Persian Gulf, through a chain

Figure 16.3 Early images of men riding equids in the Near East and Central Asia: (*top*) Akkadian seal impression from Kish, 2350–2200 BCE (after Buchanan 1966); (*middle*) seal impression of the BMAC from a looted grave in Afghanistan, 2100–1800 BCE (after Sarianidi 1986); (*bottom*) Ur III seal impression of Abbakalla, animal disburser for king Shu-Sin, 2050–2040 BCE (after Owen 1991).

of coastal kingdoms scattered from Oman to Kuwait. Harappa probably was the country referred to as "Melukkha" in the Mesopotamian cuneiform records.[3]

Akkadian armies and trade networks reached far and wide, but inside Akkad was an enemy it could not conquer with arms: crop failure. During the Akkadian era the climate became cooler and drier, and the agricultural economy of the empire suffered. Harvey Weiss of Yale has argued that some northern Akkadian cities were entirely abandoned, and their populations might have moved south into the irrigated floodplains of southern Mesopotamia.[4] The Gutians, a coalition of chiefs from the western Iranian uplands (perhaps Azerbaijan?) defeated the Akkadian army and overran the city of Akkad in 2170 BCE. Its ruins have never been found.

About 2100 BCE the first king of the Third Dynasty of Ur, even then an ancient Sumerian city in what is now southern Iraq, expelled the Gutians and reestablished the power of southern Mesopotamia. The brief Ur III period, 2100–2000 BCE, was the last time that Sumerian, the language of the first cities, was a language of royal administration. A century of bitter wars erupted between the Sumerian Ur III kings and the Elamite city-states of the Iranian plateau, occasionally interrupted by negotiations and marriage exchanges. King Shu-Sin of Ur bragged that he conquered a path across Elam and through Shimashki until his armies finally were stopped only by the Caspian Sea.

During this period of struggle and empire, 2100–2000 BCE, the bones of horses appeared for the first time at important sites on the Iranian plateau such as the large city of Malyan in Fars and the fortified administrative center at Godin Tepe in western Iran. Bit wear made with a hard bit, probably metal, appeared on the teeth of some of the equids (both mules and horses) from Malyan. Excavated by Bill Sumner and brought by Mindy Zeder to the collections of the Smithsonian Museum of Natural History in Washington, D.C. these teeth were the first archaeological specimens that we examined when we started our bit wear project in 1985. Now we know what then we only suspected: the horses and mules of the Kaftari phase at Malyan were bitted with hard bits. Bits were a new technology for controlling equids in Iran, different from the lip- and nose-rings that had appeared before this in Mesopotamian works of art. Of course bits and bit-wear were very old in the steppes by 2000 BCE.[5]

Horses also appeared in significant numbers in the cities of Mesopotamia for the first time during the Ur III period; this was when the word for *horse* first appeared in written records. It meant "ass of the mountains," showing that horses were flowing into Mesopotamia from western Iran and eastern Anatolia. The Ur III kings fed horses to lions for exotic entertainment. They did not use horse-drawn chariots, which had not yet appeared in Near Eastern warfare. But they did have solid-wheeled battle wagons and battle carts armed with javelins, pulled by teams of their smaller native equids—asses, which were manageable but small, and onagers or hemiones, which were almost untamable but larger. Ass-onager hybrids probably pulled Sumerian battle carts and battle wagons. Horses could have been used initially as breeding stock to make a larger, stronger ass-horse hybrid—a mule. Mules were bitted at Malyan.

The Sumerians recognized in horses an arched-neck pride that asses and onagers simply did not possess. King Shulgi compared himself in one inscription to "a horse of the highway that swishes his tail." We are not sure exactly what horses were doing on Ur III highways, but a seal impression of one Abbakalla, the royal animal disburser for king Shu-Sin, showed a man riding a galloping equid that looks like a horse (see figure 16.3).[6] Ceramic figurines of the same age showed humans astride schematic animals that have equine proportions; and ceramic plaques dated at the time of Ur III or just afterward showed men astride equids that probably were horses, some riding in awkward poses on the rump and others in more natural forward seats. No Ur III images showed a chariot, so the first clear images of horses in Mesopotamia show men riding them.[7]

About 2000 BCE an Elamite and Shimashki alliance defeated the last of the Ur III kings, Ibbi-Sin, and dragged him to Elam in chains. After this stunning event the kings of Elam and Shimashki played a controlling role in Mesopotamian politics for several centuries. Between 2000 and 1700 BCE the power, independence, and wealth of the Old Elamite (Malyan) and Shimashkian (Hissar? Godin?) overlords of the Iranian plateau was at its height. The treaties they negotiated for the Ur III wars were sealed by gifts and trade agreements that channeled lapis lazuli, carved steatite vessels, copper, tin, and horses from one prince to another. The Sintashta culture appeared at just the same time, but showed up 2000 km to the north in the remote grasslands of the Ural-Tobol steppes. The metal trade and the horse trade might have tied the two worlds together. Could the Elamite defeat of Ibbi-Sin have been aided by chariot-driving Sintashta mercenaries from the steppes? It is possible. Vehicles like chariots, with two spoked wheels and a standing driver, but guided by equids with lip- or nose-rings, began to appear on seal images in Anatolia just after the defeat of Ibbi-Sin. They were not yet common, but that was about to change.

The metal trade might have provided the initial incentive for prospectors to explore across the Central Asian deserts that had previously separated the northern Eurasian steppe cultures from those of Iran. Vast amounts of metal were demanded by Near Eastern merchants during the heyday of the Old Elamite kings. Zimri-Lim, king of the powerful city-state of Mari in northern Syria between 1776 and 1761 BCE, distributed gifts totaling more than 410 kg (905 lb) of tin—not bronze, but tin—to his allies during a single tour in his eighth year. Zimri-Lim also was chided by an adviser for riding a horse in public, an activity still considered insulting to the honor of an Assyrian king:[8]

> May my lord honor his kingship. You may be the king of the Hane-ans, but you are also the king of the Akkadians. May my Lord not ride horses; (instead) let him ride either a chariot or *kudanu*-mule so that he would honor his kingship.

Zimri-Lim's advisers accepted the fact that kings could ride in chariots—Near Eastern monarchs had by then ridden in wheeled vehicles of other kinds for more than a thousand years. But only rude barbarians actually rode on the backs of the large, sweaty, smelly animals that pulled them. Horses, in Zimri-Lim's day, were still exotic animals associated with crude foreigners. A steady supply of horses first began between 2100 and 2000 BCE. Chariots appeared across the Near East after 2000 BCE. How?

The Tin Trade and the Gateway to the North

Tin was the most important trade commodity in the Bronze Age Near East. In the palace records of Mari it was said to be worth ten times its weight in silver. A copper-tin alloy was easier for the metal smith to cast, and it made a harder, lighter-colored metal than either pure copper or arsenical bronze, the older alternatives. But the source of Near Eastern tin remains an enigma. Large tin deposits existed in England and Malaysia, but these places were far beyond the reach of Near Eastern traders in the Bronze Age. There were small tin deposits in western Serbia—and a scatter of Old European copper objects from the Danube valley contained elevated tin, perhaps derived from this source—but no ancient mines have been found there. Ancient mines in eastern Anatolia near Goltepe might have supplied a trickle of tin before 2000 BCE, but their proven tin content is very low, and tin was *imported* at great cost to Anatolia from northern Syria after 2000 BCE. It was imported into northern Syria from somewhere far to the east. The letters of king Zimri-Lim of Mari said flatly that he acquired his tin from Elam, through merchants at Malyan (Anshan) and Susa. An inscription on a statue of Gudea of Lagash, ca.

2100 BCE, was thought to refer to the "tin of Melukkha," implying that tin came up the Arabian Gulf in ships sent by Harappan merchants; but the passage might have been mistranslated. Intentional tin-bronze alloys occurred in about 30% of the objects tested from the Indus-valley cities of Mohenjo-Daro and Harappa, although most had such a low tin content (70% of them had only 1% tin, 99% copper) that it seems the best recipe for tin bronze (8–12% tin, 92–88% copper) was not yet known in Harappa. Still, "Melukkha" could have been one source of Mesopotamian tin. Tin-bronzes have been found in sites in Oman, at the entrance to the Arabian Gulf, in association with imported pottery and beads from Harappa and bone combs and seals made in Bactria. Oman had no tin of its own but could have been a coastal port and trans-shipment point for tin that came from the Indus valley.[9]

Where were the tin mines? Could the tin exported by the Elamite kings and by Harappan merchants have come from the same sources? Quite possibly. The most probable sources were in western and northern Afghanistan, where tin ore has been found by modern mineral surveyors, although no ancient mines have been found there, and also in the Zeravshan River valley, where the oldest tin mines in the ancient world have been found near the site of Sarazm. Sarazm also was the portal through which horses, chariots, and steppe cultures first arrived at the edges of Central Asia.

Sarazm was founded before 3500 BCE (4880±30 BP, 4940±30 BP for phase I) as a northern colony of the Namazga I–II culture. The Namazga home settlements (Namazga, Anau, Altyn-Depe, Geoksur) were farming towns situated on alluvial fans where the rivers that flowed off the Iranian plateau emerged into the Central Asian deserts. Perhaps the lure that enticed Namazga farmers to venture north across the Kara Kum desert to Sarazm was the turquoise that outcropped in the desert near the lower Zeravshan River, a source they could have learned about from Kelteminar foragers. Sarazm probably was founded as a collection point for turquoise. It was situated on the middle Zeravshan more than 100 km upstream from the turquoise deposits at an elevation where the valley was lush and green and crops could be grown. It grew to a large town, eventually covering more than 30 ha (74 acres). Its people were buried with ornaments of turquoise, carnelian, silver, copper, and lapis lazuli. Late Kelteminar pottery was found at Sarazm in its phase II, dated about 3000–2600 BCE (4230±40BP), and turquoise workshops have been found in the late Kelteminar camps of Kaptarnikum and Lyavlyakan in the desert near the lower Zeravshan. Turquoise from the Zeravshan and from a second source near Nishapur in northeastern Iran was traded into Mesopotamia, the Indus valley, and perhaps even to Maikop (the Maikop chieftain was buried with a necklace of turquoise beads). But the Zeravshan also contained polymetallic deposits of copper, lead, silver—and tin.

Oddly, no tin has been found at Sarazm itself. Crucibles, slag, and smelting furnaces appeared at Sarazm at least as early as the phase III settlement (radiocarbon dated 2400–2000 BCE), probably for processing the rich copper deposits in the Zeravshan valley. Sarazm III yielded a variety of copper knives, daggers, mirrors, fishhooks, awls, and broad-headed pins. Most were made of pure copper, but a few objects contained 1.8–2.7% arsenic, probably an intentional arsenical bronze. Tin-bronzes began to appear in small amounts in the Kopet Dag home region, in Altyn-Depe and Namazga, during the Namazga IV period, equivalent to late Sarazm II and III. A small amount of tin, perhaps just placer minerals retrieved from the river, probably came from the Zeravshan before 2000 BCE, even if we cannot see it at Sarazm.[10]

The tin mines of the Zeravshan River valley were found and investigated by N. Boroffka and H. Parzinger between 1997 and 1999.[11] Two tin mines with Bronze Age workings were excavated. The largest was in the desert on the lower Zeravshan at Karnab (Uzbekistan), about 170 km west of Sarazm, exploiting cassiterite ores with a moderate tin content—probably ordinarily about 3%, although some samples yielded as much as 22% tin. The pottery and radiocarbon dates show that the Karnab mine was worked by people from the northern steppes, connected with the Andronovo horizon (see below). Dates ranged from 1900 to 1300 BCE (the oldest was Bln 5127, 3476±32 BP, or 1900–1750 BCE; see table 16.1). A few pieces of Namazga V/VI pottery were found in the Andronovo mining camp at Karnab. The other mining complex was at Mushiston in the upper Zeravshan (Tajikistan), just 40 km east of Sarazm, working stannite, cassiterite and copper ores with a very high tin content (maximum 34%). Andronovo miners also left their pottery at Mushiston, where wood beams produced radiocarbon dates as old as Karnab. Sarazm probably was abandoned when these Andronovo mining operations began. Whether the Zeravshan tin mines were worked before the steppe cultures arrived is unknown.

Sarazm probably was abandoned around 2000 BCE, just at the Namazga V/VI transition. On the lower Zeravshan, the smaller villages of the Zaman Baba culture probably were abandoned about the same time as Sarazm.[12] The Zaman Baba culture had established small villages of pit-houses supported by irrigation agriculture in the large oasis in the lower

Zeravshan delta just a couple of centuries earlier. Zaman Baba and Sarazm were abandoned when people from the northern steppes arrived in the Zeravshan.[13]

Sarazm exported both copper and turquoise southward during the Akkadian and Ur III periods. Could it have pulled steppe copper miners and horse traders into the chain of supply for the urban trade? Could that explain the sudden intensification of copper production in Sintashta settlements and the simultaneous appearance of horses in Iran and Mesopotamia beginning about 2100 BCE? The answer lies among the ruins of walled cities in Central Asia south of Sarazm, cities that interacted with the cultures of the northern steppes before the Andronovo tin miners appeared on the Zeravshan frontier.

The Bactria-Margiana Archaeological Complex

Around 2100 BCE a substantial population colonized the Murgab River delta north of the Iranian plateau. The Murgab River flowed down from the mountains of western Afghanistan, snaked across 180 km of desert, then fanned out into the sands, dropping deep loads of silt and creating a fertile island of vegetation about 80 by 100 km in size. This was Margiana, a region that quickly became and remained one of the richest oases in Central Asia. The immigrants built new walled towns, temples, and palaces (Gonur, Togolok) on virgin soil during the late Namazga V period, at the end of the regional Middle Bronze Age (figure 16.4). They might have been escaping from the military conflicts that raged periodically across the Iranian plateau, or they might have relocated to a larger river system with more reliable flows in a period of intensifying drought. Anthropological studies of their skeletons show that they came from the Iranian plateau, and their pottery types seem to have been derived from the Namazga V-type towns of the Kopet Dag.[14]

The colonization phase in Margiana, 2100–2000 BCE, was followed by a much richer period, 2000–1800 BCE, during Namazga VI, the beginning of the regional Late Bronze Age. New walled towns now spread to the upper Amu Darya valley, ancient Bactria, where Sapalli-Tepe, Dashly-3, and Djarkutan were erected on virgin soil. The towns of Bactria and Margiana shared a distinctive set of seal types, architectural styles, brick-lined tomb types, and pottery. The LBA civilization of Bactria and Margiana is called the Bactria-Margiana Archaeological Complex (BMAC). The irrigated countryside was dominated by large towns surrounded by thick yellow-brick walls with narrow gates and high corner towers. At the

TABLE 16.1

Selected Radiocarbon Dates from Earlier Late Bronze Age Cultures in the Steppes

Lab Number	BP Date	Kurgan	Grave	Mean Intercept BCE	BCE
1. Krasnosamarskoe kurgan cemetery IV, Samara oblast, LBA Pokrovka and Srubnaya graves					
AA37038	3490±57	kurgan 3	1	1859, 1847, 1772	1881–1740
AA37039	3411±46	kurgan 3	6	1731, 1727, 1686	1747–1631
AA37042	3594±45	kurgan 3	10	1931	1981–1880
AA37043	3416±57	kurgan 3	11	1733, 1724, 1688	1769–1623
AA37044	3407±46	kurgan 3	13	1670, 1668, 1632	1685–1529
AA37045	3407±46	kurgan 3	16	1730, 1685	1744–1631
AA37046	3545±65	kurgan 3	17	1883	1940–1766
AA37047	3425±52	kurgan 3	23	1735, 1718, 1693	1772–1671

2. Krasnosamarskoe settlement, Samara oblast

Structure floor and cultural level outside structure, Pokrovka and Srubnaya occupations

Lab Number	BP Date	Square/quad		level	Mean Intercept BCE	BCE
AA41022	3531±43	L5	2	3	1879, 1832, 1826, 1790	1899–1771
AA41023	3445±51	M5	1	7	1741	1871–1678
AA41024	3453±43	M6	3	7	1743	1867–1685
AA41025	3469±45	N3	3	7	1748	1874–1690
AA41026	3491±52	N4	2	6	1860, 1846, 1772	1879–1743
AA41027	3460±52	O4	1	7	1745	1873–1685
AA41028	3450±57	O4	2	5	1742	1874–1679
AA41029	3470±43	P1	4	6	1748	1783–1735
AA41030	3477±39	S2	3	4	1752	1785–1738
AA41031	3476±38	R1	2	5	1750	1875–1706
AA41032	3448±47	N2	2	4	1742	1858–1685
AA47790	3311±54	O5	3	3	1598, 1567, 1530	1636–1518
AA47796	3416±59	Y2	2	4	1736, 1713, 1692	1857–1637
AA47797	3450±50	Y1	3	5	1742	1779–1681

Waterlogged Pokrovka artifacts from deep pit interpreted as a well inside the structure

Lab Number	BP Date	Square/quad		level	Mean Intercept BCE	BCE
AA47793	3615±41	M2	4	−276	1948	1984–1899
AA47794	3492±55	M2	4	−280	1860, 1846, 1773	1829–1742
AA47795	3550±54	M2	4	−300	1884	1946–1776

TABLE 16.1 (*continued*)

Lab Number	BP Date	Kurgan		Grave	Mean Intercept BCE	BCE
Srubnaya and Pokrovka artifacts from eroded part of settlement on the lake bottom						
AA47791	3494±56	Lake find 1		0	1862, 1845, 1774	1881–1742
AA47792	3492±55	Lake find 2		0	1860, 1846, 1773	1829–1742
Srubnaya herding camp at PD1 in the Peschanyi Dol valley						
AA47798	3480±52	A 16	3	3	1758	1789–1737
AA47799	3565±55	I 18	2	2	1889	1964–1872
3. Karnab mining camp, Zeravshan valley, Uzbekistan, Andronovo–Alakul occupation						
Bln–5127	3476±32					1880–1740
Bln–141274	3280±40					1620–1510
Bln–141275	3170±50					1520–1400
Bln–5126	3130±44					1490–1310
4. Alakul–Andronovo settlements and kurgan graves						
Alakul kurgan 15, grave 1						
Le–924	3360±50	charcoal				1740–1530
Subbotino kurgan 17, grave 3						
Le–1126	3460±50	wood				1880–1690
Subbotino kurgan 18, central grave						
Le–1196	3000±50	wood				1680–1510
Tasty–Butak settlement						
Rul–614	3550±65	wood, pit 14				2010–1770
Le–213	3190±80	wood, pit 11				1600–1320

center of the larger towns were walled palaces or citadels that contained temples. The brick houses and streets of Djarkutan covered almost 100 ha, commanded by a high-walled citadel about 100 by 100 m. Local lords ruled from smaller strongholds such as Togolok 1, just .5 ha (1.2 acres) in size but heavily walled with large corner turrets. Trade and crafts flourished in the crowded houses and alleys of these Central Asian walled towns and fortresses. Their rulers had relations with the civilizations of Mesopotamia, Elam, Harappa, and the Arabian Gulf.

Between 2000 and 1800 BCE, BMAC styles and exported objects (notably small jars made of carved steatite) appeared in many sites and

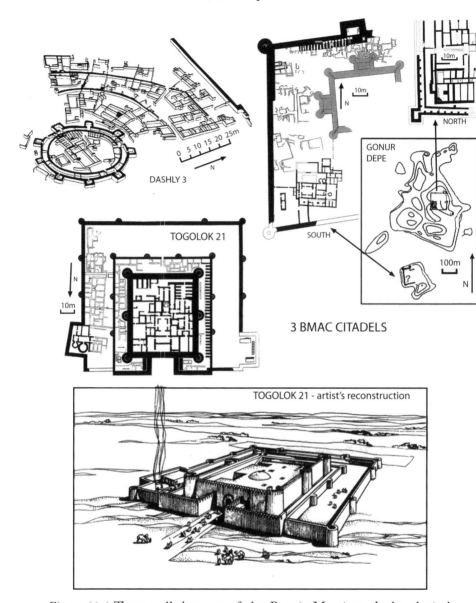

Figure 16.4 Three walled towns of the Bactria-Margiana Archaeological Complex (BMAC) in Central Asia, 2100–1800 BCE. Wall foundations of the central circular citadel/temple and town at Dashly 3, Bactria (after Sarianidi 1977, figure 13); wall foundations at Gonur Depe, Margiana (combined from Hiebert 1994; and Sarianidi 1995); wall foundations and artist's reconstruction of Togolok 21, Margiana (after Hiebert 1994; and Sarianidi 1987).

cemeteries across the Iranian plateau. Crested axes like those of the BMAC appeared at Shadad and other sites in eastern and central Iran. A cemetery at Mehrgarh VIII in Baluchistan, on the border between the Harappan and Elamite civilizations, contained so many BMAC artifacts that it suggests an actual movement of BMAC people into Baluchistan. BMAC-style sealings, ivory combs, steatite vessels, and pottery goblets appeared in the Arabian Gulf from Umm-al-Nar on the Oman peninsula up the Arabian coast to Falaika island in Kuwait. Beadmakers in BMAC towns used shells obtained from both the Indian Ocean (*Engina medicaria, Lambis truncate sebae*) and the Mediterranean Sea (*Nassarius gibbosulus*), as well as steatite, alabaster, lapis lazuli, turquoise, silver, and gold.[15]

The metalsmiths of the BMAC made beautiful objects of bronze, lead, silver, and gold. They cast delicate metal figures by the lost-wax process, which made it possible to cast very detailed metal objects. They made crested bronze shaft-hole axes with distinctive down-curved blades, tanged daggers, mirrors, pins decorated with cast animal and human figures, and a variety of distinctive metal compartmented seals (figure 16.5). The metals used in the first colonization period, late Namazga V, were unalloyed copper, arsenical bronze, and a copper-lead alloy with up to 8–10% lead.

About 2000 BCE, during the Namazga VI/BMAC period, tin-bronze suddenly appeared prominently in sites of the BMAC. Tin-bronzes were common at two BMAC sites, Sapalli and Djarkutan, reaching more than 50% of objects, although at neighboring Dashly-3, also in Bactria, tin-bronzes were just 9% of metal objects. Tin-bronzes were rare in Margiana (less than 10% of metal objects at Gonur, none at all at Togolok). Tin-bronze was abundant only in Bactria, closer to the Zeravshan. It looks like the tin mines of the Zeravshan were established or greatly expanded at the beginning of the mature BMAC period, about 2000 BCE.[16]

There were no wild horses in Central Asia. The native equids were onagers. Wild horses had not previously strayed south of what is today central Kazakhstan. Any horses found in BMAC sites must have been traded in from the steppes far off to the north. The animal bones discarded in and near BMAC settlements contained no horse bones. Hunters occasionally killed wild onagers but not horses. Most of the bones recovered from the settlement trash deposits were from sheep or goats. Asian zebu cattle and domesticated Bactrian camels also appeared. They were shown pulling wagons and carts in BMAC artwork. Small funeral wagons with solid wooden-plank wheels and bronze-studded tires were buried in royal graves associated with the first building phase, dated about 2100–2000 BCE, at

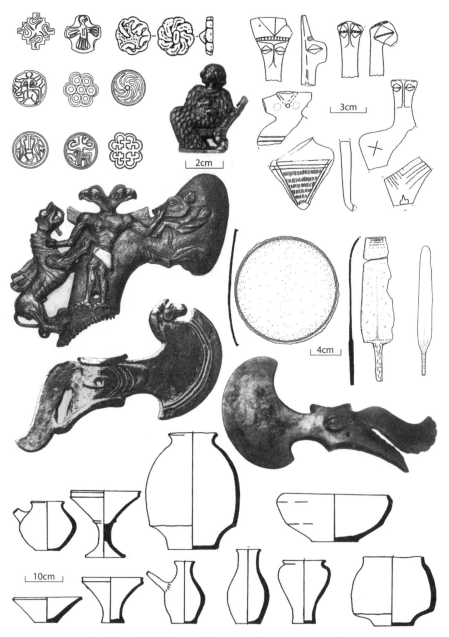

Figure 16.5 Artifacts of the Bactria-Margiana Archaeological Complex, 2100–1800 BCE: (*top left*) a sample of BMAC stamp seals, adapted after Salvatori 2000, and Hiebert 1994; (*top center*) cast silver pin head from Gonur North showing a goddess in a ritual dress, after Klochkov 1998, figure 3; (*top right*) ceramic female figurines from Gonur North, after Hiebert 1994; (*center left*)

Gonur in Margiana (called Gonur North, because the oldest phase was found at the northern end of the modern ruins).

In these graves at Gonur, associated with the early settlement of Gonur North, one horse was found. A brick-lined grave pit contained the contorted bodies of ten adult humans who were apparently killed in the grave itself, one of whom fell across a small funeral wagon with solid wooden wheels. The grave also contained a whole dog, a whole camel, and the decapitated body of a horse foal (the reverse of an Aryan horse sacrifice). This grave is thought to have been a sacrificial offering that accompanied a nearby "royal" tomb. The royal tomb contained funeral gifts that included a bronze image of a horse head, probably a pommel decoration on a wooden staff. Another horse head image appeared as a decoration on a crested copper axe of the BMAC type, unfortunately obtained on the art market and now housed in the Louvre. Finally, a BMAC-style seal probably looted from a BMAC cemetery in Bactria (Afghanistan) showed a man riding a galloping equid that looks very much like a horse (see figure 16.3). The design was similar to the contemporary galloping-horse-and-rider image on the Ur III seal of Abbakalla, dated 2040–2050 BCE. Both seals showed a galloping horse, a rider with a hair-knot on the back of his head, and a man walking.

These finds suggest that horses began to appear in Central Asia about 2100–2000 BCE but never were used for food. They appeared only as decorative symbols on high-status objects and, in one case, in a funeral sacrifice. Given their simultaneous appearance across Iran and Mesopotamia, and the position of BMAC between the steppes and the southern civilizations, horses were probably a trade commodity. After chariots were introduced to the princes of the BMAC, Iran, and the Near East around 2000–1900 BCE, the demand for horses could easily have been on the order of tens of thousands of animals annually.[17]

Steppe Immigrants in Central Asia

Fred Hiebert's excavations at the walled town of Gonur North in Margiana, dated 2100–2000 BCE, turned up a few sherds of strange pottery,

---

Figure 16.5 (continued) crested shaft-hole axes from the art market, probably from BMAC sites, with a possible horse-head on the lower one, after Aruz 1998, figure 24; and Amiet 1986, figure 167; (*center right*) a crested axe with eye amulet, and a copper mirror and dagger excavated from Gonur North, after Hiebert 1994; and Sarianidi 1995, figure 22; (*bottom*) ceramic vessel shapes from Gonur, after Hiebert 1994.

unlike any other pottery at Gonur. It was made with a paddle-and-anvil technique on a cloth-lined form—the clay was pounded over an upright cloth-covered pot to make the basic shape, and then was removed and finished. This is how Sintashta pottery was made. These strange sherds were imported from the steppe. At this stage (equivalent to early Sintashta) there was very little steppe pottery at Gonur, but it was there, at the same time a horse foal was thrown into a sacrificial pit in the Gonur North cemetery. Another possible trace of this early phase of contact were "Abashevo-like" pottery sherds decorated with horizontal channels, found at the tin miners' camp at Karnab on the lower Zeravshan. Late Abashevo was contemporary with Sintashta.

During the classic phase of the BMAC, 2000–1800 BCE, contact with steppe people became much more visible. Steppe pots were brought into the rural stronghold at Togolok 1 in Margiana, inside the larger palace/temple at Togolok 21, inside the central citadel at Gonur South, and inside the walled palace/temple at Djarkutan in Bactria (figure 16.6). These sherds were clearly from steppe cultures. Similar designs can be found on Sintashta pots at Krivoe Ozero (k. 9, gr. 3; k. 10, gr. 13) but were more common on pottery of early Andronovo (Alakul variant) type, dated after 1900–1800 BCE—pottery like that used by the Andronovo miners at Karnab. Although the amount of steppe pottery in classic BMAC sites is small, it is widespread, and there is no doubt that it derived from northern steppe cultures. In these contexts, dated 2000–1800 BCE, the most likely steppe sources were the Petrovka culture at Tugai or the first Alakul-Andronovo tin miners at Karnab, both located in the Zeravshan valley.[18]

The Petrovka settlement at Tugai appeared just 27 km downstream (west) of Sarazm, not far from the later site of Samarkand, the greatest caravan trading city of medieval Central Asia. Perhaps Tugai had a similar, if more modest, function in an early north-south trade network. The Petrovka culture (see below) was an eastern offshoot of Sintashta. The Petrovka people at Tugai constructed two copper-smelting ovens, crucibles with copper slag, and at least one dwelling. Their pottery included at least twenty-two pots made with the paddle-and-anvil technique on a cloth-lined form. Most of them were made of clay tempered with crushed shell, the standard mixture for Petrovka potters, but two were tempered with crushed talc/steatite minerals. Talc-tempered clays were typical of Sintashta, Abashevo, and even forest-zone pottery of Ural forager cultures, so these two pots probably were carried to the

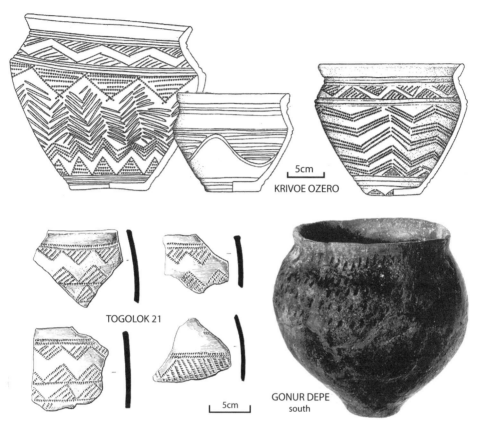

Figure 16.6 A whole steppe pot found inside the walls of the Gonur South town, after Hiebert 1994; steppe sherds with zig-zag decoration found inside the walls of Togolok 1, after Kuzmina 2003; and similar motifs on Sintashta sherds from graves at Krivoe Ozero, Ural steppes, after Vinogradov 2003, figures 39 and 74.

Zeravshan from the Ural steppes. The pottery shapes and impressed designs were classic early Petrovka (figure 16.7). A substantial group of Petrovka people apparently moved from the Ural-Ishim steppes to Tugai, probably in wagons loaded with pottery and other possessions. They left garbage middens with the bones of cattle, sheep, and goats, but they did not eat horses—although their Petrovka relatives in the northern steppes did. Tugai also contained sherds of wheel-made cups in red-polished and black-polished fabrics typical of the latest phase at Sarazm (IV). The

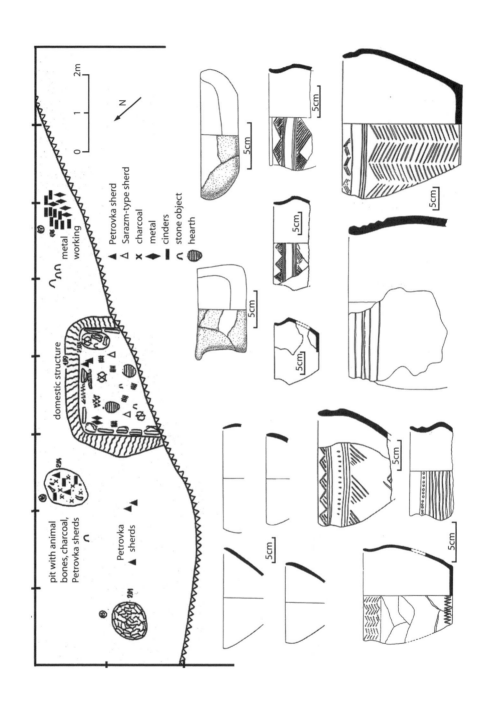

metal working

Petrovka sherd
Sarazm-type sherd
charcoal
metal
cinders
stone object
hearth

domestic structure

pit with animal
bones, charcoal,
Petrovka sherds

Petrovka
sherds

N

0   1   2m

5cm

principal activity identified in the small excavated area was copper smelting.[19]

The steppe immigrants at Tugai brought chariots with them. A grave at Zardcha-Khalifa 1 km east of Sarazm contained a male buried in a contracted pose on his right side, head to the northwest, in a large oval pit, 3.2 m by 2.1 m, with the skeleton of a ram.[20] The grave gifts included three wheel-made Namazga VI ceramic pots, typical of the wares made in Bactrian sites of the BMAC such as Sappali and Dzharkutan; a trough-spouted bronze vessel (typical of BMAC) and fragments of two others; a pair of gold trumpet-shaped earrings; a gold button; a bronze straight-pin with a small cast horse on one end; a stone pestle; two bronze bar bits with looped ends; and two largely complete bone disc-shaped cheekpieces of the Sintashta type, with fragments of two others (figure 16.8). The two bronze bar bits are the oldest known metal bits anywhere. With the four cheekpieces they suggest equipment for a chariot team. The cheekpieces were a specific Sintashta type (the raised bump around the central hole is the key typological detail), though disc-shaped studded cheekpieces also appeared in many Petrovka graves. Stone pestles also frequently appeared in Sintashta and Petrovka graves. The Zardcha-Khalifa grave probably was that of an immigrant from the north who had acquired many BMAC luxury objects. He was buried with the only known BMAC-made pin with the figure of a horse—perhaps made just for him. The Zardcha-Khalifa chief may have been a horse dealer. The Zeravshan valley and the Ferghana valley just to the north might have become the breeding ground at this time for the fine horses for which they were known in later antiquity.

The fabric-impressed pottery and the sacrificed horse foal at Gonur North and perhaps the Abashevo (?) sherds at Karnab represent the exploratory phase of contact and trade between the northern steppes and the southern urban civilizations about 2100–2000 BCE, during the period when the kings of Ur III still dominated Elam. Information and perhaps even cult practices from the south flowed back to early Sintashta societies. On the eastern frontier in Kazakhstan, where Petrovka was budding off from Sintashta, the lure of the south prompted a migration across more

---

Figure 16.7 The Petrovka settlement at Tugai on the Zeravshan River: (*top*) plan of excavation; (*center left*) imported redware pottery like that of Sarazm IV; (*center right*) two coarse ceramic crucibles from the metal-working area; (*bottom*) Petrovka pottery. Adapted from Avanessova 1996.

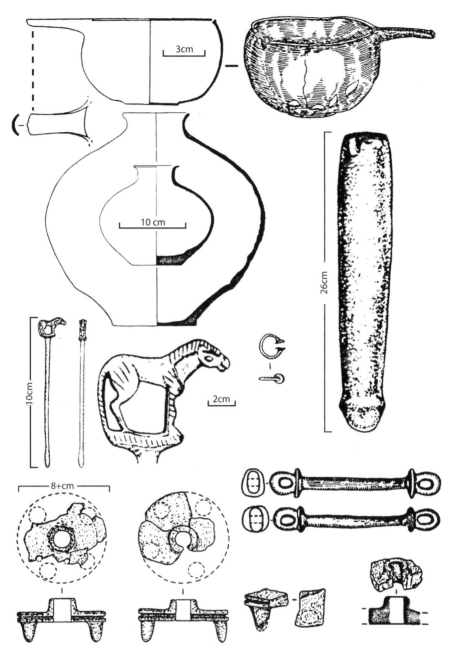

Figure 16.8 Objects from the grave at Zardcha-Khalifa on the Zeravshan River. The trough-spouted bronze vessel and ceramic pots are typical of the BMAC, 2000–1800 BCE; the cast copper horse pin shows BMAC casting methods; the bronze bar bits are the first ones dated this early; and the stone pestle, trumpet-shaped earring, and bone cheekpieces are steppe types. After Bobomulloev 1997, figures 2, 3, and 4.

than a thousand kilometers of hostile desert. The establishment of the Petrovka metal-working colony at Tugai, probably around 1900 BCE, was the beginning of the second phase, marked by the actual migration of chariot-driving tribes from the north into Central Asia. Sarazm and the irrigation-fed Zaman-Baba villages were abandoned about when the Petrovka miners arrived at Tugai. The steppe tribes quickly appropriated the ore sources of the Zeravshan, and their horses and chariots might have made it impossible for the men of Sarazm to defend themselves.

Central Asian Trade Goods in the Steppes

Did any BMAC products appear in Sintashta or Petrovka settlements? Only a few hints of a return trade can be identified. One intriguing innovation was a new design motif, the stepped pyramid or crenellation. Stepped pyramids or crenellations appeared on the pottery of Sintashta, Potapovka, and Petrovka. The stepped pyramid was the basic element in the decorative artwork on Namazga, Sarazm, and BMAC pottery, jewelry, metalwork, and even in a mural painted on the Proto-Elamite palace wall at Malyan (figure 16.9, bottom). Repeated horizontally, the stepped pyramid became a line of crenellated designs; repeated on four sides, it became a stepped cross. This motif had not appeared in any earlier pottery in the steppes, neither in the Bronze Age nor the Eneolithic. Charts of design motifs are regularly published in Russian archaeological ceramic studies. I have scanned these charts for years and have not found the stepped pyramid in any assemblage earlier than Sintashta. Stepped pyramids appeared for the first time on northern steppe pottery just when northern steppe pottery first showed up in BMAC sites. It was seen first on a small percentage (<5%) of Potapovka pottery on the middle Volga (single vessels in Potapovka kurgans 1, 2, 3, and 5) and at about the same frequency on Sintashta pottery in the Ural-Tobol steppes; later it became a standard design element in Petrovka and Andronovo pottery (but not in Srubnaya pottery, west of the Urals). Although no Sarazm or BMAC pottery has been found in Sintashta contexts, the design could have been conveyed to the northern steppes on textiles—perhaps the commodity exchanged for northern metal. I would guess that Sintashta potters copied the design from imported BMAC textiles.

There are other indications of contact. A lead wire made of two braided strands was found among the metal objects in the Sintashta settlement of Kuisak. Lead had never before appeared in the northern steppes as a pure metal, whereas a single ingot of lead weighing 10 kg was found at Sarazm.

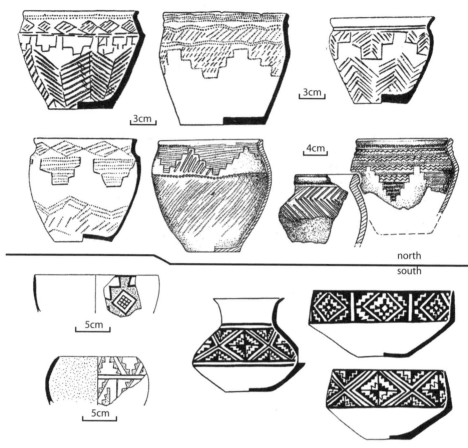

Figure 16.9 Stepped pyramid or crenellation motifs on steppe pottery and on Central Asian pottery: (*top row and left pot in second row*) Potapovka graves, middle Volga region, 2100–1800 BCE, after Vasiliev, Kuznetsov, and Semenova 1994, figures 20 and 22; (*middle row, remaining pots*) Sintashta SII cemetery, grave 1, after Gening, Zdanovich and Gening 1992, figure 172; (*bottom left*) Sarazm, level II, 3000–2500 BCE, after Lyonnet 1996, figures 4 and 12; (*bottom right*) Altyn-Depe, excavation 1, burial 296, after Masson 1988, plate 27.

The Kuishak lead wire probably was an import from the Zeravshan. A lapis lazuli bead from Afghanistan was found at Sintashta. A Bactrian-handled bronze mirror was found in a Sintashta grave at Krasnoe Znamya.[21] Finally, the technique of lost-wax metal casting first appeared in the north during the Sintashta period, in metal objects of Seima-Turbino type (described in more detail below). Lost-wax casting was familiar to BMAC metalsmiths. Southern decorative motifs (stepped pyramids), raw materials (lead and lapis lazuli), one mirror, and metal-working techniques (lost-wax casting) appeared in the north just when northern pottery, chariot-driving cheekpieces, bit wear, and horse bones appeared in the south.

The sudden shift to large-scale copper production that began about 2100–2000 BCE in the earliest Sintashta settlements must have been stimulated by a sharp increase in demand. Central Asia is the most likely source. The increase in metal production deeply affected the internal politics of northern steppe societies, which quickly became accustomed to using and consuming large quantities of bronze. Although the northern steppe producers probably had direct contact with the Central Asian market only for a short time, internal demand in the steppes remained high throughout the LBA. Once the metallurgical pump was primed, so to speak, it continued to flow. The priming happened because of contact with urban markets, but the flow after that raised the usage of metal in the steppes and in the forest zone to the north, starting an internal European cycle of exchange that would lead to a metal boom in the Eurasian steppes after 2100 BCE.

After 1900 BCE a contact zone developed in the Zeravshan valley and extended southward to include the central citadels in the BMAC towns. In the Zeravshan, migrants from the northern steppes mixed with late Kelteminar and BMAC-derived populations. The Old Indic dialects probably evolved and separated from the developing Iranian dialects in this setting. To understand how the Zeravshan-Bactrian contact zone separated itself from the northern steppes, we need to examine what happened in the northern steppes after the end of the Sintashta culture.

The Opening of the Eurasian Steppes

The Srubnaya (or Timber-Grave) culture was the most important LBA culture of the western steppes, from the Urals to the Dnieper (figure 16.10). The Andronovo horizon was the primary LBA complex of the eastern steppes, from the Urals to the Altai and the Tien Shan. Both grew from the Potapovka-Sintashta complex between the middle Volga and the Tobol. With the appearance of Srubnaya and Andronovo between about 1900 and 1800 BCE, for the first time in history a chain of broadly similar cultures extended from the edges of China to the frontiers of Europe. Innovations and raw materials began to move across the continent. The steppe world was not just a conduit, it also became an innovating

Figure 16.10 The Late Bronze Age cultures of the Eurasian steppes, 1900–1500 BCE.

center, particularly in bronze metallurgy and chariot warfare. The chariot-driving Shang kings of China and the Mycenaean princes of Greece, contemporaries at opposite ends of the ancient world at about 1500 BCE, shared a common technological debt to the LBA herders of the Eurasian steppes.

The Srubnaya Culture: Herding and Gathering in the Western Steppes

West of the Ural Mountains, the Potapovka and late Abashevo groups of the middle Volga region developed into the Pokrovka complex, dated about 1900–1750 BCE. Pokrovka was a proto-Srubnaya phase that rapidly developed directly into the Srubnaya (or Timber-Grave) culture (1800–1200 BCE). Srubnaya material culture spread as far west as the Dnieper valley. One of the most prominent features of the Srubnaya culture was the appearance of hundreds of small settlement sites, most of them containing just a few houses, across the northern steppe and the southern forest-steppe, from the Urals to the Dnieper. Although settlements had reappeared in a few places east of the Don River during the late Catacomb culture, 2400–2100 BCE, and were even more numerous in Ukraine west of the Don during the Mnogovalikovaya (MVK) period (2100–1800 BCE), the Srubnaya period was the first time since the Eneolithic that settlements appeared across the entire northern steppe zone from the Dnieper to the southern Urals and beyond into northern Kazakhstan.

The reason for this shift back to living in permanent homes is unclear. Most Srubnaya settlements were not fortified or defended. Most were small individual homesteads or extended family ranches rather than nucleated villages. The herding pattern seems to have been localized rather than migratory. During the Samara Valley Project, in 1999–2001, we studied the local Srubnaya herding pattern by excavating a series of Srubnaya herding camps that extended up a tributary stream valley, Peschanyi Dol, from the Srubnaya settlement at Barinovka, near the mouth of the valley on the Samara (figure 16.11). The largest herding camps (PD1 and 2) were those closest to the home settlement, within 4–6 km of Barinovka. Farther upstream the Srubnaya camps were smaller with fewer pottery sherds, and beyond about 10–12 km upstream from Barinovka we found no LBA herding camps at all, not even around the springs that fed the stream at its source, where there was plenty of water and good pastures. So the herding system seems to have been localized, like the new residence pattern.

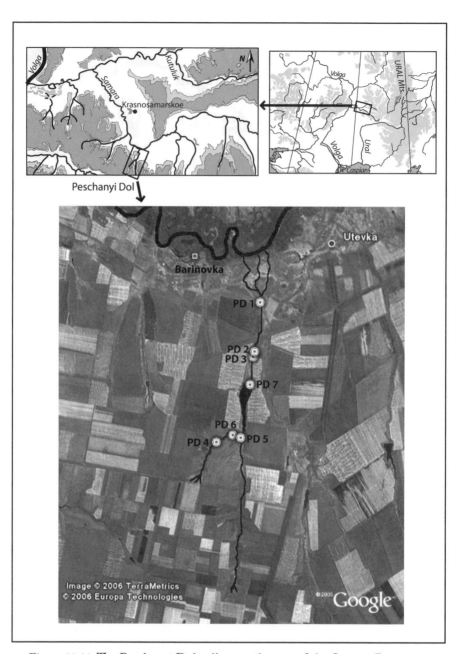

Figure 16.11 The Peschanyi Dol valley, a tributary of the Samara River, surveyed to find ephemeral camps in 1995–96. PD1, 2, and 3, were Srubnaya herding camps excavated in 2000. All numbered sites yielded at least one Srubnaya ceramic sherd. Barinovka was a larger Srubnaya settlement tested in

The Srubnaya economy in the middle Volga steppes does not seem to have required long-distance migrations.

One traditional explanation for the settling-down phenomenon is that this was when agriculture was widely adopted across the northern steppes.[22] But this explanation certainly does not apply everywhere. At the settlement of Krasnosamarskoe in the Samara River valley, where the dog sacrifice was found (chapter 15), a Pokrovka component (radiocarbon dated 1900–1800 BCE) and an early Srubnaya component (dated 1800–1700 BCE) were stratified within a single structure. In the Srubnaya period the structure probably was a well-house and woodshed where a variety of domestic tasks were conducted and food garbage was buried in pits. It was used during all seasons of the year. Anne Pike-Tay's analysis of seasonal bands in the roots of animal teeth established that the cattle and sheep were butchered in all seasons. But there was no agriculture. Laura Popova found no seeds, pollen, or phytoliths of cultivated cereals associated with the LBA occupation, only wild *Chenopodium* and *Amaranthus* seeds. The skeletons of 192 adults from twelve Srubnaya cemeteries in the Samara oblast were examined by Eileen Murray and A. Khokhlov. They showed almost no dental decay. The complete absence of caries usually is associated with a low-starch, low-carbohydrate diet, typical for foragers and quite atypical for bread eaters (figure 16.12). The dental evidence confirmed the botanical evidence. Bread was not eaten much, if at all, in the northern steppes.

In pits at Krasnosamarskoe we found an abundance of carbonized wild seeds, including *Chenopodium album* and *Amaranthus*. Modern wild *Chenopodium* (also known as goosefoot) is a weed that grows in dense stands that can produce seed yields in the range of 500–1000 kg/ha, about the same as einkorn wheat, which yields 645-835 kg/ha.[23] *Amaranthus* is equally prolific. With meat and milk from cattle, sheep, and horses, this was a sufficient diet. Although clear evidence of cereal agriculture has been found in Srubnaya settlements west of the Don in Ukraine, it is possible that agriculture was much less important east of the Don than has often been assumed. Herding and gathering was the basis for the northern steppe economy in at least some regions east of the Don as late as the LBA.[24]

Figure 16.11 (continued) 1996 but found to be badly disturbed by a historic settlement. Author's excavation. Bottom image is a Google Earth™ image, © 2006 Terra Metrics, 2006 Europa Technologies.

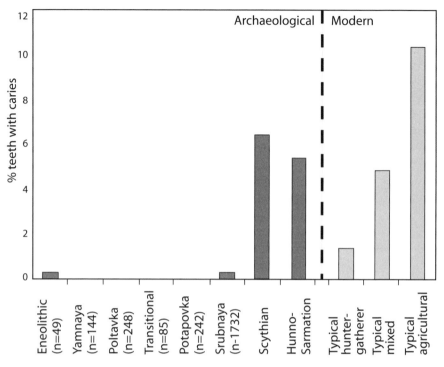

Figure 16.12 Graph of the frequency of dental caries (cavities) in populations with different kinds of food economies (*right*), in Scythian and Sarmatian cemeteries in Tuva (*center*), and in prehistoric populations in the Samara oblast, middle Volga region (*left six bars*). Bread apparently was not part of the diet in the Samara oblast. After Murphy 2003; and Murphy and Khokhlov 2001.

So if agriculture does not provide an answer, then why did people settle down during the MBA/LBA transition in the northern steppes, including the earlier episode at Sintashta? As explained in chapter 15, climate change might have been the principal cause. A cool, arid climate affected the Eurasian steppes between about 2500–2000 BCE. This was the same event that struck Akkadian agriculture and weakened the Harappan civilization. The late MBA/early LBA settling-down phenomenon, including the earliest episodes at Sintashta and Arkaim, can be interpreted as a way to maintain control over the richest winter forage areas for herds, particularly if grazing animals were the principal source of food in an economy that, in many regions, did not include agriculture. Early LBA Krasnosamarskoe overlooked one of the largest marshes on the lower Samara River.

Some permanent settlements also developed near copper mines. Cattle forage was not the only critical resource in the northern steppes. Mining and bronze working became important industries across the steppes during the LBA. A vast Srubnaya mining center operated at Kargaly near Orenburg in the South Urals, and other enormous copper mines operated near Karaganda in central Kazakhstan. Smaller mining camps were established at many small copper outcrops, like the Srubnaya mining camp at Mikhailovka Ovsianka in the southern Samara oblast.[25]

EAST OF THE URALS, PHASE I: THE PETROVKA CULTURE

The first culture of the LBA east of the Urals was the Petrovka culture, an eastern offshoot of Sintashta dated about 1900–1750 BCE. Petrovka was so similar to Sintashta in its material culture and mortuary rituals that many archaeologists (including me) have used the combined term Sintashta-Petrovka to refer to both. But Petrovka ceramics show some distinctive variations in shape and decoration, and are stratified above Sintashta deposits at several sites, so it is clear that Petrovka grew out of and was generally later than Sintashta. The oldest Petrovka sites, like the type site, Petrovka II, were settlements on the Ishim River in the steppes of northern Kazahstan (figure 16.13). The Petrovka culture probably absorbed some people who had roots in the older post-Botai horse-centered cultures of the Ishim steppes, like Sergeivka, but they were materially (and probably linguistically) almost invisible. Petrovka-style pottery then replaced Sintashta ceramics at several Sintashta fortified sites, as at Ust'ye, where the Sintashta settlement was burned and replaced by a Petrovka settlement built on a different plan. Petrovka graves were dug into older Sintashta kurgans at Krivoe Ozero and Kamenny Ambar.[26]

The settlement of Petrovka II was surrounded by a narrow ditch less than 1 m deep, perhaps for drainage. The twenty-four large houses had dug-out floors and measured from 6 by 10 m to about 8 by 18 m. They were built close together on a terrace overlooking the floodplain, a nucleated village pattern quite different from the scattered homesteads of the Srubnaya culture. Petrovka II was reoccupied by people who made classic Andronovo-horizon ceramics of both the Alakul and Federovo types, stratified above the Petrovka layer, and the Andronovo town was succeeded by a "final-LBA" settlement with Sargar ceramics. This stratified sequence made Petrovka II an important yardstick for the LBA chronology of the Kazakh steppes. Chariots continued to be buried in a few early Petrovka graves at Berlyk II and Krivoe Ozero, and many bone disk-shaped cheekpieces have

Petrovka settlement plan

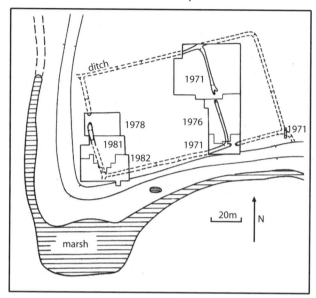

1971 excavation detail

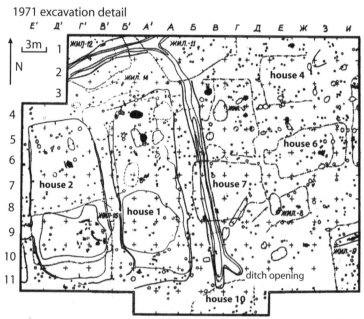

Figure 16.13 The Petrovka settlement, type site for the Petrovka culture, ca. 1900–1750 BCE: (*top*) general plan of the original ditch around the settlement, with a later enlargement at the east end, after Zdanovich 1988, Figure 12; (*bottom*) detail of overlapping rebuilt house floors in the northeast corner of the original settlement, with new houses built over the original eastern ditch, after

come from Petrovka sites. During the Petrovka period, however, chariot burials gradually ceased, the size and number of mortuary animal sacrifices also declined, and large-scale Sintashta-type fortifications were no longer built around settlements in the northern steppes.

Petrovka settlements and kurgan cemeteries spread southward into the arid steppes of central Kazkahstan, and from there to Tugai on the Zeravshan, more than 1,200 km south of central Kazakhstan. Petrovka probably also was in touch with the Okunevo culture in the western Altai, the successor of late Afanasievo. The permanent nucleated settlements of the Petrovka culture do not resemble the temporary camps of nomadic herders, so it is unlikely that the Petrovka economy depended on annual long-distance migrations. Early historic nomads, who did not live in permanent nucleated villages, wintered in the Syr Darya marshes and summered in the north Kazakh steppes, a cycle of annual movements that brought them to the doorstep of Central Asia civilizations each winter. But the Petrovka economy seems to have been less nomadic. If the Petrovka people did *not* engage in long-distance herd migrations, then their movement south to the Zeravshan was not an accidental by-product of annual herding patterns (as is often presumed) but instead was intentional, motivated by the desire for trade, loot, or glory. The later annual migration pattern does at least show that in the spring and fall it was possible to drive herds of animals across the intervening desert and semi-desert.[27]

Petrovka settlements commonly contained two-part furnaces, slag, and abundant evidence of copper smelting, like Sintashta settlements. But, unlike Sintashta, most Petrovka metal objects were made of tin-bronze.[28] A possible source for the tin in Petrovka tin-bronzes, in addition to the Zeravshan valley, was in the western foothills of the Altai Mountains. A remarkable shift occurred in the forest-steppe zone north of the Petrovka territory during the early Petrovka phase.

The Seima-Turbino Horizon in the Forest-Steppe Zone

The Seima-Turbino horizon marks the entry of the forest-steppe and forest-zone foragers into the cycle of elite competition, trade, and warfare that had erupted earlier in the northern steppes. The tin-bronze spears, daggers, and axes of the Seima-Turbino horizon were among the most

---

Figure 16.13 (continued) Maliutina 1991, Figure 14. The stratigraphic complexity of these settlements contributes to arguments about phases and chronology.

technically and aesthetically refined weapons in the ancient world, but they were made by forest and forest-steppe societies that in some places (Tashkovo II) still depended on hunting and fishing. These very high-quality tin-bronze objects first appeared among the Elunino and Krotovo cultures located on the upper and middle Irtysh and the upper Ob in the western foothills of the Altai Mountains, a surprisingly remote region for such a remarkable exhibition of metallurgical skill. But tin, copper, and gold ores all could be found on the upper Irtysh, near the confluence of the Irtysh and the Bukhtarta rivers about 600 km east of Karaganda. The exploitation of these ore sources apparently was accompanied by an explosion of new metallurgical skills.

One of the earliest and most important Seima-Turbino cemeteries was at Rostovka in the Omsk oblast on the middle Irtysh (figure 16.14). Although skeletal preservation was poor, many of the thirty-eight graves seem to have contained no human bones at all or just a few fragments of a skeleton. In the graves with whole bodies the skeleton was supine with the legs and arms extended. Grave gifts were offered both in the graves and in ritual deposits at the edge of graves. Both kinds of offerings included tin-bronze socketed spearheads, single-edged curved knives with cast figures on the pommel, and hollow-core bronze axes decorated with triangles and lozenges. Grave 21 contained bivalve molds for making all three of these weapon types. Offerings also included stemmed flint projectile points of the same types that appeared in Sintashta graves, bone plates pierced to make plate armor, and nineteen hundred sherds of Krotovo pottery (figure 16.14). One grave (gr. 2) contained a lapis lazuli bead from Afghanistan, probably traded through the BMAC, strung with beads of nephrite, probably from the Baikal region.[29]

Seima-Turbino metalsmiths were, with Petrovka metalsmiths, the first north of Central Asia to regularly use a tin-bronze alloy. But Seima-Turbino metalsmiths were unique in their mastery of lost-wax casting (for decorative figures on dagger handles) and thin-walled hollow-mold casting (for socketed spears and hollow axes). Socketed spearheads were made on Sintashta anvils by bending a bronze sheet around a socket form and then forging the seam (figure 16.15). Seima-Turbino socketed spearheads were made by pouring molten metal into a mold that created a seamless cast socket around a suspended core, making a hollow interior, a much more sophisticated operation, and easier to do with tin-bronze than with arsenical bronze. Axes were made in a similar way, tin-bronze with a hollow interior, cast around a suspended core. Lost-wax and hollow-mold casting methods probably were learned from the BMAC civilization, the only reasonably nearby source (perhaps through a skilled captive?).

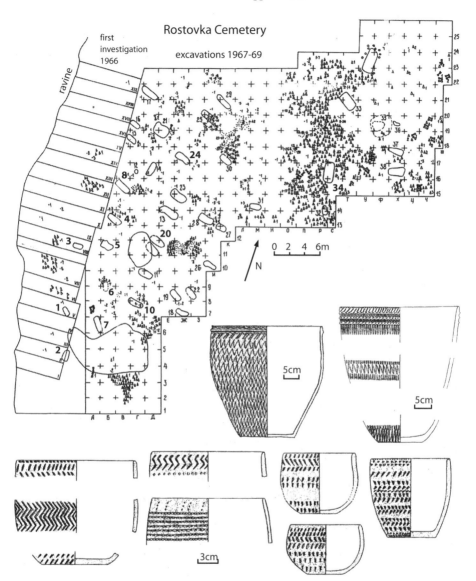

Figure 16.14 The Rostovka cemetery near Omsk, one of the most important sites of the Seima-Turbino culture. Graves are numbered. Black dots represent ceramics, metal objects, and other artifacts deposited above and beside the graves. All the pots conform to the Krotova type. After Matiushchenko and Sinitsyna 1988, figures 4, 81, 82, and 83.

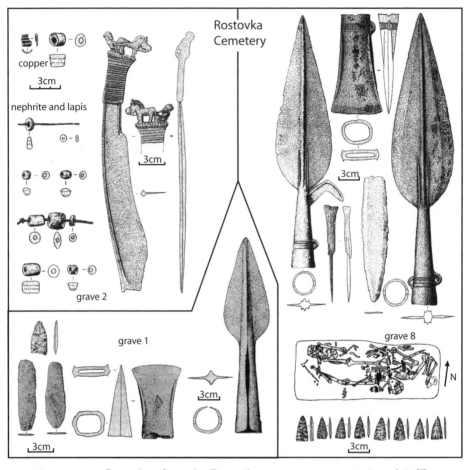

Figure 16.15 Grave lots from the Rostovka cemetery, graves 1, 2, and 8. The lost-wax cast figure of a man roping a horse and the hollow-mold casting of spears and axes were technical innovations probably learned from BMAC metalsmiths. Grave 1 contained beads made of both lapis lazuli from Afghanistan and nephrite probably from the near Lake Baikal. After Matiushchenko and Sinitsyna 1988, figures 6, 7, 17, and 18.

Beyond the western Altai/middle Irtysh core area the Seima-Turbino horizon was not a culture. It did not have a standard ceramic type, settlement type, or even a standard mortuary rite. Rather, Seima-Turbino metal-working techniques were adopted by emerging elites across the southern Siberian forest-steppe zone, perhaps in reaction to and competing with the Sintashta and Petrovka elites in the northern steppes. A series of original and distinctive new metal types quickly diffused through the forest-steppe zone from the east to the west, appearing in late Abashevo and Chirkovskaya cemeteries west of the Urals almost at the same time that they first appeared east of the Urals, beginning about 1900 BCE. The rapidity and reach of this phenomenon in the forest zone is surprising. The new metal styles probably spread more by emulation than by migration, along with fast-moving political changes in the structure of power. Seima-Turbino spearheads, daggers, and axes were displayed at the Turbino cemetery in the forests of the lower Kama, southward up the Oka, and as far south as the Borodino hoard in Moldova, in the East Carpathian foothills. East of the Urals, most Seima-Turbino bronzes were tin-bronzes, and west of the Urals, they were mostly arsenical bronzes. The source of the tin was in the east, but the styles and methods of Seima-Turbino metallurgy were diffused across the forest-steppe and forest zones from the Altai to the Carpathians. The Borodino hoard contained a nephrite axe probably made of stone quarried near Lake Baikal. In the eastern direction, Seima-Turbino metal types (hollow-cast socketed spearheads with a side hook, hollow-cast axes) appeared also in sites on the northwestern edges of the evolving archaic Chinese state, probably through a network of trading trails that passed north of the Tien Shan through Dzungaria.[30]

The dating of the Seima-Turbino horizon has changed significantly in recent years. Similarities between Seima-Turbino socketed spearheads and daggers and parallel objects in Mycenaean tombs were once used to date the Seima-Turbino horizon to a period after 1650 BCE. It is clear now, however, that Mycenaean socketed spearheads, like studded disk cheekpieces, were derived from the east and not the other way around. Seima-Turbino and Sintashta were partly contemporary, so Seima-Turbino probably began before 1900 BCE.[31] Seima-Turbino and Sintasha graves had the same kinds of flint projectile points. Sintashta forged socketed spearheads probably were the simpler predecessors of the more refined hollow-cast Seima-Turbino socketed spearheads. A hollow-cast spearhead of Seima-Turbino type was deposited in a Petrovka-culture chariot grave at Krivoe Ozero (k. 2, gr. 1); and a Sintashta bent and forged spearhead appeared in the Seima-Turbino cemetery at Rostovka (gr. 1) (see figure 16.15).

The metal-working techniques of the northern steppes (Sintashta and Petrovka) and the forest-steppe zone (Seima-Turbino) remained separate and distinct for perhaps one hundred to two hundred years. But by the beginning of the Andronovo period they merged, and some important

Seima-Turbino metal types, such as cast single-edged knives with a ring-pommel, became widely popular in Andronovo communities.

EAST OF THE URALS, PHASE II: THE ANDRONOVO HORIZON

The Andronovo horizon was the principal LBA archaeological complex in the steppes east of the Urals, the sister of the Srubnaya horizon west of the Urals, between about 1800 and 1200 BCE. Andronovo sites extended from the Ural steppes eastward to the steppes on the upper Yenisei River in the Altai, and from the southern forest zone southward to the Amu Darya River in Central Asia. Andronovo contained two principal subgroups, Alakul and Federovo. The earliest of these, the Alakul complex, appeared in some places by about 1900–1800 BCE. It grew directly out of the Petrovka culture by small modifications of ceramic decorations and vessel shapes. The Federovo style might have developed from a southern or eastern stylistic variant of Alakul, although some specialists insist that it had completely independent origins. Andronovo continued many of the customs and styles inherited through Sintashta and Petrovka: small family kurgan cemeteries, settlements containing ten to forty houses built close together, similar spear and dagger types, similar ornaments, and even the same decorative motifs on pottery: meanders, hanging triangles, "pine-tree" figures, stepped pyramids, and zig-zags. But chariots were no longer buried.

Alakul and Federovo are described as separate cultures within the Andronovo horizon, but to this observer, admittedly not an expert in the details of LBA ceramic typology, the Alakul and Federovo ceramic styles seem similar. Pot shapes varied only slightly (Federovo pots usually had a more indented, undercut lower profile) and decorative motifs also varied around common themes (some Federovo motifs were "italicized" or forward-slanted versions of Alakul motifs). Pots and potsherds of these two ceramic styles are found in the same sites from the Ural-Tobol steppes southeastward to central Kazkahstan, often in the same house and pit features, and in adjoining kurgans in the same cemeteries. Some pots are described as Alakul with Federovo elements, so the two varieties can appear on the same pot (figure 16.16). Alakul pottery is stratified beneath Federovo pottery in a few key features at some sites (at Novonikol'skoe and Petrovka II in the Ishim steppes and Atasu 1 in central Kazakhstan), but Federovo pottery has never been found stratified beneath Alakul. The earliest Alakul radiocarbon dates (1900–1700 BCE) are a little older than the earliest Federovo dates (1800–1600 BCE), so Alakul

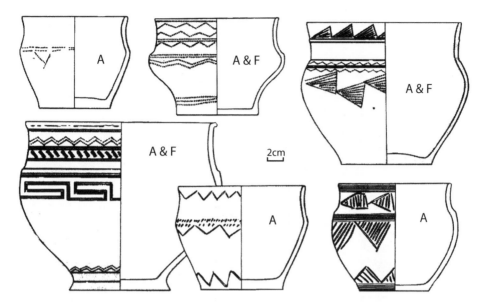

Figure 16.16 Andronovo pots that are described as typical Alkakul (A) or Alakul with Federovo traits (A+F) from the Priplodyi Log kurgan cemetery I on the Ui River, Chelyabinsk oblast, Russia. Traits of both styles can appear on the same pot. After Maliutina 1984, figure 4.

probably began a century or two earlier, although in many settlements the two are thoroughly mixed. Kurgans containing Federovo pots often had larger, more complex stone constructions around the grave and the dead were cremated, whereas kurgans with Alakul pots were simpler and the dead usually were buried in the flesh. Since the two ceramic styles occurred in the same settlements and cemeteries, and even in the same house and pit features, they cannot easily be interpreted as distinct ethnic groups.[32]

The spread of the Andronovo horizon represented the maturation and consolidation of an economy based on cattle and sheep herding almost everywhere in the grasslands east of the Urals. Permanent settlements appeared in every region, occupied by 50 to 250 people who lived in large houses. Wells provided water through the winter. Some settlements had elaborate copper-smelting ovens. Small-scale agriculture might have played a minor role in some places, but there is no direct evidence for it. In the northern steppes cattle were more important than sheep (cattle 40% of bones, sheep/goat 37%, horses 17% in the Ishim steppes), whereas in central Kazakhstan there were more sheep than cattle, and more horses as well (sheep/goat 46%, cattle 29%, horse 24%).[33]

Although it is common in long-established tribal culture areas for a relatively homogeneous material culture to mask multiple languages, the link between language and material culture often is strong among the early generations of long-distance migrants. The source of the Andronovo horizon can be identified in an extraordinary burst of economic, military, and ritual innovations by a single culture—the Sintashta culture. Many of its customs were retained by its eastern daughter, the Petrovka culture. The language spoken in Sintashta strongholds very likely was an older form of the language spoken by the Petrovka and Andronovo people. Indo-Iranian and Proto-Iranian dialects probably spread with Andronovo material culture.

Most Andronovo metals, like Petrovka metals, were tin-bronzes. Andronovo miners mined tin in the Zeravshan and probably on the upper Irtysh. Andronovo copper mines were active in two principal regions: one was south of Karaganda near Uspenskyi, working malachite and azurite oxide ores; and the other was to the west in the southern Ulutau Hills near Dzhezkazgan, working sulfide ores. (Marked on figure 15.9.) One mine of at least seven known in the Dzhezkazgan region was 1,500 m long, 500 m wide, and 15 m deep. Ore was transported from the Uspenskyi mine to copper-smelting settlements such as Atasu 1, where excavation revealed three key-shaped smelting ovens with 4 m-long stone-lined air shafts feeding into two-level circular ovens. The Karaganda-region copper mines are estimated to have produced 30 to 50,000 metric tons of smelted copper during the Bronze Age.[34] The labor and facilities at these places suggest enterprises organized for export.

Trade with and perhaps looting raids into Central Asia left clear evidence surprisingly far north in the steppes. Wheel-made Namamzga VI pottery was found in the Andronovo settlement of Pavlovka, in northern Kazakhstan near Kokchetav, 2,000 km north of Bactria. It was 12% of the pottery on two house floors. The remainder was Andronovo pottery of the Federovo type.[35] The imported Central Asian pots were made with very fine white or red clay fabrics, largely undecorated, and in forms such as pedestaled dishes that were typical of Namazaga VI (figure 16.17). Pavlovka was a settlement of about 5 ha with both Petrovka and Federovo pottery. The Central Asian pottery is said to have been associated with the Federovo component.

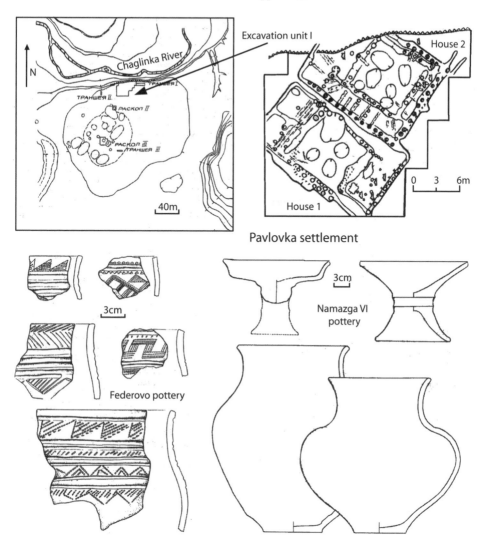

Figure 16.17 Pavlovka, an Alakul-Federovo settlement in the Kokchetav region of northern Kazakhstan, with imported Namazga VI pottery constituting more than 10% of the sherds on two house floors. After Maliutina 1991, figures 4 and 5.

Proto-Vedic Cultures in the Central Asian Contact Zone

By about 1900 BCE Petrovka migrants had started to mine copper in the Zeravshan valley at Tugai. They were followed by larger contingents of Andronovo people who mined tin at Karnab and Mushiston. After 1800 BCE Andronovo mining camps, kurgan cemeteries, and pastoral camps spread into the middle and upper Zeravshan valley. Other Andronovo groups moved into the lower Zeravshan and the delta of the lower Amu Darya (now located in the desert east of the modern delta) and became settled irrigation farmers, known as the Tazabagyab variant of the Andronovo culture. They lived in small settlements of a few large dug-out houses, much like Andronovo houses; used Andronovo pottery and Andronovo-style curved bronze knives and twisted earrings; conducted in-settlement copper smelting as at many Andronovo settlements; but buried their dead in large flat-grave cemeteries like the one at Kokcha 3, with more than 120 graves, rather than in kurgan cemeteries (figure 16.18).[36]

About 1800 BCE the walled BMAC centers decreased sharply in size, each oasis developed its own types of pottery and other objects, and Andronovo-Tazabagyab pottery appeared widely in the Bactrian and Margian countryside. Fred Hiebert termed this the *post-BMAC* period to emphasize the scale of the change, although occupation continued at many BMAC strongholds and Namazga VI–style pottery still was made inside them.[37] But Andronovo-Tazabagyab coarse incised pottery occurred both within post-BMAC fortifications and in occasional pastoral camps located outside the mudbrick walls. Italian survey teams exposed a small Andronovo-Tazabagyab dug-out house southeast of the post-BMAC walled fortress at Takhirbai 3, and American excavations found a similar occupation outside the walls of a partly abandoned Gonur. By this time the people living just outside the crumbling walls and at least some of those now living inside were probably closely related. To the east, in Bactria, people making similar incised coarse ware camped atop the vast ruins (100 ha) of the Djarkutan city. Some walled centers such as Mollali-Tepe continued to be occupied but at a smaller scale. In the highlands above the Bactrian oases in modern Tajikistan, kurgan cemeteries of the Vaksh and Bishkent type appeared with pottery that mixed elements of the late BMAC and Andronovo-Tazabagyab traditions.[38]

Between about 1800 and 1600 BCE, control over the trade in minerals (copper, tin, turquoise) and pastoral products (horses, dairy, leather) gave the Andronovo-Tazabagyab pastoralists great economic power in the old

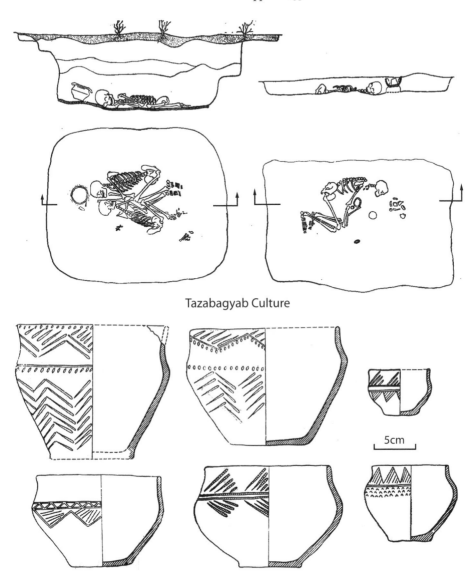

Tazabagyab Culture

5cm

Figure 16.18 Graves of the Tazabagyab-Andronovo culture at the Kokcha 3 cemetery on the old course of the lower Amu-Darya River. Pottery like this was widespread in the final phase of occupation in the declining BMAC walled towns of Central Asia, 1700–1500 BCE. After Tolstov and Kes' 1960, figure 55.

BMAC oasis towns and strongholds, and chariot warfare gave them military control. Social, political, and even military integration probably followed. Eventually the simple incised pottery of the steppes gave way to new ceramic traditions, principally gray polished wares in Margiana and the Kopet Dag, and painted wares in Bactria and eastward into Tajikistan.

By 1600 BCE all the old trading towns, cities, and brick-built fortified estates of eastern Iran and the former BMAC region in Central Asia were abandoned. Malyan, the largest city on the Iranian plateau, was reduced to a small walled compound and tower occupied within a vast ruin, where elite administrators, probably representatives of the Elamite kings, still resided atop the former city. Pastoral economies spread across Iran and into Baluchistan, where clay images of riders on horseback appeared at Pirak about 1700 BCE. Chariot corps appeared across the Near East as a new military technology. An Old Indic-speaking group of chariot warriors took control of a Hurrian-speaking kingdom in north Syria about 1500 BCE. Their oaths referred to deites (Indra, Varuna, Mithra, and the Nasatyas) and concepts (*r'ta*) that were the central deities and concepts in the *Rig Veda*, and the language they spoke was a dialect of the Old Indic Sanskrit of the *Rig Veda*.[39] The Mitanni dynasts came from the same ethnolinguistic population as the more famous Old Indic–speakers who simultaneously pushed eastward into the Punjab, where, according to many Vedic scholars, the *Rig Veda* was compiled about 1500–1300 BCE. Both groups probably originated in the hybrid cultures of the Andronovo/ Tazabagyab/ coarse-incised-ware type in Bactria and Margiana.[40]

The language of the *Rig Veda* contained many traces of its syncretic origins. The deity name *Indra* and the drug-deity name *Soma*, the two central elements of the religion of the *Rig Veda*, were non–Indo-Iranian words borrowed in the contact zone. Many of the qualities of the Indo-Iranian god of might/victory, Verethraghna, were transferred to the adopted god Indra, who became the central deity of the developing Old Indic culture.[41] Indra was the subject of 250 hymns, a quarter of the *Rig Veda*. He was associated more than any other deity with *Soma*, a stimulant drug (perhaps derived from *Ephedra*) probably borrowed from the BMAC religion. His rise to prominence was a peculiar trait of the Old Indic speakers. Indra was regarded in later Avestan Iranian texts as a minor demon. Iranian dialects probably developed in the northern steppes among Andronovo and Srubnaya people who had kept their distance from the southern civilizations. Old Indic languages and rituals developed in the contact zone of Central Asia.[42]

Loan Words Borrowed into Indo-Iranian and Vedic Sanskrit

The Old Indic of the *Rig Veda* contained at least 383 non–Indo-European words borrowed from a source belonging to a different language family. Alexander Lubotsky has shown that common Indo-Iranian, the parent of both Old Indic and Iranian, probably had already borrowed words from the *same* non–Indo-European language that later enriched Old Indic. He compiled a list of 55 non–Indo-European words that were borrowed into common Indo-Iranian *before* Old Indic or Avestan evolved, and then later were inherited into one or both of the daughters from common Indo-Iranian. The speakers of common Indo-Iranian were in touch with and borrowed terms from *the same foreign language group* that later was the source from which Old Indic speakers borrowed even more terms. This discovery carries significant implications for the geographic locations of common Indo-Iranian and formative Old Indic—they must have been able to interact with the same foreign-language group.

Among the fifty-five terms borrowed into common Indo-Iranian were the words for bread (**nagna-*), ploughshare (*spʰāra*), canal (**iaviā*), brick (**išt(i)a-*, camel (**Huštra-*), ass (**kʰara-*) sacrificing priest (**ućig-*), soma (**anću-*), and Indra (**indra-*). The BMAC fortresses and cities are an excellent source for the vocabulary related to irrigation agriculture, bricks, camels, and donkeys; and the phonology of the religious terms is the same, so probably came from the same source. The religious loans suggest a close cultural relationship between some people who spoke common Indo-Iranian and the occupants of the BMAC fortresses. These borrowed southern cults might possibly have been one of the features that distinguished the Petrovka culture from Sintashta. Petrovka people were the first to migrate from the northern steppes to Tugai on the northern edge of Central Asia.

Lubotsky suggested that Old Indic developed as a vanguard language south of Indo-Iranian, closer to the source of the loans. The archaeological evidence supports Lubotsky's suggestion. The earliest Old Indic dialects probably developed about 1800–1600 BCE in the contact zone south of the Zeravshan among northern-derived immigrants who were integrated with and perhaps ruled over the declining fortunes of the post-BMAC citadels. They retained a decidedly pastoral set of values. In the *Rig Veda* the clouds were compared to dappled cows full of milk; milk and butter were the symbols of prosperity; milk, butter, cattle, and horses were the proper offerings to the gods; Indra was compared to a mighty bull; and wealth was counted in fat cattle and swift horses. Agricultural products were never offered to the gods. The people of the *Rig Veda* did not live in brick houses and had no cities, although their enemies, the *Dasyus*, did live in walled strongholds. Chariots were used in races and war; the gods drove chariots across the sky. Almost all important deities were masculine. The only important female deity was Dawn, and she was less powerful than Indra, Varuna, Mithra, Agni, or the Divine Twins. Funerals included both cremation (as in Federovo graves) and inhumation (as in Andronovo and Tazabagyab graves). Steppe cultures are an acceptable source for all these details of belief and practice, whereas the culture of the BMAC, with its female deity in a flounced skirt, brick fortresses, and irrigation agriculture, clearly is not.

During the initial phase of contact, the Sintashta or the Petrovka cultures or both borrowed some vocabulary and rituals from the BMAC, accounting for the fifty-five terms in common Indo-Iranian. These included the drug *soma*, which remained in Iranian ritual usage as *haoma*. In the second phase of contact, the speakers of Old Indic borrowed much more heavily from the same language when they lived in the shadows of the old BMAC settlements and began to explore southward into Afghanistan and Iran. Archaeology shows a pattern quite compatible with that suggested by the linguistic evidence.

The Steppes Become a Bridge across Eurasia

The Eurasian steppe is often regarded as a remote and austere place, poor in resources and far from the centers of the civilized world. But during the Late Bronze Age the steppes became a bridge between the civilizations that developed on the edges of the continent in Greece, the Near East, Iran, the Indian subcontinent, and China. Chariot technology, horses and horseback riding, bronze metallurgy, and a strategic location gave steppe societies an importance they never before had possessed. Nephrite from Lake Baikal appeared in the Carpathian foothills in the Borodino hoard; horses and tin from the steppes appeared in Iran; pottery from Bactria appeared in a Federovo settlement in northern Kazakhstan; and chariots appeared across the ancient world from Greece to China. The road from the steppes to China led through the eastern end of the Tarim Basin, where desert-edge cemeteries preserved the dessicated mummies of brown-haired, white-skinned, wool-wearing people dated as early as 1800 BCE. In Gansu, on the border between China and the Tarim Basin, the Qijia culture acquired horses, trumpet-shaped earrings, cast bronze ring-pommel single-edged knives and axes in steppe styles between about 2000 and 1600 BCE.[45] By the time the first Chinese state emerged, beginning about 1800 BCE, it was exchanging innovations with the West. The Srubnaya and Andronovo horizons had transformed the steppes from a series of isolated cultural ponds to a corridor of communication. That transformation permanently altered the dynamics of Eurasian history.

CHAPTER SEVENTEEN

Words and Deeds

The Indo-European problem can be solved today because archaeological discoveries and advances in linguistics have eaten away at problems that remained insoluble as recently as fifteen years ago. The lifting of the Iron Curtain after 1991 made the results of steppe research more easily available to Western scholars and created new cooperative archaeological projects and radiocarbon dating programs. Linguists like Johanna Nichols, Sarah Thomason, and Terrence Kaufman came up with new ways of understanding language spread and convergence. The publication of the Khvalynsk cemetery and the Sintashta chariot burials revealed unsuspected richness in steppe prehistory. Linguistic and archaeological discoveries now converge on the probability that Proto-Indo-European was spoken in the Pontic-Caspian steppes between 4500 and 2500 BCE, and alternative possibilities are increasingly difficult to square with new evidence. Gimbutas and Mallory preceded me in arguing this case. I began this book by trying to answer questions that still bothered many reasonable observers.

One question was whether prehistoric language borders could be detected in prehistoric material culture. I suggested that they were correlated at persistent frontiers, a generally rare phenomenon that was surprisingly common among the prehistoric cultures of the Pontic-Caspian steppes. Another problem was the reluctance of Western archaeologists and the overenthusiasm of Eastern European archaeologists to use migration as an explanation for prehistoric culture change, a divergence in approach that produced Eastern interpretations that Western archaeologists would not take seriously. I introduced models from demographics, sociology, and anthropology that describe how migration works as a predictable, regular human behavior in an attempt to bring both sides to the middle. The most divisive problem was the absence of convincing evidence indicating when horse domestication and horseback riding began. Bit wear might settle the issue through the presence or absence of a clear riding-related pathology on horse teeth. A separate but related debate swirled around the question of whether pastoral nomadism was possible as early as the Yamnaya horizon, or if it depended on later horseback riding, which in this argument only began in the Iron Age; or perhaps it depended on state economies, which also appeared on the steppe border during the Iron Age. The Samara Valley Project examined the botanical and seasonal aspects of a Bronze Age steppe pastoral economy and found that it did not rely on cultivated grain even in year-round permanent settlements. Steppe pastoralism was entirely self-sustaining and independent in the Bronze Age; wild seed plants were plentiful, and wild seeds were eaten where grain was not cultivated. Pastoral nomadism did not depend for its food supply on Iron Age states. Finally, the narrative culture history of the western steppes was impenetrable to most Western linguists and archaeologists. Much of this book is devoted to my efforts to cut a path through the tangle of arguments about chronology, culture groups, origins, migrations, and influences. I have tried to reduce my areas of ignorance about steppe archaeology, but am mindful of the few years I spent doing federally funded archaeology in Massachusetts, less than half the size of the single Samara oblast on the Volga, and how we all thought it an impossible task to try to learn the archaeology of Massachusetts *and* neighboring Rhode Island—one-tenth the size of Samara oblast. Nevertheless, I have found a path that makes sense through what I have read and seen. Debate will continue on all these subjects, but I sense that a chord is emerging from the different notes.

THE HORSE AND THE WHEEL

Innovations in transportation technology are among the most powerful causes of change in human social and political life. The introduction of the private automobile created suburbs, malls, and superhighways; transformed heavy industry; generated a vast market for oil; polluted the atmosphere; scattered families across the map; provided a rolling, heated space in which young people could escape and have sex; and fashioned a powerful new way to express personal status and identity. The beginning of horseback riding, the invention of the heavy wagon and cart, and the development of the spoke-wheeled chariot had cumulative effects that unfolded more slowly but eventually were equally profound. One of those effects was to transform Eurasia from a series of unconnected cultures into a single interacting system. How that happened is a principal focus of this book.

Most historians think of war when they begin to list the changes caused by horseback riding and the earliest wheeled vehicles. But horses were first domesticated by people who thought of them as food. They were a cheap source of winter meat; they could feed themselves through the steppe winter, when cattle and sheep needed to be supplied with water and fodder. After people were familiar with horses as domesticated animals, perhaps after a relatively docile male bloodline was established, someone found a particularly submissive horse and rode on it, perhaps as a joke. But riding soon found its first serious use in the management of herds of domesticated cattle, sheep, and horses. In this capacity alone it was an important improvement that enabled fewer people to manage larger herds and move them more efficiently, something that really mattered in a world where domesticated animals were the principal source of food and clothing. By 4800–4600 BCE horses were included with obviously domesticated animals in human funeral rituals at Khvalysnk on the middle Volga.

By about 4200–4000 BCE people living in the Pontic-Caspian steppes probably were beginning to ride horses to advance to and retreat from raids. Once they began to ride, there was nothing to prevent them from riding into tribal conflicts. Organic bits functioned perfectly well, Eneolithic steppe horses were big enough to ride (13–14 hands), and the leaders of steppe tribes began to carry stone maces as soon as they began to keep herds of cattle and sheep, around 5200–4800 BCE. By 4200 BCE people had become more mobile, their single graves emphasized individual status and personal glory unlike the older communal funerals, high-status graves contained stone maces shaped like horse heads and other weapons, and raiding parties migrated hundreds of kilometers to enrich themselves with Balkan copper, which they traded or gifted back to their relatives in the Dnieper-Azov steppes. The collapse of Old Europe about 4200–4000 BCE probably was at least partly their doing.

The relationship between mounted steppe pastoralists and sedentary agricultural societies has usually been seen by historians as either violent, like the Suvorovo confrontation with Old Europe, or parasitic, or both. "Barbaric" pastoral societies, hungry for grain, metals, and wealth, none of which they could produce themselves, preyed upon their "civilized" neighbors, without whom they could not survive. But these ideas are inaccurate and incomplete even for the historical period, as the Soviet ethnographer Sergei Vainshtein, the Western historian Nicola DiCosmo, and our own botanical studies have shown. Pastoralism produced plenty of food—the average nomad probably ate better than the average agricultural peasant in Medieval China or Europe. Steppe miners and craftsmen mined their own abundant ores and made their own metal tools and weapons; in fact, the enormous copper mines of Russia and Kazakhstan and the tin mines of the Zeravshan show that the Bronze Age civilizations of the Near East depended on *them*. For the prehistoric era covered in this book, any model based on relationships between the militarized nomads of the steppes and the medieval civilizations of China or Persia is anachronistic. Although the steppe societies of the Suvorovo-Novodanilovka period did seem to prey upon their neighbors in the lower Danube valley, they were clearly more integrated and apparently had peaceful relationships with their Cucuteni-Tripolye neighbors at the same time. Maikop traders seem to have visited steppe settlements on the lower Don and even perhaps brought weavers there. The institutions that regulated peaceful exchange and cross-cultural relationships were just as important as the institution of the raid.

The reconstructed Proto-Indo-European vocabulary and comparative Indo-European mythology reveal what two of those important integrative institutions were: the oath-bound relationship between patrons and clients, which regulated the reciprocal obligations between the strong and the weak, between gods and humans; and the guest-host relationship, which extended these and other protections to people outside the ordinary social circle. The first institution, legalizing inequality, probably was very old, going back to the initial acceptance of the herding economy, about 5200–5000 BCE, and the first appearance of pronounced differences in wealth. The second might have developed to regulate migrations into unregulated geographic and social space at the beginning of the Yamnaya horizon.

When wheeled vehicles were introduced into the steppes, probably about 3300 BCE, they again found their first use in the herding economy. Early wagons and carts were slow, solid-wheeled vehicles probably pulled by oxen and covered by arched roofs made of reed mats plaited together, perhaps originally attached to a felt backing. Yamnaya-era graves often contain remnants of reed mats with other decayed organic material. On some occasions the mats were painted in red, black, and white stripes and curved designs, certainly at funerals. Wagons permitted herders to migrate with their herds into the deep steppes between the river valleys for weeks or months at a time, relying on the tents, food, and water carried in their wagons. Even if the normal annual range of movement was less than 50 km, which seems likely for Yamnaya herders, the combination of bulk wagon transport with rapid horseback transport revolutionized steppe economies, opening the majority of the Eurasian steppe zone to efficient exploitation. The steppes, largely wild and unused before, were domesticated. The Yamnaya horizon exploded across the Pontic-Caspian steppes about 3300 BCE. With it probably went Proto-Indo-European, its dialects scattering as its speakers moved apart, their migrations sowing the seeds of Germanic, Baltic, Slavic, Italic, Celtic, Armenian, and Phrygian.

The chariot, the first wheeled vehicle designed entirely for speed, first appeared in the graves of the Sintashta culture, in the southern Ural steppes, about 2100 BCE. It was meant to intimidate. A chariot was incredibly difficult to build, a marvel of carpentry and bent-wood joinery. It required a specially trained team of fast, strong horses. To drive it through a turn, you had to rein each horse independently while keeping a backless, bouncing car level by leaning your weight into each bounce. It was even more difficult to throw a javelin accurately at a target while driving a speeding chariot, but the evidence from the Sintashta chariot graves suggests that this is precisely what they did. Only men with a lot of time and resources, as well as balance and courage, could learn to fight from a chariot. When a squadron of javelin-hurling chariot warriors wheeled onto the field of battle, supported by clients and supporters on foot and horseback with axes, spears, and daggers, it was a new, lethal style of fighting that had never been seen before, something that even urban kings soon learned to admire.

This heroic world of chariot-driving warriors was dimly remembered in the poetry of the *Iliad* and the *Rig Veda*. It was introduced to the civilizations of Central Asia and Iran about 2100 BCE, when exotic Sintashta or Petrovka strangers first appeared on the banks of the Zeravshan, probably bouncing along on the backs of the new kinds of equids from the north. At first, this odd way of moving around probably was amusing to the local people of Sarazm and Zaman Baba. Very soon, however, both places were abandoned. Between 2000 and 1800 BCE first Petrovka and then Alakul-Andronovo groups settled in the Zeravshan valley and began mining copper and tin. Horses and chariots appeared across the Near East, and the warfare of cities became dependent, for the first time, on well-trained horses. The Old Indic religion probably emerged among northern-derived immigrants in the contact zone between the Zeravshan and Iran as a syncretic mixture of old Central Asian and new Indo-European elements. From this time forward the people of the Eurasian steppes remained directly connected with the civilizations of Central Asia, South Asia, and Iran, and, through intermediaries, with China. The arid lands that occupied the center of the Eurasian continent began to play a role in transcontinental economies and politics.

Jared Diamond, in *Guns, Germs, and Steel*, suggested that the cultures of Eurasia enjoyed an environmental advantage over those of Africa or the Americas partly because the Eurasian continent is oriented in an east-west direction, making it easier for innovations like farming, herding, and wheeled vehicles to spread rapidly between environments that were basically similar because they were on about the same latitude.[1] But persistent cultural borders like the Ural frontier delayed the transmission of those innovations by thousands of years even within the single ecological zone of the steppes. A herding economy was accepted on the middle Ural River, near the headwaters of the Samara River, by 4800 BCE. Hunters and gatherers in the neighboring steppes of northern Kazakhstan, at the same latitude, refused domesticated cattle and sheep for the next two thousand years (although they did begin to ride horses by 3700–3500 BCE). The potential geographic advantage Diamond described was frustrated for millennia, not a short time, by human distrust of foreign ways of doing things and admiration for the familiar ways. This tendency was hyperdeveloped when two very different cultures were brought into contact through long-distance migrations or at an ecological border. In the case of the Ural frontier, the Khvalynian Sea separated the populations east and west of the Ural Mountains for millennia, and the saline desert-steppe that replaced it (chapter 8) probably remained a significant ecological barrier for pedestrian foragers. Places like the Ural River frontier became borders where deep-rooted, intransigent traditions of opposition persisted.

These long-lasting, robust kinds of frontiers seem to have been rare in the prehistoric world of tribal politics. We have grown accustomed to them now only because the modern nation-state has made it the standard kind of border everywhere around the world, encouraging patriotism, jingoism, and the suspicion of other nations across sharply defined boundaries. In the tribal past, the long-term survival of sharp, bundled oppositions was unusual. The Pontic-Caspian steppes, however, witnessed an unusual number of persistent tribal frontiers because sharp environmental ecotones ran across it and it had a complex history of long-distance migrations, two important factors in the creation and maintenance of such frontiers.

ARCHAEOLOGY AND LANGUAGE

Indo-European languages replaced non–Indo-European languages in a multi-staged, uneven process that continues today, with the worldwide spread of English. No single factor explains every event in that complicated and drawn-out history—not race, demographics, population pressure, or imagined spiritual qualities. The three most important steps in the spread of Indo-European languages in the last two thousand years were the rise of the Latin-speaking Roman Empire (an event almost prevented by Hannibal); the expansion of Spanish, English, Russian, and French colonial powers in Asia, America, and Africa; and the recent triumph of the English-speaking Western capitalist trade system, in which American-business English has piggybacked onto British-colonial English. No historian would suggest that these events shared a single root cause. If we can draw any lessons about language expansion from them, it is perhaps only that an initial expansion can make later expansions easier (the *lingua franca* effect), and that language generally follows military and economic power (the *elite dominance* effect, so named by Renfrew). The earliest Indo-European expansions described in this book laid a foundation of sorts for later expansions by increasing the territorial extent of the Indo-European languages, but their continued spread never was inevitable, and each expansion had its own local causes and effects. These local events are much more important and meaningful than any imagined spiritual cause.

It is not likely that the initial spread of the Proto-Indo-European dialects into regions outside the Pontic-Caspian steppes was caused primarily by an organized invasion or a series of military conquests. As I suggested in chapter 14, the initial spread of Proto-Indo-European dialects probably was more like a franchising operation than an invasion. At least a few steppe chiefs must have moved into each new region, and their initial arrival might well have been accompanied by cattle raiding and violence. But equally important to their ultimate success were the advantages they enjoyed in institutions (patron-client systems and guest-host agreements that incorporated outsiders as individuals with rights and protections) and perhaps in the public performances associated with Indo-European rituals. Their social system was maintained by myths, rituals, and institutions that were adopted by others, along with the poetic language that conveyed their prayers to the gods and ancestors. Long after the genetic imprint of the original immigrant chiefs faded away, the system of alliances, obligations, myths, and rituals that they introduced was still being passed on from generation to generation. Ultimately the last remnant of this inheritance is the expanding echo of a once-shared language that survives as the Indo-European language family.

Understanding the people who lived before us is difficult, particularly the people who lived in the prehistoric tribal past. Archaeology throws a bright light on some aspects of their lives but leaves much in the dark. Historical linguistics can illuminate a few of those dark corners. But the combination of prehistoric archaeology with historical linguistics has a bad history. The opportunities for imaginative fantasies of many kinds, both innocent and malevolent, seem dangerously increased when these two very different kinds of evidence are mixed. There is no way to stop that from happening—as Eric Hobsbawm once remarked, historians are doomed to provide the raw material for bigotry and nationalism.[2] But he did not let that stop him from doing history.

For Indo-European archaeology, the errors of the past cannot be repeated as easily today. When the nineteenth-century fantasy of the Aryans began there were no material remains, no archaeological findings, to constrain the imagination. The Aryans of Madison Grant were concocted from spare linguistic evidence (and even that was twisted to his purpose), a large dose of racism, a cover of ideals derived from the Classical literature of Greece and Rome, and the grim zero-sum politics of social Darwinism. Archaeology really played no role. The scattered archaeological discoveries of the first half of the twentieth century could still be forced into this previously established imaginary mold. But that is not so easy today. A convincing narrative about the speakers of Proto-Indo-European must today be pegged to a vast array of archaeological facts, and it must remain un-contradicted by the facts that stand outside the chosen narrative path. I have used a lot of archaeological detail in this account, because the more places a narrative is pegged to the facts, and the more different kinds of facts from different sources are employed as pegs, the less likely it is that the narrative is false. As both the density of the archaeological facts and the quality of the linguistic evidence improve, advances in each field should act as independent checks on the worst abuses. Although I have used linguistic reconstructions for which there is little direct archaeological evidence (importantly patron-client and guest-host relationships), at least both would be compatible with the kinds of societies indicated by the archaeological evidence.

On the positive side, the combination of archaeological evidence and the reconstructed Proto-Indo-European vocabulary can reveal entirely new kinds of information about the prehistoric past. That promise keeps pushing the project forward both for linguists and archaeologists. At many critical points the interpretations presented here have been guided by institutions, rituals, and words that I found in reconstructed Indo-European and applied to archaeological settings. But I have barely scratched the surface of what might be accomplished by pulling material out of Proto-Indo-European and using it as a lens through which to examine archaeological evidence. Reciprocally, archaeological data add real-life complexities and contradictions to the idealized Indo-European social world of the linguists. We might not be able to retrieve the names or the personal accomplishments of the Yamnaya chiefs who migrated into the Danube valley around 3000 BCE, but, with the help of reconstructed

Proto-Indo-European language and mythology, we can say something about their values, religious beliefs, initiation rituals, kinship systems, and the political ideals they admired. Similarly, when we try to understand the personal, human motivation for the enormous animal sacrifices that accompanied the funerals of Sintashta chiefs around 2000 BCE, reading the *Rig Veda* gives us a new way of understanding the value attached to public generosity (RV 10.117):

> That man is no friend who does not give of his own nourishment to his friend, the companion at his side. Let the friend turn away from him; this is not his dwelling-place. Let him find another man who gives freely, even if he be a stranger. Let the stronger man give to the man whose need is greater; let him gaze upon the lengthening path. For riches roll like the wheels of a chariot, turning from one to another.[3]

Archaeologists are conscious of many historical ironies: wooden structures are preserved by burning, garbage pits survive longer than temples and palaces, and the decay of metals leads to the preservation of textiles buried with them. But there is another irony rarely appreciated: that in the invisible and fleeting sounds of our speech we preserve for a future generation of linguists many details of our present world.

Appendix

Author's Note on Radiocarbon Dates

All dates in this book are given as BCE (Before the Common Era) and CE (Common Era), the international equivalent of BC and AD.

All BCE dates in this book are based on calibrated radiocarbon dates. Radiocarbon dates measure the time that has passed since an organic substance (commonly wood or bone) died, by counting the amount of ^{14}C that remains in it. Early radiocarbon scientists thought that the concentration of ^{14}C in the atmosphere, and therefore in all living things, was a constant, and they also knew that the decay rate was a constant; these two factors established the basis for determining how long the ^{14}C in a dead organic substance had been decaying. But later investigations showed that the concentration of ^{14}C in the atmosphere varied, probably with sunspot activity. Organisms that lived at different times had different amounts of ^{14}C in their tissues, so the baseline for counting the amount of ^{14}C in the tissues moved up and down with time. This up-and-down variation in ^{14}C concentrations has been measured in tree rings of known age taken from oaks and bristlecone pines in Europe and North America. The tree-ring sequence is used to calibrate radiocarbon dates or, more precisely, to convert raw radiocarbon dates into real dates by correcting for the initial variation in ^{14}C concentrations as measured in a continuous sequence of annual tree rings. Uncalibrated radiocarbon dates are given here with the designation BP (before present); calibrated dates are given as BCE. Calibrated dates are "real" dates, measured in "real" years. The program used to convert BP to BCE dates is OxCal, which is accessible free for anyone at the website of the Oxford Radiocarbon Accelerator Unit.

Another kind of calibration seems to be necessary for radiocarbon dates taken on human bones, *if the humans ate a lot of fish*. It has long been recognized that in salt-water seas, organic substances like shell or fish bones absorb old carbon that is in solution in the water, which makes radiocarbon dates on shell and fish come out too old. This is called the "reservoir effect" because seas act as a reservoir of old carbon. Recent studies have indicated that the same problem can affect organisms that lived in fresh water, and most important among these were fish. Fish absorb old carbon in solution in fresh water, and people who eat a lot of fish will digest that old carbon

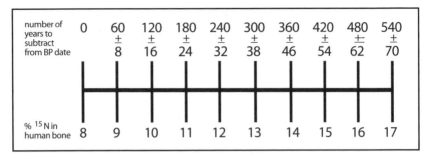

Figure A1. A proposed linear correlation between the % of ^{15}N in dared human bone (*bottom*) and the number of radiocarbon years that should be subtracked from radiocarbon dares (*top*) before they are calibrated.

and use it to build their bones. Radiocarbon dates on their bones will come out too old. Dates measured on charcoal or the bones of horses and sheep are not affected, because wood and grazing animals do not absorb carbon directly from water like fish do, and they do not eat fish. Dates on human bone can come out centuries older than dates measured on animal bone or charcoal *taken from the same grave* (this is how the problem was recognized) if the human ate a lot of fish. The size of the error depends on how much fish the human ate and how much old carbon was in solution in the groundwater where he or she went fishing. Old carbon content in groundwater seems to vary from region to region, although the amount of regional variation is not at all well understood at this time. The amount of fish in the diet can be estimated on the basis of ^{15}N levels in bone. Fish have much higher percentages of ^{15}N in their tissues than does any other animal, so humans with high ^{15}N in their bones probably ate a lot of fish. High ^{15}N in human bones is a signal that radiocarbon dates from those bones probably will yield ages that are too old.

Research to correct for reservoir effects in the steppes is just beginning as I write this, so I cannot solve the problem. But many of the radiocarbon dates from steppe archaeology are from cemeteries, and the dated material often is human bone. Widespread tests of the ^{15}N in human bone from many different steppe cemeteries, from Kazakhstan to Ukraine, indicate that fish was a very important part of most ancient steppe diets, often accounting for 50% of the meat consumed. Because I did not want to introduce dates that were probably wrong, I used an approach discussed by Bonsall, Cook, and others, and described by them as *preliminary* and *speculative*. They studied five graves in the lower Danube valley where human bone and animal bone in the same grave yielded different ages (see chapter 7 for references). Data from these graves suggested a correction method. The average level of ^{15}N in the human skeletons (15.1%) was equated with an average radiocarbon error (425±55) that should be *subtracted* prior to calibrating those dates. These averages could be placed on a scale between the known minimum and maximum levels of ^{15}N found in human bone, and, speculatively, a given level of ^{15}N could be equated with an average error in radiocarbon years. The scale shown in figure A.1 was constructed in this way. It seems to yield results that solve some long-problematic dating offsets in steppe chronology (see ch. 9, notes 4, 16, and 22; and ch. 12, note 30). When I use it—when dates are based principally on human bone—I warn readers in the text. Whatever errors it introduces probably are smaller than those caused by ignoring the problem. All the radiocarbon dates listed in the tables in this book are regular BP and calibrated BCE dates, without any correction for the reservoir effect.

Figure A.1 shows the correction scale I used to revise dates that were measured from human bone in regions where I knew the average ^{15}N levels in human bone. The top number is the number of years that should be

TABLE A.1

The average ^{13}C and ^{15}N% in human bone from seventy-two individuals excavated from graves in the Samara oblast, by time period.

Time Period	Sample Size	C13	N15	Years to Subtract
MESOLITHIC	5	−20.6	13.5	−330±42
NEOLITHIC	8	−22.3	11.8	−228±30
EARLY ENEOL	6	−20.9	14.8	−408±52
LATE ENEOL	6	−21.0	13.1	−306±39
EBA	11	−18.7	11.7	−222±30
MBA	11	−19.0	12.0	−240±32
POTAPOVKA	9	−19.1	11.3	−198±26
EARLY LBA	7	−19.1	11.4	−204±27
LATE LBA	9	−18.9	11.2	−192±26

subtracted from the BP radiocarbon date; the bottom number is the ^{15}N level associated with specific subtraction numbers.

Table A.1, based on our own studies in the Samara oblast, shows the average ^{15}N content in human bone for different periods, taken from measurements on seventy-two individuals.

NOTES

CHAPTER 1. THE PROMISE AND POLITICS OF THE MOTHER TONGUE

1. Bloch 1998:109.

2. See Sapir 1912:228.

3. Cannon 1995:28–29.

4. Poliakov 1974:188–214.

5. Veit 1989:38.

6. Grant 1916.

7. For "external origin" passages in the *Rig Veda*, see Witzel 1995. For "indigenous origin" arguments, see N. Kazanas's discussions in the *Journal of Indo-European Studies* 30, nos. 3–4 (2002); and 31, nos. 1–2 (2003).

8. For the Nazi pursuit of Aryan archaeology, see Arnold 1990.

9. For goddesses and Indo-Europeans, see Anthony 1995b; Eisler 1987, 1990; and Gimbutas 1989, 1991. For Aryan-identity politics in Russia, see Shnirelman 1998, 1999.

10. Heidegger 1959:37–51, contrasted to Boaz 1911. For the non-Aryan element in the *Rig Veda*, see Kuiper 1948, 1991.

11. Harding and Sokal 1988.

12. The *American Heritage Dictionary* has thirteen hundred unique Proto-Indo-European roots listed in its appendix. But multiple reconstructed words are derived from the same root morphemes. The number of reconstructed words with distinct meanings is much greater than the number of unique roots.

13. For doubts about proto-languages and tree diagrams, see Lincoln 1991; and Hall 1997. For a more nuanced view of tree diagrams, see Stewart 1976. For "creolization" and convergence creating Proto–Indo–European, see Renfrew 1987:78–86; Robb 1991; and Sherratt and Sherratt 1988.

14. For framing, see Lakoff 1987:328–37.

CHAPTER 2. HOW TO RECONSTRUCT A DEAD LANGUAGE

1. Here is the text of the tale:

A sheep, shorn of its wool, saw some horses, one moving a heavy cart, another carrying a big load, a third carrying a human speedily. The sheep said to the horses: "It pains me [literally, "the heart narrows itself for me"] to see human driving horses." The horses said: "Listen sheep, it pains us to see that human, the master, makes the wool of the sheep into a warm garment for himself and the sheep no longer has any wool!" On hearing that the sheep ran off into the fields.

It is impossible to construct whole sentences like this with confidence in a language known only in fragments. Proto-Indo-European tense markers in the verbs are debated, the form of the relative pronoun is uncertain, and the exact construction of a Proto-Indo-European complement (sheep saw horse carrying load) is unknown. Linguists still see it as a classic challenge. See Bynon 1977:73–74; and Mallory 1989:16–17.

2. This chapter is generally based on four basic textbooks (Bynon 1977; Beekes 1995; Hock and Joseph 1996; and Fortson 2004), and on various encyclopedia entries in Mallory and Adams 1997.

3. Embleton 1991.

4. Pinker 1994.

5. An example of a change in phonology, or pronunciation, that caused shifts in morphology, or grammar, can be seen in English. German has a complex system of noun and pronoun case endings to identify subjects, objects, and other agents, and verb endings that English lacks. English has lost these features because a particular dialect of Middle English, Old Northumbrian, lost them, and people who spoke the Old Northumbrian dialect, probably rich wool merchants, had a powerful effect on the speech of Medieval London, which happened to give us Modern English. The speakers of Old Northumbrian dropped the Germanic word-final *n* and *m* in most suffixes (*esse'*, not *essen*, for "to eat"). In late Old English the pronunciation of many short vowels (like the final *-e* that resulted here) was already merging into one vowel (the [uh] in so*fa*, called *schwa* by linguists). These two shifts in pronunciation meant that many nouns no longer had distinctive endings, and neither the infinitive nor the subjunctive plural verb had a distinct ending. Later, between 1250 and 1300, the word-final *schwa* began to be dropped from most English speech, which wiped out the distinction between two more grammatical categories. Word order became fixed, as few other guides indicated the difference between subject and object, and auxiliary particles like *to*, *of*, or *by* were employed to distinguish infinitives and other forms. Three shifts in pronunciation were responsible for much of the grammatical simplification of modern English. See Thomason and Kaufman 1988:265–275.

6. For Grimm's Law, see Fortson 2004:300–304.

7. Some linguists argue that the Proto-Indo-European root did not begin with *k* but rather with a palato-velar, a *kh*-type sound, which would require that the first consonant was moved back in the *centum* languages rather than forward in the *satem* languages. See Melchert 1994:251–252. Thanks to Bill Darden for pointing this out.

8. Hock and Joseph 1996:38.

9. For pessimistic views on the "reality" of reconstructed Proto-Indo-European, see Bynon 1977; and Zimmer 1990. For optimistic views, see Hock and Joseph 1996:532–534; and Fortson 2004:12–14.

10. Hall 1950, 1976.

11. Bynon 1977:72. Mycenaean was in a transitional state in 1350 BC, when it was recorded. Some Proto-Indo-European words with k^w had already shifted to k in Mycenaean. The alternation between $*k^w$ and $*p$ probably was already present in some dialects of Proto-Indo-European.

12. For doubts on reconstructed meanings, see Renfrew 1987:80, 82, 260. For the argument that comparing cognates requires that the meanings of the compared terms are subjected to fairly strict limits, see Nichols 1997b.

CHAPTER 3. LANGUAGE AND TIME 1

1. See Swadesh 1952, 1955; and Lees 1953.

2. The replacement rate cited here compares the core vocabulary in Modern English to the core vocabulary in Old English, or Anglo-Saxon. Much of the Old English core vocabulary was replaced by Norse but, since Norse was another Germanic language, most of the core vocabulary remains Germanic. That is why we can say that 96% of the core vocabulary remains Germanic, and at the same time say that the replacement rate in the core vocabulary was a high 26%.

3. Much of the information in this section came from Embleton 1991, 1986. See also McMahon and McMahon 2003; and Dyen, Kruskal, and Black 1992. Many linguists are hostile to any claim that a cross-cultural core vocabulary can be identified. The Australian aboriginal languages, for example, do not seem to have a core vocabulary—all vocabulary items are equally vulnerable to replacement. We do not understand why. Both sides of the debate are represented in Renfrew, McMahon, and Trask 2000.

4. Meid 1975; Winfred 1989; and Gamkrelidze and Ivanov 1984:267–319.

5. Ivanov derived Hittite (Northern Anatolian) and Luwian (Southern Anatolian) separately and directly from Proto-Indo-European, without an intervening proto-language, making them as different as Celtic and Greek. Most other linguists derive all the Anatolian languages from a common source, Proto-Anatolian; see Melchert 2001 and Diakonoff 1985. Lydian, spoken on the western coast of Anatolia in the Classical era, might have descended from the same dialect group as Hittite. Lycian, spoken on the southwestern coast, probably descended from the same dialects as Luwian. Both became extinct in the Classical era. For all these topics, see Drews 2001.

6. For the Anatolian languages, see Fortson 2004:154–179; Houwink Ten Cate 1995; Veenhof 1995; and Puhvel 1991, 1994. For the glottalic perspective, see Gamkrelidze and Ivanov 1995.

7. Wiluša was a city west of the Hittite realm. It is very possible that Wilusa was Troy and that the Trojans spoke Luwian. See Watkins 1995:145–150; and Latacz 2004.

8. The non–Indo-European substrate effect on Luwian was described by Jaan Puhvel (1994:261–262) as "agglutinative creolization . . . What has happened to Anatolian here is reminiscent of what became of French in places like Haiti." Hittite showed similar non–Indo-European substrate effects and had few speakers, causing Zimmer (1990:325) to note that, "on the whole, the Indo-Europeanization of Anatolia failed."

9. Melchert 2001.

10. Forster 2004; Baldi 1983:156–159.

11. Lehrman 2001. The ten innovations that Lehrman identified as distinctive of Proto-Indo-European included two phonological traits (e.g., loss of the laryngeals), three morphological traits in nouns (e.g., addition of the feminine gender), and five morphological traits in verbs.

12. See Sturtevant 1962 for the Indo-Hittite hypothesis. For Anatolian as a daughter of very early Proto-Indo-European, see Puhvel 1991. Lehrman (2001) pointed out that Anatolian had a different word from Proto-Indo-European for *man*, usually considered part of the core vocabulary. The Anatolian term (**pāsna-*) used a root that also meant "penis," and the Proto-Indo-European term (**wīro-*) used a root that also meant "strength." Proto-Anatolian and Proto-Indo-European did, however, share cognate terms for *grandfather* and *daughter*, so their kinship vocabularies overlapped. Classic Proto-Indo-European and Anatolian probably emerged from different places and different times in the Pre-Proto-Indo-European dialect chain.

13. For Pre-Greek language(s) of Greece, see Hainsworth 1972; and Francis 1992.

14. For the oldest language in the Indic branch I use the term *Old Indic* instead of *Indo-Aryan*. The standard nomenclature today is *Indo-Iranian* for the parent, *Avestan Iranian* for the oldest Iranian daughter, and *Indo-Aryan* for the oldest Indic daughter. But the designation *Aryan* for Indic is unnecessary; they were all Aryan. For the language and history of the *Rig-Veda*, see Erdosy 1995.

15. For Old Indic terms among the Mitanni, see Thieme 1960; Burrow 1973; and Wilhelm 1995. I thank Michael Witzel for his comments on Mitanni names. Any errors are my own.

16. For a date for Zarathustra before 1000 BCE, see Boyce 1975; and Skjærvø 1995. For the "traditional" date promulgated by ancient Greek sources, five hundred years later, see Malandra 1983.

17. Clackson (1994) and Hamp (1998) argued that Pre-Armenian was linked to the Greek-Indo-Iranian block. See also the isogloss map in Antilla 1972, figure 15.2. Many of the shared lexical items are discussed and described in Mallory and Adams 1997. I am grateful to Richard Diebold for his analysis of Greek/Indo-Iranian relations in a long letter of October 1994, where he pointed out that the shared innovations link Greek and Iranian closely, and Greek and Indic somewhat less.

18. See Rijksbaron 1988 and Drinka 1995 for the shared poetic functions of the imperfect. Poetics, shared phrases, and weapon terms are reviewed in Watkins 1995, chap. 2, 435–436.

19. See Ringe et al. 1998; and also Ringe, Warnow, and Taylor 2002. Similar cladistic methods were applied to a purely lexical data set in Rexová, Frynta, and Zrzavý 2003.

CHAPTER 4. LANGUAGE AND TIME 2

1. See Darden 2001, esp. 201–204, for the etymology of the term *wool*. For the actual textiles, see Barber 2001, 1991; and Good 1998.

2. The "unspinnable" quotation is from Barber 2001:2. The mitochondrial DNA in modern domesticated sheep indicates that all are descended from two ancient episodes of domestication. One cluster (B), including all European and Near Eastern sheep, is descended from the wild *Ovis orientalis* of eastern Anatolia or western Iran. The other cluster (A) is descended from another *Ovis orientalis* population, probably in north-central Iran. Other wild Old World ovicaprids, *Ovis ammon* and *Ovis vignei,* did not contribute to the genes of domesticated sheep. See Hiendleder et al. 2002. For a general discussion of sheep domestication, see Davis 1987; and Harris 1996.

3. In the Ianna temple of Uruk IV (3400–3100 BCE) artists depicted women making textiles. The later Sumerian names for some months incorporated the term for plucking sheep. The zoological evidence suggests that the months were named this way during the Late Uruk period or afterward, not before.

4. Zoological evidence for wool production in the Near East is reviewed by Pollack (1999:140–147). For Arslantepe, see Bökönyi 1983. An earlier date for wool sheep could be indicated by a couple of isolated pieces of evidence. The phase A occupation at Hacinebi on the Euphrates, dated 4100–3800 BCE, had spindle-whorls that seemed the right weight for spinning wool, which requires a light spindle; see Keith 1998. A clay sheep figurine from Tepe Sarab in western Iran (Kermanshah) seems to show a wooly fleece, from a level dated about 5000 BCE. For a broader discussion, see Good 2001.

5. For the caprids (sheep and/or goats) at Khvalynsk, see Petrenko 1984. Petrenko did not report the age at death for all the caprids in the Khvalynsk graves, but six of the twelve with reported ages were adults. Sacrificial deposit #11 contained 139 bones of caprids representing four adults and five sub-adults, and the *average* adult withers height was 78 cm, almost 15 cm taller than other European Neolithic caprids. For Svobodnoe sheep, see Nekhaev 1992:81. For sheep in Hungary, see Bökönyi 1979:101–116. For sheep in Poland, see Milisauskas 2002:202.

6. For wool at Novosvobodnaya, see Shishlina, Orfinskaya, and Golikov 2003. For evidence of Catacomb-period wool (dated ca. 2800–2200 BCE) in the North Caucasian steppes, see Shishlina 1999. Sherratt's updated comments on wool are included in the revised text of an older article in Sherratt 1997a.

7. The term for hub or nave, which is often included in other lists, also meant "navel" in Proto-Indo-European, so its exact meaning is unclear. For the wheel-wagon vocabulary, see Specht 1944. Three influential updates were Gamkrelidze and Ivanov 1984:718–738; Meid 1994; and Häusler 1994. I first published on the topic in Anthony and Wailes 1988; and also in Anthony 1991a, 1995a. As with most of the topics covered in this book, there is an excellent review of the Indo-European wheel vocabulary in Mallory and Adams 1997.

8. Don Ringe communicated the argument against *hurki-* to me in a letter in 1997. Bill Darden discussed the Anatolian terms in Darden 2001.

9. I am indebted to Mary Littauer for alerting me to draft experiments carried out in 1838–40 with wagons and carts on different road surfaces, where it was determined that the draft of a wagon was 1.6 times greater than that of a cart of the same weight. See Ryder 1987.

10. For the earliest wheeled vehicles, see Bakker et al. 1999; and Piggott 1983. For European wheels, see Häusler 1992; and Hayen 1989. For Mesopotamia, see Littauer and Crouwel 1979; and Oates 2001. The most comprehensive anlysis of the steppe vehicle burials, still unpublished, is by Izbitser 1993, a thesis for the Institute of the History of Material Culture in

St. Petersburg. Izbitser is working on an English-language update from her post in the New York Metropolitan Museum. Other key steppe accounts are in Mel'nik and Serdiukova 1988, and the section on wagons in Gei 2000:175–192.

11. Sherratt's essays were compiled and amended in Sherratt 1997. He continued to suggest that horseback riding in the steppes was inspired by Near Eastern donkey riding; see 1997:217. An early critical response to the SPR is Chapman 1983.

12. For Neolithic sleds in Russia, see Burov 1997. Most of them were joined with mortice-and-tenon joints, and equipped with bent-wood curved runners. These are the same carpentry skills needed to make wheels and wooden-slat tires.

13. The version of the Renfrew hypothesis I use here was published as Renfrew 2001. For assenting views among archaeologists, see Zvelebil and Zvelebil 1988; Zvelebil 1995; and Robb 1991, 1993. Robert Drews (2001) began in a different place but ended up supporting Renfrew.

14. For the north Syrian origin of the Anatolian Neolithic population, see Bar-Yosef 2002; for the likely Afro-Asiatic linguistic affiliation of these first farmers, see Militarev 2002.

15. See Gray and Atkinson 2003, reviewed by Balter 2003. The linguist L. Trask criticized Gray and Atkinson's methods, and Gray responded on his homepage, updated March 2004, at http:// www.psych.auckland.ac.nz/psych/research/Evolution/GrayRes.htm.

16. Buck 1949:664, with Indo-European terms for *turn, turn around, wind,* and *roll.* Gray's argument for a natural independent development of the term *wheel* from *to turn* (wheel = the turner) is further complicated by the fact that there are two reconstructed Proto-Indo-European terms for *wheel,* and the other one was based on the Proto-Indo-European verb **reth-* 'run' (wheel = the runner), a different semantic development.

17. Renfrew 2001:40–45; 2000. Renfrew's hypothesis of a very long-lived Proto-Indo-European phase, surviving for many millennia, is supported by some linguists. For a view that Proto-Indo-European was spoken from the Mesolithic through the end of the Corded Ware period, or about 6000–2200 BCE, see Kitson 1997, esp. 198–202.

18. Childe 1957:394.

19. Mallory 1989:145–146; and Anthony 1991a. For Africa, see Nettles 1996.

Chapter 5. Language and Place

1. For homeland theories, see Mallory 1989, chap. 6. For political uses of the past in the Soviet Union, see Shnirelman 1995, 1999; Chernykh 1995; and Kohl and Tsetskhladze 1995. For the belief in an Aryan-European "race," see Kühl 1994; and Poliakov 1974.

2. The Pontic-Caspian steppe homeland hypothesis was defended in English most clearly by Gimbutas 1970, 1977, 1991; and Mallory 1989, updated in Mallory and Mair 2000. Although I agree with Gimbutas's homeland solution, I disagree with her chronology, her suggested causes for the expansion, and her concept of Kurgan-culture migrations, as I explained in detail in Anthony 1986.

3. See Dixon 1997:43–45. Similarly for Zimmer 1990:312–313, "reconstructions are pure abstracts incapable of being located or dated . . . no philological interpretation of the reconstructed items is possible."

4. The tree model does not exclude or deny some areal convergence. All languages contain elements based on both branching structures and convergence with neighbors. On areal borrowing, see Nichols 1992.

5. See Thomason and Kaufman 1992; Nichols 1992; and Dixon 1997. All support the derivation of the Indo-European languages from Proto-Indo-European. Dixon (1997:31), although a critic of the criteria used to create some family tree models, stated: "The genetic relatedness of the Indo-European languages, in a family tree model, has of course been eminently proved." A good brief review of various approaches to convergence can be found in Hock and Joseph 1996:388–445.

6. Gradual convergence between neighboring languages can result in several different kinds of similarities, depending on the social circumstances. The range of possibilities includes *trade jargons*, crude combinations of words from neighboring languages barely sufficient to communicate for purposes of trade or barter; *pidgins,* which evolve from trade jargons or from a multitude of partially known languages in a colonial encounter where a colonial target language supplies much of the content of the pidgin; and *creoles,* which can evolve from pidgins or can arise abruptly in multiethnic forced labor communities where again a colonial target language supplies much of the content. Unlike pidgins, creoles contain the essential grammatical structures of a natural language, but in a reduced and simple form. They can, of course, be as expressive in song, poetry, and metaphor as any natural language, so the fact that they are grammatically simple is not a value statement. All these ways of speaking pass through a bottleneck of great grammatical simplification. Indo-European grammar is not at all like a creole grammar. See Bickerton 1988; and Thomason and Kaufman 1988.

7. Pulgram, in 1959, suggested that the comparative method, applied to the modern Romance words for *coffee*, would produce a false Latin root for *coffee* in Classical Latin. But Pulgram's claim was rebutted by Hall (1960, 1976). Pulgram's argument was cited in Renfrew (1987:84–86) but corrected in Diakonov (1988: n. 2).

8. For Pre–Indo-European substrate terms in Balto-Slavic, see Andersen 2003. For Greek and pre-Greek place-names, see Hester 1957; Hainsworth 1972; and Renfrew 1998. In northern Europe, at least three different extinct non–Indo-European languages have been identified: (1) the "language of Old European hydronymy," preserved principally in non–Indo-European river names; (2) the "language of bird names," preserved in the names of several kinds of birds, including the blackbird, lark, and heron, and also in other terms borrowed into early Germanic, Celtic, and Latin, including the terms for *ore* and *lightning*; and (3) the "language of geminates," which survives only in a few odd sounds quite atypical for Indo-European, borrowed principally into Germanic but also into a few Celtic words, including doubled final consonants and the word-initial [kn-], as in *knob*. See Schrijver 2001; Venneman 1994; Huld 1990; Polomé 1990; and Krahe 1954.

≠ 9. For *beech* and *salmon* as terms that limited Proto-Indo-European to northern Europe, see Thieme 1958. Friedrich 1970 showed that the *beech* root referred variously to beech, oak, and elder trees in several branches, and that in any case the common beech grew in the Caucasus Mountains, making it useless as a diagnostic northern European tree word. Diebold 1985 summarized the evidence against salmon as a limiting geographic term. For the honeybee argument, see the excellent study by Carpelan and Parpola 2001. See also the articles on salmon and beech in Mallory and Adams 1997.

10. This interpretation of Proto-Indo-European **peku* is that of Benveniste 1973:40–51.

11. This reconstruction of Proto-Indo-European society is based on Benveniste 1973, numerous entries in Mallory and Adams 1997, and Gamkrelidze and Ivanov 1995.

12. For Proto-Uralic linkages with Proto-Indo-European, see Carpelan, Parpola, and Koskikallio 2001, particularly the articles by Koivulehto and Kallio. See also Janhunen 2000; Sinor 1988; and Ringe 1997.

13. For a Yeniseian homeland, see Napol'skikh 1997.

14. Koivulehto 2001.

15. Janhunen (2000) has somewhat different forms for some of the pronouns. Nichols pointed out in a note to me that the *-m* and *-n* shared inflections are not very telling; only a whole paradigm of shared inflections is diagnostic. Also, nasal consonants occur in high frequencies and apparently are prone to occur in grammatical endings, and so it is the pronouns that are really important here.

16. Nichols 1997a.

17. For the glotallic theory, see Gamkrelidze and Ivanov 1973; see also Hopper 1973. For their current views, see Gamkrelidze and Ivanov 1995.

18. For discussions of the glottalic theory, see Diakonov 1985; Salmons 1993; and Szemerényi 1989.

19. For critical discussions of the Semitic-Proto-Indo-European and Kartvelian-Semitic-Proto-Indo-European loan words, see Diakonov 1985:122–140; and Nichols 1997a appendix. On the chronology of the Proto-Kartvelian dispersal or breakup, see Harris 1991.

Chapter 6. The Archaeology of Language

1. My definitions are adapted from Prescott 1987. A different set of definitions was suggested by Parker 2006. He suggested *boundary* as the general term (what I am calling borders) and *border* as a specific term for a political or military boundary (more or less what I am calling a boundary). Parker tried to base his definitions partly on vernacular understandings of how these words are normally used, a noble goal; but I disagree that there is any consistency of usage in the vernacular, and prefer to use established definitions. In their review of the borderland literature, Donnan and Wilson (1999:45–46) followed Prescott in using *border* as the general or unspecialized term. The classic work to which I owe a great deal of my thinking is Barth 1969. For archaeological treatments of ethnic borders, see Shennan 1989, and Stark 1998.

2. For the growth of Medieval European regional identities, see Russell 1972; and Bartlett 1993. For the anthropological deconstruction of tribes and bounded cultures, see Fried 1975; and Wolf 1982, 1984. See also Hill 1992; and Moore 2001. For good archaeological uses of this border-deconstructing approach to ethnicity see Wells 2001; Florin 2001; MacEachern 2000; and James 1999.

3. See Hobsbawm 1990; Giddens 1985; and Gellner 1973. Giddens (1985:120) famously referred to the nation-state as a "bordered power-container." For a different interpretation of ancient tribes and borders, see Smith 1998. He is accused of being a "primordialist"; see his defense in chapter 7. Also see Armstrong 1982.

4. For projectile points and language families in South Africa, see Weissner 1983. For a good review of material culture and ethnicity, see Jones 1997, esp. chap. 6.

5. For New Guinea, see Terrell 2001; see also Terrell, Hunt, and Godsen 1997. For the original argument that biology, culture, and language were separate and independent, see the introduction to Boaz 1911. For California, see Jordan and Shennan 2003. For the other examples, see Silver and Miller 1997:79–98.

6. Persistent frontiers were the subject of a flurry of studies in the 1970s; see Spicer 1971 and a volume dedicated to Spicer by Castile and Kushner 1981. The focus in these papers was the maintenance of stigmatized minority identities. In archaeology, the long-term persistence of prehistoric "culture areas" was discussed long ago in Ehrich 1961. The subject was revisited by Kuna 1991; and Neustupny 1991. My first paper on the subject was Anthony 2001.

7. For the persistence of the Hudson-Valley Iroquoian/Algonkian frontier, see Chilton 1998. For the Linear Pottery frontier, see Zvelebil 2002. For the Jastorf/Halstatt frontier, see Wells 1999.

8. Emberling (1997) used the term *redundant* rather than *robust* for material-culture borders that were marked in multiple categories of material culture, and he recognized that this redundancy suggested that these borders were particularly important socially.

9. For Wales, see Mytum 1994; and John 1972. For the genetic border at the Welsh/English frontier, see Weale et al. 2002. For the border near Basle, see Gallusser 1991. On Breton culture, see Jackson 1994; and Segalen 1991. For the German/Romansh frontier in Italy, see Cole and Wolf 1974.

10. For the Ucayali quotation, see DeBoer 1990:102. For language and genetic correlations, see Jones 2003.

11. For the Iroquois, see Wolf 1982:167; 1984:394; and, in contrast, see Tuck 1978; Snow 1994; and Richter 1992. Moore (2001:43) also used intermarriages between Amerindian tribes as an index of *general* cultural and linguistic mixing: "These [marriage] data show a continual movement of people, and hence their genes, *language, and culture,* from society to society" (emphasis mine).

12. For the borders of functional zones, see Labov 1994. For functional zones, see Chambers and Trudgill 1998; and Britain 2002.

13. See Cole and Wolf 1974:81–282; see also Barth 1969. Cole and Wolf wrote a perceptive analysis of a persistent frontier in Italy, and then in 1982 Wolf published his best-known book, which suggested that tribal borders outside Europe were much more porous and changeable. In making this argument he seems, in my view, to have made some statements contradicted by his own earlier field work.

14. For the billiard-ball analogy, see Wolf 1982:6, 14. On migration processes generally, see Anthony 1990, 1997. Archaeologists of the American Southwest have pushed migration theory further than those of any other region. For a sampling see Spielmann 1998. For migration theory in Iroquoian archaeology, see Sutton 1996.

15. For the four Colonial cultural provinces, see Fischer 1989; Glassie 1965; and Zelinsky 1973. Although anthropology veered away from cultural geography in the 1980s and 1990s, historians and folklorists continued to study it. See Upton and Vlach 1986; and Noble 1992. For a review of the historians' interest in cultural geography in North America, see Nash 1984.

16. Clark 1994.

17. Kopytoff 1987.

18. For the Nuer, see Kelley 1985. For the effect of changes in bride-price currencies on basic subsistence economies, see Cronk 1989.

19. On dialect leveling among colonists, see Siegel 1985; Trudgill 1986; and Britain 2004. The degree of leveling depends on a number of social, economic, and linguistic factors; see Mufwene 2001. For Spanish leveling in the Americas, see Penny 2000. On the history of American English dialects, see Fischer 1989.

20. For charter groups, see Porter 1965; and Breen 1984. On German immigrants in Ohio, see Wilhelm 1992. On Puritan charter groups in new England, see Fischer 1989:57–68. On the Maya, see Fox 1987, although now there are criticisms of Fox's migration-based history; on apex families, see Alvarez 1987; and on the Pueblo, see Schlegel 1992.

21. On leveling and simplification in material culture among colonists, see Noble 1992; and Upton and Vlach 1986. Burmeister (2000) noted that the external form of residential architecture tends to conform to broad norms, whereas ethnicity is expressed in internal details of decoration and ornament.

22. The Boasian approach to borders is reviewed in Bashkow 2004.

23. On the provinces of France, see Chambers and Trudgill 1998:109–123; on the Maasai, see Spear and Waller 1993; on Burma, see Leach 1968, 1960; and for a different interpretation of Burma, see Lehman 1989.

24. On language and ecology, see Hill 1996; and Nettles 1996. Hill's paper was published later in Terrell 2001:257–282. Also see Milroy 1992.

25. The concept of ecologically determined "spread zones" for languages came from Nichols 1992. Similar ideas about arid zones and language expansion can be found in Silver and Miller 1997:79–83. Renfrew (2002) applied the term *spread zone* to any region of rapid language spread, particularly any expansion of pioneer farmers, regardless of ecology. Campbell (2002), however, warned against mixing these definitions.

26. For China, see DiCosmo 2002; and Lattimore 1940.

27. For Acholi origins, see Atkinson 1989, 1994.

28. A similar model for the growth of Bronze Age chiefdoms, described long before Atkinson's case study was published, was by Gilman 1981.

29. For the Pathan-Baluch shift, see Mallory 1992; Barth 1972; and Noelle 1997.

CHAPTER 7. HOW TO RECONSTRUCT A DEAD CULTURE

1. For the history of Christian J. Thomsen's Three-Age System, see Bibby 1956.

2. I generally follow the Eneolithic and Bronze Age chronology of Victor Trifonov at the Institute of the History of Material Culture in St. Petersburg; see Trifonov 2001.

3. For the impact of radiocarbon dating on our understanding of European prehistory, see Renfrew 1973.

4. The old carbon problem in freshwater fish is explained in Cook et al. 2002; and in Bonsall et al. 2004. I used their method to create the correction scale that appears in the appendix.

5. A good historical review of radiocarbon dating in Russian archaeology is in Zaitseva, Timofeev, and Sementsov 1999.

6. For a good example of cultural identity shifting in response to changing historical situations, see Haley and Wilcoxon 2005. For Eric Wolf's and Anthony Smith's comments on situational politics alone being insufficient to explain emotional ties to a cultural identity see Cole and Wolf 1974:281–282; and Smith 1998, chap. 7.

7. For technological style and cultural borders, see Stark 1998.

CHAPTER 8. FIRST FARMERS AND HERDERS

1. The three sky gods named here almost certainly can be ascribed to Proto-Indo-European. *Dyeus Pater*, or Sky/Heaven Father, is the most certain. The Thunder/War god was named differently in different dialects but in each branch was associated with the thunderbolt, the hammer or club, and war. The Divine Twins likewise were named differently in the different branches—the Nāsatyas in Indic, Kastōr and Polydeukēs in Greek, and the Dieva Dēli in Baltic. They were associated with good luck, and often were represented as twin horses, the offspring of a divine mare. For Trita, see Watkins 1995; and Lincoln 1981:103–124. More recently, see Lincoln 1991, chap. 1. For the twins, see Puhvel 1975; and Mallory and Adams 1997:161–165.

2. For the tripartition of Indo-European society, see Dumezil 1958; and Littleton 1982. There is a good review in Mallory 1989:128–142. For an impressive example of the interweaving of three's and two's in Indo-European poetry, see Calvert Watkin's analysis of a traditional Latin poem preserved by Cato in 160 BCE, the "Lustration of the Fields." The structure is tripartite, expressed in a series of doubles. See Watkins 1995:202–204.

3. Przewalkski horses are named after the Polish colonel who first formally described them in 1881. A Russian noble, Frederic von Falz-fein, and a German animal collector, Carl Hagenbeck, captured dozens of them in Mongolia, in 1899 and 1901. All modern Przewalski's are descended from about 15 of these animals. Their wild cousins were hunted to extinction after World War II; the last ones were sighted in Mongolia in 1969. Zoo-bred populations were reintroduced to two preserves in Mongolia in 1992, where once again they are thriving.

4. For differences between east-Ural and west-Ural Upper Paleolithic cultures, see Boriskovskii 1993, and Lisitsyn 1996.

5. For a wide-ranging study of the Ice Age Caspian, the Khvalynian Sea, and the Black Sea, including the "Noah's Flood" hypothesis, see Yanko-Hombach et al. 2006.

6. For the decline of matriliny among cattle herders, see Holden and Mace 2003.

7. For Y-chromosome data on early European cattle, see Gotherstrom et al. 2005. For MtDNA, see Troy et al. 2001; and Bradley et al. 1996.

8. For agricultural frontier demography, see Lefferts 1977; and Simkins and Wernstedt 1971.

9. For the oldest Criş site in the lower Danube valley, see Nica 1977. For a Starcevo settlement in the plains north of Belgrade, see Greenfield 1994.

10. For Criş immigrants in the East Carpathians, see Dergachev, Sherratt, and Larina 1991; Kuzminova, Dergachev, and Larina 1998; Telegin 1996; and Ursulescu 1984. The count of thirty sites refers to excavated sites. Criş pottery is known in unexcavated surface exposures at many more sites listed in Ursulescu 1984. For the Criş economy in eastern Hungary, see Vörös 1980.

11. For Neolithic bread, see Währen 1989. Criş people cultivated gardens containing four varieties of domesticated wheat: *Triticum monococcum*, *T. dicoccum* Shrank, *T. spelta*, *T. aestivo-compactum* Schieman; as well as barley (*Hordeum*), millet (*Panicum miliaceum*), and peas (*Pisum*)—all foreign to eastern Europe. On the plant evidence, see Yanushevich 1989; and Pashkevich 1992.

12. Markevich 1974:14.

13. For the possible role of acculturated foragers in the origin of the East Carpathian Criş culture, see Dergachev, Sherratt, and Larina 1991; and, more emphatically, Zvelebil and Lillie 2000.

14. On pioneer farmers and language dispersal, see Bellwood and Renfrew 2002; Bellwood 2001; Renfrew 1996; and Nichols 1994. On the symbolic opposition of wild and domesticated animals, see Hodder 1990.

15. Most archaeologists have accepted the argument made by Perles (2001) that the Greek Neolithic began with a migration of farmers from Anatolia. For the initial spread from Greece into the Balkans, see Fiedel and Anthony 2003. Also see Zvelebil and Lillie 2000; and van Andel and Runnels 1995. The practical logistics of a Neolithic open-boat crossing of the Aegean are discussed in Broodbank and Strasser 1991.

16. For **tawro-s*, see Nichols 1997a: appendixes. For the association of Afro-Asiatic with the initial Neolithic, see Militarev 2003.

17. The classic Russian-language works on the Bug-Dniester culture are in Markevich 1974; and Danilenko 1971; the classic discussion in English is in Tringham 1971. More recently, see Telegin 1977, 1982, and 1996; and Wechler, Dergachev, and Larina 1998.

18. For the Mesolithic groups around the Black Sea, see Telegin 1982; and Kol'tsov 1989. On the Dobrujan Mesolithic, see Paunescu 1987. For zoological analyses, see Benecke 1997.

19. Most of the dates for the earliest Elshanka sites are on shell, which might need correction for old carbon. Corrected, Elshanka dates might come down as low as 6500–6200 BCE. See Mamonov 1995, and other articles in the same edited volume. For radiocarbon dates, see Timofeev and Zaitseva 1997. For the technology and manufacture of this silt/mud/clay pottery, see Bobrinskii and Vasilieva 1998.

20. For the dates from Rakushechni Yar, see Zaitseva, Timofeev, and Sementsov 1999. For the excavations at Rakushechni Yar, see Belanovskaya 1995. Rakushechni Yar was a deeply stratified dune site. Telegin (1981) described sedimentary stratum 14 as the oldest cultural occupation. A series of new radiocarbon dates, which I ignore here, have been taken from organic residues that adhered to pottery vessels said to derive from levels 9 to 20. Levels 15 to 20 would have been beneath the oldest cultural level, so I am unsure about the context of the pottery. These dates were in the calibrated range of 7200–5800 BCE (7930±130 to 6825±100 BP). If they are correct, then this pottery is fifteen hundred years older than the other pottery like it, *and* domesticated sheep appeared in the lower Don valley by 7000 BCE. All domesticated sheep are genetically proven to have come from a maternal gene pool in the mountains of eastern Turkey, northern Syria, and Iraq about 8000–7500 BCE, and no domesticated sheep appeared in the Caucasus, northwestern Anatolia, or anywhere else in Europe in any site dated as early as 7000 BCE. The earliest dates *on charcoal* from Rakushechni Yar (6070+100 BP, 5890+105 BP for level 8) come out about 5200–4800 BCE, in agreement with other dates for the earliest domesticated animals in the steppes. If the dated organic residue was full of boiled fish, it could need a correction of five hundred radiocarbon years, which would bring the earliest dates down to about 6400–6200 BCE—somewhat more reasonable. I think the dates are probably contaminated and the sheep are mixed down from upper levels.

21. For 155 Late Mesolithic and Neolithic radiocarbon dates from Ukraine, see Telegin et al. 2002, 2003.

22. On Bug-Dniester plant foods, see Yanushevich1989; and Kuzminova, Dergachev, and Larina 1998. A report of millet and barley impressions from the middle-phase site of Soroki I/level 1a is contained in Markevich 1965. Yanushevich did not include this site in her 1989 list of Bug-Dniester sites with domesticated seed imprints; it is the only Bug-Dniester site I have seen with reports of barley and millet impressions.

23. The dates here are not on human bones, so they need no correction. The bone percentages are extracted from Table 7 in Markevich 1974; and Benecke 1997. Benecke dismissed the Soviet-era claims that pigs or cattle or both were domesticated independently in the North Pontic region. Telegin (1996:44) agreed. Mullino in the southern Urals produced domesticated sheep bones supposedly dated to 7000 BCE, cited by Matiushin (1986) as evidence for migrations from Central Asia; but like the claimed sheep in deep levels at Rakushechni Yar, these sheep would have been *earlier* than their proposed parent herds at Djeitun, and the wild species was not native to Russia. The sheep bones probably came from later Eneolithic levels. Matiushin's report was criticized for stratigraphic inconsistencies. See Matiushin 1986; and, for his critics, Vasiliev, Vybornov, and Morgunova 1985; and Shorin 1993.

24. Zvelebil and Rowley-Conwy 1984.

25. For captured women and their hyper-correct stylistic behavior, see DeBoer 1986. The archaeological literature on technological style is vast, but a good introduction is in Stark 1998.

26. The Linear Pottery culture in the East Carpathian piedmont overlapped with the Criş culture around 5500–5400 BCE. This is shown at late Criş sites like Grumazeşti and Sakarovka that contained a few Linear Pottery sherds. Sakarovka also had Bug-Dniester sherds, so it shows the brief contemporaneity of all three groups.

27. There is, of course, generosity and sharing among farmers, but farmers also understand that certain potential foods are not food at all but investments. Generosity with food has practical limits in bad times among farmers; these are generally absent among foragers. See Peterson 1993; and Rosenberg 1994.

28. The classic text on the Dnieper-Donets culture is Telegin 1968. For an English-language monograph see Telegin and Potekhina. In this chapter I only discuss the first phase, Dnieper-Donets I.

29. For DDI chipped axes, see Neprina 1970; and Telegin 1968:51–54.

30. Vasilievka V was published as a Dnieper-Donets II cemetery, but its radiocarbon dates suggest that it should have dated to DD I. Vasilievka I and III were published as Late Mesolithic, broadly around 7000–6000 BCE, but have radiocarbon dates of the very Early Mesolithic, closer to 8000 BCE. Vasilievka II and Marievka were published as Neolithic but have no ceramics and Late Mesolithic radiocarbon dates, 6500–6000 BCE, and so are probably Late Mesolithic. Changes in human skeletal morphology that were thought to have occurred between the Late Mesolithic and Neolithic (Jacobs 1993) now appear to have occurred between the Early and Late Mesolithic. These revisions in chronology have not generally been acknowledged. For radiocarbon dates, see Telegin et al. 2002, 2003. See also Jacobs 1993, and my reply in Anthony 1994.

31. For Varfolomievka, see Yudin 1998, 1988.

32. The zoologist Bibikova identified domesticated animals—sheep, cattle, and horses—at Matveev Kurgan in levels dated 6400–6000 BCE. Today neither the German zoologist Benecke nor the Ukrainian archaeologist Telegin give credit to Bibikova's claims for an independent local domestication of animals in Ukraine. Matveev Kurgan (a settlement, not a kurgan) is located in the Mius River valley north of the Sea of Azov, near Mariupol. Two sites were excavated between 1968 and 1973, numbered 1 and 2. Both contained Grebenikov-type microlithic flint tools and were thought to be contemporary. Two radiocarbon dates from MK 1 average about 6400–6000 BCE, but the single date (on bone) from MK 2 was about 4400–4000 BCE.

In the latter period domesticated animals including sheep were common in the region. The artifacts from all depths were analyzed and reported as a single cultural deposit. But at MK 1 the maximum number of flint tools and animal bones was found at a depth of 40–70 cm (Krizhevskaya 1991:8), and the dwelling floor and hearths were at 80–110 cm (Krizhevskaya 1991:16). Most of the animal bones from MK 1 and 2 were from wild animals, principally horses, onagers, and wild pigs, and these probably were associated with the older dates. But the bones identified as domesticated horses, cattle, and sheep probably came from later levels associated with the later date. See Krizhevskaya 1991. Stratigraphic inconsistencies mar the reporting of all three Pontic-Ural sites with claimed very early domesticated animals—Rakushechni Yar, Mullino, and Matveev Kurgan.

CHAPTER 9. COWS, COPPER, AND CHIEFS

1. Benveniste 1973:61–63 for feasts; also see the entry for GIVE in Mallory and Adams 1997:224–225; and the brief recent review by Fortson 2004:19–21.

2. The dates defining the beginning of the Eneolithic in the steppes are principally from human bone, whereas the dates from Old Europe are not. The date of 5200–5000 BCE for the beginning of the Eneolithic Dnieper-Donets II culture incorporates a reduction of -228 ± 30 radiocarbon years prior to recalibration. There is a discussion of this below in note 16.

3. "Old Europe" was a term revived by Marija Gimbutas, perhaps originally to distinguish Neolithic European farming cultures from Near Eastern civilizations, but she also used the term to separate southeastern Europe from all other European Neolithic regions. See Gimbutas 1991, 1974. For chronologies, economy, environment, and site descriptions, see Bailey and Panayotov 1995; and Lichardus 1991. For the origin of the term *Alteuropa* see Schuchhardt 1919.

4. Most of these dates are on charcoal or animal bone and so need no correction. The earliest copper on the Volga is at Khvalynsk, which is dated by human bone that tested high in ^{15}N (mean 14.8%) and also seemed too old, from about 5200–4700 BCE, older than most of the copper in southeastern Europe, which was the apparent source of the Khvalynsk copper. I have subtracted four hundred radiocarbon years from the original radiocarbon dates to account for reservoir effects, making the Khvalynsk cemetery date 4600–4200 BCE, which accords better with the florescence of the Old European copper age and therefore makes more sense.

5. For the pathologies on cattle bones indicating they were used regularly for heavy draft, see Ghetie and Mateesco 1973; and Marinescu-Bîlcu et al. 1984.

6. For signs and notation, see Gimbutas 1989; and Winn 1981. The best book on female figurines is Pogozheva 1983.

7. Copper tools were found in Early Eneolithic Slatina in southwestern Bulgaria, and copper ornaments and pieces of copper ore (malachite) were found in Late Neolithic Hamangia IIB on the Black Sea coast in the Dobruja hills south of the Danube delta, both probably dated about 5000 BCE. For Old European metals in Bulgaria, see Pernicka et al. 1997. For the middle Danube, see Glumac and Todd 1991. For general overviews of Eneolithic metallurgy, see Chernykh 1992; and Ryndina 1998.

8. For vegetation changes during the Eneolithic, see Willis 1994; Marinescu-Bîlcu, Cârciumaru, and Muraru 1981; and Bailey et al. 2002.

9. Kremenetski et al. 1999; see also Kremenetskii 1997. For those who follow the "beech line" argument in Indo-European origin debates, these pollen studies indicate that Atlantic-period beech forests grew in the Dniester uplands and probably spread as far west as the Dnieper.

10. For the ceramic sequence, see Ellis 1984:48 and n. 3. The Pre-Cucuteni I phase was defined initially on the basis of ceramics from one site, Traian-Dealul Viei; small amounts of

similar ceramics were found later at four other sites, and so the phase probably is valid. For an overview of the Tripolye culture, see Zbenovich 1996.

11. Marinescu-Bîlcu et al. 1984.

12. Some Tripolye A settlements in the South Bug valley (Lugach, Gard 3) contained sherds of Bug-Dniester pottery, and others had a few flint microlithic blades like Bug-Dniester forms. These traces suggest that some late Bug-Dniester people were absorbed into Tripolye A villages in the South Bug valley. But late Bug-Dniester pottery was quite different in paste, temper, firing, shape, and decoration from Tripolye pottery, so the shift to using Tripolye wares would have been an obvious and meaningful act. For the absence of Bug-Dniester traits in Tripolye material culture, see Zbenovich 1980:164–167; and for Lugach and Gard 3, see Tovkailo 1990.

13. For Bernashevka, see Zbenovich 1980. For the Tripolye A settlement of Luka-Vrublevetskaya, see Bibikov 1953.

14. For the Karbuna hoard, see Dergachev 1998.

15. The Early Eneolithic cultures I describe in this section are also called Late Neolithic or Neo-Eneolithic. Telegin (1987) called the DDII cemeteries of the Mariupol-Nikol'skoe type Late Neolithic, and Yudin (1988) identified Varfolomievka levels 1 and 2 as Late Neolithic. But in the 1990s Telegin began to use the term "Neo-Eneolithic" for DDII sites, and Yudin (1993) started calling Varfolomievka an Eneolithic site. I have to accept these changes, so sites of Mariupol-Nikol'skoe (DDII) type and all sites contemporary with them, including Khvalynsk and Varfolomievka, are called Early Eneolithic. The Late Neolithic apparently has disappeared. The terminological sequence in this book is Early Neolithic (Surskii), Middle Neolithic (Bug-Dniester–DDI), Early Eneolithic (Tripolye A–DDII–Khvalynsk), and Late Eneolithic (Tripolye B, C1-Sredni Stog-Repin). For key sites in the Dnieper-Azov region, see Telegin and Potekhina 1987; and Telegin 1991. For sites on the middle Volga, see Vasiliev 1981; and Agapov, Vasiliev, and Pestrikova 1990. In the Caspian Depression, see Yudin 1988, 1993.

16. The average level of ^{15}N in DDII human bones is 11.8 percent, which suggests an average offset of about -228 ± 30 BP, according to the method described in the appendix. I subtracted 228 radiocarbon years from the BP dates for the DDII culture and calibrated them again. The unmodified dates from the earliest DDII cemeteries (Dereivka, Yasinovatka) suggested a calibrated earliest range of 5500–5300 BCE (see Table 9.1), but these dates always seemed too early. They would equate DDII with the middle Bug-Dniester and Criş cultures. But DDII came for the most part *after* Bug-Dniester, during the Tripolye A period. The modified radiocarbon dates for Dnieper-Donets II fit better with the stratigraphic data and with the Tripolye A sherds found in Dnieper-Donets II sites. For lists of dates, see Trifonov 2001; Rassamakin 1999; and Telegin et al. 2002, 2003.

17. For lists of fauna, see Benecke 1997:637–638; see also Telegin 1968:205–208. For ^{15}N in the bones, see Lillie and Richards 2000. Western readers might be confused by statements in English that the DDII economy was based on hunting and fishing (Zvelebil and Lillie 2000:77; Telegin, et al. 2003:465; and Levine 1999:33). The DDII people ate cattle and sheep in percentages between 30% and 78% of the animal bones in their garbage pits. Benecke (1997:637), a German zoologist, examined many of the North Pontic bone collections himself and concluded that domesticated animals "first became evident in faunal assemblages that are synchronized with level II of the Dnieper-Donets culture." People who kept domesticated animals were no longer hunter-gatherers.

18. Flint blades 5–14 cm long with sickle gloss are described by Telegin (1968:144). The northwestern DDII settlements with seed impressions are listed in Pashkevich 1992, and Okhrimenko and Telegin 1982. DDII dental caries are described in Lillie 1996.

19. Telegin 1968:87.

20. The Vasilievka II cemetery was recently dated by radiocarbon to the Late Mesolithic, about 7000 BCE. The cemetery was originally assigned to the DDII culture on the basis of

a few details of grave construction and burial pose. Telegin et al. 2002 extended the label "Mariupol culture" back to include Vasilievka II, but it lacks all the artifact types and many of the grave features that define DDII-Mariupol graves. The DDII cemeteries are securely dated to a period after 5400–5200 BCE. Vasilievka II is Late Mesolithic.

21. For funeral feasts, see Telegin and Potekhina 1987:35–37, 113, 130.

22. I have modified Khvalynsk dates on human bone to account for the very high average ^{15}N in human bone from Khvalynsk, which we measured at 14.8%, suggesting that an average −408±52 radiocarbon years should be subtracted from these dates before calibrating them (see Authors Note on Dating, and chapter 7). After doing this I came up with dates for the Khvalynsk cemetery of 4700/4600–4200/4100 BCE, which makes it overlap with Sredni Stog, as many Ukrainian and Russian archaeologists thought it should on stylistic and typological grounds. It also narrows the gap between late Khvalynsk on the lower Volga (now 3600–3400 BCE) and earliest Yamnaya. See Agapov, Vasiliev, and Pestrikova 1990; and Rassamakin 1999.

23. Until Khvalynsk II is published, the figure of forty three graves is conditional. I was given this figure in conversation.

24. For the enhancement of male status with herding economies, see Holden and Mace 2003.

25. In Anthony and Brown (2000) we reported a smaller number of horses, cattle, and sheep from the cemetery at Khvalynsk, based on only the twelve "ritual deposits" placed above the graves. I later compiled the complete animal bone reports from two sources: Petrenko 1984; and Agapov, Vasiliev, and Pestrikova 1990, tables 1, 2. They presented conflicting descriptions of the numbers of sheep in ritual deposits 10 and 11, and this discrepancy resulted in a total count of either fifty-two or seventy sheep MNI.

26. See Ryndina 1998:151–159, for Khvalynsk I and II metals.

27. For ornaments see Vasiliev 2003.

28. For the possibility that the first domesticated animals came across the North Caucasus from the Near East, see Shnirelman 1992; and Jacobs 1993; and, in opposition, see Anthony1994.

29. Yanushevich 1989.

30. Nalchik is described in Gimbutas 1956:51–53.

31. I found this grave referenced in Gei 2000:193.

32. The bones at Dzhangar were originally reported to contain domesticated cattle, but the zoologist Pavel Kosintsev told me, in 2001, that they were all onager and horse, with no obvious domesticates.

33. The Neolithic cultures of the North Caspian Depression, east of the Volga, were first called the Seroglazivka culture by Melent'ev (1975). Seroglazivka included some Neolithic forager camps similar to Dzhangar and later sites with domesticated animal bones like Varfolomievka. Yudin suggested in 1998 that a new label, "Orlovka culture," should be applied to the Early Eneolithic sites with domesticated animals. On Varfolomievka, see Yudin 1998, 1988. Razdorskoe was described by Kiyashko 1987. Older but still informative is Telegin 1981.

34. The Orlovka site was first described by Mamontov 1974.

35. The Samara Neolithic culture, with the cemetery of S'yezzhe, usually is placed earlier than Khvalynsk, as one S'yezzhe grave contained a boars-tusk plaque exactly like a DDII type. Radiocarbon dates now indicate that early Khvalynsk overlapped with the late Samara Neolithic (and late DDII). The Samara Neolithic settlement of Gundurovka contained Khvalynsk pottery. The Samara culture might have begun before Khvalynsk; see Vasiliev and Ovchinnikova 2000. For S'yezzhe, see Vasiliev and Matveeva 1979. For animal bones, see Petrenko 1984:149; and Kuzmina 2003.

CHAPTER 10. THE DOMESTICATION OF THE HORSE
AND THE ORIGINS OF RIDING

1. See Clayton and Lee 1984; and Clayton 1985. For a recent update, see Manfredi, Clayton, and Rosenstein 2005.

2. For early descriptions of bit wear, see Clutton-Brock 1974; and Azzaroli 1980. Doubts about the causes of this kind of wear had been expressed by Payne (1995) in a study published after long delays.

3. We were provided with horse teeth by Mindy Zeder at the Smithsonian Institution; the Large Mammal Veterinary Facility at Cornell University; the University of Pennsylvania's New Bolton Veterinary Center; the Bureau of Land Management, Winnemucca, Nevada; and Ron Keiper of Pennsylvania State University. We learned mold-making and casting procedures from Sandi Olsen and Pat Shipman, then at Johns Hopkins University. Mary Littauer gave us invaluable advice and the use of her unparalleled library. Our first steps were supported by grants from the Wenner-Gren Foundation and the American Philosophical Society.

4. On horse MtDNA, see Jansen et al. 2002; and Vilà et al. 2001. For horse Y-chromosomes, see Lindgren et al. 2004.

5. For equids in Anatolia, see Summers 2001; and online reports on the Catal Höyük project. For horses in Europe, see Benecke 1994; and Peške 1986.

6. For Mesolithic and Neolithic Pontic-Caspian horses, see Benecke 1997; Vasiliev, Vybornov, and Komarov 1996; and Vasilev 1998. For horse bones at Ivanovskaya in the Samara Neolithic, see Morgunova 1988. In the same volume, see I. Kuzmina 1988.

7. For Mongol horse keeping, see Sinor 1972; and Smith 1984. For horses and cattle in the blizzard of 1886, see Ryden 1978:160–162. For feral horses see also Berger 1986.

8. For a review of these methods, see Davis 1987. For riding-related pathologies in vertebrae, see Levine 1999b. For crib-biting, see Bahn 1980; and the critique in White 1989.

9. The graphs from Benecke and von den Driesch (2003) are combined and reprinted as figure 10.3 here. See also Bökönyi 1974. For a critical view of Dereivka, see Uerpmann 1990.

10. The ratio of females to males in a harem band, counting immature horses, should be about 2:1, but the *skeletons* of immature males cannot be assigned a sex as the canine teeth do not erupt until about four to five years of age, and the presence of erupted canines is the principal way to identify males. From the bones, a harem band would contain just one *identifiable* male.

11. A horse's age at death can be estimated from a loose molar by measuring the molar crown height, the length of the tooth from the bifurcation between the roots to the occlusal surface. This measurement decreases with age as the tooth wears down. Spinage (1972) was the first to publish crown height-versus-age statistics for equids, based on zebras; Levine (1982) published statistics for a small sample of horses using measurements from X-rays. We largely confirmed Levine's numbers with direct measurements on our larger sample. But we found that estimates based *only* on crown heights have *at best* a ±1.5 year degree of uncertainty (a three-year span). The crown height on the right and left P_2s of the same horse can vary by as much as 5 mm, which would normally be interpreted as indicating a difference in age of more than three years. See note 18, below.

12. Bibikova (1967, 1969) noted that fifteen of seventeen sexable mandibles were male. I subtracted the cult stallion, an Iron Age intrusion, making fourteen of sixteen males. Bibikova never published a complete description of the Dereivka horse bones, but she did note that the MNI was fifty-two individuals; 23% of the population was aged one to two years (probably looking at long bone fusion); fifteen of seventeen sexable jaw fragments were from males older than five, as this is when the canine teeth emerge; and there were no very old individuals.

Levine's age-at-death statistics were based on the crown heights of all the teeth kept in 1998, with an MNI of only sixteen—about two-thirds of the original collection had been lost. Only 7% of this remnant population was one to two years of age based on long-bone fusion (1999b:34) and about one-third of the surviving teeth were from the Iron-Age cult stallion. For Levine's age-at-death graphs, see Levine 1990, 1999a, 1999b.

13. The analysis of the equid P_2s from Leisey was conducted by Christian George as part of his MA Thesis in Geosciences at the University of Florida. The 1.5-million-year-old Leisey equids were *Equus "leidyi,"* possibly an eastern variant of *Equus scotti,* a common member of the Rancholabrean fauna, very similar in dentition, diet and stature to true horses. Of the 113 P2s from this site, 39 were eliminated because of age, damage, or pathologies, leaving 74 measurable P_2s from mature equids. See George 2002; Anthony, Brown, and George 2006; and Hulbert, Morgan, and Webb 1995. Our collection of P_2s was assembled through the generosity of the New Bolton Center at the University of Pennsylvania, the Cornell University College of Veterinary Medicine, the Bureau of Land Management in Winnemucca, NE; and Ron Keiper, then at Pennsylvania State University.

14. We are grateful to the National Science Foundation for supporting the riding experiment, and to the State University of New York at Cobleskill for hosting and managing it. Dr. Steve MacKenzie supervised the project, and the riding and recording was done by two students in the Horse Training and Behavior Program, Stephanie Skargensky and Michelle Beleyea. The bone bit and antler cheekpieces were made with flint tools by Paul Trotta. The hemp rope was supplied by Vagn Noeddlund of Randers Ropeworks. Mary Littauer and Sandra Olsen provided valuable suggestions on bits and mold-making. All errors were our own.

15. The pre-experiment, never-bitted mean bevel measurement for the three horses bitted with soft bits was 1.1 mm, the same as the never-bitted Pleistocene Leisey equids. The standard deviation for the three was 0.42 mm. The post-experiment mean was 2.04 mm, more than two standard deviations greater than the pre-experiment mean. Another 300 hours of riding might have created a bevel of 3 mm, our threshold for archaeological specimens.

16. The 74 never-bitted equid teeth from Leisey exhibited a greater range of variation than the 31 never-bitted modern P_2s we collected, not surprising with a larger sample. The distribution of measurements was normal, and a t-Test of the difference between the means for our bitted sample and the Leisey sample showed a significant difference. The threshold of 3 mm for identifying bit wear in archaeological specimens is supported by the Leisey data.

17. Levine outlined six problems with our bit wear studies in 1999b:11–12 and 2004:117–120. She placed it in a category she termed "false direct evidence," with so-called bridle cheekpieces whose forms vary wildly and whose function is entirely speculative. We believe Levine's criticisms are based on factual errors, distortions, and misunderstandings. For our reply to each of her six criticisms, see Anthony, Brown, and George 2006. We remain confident in our analysis of bit wear.

18. Permanent horse P2s become flattened or "tabled" by occlusion with the opposing tooth gradually between two and three years of age. Brown determined that a P_2 with a crown height greater than 5.0 mm *and* an occlusal length-to-width ratio greater than 2.1 is probably from a horse three years old or younger, so should be excluded from studies of bit wear (Brown and Anthony 1998:338–40). Brown was the first to combine the crown height and the occlusal length-width ratio to produce an age-at-death estimate this precise. If she had not done this we would have been forced to discard half of our sample to avoid using 2-3-year-old teeth. Christian George also used Brown's method to eliminate young teeth (≤ 3 yr) from the Leisey sample. It should be noted that George found one P_2 with a bevel of 3.05 mm, but it was probably from a horse less than three years old.

19. Bendrey (2007), as this book went to press, reported new bevel measurements on never-bitted Przewalski horses, from zoos in England and Prague. Bendrey measured 29 P_2s from 15 Przewalksi horses of acceptable age (>3 and <21), and found 3mm bevels on three, or 10%. We found one bevel of *almost* 3mm in 105 never-bitted P_2s, less than 1%. The Przewalski bevels all

were caused by malocclusion with the opposing upper P^2; one 3mm bevel was filed down as a veterinary treatment for underbite. Malocclusion occurred among zoo-kept Przewalskis more frequently than among Pleistocene equids or Nevada mustangs. All zoo Przewalskis are descended from about 15 captured in the wild, and these founders might have had unusually bad occlusion. Also domestic horses were bred with the founders, perhaps mixing genes for different tooth and jaw sizes.

20. Raulwing 2000:61, with references.

21. For Dereivka, see Telegin 1986. For the horse bones, see Bibikova 1967, 1970; Bökönyi 1974, 1978, 1979; and Nobis 1971.

22. For criticisms of the traditional evidence for horse domestication at Dereivka, see Anthony 1986, 1991b; and Levine 1990.

23. Our research at the Institute of Zoology in Kiev was hosted by a generous and thoughtful Natalya Belan; in Samara, Russia, by Igor Vasiliev; and in Petropavlovsk, Kazakhstan, by Victor Zaibert. In Budapest Sandor Bökönyi made us welcome in the gracious manner for which he was widely known and is widely missed. The project was supported by a grant from the National Science Foundation. For reports, see Anthony and Brown 1991; and Anthony, Telegin, and Brown 1991.

24. See Häusler 1994.

25. For the redating of the Dereivka cult stallion, see Anthony and Brown 2000; reiterated in Anthony and Brown 2003.

26. Both Botai and Tersek showed some influence in their ceramics from forager cultures of the forest-steppe zone in the southeastern Urals, known as Ayatskii, Lipchin, and Surtanda. Botai-Tersek might have originated as a southern, steppe-zone offshoot of these cultures. For a description of Botai and Tersek in English, see Kislenko and Tatarintseva 1999; in Russian, see Zaibert 1993. For discussions of the horse remains at Botai and related sites, see Olsen 2003; and Brown and Anthony 1998.

27. Our initial measurements of the horse teeth from Kozhai 1 (made in a hotel room in Petropavlovsk, Kazakhstan) produced one tooth with a 3 mm bevel. This is how we described the Kozhai results before 2006. We remeasured the twelve Kozhai 1 casts for Anthony, Brown, and George 2006, and agreed that a borderline 2.9+ measurement was actually 3 mm, resulting in two teeth with bit wear. Two other P_2s from Kozhai 1 measured 2 mm or more, an unusually high measurement among wild horses.

28. Describing the Botai horses as wild were Levine 1999a, 1999b; Benecke and von den Dreisch 2003; and Ermolova, in Akhinzhalov, Makarova, and Nurumov 1992.

29. See Olsen 2003:98–101.

30. French and Kousoulakou 2003:113.

31. The Atbasar Neolithic preceded Botai in the northern Kazakh steppes; see Kislenko and Tatarintseva 1999. Benecke and von den Dreisch (2003: table 6.3) reported that domesticated sheep and cattle bones were found in Atbasar sites in the Kazakh steppes, dated before Botai. This is true, *but* the Russian and Kazakh authors they cite described the bones of domesticated sheep and cattle as later intrusions in the Neolithic levels; they were less weathered than the bones of the wild animals. The animal bones from Atbasar sites are interpreted by Akhinzhalov, Makarova, and Nurumov as indicating a foraging economy based on wild horses, short-horned bison, saiga antelope, gazelle, red deer, and fish. Domesticated animals appeared at the end of the Botai era. For their comments on differential bone weathering in Atbasar sites, see Akhinzhalov, Makarova, and Nurumov 1992:28–29, 39.

32. Logvin (1992) and Gaiduchenko (1995) interpreted some animal bones in sites of the Eneolithic Tersek culture, centered in the Tugai steppes near Kustenai, Kazakhstan, and dated to the same period as Botai, as domesticated cattle, particularly from Kumkeshu I. Another zoologist, Makarova, had identified the Tersek bovid bones as those of wild bison (Akhinzhalov, Makarova, and Nurumov 1992:38). Some domesticated cattle might have been kept in Tersek sites, which were closer to the Pontic-Caspian herders. None appeared at Botai. For Kumkeshu I, see Logvin, Kalieva, and Gaiduchenko 1989.

33. For horses in the Caucasus I relied on the text of a conference paper by Mezhlumian (1990). A few horses might have passed through the Caucasus into northern Iran before 3000 BCE, indicated by a few probable horse teeth at the site of Qabrestan, west of Teheran (see Mashkour 2003) and a possible horse tooth at Godin Tepe (see Gilbert 1991). No definite horse remains have been identified in eastern Iran, Central Asia, or the Indian subcontinent in deposits dated earlier than 2000 BCE, claims to the contrary notwithstanding. For a review of this debate, see Meadow and Patel 1997.

34. For central European horses, see See Benecke 1994; Bökönyi 1979; and Peške 1986.

35. Khazanov 1994:32.

36. For war and the prestige trade, see Vehik 2002.

37. The American Indian analogy is described in Anthony 1986. The most detailed analysis of the effects of horseback riding and horse keeping on Plains Indian cultures is Ewers 1955.

38. One argument against riding before 1500 BCE was that steppe horses were too small to ride. This is not true. More than 70% of the horses at Dereivka and Botai stood 136–144 cm at the withers, or about 13–14 hands high, and some were 15 hands high. They were the same size as Roman cavalry horses. Another argument is that rope and leather bits were inadequate for controlling horses in battle. This is also not true, as the American Indians demonstrated. Our SUNY students at Cobleskill also had "no problem" controlling horses with rope bits. The third is that riders in the steppes rode sitting back on the rump of the horse, a manner suited only to riding donkeys, which did not exist in the steppes. We have rebutted these doubts about Eneolithic riding in Anthony, Brown, and George 2006. For the arguments against Eneolithic riding, see Sherratt 1997a:217; Drews 2004:42–50; Renfrew 2002; and E. Kuzmina 2003:213.

39. The remains of a bow found in Berezovka kurgan 3, grave 2, on the Volga, in a grave of Pokrovka type probably dated about 1900–1750 BCE, had bone plates reinforcing the shaft and bone tips at the ends—a composite bow. The surviving pieces suggest a length of 1.4–1.5 m, almost five feet from tip to tip. See Shishlina 1990; and Malov 2002. For an overview of early archery and bows, see Zutterman 2003.

40. I am indebted to Dr. Muscarella for some of these ideas about arrow points. For a discussion of the initial appearance and usage of socketed bronze arrowheads, see Derin and Muscarella 2001. For a catalogue and discussion of the early Iron Age socketed arrowheads of the Aral Sea region, see Itina and Yablonskii 1997. Socketed bronze spear points were made in the steppes as early as 2000 BCE, and smaller socketed points began to appear occasionally in steppe sites about the middle of the Late Bronze Age, around 1500 BCE, but their potential was not immediately exploited. The ideal bows, arrows, and arrowheads for mounted archery evolved slowly.

41. For tribal warfare, see Keeley 1996.

CHAPTER 11. THE END OF OLD EUROPE AND THE RISE OF THE STEPPE

1. For the gold at Varna, see Bailey 2000:203–224; Lafontaine and Jordanov 1988; and Eleure 1989.

2. Chapman 1989.

3. For off-tell settlement at Bereket, see Kalchev 1996; at Podgoritsa, see Bailey et al. 1998.

4. The decrease in solar insolation that bottomed out at 4000–3800 BCE is documented in Perry and Hsu 2000; and Bond et al. 2001. For the Piora Oscillation in the Swiss Alps, see Zöller 1977. For indicators of cooling in about 4000 BCE in the Greenland ice cores, see O'Brien et al. 1995. For climate change in Central Europe in the German oak tree rings, see Leuschner et al. 2002. For the Pontic steppes, see Kremenetski, Chichagova, and Shishlina 1999.

5. For the flooding and agricultural shifts, see Bailey et al. 2002. For overgrazing and soil erosion, see Dennell and Webley 1975.

6. For Jilava, see Comsa 1976.

7. The pollen changes are described in Marinova 2003.

8. Cast copper objects began to appear regularly in western Hungary with the Lasinja-Balaton culture at about 4000 BCE; see Bánffy 1995; also Parzinger 1992.

9. Todorova 1995:90; Chernykh 1992:52. The burning of houses might have been an intentional ritual act during the Eneolithic; see Stevanovic 1997. But the final fires that consumed the Eneolithic towns of the lower Danube valley and the Balkans about 4000 BCE were followed by region-wide abandonment and abrupt culture change. Region-wide abandonments of large settlements in the North American Southwest (1100–1400 CE) and in Late Classic Maya sites (700–900 CE) in Mesoamerica were associated with intense warfare; see Cameron and Tomka 1993. The kind of climate shift that struck the lower Danube valley about 4100–3800 BCE would not have made tell settlements uninhabitable. Warfare therefore seems a likely explanation.

10. For evidence of overgrazing and soil erosion at the end of the Karanovo VI period, see Dennell and Webley 1975; for the destruction of Eneolithic Yunatsite, see Merpert 1995; and Nikolova 2000.

11. Todorova 1995.

12. See Ellis 1984 for ceramic workshops, and Popov 1979 for flint workshops. I use the Russian spelling (Tripolye, Tomashovka) rather than the Ukrainian (Tripil'ye, Tomashivka), because many site names such as Tripolye are established in the literature outside Ukraine in their Russian spelling.

13. On the demographics, see Dergachev 2003; and Masson 1979. On the flight of Bolgrad-Aldeni refugees, see Sorokin 1989.

14. On Tripolye B1 warfare generally, see Dergachev 2003, 1998b; and Chapman 1999. On Drutsy 1, see Ryndina and Engovatova 1990. For much of the other information in this section I have relied on the review article by Chernysh 1982.

15. The Cucuteni C designation refers only to a type of shell-tempered pottery. The Cucuteni chronology ends with Cucuteni B$_2$. Cucuteni C ware appeared first in sites dated to the Cucuteni A$_3$/Tripolye B1 period and ultimately dominated ceramic assemblages. See Ellis 1984:40–48.

16. The source of the steppe influence on Cucuteni C pottery is usually identified as the early Sredni Stog culture, phase Ib, for Telegin; or the Skelya culture, for Rassamakin.

17. Shell-temper adds to the durability and impact resistance of vessels that are regularly submitted to thermal shock through reheating, and also increases the cooling effect of evaporation, making a shell-tempered pot good for cooking or storing cool drinking water. Cucuteni C ware and fine painted wares were found together both in pit-houses and large two-storied surface houses. Contextual differences in the distribution of Cucuteni C ware and fine ware in settlements have not been described. At some sites the appearance of Cucuteni C wares seems abrupt: Polivanov Yar had traditional grog-tempered coarse wares in the Tripolye B2 occupation but switched to shell-tempered C wares of different shapes and designs in Tripolye C1, whereas the fine painted wares showed clear continuity between the two phases. See Bronitsky and Hamer 1986; Gimbutas 1977; and Marinescu-Bilcu 1981.

18. For the horse-head maces see Telegin et al. 2001; Dergachev 1999; Gheorgiu 1994; and Govedarica and Kaiser 1996.

19. For the skull shapes, see Necrasov 1985; and Marcsik 1971. Gracile "Mediterranean" Tripolye skulls have been found in ritual foundation deposits at Traian (Tripolye B2).

20. For Mirnoe, see Burdo and Stanko 1981.

21. For the eastern migration, see Kruts and Rizhkov 1985.

22. The Iron Age stereotype of nomadic cavalry seems to lie behind some of the writings of Merpert (1974, 1980) and Gimbutas (1977), who were enormously influential.

23. The "awkward seat" hypothesis is based on Near Eastern images that show riders sitting awkwardly on the horse's rump, a seat more suited to donkey riding. Donkeys have low withers

and a high, broad rump. If you sit forward on a donkey and the animal lowers its head, you can easily fall forward to the ground. Donkey riders, therefore, usually sit back on the rump. Horses have high withers, so horse riders sit forward, which also permits them to hang onto the mane. You have to push and lift to get yourself onto a horse's rump, and then there's nothing to hold on to. Artistic images that show riders on horseback sitting back on the rump probably indicate only that many Near Eastern artists before 1000 BCE, particularly in Egypt, were more familiar with riding donkeys than horses. The suggestion that riders in the steppes would adopt and maintain a donkey seat on horses is inherently implausible. See Drews 2004:40–55, for this argument.

24. For mutualism and economic exchanges between Old Europe and the Eneolithic cultures of the Pontic steppe, see Rassamakin 1999:112; see also Manzura, Savva, and Bogotaya 1995; and Nikolova 2005:200. Nikolova has argued that transhumant pastoralism was already part of the Old European economy in Bulgaria, but the Yagodinska cave sites she cited are radiocarbon dated about 3900 BCE, during or just after the collapse. Upland pastoral settlements were a small and comparatively insignificant aspect of the tell economies, and only a serious crisis made them the basis for a new economy.

25. Ewers 1955:10.

26. See Benveniste 1973:53–70, for *Give* and *Take*, esp. 66–67 for the Hittite terms; for the quotation, see 53. Hittite *pai* was derived from the preverb *pe-* with **ai-*, with reflexes meaning "give" in Tocharian *ai-*. Also see the entry for *Give* in Mallory and Adams 1997:224–225.

27. See Keeley 1996. For mutualist models of the Linear Pottery frontier, see Bogucki 1988. An ethnographic case frequently cited in discussions of mutualist food exchange is that of the horticultural Pueblo Indians and the pedestrian buffalo hunters of the Plains. But a recent study by Susan Vehik suggested that the Pueblo Indians and the Plains bison hunters traded prestige commodities—flint arrowheads, painted pottery, and turquoise—not food. And during a period of increasing conflict in the Plains after 1250 CE, trade actually greatly increased; see Vehik 2002.

28. See Kershaw 2000.

29. See "bride-price" in Mallory and Adams 1997:82–83.

30. In East Africa a group of foragers and beekeepers, the Mukogodo, were forced to obtain livestock after they began to interact and intermarry with stock-raising tribes, because it became impossible for Mukogodo men to obtain wives by offering beehives when non-Mukogodo suitors offered cattle. Cattle were just more valuable. The Mukogodo became pastoralists so that they could continue to have children. See Cronk 1989, 1993.

31. Ewers 1955:185–187.

32. The Sredni Stog site had two levels, Sredni Stog 1 and 2. The lower level (Sredni Stog 1) was an Early Eneolithic DDII occupation, and the upper was the type site for the Late Eneolithic Sredni Stog culture. In older publications the Sredni Stog culture is sometimes called Sredni Stog 2 (or II) to differentiate it from Sredni Stog 1 (or I).

33. The Sredni Stog culture is defined in Telegin 1973. The principal settlement site of the Sredni Stog cultre, Dereivka, is described in English in Telegin 1986; for the Sredni Stog origin of Cucuteni C ware, see 111–112. Telegin's chronological outline is described in English in Telegin 1987.

34. The longest and most detailed version of Rassamakin's new model in English is the 123-page article, Rassamakin 1999. Telegin's four phases (Ia, Ib, IIa, IIb) of the Sredni Stog culture represented, for Rassamakin, at least three separate and successive cultures: (1) the Skelya culture, 4500–4000 BCE (named for Strilcha Skelya, a phase Ib Sredni Stog site for Telegin); (2) the Kvityana culture, 3600–3200 BCE (Kvityana was a phase Ia site for Telegin, but Rassamakin moved it to the equivalent of Telegin's *latest* phase IIb); and (3) the Dereivka culture, 3200–3000 BCE (a phase IIa site for Telegin, dated 4200–3700 BCE by radiocarbon). Telegin seemed to stick to the stratigraphy, grave associations, and radiocarbon dates, whereas Rassamakin relied on stylistic arguments.

35. For Sredni Stog ceramics, see Telegin 1986:45–63; 1973:81–101. For skeletal studies, see Potekhina 1999:149–158.

36. For the seeds at Moliukhor Bugor, see Pashkevich 1992:185. For the tools at Dereivka, see Telegin 1973:69, 43. Bibikova actually reported 2,412 horse bones and 52 horse MNI. I have edited out the mandible, skull, and two metacarpals of the "cult stallion."

37. Only four settlement animal bone samples are reported for Sredni Stog. Most of them are worryingly small (a few hundred bones) and screens were not used in excavations (still are not), so bone recovery varied between excavations. For these reasons, the published animal bone percentages can be taken only as rough guides. For an English translation of the faunal reports, see Telegin 1986.

38. Rassamakin (1999:128) assigned the Dereivka cemetery, which he called Dereivka 2, to the Skelya period, before 4000 BCE, and assigned the Dereivka settlement to the Late Eneolithic, around 3300–3000 BCE. Telegin, following the radiocarbon dates from the settlement and the Tripolye B2 bowl found in the cemetery, assigned both to the same period.

39. See Dietz 1992 for the varied interpretations of antler "cheekpieces."

40. For the Suvorovo-Novodanilovka group, see Nechitailo 1996; and Telegin et al. 2001. The metals are analyzed in Ryndina 1998:159–170; for an English summary, see 194–195. English-language discussions of the Suvorovo-Novodanilovka group are few. In addition to Rassamakin's description of the Skelya culture, which incorporates Suvorovo-Novodanilovka, see Dergachev 1999; and Manzura, Savva, and Bogotaya 1995. And there is a useful entry under "Suvorovo" in Mallory and Adams 1997.

41. Telegin 2002, 2001.

42. The physical type in Novodanilovka graves is discussed in Potekhina 1999:149–154. The types of the lower Danube valley are described by Potekhina in Telegin et al. 2001; and in Necrasov and Cristescu 1973.

43. Ryndina (1998:159–170) examined copper objects from graves at Giugiurleşti, Suvorovo, Novodanilovka, Petro-Svistunovo, and Chapli. For the copper of Varna and Gumelnitsa, see Pernicka et al. 1997. They document the end of the Balkan mines and the switch to Carpathian ores at about 4000 BCE.

44. The horse-head examples in the Volga steppes were found at Novoorsk near Orenburg and at Lebyazhinka near Samara. For the polished stone mace heads, see Kriukova 2003.

45. For Old European weapons, see Chapman 1999.

46. *Equus hydruntinus* had a special ritual status in the cemeteries of Varna and Durankulak, but was unimportant in the diet and was on the brink of extinction. Horses (*Equus caballus*) were rare or absent in the Eneolithic settlements and cemeteries of the Danube valley before the Cernavoda I period, except for sites of the Bolgrad variant. The Gumelniţa-related Bolgrad sites had about 8% horse bones. Other Old European sites in the Danube valley had few or no horses. For the Varna and Durankulak equids, see Manhart 1998.

47. See Vehik 2002 on increased warfare and long-distance trade in the Southwest. DiCosmo (1999) observed that increased warfare in the steppes encouraged organizational changes in preexisting institutions, and these changes later made large nomadic armies possible.

48. Contacts between late Tripolye A/early B1 settlements and the Bolgrad culture are summarized in Burdo 2003. Most of the contact is dated to late Tripolye A—Tripolye AIII2 and III3.

49. For Bolgrad sites, see Subbotin 1978, 1990.

50. For the intrusive cemeteries, see Dodd-Opriţescu 1978. For the gold and copper hoards, see Makkay 1976.

51. For the Suvorovo kurgan group, see Alekseeva 1976. The Kopchak kurgan is described in Beilekchi 1985.

52. Giurgiuleşti is described briefly in Haheu and Kurciatov 1993. One radiocarbon date is published from Giurgiuleşti: Ki-7037, 5380 ± 70 BP, or about 4340–4040 BCE, calibrated; I have been told that the date is misprinted in Telegin et al. 2001, 128.

53. The Novodanilovka grave, which was isolated and not in a cemetery, is described in Telegin 1973:113; for Petro-Svistunovo and Chapli, see Bodyans'kii 1968; and Dobrovol'ski 1958.

54. The region-wide abandonment of tells in about 4000–3500 BCE is observed in Coleman 2000. I do not see how this could have been the event that brought Greek speakers into Greece, because Greek shared many traits with the Indo-Iranian language branch (see the end of chapter 3), and Indo-Iranian emerged much later. The crisis of 4000 BCE probably brought Pre-Anatolian speakers into southeastern Europe.

55. See Madgearu 2001 on de-urbanization in post–Roman Bulgaria. Mace (1993) notes that if grain production falls, cattle are insurance against starvation. Cattle can be moved into a protected area during a period of conflict. Under conditions of declining agricultural yields and increasing conflict, a shift to a greater reliance on herding would make good economic sense.

56. For loot, lucre, and booty in Proto-Indo-European, see Benveniste 1973:131–137; for language shift among the Pathan, see Barth 1972.

57. For Cernavoda I, see Morintz and Roman 1968; and Roman 1978; see also Georgieva 1990; Todorova 1995; and Ilčeva 1993. A good recent summary is in Manzura 1999. For the cemetery of Ostrovul Corbului, see Nikolova 2002, 2000.

58. Sherratt 1997b, 1997c. Sherratt suggested that the drinking vessels of the period from 4000 to 2500 BCE were used to serve a beverage that included honey (the basis of mead) and grain (the source of beer), both directly attested in Early Bronze Age Bell Beaker cups. Honey, he suggested, would have been available only in small quantities, and might have been under the control of an elite who apportioned the fermented drink in ceremonies and closed gatherings open to just their inner circle. Proto-Indo-European contained a word for honey (*melit-) and a derivative term for a honey drink (*medhu-).

59. For Cernavoda I-Late Lengyel horses, see Peške 1986; and Bökönyi 1979.

60. For pastoralism, see Greenfield 1999; Bökönyi 1979; and Milisauskas 2002:202.

61. For the prayer to Sius, see Puhvel 1991.

Chapter 12. Seeds of Change on the Steppe Borders

1. Ryndina (1998:170–171) counted 79 copper objects from steppe graves for the Post-Suvorovo period, compared to 362 for Suvorovo-Novodanilovka graves.

2. See Telegin 2002, 1988, 1987; see also Nikolova and Rassamakin 1985; and Rassamakin 1999. Early reports on Mikhailovka are Lagodovskaya, Shaposhnikova, and Makarevich 1959; Shaposhnikova 1961 (this was the article where the division between lower and upper stratum 2 was noticed); and Shevchenko 1957. For the stratigraphic position of Lower Mikhailovka graves, see Cherniakov and Toshchev 1985. Radiocarbon dates for graves with Mikhailovka I pottery are reported in Videiko and Petrenko 2003. Early Mikhailovka II begins about 3500 BCE, in Kotova and Spitsyna 2003.

3. For the Maikop sherd at Mikhailovka I, see Nechitailo 1991:22. For the other pottery exchanges, see Rassamakin 1999:92; and Telegin 2002:36.

4. Pashkevich 2003.

5. The sheep of the Early Bronze Age in southeastern Europe were significantly larger than Eneolithic sheep, which Bökönyi (1987) attributed to a new breed of wool sheep that appeared after about 3500 BCE.

6. At the Cernavoda site three excavation areas yielded three successive archaeological cultures, of which the oldest was Cernavoda I, about 4000–3600 BCE; next was Cernavoda III, about 3600–3000 BCE, contemporary with Baden; and the youngest was Cernavoda II, 3000–2800 BCE. Mikhailovka I probably was contemporary with the end of Cernavoda I and the first half of Cernavoda III. See Manzura, Savva, and Bogatoya 1995.

7. For Mikhailovka I graves at Olaneshti, see Kovapenko and Fomenko 1986; and for Sokolovka, see Sharafutdinova 1980.

8. Potekhina 1999:150–151.

9. "Post-Mariupol" was the label first assigned by Kovaleva in the 1970s. See Nikolova and Rassamakin 1985; Telegin 1987; and Kovaleva 2001.

10. See Ryndina 1998:170–179, for Post-Mariupol metal types.

11. The two graves were Verkhnaya Maevka XII k. 2, gr. 10; and Samarska k.1, gr. 6 in the Orel-Samara region. See Ryndina 1998:172–173.

12. For Razdorske, see Kiyashko 1987, 1994.

13. The percentage of horse bones at Repin is often said to be 80%. Shilov (1985b) reviewed the numbers and came up with 55% horse bones, still a very high number.

14. For Repin/Yamnaya at Cherkasskaya, see Vasiliev and Siniuk 1984:124–125.

15. For Kara Khuduk and Kyzyl-Khak, see Barynkin and Vasiliev 1988; for the fauna, see I. Kuzmina 1988. Also see Ivanov and Vasiliev 1995; and Barynkin, Vasiliev, and Vybornov 1998. For the radiocarbon dates for Kyzyl Khak, see Lavrushin, Spiridonova, and Sulerzhitskii 1998:58–59. For late Khvalynsk graves on the lower Volga, see Dremov and Yudin 1992; and Klepikov 1994.

16. Kruts typed the Chapaevka ceramics as late Tripolye C1, whereas Videiko described Chapaevka as a late Tripolye B2 settlement. See Kruts 1977; and Videiko 2003. Videiko argued that ceramic craft traditions changed at different rates in different settlement groups. Tripolye B2 stylistic habits lingered longer, he suggested, in the Dnieper group (Chapaevka) than they did in the super-settlements of the South Bug group, which shifted to Tripolye C1 styles earlier. Tripolye C2 styles began on the Dniester at Usatovo about 3400–3300 BCE, but Tripolye C2 styles appeared on the Dnieper about 3100 BCE.

17. Kruts 1977:48.

18. For the super-sites, see Videiko 1990, and other articles in the same volume; also see Shmagli and Videiko 1987 and Kohl 2007.

19. At Maidanets'ke, emmer and spelt wheats were the most common cereals recovered; barley and peas also were found in one house. Cattle (35% of domesticates, MNI) were the most important source of meat, with pig (27%) and sheep (26%) as secondary sources; the remaining 11% was equally divided between dogs and horses. About 15% of the animals were red deer, wild boar, bison, hare, and birds. The cattle, pigs, and abundant wild animals indicate substantial forest near the settlement. A forest of about 20 <km² would have provided sufficient firewood for the town, figuring about 2.2 ha of hardwood forest per family of five for a sustainable woodlot. Since ecological degradation is not obvious, the abandonment of the town perhaps was caused by warfare. See Shmagli and Videiko 1987:69, and several articles on economy in the volume cited above as Videiko 1990.

20. The Tripolye B1 settlement of Polivanov Yar on the Dniester overlooked outcrops of high-quality flint. One house was engaged heavily in flint working, with all stages of the tool-making process. In the later Tripolye C1 settlement, all six excavated structures were engaged in flint working, the initial shaping occurred elsewhere, and new products were made (heavy flint axes and chisels about 10 cm long). The Tripolye C1 settlement had become a specialized village of flint workers. Maidanets'ke imported finished flint tools of Dniester flint, probably from Polivanov Yar. At Veseli Kut (150 ha), a Tripolye B2 town east of the South Bug valley, two structures were identified as ceramic workshops. Eight buildings dedicated to ceramic production were found at Varvarovka VIII (40 ha and 200 houses—the largest town in its region), and a similar ceramic factory appeared at Petreni on the Dniester, again the largest town in its area. At Maidanets'ke, eight houses in a row contained looms (indicated by clusters of up to seventy ceramic loom weights) and some had two looms, perhaps a specialized weaver's quarter. For Polivanov Yar, see Popova 1979; for ceramic workshops, see Ellis 1984.

21. For the Uruk expansion, see Algaze 1989; Stein 1999; and Rothman 2001. For copper production at Hacinebi, see Özbal, Adriaens, and Earl 2000; for the copper of Iran, see Matthews and Fazeli 2004. For the wool sheep, see Bökönyi 1983; and Pollack 1999.

22. For Sos and Berikldeebi, see Kiguradze and Sagona 2003; and Rothman 2003.

23. The Maikop-like pottery was found in pre-Kura-Araxes levels at Berikldeebi. Early Maikop began before the Early Transcaucasian Culture. See Glonti and Dzhavakhishvili 1987.

24. For pre-Maikop Svobodnoe, see Nekhaev 1992; and Trifonov 1991. For steppe-Svobodnoe exchanges, see Nekhaev 1992; and Rassamakin 2002.

25. The poses of those buried in the Maikop chieftain's grave were not clear. For an English-language description of the Maikop culture, see Chernykh 1992:67–83. Quite dated accounts are Childe 1936; and Gimbutas 1956:56–62. A long, detailed description in Russian is in Munchaev 1994. For the Novosvobodnaya graves, see Rezepkin 2000. For the archaeological culture history in the North Caucasus, see Trifonov 1991.

26. For the silver and gold staff casings with bulls, see Chernopitskii 1987. The 47-cm length of the riveted copper blade is emphasized in Munchaev 1994:199.

27. Rostovtseff (1922:18–32) argued that Maikop was a Copper Age or, in Anatolian terms, a Late Chalcolithic culture. But Maikop became established as a North Caucasian Bronze Age culture, so it begins somewhat earlier than the Anatolian Bronze Age to which it was originally linked. Some Russian archaeologists now suggest an early Maikop phase that would be Late Eneolithic, whereas later Maikop would remain Early Bronze Age. For Maikop chronology, see Trifonov 1991, 2001. For my own mistaken chronology, see Glumac and Anthony 1992. I should have believed Rostovtseff.

28. For the east Anatolian seal, see Nekhaev 1986; and Munchaev 1994:169, table 49:1–4.

29. For Galugai, see Korenevskii 1993, 1995; the fauna is described in 1995:82. Korenevskii considered Galugai a pioneer settlement by migrants from Arslantepe VIA. For Maikop horses, see Chernykh 1992:59.

30. Rezepkin (1991, 2000) argued that Maikop and Novosvobodnaya were separate and contemporary cultures. Similar radiocarbon dates from Galugai (Maikop) and Klady (Novos-vobodnaya) suggested this. But the radiocarbon dates for Galugai are on charcoal and those from Klady are on human bone, which might be affected by old carbon in fish if the Klady people ate a lot of fish. Adjusted for a ^{15}N content of 11%, which would be at the low end of the levels known in the steppes, the *oldest* Klady dates might drop from about 3700–3500 to about 3500–3350 BCE. I follow the traditional view and represent Novosvobodnaya as an outgrowth of Maikop. Rezepkin compared Novosvobodnaya pottery to TRB or Funnel Beaker pottery from Poland, and megalithic porthole graves at Klady to TRB dolmen porthole graves. He sug-gested that Novosvobodnaya began with a migration from Poland. Sergei Korenevskii (1993) tried to bring the two phases back into a single culture. Black burnished pottery is found in central Anatolia at Late Chalcolithic and at EBI sites such as Kösk Höyük and Pinarbişi, a closer alternative source.

31. Shishlina, Orfinskaya, and Golikov 2003.

32. See Kiguradze and Sagona 2003:89, for the beads at Alikemek Tepesi.

33. The Maikop-Novosvobodnaya connections of the Sé Girdan kurgans were noticed by A. D. Rezepkin and B. A. Trifonov; both published Russian-language articles describing these connections in 2000. These were brought to Muscarella's attention in 2002 by Elena Izbitser at the Metropolitan Museum of Art in New York. Muscarella (2003) reviewed this history.

34. For the symbolic power of long-distance trade, see Helms 1992. For primitive valuables, see Dalton 1977; and Appadurai 1986.

35. For the Novosvobodnaya wagon grave, see Rezepkin and Kondrashov 1988:52.

36. Shilov and Bagautdinov 1998.

37. See Nechitailo 1991, for Maikop-steppe contacts. Rassamakin (2002) suggested that Late Tripolye migrants of the Kasperovka type influenced the formation of the Novosvobod-naya culture.

38. Cannabis might have been traded from the steppes to Mesopotamia. Greek *kánnabis* and Proto-Germanic **hanipiz* seem related to Sumerian *kunibu*. Sumerian was dead as a widely

spoken language by about 1700 BCE, so the connection must have been a very ancient one, and the international trade of the Late Uruk period provides a suitable context; see Sherratt 2003, 1997c. Wine could have been a linked commodity; the Greek, Latin, Armenian, and Hittite roots for "wine" are cognates, and some linguists feel that the root was of Semitic or Afro-Asiatic origin. See Hock and Joseph 1996:513.

39. For Caucasian horses, see Munchaev 1982; Mezhlumian 1990; and Chernykh 1992:59. For Norşuntepe and Anatolia, see Bökönyi 1991.

CHAPTER 13. WAGON DWELLERS OF THE STEPPE

1. For climate change at the beginning of the Yamnaya period, see Kremenetski 1997b, 2002.

2. The *ghos-ti-* root survived only in Italic, Germanic, and Slavic, but the institution was more widespread. See Benveniste 1973:273–288 on *Phílos*, and entries in Mallory and Adams 1997 on *guest* and *friend*. Ivanov suggested that Luwian *kaši-* 'visit' might possibly be cognate with Proto-Indo-European *ghos-ti-*, but the relationship was unclear. See Gamkrelidze and Ivanov 1995:657–658, for their discussion of *hospitality*. In later Indo-European societies, this institution was critical for the protection of merchants and visiting elites or nobles; see Kristiansen and Larsson 2005:236–240. See also Rowlands 1980.

3. As Mallory has noted, the eastern Indo-European branches did have some agricultural vocabulary. The eastern Indo-Europeans talked about plowed fields, grain, and chaff. The archaeological contrast between east and west is more extreme than the linguistic one, which perhaps reflects the difference between what people knew and could talk about (language) and how they actually behaved most of the time (archaeology). See entries on *agriculture, field,* and *plow* in Mallory and Adams 1997.

4. For the feminine gender as one of the ten innovations distinguishing classic Proto-Indo-European from the archaic form preserved in Anatolian, see Lehrman 2001. For the Afro-Asiatic loans in western Indo-European, see Hock and Joseph 1996:513. For Rudra's female consorts, see Kershaw 2000:212

5. Gimbutas 1956:70ff. I would never have thought it possible to penetrate the archaeology of Eastern Europe had it not been for this pioneering English-language synthesis, which opened the door. Nevertheless, I soon began to disagree with her; see Anthony 1986. I was very pleased to spend a few days with her in 1991 at a the National Endowment for the Humanities conference in Austin, Texas, organized by Edgar Polomé.

6. The hundred-year anniversary of Gorodtsov's 1903 archaeological expedition on the Northern Donets River was celebrated by three conferences on the Bronze Age (or at least three were planned). The first conference was in Samara in 2001, and the proceedings make a valuable primer on the Bronze Age cultures of the steppes. See Kolev et al. 2001.

7. See Merpert 1974:123–146, for the Yamnaya "cultural-historical community."

8. This steppe-pine-forest vegetation community is designated number 19 in the Atlas SSSR, 1962, edited by S. N. Teplova, 88–89. It occurs both in the lowland and mountain steppe environments.

9. Afanasievo radiocarbon dates are listed in table 13.3. Most of the Afanasievo dates appear to be on wood from the graves, but some are on human bone. Although I have not seen ^{15}N measurements for Afanasievo individuals, later skeletons from graves in the Altai had ^{15}N levels of 10.2 to 14.3%. Applying the correction scale I am using in this book, the Afanasievo dates taken on bone might be too old by 130 to 375 radiocarbon years. I have not corrected them, because, as I said, most appear to have been measured on samples of wood taken from graves, not human bone.

10. V. N. Logvin (1995) noted that some undated flat-grave cemeteries in northern Kazakhstan might represent a short-lived mixture of early Yamnaya or Repin and Botai-Tersek people. For the Karagash kurgan, see Evdokimov and Loman 1989.

11. The pottery in the earliest Yamnaya graves in the Volga-Ural region (Pokrovka cemetery I, k. 15, gr. 2; Lopatino k. 1, gr. 31; Gerasimovka II, k. 4, gr. 2) was Repin-influenced; and the pottery in the earliest Afanasievo kurgans (Bertek 33, Karakol) in the Gorny-Altai region also looks Repin-influenced.

12. For Afanasievo, see Molodin 1997; and Kubarev 1988. On the craniometrics, see Hemphill and Mallory 2003; and Hemphill, Christensen, and Mustafakulov 1997. For the faunal remains from Balyktyul, see Alekhin and Gal'chenko 1995.

13. On the local cultures, see Weber, Link, and Katzenberg 2002; also Bobrov 1988.

14. Chernykh 1992:88; Chernykh, Kuz'minykh, and Orlovskaya 2004.

15. For Tocharian linkages to Afanasievo, see Mallory and Mair 2000.

16. See Gei 2000:176, for the count of all steppe vehicle graves, and for the wagons of the Novotitorovskaya culture. For the Yamnaya wagon grave at Balki kurgan, see Lyashko and Otroshchenko 1988. For the Yamnaya vehicle at Lukyanovka, see Mel'nik and Serdyukova 1988. For the Yamnaya vehicle graves north of the Danube delta, see Gudkova and Chernyakov 1981. The Yamnaya vehicle graves at Shumaevo cemetery II, kurgans 2 and 6, were the first wagon graves found in the Volga-Ural region in decades, excavated by M. A. Turetskii and N. L. Morgunova in 2001–2002. One wheel was recognized in kurgans 6 and three in kurgan 2; see Morgunova and Turetskii 2003. For early wheeled vehicles in general, see Bakker, et al. 1999.

17. Mel'nik and Serdiukova (1988:123) suggested that Yamnaya wagons had no practical use but were purely ritual imitations of vehicles used in the cults of Near Eastern kings. This ascribes to the Yamnaya people more veneration of distant Near Eastern symbols and less practical sense than seems likely to me. It also leaves unexplained the Yamnaya shift to an economy based on mobility. Even if some of the wagons placed in graves *were* lightly built funeral objects, that does not mean that sturdier originals did not exist.

18. Izbitser (1993) asserted that all these steppe vehicles, including those in graves where only two wheels were found, were four-wheeled wagons. Her opinion has been cited in arguments over the origin of the chariot to suggest that the steppe cultures perhaps had no experience making two-wheeled vehicles; see Littauer and Crouwel 1996:936. But many graves contain just two wheels, including Bal'ki kurgan, grave 57. The image on the Novosvobodnaya cauldron at Evdik looks like a cart. Ceramic cart models associated with the Catacomb culture (2800–2200 BCE) and in the North Caucasus at the Badaani site of the ETC or Kura-Araxes culture (3500–2500 BCE) are interpreted by Izbitser as portraying something other than vehicles. Gei, on the other hand, sees evidence for both carts and wagons, as do I. See Gei 2000:186.

19. The Dnieper region of Merpret 1974 was divided into no fewer than six microregions by Syvolap 2001.

20. Telegin, Pustalov, and Kovalyukh 2003.

21. See Sinitsyn 1959; Merpert 1974; and Mallory 1977. For reconsiderations of Merpert's scheme in the light of the discovery of the Khvalynsk culture, see Dremov and Yudin 1992; and Klepikov 1994. For a review of all the early Yamnaya variants in the Volga-Don-Caucasus region, and their chronology, see Vasiliev, Kuznetsov, and Turetskii 2000.

22. Whereas Mikhailovka I produced 1,166 animal bones, Mikhailovka II and III together yielded 52,540 bones.

23. For Yamnaya seed imprints, see Pashkevich 2003. Pashkevich identifies Mikhailovka II as a settlement of the Repin culture, reflecting the debate about its ceramic affiliation referred to in the text; see also Kotova and Spitsyna 2003.

24. For Yamnaya and Catacomb chronology, see Trifonov 2001; Gei 2000; and Telegin, Pustalov, and Kovalyukh 2003. For western Yamnaya and Catacomb dates, see Kośko and Klochko 2003.

25. These views were well stated by Khazanov (1994) and Barfield (1989).

26. For grain cultivation by steppe nomads, see Vainshtein 1980; and DiCosmo 1994. For modern nomads who ate very little grain, see Shakhanova 1989. For the growth of bodyguards into armies, see DiCosmo 1999, 2002.

27. See Shilov 1985b.

28. For a study of seasonal indicators in kurgans in the Kalmyk steppes, see Shishlina 2000. For comments on the Yamnaya herding pattern in the Dnieper steppes, see Bunyatyan 2003.

29. For Samsonova, see Gei 1979. For Liventsovka, see Bratchenko 1969. The predominance of cattle at these places is mentioned in Shilov 1985b:30.

30. Surface scatters of Yamnaya lithics and ceramics in the Manych Depression in Kalmykia are mentioned by Shishlina and Bulatov 2000; and in the lower Volga and North Caspian steppes by Sinitsyn 1959:184. Desert or semi-desert conditions in these places make surface sites more visible than they are in the northern steppes, where the sod hides the ground. In the Samara oblast we found LBA occupations 20–30 cm beneath the modern ground surface; see Anthony et al. 2006. The winter camps of the Blackfeet are described in Ewers 1955:124–126: "Green Grass Bull said that bands whose members owned large horse herds had to move camp several times each winter. . . . However, a short journey of less than a day's march might bring them to a new site possessing adequate resources for another winter camp . . . Demands on fuel and grass were too great to allow all the members of a tribe to winter in one large village." This kind of behavior might make Yamnaya camps hard to find.

31. The Tsa-Tsa grave is described in Shilov 1985a.

32. Yamnaya dental pathologies in the middle Volga region with comparative data from Hsiung-Nu and other cemeteries were studied by Eileen Murphy at Queen's University Belfast as part of the Samara Valley Project. The unpublished internal report is in Murphy and Khokhlov 2004; see also Anthony et al. 2006. For caries in different populations, see Lukacs 1989.

33. For phytoliths in Yamnaya graves, see Shishlina 2000. The yields of *Chenopodium* and einkorn wheat were compared by Smith 1989. *Amaranthus* has 22% more protein (g/kg) than bread wheat, and *Chenopodium* has 34% more; wheat is higher in carbohydrates than either. For nutrient comparisons, see Gremillion 2004.

34. For the high incidence of curbitra orbitalis among Yamnaya skeletons, see Murphy and Khokhlov 2004; and Anthony et al. 2006.

35. For lactose tolerance, see Enattah 2005.

36. See Vainshtein 1980:59, 72, for comments on cows, milk foods, and poverty.

37. Mallory 1990.

38. On genders in Yamnaya graves, see Murphy and Khokhlov 2004; Gei 1990; Häusler 1974; and Mallory 1990.

39. On "Amazon" graves, see Davis-Kimball 1997; and Guliaev 2003.

40. Alexander Gei (1990) estimated a population density of 8–12 people per 100 km[2] in the EBA Novotitorovskaya and 12–14 per 100 km[2] in the MBA Catacomb periods in the Kuban steppes. But kurgans were erected only for a small percentage of those who died, so Gei's figures undercount the actual population density by an order of magnitude. At ten times his grave-based estimate, or about 120 people per 100 km[2], the population density would have been like that of modern Mongolia, where pastoralism is the dominant element in the economy.

41. Golyeva 2000.

42. For the equation between the status and man-days invested in the funeral, see Binford 1971. See also Dovchenko and Rychkov 1988; Mallory's analysis of their study in Mallory 1990; and Morgunova 1995.

43. The granulated decoration on the two golden rings from Utyevka I, kurgan 1, grave 1, is surprising, since the technique of making and applying golden granulation requires very specific skills that first appeared about 2500 BCE (Troy II, Early Dynastic III). The middle Volga was apparently connected with the Troad through some kind of network at this time. The axe in the Utyevka grave is an early type, similar to the axes of Novosvobodnaya and Yamnaya, and that implies a very early Poltavka date. The grave form and artifact assemblage taken together suggested to Vasiliev a date at the late Yamnaya–early Poltavka transition, so probably about 2800 BCE. The grave has not been dated by radiocarbon. For Utyevka I and its analogies, see

Vasiliev 1980. For the Kutuluk grave with the mace, see Kuznetsov 1991, 2005. For an overview, see Chernykh 1992:83–92.

44. Chernykh 1992:83–92.

45. For the Yamnaya grave at Pershin, see Chernykh; and Isto 2002. For the "clean" copper on the Volga, see Korenevskii 1980.

46. For the Post-Mariupol graves, see Ryndina 1998:170–179; for Lebedi, see Chernykh 1992:79–83; and for Voroshilovgrad, see Berezanskaya 1979.

47. For the iron blade, see Shramko and Mashkarov 1993.

48. Oared longboats are not actually portrayed in surviving art until Early Cycladic II, after 2900–2800 BCE, but the number of settled Cycladic Islands jumped from 10% to 90% for the first time in Early Cycladic I, beginning about 3300 BCE. This was possible only with a reliable form of seagoing transport. Longboats capable of holding twenty to forty oarsmen probably appeared earlier than ECII. See Broodbank 1989.

49. For Kemi-Oba graves in the Odessa oblast, see Subbotin 1995. For stone stelae in the North Pontic steppes generally, see Telegin and Mallory 1994.

Chapter 14. The Western Indo-European Languages

1. For a good essay on the subject of language shift, see the introduction in Kulick 1992. For Scots Gaelic, see Dorian 1981; see also Gal 1978.

2. For the Galgenberg site of the Cham culture, see Ottaway 1999. Bökönyi saw the statistical source of the larger horses that appeared in Central Europe in the horse population at Dereivka; Benecke suggested that the horses of Late Mesolithic Mirnoe in the steppes north of the Danube delta were a closer match. But both agreed that the source of the new larger breeds was in the steppes. See Benecke 1994:73–74; and Bökönyi 1974.

3. For the Bukhara horse trade, see Levi 2002. I am indebted to Peter Golden and Ranabir Chakravarti for calling my attention to it.

4. Polomé 1991. For the translation of the *Rig Veda* passage, see O'Flaherty 1981:92.

5. See Kristiansen and Larsson 2005:238.

6. See Benveniste 1973:61–63 for feasts; also see the entry for GIVE in Mallory and Adams 1997:224–225; and Markey 1990. For poets, see Watkins 1995:73–84. For the general importance of feasting in tribal societies, see Dietler and Hayden 2001. For an ethnographic parallel where chiefs and poets were mutually dependent, see Lehman 1989.

7. Mallory (1998) referred to this process using the wry metaphor of the *Kulturkugel*, a bullet of language and culture that acquired a new cultural skin after penetrating a target culture, but retained its linguistic core.

8. A broad scatter of kurgan graves in the steppes contained imported Tripolye C2 pots (among other imported pot types) and a few, like Serezlievka, also contained Tripolye-like schematic rod-headed figurines. The Serezlievka-type graves in the South Bug valley probably were contemporary with Yamnaya graves of the Zhivotilovka-Volchansk group in the Dnieper-Azov steppes that also contained imported Tripolye C2 pots, dated by radiocarbon about 2900–2800 BCE. Rassamakin (1999, 2002) thought that Zhivotilovka-Volchansk graves represented a migration of Tripolye C2 people from the forested upper Dniester deep into the steppes east of the Dnieper. But a Tripolye pot in a Yamnaya grave is most simply interpreted as a souvenir, gift, or acquisition rather than as a migrant Tripolye person. Yamnaya graves rarely contained any pots. Cotsofeni pots filled that customary void in the Yamnaya graves of the Danube valley, just as pottery of the Tripolye C2, late Maikop, and Globular Amphorae types did in the Ukrainian steppes.

9. For the Usatovo culture see Zbenovich 1974; Dergachev 1980; Chernysh 1982; and Patovka et al. 1989. For a history of excavations at Usatovo, see Patovka 1976. The Cernavoda I affiliations of pre-Usatovo coastal steppe kurgans are discussed in Manzura, Savva and Boga-

toya 1995. A Cernavoda I feature in Usatovo is described in Boltenko 1957:42. Recent radiocarbon dates are discussed in Videiko 1999.

10. For Usatovo fauna see Zbenovich 1974: 111–115.

11. For spindle whorls, see Dergachev 1980:106.

12. See Kuz'minova 1990, for Usatovo paleobotany.

13. For Usatovo ceramics, see Zbenovich 1968, with a brief notice of the orange-slipped grey wares on page 54.

14. For trade between Usatovo, late Cernavoda III, and late Maikop, see Zbenovich 1974:103, 141. The single glass bead at Usatovo was colored white by the inclusion of phosphorus. It was in a grave pit covered by a stone lid, a stone cairn, and then by the kurgan. The pear-shaped bead measured 9 mm in diameter, had a hole 5 mm in diameter, and had slightly darker spiraling on its surface. Two cylindrical glass beads, colored with copper (green-blue) were recovered from the Tripolye C2 grave 125 at Sofievka on the Dnieper near Kiev, dated a century or two later, about 3000–2800 BCE (4320+70 BP, 4270+90 BP, 4300+45 BP, from three other graves at Sofievka). Two other glass beads were found on the surface near this grave but certainly were not from it. The glass in both Sofievka and Usatovo was made with ash as an alkali, not soda. An ash recipe was used in the Near East. For analyses, see Ostroverkhov 1985. For the radiocarbon dates from Sofievka and the amber beads from Zavalovka, see Videiko 1999.

15. For the daggers, see Anthony 1996. For oared longboats, see the end of the last chapter of this volume, and Broodbank 1989.

16. For the ochre-painted skulls, see Zin'kovskii and Petrenko 1987.

17. For Zimnea, see Bronicki, Kadrow, and Zakościelna 2003; see also Movsha 1985; and Kośko 1999.

18. For fortifications, see Chernysh 1982:222.

19. See Boyadziev 1995, for the dating of the migration.

20. For the large cluster in Hungary, see Ecsedy 1979, 1994. For the cluster in Oltenia, see Dumitrescu 1980. For the cluster in northern Serbia, see Jovanovich 1975. For Bulgaria, see Panayotov 1989. For overviews see, Nikolova 2000, 1994. For relative chronologies at the time of the migration event in southeastern Europe generally, see Parzinger 1993. For the wagon grave at Plachidol, see Sherratt 1986. For the stone stelae, see Telegin and Mallory 1994. Ecsedy mentions that undecorated stone stelae were found near Yamnaya kurgans in Hungary.

21. The graves in Hungary could possibly have been the result of a separate migration stream that passed directly over the Carpathians through Late Tripolye territory rather than being a continuation of the lower Danube valley stream.

22. Most of the radiocarbon dates for Yamnaya graves in the Odessa oblast, the heart of the Dniester steppes, are quite late, beginning about 2800–2600 BCE, by which time the Usatovo culture was gone. There are a few earlier radiocarbon dates (Semenovskii, k.11, 14; Liman, k.2; Novoseltsy, k.19), but in both of the Semenovskii kurgans the primary grave for which the kurgan was raised was an Usatovo grave, and all the Yamnaya graves were secondary. The stratigraphy makes me wonder about the early radiocarbon dates. Yamnaya seems to have taken over the Odessa oblast steppes after the Usatovo culture. See Gudkova and Chernyakov 1981; and Subbotin 1985.

23. Kershaw 2000; see also entries on *korios* and warfare in Mallory and Adams 1997. The cattle raid, a related institution, is discussed in Walcot 1979.

24. For Yamnaya dog-tooth ornaments on the Ingul, see Bondar and Nechitailo 1980.

25. For the stelae of the steppes, see Telegin and Mallory 1994. For the symbolic importance of belts, see Kershaw 2000:202–203; and Falk 1986:22–23.

26. Kalchev 1996.

27. Nikolova 1996.

28. Alexandrov 1995.

29. Panayotov 1989:84–93.

30. Barth 1965:69.

31. Bell Beaker decorated cup styles, domestic pot types, and grave and dagger types from the middle Danube were adopted about 2600 BCE in Moravia and Southern Germany. This material network could have been the bridge through which pre-Celtic dialects spread into Germany. See Heyd, Husty, and Kreiner 2004, especially the final section by Volker Heyd.

32. See Hamp 1998; and Schmidt 1991, for connections between Italic and Celtic.

33. For the effects of wheeled vehicles, see Maran 2001.

34. See Szmyt 1999, esp. 178–188.

35. On the Slavic homeland, see Darden 2004.

36. Coleman (2000) argued that Greek speakers entered Greece during the Final Neolithic/Bronze Age transition, about 3200 BCE. If an Indo-European language spread into Greece this early I think it was more likely an Anatolian-type language. For a northern steppe origin for Greek, but in a later era more amenable to my scenario, see Lichardus and Vladar 1996; and Penner 1998. The same evidence is marshaled for another purpose in Makkay 2000, and in detail by Kristiansen and Larsson 2005. Another argument for a northern connection of the Shaft Grave princes is presented in Davis 1983. Connections between southeastern Europe and Greece are outlined in Hänsel 1982. Robert Drews (1988) also argued that the Shaft Grave princes were an immigrant dynasty from the north, although he derived them from Anatolia.

37. Mallory 1998:180.

Chapter 15. Chariot Warriors of the Northern Steppes

1. See Gening, Zdanovich, and Gening 1992, for the original report on Sintashta.

2. The Sintashta culture remained unrecognized as recently as 1992. Chernykh (1992:210–234) discussed Sintashta-type metals as part of the "Andronovo historico-cultural community," assigning it to about 1600–1500 BCE. Dorcas Brown and I visited Nikolai Vinogradov in 1992, and I was permitted to take bone samples from the chariot grave at Krivoe Ozero for radiocarbon dating. This resulted in two articles: Anthony 1995a; and Anthony and Vinogradov 1995. See Vinogradov 2003, for the complete report on the Krivoe Ozero cemetery. For the settlement and cemeteries at Arkaim, see Zdanovich 1995; and Kovaleva and Zdanovich 2002. For the Sintashta cemetery at Kammeny Ambar, see Epimakhov 2002. For a wide-ranging overview, see Grigoriev 2002, marred by the assumption that the Sintashta culture and many other steppe cultures originated from a series of south-to-north folk migrations from Anatolia and Syria, where he argued that the Indo-European homeland was located. See Lamberg-Karlovsky 2002, for connections to Central Asia. For conference proceedings, see Jones-Bley and Zdanovich 2002; Boyle, Renfrew, and Levine 2002; and Levine, Renfrew, and Boyle 2003.

3. I use the term *Aryan* here as it is defined it in chapter 1, as the self-designation of the people who composed the hymns and poems of the *Rig Veda* and *Avesta* and their immediate Indo-Iranian ancestors.

4. For the contact zone between Corded Ware, Globular Amphorae, and Yamnaya at about 2800–2600 BCE, see Szmyt 1999, esp. pp. 178–188. Also see Machnik 1999; and Klochko, Kośko, and Szmyt 2003. A classic review of the archaeological evidence for mixed Yamnaya, late Tripolye (Chapaevka), and Corded Ware elements in Middle Dnieper origins is Bondar 1974. A recent review emphasizes the Yamnaya influence on the Middle Dnieper culture, in Telegin 2005.

5. For Middle Dnieper chronology, see Kryvaltsevich and Kovalyukh 1999; and Yazepenka and Kośko 2003.

6. Machnik 1999.

7. Before the Middle Dnieper culture appeared, the east side of the river near Kiev had been occupied between about 3000 and 2800 BCE by the mixed-origin late Tripolye C2 Sofievka group, which cremated its dead, used riveted daggers like those at Usatovo, and made

pottery that showed both cord-impressed steppe elements and late Tripolye elements. For the Sofievka settlement, see Kruts 1977:109–138; for radiocarbon dates, see Videiko 1999.

8. See Carpelan and Parpola 2001. This almost monograph-length article covers much of the subject matter discussed in this chapter. For Corded Ware migrations from the genetic point of view, see Kasperavičiūtė, Kučinskas, and Stoneking 2004.

9. For Balanovo, Abashevo, and Volosovo, see Bol'shov 1995. For Abashevo ceramics, see Kuzmina 1999. The classic work on Abashevo is Pryakhin 1976, updated in Pryakhin 1980. For an English account, in addition to Carpelan Parpola and 2001, see Chernykh 1992:200–204 and Koryakova and Epimakhov 2007.

10. For the Volosovo culture, see Korolev 1999; Vybornov and Tretyakov 1991; and Bakharev and Obchinnikova 1991.

11. For Abashevo and Indo-Iranian linkages, see Carpelan and Parpola 2001; and Pryakhin 1980.

12. For the headbands, see Bol'shov 1995.

13. See Keeley 1996, on tribal war.

14. See Koivulehto 2001; and Carpelan and Parpola 2001.

15. See Ivanova 1995:175–176, for the Aleksandrovska IV kurgan cemetery.

16. For Kuisak settlement, see Maliutina and Zdanovich 1995.

17. In Table 1, sample AA 47803, dated ca. 2900–2600 BCE, was from a human skeleton of the Poltavka period that was later cut through and decapitated by a much deeper Potapovka grave pit. A horse sacrifice above the Potapovka grave is dated by sample AA 47802 to about 1900–1800 BCE. Although they were almost a thousand years apart, they looked, on excavation, like they were deposited together, with the Potapovka horse skull lying above the shoulders of the decapitated Poltavka human. Before dates were obtained on both the horse and the skeleton this deposit was interpreted as a "centaur"—a decapitated human with his head replaced by the head of a horse, an important combination in Indo-Iranian mythology. But Nerissa Russell and Eileen Murphy found that both the horse and the human were female, and the dates show that they were buried a thousand years apart. Similarly sample AA-12569 was from an older Poltavka-period dog sacrifice found on the ancient ground surface at the edge of Potapovka grave 6 under kurgan 5 at the same cemetery. Older Poltavka sacrifices and graves were discovered under both kurgans 3 and 5 at Potapovka cemetery I. The Poltavka funeral deposits were so disturbed by the Potapovka grave diggers that they remained unrecognized until the radiocarbon dates made us take a second look. The "centaur" possibility was mentioned in Anthony and Vinogradov 1995, five or six years before the two pieces were dated. Of course, it now must be abandoned.

18. For Sarazm, see Isakov 1994.

19. For Kelteminar, see Dolukhanov 1986; and Kohl, Francfort, and Gardin 1984. The classic work on Kelteminar is Vinogradov 1981.

20. For a radiocarbon date from Sergeivka, see Levine and Kislenko 2002, but note that their discussion mistakenly assigns it to the Andronovo period, 1900–1700 BCE. See also Kislenko and Tatarintseva 1990. Another transitional forager-herder group influenced by Poltavka was the Vishnevka 1 pottery group in the forest-steppe on the northern Ishim; see Tatarintseva 1984. For Sergeivka sherds at the Poltavka cemetery of Aleksandrovka, see Maliutina and Zdanovich 1995:105.

21. For climate deterioration, see Blyakharchuk et al. 2004; and Kremenetski 2002, 1997a, 1997b.

22. Rosenberg 1998.

23. For the Mesopotamian metal trade, see Muhly 1995; Potts 1999:168–171, 186.

24. For metals and mining, see Grigoriev 2002:84; and Zaikov, Zdanovich, and Yuminov 1995. See also Kovaleva and Zdanovich 2002. Grigoriev suggested that the amount of slag found in each house was so small that it could represent household production. However, slag is often found in small amounts even at industrial sites, and that all houses contained slag and

production facilities (ovens with attached wells that aided in the updraft) shows an intensity of metal production that was unprecedented in the steppes.

25. See DiCosmo 1999, 2002; and Vehik 2002.

26. Ust'e, like Chernorech'e III, was excavated by Nikolai Vinogrado. Vinogradov was kind enough to show me his plans and photographs from Ust'e, where Sintashta houses are clearly stratified beneath a Petrovka occupation.

27. See Epimakhov 2002:124–132 for the artifact catalogue.

28. For the ballistics of flint projectile points, see Knecht 1997; and Van Buren 1974. For javelins in Greek chariot warfare, see Littauer 1972; and Littauer and Crouwel 1983.

29. For the chariot petroglyphs, see Littauer 1977; Samashev 1993; and Jacobsen-Tepfer 1993. On the derivation of steppe cheekpieces from Mycenaean cheekpieces, see E. Kuzmina 1980. For a review of European cheekpieces, see Hüttel 1992. Littauer and Crouwel (1979) argued persuasively for the Near Eastern origin of the chariot, overthrowing pre-World War II suggestions that the chariot was a super-weapon of the steppe Aryans. Piggott (1983, 1992) began to challenge the Near Eastern origin hypothesis almost immediately. Moorey (1986) also supported a multiregional invention of the various elements combined in the chariot.

30. See Epimakhov 2002:124–132 for a grave inventory that totals sixteen chariot graves; see Kuzmina 2001:12 for an estimate of twenty. The sites Kuzmina lists include Sintashta (seven chariot graves), Kamenny Ambar (two), Solntse II (three), Krivoe Ozero (three), and, in northern Kazakhstan, in Petrovka graves, Ulybai (one), Kenes (one), Berlyk II (two), and Satan (one).

31. For arguments against the functionality of steppe chariots, see Littauer and Crouwel 1996; Jones-Bley 2000; and Vinogradov 2003:264, 274. For arguments in favor of the steppe chariots as effective instruments of war, see Anthony and Vinogradov 1995; and Nefedkin 2001.

32. For English descriptions of the narrow-gauge chariots, see Gening 1979; Anthony and Vinogradov 1995; and Anthony 1995a. For two critical replies, see Littauer and Crouwel 1996; and Jones-Bley 2000. For the limitations of the chariot in battle, see Littauer 1972; and Littauer and Crouwel 1983.

33. For Bronze Age steppe bows, see Grigoriev 2002:59–60; Shishlina 1990; Malov 2002; and Bratchenko 2003:199. For ancient bows of the Near East and Iran, see Zutterman 2003.

34. See Littauer 1968.

35. For the disk cheekpieces, see Priakhin and Besedin 1999; Usachuk 2002; and Kuzmina 2003, 1980. For left and right side differences, see Priakhin and Besedin 1999:43–44. For chariots in the *Rig Veda*, see Sparreboom 1985. For the metal examples in the Levant, see Littauer and Crouwel 1986, 2001. This type of cheekpiece probably spread into Mycenaean Greece from southeastern Europe, where it appeared in Otomani, Monteoru, and Vatin contexts. For radiocarbon dates for these cultures, see Forenbaher 1993, and for disk-shaped cheekpieces in those contexts, see Boroffka 1998, and Hüttel 1994. The European origin of Mycenaean chariotry might explain why Mycenaean chariot warriors, like the early charioteers of the northern steppes, sometimes carried spears or javelins. For chariots in Greece, see Crouwel 1981.

36. For a review of the Near Eastern evidence for chariots, see Oates 2003; for older studies, see Moorey 1986, and Littauer and Crouwel 1979. For vehicles at Tell Brak, see Oates 2001:141–154. If we were to accept the "low" chronology, which seems increasingly likely, the date for the end of Ur III and the earliest proto-chariots would shift down from 2000 to 1900 BCE. See Reade 2001.

37. See Stillman and Tallis 1984:25 for Mitanni chariot squadrons; for Chinese chariot squadrons, see Sawyer 1993:5.

38. See Appuradai 1986:21 for the "tournament of values."

39. For human pathologies, see Lindstrom 2002, who notes the complete absence of dental caries, even in the oldest individuals (161). Lindstrom was the first Western archaeologist to participate in excavations at a Sintashta site.

40. Igor Ivanov, a geomorphologist at Arkaim, told me in 2000 that the reports of irrigation channels at Arkaim were mistaken, that these were natural features.

41. See Gening, Zdanovich, and Gening 1992:234–235 for Sacrificial Complex 1, and page 370 for the man-days for the SB kurgan.

42. For feasting in tribal societies, see Hayden 2001.

43. For the fauna, see Kosintsev 2001; and Gaiduchenko 1995. For $^{N}15$ isotopes in human and animal bones, see Privat 2002.

44. For doubts about social hierarchy in Sintashta society, see Epimakhov 2000:57–60.

45. Witzel 1995:109, citing Kuiper 1991.

46. For various theories on how to link Sintashta and the Indo-Iranians, see Parpola 1988, 2004–2005; E. Kuzmina 1994, 2001; and Witzel 2003.

47. All quotations are from O'Flaherty 1981.

48. For the Indo-European dog sacrifice and New Year initiation ceremony, see Kershaw 2000; and Kuiper 1991, 1960.

49. Epimakhov 2002; and Anthony et al. 2005.

CHAPTER 16. THE OPENING OF THE EURASIAN STEPPES

1. For exotic knowledge and power, see Helms 1992.

2. For Indic terms among the Mitanni, see chapter 3; Thieme 1960; and Burrow 1973.

3. Elamite was a non–Indo-European language of uncertain affiliations. As Dan Potts stressed, the people of the western Iranian highlands never used this or any other common term as a blanket ethnic designation for themselves. They did not even all speak Elamite. See Potts 1999:2–4. For the appearance of horses, see Oates 2003.

4. See Weiss 2000; also Perry and Hsu 2000.

5. At Godin Tepe, onagers were 94% of the equid bones. A cheektooth and a metacarpal from Godin IV, dated about 3000–2800 BCE, might be horse. The first clear and unambiguous horse bones at Godin appeared in period III, dated 2100–1900 BCE; see Gilbert 1991. On horses and mules at Malyan, see Zeder 1986. The bit wear at Malyan is the earliest unambiguous bit wear in the Near East. Copper stains reported on the P_2s of asses from Tell Brak, dated 2300–2000 BCE, might have had another cause (perhaps corroded lip rings). See Clutton-Brock 2003.

6. Owen 1991.

7. The phrase *Fahren und Reiten,* or "To drive and to ride," appeared between 1939 and 1968 in the titles of three influential publications by Joseph Weisner, and the order of terms in this phrase—driving *before* riding—has become a form of shorthand referring to the historical priority of the chariot over the ridden horse in the Bronze Age civilizations of the Near East. Certainly wheeled vehicles preceded horseback riding in the Near East, and horse-drawn chariots dominated Near Eastern warfare long before cavalry, but this was not because riding was invented after chariotry (see chapter 10). If images of horseback riding can now be dated before 1800 BCE, as seems to be the case, they preceded the appearance of horses with chariots in Near Eastern art. See Weisner 1939, 1968; Drews 2004:33–41, 52; and Oates 2003.

8. For Zimri-Lim's adviser's advice, see Owen 1991; n. 12.

9. For tin sources, see Muhly 1995:1501–1519; Yener 1995; and Potts 1999:168–171, 186. For Eneolithic Serbian tin-copper alloys, see Glumac and Todd 1991. For the possible mistranslation of the Gudea inscription I am indebted to Chris Thornton, and, through him, to Greg Possehl and Steven Tinney. For the seaborne tin trade in the Arabian Gulf, see Weeks 1999; and for the Bactrian comb at Umm-al-Nar, see Potts 2000:126. For Harappan metals, see Agrawal 1984.

10. The polymetallic ores of the Zeravshan probably produced the metals of Ilgynly-Depe, near Anau, during the fourth millennium BCE. At Ilgynly, among sixty-two copper artifacts, primarily tanged knives, one object contained traces of tin; see Solovyova et al. 1994. For tin

bronzes in early third-millennium Namazga IV, see Salvatori et al. 2002. For Sarazm, see Isakov 1994; for its radiocarbon dates and metals, see Isakov, et al. 1987.

11. For the tin mines of the Zeravshan, see Boroffka et al. 2002; and Parzinger and Boroffka 2003.

12. Zaman Baba graves have been seen as a hybrid between Kelteminar and Namazga V/VI-type cultures, see Vinogradov 1960:80–81; and as a hybrid with Catacomb cultures on the supposition that Catacomb-culture people migrated to Central Asia, see Klejn 1984. I support the former. For recent debates over Zaman Baba, see E. Kuzmina 2003:215–216.

13. Lyonnet (1996) sees Sarazm IV ending during Namazga IV, or during the middle of the third millennium BCE. I see Sarazm ending in late Namazga V/early VI, based on the co-occurrence of Petrovka and late Sarazm pottery at Tugai, and on radiocarbon dates indicating that Sarazm III was occupied in 2400–2000 BCE, so Sarazm IV had to be later.

14. For skull type affiliations, see Christensen, Hemphill, and Mustafakulov 1996.

15. For BMAC, see Hiebert 1994, 2002. Salvatori (2000) disagreed with Hiebert, suggesting that BMAC began much earlier than 2100 BCE, and grew from local roots, not from an intrusion from the south, making the growth of BMAC more gradual. For the BMAC graves at Mehrgarh VIII, see Jarrige 1994. For BMAC materials in the Arabian Gulf, see Potts 2000, During Caspers 1998; and Winckelmann 2000.

16. For tin-bronzes in Bactria and lead-copper alloys in Margiana, see Chernykh 1992:176–182; and Salvatori et al. 2002. For the lead ingot at Sarazm, see Isakov 1994:8. For the Iranian background, see Thornton and Lamberg-Karlovsky 2004.

17. For horse bones in BMAC, see Salvatori 2003; and Sarianidi 2002. For the BMAC seal with the rider, see Sarianidi 1986. A few horses might have passed through the Caucasus into western Iran before 3000 BCE, indicated by a few probable horse teeth at the site of Qabrestan, west of Teheran; see Mashkour 2003. No definite horse remains have been identified in eastern Iran or the Indian subcontinent dated earlier than 2000 BCE. See Meadow and Patel 1997.

18. For the steppe sherds in BMAC sites, see Hiebert 2002. For the "Abashevo-like"sherds at Karnab, see Parzinger and Boroffka 2003:72, and Figure 49.

19. For Tugai, see Hiebert 2002; E. Kuzmina 2003; and the original report, Avanessova 1996. The talc temper in two pots, an indication that they were made in the South Ural steppes, is described in Avanessova 1996:122.

20. For Zardcha Khalifa, see Bobomulloev 1997; and E. Kuzmina 2001, 2003:224–225.

21. For the lead wires at Kuisak, see Maliutina and Zdanovich 1995:103. For the lapis bead and the grave at Krasnoe Znamya, see E. Kuzmina 2001:20.

22. For Srubnaya subsistence, see Bunyatyan 2003; and Ostroshchenko 2003.

23. For Chenopodium yields, see Smith 1989:1569.

24. For the Samara Valley Project, see Anthony et al. 2006. The results obtained here were replicated at Kibit, another Srubnaya settlement in Samara Oblast, excavated by L. Popova and D. Peterson, where there was no cultivated grain and many seeds of *Chenopodium*.

25. For the enormous Srubnaya mining center at Kargaly, see Chernykh 1997, 2004. For the mining center in Kazakhstan near Atasu, see Kadyrbaev and Kurmankulov 1992.

26. For stratigraphic relationships between Sintashta and Petrovka, see Vinogradov 2003; and Kuzmina 2001:9. The Petrovka culture was a transitional culture marking the beginning of the LBA. For Petrovka and its stratigraphic relationships to Alakul and Federovo, see Maliutina 1991. I would like to acknowledge the difficulty of keeping all these P-k cultures straight: on the middle Volga the MBA Poltavka culture evolved into final MBA Potapovka and then into early LBA Pokrovka, which was contemporary with early LBA Petrovka in Kazakhstan.

27. For the north-south movements of nomads in Kazakhstan, see Gorbunova 1993/94.

28. See Grigoriev 2002:78–84, for Petrovka metals.

29. For the Rostovka cemetery, see Matiushchenko and Sinitsyna 1988. For general discussions in English, see Chernykh 1992:215–234; and Grigoriev 2002:192–205.

30. For Seima-Turbino hollow-cast bronze casting and its influence on early China through the Qijia culture of Gansu province, see Mei 2003a, 2003b; and Li 2002. See also Fitzgerald-Huber 1995 and Linduff, Han, and Sun 2000.

31. See Epimakhov, Hanks, and Renfrew 2005 for dates. Seima-Turbino might possibly have begun west of the Urals and spread eastward. Sintashta fortifications might then be seen as a reaction to the emergence of Seima-Turbino warrior bands in the forest zone, but this is a minority position; see Kuznetsov 2001.

32. For Alakul and Federovo elements on the same pot, see Maliutina 1984; for the stratigraphic relations between the two, see Maliutina 1991. For radiocarbon dates, see Parzinger and Boroffka 2003:228.

33. E. Kuzmina 1994:207–208.

34. For Andronovo mines near Karaganda, see Kadyrbaev and Kurmankulov 1992; for mines near Dzhezkazgan, see Zhauymbaev 1984. For the estimate of copper production, see Chernykh 1992:212

35. For the Namazga VI pottery at Pavlovka, see Maliutina 1991:151–159.

36. For Andronovo sites in the Zeravshan, see Boroffka et al. 2002. For Tazabagyab sites on the former Amu-Darya delta, see Tolstov and Kes' 1960:89–132.

37. Hiebert 2002.

38. For the post-BMAC pastoral groups who made coarse incised ware, see Salvatori 2003:13; also Salvatori 2002. For the Vaksh and Bishkent groups, see Litvinsky and P'yankova 1992.

39. See Witzel 1995.

40. Books 2 and 4 of the *Rig Veda* referred to places in eastern Iran and Afghanistan. Book 6 described two clans who claimed they had come from far away, crossed many rivers, and gone through narrow passages, fighting indigenous people referred to as *Dasyus*. These details suggest that the Aryans fought their way into the Indian subcontinent from eastern Iran and Afghanistan. Although some new elements such as horses can be seen moving from Central Asia into the Indian subcontinent at this time, and intrusive pottery styles can be identified here or there, no single material culture spread with the Old Indic languages. For discussions, see Parpola 2002; Mallory 1998; and Witzel 1995:315–319.

41. For *Indra* and *Soma* as loan words, see Lubotsky 2001. Indra combined attributes that originally were separate: the mace was Mithra's; some of his epithets, his martial power, and perhaps his ability to change form were Verethraghna's; and the slaying of the serpent was the feat of the hero Thrataona, the Third One. The Old Indic poets gave these Indo-Iranian traits to Indra. The most prominent aspect of Indo-Iranian Verethraghna, the god of might/victory, was his shape-shifting ability, especially his form as the Boar. See Malandra 1983:80–81.

42. V. Sarianidi proposed that the people of the BMAC spoke Iranian. Sarianidi suggested that "white rooms" inside the walled buildings at Togolok 21, Togolok 1, and Gonur were fire temples like those of the Zoroastrians, with vessels containing *Ephedra*, *Cannabis*, and poppy seeds, which he equated with *Soma* (RV) or *Haoma* (AV). But examinations of the seed and stem impressions from the "white rooms" at Gonur and Togolok 21 by paleobotanists at Helsinki and Leiden Universities proved that the vessels contained no *Cannabis* or *Ephedra*. Instead the impressions probably were made by millet seeds and stems (*Panicum miliaceum*); see Bakels 2003. The BMAC culture makes a poor match with Indo-Iranian. The BMAC people lived in brick-built fortified walled towns, depended on irrigation agriculture, worshiped a female deity who was prominent in their iconography (a goddess with a flounced skirt), had few horses, no chariots, did not build kurgan cemeteries, and did not place carefully cut horse limbs in their graves.

43. Li 2002; and Mei 2003a.

Chapter 17. Words and Deeds

1. See Diamond 1997.

2. Hobsbawm 1997:5–6: "For history is the raw material for nationalist or ethnic or fundamentalist ideologies, as poppies are the raw material for heroin addiction. . . . This state of affairs affects us in two ways. We have a responsibility for historical facts in general and for criticizing the politico-ideological abuse of history in particular."

3. O'Flaherty 1981:69.

REFERENCES

Agapov, S. A., I. B. Vasiliev, and V. I. Pestrikova. 1990. *Khvalynskii Eneoliticheskii Mogil'nik.* Saratov: Saratovskogo universiteta.

Agrawal, D. P. 1984. Metal technology of the Harappans. In *Frontiers of the Indus Civilization*, ed. B. B. Lal and S. P. Gupta, pp. 163–167. New Delhi: Books and Books, Indian Archaeological Society.

Akhinzhalov, S. M., L. A. Makarova, and T. N. Nurumov. 1992. *K Istorii Skotovodstva i Okhoty v Kazakhstane.* Alma-Ata: Akademiya nauk Kazakhskoi SSR.

Alekhin, U. P., and A. V. Gal'chenko. 1995. K voprosu o drevneishem skotovodstve Altaya. In *Rossiya i Vostok: Problemy Vzaimodeistviya*, pt. 5, bk. 1: *Kul'tury Eneolita-Bronzy Stepnoi Evrazii*, pp. 22–26. Chelyabinsk: 3-ya Mezhdunarodnaya nauchnaya konferentsiya.

Alekseeva, I. L. 1976. O drevneishhikh Eneoliticheskikh pogrebeniyakh severo-zapadnogo prichernomor'ya. In *Materialy po arkheologii severnogo prichernomor'ya* (Kiev) 8:176–186.

Alexandrov, Stefan. 1995. The early Bronze Age in western Bulgaria: Periodization and cultural definition. In *Prehistoric Bulgaria*, ed. Douglass W. Bailey and Ivan Panayotov, pp. 253–270. Monographs in World Archaeology 22. Madison, Wis.: Prehistory Press.

Algaze, G. 1989. The Uruk Expansion: Cross-cultural exchange in Early Mesopotamian civilization. *Current Anthropology* 30:571–608.

Alvarez, Robert R., Jr. 1987. *Familia: Migration and Adaptation in Baja and Alta California, 1800–1975.* Berkeley: University of California Press.

Amiet, Pierre. 1986. *L'Âge des Échanges Inter-Iraniens 3500–1700 Avant J-C.* Paris: Editions de la Réunion des Musées Nationaux.

Andersen, Henning. 2003. Slavic and the Indo-European migrations. In *Language Contacts in Prehistory: Studies in Stratigraphy*, ed. Henning Andersen, pp. 45–76. Amsterdam and Philadelphia: Benjamins.

Antilla, R. 1972. *An Introduction to Historical and Comparative Linguistics.* New York: Macmillan.

Anthony, David W. 2001. Persistent identity and Indo-European archaeology in the western steppes. In *Early Contacts between Uralic and Indo-European: Linguistic and Archaeological Considerations*, ed. Christian Carpelan, Asko Parpola, and Petteri Koskikallio, pp. 11–35. Memoires de la Société Finno-Ugrienne 242. Helsinki: Suomalais-Ugrilainen Seura.

———. 1997. "Prehistoric migration as social process." In *Migrations and Invasions in Archaeological Explanation*, ed. John Chapman and Helena Hamerow, pp. 21–32. British Archaeological Reports International Series 664. Oxford: Archeopress.

———. 1996. V. G. Childe's world system and the daggers of the Early Bronze Age. In *Craft Specialization and Social Evolution: In Memory of V. Gordon Childe*, ed. Bernard Wailes, pp. 47–66. Philadelphia: University of Pennsylvania Museum Press.

———. 1995a. Horse, wagon, and chariot: Indo-European languages and archaeology. *Antiquity* 69 (264): 554–565.

———. 1995b. Nazi and Ecofeminist prehistories: ideology and empiricism in Indo-European archaeology. In *Nationalism, Politics, and the Practice of Archaeology*, ed. Philip Kohl and Clare Fawcett, pp. 82–96. Cambridge: Cambridge University Press.

———. 1994. On subsistence change at the Mesolithic-Neolithic transition in Ukraine. *Current Anthropology* 35 (1): 49–52.

———. 1991a. The archaeology of Indo-European origins. *Journal of Indo-European Studies* 19 (3–4): 193–222.

———. 1991b. The domestication of the horse. In *Equids in the Ancient World*, vol. 2, ed. Richard H. Meadow and Hans-Peter Uerpmann, pp. 250–277. Weisbaden: Verlag.

———. 1990. Migration in archaeology: The baby and the bathwater. *American Anthropologist* 92 (4): 23–42.

———. 1986. The "Kurgan Culture," Indo-European origins, and the domestication of the horse: A reconsideration. *Current Anthropology* 27:291–313.

Anthony, David W., and Dorcas Brown. 2003. Eneolithic horse rituals and riding in the steppes: New evidence. In *Prehistoric Steppe Adaptation and the Horse*, ed. Marsha Levine, Colin Renfrew, and Katie Boyle, pp. 55–68. Cambridge: McDonald Institute for Archaeological Research.

———. 2000. Eneolithic horse exploitation in the Eurasian steppes: Diet, ritual, and riding. *Antiquity* 74:75–86.

———. 1991. The origins of horseback riding. *Antiquity* 65:22–38.

Anthony, David W., D. Brown, E. Brown, A. Goodman, A. Kokhlov, P. Kosintsev, P. Kuznetsov, O. Mochalov, E. Murphy, D. Peterson, A. Pike-Tay, L. Popova, A. Rosen, N. Russel, and A. Weisskopf. 2005. The Samara Valley Project: Late Bronze Age economy and ritual in the Russian steppes. *Eurasia Antiqua* 11:395–417.

Anthony, David W., Dorcas R. Brown, and Christian George. 2006. Early horseback riding and warfare: The importance of the magpie around the neck. In *Horses and Humans: The Evolution of the Equine-Human Relationship*, ed. Sandra Olsen, Susan Grant, Alice Choyke, and László Bartosiewicz. pp. 137–156. British Archaeological Reports International Series 1560. Oxford: Archeopress.

Anthony, David W., Dimitri Telegin, and Dorcas Brown. 1991. The origin of horseback riding. *Scientific American* 265:94–100.

Anthony, David W., and Nikolai Vinogradov. 1995. The birth of the chariot. *Archaeology* 48 (2): 36–41.

Anthony, David W., and B. Wailes. 1988. CA review of *Archaeology and Language* by Colin Renfrew. *Current Anthropology* 29 (3): 441–445.

Appadurai, Arjun. 1986. Introduction: Commodities and the politics of value. In *The Social Life of Things: Commodities in Cultural Perspective*, ed. Arjun Appadurai, pp. 3–63. Cambridge: Cambridge University Press.

Armstrong, J. A. 1982. *Nations before Nationalism*. Chapel Hill: University of North Carolina Press.

Arnold, Bettina. 1990. The past as propaganda: Totalitarian archaeology in Nazi Germany. *Antiquity* 64:464–478.

Aruz, Joan. 1998. Images of the supernatural world: Bactria-Margiana seals and relations with the Near East and the Indus. *Ancient Civilizations from Scythia to Siberia* 5 (1): 12–30.

Atkinson, R. R. 1994. *The Roots of Ethnicity: The Origins of the Acholi of Uganda before 1800*. Philadelphia: University of Pennsylvania Press.

———. 1989. The evolution of ethnicity among the Acholi of Uganda: The precolonial phase. *Ethnohistory* 36 (1): 19–43.

Avanessova, N. A. 1996. Pasteurs et agriculteurs de la vallée du Zeravshan (Ouzbekistan) au début de l'age du Bronze: relations et influences mutuelles. In B. Lyonnet, *Sarazm (Tadjikistan) Céramiques (Chalcolithique et Bronze Ancien)*, pp. 117–131. Paris: Mémoires de la Mission Archéologique Française en Asie Centrale Tome 7.

Azzaroli, Augusto. 1980. Venetic horses from Iron Age burials at Padova. *Rivista di Scienze Preistoriche* 35 (1–2): 282–308.

Bahn, Paul G. 1980. "Crib-biting: Tethered horses in the Palaeolithic?" *World Archaeology* 12:212–217.

Bailey, Douglass W. 2000. *Balkan Prehistory: Exclusion, Incorporation, and Identity*. London: Routledge.

Bailey, Douglass W., R. Andreescu, A. J. Howard, M. G. Macklin, and S. Mills. 2002. Alluvial landscapes in the temperate Balkan Neolithic: Transitions to tells. *Antiquity* 76:349–355.

Bailey, Douglass W., and Ivan Panayotov, eds. 1995. Monographs in World Archaeology 22. *Prehistoric Bulgaria*. Madison, Wis.: Prehistory Press.

Bailey, Douglass W., Ruth Tringham, Jason Bass, Mirjana Stefanović, Mike Hamilton, Heike Neumann, Ilke Angelova, and Ana Raduncheva. 1998. Expanding the dimensions of early agricultural tells: The Podgoritsa archaeological project, Bulgaria. *Journal of Field Archaeology* 25:373–396.

Bakels, C. C. 2003. The contents of ceramic vessels in the Bactria-Margiana Archaeological Complex, Turkmenistan. *Electronic Journal of Vedic Studies* 9 (1).

Bakharev, S. S., and N. V. Obchinnikova. 1991. Chesnokovskaya stoiankana na reke Sok. In *Drevnosti Vostochno-Evropeiskoi Lesotepi*, ed. V. V. Nikitin, pp. 72–93. Samara: Samarskii gosudartsvennyi pedagogicheskii institut.

Bakker, Jan Albert, Janusz Kruk, A. L. Lanting, and Sarunas Milisauskas. 1999. The earliest evidence of wheeled vehicles in Europe and the Near East. *Antiquity* 73:778–790.

Baldi, Philip. 1983. *An Introduction to the Indo-European Languages*. Carbondale: Southern Illinois University Press.

Balter, Michael. 2003. Early date for the birth of Indo-European languages. *Science* 302 (5650): 1490–1491.

Bánffy, Ester. 1995. South-west Transdanubia as a mediating area: on the cultural history of the early and middle Chalcolithic. In *Archaeology and Settlement History in the Hahót Basin, South-West Hungary*, ed. Béla Miklós Szőke. Antaeus 22. Budapest: Archaeological Institute of the Hungarian Academy of Sciences.

Bar-Yosef, Ofer. 2002. The Natufian Culture and the Early Neolithic: Social and Economic Trends in Southwestern Asia. In *Examining the Farming/Language Dispersal Hypothesis*, ed. Peter Bellwood and Colin Renfrew, pp. 113–126. Cambridge: McDonald Institute for Archaeological Research.

Barber, Elizabeth J. W. 2001. The clues in the clothes: Some independent evidence for the movement of families. In *Greater Anatolia and the Indo-Hittite Language Family*, ed. Robert Drews, pp. 1–14. Journal of Indo-European Studies Monograph 38. Washington, D.C.: Institute for the Study of Man.

———. 1991. *Prehistoric Textiles*. Princeton, N. J.: Princeton University Press.

Barfield, Thomas. 1989. *The Perilous Frontier*. Cambridge: Blackwell.

Barth, Frederik. 1972 [1964]. "Ethnic processes on the Pathan-Baluch boundary." In *Directions in Sociolinguistics: The Ethnography of Communication*, ed. John J. Gumperz and Dell Hymes, pp. 454–464. New York: Holt Rinehart.

———. 1965 [1959]. *Political Leadership among Swat Pathans*. Rev. ed. London: Athalone.

Barth, Fredrik. 1969. *Ethnic Groups and Boundaries: The Social Organization of Culture Difference*. Repr. ed. Prospect Heights: Waveland.

Bartlett, Robert. 1993. *The Making of Europe: Conquest, Colonization, and Cultural Change, 950–1350*. Princeton, N. J.: Princeton University Press.

Barynkin, P. P., and E. V. Kozin. 1998. Prirodno-kilmaticheskie i kul'turno-demograficheskie protsessy v severnom priKaspii v rannem i srednem Golotsene. In *Arkheologicheskie Kul'tury Severnogo Prikaspiya*, ed. R. S. Bagautdinov, pp. 66–83. Kuibyshev: Kuibyshevskii gosudarts-vennyi pedagogicheskii institut.

Barynkin, P. P., and I. B. Vasiliev. 1988. Stoianka Khvalynskoi eneoliticheskoi kulturi Kara-Khuduk v severnom Prikaspii. In *Arkheologicheskie Kul'tury Severnogo Prikaspiya*, ed. R. S. Bagautdinov, pp. 123–142, Kuibyshev: Kuibyshevskii gosudarstvennyi pedagogicheskii institut.

Barynkin, P. P., I. B. Vasiliev, and A. A. Vybornov. 1998. Stoianka Kyzyl-Khak II: pamyatnik epokhi rannei Bronzy severnogo prikaspiya. In *Problemy Drevnei Istorii Severnogo Prikaspiya*, ed. V. S. Gorbunov, pp. 179–192, Samara: Samarskogo gosudarstvennogo pedagogicheskogo universiteta.

Bashkow, Ira. 2004. A neo-Boasian conception of cultural boundaries. *American Anthropologist* 106 (3): 443–458.

Beekes, Robert S. P. 1995. *Comparative Indo-European Linguistics: An Introduction.* Amsterdam: John Benjamins.

Beilekchi, V. S. 1985. Raskopki kurgana 3 u s. Kopchak. *Arkheologicheskie Issledovaniya v Moldavii v 1985 g.*, pp. 34–49. Kishinev: Shtiintsa.

Belanovskaya, T. D. 1995. *Iz drevneishego proshlogo nizhnego po Don'ya.* St. Petersburg: IIMK.

Bellwood, Peter. 2001. Early agriculturalist population diasporas? Farming, language, and genes. *Annual Review of Anthropology* 30:181–207.

Bellwood, Peter, and Colin Renfrew, eds. 2002. *Examining the Farming/Language Dispersal Hypothesis.* Cambridge: McDonald Institute for Archaeological Research.

Bendrey, Robin. 2007. New methods for the identification of evidence for bitting on horse remains from archaeological sites. *Journal of Archaeological Science* 34:1036–1050.

Benecke, Norbert. 1997. Archaeozoological studies on the transition from the Mesolithic to the Neolithic in the North Pontic region. *Anthropozoologica* 25–26:631–641.

———. 1994. *Archäologische Studien zur Entwicklung der Haustierhaltung in Mitteleuropa und Südskandinavien von Anfängen bis zum Ausgehenden Mittelalter.* Berlin: Akademie Verlag.

Benecke, Norbert, and Angela von den Dreisch. 2003. Horse exploitation in the Kazakh steppes during the Eneolithic and Bronze Age. In *Prehistoric Steppe Adaptation and the Horse*, ed. Marsha Levine, Colin Renfrew, and Katie Boyle, pp. 69–82. Cambridge: McDonald Institute for Archaeological Research.

Benveniste, Emile. 1973 [1969]. *Indo-European Language and Society.* Translated by Elizabeth Palmer. Coral Gables, Fla.: University of Miami Press.

Berger, Joel. 1986. *Wild Horses of the Great Basin: Social Competition and Population Size.* Chicago: University of Chicago Press.

Berezanskaya, S. S. 1979. Pervye mastera-metallurgi na territorii Ukrainy. In *Pervobytnaya arkheologiya: poiski i nakhodki*, ed. N. N. Bondar and D. Y. Telegin, pp. 243–256. Kiev: Naukova Dumka.

Bibby, Geoffrey. 1956. *The Testimony of the Spade.* New York: Knopf.

Bibikov, S. N. 1953. *Rannetripol'skoe Poselenie Luka-Vrublevetskaya na Dnestre.* Materialy i issledovaniya po arkheologii SSR 38. Moscow: Akademii Nauk SSSR.

Bibikova, V. I. 1970. K izucheniyu drevneishikh domashnikh loshadei vostochnoi Evropy, soobshchenie 2. *Biulleten moskovskogo obshchestva ispytatlei prirodi otdel biologicheskii* 75 (5): 118–126.

———. 1967. K izucheniyu drevneishikh domashnikh loshadei vostochnoi Evropy. *Biulleten moskovskogo obshchestva ispytatelei prirodi Otdel Biologicheskii* 72 (3): 106–117.

Bickerton, D. 1988. Creole languages and the bioprogram. In *Linguistics: The Cambridge Survey*, vol. 2 ed. F. J. Newmeyer, pp. 267–284. Cambridge: Cambridge University Press.

Binford, Lewis. 1971. Mortuary practices: Their study and their potential. In *Approaches to the Social Dimensions of Mortuary Practices*, ed. James A. Brown, pp. 92–112. Memoirs No. 25. Washington, D.C.: Society for American Archaeology.

Blyakharchuk, T. A., H. E. Wright, P. S. Borodavko, W. O. van der Knaap, and B. Ammann. 2004. Late Glacial and Holocene vegetational changes on the Ulagan high-mountain plateau, Altai Mts., southern Siberia. *Palaeogeography, Paleoclimatology, and Paleoecology* 209:259–279.

Bloch, Maurice E. F. 1998. Time, narratives, and the multiplicity of representations of the past. In *How We Think They Think*, ed. Maurice E. F. Bloch, 100–113. Boulder, CO: Westview Press.

Boaz, Franz. 1911. Introduction. In *Handbook of American Indian Languages*, pt. 1, pp. 1–82. Bulletin 40. Washington, D.C.: Bureau of American Ethnology.

Bobomulloev, Saidmurad. 1997. Ein bronzezeitliches Grab aus Zardča Chalifa bei Pendžikent (Zeravšan-Tal). *Archäologische Mitteilungen aus Iran und Turan* 29:122–134.

Bobrinskii, A. A., and I. N. Vasilieva. 1998. O nekotorykh osobennostiakh plasticheskogo syr'ya v istorii goncharstva. In *Problemy drevnei istorii severnogo prikaspiya*, pp. 193–217. Samara: Institut istorii i arkheologii povolzh'ya.

Bobrov, V. V. 1988. On the problem of interethnic relations in South Siberia in the third and second millennia BC. *Arctic Anthropology* 25 (2): 30–46.

Bodyans'kii, O. V. 1968. Eneolitichnii mogil'nik bilya s. Petyro-Svistunovo. *Arkheologiya* (Kiev) 21:117–125.

Bogucki, Peter. 1988. *Forest Farmers and Stockherders*. Cambridge: Cambridge University Press.

Bökönyi, Sandor. 1991. Late Chalcolithic horses in Anatolia. In *Equids in the Ancient World*, ed., Richard Meadow and Hans-Peter Uerpmann, vol. 2, pp. 123–131. Wiesbaden: Ludwig Reichert.

———. 1987. Horses and sheep in East Europe. In *Proto-Indo-European: The Archaeology of a Linguistic Problem*, ed. Susan Skomal, pp. 136–144. Washington, D.C.: Institute for the Study of Man.

———. 1983. Late Chalcolithic and Early Bronze I animal remains from Arslantepe (Malatya), Turkey: A preliminary report. *Origini* 12 (2): 581–598.

———. 1979. Copper age vertebrate fauna from Kétegyháza. In *The People of the Pit-Grave Kurgans in Eastern Hungary*, ed. Istvan Ecsedy, pp. 101–116. Budapest: Akademiai Kiado.

———. 1978. The earliest waves of domestic horses in East Europe. *Journal of Indo-European Studies* 6 (1/2): 17–76.

———. 1974. *History of Domestic Animals in Central and Eastern Europe*. Budapest: Akademiai Kiado.

Bol'shov, S. V. 1995. Problemy kulturogeneza v lesnoi polose srednego povolzh'ya v Abashevskoe vremya. In *Drevnie IndoIranskie Kul'tury Volgo-Ural'ya*, ed. I. B. Vasilev and O. V. Kuz'mina, pp. 141–156. Samara: Samara Gosudarstvennogo Pedagogicheskogo Universiteta.

Boltenko, M. F. 1957. Stratigrafiya i khronologiya Bol'shogo Kulial'nika. *Materiali i issledovaniya po arkheologii severnogo prichernomoriya (Kiev)* 1:21–46.

Bond, G., Kromer, B., Beer, J., Muscheler, R., Evans, M. N., Showers, W., Hoffmann, S., Lotti-Bond, R., Hajdas, I. and Bonani, G., 2001. Persistent solar influence on North Atlantic climate during the Holocene. *Science* 294:2130–2136.

Bondar, N. N. and Nechitailo, A. L., eds. 1980. *Arkheologicheskie pamyatniki po ingul'ya*. Kiev: Naukova Dumka.

Bondar, N. N. 1974. K voprosu o proiskhozhdenii serdnedneprovskoi kul'tury. *Zborník Filozofickej Fakulty Univerzity Komenského Musaica (Bratislava)* 14:37–53.

Bonsall, C., G. T. Cook, R. E. M. Hedges, T. F. G. Higham, C. Pickard, and I. Radovanovic. 2004. Radiocarbon and stable isotope evidence of dietary change from the Mesolithic to the Middle Ages in the Iron Gates: New results from Lepenski Vir. *Radiocarbon* 46 (1): 293–300.

Boriskovskii, Pavel I. 1993. Determining Upper Paleolithic historico-cultural regions. In *From Kostienki to Clovis, Upper Paleolithic: Paleo-Indian Adaptations*, ed. Olga Soffer and N. D. Praslov, pp. 143–147. New York: Plenum.

Boroffka, Nikolaus. 1998. Bronze- und früheizenzeitliche Geweihtrensenknebel aus Rumänien und ihre Beziehungen. *Eurasia Antiqua* (Berlin) 4:81–135.

Boroffka, Nikolaus, Jan Cierny, Joachim Lutz, Hermann Parzinger, Ernst Pernicka, and Gerd Weisberger, 2002. Bronze Age tin from central Asia: Preliminary notes. In *Ancient Interactions: East and West in Eurasia*, ed. Katie Boyle, Colin Renfrew, and Marsha Levine, pp. 135–159, Cambridge: McDonald Institute for Archaeological Research.

Boyadziev, Yavor D. 1995. Chronology of the prehistoric cultures in Bulgaria. In *Prehistoric Bulgaria*, ed. Douglass W. Bailey and Ivan Panayotov, pp. 149–191. Monographs in World Archaeology 22. Madison, Wis.: Prehistory Press.

Boyce, Mary. 1975. *A History of Zoroastrianism*. Vol. 1. Leiden: Brill.

Britain, David. 2002. Space and spatial diffusion. In *The Handbook of Language Variation and Change*, ed. J. Chambers, P. Trudgill, and N. Schilling-Estes, pp. 603–637. Oxford: Blackwell.

Boyle, Katie, Colin Renfrew, and Marsha Levine, eds. 2002. *Ancient Interactions: East and West in Eurasia*. Cambridge: McDonald Institute for Archaeological Research.

Bradley D. G., D. E. MacHugh, P. Cunningham, and R. T. Loftus. 1996. Mitochondrial diversity and the origins of African and European cattle. *Proceedings of the National Academy of Sciences* 93 (10): 5131–5135.

Bratchenko, S. N. 2003. Radiocarbon chronology of the Early Bronze Age of the Middle Don, Svatove, Luhansk region. *Baltic-Pontic Studies* 12:185–208.

———. 1976. *Nizhnee Podone v Epokhu Srednei Bronzy*. Kiev: Naukovo Dumka.

———. 1969. Bagatosha rove poselennya Liventsivka I na Donu. *Arkheologiia* (Kiev) 22:210–231.

Breen, T. H. 1984. Creative adaptations: Peoples and cultures. In *Colonial British America*, ed. Jack P. Green and J. R. Pole, pp. 195–232. Baltimore, Md.: Johns Hopkins University Press.

Britain, David. 2004. Geolinguistics—Diffusion of Language. In *Sociolinguistics: International Handbook of the Science of Language and Society* vol. 1, ed. Ulrich Ammon, Norbert Dittmar, Klaus J. Mattheier, and Peter Trudgill, pp. 34–48, Berlin: Mouton de Gruyter.

Bronicki, Andrzej, Sławomir Kadrow, and Anna Zakościelna. 2003. Radiocarbon dating of the Neolithic settlement in Zimne, Volhynia. *Baltic-Pontic Studies* 12:22–66.

Bronitsky, G., and R. Hamer. 1986. Experiments in ceramic technology: The effects of various tempering material on impact and thermal-shock resistance. *American Antiquity* 51 (1): 89–101.

Broodbank, Cyprian. 1989. The longboat and society in the Cyclades in the Keros-Syros culture. *American Journal of Archaeology* 85:318–337.

Broodbank, Cyprian, and T. F. Strasser. 1991. Migrant farmers and the colonization of Crete. *Antiquity* 65:233–245.

Brown, D. R., and David W. Anthony. 1998. Bit wear, horseback riding, and the Botai site in Kazakstan. *Journal of Archaeological Science* 25:331–347.

Bryce, T. 1998. *The Kingdom of the Hittites*. Oxford: Clarendon.

Buchanan, Briggs. 1966. *Catalogue of Ancient Near Eastern Seals in the Ashmolean Museum*. Vol. 1, *Cylinder Seals*. Oxford: Clarendon.

Buck, Carl Darling. 1949. *A Dictionary of Selected Synonyms in the Principal Indo-European Languages*. Chicago: University of Chicago Press.

Bunyatyan, Katerina P. 2003. Correlations between agriculture and pastoralism in the northern Pontic steppe area during the Bronze Age. In *Prehistoric Steppe Adaptation and the Horse*, ed. Marsha Levine, Colin Renfrew, and Katie Boyle, pp. 269–286. Cambridge: McDonald Institute for Archaeological Research.

Burdo, Natalia B. 2003. Cultural contacts of early Tripolye tribes. Paper delivered at the Ninth Annual Conference of the European Association of Archaeologists. St Petersburg, Russia.

Burdo, Natalia B., and V. N. Stanko. 1981. Eneoliticheskie nakhodki na stoianke Mirnoe. In *Drevnosti severo-zapadnogo prichernomor'ya*, pp. 17–22. Kiev: Naukovo Dumka.

Burmeister, Stefan. 2000. Archaeology and migration: Approaches to an archaeological proof of migration. *Current Anthropology* 41 (4): 554–555.

Burov, G. M. 1997. Zimnii transport severnoi Evropy i Zaural'ya v epokhu Neolita i rannego metalla. *Rossiskaya arkheologiya* 4:42–53.

Burrow, T. 1973. The Proto-Indoaryans. *Journal of the Royal Asiatic Society* (n. 5.) 2:123–40.

Bynon, Theodora. 1977. *Historical Linguistics*. Cambridge: Cambridge University Press.

Cameron, Catherine, and Steve A. Tomka, eds. 1993. *Abandonment of Settlements and Regions: Ethnoarchaeological and Archaeological Approaches*. Cambridge: Cambridge University Press.

Campbell, Lyle. 2002. What drives linguistic diversification and language spread? In *Examining the Farming/Language Dispersal Hypothesis*, ed. Peter Bellwood and Colin Renfrew, pp. 49–63. Cambridge: McDonald Institute for Archaeological Research.

Cannon, Garland. 1995. "Oriental Jones: Scholarship, Literature, Multiculturalism, and Humankind." In *Objects of Enquiry: The Life, Contributions, and Influences of Sir William Jones*, pp. 25–50. New York: New York University Press.

Carpelan, Christian, and Asko Parpola. 2001. Emergence, contacts and dispersal of Proto-Indo-European, proto-Uralic and proto-Aryan in archaeological perspective. In *Early Contacts between Uralic and Indo-European: Linguistic and Archaeological Considerations*, ed. Christian Carpelan, Asko Parpola, and Petteri Koskikallio, pp. 55–150. Memoires de la Société Finno-Ugrienne 242. Helsinki: Suomalais-Ugrilainen Seura.

Castile, George Pierre, and Gilbert Kushner, eds. 1981. *Persistent Peoples: Cultural Enclaves in Perspective*. Tucson: University of Arizona Press.

Chambers, Jack, and Peter Trudgill. 1998. *Dialectology*. Cambridge: Cambridge University Press.

Chapman, John C. 1999. The origins of warfare in the prehistory of Eastern and central Europe. In *Ancient Warfare: Archaeological Perspectives*, ed. John Carman and Anthony Harding, pp. 101–142. Phoenix Mill: Sutton.

———. 1989. The early Balkan village. In *The Neolithic of Southeastern Europe and Its Near Eastern Connections*, ed. Sándor Bökönyi, pp. 33–53. Budapest: Varia Archaeologica Hungarica II.

———. 1983. The Secondary Products Revolution and the limitations of the Neolithic. *Bulletin of the Institute of Archaeology* (London) 19:107–122.

Cherniakov, I. T., and G. N. Toshchev. 1985. Kul'turno-khronologicheskie osobennosti kurgannykh pogrebenii epokhi Bronzy nizhnego Dunaya. In *Novye Materialy po Arkheologii Severnogo-Zapadnogo Prichernomor'ya*, ed. V. N. Stanko, pp. 5–45, Kiev: Naukovo Dumka.

Chernopitskii, M. P. 1987. Maikopskii "baldachin." *Kratkie soobshcheniya institut arkheologii* 192:33–40.

Chernykh, E. N., ed. 2004. *Kargaly*. Vol. 3, *Arkheologicheskie materialy, tekhnologiya gornometallurgicheskogo proizvodstva, arkheobiologicheskie issledovaniya*. Moscow: Yaziki slavyanskoi kul'tury.

———. 1997. *Kargaly: Zabytyi Mir*. Moscow: NOX.

———. 1995. Postscript: Russian archaeology after the collapse of the USSR: Infrastructural crisis and the resurgence of old and new nationalisms. In *Nationalism, Politics, and the Practice of Archaeology*, ed. Philip L. Kohl and Clare Fawcett, pp. 139–148, Cambridge: Cambridge University Press.

———. 1992. *Ancient Metallurgy in the USSR*. Cambridge: Cambridge University Press.

Chernykh, E. N., and K. D. Isto. 2002. Nachalo ekspluatsii Kargalov: Radiouglerodnyi daty. *Rossiiskaya arkheologiya* 2: 44–55.

Chernykh, E.N., E. V. Kuz'minykh, and L. B. Orlovskaya. 2004. Ancient metallurgy of northeast Asia: From the Urals to the Saiano-Altai. In *Metallurgy in Ancient Eastern Eurasia from the Urals to the Yellow River*, ed. Katheryn M. Linduff, pp. 15–36. Lewiston, Me.: Edwin Mellen.

Chernysh, E. K. 1982. Eneolit pravoberezhnoi Ukrainy i Moldavii. In *Eneolit SSSR*, ed. V. M. Masson and N. Y. Merpert, pp. 165–320. Moscow: Nauka.

Childe, V. Gordon. 1957. *The Dawn of European Civilization*. 6th ed. London: Routledge Kegan Paul.

———. 1936. The axes from Maikop and Caucasian metallurgy. *Annals of Archaeology and Anthropology* (Liverpool) 23:113–119.

Chilton, Elizabeth S. 1998. The cultural origins of technical choice: Unraveling Algonquian and Iroquoian ceramic traditions in the Northeast. In *The Archaeology of Social Boundaries*, ed. Miriam Stark, pp 132–160. Washington, D.C.: Smithsonian Institution Press.

Chretien, C. D. 1962. The mathematical models of glottochronology. *Language* 38:11–37.

Christensen, A. F., Brian E. Hemphill, and Samar I. Mustafakulov. 1996. Bactrian relationships to Russian and Central Asian populations: A craniometric assessment. *American Journal of Physical Anthropology* 22:84–85.

Clackson, James. 1994. *The Linguistic Relationship between Greek and Armenian*. Oxford: Blackwell.

Clark, Geoffry. 1994. Migration as an explanatory concept in Paleolithic archaeology. *Journal of Archaeological Method and Theory* 1 (4): 305–343.

Clark, Grahame. 1941. Horses and battle-axes. *Antiquity* 15 (57): 50–69.

Clayton, Hilary. 1985. A fluoroscopic study of the position and action of different bits in the horse's mouth. *Equine Veterinary Science* 5 (2): 68–77.

Clayton, Hilary M., and R. Lee. 1984. A fluoroscopic study of the position and action of the jointed snaffle bit in the horse's mouth. *Equine Veterinary Science* 4 (5): 193–196.

Clutton-Brock, Juliet. 2003. Were the donkeys of Tell Brak harnessed with a bit? In *Prehistoric Steppe Adaptation and the Horse*, ed. Marsha Levine, Colin Renfrew, and Katie Boyle, pp. 126–127. Cambridge: McDonald Institute for Archaeological Research.

———. 1974. The Buhen horse. *Journal of Archaeological Science* 1:89–100.

Cole, John W., and Eric Wolf. 1974. *The Hidden Frontier: Ecology and Ethnicity in an Alpine Valley*. New York: Academic Press.

Coleman, John. 2000. An archaeological scenario for the "Coming of the Greeks" ca. 3200 BC." *Journal of Indo-European Studies* 28 (1–2): 101–153.

Comsa, Eugen. 1976. Quelques considerations sur la culture Gumelnitsa. *Dacia* 20:105–127.

Cook, G. T., C. Bonsall, R. E. M. Hedges, K. McSweeney, V. Boroneanţ, L. Bartosiewicz, and P. B. Pettitt, 2002. Problems of dating human bones from the Iron Gates. *Antiquity* 76:77–85.

Cronk, Lee. 1993. CA comment on transitions between cultivation and pastoralism in Sub-Saharan Africa. *Current Anthropology* 34 (4): 374.

———. 1989. From hunters to herders: Subsistence change as a reproductive strategy. *Current Anthropology* 30:224–34.

Crouwel, Joost H. 1981. *Chariots and Other Means of Land Transport in Bronze Age Greece*. Allard Pierson Series 3. Amsterdam: Allard Pierson Museum.

Dalton, G. 1977. Aboriginal economies in stateless societies. In *Exchange Systems in Prehistory*, ed. Timothy Earle and J. Ericson, pp. 191–212, New York: Academic Press.

Danilenko, V. M. 1971. *Bugo-Dnistrovs'ka Kul'tura*. Kiev: Dumka.

Darden, Bill J. 2001. On the question of the Anatolian origin of Indo-Hittite. In *Greater Anatolia and the Indo-Hittite Language Family*, ed. Robert Drews, pp. 184–228. Journal of Indo-European Studies Monograph 38. Washington, D.C.: Institute for the Study of Man.

———. 2004. Who were the Sclaveni and where did they come from? *Byzantinische Forschungen* 28:133–157.

Davis, E. M. 1983. The gold of the shaft graves: The Transylvanian connection. *Temple University Aegean Symposium* 8:32–38.

Davis, Simon J. M. 1987. *The Archaeology of Animals*. New Haven: Yale University Press.

Davis-Kimball, Jeannine. 1997. Warrior women of the Eurasian steppes. *Archaeology* 50 (1): 44–49.

DeBoer, Warren. 1990. Interaction, imitation, and communication as expressed in style: The Ucayali experience. In *The Uses of Style in Archaeology*, ed. M. Conkey and Christine Hastorf, pp. 82–104. Cambridge: Cambridge University Press.

———. 1986. Pillage and production in the Amazon: A view through the Conibo of the Ucayali Basin, eastern Peru. *World Archaeology* 18 (2): 231–246.

Dennell, R. W., and D. Webley. 1975. Prehistoric settlement and land use in southern Bulgaria. In *Palaeoeconomy*, ed. E. S. Higgs, pp. 97–110. Cambridge: Cambridge University Press.

Dergachev, Valentin A. 2003. Two studies in defense of the migration concept. In *Ancient Interactions: East and West in Eurasia*, ed. Katie Boyle, Colin Renfrew, and Marsha Levine, pp. 93–112. McDonald Institute Monographs. Cambridge: University of Cambridge Press.

————. 1999. Cultural-historical dialogue between the Balkans and Eastern Europe (Neolithic-Bronze Age). *Thraco-Dacica* 20 (1–2): 33–78.

————. 1998a. *Karbunskii Klad.* Kishinev: Academiei Ştiinţe.

————. 1998b. Kulturell und historische Entwicklungen im Raum zwischen Karpaten und Dnepr. In *Das Karpatenbecken und Die Osteuropäische Steppe*, ed. Bernhard Hänsel and Jan Machnik, pp. 27–64. München: Südosteuropa-Schriften Band 20, Verlag Marie Leidorf GmbH.

————. 1980. *Pamyatniki Pozdnego Tripol'ya.* Kishinev: Shtiintsa.

Dergachev, V., A. Sherratt, and O. Larina. 1991. Recent results of Neolithic research in Moldavia (USSR). *Oxford Journal of Prehistory* 10 (1): 1–16.

Derin, Z., and Oscar W. Muscarella. 2001. Iron and bronze arrows. In *Ayanis I. Ten Years' Excavations at Rusahinili Eiduru-kai 1989–1998*, ed. A. Çilingiroğlu and M. Salvini, pp. 189–217. Roma: Documenta Asiana VI ISMEA.

Diakonov, I. M. 1988. Review of *Archaeology and Language. Annual of Armenian Linguistics* 9:79–87.

————. 1985. On the original home of the speakers of Indo-European. *Journal of Indo-European Studies* 13 (1–2): 93–173.

Diamond, Jared. 1997. *Guns, Germs, and Steel: The Fates of Human Societies.* New York: Norton.

DiCosmo, Nicola. 2002. *Ancient China and Its Enemies: The Rise of Nomadic Power in East Asian History.* Cambridge: Cambridge University Press.

————. 1999. State Formation and periodization in Inner Asian prehistory. *Journal of World History* 10 (1): 1–40.

————. 1994. Ancient Inner Asian Nomads: Their Economic basis and its significance in Chinese history. *Journal of Asian Studies* 53 (4): 1092–1126.

Diebold, Richard. 1985. *The Evolution of the Nomenclature for the Salmonid Fish: The Case of "huchen" (Hucho spp.).* Journal of Indo-European Studies Monograph 5. Washington, D.C.: Institute for the Study of Man.

Dietler, Michael, and Brian Hayden, eds. 2001. *Feasts.* Washington, D.C.: Smithsonian Institution Press.

Dietz, Ute Luise. 1992. Zur frage vorbronzezeitlicher Trensenbelege in Europa. *Germania* 70 (1): 17–36.

Dixon, R. M. W. 1997. *The Rise and Fall of Languages.* Cambridge: Cambridge University Press.

Dobrovol'skii, A. V. 1958. Mogil'nik vs. Chapli. *Arkheologiya* (Kiev) 9:106–118.

Dodd-Opriţescu, 1978, Les elements steppiques dans l'Énéolithique de Transylvanie. *Dacia* 22:87–97.

Dolukhanov, P. M. 1986. Foragers and farmers in west-Central Asia. In *Hunters in Transition*, ed. Marek Zvelebil, pp. 121–132. Cambridge: Cambridge University Press.

Donnan, Hastings, and Thomas M. Wilson. 1999. *Borders: Frontiers of Identity, Nation, and State.* Oxford: Berg.

Dorian, N. 1981. *Language Death: The Life Cycle of a Scottish Gaelic Dialect.* Philadelphia: University of Pennsylvania Press.

Dovchenko, N. D., and N. A. Rychkov. 1988. K probleme sotsial'noi stratigrafikatsii plemen Yamnoi kul'turno-istoricheskoi obshchnosti. In *Novye Pamyatniki Yamnoi Kul'tury Stepnoi Zony Ukrainy*, pp. 27–40. Kiev: Naukova Dumka.

Dremov, I. I., and A. I. Yudin. 1992. Drevneishie podkurgannye zakhoroneniya stepnogo zaVolzh'ya. *Rossiskaya arkheologiya* 4:18–31.

Drews, Robert. 2004. *Early Riders.* London: Routledge.

————, ed. 2001. *Greater Anatolia and the Indo-Hittite Language Family.* Journal of Indo-European Studies Monograph 38. Washington, D.C.: Institute for the Study of Man.

————. 1988. *The Coming of the Greeks: Indo-European Conquests in the Aegean and the Ancient Near East.* Princeton, N. J.: Princeton University Press.

Drinka, Bridget. 1995. Areal linguistics in prehistory: Evidence from Indo-European aspect. In *Historical Linguistics 1993*, ed. Henning Andersen, pp. 143–158. Amsterdam: John Benjamins.

Dumezil, Georges. 1958. *L'Idéologie Tripartie des Indo-Européens*. Brussels: Latomus.

Dumitrescu, Vladimir. 1980. Tumuli from the period of transition from the Eneolithic to the Bronze Age excavated near Rast. In *The Neolithic Settlement at Rast*, appendix 3, pp. 126–133. British Archaeological Reports International Series 72. Oxford: Archaeopress.

During Caspers, E. C. L. 1998. The MBAC and the Harappan script. *Ancient Civilizations from Scythia to Siberia* 5 (1): 40–58.

Dyen, I., J. B. Kruskal, and P. Black. 1992. An Indo-European classification: A lexicostatistical experiment. *Transactions of the American Philosophical Society* 82 (5): 1–132.

Ecsedy, István. 1994. "Camps for eternal rest: Some aspects of the burials by the earliest nomads of the steppes." In *The Archaeology of the Steppes: Methods and Strategies*, ed. Bruno Genito, pp. 167–176. Napo: Instituto universitario oreintale series minor 44.

———, ed. 1979. *The People of the Pit-Grave Kurgans in Eastern Hungary*. Budapest: Akadémia Kiadó.

Ehrich, Robert W. 1961. On the persistence and recurrences of culture areas and culture boundaries during the course of European prehistory, protohistory, and history. In *Berichte über den V Internationalen Kongress für Vor- und Frühgeschichte*, pp. 253–257. Berlin: Gebrüder Mann.

Eisler, Riane. 1990. The Gaia tradition and the partnership future: An ecofeminist manifesto. In *Reweaving the World*, ed. Irene Diamond and G. F. Orenstein, pp. 23–34. San Francisco: Sierra Club Books.

———. 1987. *The Chalice and the Blade*. San Francisco: Harper and Row.

Eleure, C., ed. 1989. *Le Premier Or de l'Humanité en Bulgarie 5e millénaire*. Paris: Musées Nationaux.

Ellis, Linda. 1984. *The Cucuteni-Tripolye Culture: A Study in Technology and the Origins of Complex Society*. British Archaeological Reports International Series 217. Oxford: Archaeopress.

Emberling, Geoff. 1997. Ethnicity in complex societies: Archaeological perspectives. *Journal of Archaeological Research* 5 (4): 295–344.

Embleton, Sheila. 1991. Mathematical methods of genetic classification. In *Sprung from Some Common Source: Investigations into the Prehistory of Languages*, ed. Sidney Lamb and E. Douglass Mitchell, pp. 365–388. Stanford: Stanford University Press.

———. 1986. *Statistics in Historical Linguistics*. Bochum: Brockmeyer.

Enattah, Nabil Sabri. 2005. *Molecular Genetics of Lactase Persistence*. Ph.D. dissertation, Department of Medical Genetics, Faculty of Medicine, University of Helsinki, Finland.

Epimakhov, A. V. 2002. *Iuzhnoe zaural'e v epokhu srednei bronzy*. Chelyabinsk: YUrGU.

Epimakhov, A., B. Hanks, and A. C. Renfrew. 2005. Radiocarbon dating chronology for the Bronze Age monuments in the Transurals, Russia. *Rossiiskaia Arkheologiia* 4:92–102.

Erdosy, George, ed. 1995. *The Indo-Aryans of Ancient South Asia: Language, Material Culture and Ethnicity*. Indian Philology and South Asian Studies 1. Berlin: Walter de Gruyter.

Euler, Wolfram. 1979. *Indoiranisch-griechische Gemeinsamkeiten der Nominalbildung und deren Indogermanische Grundlagen*. Innsbruck: Institut für Sprachwissenschaft der Universität Innsbruck, vol. 30.

Evdokimov, V. V., and V. G. Loman. 1989. Raskopi Yamnogo kurgana v Karagandinskoi Oblasti. In *Voprosy arkheologii tsestral'nogo i severnogo Kazakhstana*, ed. K.M. Baipakov, pp. 34–46. Karaganda: Karagandinskii gosudarstvennyi universitet.

Ewers, John C. 1955. *The Horse in Blackfoot Indian Culture*. Washington, D.C.: Smithsonian Institution Press.

Falk, Harry. 1986. *Bruderschaft und Würfelspiel*. Freiburg: Hedwig Falk.

Fiedel, Stuart, and David W. Anthony. 2003. Deerslayers, pathfinders, and icemen: Origins of the European Neolithic as seen from the frontier. In *The Colonization of Unfamiliar Landscapes*, ed. Marcy Rockman and James Steele, pp. 144–168. London: Routledge.

Fischer, David Hackett. 1989. *Albion's Seed: Four British Folkways in America*. New York: Oxford University Press.

Fitzgerald-Huber, Louise G. 1995. Qijia and Erlitou: The question of contacts with distant cultures. *Early China* 20:17–67.

Florin, Curta. 2001. *The Making of the Slavs*. Oxford: Oxford University Press.

Forenbaher, S. 1993. Radiocarbon dates and absolute chronology of the central European Early Bronze Age. *Antiquity* 67:218–256.

Forsén, J. 1992. *The Twilight of the Early Helladics: A Study of the Disturbances in East-Central and Southern Greece toward the End of the Early Bronze Age*. Jonsered, Sweden: P. Åströms Förlag.

Fortson, Benjamin W., IV. 2004. *Indo-European Language and Culture: An Introduction*. Oxford: Blackwell.

Fox, John W. 1987. *Maya Postclassic State Formation*. Cambridge: Cambridge University Press.

Francis, E. D. 1992. The impact of non-Indo-European languages on Greek and Mycenaean. In *Reconstructing Languages and Cultures*, ed. E. Polome and W. Winter, pp. 469–506. Trends in Linguistics: Studies and Monographs 58. Berlin: Mouton de Gruyter.

French, Charly, and Maria Kousoulakou. 2003. Geomorphological and micro-morphological investigations of paleosols, valley sediments and a sunken-floored dwelling at Botai, Kazakstan. In *Prehistoric Steppe Adaptation and the Horse*, ed. Marsha Levine, Colin Renfrew, and Katie Boyle, pp. 105–114. Cambridge: McDonald Institute for Archaeological Research.

Fried, Morton H. 1975. *The Notion of Tribe*. Menlo Park, Calif.: Cummings.

Friedrich, Paul. 1970. *Proto-Indo-European Trees*. Chicago: University of Chicago Press.

Gaiduchenko, L. L. 1995. Mesto i znachenie Iuzhnogo Urala v eksportno-importnikh operatsiyakh po napravleniu vostok-zapad v eopkhu bronzy. In *Rossiya i vostok: Problemy vzaimodeistviya*, pt. 5, bk. 1: *Kul'tury eneolita-bronzy stepnoi evrazii*, pp. 110–115. Chelyabinsk: 3-ya Mezhdunarodnaya nauchnaya konferentsiya.

Gal, S. 1978. *Language Shift: Social Determinants of Linguistic Change in Bilingual Austria*. New York: Academic Press.

Gallusser, W. A. 1991. Geographical investigations in boundary areas of the Basle region ("Regio"). In *The Geography of Border Landscapes*, ed. D. Rumley and J. V. Minghi, pp. 32–42. London: Routledge.

Gamkrelidze, Thomas V., and Vyacheslav Ivanov. 1995. *Indo-European and the Indo-Europeans: A Reconstruction and Historical Analysis of a Proto-Language and a Proto-Culture*. Vol. 1. Translated by Johanna Nichols. Edited by Werner Winter. Trends in Linguistics: Studies and Monographs 80. Berlin: Mouton de Gruyter.

———. 1984. *Indoevropeiskii iazyk i indoevropeitsy*. Tiflis: Tbilisskogo Universiteta.

———. 1973. Sprachtypologie und die Rekonstruktion der gemeinindogermanischen Verschlüsse. *Phonetica* 27:150–156.

Gei, A. N. 2000. *Novotitorovskaya kul'tura*. Moscow: Institut Arkheologii.

———. 1990. Poyt paleodemograficheskogo analiza obshchestva stepnykh skotovodov epokhi bronzy: po pogrebal'nym pamyatkikam prikuban'ya. *Kratkie Soobshcheniya Institut Arkheologii* 201:78–87.

———. 1986. Pogrebenie liteishchika Novotitorovskoi kul'tury iz nizhnego pri kuban'ya. In *Arkheologicheskie Otkrytiya na Novostroikakh: Drevnosti severnogo kavkaza* (Moscow) 1:13–32.

———. 1979. Samsonovskoe mnogosloinoe poselenie na Donu. *Sovietskaya arkheologiya* (2): 119–131.

Gellner, Ernest. 1973. *Nations and Nationalism*. Ithaca, N.Y.: Cornell University Press.

Gening, V. F. 1979. The cemetery at Sintashta and the early Indo-Iranian peoples. *Journal of Indo-European Studies* 7:1–29.

Gening, V. F., G. B. Zdanovich, and V. V. Gening. 1992. *Sintashta*. Chelyabinsk: Iuzhno-ural'skoe knizhnoe izdatel'stvo.

George, Christian. 2002. *Quantification of Wear in Equus Teeth from Florida*. MA thesis, Department of Geological Sciences, University of Florida, Gainesville.

Georgieva, P. 1990. Ethnocultural and socio-economic changes during the transitional period from Eneolithic to Bronze Age in the region of the lower Danube. *Glasnik Centara za Balkanoloških Ispitavanja* 26:123–154.

Gheorgiu, Drago. 1994. Horse-head scepters: First images of yoked horses. *Journal of Indo-European Studies* 22 (3–4): 221–250.

Ghetie, B., and C. N. Mateesco. 1973. L'utilisation des bovines a la tracation dans le Neo-lithique Moyen. *International Conference of Prehistoric and Protohistoric Sciences* (Belgrade) 10:454–461.

Giddens, Anthony. 1985. *The Nation-state and Violence*, Cambridge: Polity.

Gilbert, Allan S. 1991. Equid remains from Godin Tepe, western Iran: An interim summary and interpretation, with notes on the introduction of the horse into Southwest Asia. In *Equids in the Ancient World*, vol. 2, ed. Richard H. Meadow and Hans-Peter Uerpmann, pp. 75–122. Wiesbaden: Reichert.

Gilman, Antonio. 1981. The development of social stratification in Bronze Age Europe. *Current Anthropology* 22 (1): 1–23.

Gimbutas, Marija. 1991. *The Civilization of the Goddess*. San Francisco: Harper.

———. 1989a. *The Language of the Goddess*. London: Thames and Hudson.

———. 1989b. Women and culture in Goddess-oriented Old Europe. In *Weaving the Visions*, ed. Judith Plaskow, and C. C. Christ, pp. 63–71. San Francisco: Harper and Row.

———. 1977. The first wave of Eurasian steppe pastoralists into Copper Age Europe. *Journal of Indo-European Studies* 5 (4): 277–338.

———. 1974. *The Goddesses and Gods of Old Europe: Myths and Cult Images (6500–3500 B.C.)*, London: Thames and Hudson.

———. 1970. Proto-Indo-European culture: The Kurgan Culture during the fifth, fourth, and third millennia B.C. In *Indo-European and the Indo-Europeans*, ed. George Cardona, Henry Hoenigswald, and Alfred Senn, pp. 155–198. Philadelphia: University of Pennsylvania Press.

———. 1956. *The Prehistory of Eastern Europe, Part 1*. Cambridge: American School of Prehistoric Research Bulletin 20.

Glassie, Henry. 1965. *Pattern in the Material Folk Culture of the Eastern United States*. Philadelphia: University of Pennsylvania Press.

Glonti, L. I. and A. I. Dzhavakhishvili. 1987. Novye dannye o mnogosloinom pamyatniki ep-okhi Eneolita-Pozdnei Bronzy v shida Kartli-Berkldeebi. *Kratkie Soobshcheniya Institut Arkheologii* 192:80–87.

Glumac, P. D., and J. A. Todd. 1991. Eneolithic copper smelting slags from the Middle Danube basin. In *Archaeometry '90*, ed. Ernst Pernicka and Günther A. Wagner, pp. 155–164. Basel: Birkhäuser Verlag.

Glumac, Petar, and David W. Anthony. 1992. Culture and environment in the prehistoric Caucasus, Neolithic to Early Bronze Age. In *Chronologies in Old World Archaeology, 3rd ed.*, ed. Robert Ehrich, pp. 196–206. Chicago: Aldine.

Golyeva, A. A. 2000. Vzaimodeistvie cheloveka i prirody v severo-zapadnom Prikaspii v epokhu Bronzy. In *Sezonnyi ekonomicheskii tsikl naseleniya severo-zapadnogo Prikaspiya v Bron-zovom Veke*, vol. 120, ed. N. I. Shishlina, pp. 10–29. Moscow: Trudy gosudarstvennogo is-toricheskogo muzeya.

Good, Irene. 2001. Archaeological textiles: A review of current research. *Annual Review of Anthropology* 30:209–226.

———. 1998. Bronze Age cloth and clothing of the Tarim Basin: The Chärchän evidence. In *The Bronze Age and Early Iron Age Peoples of Eastern Central Asia*, ed. Victor Mair, vol. 2, pp. 656–668. Journal of Indo European Studies Monograph 26. Washington, D.C.: Institute for the Study of Man.

Gorbunova, Natalya G. 1993/94. Traditional movements of nomadic pastoralists and the role of seasonal migrations in the formation of ancient trade routes in Central Asia. *Silk Road Art and Archaeology* 3:1–10.

Gotherstrom, A., C. Anderung, L. Hellborg, R. Elburg, C. Smith, D. G. Bradley, H. Ellegren 2005. Cattle domestication in the Near East was followed by hybridization with aurochs bulls in Europe. *Proceedings of Biological Sciences* 272 (1579): 2337–44.

Govedarica, B., and E. Kaiser. 1996. Die äneolithischen abstrakten und zoomorphen Steinzepter Südostund Europas. *Eurasia Antiqua* 2:59–103.

Grant, Madison. 1916. *The Passing of the Great Race; or, The Racial Basis of European History.* New York: Scribner's.

Gray, Russell D., and Quentin D. Atkinson. 2003. Language-tree divergence times support the Anatolian theory of Indo-European origin. *Nature* 426 (6965): 435–439.

Greenfield, Haskell. 1994. Preliminary report on the 1992 excavations at Foeni-Sălaş: An early Neolithic Starčevo-Criş settlement in the Romanian Banat. *Analele Banatului* 3:45–93.

———. 1999. The advent of transhumant pastoralism in temperate southeast Europe: A zooarchaeological perspective from the central Balkans. In *Transhumant Pastoralism in Southern Europe*, ed. L. Bartosiewicz and Haskell Greenfield, pp. 15–36. Budapest: Archaeolingua.

Gremillion, Kristen J. 2004. Seed processing and the origins of food production in eastern North America. *American Antiquity* 69 (2): 215–233.

Grigoriev, Stanislav A. 2002. *Ancient Indo-Europeans.* Chelyabinsk: RIFEI.

Gudkova, A. V., and I. T. Chernyakov. 1981. Yamnye pogebeniya s kolesami u s. Kholmskoe. In *Drevnosti severo-zapanogo prichernomor'ya*, pp. 38–50. Kiev: Naukovo Dumka.

Guliaev, V. I. 2003. Amazons in the Scythia: New finds at the Middle Don, Southern Russia. *World Archaeology* 35 (1): 112–125.

Haheu, Vasile, and Serghei Kurciatov. 1993. Cimitirul plan Eneolitic de lingă satul Giurgiuleşti. *Revista Arkheologică* (Kishinev) 1:101–114.

Hainsworth, J. B. 1972. Some observations on the Indo-European placenames of Greece. In *Acta of the 2nd International Colloquium on Aegean Prehistory*, pp. 39–42. Athens: Ministry of Culture and Sciences.

Haley, Brian D., and Larry R. Wilcoxon. 2005. How Spaniards became Chumash and other tales of ethnogenesis. *American Anthropologist* 107 (3): 432–445.

Hall, Jonathan M. 1997. *Ethnic Identity in Greek Antiquity.* Cambridge: Cambridge University Press.

Hall, Robert A., Jr. 1976. *Proto-Romance Phonology.* New York: Elsevier.

———. 1960. On realism in reconstruction. *Language* 36:203–206.

———. 1950. The reconstruction of Proto-Romance. *Language* 26:6–27.

Hamp, Eric. 1998. Whose were the Tocharians? In *The Bronze Age and Early Iron Age Peoples of Eastern Central Asia*, ed. Victor H. Mair, vol. 1, pp. 307–346. Journal of Indo-European Studies Monograph 26. Washington, D.C.: Institute for the Study of Man.

Hänsel, B. 1982. Südosteuropa zwischen 1600 und 1000 V. Chr. In *Südosteuropa zwischen 1600 und 1000 V. Chr.*, ed. B. Hänsel, pp. 1–38. Berlin: Moreland Editions.

Harding, R. M., and R. R. Sokal. 1988. Classification of the European language families by genetic distance. *Proceedings of the National Academy of Sciences* 85:9370–9372.

Harris, Alice C. 1991. Overview on the history of the Kartvelian languages. In *The Indigenous Languages of the Caucasus*, vol. 1, *The Kartvelian Languages*, ed. Alice C. Harris, pp. 7–83. Delmar, N.Y.: Caravan Books.

Harris, D. R., ed. 1996. *The Origins and Spread of Agriculture and Pastoralism in Eurasia.* London: University College.

Häusler, A. 1994. Archäologische Zeugnisse für Pferd und Wagen in Ost- und Mitteleuropa. In *Die Indogermanen und das Pferd: Festschrift für Bernfried Schlerath*, ed. B. Hänsel and S. Zimmer, pp. 217–257. Budapest: Archaeolingua.

———. 1992. "Der ursprung der Wagens in der Diskussion der gegenwart." *Archäologische Mitteilungen aus Nordwestdeutschland* 15:179–190.

———. 1974. *Die Gräber der älteren Ockergrabkultur zwischen Dnepr und Karpaten.* Berlin: Akadmie-Verlag.

Hayden, Brian. 2001. Fabulous feasts: A prolegomenon to the importance of feasting. In *Feasts*, ed. M. Dietler, and Brian Hayden, pp.23–64. Washington, D.C.: Smithsonian Institution Press.

Hayen, Hajo. 1989. Früheste Nachweise des Wagens und die Entwicklung der Transport-Hilfsmittel. *Mitteilungen der Berliner Gesellschaft für Anthropologie, Ethnologie und Urgeschichte* 10:31–49.

Heidegger, Martin. 1959. *An Introduction to Metaphysics.* 1953 [1935]. Translated by Ralph Manheim. New Haven: Yale University Press.

Helms, Mary. 1992. Long-distance contacts, elite aspirations, and the age of discovery. In *Resources, Power, and Inter-regional Interaction*, ed. Edward M. Schortman and Patricia A. Urban, pp. 157–174. New York: Plenum.

Hemphill, Brian E., A. F. Christensen, and Samar I. Mustafakulov. 1997. Trade or travel: An assessment of interpopulational dynamics among Bronze-Age Indo-Iranian populations. *South Asian Archaeology, 1995: Proceedings of the 13th Meeting of the South Asian Archaeologists of Europe, Cambridge, UK*, ed. Bridget Allchin, pp. 863–879, Oxford: IBH.

Hemphill, Brian E., and J. P. Mallory. 2003. Horse-mounted invaders from the Russo-Kazakh steppe or agricultural colonists from western Central Asia? A craniometric investigation of the Bronze Age settlements of Xinjiang. *American Journal of Physical Anthropology* 124 (3): 199–222.

Hester, D. A. 1957. Pre-Greek placenames in Greece and Asia Minor. *Revue Hittite et Asianique* 15:107–119.

Heyd, V., L. Husty, and L. Kreiner. 2004. *Siedlungen der Glockenbecherkultur in Süddeutschland und Mitteleuropa.* Büchenbach: Arbeiten zur Archäologie Süddeutschlands 17 (Dr. Faustus Verlag).

Hiebert, Frederik T. 2002. Bronze age interaction between the Eurasian steppe and Central Asia. In *Ancient Interactions: East and West in Eurasia*, ed. Katie Boyle, Colin Renfrew, and Marsha Levine, pp. 237–248, Cambridge: McDonald Institute for Archaeological Research.

———. 1994. *Origins of the Bronze Age Oasis Civilizations of Central Asia.* Bulletin of the American School of Prehistoric Research 42. Cambridge, Mass.: Peabody Museum of Archaeology and Ethnology, Harvard University.

Hiendleder, Stefan, Bernhard Kaupe, Rudolf Wassmuth, and Axel Janke. 2002. Molecular analysis of wild and domestic sheep. *Proceedings of the Royal Society of London* 269:893–904.

Hill, Jane. 1996. Languages on the land: Toward an anthropological dialectology. In *David Skomp Distinguished Lectures in Dialectology.* Bloomington: Indiana University Press.

Hill, Jonathon D. 1992. Contested pasts and the practice of archaeology: Overview. *American Anthropologist* 94 (4): 809–815.

Hobsbawm, Eric. 1997. *On History.* New York: New Press.

———. 1990. *Nations and Nationalism since 1780.* Cambridge: Cambridge University Press.

Hock, Hans Henrich, and Brian D. Joseph. 1996. *Language History, Language Change, and Language Relationship: An Introduction to Historical and Comparative Linguistics.* Berlin: Mouton de Gruyter.

Hodder, Ian. 1990. *The Domestication of Europe: Structure and Contingency in Neolithic Societies.* Cambridge: Cambridge University Press.

Holden, Clare, and Ruth Mace. 2003. Spread of cattle led to the loss of matriliny in Africa: A co-evolutionary analysis. *Proceedings of the Royal Society B* 270:2425–2433.

Hopper, Paul. 1973. Glottalized and murmured occlusives in Indo-European. *Glossa* 7:141–166.

Houwink Ten Cate, P. H. J. 1995. Ethnic diversity and population movement in Anatolia. In *Civilizations of the Ancient Near East*, ed. Jack M. Sasson, John Baines, Gary Beckman, and Karen R. Rubinson, vol. 1, pp. 259–270, New York: Scribner's.

Hulbert, R. C., G. S. Morgan, and S. D. Webb, eds. 1995. Paleontology and Geology of the Leisey Shell Pits, Early Pleistocene of Florida. *Bulletin of the Florida Museum of Natural History* 37 (1–10).

Huld, Martin E. 2002. "Linguistic science, truth, and the Indocentric hypothesis." *Journal of Indo-European Studies* 30 (3–4): 353–364.

———. 1990. "The linguistic typology of Old European substrata in north central Europe." *Journal of Indo-European Studies* 18:389–417.

Hüttel, Hans-Georg. 1992. "Zur archäologischen Evidenz der Pfredenutzung in der Kupfer- und Bronzezeit." In *Die Indogermanen und das Pferd: Festschrift für Bernfried Schlerath*, ed. B. Hänsel and S. Zimmer, pp. 197–215. Archaeolingua 4. Budapest: Archaeolingua Foundation.

Ilčeva, V. 1993. Localités de periode de transition de l'énéolithique a l'âdu bronze dans la region de Veliko Tirnovo. In *The Fourth Millennium B.C.*, ed. Petya Georgieva, pp. 82–98. Sofia: New Bulgarian University.

Isakov, A. I. 1994. Sarazm: An agricultural center of ancient Sogdiana. *Bulletin of the Asia Institute* (n. s.) 8:1–12.

Isakov, A. I., Philip L. Kohl, C. C. Lamberg-Karlovsky, and R. Maddin. 1987. Metallurgical analysis from Sarazm, Tadjikistan SSR. *Archaeometry* 29 (1): 90–102.

Itina, M. A., and L. T. Yablonskii. 1997. *Saki Nizhnei Syrdar'i*. Moscow: Rosspen.

Ivanov, I. V., and I. B. Vasiliev. 1995. *Chelovek, Priroda i Pochvy Ryn-Peskov Volgo-Ural'skogo Mezhdurech'ya v Golotsene*. Moscow: Intelleckt.

Ivanova, N. O. 1995. Arkheologicheskaya karta zapovednika Arkaim: Istotiya izucheniya arkheologicheskikh pamyatnikov. In *Arkaim*, ed. G. B. Zdanovich, pp. 159–195. Chelyabinsk: "Kammennyi Poyas."

Izbitser, Elena. 1993. Wheeled vehicle burials of the steppe zone of Eastern Europe and the Northern Caucasus, 3rd to 2nd millennium B.C. Doctoral Thesis, Institute of the History of Material Culture, St. Petersburg, Russia.

Jackson, Kenneth H. 1994. *Language and History in Early Britain*. Dublin: Four Courts.

Jacobs, Kenneth. 1993. Human postcranial variation in the Ukrainian Mesolithic-Neolithic. *Current Anthropology* 34 (3): 311–324.

Jacobsen-Tepfer, Esther. 1993. *The Deer-Goddess of Ancient Siberia: A Study in the Ecology of Belief*. Leiden: Brill.

James, Simon. 1999. *The Atlantic Celts: Ancient People or Modern Invention?* London: British Musem Press.

Janhunen, Juha. 2001. "Indo-Uralic and Ural-Altaic: On the diachronic implications of areal typology." In *Early Contacts between Uralic and Indo-European: Linguistic and Archaeological Considerations*, ed. Christian Carpelan, Asko Parpola, and Petteri Koskikallio, pp. 207–220. Memoires de la Société Finno-Ugrienne 242. Helsinki: Suomalais-Ugrilainen Seura.

———. 2000. Reconstructing Pre-Proto-Uralic typology: Spanning the millennia of linguistic evolution. In *Congressus Nonus Internationalis Fenno-Ugristarum*, pt. 1: *Orationes Plenariae & Orationes Publicae*, ed. Anu Nurk, Triinu Palo, and Tõnu Seilenthal, pp. 59–76. Tartu: CIFU.

Jansen, Thomas, Peter Forster, Marsha A. Levine, Hardy Oelke, Matthew Hurles, Colin Renfrew, Jürgen Weber, and Klaus Olek. 2002. Mitochondrial DNA and the origins of the domestic horse. *Proceedings of the National Academy of Sciences* 99:10905–10910.

Jarrige, Jean-Francois. 1994. The final phase of the Indus occupation at Nausharo and its connection with the following cultural complex of Mehrgarh VIII. *South Asian Archaeology* 1993 (1): 295–313.

John, B. S. 1972. The linguistic significance of the Pembrokeshire Landsker. *The Pembrokeshire Historian* 4:7–29.

Jones, Doug. 2003. Kinship and deep history: Exploring connections between culture areas, genes, and languages. *American Anthropologist* 105 (3): 501–514.

Jones, Siân. 1997. *The Archaeology of Ethnicity: Constructing Identities in the Past and Present.* London: Routledge.

Jones-Bley, Karlene. 2000. The Sintashta "chariots." In *Kurgans, Ritual Sites, and Settlements: Eurasian Bronze and Iron Age,* ed. Jeannine Davis-Kimball, Eileen Murphy, Ludmila Koryakova, and Leonid Yablonsky, pp. 135–140. BAR International Series 89. Oxford: Archeopress.

Jones-Bley, Karlene, and D. G. Zdanovich, eds. 2002. *Complex Societies of Central Eurasia from the 3rd to the 1st Millennium BC.* Vols. 1 and 2. Journal of Indo-European Studies Monograph 45. Washington, D.C.: Institute for the Study of Man.

Jordan, Peter, and Stephen Shennan. 2003. Cultural transmission, language, and basketry traditions amongst the California Indians. *Journal of Anthropological Archaeology* 22:42–74.

Jovanovich, B. 1975. Tumuli stepske culture grobova jama u Padunavlu," *Starinar* 26:9–24.

Kadyrbaev, M. K., and Z. Kurmankulov. 1992. *Kul'tura Drevnikh Skotobodov i Metallurgov Sary-Arki.* Alma-Ata: Gylym.

Kalchev, Petar. 1996. Funeral rites of the Early Bronze Age flat necropolis near the Bereket tell, Stara Zagora." In *Early Bronze Age Settlement Patterns in the Balkans,* pt. 2. Reports of Prehistoric Research Projects 1 (2–4): 215–225. Sofia: Agatho Publishers, Prehistory Foundation.

Kallio, Petri. 2001. Phonetic Uralisms in Indo-European? In *Early Contacts between Uralic and Indo-European: Linguistic and Archaeological Considerations,* ed. Christian Carpelan, Asko Parpola, and Petteri Koskikallio, pp. 221–234. Memoires de la Société Finno-Ugrienne 242. Helsinki: Suomalais-Ugrilainen Seura.

Kasperavičiūtė, D., V. Kučinskas, and M. Stoneking. 2004. Y chromosome and mitochondrial DNA variation in Lithuanians. *Annals of Human Genetics* 68:438–452.

Keeley, Lawrence, H. 1996. *War before Civilization.* New York: Oxford University Press.

Keith, Kathryn. 1998. Spindle whorls, gender, and ethnicity at Late Chalcolithic Hacinebi Tepe. *Journal of Field Archaeology* 25:497–515.

Kelley, Raymond C. 1985. *The Nuer Conquest.* Ann Arbor: University of Michigan Press.

Kershaw, Kris. 2000. *The One-Eyed God: Odin and the Indo-Germanic Männerbünde.* Journal of Indo-European Studies Monograph 36. Washington, D.C.: Institute for the Study of Man.

Khazanov, Anatoly. 1994 [1983]. *Nomads and the Outside World.* Rev. ed. Madison: University of Wisconsin Press.

Kiguradze, Tamaz, and Antonio Sagona. 2003. On the origins of the Kura-Araxes cultural complex. In *Archaeology in the Borderlands,* ed. Adam T. Smith and Karen Rubinson, pp. 38–94. Los Angeles: Cotsen Institute.

Kislenko, Aleksandr, and N. Tatarintseva. 1999. The eastern Ural steppe at the end of the Stone Age. In *Late Prehistoric Exploitation of the Eurasian Steppe,* ed. Marsha Levine, Yuri Rassamakin, A. Kislenko, and N. Tatarintseva, pp. 183–216. Cambridge: McDonald Institute for Archaeological Research.

Kitson, Peter R. 1997. Reconstruction, typology, and the "original homeland" of the Indo-Europeans. In *Linguistic Reconstruction and Typology,* ed. Jacek Fisiak, pp. 183–239, esp. pp. 198–202. Berlin: Mouton de Gruyter.

Kiyashko, V. Y. 1994. *Mezhdu Kamnem i Bronzoi.* Vol. 3. Azov: Donskie drevnosti.

———. 1987. Mnogosloinoe poselenie Razdorskoe i na Nizhnem Donu. *Kratkie soobschcheniya institut arkheologii* 192:73–79.

Klejn, L. 1984. The coming of the Aryans: Who and whence? *Bulletin of the Deccan College Research Institute* 43:57–69.

Klepikov, V. M. 1994. Pogrebeniya pozdneneoliticheskkogo vremeni u Khutora Shlyakhovskii v nizhnem Povolzh'e. *Rossiskaya arkheologiya* (3): 97–102.

Klochko, Viktor I., Aleksandr Kośko, and Marzena Szmyt. 2003. A comparative chronology of the prehistory of the area between the Vistula and the Dnieper: 4000–1000 BC. *Baltic-Pontic Studies* 12:396–414.

Knecht, Heidi, ed. 1997. *Projectile Technology*. New York: Plenum.

Kniffen, F. B. 1986. Folk housing: Key to diffusion. In *Common Places: Readings in American Vernacular Architecture*, ed. Dell V. Upton and John M. Vlach, pp. 3–23. Athens: University of Georgia Press.

Kohl, Philip, 2007. *The Making of Bronze Age Eurasia*. Cambridge: Cambridge University Press.

Kohl, Philip L., and Gocha R. Tsetskhladze. 1995. Nationalism, politics, and the practice of archaeology in the Caucasus. In *Nationalism, Politics, and the Practice of Archaeology*, ed. Philip L. Kohl and Clare Fawcett, pp. 149–174. Cambridge: Cambridge University Press.

Kohl, Philip L., Henri-Paul Francfort, and Jean-Claude Gardin. 1984. *Central Asia Palaeolithic Beginnings to the Iron Age*. Paris: Editions recherche sur les civilisations.

Koivulehto, Jorma. 2001. The earliest contacts between Indo-European and Uralic speakers in the light of lexical loans. In *Early Contacts between Uralic and Indo-European: Linguistic and Archaeological Considerations*, ed. Christian Carpelan, Asko Parpola, and Petteri Koskikallio, pp. 235–263. Memoires de la Société Finno-Ugrienne 242. Helsinki: Suomalais-Ugrilainen Seura.

Kolev, U. I., Kuznetsov, P. F., Kuz'mina, O. V., Semenova, A. P., Turetskii, M. A., and Aguzarov, B. A., eds. 2001. *Bronzovyi Vek Vostochnoi Evropy: Kharaketristika Kul'tur, Khronologiia i Periodizatsiya*. Samara: Samarskii gosudarstvennyi pedagogicheskii universitet.

Kol'tsov, L. V., ed. 1989. *Mezolit SSSR*. Moscow: Nauka.

Kopytoff, Igor. 1987. The internal African frontier: The making of African political culture. In *The African Frontier: The Reproduction of Traditional African Societies*, ed. Igor Kopytoff, pp. 3–84, Bloomington: Indiana University Press.

Korenevskii, S. N. 1995. *Galiugai I, poselenie Maikopskoi kul'tury*. Moscow: Biblioteka rossiskogo etnografa.

———. 1993. *Drevneishee osedloe naselenie na srednem Tereke*. Moscow: Stemi.

———. 1980. O metallicheskikh veshchakh i Utyevskogo mogil'nika. In *Arkheologiya Vostochno-Evropeiskoi Lesostepi*, ed. A. D. Pryakhin, pp. 59–66. Voronezh: Vorenezhskogo universiteta.

Korpusova, V. N., and S. N. Lyashko. 1990. Katakombnoe porgebenie s pshenitsei v Krimu. *Sovietskaya Arkheologiia* 3:166–175.

Korolev, A. I. 1999. Materialy po khronologii Eneolita pri Mokshan'ya. In *Voprosy Arkheologii Povolzh'ya, Sbornik Statei*, Vol. 1, ed. A. A. Vybornov and V. N. Myshkin, pp. 106–115. Samara: Samarskii gosudarstvennyi pedagogicheskii universitet.

Koryakova, L., and A. D. Epimakhov, 2007. *The Urals and Western Siberia in the Bronze and Iron Ages*. Cambridge: Cambridge University Press.

Kosintsev, Pavel. 2001. Kompleks kostnykh ostatkov domashnikh zhivotnykh iz poselenii i mogilnikov epokhi Bronzy Volgo-Ural'ya i ZaUral'ya. In *Bronzovyi Vek Vostochnoi Evropy: Kharakteristika Kul'tur, Khronologiya i Periodizatsiya*, ed. Y. I. Kolev, pp. 363–367. Samara: Samarskii gosudarstvennyi pedagogicheskii universitet.

Kośko, Aleksander, ed. 1999. *The Western Border Area of the Tripolye Culture*. Baltic-Pontic Studies 9. Poznań: Adam Mickiewicz University.

Kośko, Aleksandr, and Viktor I. Klochko, eds. 2003. *The Foundations of Radiocarbon Chronology of Cultures between the Vistula and Dnieper, 4000–1000 BC*. Baltic-Pontic Studies 12. Poznán: Adam Mickiewicz University.

Kotova, Nadezhda, and L. A. Spitsyna. 2003. Radiocarbon chronology of the middle layer of the Mikhailivka settlement. *Baltic-Pontic Studies* 12:121–131.

Kovaleva, I. F. 2001. "Vityanutye" pogrebeniya iz raskopok V. A. Gorodtsovym kurganov Donetchiny v kontekste Postmariupol'skoi kul'tury. In *Bronzovy Vek v Vostochnoi Evropy: Kharakteristika Kul'tur', Khronologiya i Periodizatsiya*, ed. Y. U. Kolev, pp. 20–24. Samara: Samara gosudarstvennyi pedagogicheskii universitet.

Kovaleva, V. T., and Zdanovich, G. B., eds. 2002. *Arkaim: Nekropol (po materialam kurgana 25 Bol'she Karaganskoe Mogil'nika)*. Chelyabinsk: Yuzhno-Ural'skoe knizhnoe izdatel'stvo.

Kovapenko, G. T., and V. M. Fomenko. 1986. Pokhovannya dobi Eneolitu-ranni Bronzi na pravoberezhzhi Pivdennogo Bugu. *Arkheologiya* (Kiev) 55:10–25.

Krahe, Hans. 1954. *Sprach und Vorzeit*. Heidelberg: Quelle und Meyer.

Kremenetski, C. V. 2002. Steppe and forest-steppe belt of Eurasia: Holocene environmental history. In *Prehistoric Steppe Adaptation and the Horse*, ed. M. Levine, C. Renfrew, and K. Boyle, pp. 11–27. Cambridge: Cambridge University Press.

———. 1997a. Human impact on the Holocene vegetation of the South Russian plain. In *Landscapes in Flux: Central and Eastern Europe in Antiquity*, ed. John Chapman and Pavel Dolukhanov, pp. 275–287. London: Oxbow Books.

———. 1997b. The Late Holocene environment and climate shift in Russia and surrounding lands. In *Climate Change in the Third Millennium BC*, ed. H. Dalfes, G. Kukla, and H. Weiss, pp. 351–370. New York: Springer.

Kremenetski, C. V., T. Böttger, F. W. Junge, A. G. Tarasov. 1999. Late- and postglacial environment of the Buzuluk area, middle Volga region, Russia. *Quaternary Science Reviews* 18:1185–1203.

Kremenetski, C. V., O. A. Chichagova, and N. I. Shishlina. 1999. Palaeoecological evidence for Holocene vegetation, climate and land-use change in the low Don basin and Kalmuk area, southern Russia. *Vegetation History and Archaeology* 8 (4): 233–246.

Kristiansen, Kristian, and Thomas Larsson. 2005. *The Rise of Bronze Age Society: Travels, Transmissions, and Transformations*. Cambridge: Cambridge University Press.

Kriukova, E. A. 2003. Obraz loshadi v iskusstve stepnogo naseleniya epokhi Eneolita-Rannei Bronzy. In *Voprosy Arkheologii Povolzh'ya*, pp. 134–143. Samara: Samarskii nauchnyi tsentr RAN.

Krizhevskaya, L. Y. 1991. *Nachalo Neolita v stepyakh severnogo Priochernomor'ya*. St. Petersburg: Institut istorii material'noi kul'tury Akademii Nauk SSSR.

Kruts, V. O. 1977. *Pozdnetripol'skie pamyatniki srednego Podneprov'ya*. Kiev: Naukovo Dumka.

Kruts, V. O., and S. M. Rizhkov, 1985, Fazi rozvitku pam'yatok Tomashivs'ko-Syshkivs'koi grupi. *Arkheologiya* (Kiev) 51:45–56.

Kryvaltsevich, Mikola M., and Nikolai Kovalyukh. 1999. Radiocarbon dating of the Middle Dnieper culture from Belarus. *Baltic Pontic Studies* 7:151–162.

Kubarev, V. D. 1988. *Drevnie rospisi Karakola*. Novosibirsk: Nauka.

Kühl, Stefan. 1994. *The Nazi Connection: Eugenics, American Racism, and German National Socialism*. New York: Oxford University Press.

Kuiper, F. B. J. 1991. *Aryans in the Rig-Veda*. Amsterdam: Rodopi.

———. 1960. The ancient Aryan verbal contest. *Indo-Iranian Journal* 4:217–281.

———. 1955. Rig Vedic Loanwords. *Studia Indologica* (Festschrift für Willibaldkirfel), pp. 137–185. Bonn: Selbst Verlag des Orientalishen Seminars des Universität.

———. 1948. *Proto-Munda Words in Sanskrit*. Amsterdam: Noord-Hollandische Uitgevers Maatschappij.

Kulick, Don. 1992. *Language Shift and Cultural Reproduction: Socialization, Self, and Syncretism in a Papuan New Guineau Village*. Cambridge: Cambridge University Press.

Kuna, Martin. 1991. The structuring of the prehistoric landscape. *Antiquity* 65:332–347.

Kuzmina, I. E. 1988. Mlekopitayushchie severnogo pri Kaspiya v Golotsene. *Arkheologocheskie Kul'tury Severnogo Prikaspiya*, ed. R. S. Bagautdinov, pp. 173–188. Kuibyshev: Samarskii gosudarstvennyi pedagogicheskii universitet.

Kuzmina, Elena E. 2003. Origins of pastoralism in the Eurasian steppes. In *Prehistoric Steppe Adaptation and the Horse*, ed. Marsha Levine, Colin Renfrew, and Katie Boyle, pp. 203–232. Cambridge: McDonald Institute for Archaeological Research.

———. 2001. The first migration wave of Indo-Iranians to the south. *Journal of Indo-European Studies* 29 (1–2): 1–40.

————. 1994. *Otkuda prishli indoarii?* Moscow: MGP "Kalina" VINITI RAN.

————. 1980. Eshche raz o diskovidniykh psaliakh Evraziiskikh stepei. *Kratkie Soobshcheniya Institut Arkheologii* 161:8–21.

Kuzmina, O. V. 1999. Keramika Abashevskoi kul'tury. In *Voprosy Arkheologii Povolzh'ya, Sbornik Statei*, vol. 1, ed. A. A. Vybornov and V. N. Myshkin, pp. 154–205. Samara: Samarskii gosudarstvennyi pedagogicheskii universitet.

Kuzminova, N. N. 1990. Paleoetnobotanicheskii i palinologicheskii analizy materialov iz kurganov nizhnego podnestrov'ya. In *Kurgany Eneolita-Eopkhi Bronzy Nizhnego Podnestrov'ya*, ed. E. V. Yarovoi, pp. 259–267. Kishinev: Shtiintsa.

Kuzminova, N. N., V. A. Dergachev, and O. V. Larina. 1998. Paleoetnobotanicheskie issledovaniya na poselenii Sakarovka I. *Revista Arheologică* (Kishinev) (2): 166–182.

Kuznetsov, Pavel. 2005. An Indo-European symbol of power in the earliest steppe kurgans. *Journal of Indo-European Studies* 33 (3–4): 325–338.

————. 2001. Territorial'nye osobennosti i vremennye ramki perekhodnogo perioda k epokhe Pozdnei Bronzy Vostochnoi Evropy. In *Bronzovyi Vek Vostochnoi Evropy: Kharakteristika Kul'tur, Khronologiya i Periodizatsiya*, ed. Y. I. Kolev et al., pp. 71–82. Samara: Samarskii gosudarstvennyi pedagogicheskii universitet.

————. 1991. Unikalnoe pogrebenie epokhi rannei Bronzy na r. Kutuluk. In *Drevnosti Vostochno-Evropeiskoi Lesostepi*, ed. V. V. Nikitin, pp. 137–139. Samara: Samarskii gosudarstvennyi pedagogicheskii institut. Labov, William. 1994. *Principles of Linguistic Change: Internal Factors*. Oxford: Blackwell.

Lafontaine, Oskar, and Georgi Jordanov, eds. 1988. *Macht, Herrschaft und Gold: Das Gräberfeld von Varna (Bulgarien) und die Anfänge Einer Neuen Europäischen Zivilisation*. Saarbrücken: Moderne Galerie des Saarland-Museums.

Lagodovskaya, E. F., O. G. Shaposhnikova, and M. L. Makarevich. 1959. Osnovnye itogi issledovaniya Mikhailovskogo poseleniya. *Kratkie soobshcheniya institut arkheologii* 9:21–28.

Lakoff, George. 1987. *Women, Fire and Dangerous Things: What Categories Reveal about the Mind*. Chicago: University of Chicago Press.

Lam, Andrew. 2006. *Learning a Language, Inventing a Future*. Commentary on National Public Radio, May 1, 2006.

————. 2005. *Perfume Dreams: Reflections on the Vietnamese Diaspora*. Foreword by Richard Rodriguez. Berkeley: Heyday Books.

Lamberg-Karlovsky, C. C. 2002. Archaeology and language: The Indo-Iranians. *Current Anthropology* 43 (1): 63–88.

Latacz, Joachim. 2004. *Troy and Homer: Toward a Solution of an Old Mystery*. Oxford: Oxford University Press.

Lattimore, Owen. 1940. *Inner Asian Frontiers of China*. Boston: Beacon.

Lavrushin, Y. A., E. A. Spiridonova, and L. L. Sulerzhitskii. 1998. Geologo-paleoekologocheskie sobytiya severa aridnoi zony v poslednie 10- tys. let. In *Problemy Drevnei Istorii Severnogo Prikaspiya*, ed. V. S. Gorbunov, pp. 40–65. Samara: Samarskogo gosudarstvennogo pedagogicheskogo universiteta.

Leach, Edmund R. 1968. *Political Systems of Highland Burma*. Boston: Beacon.

————. 1960. The frontiers of Burma. *Comparative Studies in Society and History* 3 (1): 49–68.

Lees, Robert. 1953. The basis of glottochronology. *Language* 29 (2): 113–127.

Lefferts, H. L., Jr. 1977. Frontier demography: An introduction. In *The Frontier, Comparative Studies*, ed. D. H. Miller and J. O. Steffen, pp. 33–56. Norman: University of Oklahoma Press.

Legge, Tony. 1996. The beginning of caprine domestication in southwest Asia. In *The Origins and Spread of Agriculture and Pastoralism in Eurasia*, ed. David R. Harris, pp. 238–262. London: University College London Press.

Lehman, F. K. 1989. Internal inflationary pressures in the prestige economy of the Feast of Merit complex: The Chin and Kachin cases from upper Burma. in *Ritual, Power and Economy:*

Upland-Lowland Contrasts in Mainland Southeast Asia, ed. Susan D. Russell, pp. 89–101. Occasional Paper 14. DeKalb, Ill.: Center for Southeast Asian Studies.

Lehmann, Winfred. 1989. Earlier stages of Proto-Indo-European. In *Indogermanica Europaea*, ed. K. Heller, O. Panagi, and J. Tischler, pp. 109–131. Grazer Linguistische Monographien 4. Graz: Institut für Sprachwissenschaft der Universität Graz.

Lehrman, Alexander. 2001. Reconstructing Proto-Hittite. In *Greater Anatolia and the Indo-Hittite Language Family*, ed. Robert Drews, pp. 106–130. Journal of Indo-European Studies Monograph 38. Washington, D.C.: Institute for the Study of Man.

Leuschner, Hans Hubert, Ute Sass-Klaassen, Esther Jansma, Michael Baillie, and Marco Spurk. 2002. Subfossil European bog oaks: Population dynamics and long-term growth depressions as indicators of changes in the Holocene hydro-regime and climate. *The Holocene* 12 (6): 695–706.

Levi, Scott C. 2002. *The Indian Diaspora in Central Asia and Its Trade, 1550–1900*. Leiden: Brill.

Levine, Marsha. 2004. Exploring the criteria for early horse domestication. In *Traces of Ancestry: Studies in Honor of Colin Renfrew*, ed. Martin Jones, pp. 115–126. Cambridge: McDonald Institute for Archaeological Research.

———. 2003. Focusing on Central Eurasian archaeology: East meets west. In *Prehistoric Steppe Adaptation and the Horse*, ed. Marsha Levine, Colin Renfrew, and Katie Boyle, pp. 1–7. Cambridge: McDonald Institute for Archaeological Research.

———. 1999a. Botai and the origins of horse domestication. *Journal of Anthropological Archaeology* 18:29–78.

———. 1999b. The origins of horse husbandry on the Eurasian steppe. In *Late Prehistoric Exploitation of the Eurasian Steppe*, ed. Marsha Levine, Yuri Rassamakin, Aleksandr Kislenko, and Nataliya Tatarintseva, pp. 5–58. Cambridge: McDonald Institute for Archaeological Research.

———. 1990. Dereivka and the problem of horse domestication. *Antiquity* 64:727–740.

———. 1982. The use of crown height measurements and eruption-wear sequences to age horse teeth. In *Ageing and Sexing Animal Bones from Archaeological Sites*, ed. B. Wilson, C. Grigson, and S. Payne, pp. 223–250. British Archaeological Reports, British Series 109. Oxford: Archaeopress.

Levine, Marsha, and A. M. Kislenko. 2002. New Eneolithic and Early Bronze Age radiocarbon dates for northern Kazakhstan and south Siberia. In *Ancient Interactions: East and West in Eurasia*, ed. Katie Boyle, Colin Renfrew, and Marsha Levine, pp. 131–134, Cambridge: McDonald Institute for Archaeological Research.

Levine, Marsha, Colin Renfrew, and Katie Boyle, eds. 2003. *Prehistoric Steppe Adaptation and the Horse*. Cambridge: McDonald Institute for Archaeological Research.

Li, Shuicheng. 2002. The interaction between northwest China and Central Asia during the second millennium BC: An archaeological perspective. In *Ancient Interactions: East and West in Eurasia*, ed. Katie Boyle, Colin Renfrew, and Marsha Levine, pp. 171–182. Cambridge: McDonald Institute for Archaeological Research.

Lichardus, Jan, ed. 1991. *Die Kupferzeit als historische Epoche*. Bonn: Dr. Rudolf Hebelt GMBH.

Lichardus, Jan, and Josef Vladar. 1996. Karpatenbecken-Sintashta-Mykene: ein Beitrag zur Definition der Bronzezeit als Historischer Epoche. *Slovenska Archeologia* 44 (1): 25–93.

Lillie, Malcolm C. 1996. Mesolithic and Neolithic populations in Ukraine: Indications of diet from dental pathology. *Current Anthropology* 37 (1): 135–142.

Lillie, Malcolm C., and M. P. Richards. 2000. Stable isotope analysis and dental evidence of diet at the Mesolithic-Neolithic transition in Ukraine. *Journal of Archaeological Science* 27:965–972.

Lincoln, Bruce. 1981. *Priests, Warriors, and Cattle: A Study in the Ecology of Religions*. Berkeley: University of California Press.

———. 1991. *Death, War and Sacrifice: Studies in Ideology and Practice*. Chicago: University of Chicago Press.

Lindgren, G., N. Backström, J. Swinburne, L. Hellborg, A. Einarsson, K. Sandberg, G. Co-thran, Carles Vilà, M. Binns, and H. Ellegren. 2004. Limited number of patrilines in horse domestication. *Nature Genetics* 36 (3): 335–336.

Lindstrom, Richard W. 2002. Anthropological characteristics of the population of the Bol-shekaragansky cemetery, kurgan 25. In *Arkaim: Nekropol (po materialam kurgana 25 Bol'she Karaganskoe Mogil'nika)*, ed. V. T. Kovaleva and G.B. Zdanovich, pp. 159–166, Chelyabinsk: Yuzhno-Ural'skoe knizhnoe izdatel'stvo.

Linduff, Katheryn M., Han Rubin, and Sun Shuyun, eds. 2000. *The Beginnings of Metallurgy in China*. New York: Edwin Mellen Press.

Lisitsyn, N. F. 1996. Srednii etap pozdnego Paleolita Sibiri. *Rossiskaya arkheologiya* (4): 5–17.

Littauer, Mary A. 1977. Rock carvings of chariots in Transcaucasia, Central Asia, and Outer Mongolia. *Proceedings of the Prehistoric Society* 43:243–262.

———. 1972. The military use of the chariot in the Aegean in the Late Bronze Age. *American Journal of Archaeology* 76:145–157.

———. 1968. A 19th and 20th dynasty heroic motif on Attic black-figured vases? *American Journal of Archaeology* 72:150–152.

Littauer, Mary A., and Joost H. Crouwel. 1996. The origin of the true chariot. *Antiquity* 70:934–939.

———. 1986. A Near Eastern bridle bit of the second millennium BC in New York. *Levant* 18:163–167.

———. 1983. Chariots in Late Bronze Age Greece. *Antiquity* 57:187–192.

———. 1979. *Wheeled Vehicles and Ridden Animals in the Ancient Near East*. Leiden: Brill.

Littleton, C. S. 1982. *The New Comparative Mythology*. Berkeley: University of California Press.

Litvinsky, B. A., and L. T. P'yankova. 1992. Pastoral tribes of the Bronze Age in the Oxus valley (Bactria). In *History of the Civilizations of Central Asia*, ed. A. H. Dani and V. M. Masson, vol. 1, pp. 379–394. Paris: UNESCO.

Logvin, V. N. 1995. K probleme stanovleniya Sintashtinsko-Petrovskikh drevnostei. In *Rossiya i Vostok: Problemy Vzaimodeistviya*, pt. 5, bk. 1: *Kul'tury Eneolita-Bronzy Stepnoi Evrazii*, pp. 88–95. Chelyabinsk: 3-ya Mezhdunarodnaya nauchnaya konferentsiya.

———. 1992. Poseleniya Tersekskogo tipa Solenoe Ozero I. *Rossiskaya arkheologiya* (1): 110–120.

Logvin, V. N., S. S. Kalieva, and L. L. Gaiduchenko. 1989. O nomadizme v stepyakh Kazakh-stana v III tys. do n. e. In *Margulanovskie chteniya*, pp. 78–81. Alma-Ata: Akademie Nauk Kazakhskoi SSR.

Lubotsky, Alexsander. 2001. The Indo-Iranian substratum. In *Early Contacts between Uralic and Indo-European: Linguistic and Archaeological Considerations*, ed. Christian Carpelan, Asko Parpola, and Petteri Koskikallio, pp. 301–317. Helsinki: Suomalais-Ugrilainen Seura.

Lukacs, J. R. 1989. Dental paleopathology: Methods for reconstructing dietary patterns. In *Reconstruction of Life From the Skeleton*, ed. M. Y. Iscan, and K. A. R. Kennedy, pp. 261–286. New York: Alan Liss.

Lyashko, S. N., and V. V. Otroshchenko. 1988. Balkovskii kurgan. In *Novye pamyatniki yam-noi kul'tury stepnoi zony Ukrainy*, ed. A. A. Zolotareva, pp. 40–63. Kiev: Naukovo Dumka.

Lyonnet, B., ed. 1996. *Sarazm (Tajikistan). Céramiques (Chalcolithiques et Bronze Ancien)*. Mémoire de la Mission Archéologique Française en Asie Centrale 7. Paris: De Boccard.

Mace, Ruth. 1993. Transitions between cultivation and pastoralism in sub-Saharan Africa. *Current Anthropology* 34 (4): 363–382.

MacEachern, Scott. 2000. Genes, tribes, and African history. *Current Anthropology* 41 (3): 357–384.

Machnik, Jan. 1999. Radiocarbon chronology of the Corded Ware culture on Grzeda Sokalska: A Middle Dnieper traits perspective. *Baltic-Pontic Studies* 7:221–250.

Madgearu, Alexandru. 2001. The end of town life in Scythia Minor. *Oxford Journal of Archaeology* 20 (2): 207–217.

Makkay, Janos. 2000. *The Early Mycenaean Rulers and the Contemporary Early Iranians of the Northeast*. Tractata Miniscula 22. Budapest: szerzo kiadása.

———. 1976. Problems concerning Copper Age chronology in the Carpathian Basin: Copper Age gold pendants and gold discs in central and south-east Europe. *Acta Archaeologica Hungarica* 28:251–300.

Malandra, William. 1983. *An Introduction to Ancient Iranian Religion*. Minneapolis: University of Minnesota Press.

Maliutina, T. S. 1991. Stratigraficheskaya pozitsiya materilaov Fedeorovskoi kul'tury na mnogosloinikh poseleniyakh Kazakhstanskikh stepei. In *Drevnosti Vostochno-Evropeiskoi Lesostepi*, ed. V. V. Nikitin, pp. 141–162. Samara: Samarskii gosudarstvennyi pedagogicheskii institut.

———. 1984. Mogil'nik Priplodnyi Log 1. In *Bronzovyi Vek Uralo-Irtyshskogo Mezhdurech'ya*, pp. 58–79. Chelyabinsk: Chelyabinskii gosudarstvennyi universitet.

Maliutina, T. S., and G. B. Zdanovich. 1995. Kuisak—ukreplennoe poselenie protogorodskoi tsivilizatsii iuzhnogo zaUral'ya. In *Rossiya i Vostok: Problemy Vzaimodeistviya*, pt. 5, bk. 1: *Kul'tury Eneolita-Bronzy Stepnoi Evrazii*, pp. 100–106. Chelyabinsk: 3-ya Mezhdunarodnaya nauchnaya konferentsiya.

Mallory, J. P. 1998. A European perspective on Indo-Europeans in Asia. In *The Bronze Age and Early Iron Age Peoples of Eastern Central Asia*, ed. Victor H. Mair, vol. 1, pp. 175–201. Philadelphia: University of Pennsylvania Press.

———. 1992. Migration and language change. *Peregrinatio Gothica III, Universitetets Oldsaksamlings Skrifter Ny Rekke* (Oslo) 14:145–153.

———. 1990. Social structure in the Pontic-Caspian Eneolithic: A preliminary review. *Journal of Indo-European Studies* 18 (1–2): 15–57.

———. 1989. *In Search of the Indo-Europeans*. London: Thames and Hudson.

———. 1977. The chronology of the early Kurgan tradition. *Journal of Indo-European Studies* 5:339–368.

Mallory, J. P., and Douglas Q. Adams. 1997. *Encyclopedia of Indo-European Culture*. London: Fitzroy Dearborn.

Mallory, J. P., and Victor H. Mair. 2000. *The Tarim Mummies: Ancient China and the Mystery of the Earliest Peoples from the West*. London: Thames and Hudson.

Malov, N. M. 2002. Spears: Signs of archaic leaders of the Pokrovsk archaeological cultures. In *Complex Societies of Central Eurasia from the 3rd to the 1st Millennium BC*, vols. 1 and 2, ed. Karlene Jones-Bley and D. G. Zdanovich, pp. 314–336. Journal of Indo-European Studies Monograph 45. Washington, D.C.: Institute for the Study of Man.

Mamonov, A. E. 1995. Elshanskii kompleks stoianki Chekalino IV. In *Drevnie kul'tury lesostepnogo povolzh'ya*, pp. 3–25. Samara: Samarskogo gosudarstvennogo pedagogicheskogo universiteta.

Mamontov, V. I. 1974. Pozdneneoliticheskaya stoianka Orlovka. *Sovietskaya arkheologiya* (4): 254–258.

Manfredi, J., Hilary M. Clayton, and D. Rosenstein. 2005. Radiographic study of bit position within the horse's oral cavity. *Equine and Comparative Exercise Physiology* 2 (3): 195–201.

Manhart, H. 1998. Die vorgeschichtliche Tierwelt von Koprivec und Durankulak und anderen prähistorischen Fundplätzen in Bulgarien aufgrund von Knochenfunden aus archäologischen Ausgrabungen. *Documenta Naturae* (München) 116:1–353.

Manzura, I. 1999. The Cernavoda I culture. In *The Balkans in Later Prehistory*, ed. Lolita Nikolova, pp. 95–174. British Archaeological Reports, International Series 791. Oxford: Archaeopress.

Manzura, I., E. Savva, and L. Bogotaya. 1995. East-west interactions in the Eneolithic and Bronze Age cultures of the north-west Pontic region. *Journal of Indo-European Studies* 23 (1–2): 1–51.

Maran, Joseph. 2001. Zur Westausbreitung von Boleráz-Elementen in Mitteleuropa. In *Cernavoda III-Boleráz, Ein vorgeschichtliches Phänomen zwischen dem Oberrhein und der unteren Donau*, ed. P. Roman, and S. Diamandi, pp. 733–752. Bucharest: Studia Danubiana.

————. 1998. *Kulturwandel auf dem Griechischen Festland und den Kykladen im späten 3. Jahrtausend v. Chr.* Bonn: Habelt.

Marcsik, Antónia. 1971. Data of the Copper Age anthropological find of Bárdos-Farmstead at Csongrád-Kettöshalom. *A Móra Ferenc Múzeum Évkönyve* (2): 19–27.

Marinescu-Bîlcu, S. 1981. Tirpeşti: From prehistory to history in Eastern Romania. British Archaeological Reports, International Series 107. Oxford: Archeopress.

Marinescu-Bîlcu, S., Alexandra Bolomey, Marin Cârciumâru, and Adrian Muraru. 1984. Ecological, economic and behavioral aspects of the Cucuteni A4 community at Draguşeni. *Dacia* 28 (1–2): 41–46.

Marinesu-Bîlcu, Silvia, M. Cârciumaru, and A. Muraru. 1981. Contributions to the ecology of pre- and proto-historic habitations at Tîrpeşti. *Dacia* 25:7–31.

Marinova, Elena. 2003. The new pollen core Lake Durankulak-3: The vegetation history and human impact in Northeastern Bulgaria. In *Aspects of Palynology and Paleontology*, ed. S. Tonkov, pp. 279–288. Sofia: Pensoft.

Markevich, V. I. 1974. *Bugo-Dnestrovskaya kul'tura na territorii Moldavii.* Kishinev: Shtintsa.

————. 1965. Issledovaniia Neolita na srednem Dnestre. *Kratkie soobshcheniya institut arkheologii* 105:85–90.

Markey, T. L. 1990. Gift, payment, and reward revisited. In *When Worlds Collide: The Indo-Europeans and the Pre-Indo-Europeans*, ed. T. L. Markey and John Grippin, pp. 345–362. Ann Arbor, Mich.: Karoma.

Markovin, V. I. 1980. O nekotorykh voprosakh interpretatsii dol'mennykh i drugikh arkheologicheskikh pamyatnikov Kavkaza. *Kratkie soobshchenniya institut arkheologii* 161:36–45.

Mashkour, Marjan. 2003. Equids in the northern part of the Iranian central plateau from the Neolithic to the Iron Age: New zoogeographic evidence. In *Prehistoric Steppe Adaptation and the Horse*, ed. Marsha Levine, Colin Renfrew, and Katie Boyle, pp. 129–138. Cambridge: McDonald Institute for Archaeological Research.

Masson, V. M. 1988. *Altyn-Depe.* Translated by Henry N. Michael. University Museum Monograph 55. Philadelphia: University of Pennsylvania Press.

————. 1979. Dinamika razvitiya Tripol'skogo obshchestva v svete paleo-demograficheskikh otsenok. In *Pervobytnaya Arkheologiya, Poiski i Nakhodki*, ed. N. N. Bondar and D. Y. Telegin, pp. 204–212. Kiev: Naukovo Dumka.

Matiushchenko, V. I., and G. V. Sinitsyna. 1988. *Mogil'nik u d. Rostovka Vblizi Omska.* Tomsk: Tomskogo universiteta.

Matiushin, G. N. 1986. The Mesolithic and Neolithic in the southern Urals and Central Asia. In *Hunters in Transition: Mesolithic Societies of Temperate Eurasia and Their Transition to Farming*, ed. M. Zvelebil, pp. 133–150. Cambridge: Cambridge University Press.

Matthews, Roger, and Hassan Fazeli. 2004. Copper and complexity: Iran and Mesopotamia in the fourth millennium BC. *Iran* 42:61–75.

McMahon, April, and Robert McMahon. 2003. Finding families: Quantitative methods in language classification. *Transactions of the Philological Society* 10:7–55.

Meadow, Richard H., and Ajita Patel. 1997. A comment on "Horse Remains from Surkotada" by Sándor Bökönyi. *South Asian Studies* 13:308–315.

Mei, Jianjun. 2003a. Cultural interaction between China and Central Asia during the Bronze Age. *Proceedings of the British Academy* 121:1–39.

————. 2003b. Qijia and Seima-Turbino: The question of early contacts between northwest China and the Eurasian steppe. *Bulletin of the Museum of Far Eastern Antiquities* 75: 31–54.

Meid, Wolfgang. 1994. Die Terminologie von Pferd und Wagen im Indogermanischen. In *Die Indogermanen und das Pferd*, ed. B. Hänsel and S. Zimmer, pp. 53–65. Budapest: Archaeolingua.

————. 1975. Probleme der räumlichen und zeitlichen Gliederung des Indogermanischen. In *Flexion und Wortbildung*, ed. Helmut Rix, pp. 204–219. Weisbaden: Reichert.

Melchert, Craig. 2001. Critical responses. In *Greater Anatolia and the Indo-Hittite Language Family*, ed. Robert Drews, pp. 229–235. Journal of Indo-European Studies Monograph 38. Washington, D.C.: Institute for the Study of Man.

———. 1994. *Anatolian Historical Phonology*. Amsterdam: Rodopi.

Melent'ev, A. N. 1975. Pamyatniki seroglazivskoi kul'tury (neolit Severnogo Prikaspiya). *Kratkie soobshcheniya institut arkheologii* (Moscow) 141:112–118.

Mel'nik, A. A., and I. L. Serdiukova. 1988. Rekonstruktsiya pogrebal'noi povozki Yamnoi kul'tury. In *Novye pamyatniki yamnoi kul'tury stepnoi zony Ukrainy*, ed. N. N. Bondar and D. Y. Telegin, pp. 118–124. Kiev: Dumka.

Merpert, N. Y. 1995. Bulgaro-Russian archaeological investigations in the Balkans. *Ancient Civilizations from Scythia to Siberia* 2 (3): 364–383.

———. 1980. Rannie skotovody vostochnoi Evropy i sudby drevneishikh tsivilizatsii. *Studia Praehistorica* 3:65–90.

———. 1974. *Drevneishie Skotovody Volzhsko-Uralskogo Mezhdurechya*. Moscow: Nauka.

Mezhlumian, S. K. 1990. Domestic horse in Armenia. Paper delivered at the International Conference on Archaeozoology, Washington, D.C.

Milisauskas, Sarunas. 2002. *European Prehistory: A Survey*. New York: Kluwer.

Militarev, Alexander. 2002. The prehistory of a dispersal: The Proto-Afrasian (Afroasiatic) farming lexicon. In *Examining the Farming/Language Dispersal Hypothesis*, ed. Peter Bellwood and Colin Renfrew, pp. 135–150. Cambridge: McDonald Institute for Archaeological Research.

Milroy, James. 1992. *Linguistic Variation and Change*. Oxford: Blackwell.

Molleson, Theya, and Joel Blondiaux. 1994. Riders' bones from Kish, Iraq. *Cambridge Archaeological Journal* 4 (2): 312–316.

Molodin, V. I. 1997. Nekotoriye itogi arkheologicheskikh isseldovanii na Iuge Gornogo Altaya. *Rossiiskaya arkheologiya* (1): 37–49.

Moore, John. 2001. Ethnogenetic patterns in Native North America. In *Archaeology, Language, and History*, ed. John E. Terrell, pp. 31–56. Westport, Conn.: Bergin and Garvey.

Moorey, P. R. S. 1986. The emergence of the light, horse-drawn chariot in the Near East, c. 2000–1500 BC. *World Archaeology* 18 (2): 196–215.

Morgunova, N. L. 1995. Elitnye kurgany eopkhi rannei I srednei bronzy v stepnom Orenburzh'e. In *Rossiya i Vostok: Problemy Vzaimodeistviya*, pt. 5, bk. 1, *Kul'tury Eneolita-Bronzy Stepnoi Evrazii*, pp. 120–123. Chelyabinsk: 3-ya Mezhdunarodnaya nauchnaya konferentsiya.

———. 1988. Ivanovskaya stoianka v Orenburgskoi oblasti. In *Arkheologocheskie kul'tury severnogo prikaspiya*, ed. R. S. Bagautdinov, pp. 106–122. Kuibyshev: Samarskii gosudarstvennyi pedagogicheskii universitet.

Morgunova, N. L., and M. A. Turetskii. 2003. Yamnye pamyatniki u s. Shumaevo: novye dannye o kolesnom transporte u naseleniya zapadnogo Orenburzh'ya v epokha rannego metalla. In *Voprosy arkheologii povozh'ya*, vol. 3, pp. 144–159. Samara: Samarskii nauchnyi tsentr RAN.

Morintz, Sebastian, and Petre Roman. 1968. Aspekte des Ausgangs des Äneolithikums und der Übergangsstufe zur Bronzezeit im Raum der Niederdonau. *Dacia* 12:45–128.

Movsha, T. G. 1985. Bzaemovidnosini Tripillya-Kukuteni z sinkhronimi kul'turami Tsentral'noi Evropi. *Arkheologiia* (Kiev) 51:22–31.

Mufwene, Salikoko. 2001. *The Ecology of Language Evolution*. Cambridge: Cambridge University Press.

Muhly, J. D. 1995. Mining and Metalwork in Ancient Western Asia. In *Civilizations of the Ancient Near East*, ed. Jack M. Sasson, John Baines, Gary Beckman, and Karen R. Rubinson, vol. 3, pp. 1501–1519. New York: Scribner's.

Munchaev, R. M. 1994. Maikopskaya kul'tura. In *Epokha Bronzy Kavkaza i Srednei Azii: Rannyaya i Srednyaya Bronza Kavkaza*, ed. K. X. Kushnareva and V. I. Markovin, pp. 158–225. Moscow: Nauka.

————. 1982. Voprosy khozyaistva i obshchestvennogo stroya Eneoliticheskikh plemen Kavkaza. In *Eneolit SSSR*, ed. V. M. Masson and N. Y. Merpert, pp. 132–137. Moscow: Akademiya nauk.

Murphy, Eileen. 2003. *Iron Age Archaeology and Trauma from Aymyrlyg, South Siberia*. British Archaeological Reports International Series 1152. Oxford: Archeopress.

Murphy, Eileen, and Aleksandr Kokhlov. 2004. Osteological and paleopathological analysis of Volga populations from the Eneolithic to the Srubnaya periods. Samara Valley Project Interim Reports, private manuscript.

Muscarella, Oscar W. 2003. The chronology and culture of Se Girdan: Phase III. *Ancient Civilizations* 9 (1–2): 117–131.

Mytum, Harold. 1994. Language as symbol in churchyard monuments: the use of Welsh in nineteenth and twentieth-century Pembrokeshire. *World Archaeology* 26 (2): 252–267.

Napol'skikh, V. V. 1997. *Vvedenie v Istoricheskuiu Uralistiku*. Izhevsk: Udmurtskii institut istorii, yazika i literatury.

Nash, Gary. 1984. Social development. In *Colonial British America*, ed. Jack P. Green and J. R. Pole, pp. 233–261. Baltimore, Md.: Johns Hopkins University Press.

Nechitailo, A. P. 1996. Evropeiskaya stepnaya obshchnost' v epokhu Eneolita. *Rossiiskaya arkheologiya* (4): 18–30.

————. 1991. *Svyazi naseleniya stepnoi Ukrainy i severnogo Kavkaza v epokhy Bronzy*. Kiev: Nauknovo Dumka.

Necrasov, Olga. 1985. Doneés anthropologiques concernant la population du complexe culturel Cucuteni-Ariuşd-Tripolié: Phases Cucuteni et Ariuşd. *Annuaire Roumain D'Anthropologie (Bucarest)* 22:17–23.

Necrasov, Olga, and M. Cristescu. 1973. Structure anthropologique des tribus Neo-Eneolithiques et de l'age du Bronze de la Roumanie. In *Die Anfänge des Neolithikums vom Orient bis Nordeuropa VIIIa, Fundamenta*, vol. 3, pp. 137–152. Cologne: Institut für Ur-und Frügeschichte der Universität zu Köln.

Nefedkin, A. K. and E. D. Frolov. 2001. *Boevye kolesnitsy i kolesnichie drevnikh Grekov (XVI–I vv. do n.e.)*. St. Petersburg: Peterburgskoe Vostokovedenie.

Nekhaev, A. A. 1992. Domakiopskaya kul'tura severnogo Kavkaza. *Arkheologicheskie vesti* 1:76–96.

————. 1986. Pogrebenie Maikopskoi kul'tury iz kurgana u s. Krasnogvardeiskoe. *Sovietskaya arkheologiya* (1): 244–248.

Neprina, V. I. 1970. Neolitichne poselenniya v Girli r. Gnilop'yati. *Arkheologiya* (Kiev) 24:100–111.

Nettles, Daniel. 1996. Language diversity in West Africa: An ecological approach. *Journal of Anthropological Archaeology* 15:403–438.

Neustupny, E. 1991. Community areas of prehistoric farmers in Bohemia. *Antiquity* 65:326–331.

Nica, Marin. 1977. Cîrcea, cea mai veche aşezare neolită de la sud de carpaţi. *Studii si Cercetări de Istore Veche şi Arheologie* 27 (4): 4, 435–463.

Nichols, Johanna. 1997a. The epicentre of the Indo-European linguistic spread. In *Archaeology and Language, I vol. 1, Theoretical and Methodological Orientations*, ed. Roger Blench, and Matthew Spriggs, pp. 122–148. London: Routledge.

————. 1997b. Modeling ancient population structures and movement in linguistics. *Annual Review of Anthropology* 26:359–384.

————. 1994. The spread of languages around the Pacific rim. *Evolutionary Anthropology* 3:206–215.

————. 1992. *Linguistic Diversity in Space and Time*. Chicago: University of Chicago Press.

Nikolova, A. V., and Y. Y. Rassamakin. 1985. O pozdneeneoliticheskie pamyatnikakh pravoberezh'ya Dnepra. *Sovietskaya arkheologiya* (3):37–56.

Nikolova, Lolita. 2005. Social changes and cultural interactions in later Balkan prehistory (later fifth and fourth millennia calBC). *Reports of Prehistoric Research Projects* 6–7:87-96. Salt Lake City, Utah: International Institute of Anthropology.

————. 2002. Diversity of prehistoric burial customs. In *Material Evidence and Cultural Pattern in Prehistory*, ed. L. Nikolova, pp. 53–87. Salt Lake City: International Institute of Anthropology.

————. 2000. Social transformations and evolution in the Balkans in the fourth and third millennia BC. In *Analyzing the Bronze Age*, ed. L. Nikolova, pp. 1–8. Sofia: Prehistory Foundation.

————. 1996. Settlements and ceramics: The experience of Early Bronze Age in Bulgaria. In *Early Bronze Age Settlement Patterns in the Balkans*, pt. 2, ed. Lolita Nikolova, pp. 145–186. Sofia: Reports of Prehistoric Research Projects 1 (2–4).

————. 1994. On the Pit-Grave culture in northeastern Bulgaria. *Helis* (Sofia) 3:27–42.

Nobis, G. 1971. *Vom Wildpferd zum Hauspferd*. Fundamenta Reihe B, vol. 6. Cologne: Bohlau-Verlag.

Noble, Allen G. 1992. Migration to North America: Before, during, and after the nineteenth century. In *To Build in a New Land: Ethnic Landscapes in North America*, ed. Allen G. Noble, pp. 3–24. Baltimore, Md.: Johns Hopkins University Press.

Noelle, Christine. 1997. *State and Tribe in Nineteenth-Century Afghanistan: The reign of Amir Dost Muhammad Khan (1826–1863)*. Richmond, Surrey: Curzon.

Oates, Joan. 2003. A note on the early evidence for horse and the riding of equids in Western Asia. In *Prehistoric Steppe Adaptation and the Horse*, ed. Marsha Levine, Colin Renfrew, and Katie Boyle, pp. 115–125. Cambridge: McDonald Institute for Archaeological Research.

————. 2001. Equid figurines and "chariot" models. In *Excavations at Tell Brak*, ed. David Oates, Joan Oates, and Helen McDonald, vol. 2, pp. 279–293. Cambridge: McDonald Institute for Archaeological Research.

O'Brien, S. R., P. A. Mayewski, L. D. Meeker, D. A. Meese, M. S. Twickler, and S. I. Whitlow. 1995. Complexity of Holocene climate as reconstructed from a Greenland ice core. *Science* 270:1962–1964.

O'Flaherty, Wendy Doniger. 1981. *The Rig Veda: An Anthology*. London: Penguin.

Olsen, Sandra. 2003. The exploitation of horses at Botai, Kazakhstan. In *Prehistoric Steppe Adaptation and the Horse*, ed. Marsha Levine, Colin Renfrew, and Katie Boyle, pp. 83–104. Cambridge: McDonald Institute for Archaeological Research.

Okhrimenko, G. V., and D. Y. Telegin. 1982. Novi pam'yatki mezolitu ta neolitu Volini. *Arkheologiya* (Kiev) 39:64–77.

Ostroshchenko, V. V. 2003. The economic preculiarities of the Srubnaya cultural-historical entity. In *Prehistoric Steppe Adaptation and the Horse*, ed. Marsha Levine, Colin Renfrew, and Katie Boyle, pp. 319–328. Cambridge: McDonald Institute for Archaeological Research.

Ostroverkhov, A. S. 1985. Steklyannye busy v pamyatnikakh pozdnego Tripolya. In *Novye materialy po arkheologii severo-zapadnogo prichernomorya*, ed. V. N. Stanko, pp. 174–180. Kiev: Naukovo Dumka.

Ottaway, Barbara S., ed. 1999. *A Changing Place: The Galgenberg in Lower Bavaria from the Fifth to the First Millennium BC*. British Archaeological Reports, n.s. 752. Oxford: Archeopress.

Owen, David I. 1991. The first equestrian: An UrIII glyptic scene. *Acta Sumerologica* 13:259–273.

Özbal, H., A. Adriaens, and B. Earl. 2000. Hacinebi metal production and exchange. *Paleorient* 25 (1): 57–65.

Panayotov, Ivan. 1989. *Yamnata Kultuyra v B'lgarskite Zemi*. Vol. 21. Sofia: Razkopki i Prouchvaniya.

Parker, Bradley. 2006. Toward an understanding of borderland processes. *American Antiquity* 71 (1): 77–100.

Parpola, Asko. 2004–2005. The Nāsatyas, the chariot, and Proto-Aryan religion. *Journal of Indological Studies* 16, 17:1–63.

————. 2002. From the dialects of Old Indo-Aryan to Proto-Indo-Aryan and Proto-Iranian. In *Indo-Iranian Languages and Peoples*, ed. N. Sims-Williams, pp. 43–102. London: Oxford University Press.

————. 1988. The coming of the Aryans to Iran and India and the cultural and ethnic identity of the Dāsas. *Studia Orientalia* (Helsinki) 64:195–302.

Parzinger, H. 2002. Germanskii Arkheologogicheskii Institut: zadachi i perspektivy arkheologicheskogo izucheniya Evrazii. *Rossiiskaya arkheologiya* (3): 59–78.

————. 1993. *Studien zur Chronologie und Kulturgeschichte der Jungstein, Kupfer- und Frühbronzezeit Zwischen Karpaten und Mittelerem Taurus.* Mainz am Rhein: Römish-Germanische Forschungen B 52.

————. 1992. Hornstaad-Hlinskoe-Stollhof: Zur absoluten datierung eines vor-Badenzeitlichen Horizontes. *Germania* 70:241–250.

Parzinger, Hermann, and Nikolaus Boroffka. 2003. *Das Zinn der Bronzezeit in Mittelasien.* Vol. 1, *Die siedlungsarchäologischen Forschgungen im Umfeld der Zinnlagerstätten.* Archäologie in Iran und Turan, Band 5. Mainz am Rhein: Philipp von Zabern.

Pashkevich. G. O. 2003. Paleoethnobotanical evidence of agriculture in the steppe and the forest-steppe of east Europe in the late Neolithic and the Bronze Age. In *Prehistoric Steppe Adaptation and the Horse,* ed. Marsha Levine, Colin Renfrew, and Katie Boyle, pp. 287–297. Cambridge: McDonald Institute for Archaeological Research.

————. 1992. Do rekonstruktsii asortmentu kul'turnikh roslin epokhi Neolitu-Bronzi na territorii Ukraini. In *Starodavne Vibornitstvo na Teritorii Ukraini,* ed. S. V. Pan'kov and G. O. Voznesens'ka, pp. 179–194. Kiev: Naukovo Dumka.

Patovka, E. F. 1976. Usatovo: iz istorii issledovaniya. *Materiali i issledovaniya po arkheologii severnogo prichernomoriya* (Kiev) 8:49–60.

Patovka, E. F., et al. 1989. *Pamyatniki tripol'skoi kul'tury v severo-zapadnom prichernomor'ye.* Kiev: Naukovo Dumka.

Payne, Sebastian. 1995. Appendix B. In *The Gordion Excavations (1950–1973) Final Reports,* vol. 2, pt. 1, *The Lesser Phrygian Tumuli: The Inhumations,* ed. Ellen L. Kohler. Philadelphia: University Museum Press.

Paunescu, Alexandru. 1987. Tardenoasianul din Dobrogea. *Studii şi Cercetări de Istorie Veche şi Arheologie* 38 (1): 3–22.

Penner, Sylvia. *Schliemanns Schachtgräberund und der Europäische Nordosten: Studien zur Herkunft der frühmykenischen Streitwagenausstattung.* Vol. 60. Bonn: Saarbrücker Beiträge zur Alterumskunde.

Penny, Ralph. 2000. *Variation and Change in Spanish.* Cambridge: Cambridge University Press.

Perles, Catherine. 2001. *Early Neolithic Greece.* Cambridge: Cambridge University Press.

Pernicka, Ernst, et al. 1997. Prehistoric copper in Bulgaria. *Eurasia Antiqua* 3:41–179.

Perry, C. A., and K. J. Hsu. 2000. Geophysical, archaeological, and historical evidence support a solar-output model for climate change. *Proceedings of the National Academy of Sciences* 7 (23): 12,433–12,438.

Peške, Lubomir. 1986. Domesticated horses in the Lengyel culture? In *Internationales Symposium Über die Lengyel-Kultur,* pp. 221–226. Nitra-Wien: Archäologisches Institut der Slowakischen Akademie der Wissenschaften in Nitra.

Peterson, Nicholas. 1993. Demand sharing: Reciprocity and the pressure for generosity among foragers. *American Anthropologist* 95 (4): 860–874.

Petrenko, A. G. 1984. *Drevnee i srednevekovoe zhivotnovodstvo srednego povolzh'ya i predural'ya.* Moscow: Nauka.

Piggott, Stuart. 1992. *Wagon, Chariot and Carriage: Symbol and Status in the History of Transport.* London: Thames and Hudson.

————. 1983. *The Earliest Wheeled Transport: From the Atlantic Coast to the Caspian Sea.* New York: Cornell University Press.

————. 1974. Chariots in the Caucasus and China. *Antiquity* 48:16–24.

————. 1962. Heads and hoofs. *Antiquity* 36 (142): 110–118.

Pinker, Steven. 1994. *The Language Instinct.* New York: William Morrow.

Pogozheva, A. P. 1983. *Antropomorfnaya Plastika Tripol'ya*. Novosibirsk: Akademiia nauk, Sibirskoe otdelenie.

Poliakov, Leon. 1974. *The Aryan Myth: A History of Racist and Nationalist Ideas in Europe*. Translated by Edmund Howard. New York: Basic Books.

Pollack, Susan. 1999. *Ancient Mesopotamia*. Cambridge: Cambridge University Press.

Polomé, Edgar C. 1991. Indo-European religion and the Indo-European religious vocabulary. In *Sprung from Some Common Source: Investigations into the Prehistory of Languages*, ed. S. M. Lamb and E. D. Mitchell, pp. 67–88. Stanford: Stanford University Press.

————. 1990. Types of linguistic evidence for early contact: Indo-Europeans and non-Indo-Europeans. In *When Worlds Collide: Indo-Europeans and the Pre-Indo-Europeans*, ed. T. L. Markey, and John A. C. Greppin, pp. 267–289. Ann Arbor, Mich.: Karoma.

Popova, T. A. 1979. Kremneobrabatyvaiushchee proizvodstvo Tripol'skikh plemen. In *Pervobytnaya Arkheologiya, Poiski i Nakhodki*, ed. N. N. Bondar and D. Y. Telegin, pp. 145–163. Kiev: Nauknovo Dumka.

Porter, John. 1965. *The Vertical Mosaic: An Analysis of Social Class and Power in Canada*. Toronto: University of Toronto Press.

Potekhina, I. D. 1999. *Naselenie Ukrainy v Epokhi Neolita i Rannego Eneolita*. Kiev: Insitut arkheologii.

Potts, Dan T. 2000. *Ancient Magan: The Secrets of Tell Abraq*. London: Trident.

————. 1999. *The Archaeology of Elam*. Cambridge: Cambridge University Press.

Prescott, J. R. V. 1987. *Political Frontiers and Boundaries*. London: Unwin Hyman.

Privat, Karen. 2002. Preliminary report of paleodietary analysis of human and faunal remains from Bolshekaragansky kurgan 25. In *Arkaim: Nekropol (po materialam kurgana 25 Bol'she Karaganskoe Mogil'nika)*, ed. V. T. Kovaleva, and G. B. Zdanovich, pp. 166–171. Chelyabinsk: Yuzhno-Ural'skoe knizhnoe izdatel'stvo.

Pryakhin, A. D., 1980. Abashevskaya kul'turno-istoricheskaya obshchnost' epokhi bronzy i lesostepe. In *Arkheologiya Vostochno-Evropeiskoi Lesostepi*, ed. A. D. Pryakhin, pp. 7–32. Voronezh: Voronezhskogo universiteta.

————. 1976. *Poseleniya Abashevskoi Obshchnosti*. Voronezh: Voronezhskogo universiteta.

Pryakhin, A. D., and V. I. Besedin. 1999. The horse bridle of the Middle Bronze Age in the East European forest-steppe and the steppe. *Anthropology and Archaeology of Eurasia* 38 (1): 39–59.

Puhvel, Jaan. 1994. Anatolian: Autochthonous or interloper? *Journal of Indo-European Studies* 22:251–263.

————. 1991. Whence the Hittite, whither the Jonesian vision? In *Sprung from Some Common Source*, ed. Sydney M. Lamb and E. D. Mitchell, pp. 52–66. Stanford: Stanford University Press.

————. 1975. Remus et Frater. *History of Religions* 15:146–157.

Pulgram, E. 1959. Proto-Indo-European reality and reconstruction. *Language* 35:421–426

Rassamakin, Yuri. 2002. Aspects of Pontic steppe development (4550–3000 BC) in the light of the new cultural-chronological model. In *Ancient Interactions: East and West in Eurasia*, ed. Katie Boyle, Colin Renfrew, and Marsha Levine, pp. 49–74. Cambridge: McDonald Institute for Archaeological Research.

————. 1999. The Eneolithic of the Black Sea steppe: dynamics of cultural and economic development, 4500–2300 BC. In *Late Prehistoric Exploitation of the Eurasian Steppe*, ed. Marsha Levine, Yuri Rassamakin, Aleksandr Kislenko, and Nataliya Tatarintseva, pp. 59–182. Cambridge: McDonald Institute for Archaeological Research.

Raulwing, Peter. 2000. *Horses, Chariots and Indo-Europeans*. Archaeolingua Series Minor 13. Budapest: Archaeolingua Foundation.

Reade, Julian. 2001. Assyrian king-lists, the royal tombs of Ur, and Indus Origins. *Journal of Near Eastern Studies* 60 (1): 1–29.

Renfrew, Colin. 2002a. Pastoralism and interaction: Some introductory questions. In *Ancient Interactions: East and West in Eurasia*, ed. Katie Boyle, Colin Renfrew, and Marsha Levine, pp. 1–12. Cambridge: McDonald Institute for Archaeological Research.

———. 2002b. The emerging synthesis: The archaeogenetics of farming/language dispersals and other spread zones. In *Examining the Farming/Language Dispersal Hypothesis*, ed. Peter Bellwood and Colin Renfrew, pp. 3–16. Cambridge: McDonald Institute for Archaeological Research.

———. 2001. The Anatolian origins of Proto-Indo-European and the autochthony of the Hittites. In *Greater Anatolia and the Indo-Hittite Language Family*, ed. Robert Drews, pp. 36–63. Journal of Indo-European Studies Monograph 38. Washington, D.C.: Institute for the Study of Man.

———. 2000. At the edge of knowability: Towards a prehistory of languages. *Cambridge Archaeological Journal* 10 (1): 7–34.

———. 1998. Word of Minos: The Minoan contribution to Mycenaean Greek and the linguistic geography of the Bronze Age Aegean. *Cambridge Archaeological Journal* 8 (2): 239–264.

———. 1996. Language families and the spread of farming. In *The Origins and Spread of Agriculture and Pastoralism in Eurasia*, ed. David Harris, pp. 70–92. Washington, D.C.: Smithsonian Institution Press.

———. 1987. *Archaeology and Language: The Puzzle of Indo-European Origins*. London: Jonathan Cape.

———. 1973. *Before Civilization: The Radiocarbon Revolution and Prehistoric Europe*. London: Jonathon Cape.

Renfrew, Colin, April McMahon, and Larry Trask, eds. 2000. *Time Depth in Historical Linguistics*. Cambridge: McDonald Institute for Archaeological Research.

Rexová, Katerina, Daniel Frynta, and Jan Zrzavý. 2003. Cladistic analysis of languages: Indo-European classification based on lexicostatistical data. *Cladistics* 19 (2): 120–127.

Rezepkin, A. D. 2000. *Das Frühbronzezeitliche Gräberfeld von Klady und die Majkop-Kultur in Nordwestkaukasien*. Archäologie in Eurasien 10. Rahden: Verlag Marie Leidorf.

———. 1991. Kurgan 31 mogil'nika Klady problemy genezisa i khronologii Maikopskoi kul'tury. In *Drevnie kul'tury prikuban'ya*, ed. V. M. Masson, pp. 167–197. Leningrad: Nauka.

Rezepkin, A. D. and A. V. Kondrashov. 1988. Novosvobodnenskoe pogrebenie s povozkoy. *Kratkie soobshcheniya instituta arkheologii AN SSSR* 193:91–97.

Richter, Daniel K. 1992. *The Ordeal of the Longhouse: The Peoples of the Iroquois League in the Era of European Colonization*. Chapel Hill: University of North Carolina Press.

Rijksbaron, A. 1988. The discourse function of the imperfect. In *In the Footsteps of Raphael Kühner*, ed. A. Rijksbaron, H. A. Mulder, and G. C. Wakker, pp. 237–254. Amsterdam: J. C. Geiben,

Ringe, Don. 1997. A probabilistic evaluation of Indo-Uralic. In *Nostratic: Sifting the Evidence*, ed. B. Joseph and J. Salmons, pp. 153–197. Philadelphia: Benjamins.

Ringe, Don, Tandy Warnow, and Ann Taylor. 2002. Indo-European and computational cladistics. *Transactions of the Philological Society* 100:59–129.

Ringe, Don, Tandy Warnow, Ann Taylor, A. Michailov, and Libby Levison. 1998. Computational cladistics and the position of Tocharian. In *The Bronze Age and Early Iron Age Peoples of Eastern Central Asia*, ed. Victor Mair, pp. 391–414. Washington, D.C.: Institute for the Study of Man.

Robb, J. 1993. A social prehistory of European languages. *Antiquity* 67:747–760.

———. 1991. Random causes with directed effects: The Indo-European language spread and the stochastic loss of lineages. *Antiquity* 65:287–291.

Roman, Petre. 1978. Modificări în tabelul sincronismelor privind eneoliticul Tîrziu. *Studii si Cercetări de Istorie Veche și Arheologie* (Bucharest) 29 (2): 215–221.

Rosenberg, Michael. 1998. Cheating at musical chairs: Territoriality and sedentism in an evolutionary context. *Current Anthropology* 39 (5): 653–681.

———. 1994. Agricultural origins in the American Midwest: A reply to Charles. *American Anthropologist* 96 (1): 161–164.

Rostovtseff, M. 1922. *Iranians and Greeks in South Russia.* Oxford: Clarendon.

Rothman, Mitchell S. 2003. Ripples in the stream: Transcaucasia-Anatolian interaction in the Murat/Euphrates basin at the beginning of the third millennium BC. In *Archaeology in the Borderlands*, ed. Adam T. Smith and Karen Rubinson, pp. 95–110. Los Angeles: Cotsen Institute.

———. 2001. *Uruk Mesopotamia and Its Neighbors: Cross-cultural Interactions in the Era of State Formation.* Santa Fe: SAR.

Russell, Josiah Cox. 1972. *Medieval Regions and Their Cities.* Bloomington: Indiana University Press.

Rutter, Jeremy. 1993. Review of Aegean prehistory II: The prepalatial Bronze Age of the southern and central Greek mainland. *American Journal of Archaeology* 97:745–797.

Ryden, Hope. 1978. *America's Last Wild Horses.* New York: Dutton.

Ryder, Tom. 1987. Questions and Answers. *The Carriage Journal* 24 (4): 200–201.

Ryndina, N. V. 1998. *Dreneishee Metallo-obrabatyvaiushchee Proizvodstvo Iugo-Vostochnoi Evropy.* Moscow: Editorial.

Ryndina, N. V. and A. V. Engovatova. 1990. Opyt planigraficheskogo analiza kremnevykh orudii Tripol'skogo poseleniya Drutsy 1. In *Rannezemledel'cheskie Poseleniya-Giganty Tripol'skoi Kul'tury na Ukraine*, ed. I. T. Chernyakov, pp. 108–114. Tal'yanki: Institut arkheologii akademii nauk USSR.

Salminen, Tapani. 2001. The rise of the Finno-Ugric language family. In *Early Contacts between Uralic and Indo-European: Linguistic and Archaeological Considerations*, ed. Christian Carpelan, Asko Parpola, and Petteri Koskikallio, pp. 385–395. Memoires de la Société Finno-Ugrienne 242. Helsinki: Suomalais-Ugrilainen Seura.

Salmons, Joe. 1993. *The Glottalic Theory: Survey and Synthesis.* Journal of Indo-European Studies Monograph 10. Washington, D.C.: Institute for the Study of Man.

Salvatori, Sandro. 2003. Pots and peoples: The "Pandora's Jar" of Central Asian archaeological research: On two recent books on Gonur graveyard excavations. *Rivista di Archeologia* 27:5–20.

———. 2002. Project "Archaeological map of the Murghab Delta" (Turkmenistan): Test trenches at the sites of Adzhi Kui 1 and 9. *Ancient Civilizations from Scythia to Siberia* 8 (1–2): 107–178.

———. 2000. Bactria and Margiana seals: A new assessment of their chronological position and a typological survey. *East and West* 50 (1–4): 97–145.

Salvatori, Sandro, Massimo Vidale, Giuseppe Guida, and Giovanni Gigante. 2002. A glimpse on copper and lead metalworking at Altyn-Depe (Turkmenistan) in the 3rd millennium BC. *Ancient Civilizations from Scythia to Siberia* 8:69–101.

Samashev, Z. 1993. *Petroglyphs of the East Kazakhstan as a Historical Source.* Almaty: Rakurs.

Sapir, Edward, 1912. Language and environment. *American Anthropologist* 14(2): 226–42.

Sarianidi, V. I. 2002. *Margush: Drevnevostochnoe tsarstvo v staroi del'te reki Murgab.* Ashgabat: Turkmendöwletthebarlary.

———. 1995. New discoveries at ancient Gonur. *Ancient Civilizations from Scythia to Siberia* 2 (3): 289–310.

———. 1987. Southwest Asia: Migrations, the Aryans, and Zoroastrians. *Information Bulletin, International Association for the Study of the Cultures of Central Asia* (Moscow) 13:44–56.

———. 1986. Mesopotamiia i Baktriia vo ii tys. do n.e. *Sovietskaia Arkheologiia* (2): 34–46.

———. 1977. *Drevnie Zemledel'tsy Afganistana: Materialy Sovetsko-Afganskoi Ekspeditsii 1969–1974 gg.* Moscow: Akademiia Nauka.

Sawyer, Ralph D. 1993. *The Seven Military Classics of Ancient China.* Boulder, Colo.: Westview.

Schlegel, Alice. 1992. African political models in the American Southwest: Hopi as an internal frontier society. *American Anthropologist* 94 (2): 376–97.

Schmidt, Karl Horst. 1991. Latin and Celtic: Genetic relationship and areal contacts. *Bulletin of the Board of Celtic Studies* 38:1–19.

Schrijver, Peter. 2001. Lost languages in northern Europe. In *Early Contacts between Uralic and Indo-European: Linguistic and Archaeological Considerations*, ed. Christian Carpelan, Asko Parpola, and Petteri Koskikallio, pp. 417–425. Memories de la Société Finno-Ugrienne 242. Helsinki: Suomalais-Ugrilainen Seura.

Schuchhardt, C. 1919. *Alteuropa in seiner Kultur- und Stilentwicklung*. Berlin: Walter de Gruyter.

Segalen, Martine. 1991. *Fifteen Generations of Bretons: Kinship and Society in Lower Brittany, 1720–1980*. Cambridge: Cambridge University Press.

Shakhanova, N. 1989. The system of nourishment among the Eurasian nomads: The Kazakh example. In *Ecology and Empire: Nomads in the Cultural Evolution of the Old World*, pp. 111–117. Los Angeles: University of Southern California Ethnographics Press.

Shaposhnikova, O. G. 1961. Novye dannye o Mikhailovskom poselenii. *Kratkie soobshcheniya institut arkheologii* 11:38–42.

Sharafutdinova, I. N. 1980. Severnaya kurgannaya grupa u s. Sokolovka. In *Arkheologicheskie pamyatniki poingul'ya*, pp. 71–123. Kiev: Naukovo Dumka.

Shaughnessy, Edward L. 1988. Historical perspectives on the introduction of the chariot into China. *Harvard Journal of Asian Studies* 48:189–237.

Shennan, Stephen J., ed. 1989. *Archaeological Approaches to Cultural Identity*. London: Routledge.

Sherratt, Andrew. 2003. The horse and the wheel: The dialectics of change in the circum-Pontic and adjacent areas, 4500–1500 BC. In *Prehistoric Steppe Adaptation and the Horse*, ed. Marsha Levine, C. Renfrew, and K. Boyle, pp. 233–252. McDonald Institute Monographs. Cambridge: University of Cambridge Press.

——. 1997a [1983]. The secondary exploitation of animals in the Old World. In *Economy and Society in Prehistoric Europe: Changing Perspectives*, rev. ed., ed. Andrew Sherratt, pp.199–228. Princeton, N.J.: Princeton University Press.

——. 1997b. The introduction of alcohol to prehistoric Europe. In *Economy and Society in Prehistoric Europe*, ed. Andrew Sherratt, pp. 376–402. Princeton, N.J.: Princeton University Press.

——. 1997c [1991]. Sacred and profane substances: The ritual use of narcotics in later Neolithic Europe. In *Economy and Society in Prehistoric Europe*, ed. Andrew Sherratt, rev. ed. pp. 403–430. Princeton, N.J.: Princeton University Press.

——. 1986. Two new finds of wooden wheels from Later Neolithic and Early Bronze Age Europe. *Oxford Journal of Archaeology* 5:243–248.

Sherratt, Andrew, and E. S. Sherratt. 1988. The archaeology of Indo-European: An alternative view. *Antiquity* 62 (236): 584–595.

Shevchenko, A. I., 1957. Fauna poseleniya epokhi bronzy v s. Mikhailovke na nizhnem Dnepre. *Kratkie soobshcheniya institut arkheologii* 7:36–37.

Shilov, V. P. 1985a. Kurgannyi mogil'nik y s. Tsatsa. In *Drevnosti Kalmykii*, pp. 94–157. Elista: Kalmytskii nauchno-issledovatel'skii institut istorii, filogii i ekonomiki.

——. 1985b. Problemy proiskhozhdeniya kochevogo skotovodstva v vostochnoi Evrope. In *Drevnosti kalmykii*, pp. 23–33. Elista: Kalmytskii nauchno-issledovatel'skii institut istorii, filogii i ekonomiki.

Shilov, V. P., and R. S. Bagautdinov. 1998. Pogebeniya Eneolita-rannei Bronzy mogil'nika Evdyk. In *Problemy drevnei istorii severnogo prikaspiya*, ed. I. B. Vasiliev, pp. 160–178. Samara: Samarskii gosudarstvenyi pedagogicheskii universitet.

Shishlina, N. I., ed. 2000. *Sezonnyi ekonomicheskii tsikl naseleniya severo-zapadnogo Prikaspiya v Bronzovom Veke*. Vol. 120. Moscow: Trudy gosudarstvennogo istoricheskogo muzeya.

——, ed. 1999. *Tekstil' epokhi Bronzy Evraziiskikh stepei*. Vol. 109. Moscow: Trudy gosudarstvennogo istoricheskogo muzeya.

———. 1990. O slozhnom luke Srubnoi kul'tury. In *Problemy arkheologii evrazii*, ed. S. V. Studzitskaya, vol. 74, pp. 23–37. Moscow: Trudy gosudarstvennogo oedena Lenina istoricheskogo muzeya.

Shishlina, N. I., and V. E. Bulatov. 2000. K voprosu o sezonnoi sisteme ispol'zovaniya pastbishch nositelyami Yamnoi kul'tury Prikaspiiskikh stepei v III tys. do n.e. In *Sezonnyi Ekonomicheskii Tsikl Naseleniya Severo-Zapadnogo Prikaspiya v Bronzovom Veke*, ed. N. I. Shishlina, vol. 120, pp. 43–53. Moscow: Trudy gosudarstvennogo istoricheskogo muzeya.

Shishlina, N. I., O. V. Orfinskaya, and V. P. Golikov. 2003. Bronze Age textiles from the North Caucasus: New evidence of fourth millennium BC fibres and fabrics. *Oxford Journal of Archaeology* 22 (4): 331–344.

Shmagli, M. M., and M. Y. Videiko. 1987. Piznotripil'ske poseleniya poblizu s. Maidanets'kogo na Cherkashchini. *Arkheologiya* (Kiev) 60:58–71.

Shnirelman, Victor, A. 1999. Passions about Arkaim: Russian nationalism, the Aryans, and the politics of archaeology. *Inner Asia* 1:267–282.

———. 1998. Archaeology and ethnic politics: The discovery of Arkaim. *Museum International* 50 (2): 33–39.

———. 1995. Soviet archaeology in the 1940s. In *Nationalism, Politics, and the Practice of Archaeology*, ed. Philip L. Kohl and Clare Fawcett, pp. 120–138. Cambridge: Cambridge University Press.

———. 1992. The emergence of food-producing economy in the steppe and forest-steppe zones of Eastern Europe. *Journal of Indo-European Studies* 20:123–143.

Shorin, A. F. 1993. O za Uralskoi oblasti areala lesnikh Eneoliticheskikh kul'tur grebenchatoi keramiki. In *Voprosy arkheologii Urala*, pp. 84–93. Ekaterinburg: Uralskii gosudarstvenyi universitet.

Shramko, B. A., and Y. A. Mashkarov. 1993. Issledovanie bimetallicheskogo nozha iz pogrebeniya Katakombnoi kul'tury. *Rossiskaya arkheologiya* (2): 163–170.

Siegel, Jeff. 1985. Koines and koineisation. *Language in Society* 14:357–378.

Silver, Shirley, and Wick R. Miller. 1997. *American Indian Languages: Cultural and Social Contexts*. Tucson: University of Arizona Press.

Simkins, P. D., and F. L. Wernstedt. 1971. *Philippines Migration: Settlement of the Digos–Padada Valley, Padao Province*. Southeast Asia Studies 16. New Haven: Yale University Press.

Sinitsyn, I. V. 1959. Arkheologicheskie issledovaniya Zavolzhskogo otriada (1951–1953). *Materialy i issledovaniya Institut arkheologii* (Moscow) 60:39–205.

Siniuk, A. T., and I. A. Kozmirchuk. 1995. Nekotorye aspekti izucheniya Abashevskoi kul'tury v basseine Dona. In *Drevnie IndoIranskie Kul'tury Volgo-Ural'ya*, ed. V. S. Gorbunov, pp. 37–72. Samara: Samarskogo gosudarstvennogo pedagogicheskogo universiteta.

Sinor, Dennis, ed. 1988. *The Uralic Languages*. Leiden: Brill.

———. 1972. Horse and pasture in Inner Asian history. *Oriens Extremus* 19:171–183.

Skjærvø, P. Oktor. 1995. The Avesta as a source for the early history of the Iranians. In *The Indo-Aryans of Ancient South Asia: Language, Material Culture and Ethnicity*, ed. George Erdosy, pp. 155–176. Indian Philology and South Asian Studies 1. Berlin: Walter de Gruyter.

Smith, Anthony D. 1998. *Nationalism and Modernism*. London: Routledge.

Smith, Bruce. 1989. Origins of agriculture in eastern North America. *Science* 246 (4937): 1,566–1,571.

Smith, John Masson. 1984. Mongol campaign rations: Milk, marmots, and blood? In *Turks, Hungarians, and Kipchaks: A Festschrift in Honor of Tibor Halasi-Kun*, ed. şinasi Tekin and Gönül Alpay Tekin. Journal of Turkish Studies 8:223–228. Cambridge, Mass.: Harvard University Print Office.

Snow, Dean. 1994. *The Iroquois*. Oxford: Blackwell.

Solovyova, N. F., A. N. Yegor'kov, V. A. Galibin, and Y. E. Berezkin. 1994. Metal artifacts from Ilgynly-Depe, Turkmenistan. In *New Archaeological Discoveries in Asiatic Russia and*

Central Asia, ed. A. G. Kozintsev, V. M. Masson, N. F. Solovyova, and V. Y. Zuyev, pp. 31–35. Archaeological Studies 16. St. Petersburg: Institute of the History of Material Culture.

Sorokin, V. Y. 1989. Kulturno-istoricheski problemy plemen srednogo Triploya Dnestrovsko-Prutskogo mezhdurechya. *Izvestiya Akademii Nauk Moldavskoi SSR* 3:45–54.

Southworth, Franklin. 1995. Reconstructing social context from language: Indo-Aryan and Dravidian prehistory. In *The Indo-Aryans of Ancient South Asia: Language, Material Culture and Ethnicity*, ed. George Erdosy, pp. 258–277. Indian Philology and South Asian Studies 1. Berlin: Walter de Gruyter.

Sparreboom, M. 1985. *Chariots in the Vedas*. Edited by J. C. Heesterman and E. J. M. Witzel. Memoirs of the Kern Institute 3. Leiden: Brill.

Spear, Thomas, and Richard Waller, eds. 1993. *Being Maasai: Ethnicity and Identity in East Africa*. Oxford: James Currey.

Specht, F. 1944. *Der Ursprung der Indogermanischen Deklination*. Göttingen: Vandenhoeck and Ruprecht.

Spicer, Edward. 1971. Persistent cultural systems: A comparative study of identity systems that can adapt to contrasting environments. *Science* 174:795–800.

Spielmann, Katherine A., ed. 1998. *Migration and Reorganization: The Pueblo IV Period in the American Southwest*. Anthropological Research Papers 51. Tempe: Arizona State University Press.

Spinage, C. A. 1972. Age estimation of zebra. *East African Wildlife Journal* 10:273–277.

Stark, Miriam T., ed. 1998. *The Archaeology of Social Boundaries*. Washington, D.C.: Smithsonian Institution Press.

Stein, Gil. 1999. *Rethinking World Systems: Diasporas, Colonies, and Interaction in Uruk Mesopotamia*. Tucson: University of Arizona Press.

Stevanovic, Mirjana. 1997. The Age of Clay: The Social Dynamics of House Destruction. *Journal of Anthropological Archaeology* 16:334–395.

Stewart, Ann H. 1976. *Graphic Representation of Models in Linguistic Theory*. Bloomington: Indiana University Press.

Stillman, Nigel, and Nigel Tallis. 1984. *Armies of the Ancient Near East*. Worthing, Sussex: Flexiprint.

Sturtevant, William. 1962. The Indo-Hittite hypothesis. *Language* 38:105–110.

Subbotin, L.V. 1995. Grobniki Kemi-Obinskogo tipa severo-zapadnogo Prichernomor'ya. *Rossiskaya arkheologiya* (3): 193–197.

———. 1990. Uglubennye zhilishcha kul'tury Gumelnitsa v nizhnem podunav'e. In *Rannezemledel'cheski poseleniya-giganty Tripol'skoi kul'tury na Ukraine*, ed. I. T. Chenyakov, pp. 177–182. Tal'yanki: Institut arkheologii AN USSR.

———. 1985. Semenovskii mogil'nik epokhi Eneolita-Bronzy. In *Novye material'i po arkheologii severo-zapadnogo prichernomor'ya*, ed. V. N. Stanko, pp. 45–95. Kiev: Naukovo Dumka.

———. 1978. O sinkhronizatsii pamyatnikov kul'tury Gumelnitsa v nizhnem Podunav'e. In *Arkheologicheskie issledovaniya severo-zapadnogo prichernomor'ya*, ed. V. N. Stanko, pp. 29–41. Kiev: Naukovo Dumka.

Summers, Geoffrey D. 2001. Questions raised by the identification of the Neolithic, Chalcolithic, and Early Bronze Age horse bones in Anatolia. In *Greater Anatolia and the Indo-Hittite Language Family*, ed. Robert Drews, pp. 285–292. Journal of Indo-European Studies Monograph 38. Washington, D.C.: Institute for the Study of Man.

Sutton, Richard E. 1996. The Middle Iroquoian colonization of Huronia. Ph.D. dissertation. McMaster University, Hamilton, Ontario.

Swadesh, M. 1955. Towards greater accuracy in lexicostatistic dating. *International Journal of American Linguistics* 21:121–37.

———. 1952. Lexico-statistic dating of prehistoric ethnic contacts. *Proceedings of the American Philosophical Society* 96:452–463.

Syvolap, M. P. 2001. Kratkaya kharakteristika pamyatnikov Yamnoi kul'tury srednego podneprov'ya. In *Bronzovyi vek vostochnoi Evropy: Kharakteristika kul'tur, khronologiya i periodizatsiya*, ed. Y. I. Kolev, P. F. Kuznetsov, O. V. Kuzmina, A. P. Semenova, M. A. Turetskii, and B. A. Aguzarov, pp. 109–117. Samara: Samarskii Gosudarstvennyi Pedagogicheskii Universitet.

Szemerényi, Oswald. 1989. The new sound of Indo-European. *Diachronica* 6:237–269.

Szmyt, Marzena. 1999. *Between West and East: People of the Globular Amphorae Culture in Eastern Europe, 2950–2350 BC*. Baltic-Pontic Studies 8. Poznań: Adam Mickiewicz University.

Tatarintseva, N. S. 1984. Keramika poseleniya Vishnevka 1 v lesostepnom pri Ishim'e. In *Bronzovyi Vek Uralo-Irtyshskogo Mezhdurech'ya*, pp. 104–113. Chelyabinsk: Chelyabinskii gosudarstvennyi universitet.

Telegin, D. Y. 2005. The Yamna culture and the Indo-European homeland problem. *Journal of Indo-European Studies* 33 (3–4): 339–358.

———. 2002. A discussion on some of the problems arising from the study of Neolithic and Eneolithic cultures in the Azov-Black Sea region. In *Ancient Interactions: East and West in Eurasia*, ed. Katie Boyle, Colin Renfrew, and Marsha Levine, pp. 25–47. Cambridge: McDonald Institute for Archaeological Research.

———. 1996. Yugo-zapad vostochnoi Evropy; and Yug vostochnoi Evropy. In *Neolit severnoi evrazii*, ed. S. V. Oshibkina, pp. 19–86. Moscow: Nauka.

———. 1991. *Neoliticheskie mogil'niki mariupol'skogo tipa*. Kiev: Naukovo Dumka.

———. 1988. Keramika rannogo Eneolitu tipu Zasukhi v lisostepovomu liboberezhzhi Ukriani. *Arkheologiya* (Kiev) 64:73–84.

———. 1987. Neolithic cultures of the Ukraine and adjacent areas and their chronology. *Journal of World Prehistory* 1 (3): 307–331.

———. 1986. *Dereivka: A Settlement and Cemetery of Copper Age Horse Keepers on the Middle Dnieper*. Edited by J. P. Mallory. Translated by V. K. Pyatkovskiy. British Archaeological Reports International Series 287. Oxford: Archeopress.

———. 1982. *Mezolitichni pam'yatki Ukraini*. Kiev: Naukovo Dumka.

———. 1981. Pro neolitichni pam'yatki Podonnya i steponogo Povolzhya. *Arkheologiya* (Kiev) 36:3–19.

———. 1977. Review of Markevich, V. I., 1974. *Bugo-Dnestrovskaya kul'tura na territorii Moldavii*. *Arkheologiia* (Kiev) 23:88–91.

———. 1973. *Seredno-Stogivs'ka kul'tura Epokha Midi*. Kiev: Naukovo Dumka.

———. 1968. *Dnipro-Donets'ka kul'tura*. Kiev: Naukovo Dumka.

Telegin, D. Y., and James P. Mallory. 1994. *The Anthropomorphic Stelae of the Ukraine: The Early Iconography of the Indo-Europeans*. Journal of Indo-European Studies Monograph 11. Washington D.C.: Institute for the Study of Man.

Telegin, D. Y., A. L. Nechitailo, I. D. Potekhina, and Y. V. Panchenko. 2001. *Srednestogovskaya i novodanilovskaya kul'tury Eneolita Azovo-Chernomorskogo regiona*. Lugansk: Shlyakh.

Telegin, D. Y., and I. D. Potekhina. 1987. *Neolithic Cemeteries and Populations in the Dnieper Basin*, ed. J. P. Mallory. British Archaeological Reports International Series 383. Oxford: Archeopress.

Telegin, D. Y., I. D. Potekhina, M. Lillie, and M. M. Kovaliukh. 2003. Settlement and economy in Neolithic Ukraine: A new chronology. *Antiquity* 77 (296): 456–470.

———. 2002. The chronology of the Mariupol-type cemeteries of Ukraine revisited. *Antiquity* 76:356–363.

Telegin, D. Y., Sergei Z. Pustalov, and N. N. Kovalyukh. 2003. Relative and absolute chronology of Yamnaya and Catacomb monuments: The issue of co-existence. *Baltic-Pontic Studies* 12:132–184.

Teplova, S. N. 1962. *Atlas SSSR*. Moscow: Ministerstva geologii i okhrany nedr SSSR.

Terrell, John Edward, ed. 2001. *Archaeology, Language and History: Essays on Culture and Ethnicity*. Westport, Conn.: Bergin and Garvey.

Terrell, John Edward, T. L. Hunt, and Chris Godsen. 1997. The dimensions of social life in the Pacific: Human diversity and the myth of the primitive isolate. *Current Anthropology* 38:155–195.

Thieme, Paul. 1960. The Aryan gods of the Mitanni treaties. *Journal of the American Oriental Society* 80:310–317.

———. 1958. The Indo-European language. *Scientific American* 199 (4): 63–74.

Thomason, Sarah Gray, and Terrence Kaufman. 1988. *Language Contact, Creolization, and Genetic Linguistics*. Los Angeles: University of California Press.

Thornton, C. P., and C. C. Lamberg-Karlovsky. 2004. A new look at the prehistoric metallurgy of southeastern Iran. *Iran* 42:47–59.

Timofeev, V. I., and G. I. Zaitseva. 1997. K probleme radiouglerodnoi khronologii Neolita stepnoi i iuga lesnoi zony Evropeiskoi chasti Rossii i Sibiri. *Radiouglerod i arkheologiya* (St. Petersburg) 2:98–108.

Todorova, Henrietta. 1995. The Neolithic, Eneolithic, and Transitional in Bulgarian Prehistory. In *Prehistoric Bulgaria*, ed. Douglass W. Bailey and Ivan Panayotov, pp. 79–98. Monographs in World Archaeology 22. Madison, Wis.: Prehistory Press.

Tolstov, S. P., and A. S. Kes'. 1960. *Nizov'ya Amu-Dar'i, Sarykamysh, Uzboi: Istoriya formirovaniya i zaseleniya*. Vol. 3. Moscow: Materialy khorezmskoi ekspeditsii.

Tovkailo, M. T. 1990. Do pitannya pro vzaemini naseleniya Bugo-Dnitrovskoi ta ranne Triplil'skoi kul'tur u stepovomu po Buzhi. In *Rannezemledel'cheskie poseleniya-Giganty Tripol'skoi Kul'tury na Ukraine*, ed. V. G. Zbenovich, and I. T. Chernyakov, pp. 191–194. Tal'yanki: Institut arkheologii akademiya nauk.

Trifonov, V. A. 2001. Popravki absoliutnoi khronologii kultur epokha Eneolita-Srednei Bronzy Kavkaza, stepnoi i lesostepnoi zon vostochnoi Evropy (po dannym radiouglerodnogo datirovaniya). In *Bronzovyi vek Vostochnoi Evropy: Kharakteristika kul'tur, Khronologiya i Periodizatsiya*, ed. Y. I. Kolev, P. F. Kuznetsov, O. V. Kuzmina, A. P. Semenova, M. A. Turetskii, and B. A. Aguzarov, pp.71–82, Samara: Samarskii gosudarstvennyi pedagogicheskii universitet.

———. 1991. Stepnoe prikuban'e v epokhu Eneolita: Srednei Bronzy (periodizatsiya). In *Drevnie kul'tury Prikuban'ya*, ed. V. M. Masson, pp. 92–166. Leningrad: Nauka.

Tringham, Ruth. 1971. *Hunters, Fishers and Farmers of Eastern Europe, 6000–3000 BC*. London: Hutchinson.

Troy, C. S., D. E. MacHugh, J. F. Bailey, D. A. Magee, R. T. Loftus, P. Cunningham, A. T. Chamberlain, B. C. Sykes, and D. G. Bradley. 2001. Genetic Evidence for Near-Eastern Origins of European Cattle. *Nature* 410:1088–1091.

Trudgill, Peter. 1986. *Dialects in Contact*. Oxford: Blackwell.

Tuck, J. A. 1978. Northern Iroquoian prehistory. In *Northeast Handbook of North American Indians*, ed. Bruce G. Trigger, vol. 15, pp. 322–333. Washington, D.C.: Smithsonian Institution Press.

Uerpmann, Hans-Peter. 1990. Die Domestikation des Pferdes im Chalcolithikum West- und Mitteleuropas. *Madrider Mitteilungen* 31:109–153.

Upton, Dell, and J. M. Vlach, eds. 1986. *Common Places: Readings in American Vernacular Architecture*. Athens: University of Georgia Press.

Ursulescu, Nicolae. 1984. *Evoluția Culturii Starčevo-Criș Pe Teritoriul Moldovei*. Suceava: Muzeul Județean Suceava.

Vainshtein, Sevyan. 1980. *Nomads of South Siberia: The Pastoral Economies of Tuva*. Edited by Caroline Humphrey. Translated by M. Colenso. Cambridge: Cambridge University Press.

Van Andel, T. H., and C. N. Runnels. 1995. The earliest farmers in Europe. *Antiquity* 69:481–500.

Van Buren, G. E. 1974. *Arrowheads and Projectile Points*. Garden Grove, Calif.: Arrowhead.

Vasiliev, I. B. 2003. Khvalynskaya Eneoliticheskaya kul'tura Volgo-Ural'skoi stepi i lesostepi (nekotorye itogi issledovaniya). *Voprosy Arkeologii Povolzh'ya* v.3: 61–99. Samara: Samarskii Gosudarstvennyi Redagogieheskii Univerditet.

Vasiliev, I. B., ed. 1998. *Problemy drevnei istorii severnogo prikaspiya.* Samara: Samarskii gosu-darstvennyi pedagogicheskii universitet.

———. 1981. *Eneolit Povolzh'ya.* Kuibyshev: Kuibyshevskii gosudarstvenyi pedagogicheskii institut.

———. 1980. Mogil'nik Yamno-Poltavkinskogo veremeni u s. Utyevka v srednem Povolzh'e. In *Arkheologiya Vostochno-Evropeiskoi Lesostepi,* pp. 32–58. Voronezh: Voronezhskogo universiteta.

Vasiliev, I. B., and G. I. Matveeva. 1979. Mogil'nik u s. S'yezhee na R. Samare. *Sovietskaya arkheologiia* (4): 147–166.

Vasiliev, I. B., P. F. Kuznetsov, and A. P. Semenova. 1994. *Potapovskii Kurgannyi Mogil'nik Indoiranskikh Plemen na Volge.* Samara: Samarskii universitet.

Vasiliev, I. B., P. F. Kuznetsov, and M. A. Turetskii. 2000. Yamnaya i Poltavkinskaya kul'tura. In *Istoriya samarskogo po volzh'ya s drevneishikh vremen do nashikh dnei: Bronzovyi Vek,* ed. Y. I. Kolev, A. E. Mamontov, and M. A. Turetskii, pp. 6–64. Samara: Samarskogo nauchnogo tsentra RAN.

Vasiliev, I. B., and N. V. Ovchinnikova. 2000. Eneolit. In *Istoriya samarskogo povolzh'ya s drevneishikh vremen do nashikh dnei,* ed. A. A. Vybornov, Y. I. Kolev, and A. E. Mamonov, pp. 216–277. Samara: Integratsiya.

Vasiliev, I. B., and Siniuk, A. T. 1984. Cherkasskaya stoiyanka na Srednem Donu. In *Epokha Medi Iuga Vostochnoi Evropy,* ed. S. G. Basina and G. I. Matveeva, pp. 102–129. Kuibyshev: Kuibyshevskii gosudarstvennyi pedagogicheskii institut.

Vasiliev, I. B., A. Vybornov, and A. Komarov. 1996. *The Mesolithic of the North Caspian Sea Area.* Samara: Samara State Pedagogical University.

Vasiliev, I. B., A. A. Vybornov, and N. L. Morgunova. 1985. Review of *Eneolit iuzhnogo Urala* by G. N. Matiushin. *Sovetskaia arkheologiia* (2): 280–289.

Veenhof, Klaas R. 1995. Kanesh: An Assyrian Colony in Anatolia. In *Civilizations of the Ancient Near East,* ed. Jack M. Sasson, John Baines, Gary Beckman, and Karen R. Rubinson, vol. 1, pp. 859–871. New York: Scribner's.

Vehik, Susan. 2002. Conflict, trade, and political development on the southern Plains. *American Antiquity* 67 (1): 37–64.

Veit, Ulrich. 1989. Ethnic concepts in German prehistory: A case study on the relationship between cultural identity and archaeological objectivity." In *Archaeological Approaches to Ethnic Identity,* ed. S. J. Shennan, pp. 35–56. London: Unwin Hyman.

Venneman, Theo. 1994. Linguistic reconstruction in the context of European prehistory. *Transactions of the Philological Society* 92:215–284.

Videiko, Mihailo Y. 2003. Radiocarbon chronology of settlements of BII and CI stages of the Tripolye culture at the middle Dnieper. *Baltic-Pontic Studies* 12:7–21.

———. 1999. Radiocarbon dating chronology of the late Tripolye culture. *Baltic-Pontic Studies* 7:34–71.

———. 1990. Zhilishchno-khozyaistvennye kompleksy poseleniya Maidanetskoe i voprosy ikh interpretatsii. In *Rannezemledel'cheskie Poseleniya-Giganty Tripol'skoi kul'tury na Ukraine,* ed. I. T. Cherniakhov, pp. 115–120. Tal'yanki: Vinnitskii pedagogicheskii institut.

Videiko, Mihailo Y., and Vladislav H. Petrenko. 2003. Radiocarbon chronology of complexes of the Eneolithic–Early Bronze Age in the North Pontic region, a preliminary report. *Baltic-Pontic Studies* 12:113–120.

Vilà, Carles, J. A. Leonard, A. Götherdtröm, S. Marklund, K. Sandberg, K. Lidén, R. K. Wayne, and Hans Ellegren. 2001. Widespread origins of domestic horse lineages. *Science* 291 (5503): 474–477.

Vinogradov, A. V. 1981. *Drevnie okhotniki i rybolovy sredneaziatskogo mezhdorechya.* Vol. 10. Moscow: Materialy khorezmskoi ekspeditsii.

———. 1960. Novye Neoliticheskie nakhodki Korezmskoi ekspeditsii AN SSSR 1957 g. In *Polevye issledovaniya khorezmskoi ekspeditsii v 1957 g.,* ed. S. P. Tolstova, vol. 4, pp. 63–81. Moscow: Materialy khorezmskoi ekspeditsii.

Vinogradov, Nikolai. 2003. *Mogil'nik Bronzovogo Beka: Krivoe ozero v yuzhnom Zaural'e*. Chelyabinsk: Yuzhno-Ural'skoe knizhnoe izdatel'stvo.

Vörös, Istvan. 1980. Zoological and paleoeconomical investigations on the archaeozoological material of the Early Neolithic Körös culture. *Folia Archaeologica* 31:35–64.

Vybornov, A. A., and V. P. Tretyakov. 1991. Stoyanka Imerka VII v Primokshan. In *Drevnosti Vostochno-Evropeiskoi Lesotepi*, ed. V. V. Nikitin, pp. 42–55. Samara: Samarskii gosudartsvennyi pedagogicheskii institut.

Währen, M. 1989. Brot und Gebäck von der Jungsteinzeit bis zur Römerzeit. *Helvetia Archaeologica* 20:82–116.

Walcot, Peter. 1979. Cattle raiding, heroic tradition, and ritual: The Greek evidence. *History of Religions* 18:326–351.

Watkins, Calvert. 1995. *How to Kill a Dragon: Aspects of Indo-European Poetics*. Oxford: Oxford University Press.

Weale, Michael E., Deborah A. Weiss, Rolf F. Jager, Neil Bradman, and Mark G. Thomas. 2002. Y Chromosome Evidence for Anglo-Saxon Mass Migration. *Molecular Biology and Evolution* 19:1008–1021.

Weber, Andrzej, David W. Link, and M. Anne Katzenberg. 2002. Hunter-gatherer culture change and continuity in the Middle Holocene of the Cis-Baikal, Siberia. *Journal of Anthropological Archaeology* 21:230–299.

Wechler, Klaus-Peter, V. Dergachev, and O. Larina. 1998. Neue Forschungen zum Neolithikum Osteuropas: Ergebnisse der Moldawisch-Deutschen Geländearbeiten 1996 und 1997. *Praehistorische Zeitschrift* 73 (2): 151–166.

Weeks, L. 1999. Lead isotope analyses from Tell Abraq, United Arab Emirates: New data regarding the "tin problem" in Western Asia. *Antiquity* 73:49–64.

Weisner, Joseph. 1968. *Fahren und Reiten*. Göttingen: Vandenhoeck and Ruprecht, Archaeologia Homerica.

———. 1939. Fahren und Reiten in Alteuropa und im alten Orient. In *Der Alte Orient* Bd. 38, fascicles 2–4. Leipzig: Heinrichs Verlag.

Weiss, Harvey. 2000. Beyond the Younger Dryas: Collapse as adaptation to abrupt climate change in ancient West Asia and the Eastern Mediterranean. In *Environmental Disaster and the Archaeology of Human Response*, ed. Garth Bawden and Richard M. Reycraft, pp. 75–98. Anthropological Papers no. 7. Albuquerque: Maxwell Museum of Anthropology.

Weissner, Polly. 1983. Style and social information in Kalahari San projectile points. *American Antiquity* 48 (2): 253–275.

Wells, Peter S. 2001. *Beyond Celts, Germans and Scythians: Archaeology and Identity in Iron Age Europe*. London: Duckworth.

———. 1999. *The Barbarians Speak*. Princeton, N. J.: Princeton University Press.

White, Randall. 1989. Husbandry and herd control in the Upper Paleolithic: A critical review of the evidence. *Current Anthropology* 30 (5): 609–632.

Wilhelm, Gernot. 1995. The Kingdom of Mitanni in Second-Millennium Upper Mesopotamia. In *Civilizations of the Ancient Near East*, vol. 2, ed. Jack M. Sasson, John Baines, G. Beckman, and Karen S. Rubinson, pp. 1243–1254. New York: Scribner's.

Wilhelm, Hubert G. H. 1992. Germans in Ohio. In *To Build in a New Land: Ethnic Landscapes in North America*, ed. Allen G. Noble, pp. 60–78. Baltimore, Md.: Johns Hopkins University Press.

Willis, K. J. 1994. The vegetational history of the Balkans. *Quaternary Science Reviews* 13: 769–788.

Winckelmann, Sylvia. 2000. Intercultural relations between Iran, the Murghabo-Bactrian Archaeological Complex (BMAC), northwest India, and Falaika in the field of seals. *East and West* 50 (1–4): 43–96.

Winn, S.M.M., 1981. *Pre-Writing in Southeastern Europe: The Sign System of the Vinča Culture ca. 4000 B.C.* Calgary: Western.

Witzel, Michael. 2003. *Linguistic Evidence for Cultural Exchange in Prehistoric Western Central Asia*. Sino-Platonic Papers 129:1–70. Philadelphia: Department of Asian and Middle Eastern Languages, University of Pennsylvania.

———. 1995. Rgvedic history: Poets, chieftans, and polities. In *The Indo-Aryans of Ancient South Asia: Language, Material Culture and Ethnicity*, ed. George Erdösy, pp. 307–352. Indian Philology and South Asian Studies 1. Berlin: Walter de Gruyter.

Wolf, Eric. 1984. Culture: Panacea or problem? *American Antiquity* 49 (2): 393–400.

———. 1982. *Europe and the People without History*. Berkeley: University of California Press.

Wylie, Alison. 1995. Unification and convergence in archaeological explanation: The agricultural "wave of advance" and the origins of Indo-European languages. In *Explanation in the Human Sciences*, ed. David K. Henderson, pp. 1–30. Southern Journal of Philosophy Supplement 34. Memphis: Department of Philosophy, University of Memphis.

Yanko-Hombach, Valentina, Allan S. Gilbert, Nicolae Panin, and Pavel M. Dolukhanov. 2006. *The Black Sea Flood Question: Changes in Coastline, Climate, and Human Settlement*. NATO Science Series. Dordrecht: Springer.

Yanushevich, Zoya V. 1989. Agricultural evolution north of the Black Sea from the Neolithic to the Iron Age. In *Foraging and Farming: The Evolution of Plant Expoitation*, ed. David R. Harris and Gordon C. Hillman, pp. 607–619. London: Unwin Hyman.

Yarovoy, E. V. 1990. *Kurgany Eneolita-epokhi Bronzy nizhnego poDnestrov'ya*. Kishinev: Shtiintsa.

Yazepenka, Igor, and Aleksandr Kośko. 2003. Radiocarbon chronology of the beakers with short-wave moulding component in the development of the Middle Dnieper culture. *Baltic-Pontic Studies* 12:247–252.

Yener, A. 1995. Early Bronze Age tin processing at Göltepe and Kestel, Turkey. In *Civilizations of the Ancient Near East*, ed. Jack M. Sasson, John Baines, Gary Beckman, and Karen R. Rubinson, vol. 3, pp. 1519–1521. New York: Scribner's.

Yudin, A. I. 1998. Orlovskaya kul'tura i istoki formirovaniya stepnego Eneolita za Volzh'ya. In *Problemy Drevnei Istorii Severnogo Prikaspiya*, pp. 83–105. Samara: Samarskii gosudarstvennyi pedagogicheskii universitet.

———. 1988. Varfolomievka Neoliticheskaya stoianka. In *Arkheologicheskie kul'tury severnogo Prikaspiya*, pp. 142–172. Kuibyshev: Kuibyshevskii gosudarstvenii pedagogicheskii institut.

Zaibert, V. F. 1993. *Eneolit Uralo-Irtyshskogo Mezhdurech'ya*. Petropavlovsk: Nauka.

Zaikov, V. V., G. B. Zdanovich, and A. M. Yuminov. 1995. Mednyi rudnik Bronzogo veka "Vorovskaya Yama." In *Rossiya i Vostok: Problemy Vzaimodeistviya*, pt. 5, bk. 1: *Kul'tury Eneolita-Bronzy Stepnoi Evrazii*, pp. 157–162. Chelyabinsk: 3-ya Mezhdunarodnaya nauchnaya konferentsiya.

Zaitseva, G. I., V. I. Timofeev, and A. A. Sementsov. 1999. Radiouglerodnoe datirovanie v IIMK RAN: istoriya, sostoyanie, rezul'taty, perspektivy. *Rossiiskaya arkheologiia* (3): 5–22.

Zbenovich, V. G. 1996. The Tripolye culture: Centenary of research. *Journal of World Prehistory* 10 (2): 199–241.

———. 1980. *Poselenie Bernashevka na Dnestre (K Proiskhozhdenniu Tripol'skoi Kul'tury)*. Kiev: Naukovo Dumka.

———. 1974. *Posdnetriplos'kie plemena severnogo Prichernomor'ya*. Kiev: Naukovo Dumka.

———. 1968. Keramika usativs'kogo tipu. *Arkheologiya* (Kiev) 21:50–78.

Zdanovich, G. B., ed. 1995. *Arkaim: Issledovaniya, Poiski, Otkrytiya*. Chelyabinsk: "Kammennyi Poyas."

———. 1988. *Bronzovyi Vek Uralo-Kazakhstanskikh Stepei*. Sverdlovsk: Ural'skogo universiteta, for Berlyk II.

Zeder, Melinda. 1986. The equid remains from Tal-e Malyan, southern Iran. In *Equids in the Ancient World*, vol. 1, ed. Richard Meadow and Hans-Peter Uerpmann, pp. 366–412. Weisbaden: Reichert.

Zelinsky, W. 1973. *The Cultural Geography of the United States*. Englewood Cliffs, N.J.: Prentice-Hall.

Zhauymbaev, S. U. 1984. Drevnie mednye rudniki tsentral'nogo Kazakhstana. In *Bronzovyi Vek Uralo-Irtyshskogo Mezhdurech'ya*, pp. 113–120. Chelyabinsk: Chelyabinskii gosudarstvennyi universitet.

Zimmer, Stefan. 1990. The investigation of Proto-Indo-European history: Methods, problems, limitations. In *When Worlds Collide: Indo-Europeans and the Pre-Indo-Europeans*, ed. T. L. Markey, and John A. C. Greppin, pp. 311–344. Ann Arbor, Mich.: Karoma.

Zin'kovskii, K. V., and V. G. Petrenko. 1987. Pogrebeniya s okhroi v Usatovskikh mogil'nikakh. *Sovietskaya arkheologiya* (4): 24–39.

Zöller, H. 1977. Alter und Ausmass postgläzialer Klimaschwankungen in der Schweizer Alpen. In *Dendrochronologie und Postgläziale Klimaschwangungen in Europa*, ed. B. Frenzel, pp. 271–281. Wiesbaden: Franz Steiner Verlag.

Zutterman, Christophe. 2003. The bow in the ancient Near East, a re-evaluation of archery from the late 2nd millennium to the end of the Achaemenid empire. *Iranica Antiqua* 38: 119–165.

Zvelebil, Marek. 2002. Demography and dispersal of early farming populations at the Mesolithic/Neolithic transition: Linguistic and demographic implications. In *Examining the Farming/Language Dispersal Hypothesis*, ed. Peter Bellwood and Colin Renfrew, pp. 379–394. Cambridge: McDonald Institute for Archaeological Research.

———. 1995. Indo-European origins and the agricultural transition in Europe. *Journal of European Archaeology* 3:33–70.

Zvelebil, Marek, and Malcolm Lillie. 2000. Transition to agriculture in eastern Europe. In *Europe's First Farmers*, ed. T. Douglas Price, pp. 57–92. Cambridge: Cambridge University Press.

Zvelebil, Marek, and Peter Rowley-Conwy. 1984. Transition to farming in northern Europe: A hunter-gatherer perspective. *Norwegian Archaeological Review* 17:104–128.

Zvelebil, Marek, and K. Zvelebil. 1988. Agricultural transition and Indo-European dispersals. *Antiquity* 62:574–583.

INDEX

Franklin Pierce College Library

00172582

DATE DUE

AUG 23 2008

GAY

P9-AOA-159

1910 *1980* 1915 1920 1925 1930 1935

1915
David Sarnoff writes famous memo on the future of wireless.

1927–1934
Radio Act of 1927 withstands challenges in court.

1919
Owen T. Young negotiates the formation of RCA.

1926
Case of *U.S. v. Zenith Radio Corp.*

1933
President Roosevelt uses radio for "fireside chats" with the public.

1917
Alexanderson's alternator takes on increased importance for international communication.

1927
Radio Act of 1927 is passed. Forms five-member Federal Radio Commission (FRC).

1923
V.K. Zworykin patents the iconoscope pickup tube for television.

1933
Edwin Armstrong demonstrates FM broadcasting for RCA.

1914
Amateur radio operators form the American Radio Relay League (A.R.R.L.).

1931
Zworykin and RCA officials visit Farnsworth labs in California. RCA later enters royalty agreement with Farnsworth.

1909
Charles David Herrold's station broadcasts from San Jose.

1932
Closed Circuit ETV begins at State University of Iowa.

1922–1925
National Radio Conferences. (Four held before new legislation)

1932
Shuler case is decided.

1921
Philo Farnsworth outlines to his science teacher the concept of electronic television.

1930
Philo Farnsworth applies for permission to experiment with 300-line TV system.

1923
Case of *Hoover v. Intercity Radio.*

1934
Communications Act of 1934 is passed. Forms seven-member Federal Communications Commission (FCC). Independent regulatory body.

1912
Wireless gains publicity by aiding rescue efforts from the *Titanic.*

1926
RCA forms subsidiary NBC to operate Red and Blue networks.

1934
Mutual network begins as four-station cooperative.

1919
9XM at the University of Wisconsin, Madison, signs on the air. Becomes WHA in 1922.

1928
CBS begins when interests are purchased by Wm. S. Paley and Congress Cigar Company.

1912
Radio Act of 1912.

1930
Zworykin visits Farnsworth labs in California to examine 300-line TV scanning system.

1910
Wireless Ship Act of 1910.

1920
WWJ in Detroit begins intermittent broadcasting schedules in August.

1929
First NAB "Code of Ethics" is passed.

1920
KDKA in Pittsburgh begins regular programming in November.

1929
First broadcast rating by Crosley Radio Company.

1914–1918
Wireless used extensively in World War I.

1922–1923
National Association of Broadcasters (NAB) is formed.

1933
Press–Radio War ends with Biltmore agreement.

1922
Toll broadcasting begins at WEAF.

1931
KFKB (*Brinkley*) case is decided.

1910 1915 1920 1925 1930 1935

384.54
B625

88680

LIBRARY
College of St. Francis
JOLIET, ILL.

DEMCO

University of St. Francis
GEN 384.54 B625
Bittner, John R.,
Broadcasting :

3 0301 00073674 0

BROADCASTING

an introduction

John R. Bittner
DePauw University

LIBRARY
College of St. Francis
JOLIET, ILL.

PRENTICE-HALL, INC.
ENGLEWOOD CLIFFS, NEW JERSEY 07632

Library of Congress Cataloging in Publication Data

Bittner, John R (date)
   Broadcasting.

   Includes bibliographies and index.
   1. Broadcasting   I. Title.
PN1990.8.B5   384.54   79-21909
ISBN 0-13-083535-8

© 1980 by Prentice-Hall, Inc., Englewood Cliffs, N.J. 07632

All rights reserved. No part of this book
may be reproduced in any form or
by any means without permission in writing
from the publisher.

Printed in the United States of America

10   9   8   7   6   5   4   3   2   1

PRENTICE-HALL INTERNATIONAL, INC., *London*
PRENTICE-HALL OF AUSTRALIA PTY. LIMITED, *Sydney*
PRENTICE-HALL OF CANADA, LTD., *Toronto*
PRENTICE-HALL OF INDIA PRIVATE LIMITED, *New Delhi*
PRENTICE-HALL OF JAPAN, INC., *Tokyo*
PRENTICE-HALL OF SOUTHEAST ASIA PTE. LTD., *Singapore*
WHITEHALL BOOKS LIMITED, *Wellington, New Zealand*

384.54
B625

1-30-80 Publisher #13.05

88680

for John and Donald

contents

v

PART 4   REGULATORY CONTROL

preface and acknowledgments

This book is designed for use in single semester or term introductory broadcasting courses. It is directed toward college students who aspire to be either responsible consumers of broadcasting in society or practicing professionals in broadcasting or related fields. It also is for students who enroll in a single broadcasting course for a greater appreciation of the broadcast media or for students who take additional courses for a greater measure of expertise in the discipline.

The text goes beyond names and dates by integrating broadcasting into the wider realm of telecommunication and human interaction. The reader will find, along with chapters on the historical and contemporary aspects of radio and television, separate chapters highlighting such subjects as the broadcast audience and effects, cable, uses of satellites and microwaves, television in business and industry, educational and public broadcasting, ratings, regulation and regulatory issues, research in broadcasting, and advertising and economics.

The text treats the historical basis of broadcasting, but it is not a history book. Although there are chapters that deal with the beginning of the wireless and the development of the industry, much of the book is based on the contemporary industry.

At the beginning of each chapter is a special "focus" to create a mental receptacle in which to plug in the major points of the following chapter. Each chapter concludes with a bibliography presented as a "spotlight on further learning."

The Appendix has a special guide to help the student use library resources to learn more about broadcasting and to prepare assigned papers and reports.

A comprehensive instructor's manual accompanies the text.

Naturally, in any book the author's own experiences are part of the content and organization of the work. Since my career has been in both college teaching and the broadcasting industry, the text is a bridge joining elements of both academia and the industry.

As author, I am deeply indebted to the students in my courses at DePauw University, and equally to the staff I had when serving as general manager of a broadcasting station and, prior to that, as a radio news director and television journalist. The following acknowledgments can only begin to express my gratitude to those who have helped during the four years in which this book has been in preparation.

Chris Sterling, Robert Burdman, William Fraser, and Jay Wright provided valuable assistance at various stages of the book's development. Many other individuals and colleagues gave suggestions and encouragement. Barbara Christenberry, Eric Bernsee, G. Gail Crotts, Barbara Moran, Larry Barker, Robert Kibler, Brian Walker, George Foulkes, Jean Wachter, Frida Schubert, Richard Shaheen, Larry Taylor, and Ralph Taylor all deserve special thanks. In addition, Mike Warner, Maureen Lathers, John Busch, Judy Mortell, Bill Ketter, John Payne, Roy Mehlman, Julie Zellers, Hal Youart, Thom Brown, Pat Nugent, Dan Bronson, Barbara Moore, Greg Rice, and Walt Stewart all should be mentioned as well as Bruce Kennan, Colette Conboy, Karen Thompson, Denise Moderack and Shelia Whiting.

Staffs of ABC, CBS, NBC, and PBS all helped, as did personnel in government agencies, professional associations including NAB and NCTA, and numerous stations including WAZY, KRON, WNET, and others.

Few authors could ask for more support than I received from the entire library staff of DePauw University. Their professional courtesy, tireless help, and cooperative attitude are a most valuable resource. Few authors could have a finer staff than Jim Martindale, Cathie Bean, Catherine Bean, Patricia Renn-Scanlan, Rea Brown, Celia Lemmink, Dave Horn, Dan Smith, and Judy Meyer.

A special acknowledgment is made to Larry L. Hardesty, reference librarian at DePauw University, for his efforts in developing the library guide found in this book.

Pamela Loy and Kent Mecum provided helpful translations of foreign-language documents and sources. Enjoying an office next to Beverly Whitaker-Long kept a professional spirit alive.

Discussions with John Jakes during his many visits to DePauw provided valuable writing discipline and contagious enthusiasm.

Rea Zeiner, Mildred Hancock, Linda Rowe, Barbara Boyll, and Linette Finstad helped type the manuscript. Elinor Zeigleman and Midge Cook kept an office running smoothly.

And no one deserves more credit than Denise.

illustrations
to help you learn

Chapter 18: The Research Process

Using the Library to Learn About Broadcasting

1

broadcasting as mass communication

FOCUS After completing this chapter we should be able to

Understand the meaning of the terms broadcast and communicate.

Explain the differences among the intrapersonal, interpersonal, and mass communication processes.

Diagram the components of the basic model of communication and mass communication.

Explain how the definition of mass communication is changing.

Define the terms gatekeeper, delayed feedback, and noise.

Understand the social context of mass communication.

Discuss broadcasting as a form of mass communication.

The young boy listened with fascination as his father spun a tale of yesteryear. He could not imagine the almost primitive society of the 1980s in comparison to his world at the turn of the twenty-first century. Stories of a mere four major television networks, watching television with only a dozen channels from which to choose, and tens of radio stations crowding their frequencies together on a tiny portion of the dial just did not seem possible. Why, now there were hundreds of television channels from which to choose, and they were programmed like radio in the 1980s. Some television stations broadcast only sports or news, others only old-time movies or symphonies, and still others broadcast only political caucuses, or government meetings, or anything else you could ever want to watch.

Radio was now an even more specialized medium. Almost everyone listened to FM for music, and the AM band was mostly for talk and informational programming. Virtually every station had quadraphonic sound—the boy could not imagine listening to radio through a single speaker on the kitchen table. His house, like others, had a media room in which four huge speakers in each corner made a recording of a symphony orchestra sound like a live concert. He did not just listen to classical music, he experienced what it was like to be sitting in the middle of the orchestra.

Every city had at least two and sometimes three television networks with satellite connection feeding programs around the world. The older man reminisced about the excitement of watching the new networks pop onto the giant screen. Networks from Chicago, Montreal, Mexico City, Dallas, and Tokyo all came into the living room over a cable system that could carry as many as a thousand channels. Even now that system was still in its infancy. The youngster looked forward to the day when he would be licensed to program his own amateur television station and broadcast to other amateur operators around the world.

What really amazed him was the realization that the concept of electronic communication was only slightly more than one hundred years old—a minute speck on the yardstick of time. The electronic world he lived in had been molded first by the discovery of electricity, then electromagnetic energy, next wireless transmission, and then computers storing and processing new information with lightning speed. All of these discoveries had shaped a world of electronic wizardry capable of communicating with millions of people the world over. Even now it was

2

testing the new frontiers of communication in the stellular systems of the universe.

No longer was broadcasting simply the spark-gap experiments of 1887 when German physicist Heinrich Rudolph Hertz discovered the presence of electromagnetic waves. Nor was it the crude device that American farmer Nathan B. Stubblefield used in 1892 to send wireless messages from a Kentucky shack to an apple orchard a short distance away.[1] Guglielmo Marconi made history a few years later with his "wireless," or "radio" as it later became known, by broadcasting the first transatlantic signal code message on December 12, 1901 from Signal Hill, Newfoundland, Canada.

From that point on, the wireless transmission of messages became a passion of inventors, amateur experimenters, and large companies. As with any limited resource which the public clamors too ardently to possess, government control was inevitable. Political pressures, the need for security in time of war, and a federal directive to serve in the public's "interest, convenience, and necessity," all contributed to the development of broadcasting. These forces helped, harnessed, and harrassed. But the medium survived and prospered.

The boy who now sits listening to his father's reminiscences could have felt the same awe thirty years earlier—and thirty years earlier than that. Because broadcasting has gone through such rapid changes, there has never been a decade since its birth that has not fostered a mushroom of new technology. Messages that were once carried on a spark now play leapfrog with satellites. And a world that once gawked with amusement now looks in amazement.

what is broadcasting?

Before beginning this book, let's first determine what the term *broadcast* (Figure 1–1) means. In its most basic sense, it can mean "scattered over a wide area," or "in a scattered manner; far and wide,"[2] along with those statements, the dictionary also includes such definitions as "to make known over a wide area: broadcast rumors." Certainly a disgruntled and defeated politician might agree with that definition. Or consider the definition "to participate in a radio or television program." The guest home economist on an afternoon radio program for consumers would agree with that definition. The farmer in the 1800s who never had heard of radio or television would have agreed with the dictionary's definition that broadcast means "to sow [seed] over a wide area, especially by hand."

Consulting a thesaurus, we find that words with similar meanings to "broadcast" include "disperse, generalize, let fall, cultivate, communicate,

```
broad·cast (brôd′kăst′, -käst′) v. -cast or -casted, -casting,
-casts. —tr. 1. To transmit (a program) by radio or television.
2. To make known over a wide area: broadcast rumors. 3. To
sow (seed) over a wide area, especially by hand. —intr. 1. To
transmit a radio or television program. 2. To participate in a
radio or television program. —n. 1. Transmission of a radio or
television program or signal. 2. A radio or television program,
or the duration of such a program. 3. The act of scattering
seed. —adj. 1. Of or pertaining to transmission by radio or
television. 2. Scattered over a wide area. —adv. In a scattered
manner; far and wide. —broad′cast′er n.
```

Figure 1–1   (© 1969, 1970, 1971, 1973, 1975, 1976, Houghton Mifflin Company. Reprinted by permission from *The
American Heritage Dictionary of the English Language*.)

publish, telecommunication, oration, and waste."[3] We would not have to
travel far to encounter people who would agree with all of those
meanings. The ad executive responsible for selling a client's product
would "disperse" knowledge through broadcasting commercials. The
supporter of noncommercial public broadcasting would argue that quality
programming "cultivates" an interest in culture and the arts (Figure 1–2).
The broadcast journalist who was subpoenaed before a grand jury and
asked to divulge the source of her latest investigative report would argue
that under the First Amendment to the U.S. Constitution, broadcast
means the same as to "publish," and her rights to protect the confiden-
tiality of news sources are the same as those of newspaper reporters. The
word broadcast to the corporate executive might be more closely associated with "telecommunication." As two people sit in the corporate
boardroom, their picture is reproduced on television monitors one
continent away. There, other corporate executives talk back to the
boardroom executives via a two-way television system. For the person
highly critical of television programming, the term "waste" might be
more appropriate. The term "vast wasteland," coined by a former
chairman of the Federal Communications Commission, Newton Minow,
has become a favorite term of critics of commercial television.[4]

In the following chapters, we shall journey into the exciting worlds of
broadcasting, experiment with innovative technology, predict the future,
and reflect upon the past. Such terms as satellites, community antenna
television (CATV), Westar®, educational television (ETV), the Communications Act, business and industrial television, AM stereo, and many
more will become part of your vocabulary. You will read about Captain
Kirk, Barbara Walters, Hopalong Cassidy, Rhoda, and other figures. You
will also discover the economic factors which are part of any broadcasting
system, the political and social controls which check and balance the
system, and the audience which uses it.

Figure 1–2 Typical of quality programming cultivating an interest in culture and the arts is the Public Broadcasting Service's "Dance in America" series. Pictured is a scene from "Adorations" produced by WNET/13 New York. We shall learn more about public broadcasting in chapter 8. (WNET/13)

First we need to examine where broadcasting fits into the total communicative process. To do that, we shall explore the different types of communication.

understanding communication

What exactly does it mean to *communicate?* (Figure 1–3). To the physiologist concerned with the processes of life, it has one meaning. To a psychologist specializing in counseling, it has another. To the anthropologist studying different cultures, it has still another meaning. If we return to our dictionary, we will also find different definitions under the listing "communicate," such as "to make known, impart, transmit."[5]

Still other definitions include that of Stewart Tubbs and Sylvia Moss in their book *Human Communication,* in which they define communication as

com·mu·ni·cate (kə-myōō′nə-kāt′) v. **-cated, -cating, -cates.** —tr. **1.** To make known; impart: *communicate information.* **2.** To transmit (a disease, for example). —intr. **1.** To have an interchange, as of thoughts or ideas. **2.** To receive Communion. **3.** To be connected or form a connecting passage. [Latin *commūnicāre*, "to make common," make known, from *commūnis*, COMMON.] —**com·mu′ni·ca′tor** (-kā′tər) n.

Figure 1–3 (© 1969, 1970, 1971, 1973, 1975, 1976, Houghton Mifflin Company. Reprinted by permission from *The American Heritage Dictionary of the English Language.*)

"the process of creating meaning."[6] Kenneth Andersen in his book *Introduction to Communication Theory and Practice* defines communication as a "dynamic process in which man consciously or unconsciously affects the cognitions of another through materials or agencies used in symbolic ways."[7] Wilbur Schramm in his *The Process and Effects of Mass Communication* stresses the importance of viewing communication as "an act of sharing, rather than something someone does to someone else."[8] To understand these definitions better, let's examine three types of communication: intrapersonal, interpersonal, and mass.

intrapersonal communication

The physiologist may think of communication as *intrapersonal communication,* communication within ourselves. In human physiology, our senses, our nervous system, and our brain become the main components of the communicative process. For example, if we are watching an instructional television program about basic mathematics, our eyes and ears respond to what is on the screen. These two senses of sight and hearing send electrochemical impulses through the nervous system to the brain. After receiving the impulses, the brain then feeds back other impulses to our motor nerves, nerves which influence movement and which may cause us to pick up a pencil and paper and work the math problem. In our example, different components of the communication process come into play: the *sender* (eyes and ears), *message* (electrochemical impulses), *medium* (nervous system), *receiver* (brain), and *feedback* (electrochemical impulses). Another factor, *noise,* can interfere with the communication process. Your head may ache to the point at which you cannot think. A sickness or injury may damage your nervous system, either interrupting the passage of electrochemical impulses or interfering with your ability to respond to commands given your motor nerves by your brain. All of these are examples of one type of noise, physical noise.

6

To understand the process better, let's look at a communication model (Figure 1–4). In a sense, a communication model "freezes" the communication process so we may more leisurely examine its parts. Note where each of the components in the process of intrapersonal communication fits into the model: sender, message, medium (sometimes called channel), receiver, feedback, and noise. We will continue to use this basic communication model as we discuss how broadcasting fits into the communication process.

Remember, for human communication of any kind to occur, intrapersonal communication must be present. In examining our dictionary definition of "communicate," we see that someone must first *think* about information before it can be "made known" or "imparted." Before "an interchange" of "thoughts or ideas" can take place between two people, each person must first employ the process of intrapersonal communication in order to react to the other's message. In Kenneth Andersen's definition, intrapersonal communication must take place before "man consciously or unconsciously affects the cognitions of another. . . ."

interpersonal communication

Interpersonal communication is communication in a face-to-face situation between at least two people, often many more, such as a group discussion or a speech to a crowd. In interpersonal communication, the same components of communication apply as those to intrapersonal communication. Here, however, the components are different.

Using our example, imagine that instead of watching the instructional

Figure 1–4 Basic model of communication.

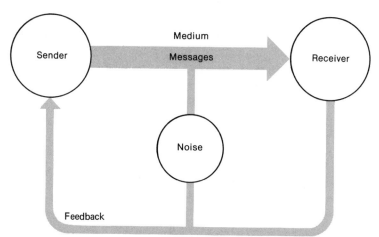

television program about mathematics, you are attending the instructor's class in person. Now the instructor becomes the sender of communication; the messages become the words spoken by the instructor; the medium is the human voice; and you are the receiver of communication. If you do not understand something the instructor is saying, you can immediately raise your hand to ask a question. Your hand being raised is a form of feedback to the instructor.

Noise can also be present. Physical noise may occur if the lights go out and you cannot see the instructor. Or a student next to you may drop a pile of books, distracting your attention. In interpersonal communication, a second type of noise can be present—*semantic* noise. Semantic noise can occur when the instructor uses a word or phrase with many different meanings, confusing you as to which meaning she wants to convey.

Using our example of the lecture, we can begin to see the reason for Wilbur Schramm's emphasis on communication as a "sharing" process. If constructive communication is to take place, you must share certain things with the instructor. First, you must understand the language, both written and oral. Second, you must know something about the subject of mathematics or else the lecture would have little value to you, and you could not begin to work the problems. You may also respect the instructor's ability to teach, perceiving her as having a genuine interest in mathematics whether or not you are able to comprehend the subject. If you have proved yourself a good student, the instructor will probably consider you as being interested in the subject, respect your ability to learn, and perceive you as having a genuine interest in learning. Kenneth Burke stresses this concept of sharing as an "identification" not only between people's experiences but between language as well. Burke is interpreted as seeing people with "common interests because of their experiences as members of groups. But if they are to function as grounds for identification, these interests must be recognized and announced symbolically for unification of attitude or action to occur."[9]

Communication researchers have used behavioral research methods to examine this concept in more detail. It has been examined by such well-known social scientists as Paul Lazarsfeld and Robert Merton. The technical term used to describe the concept is *homophily*. Homophily can best be understood as overlap. Or, as James McCroskey and Lawrence Wheeless state, "To the extent that the attitudes, beliefs, experiences, education, background, culture, and so forth, of the source (sender) and the receiver overlap, they are more likely to attempt communication with each other, and equally as important, they are more likely to be effective in their communication attempts."[10] This concept of "sharing," "interaction," and "homophily" are important to remember, because they center on how we react to all communication.

mass communication

Now that we have a basic understanding of the processes of intrapersonal and interpersonal communication, we need to understand the process of mass communication and, specifically, where broadcasting fits into the process. Mass communication is different from intrapersonal and interpersonal communication in three ways.

Defining Mass

First, as the word *mass* denotes, mass communication has the ability to reach a large number of people through a mass medium. The number of people who could attend the lecture on mathematics was determined by the size of the classroom. However, if it were televised, it could be made available to many thousands, perhaps millions of people.

The Medium

To make the lecture available to those thousands of people, it is necessary to alter our concept of medium. No longer is it just the human voice or the nervous system; we add the mass medium of television. We could just as easily add the medium of radio, books, or even newspapers, depending upon the applicability of the medium to our task. It may be somewhat difficult, although certainly not impossible, to teach our mathematics section by radio. We may even produce a series of articles for the newspaper. If we want to teach music appreciation, radio might be just as effective as television and considerably cheaper. On the other hand, if we want to teach surgical techniques, television would be far superior. Nonetheless, some mass medium would be necessary to reach the audience beyond the limitation of interpersonal communication. For our purposes, we will therefore define *mass communication* as *messages directed toward a group of people by a mass medium.*

Altering the Definition of Mass

At first glance, it may seem as if the appropriate wording of our definition should be messages directed toward a *mass audience,* or large number of people, by a mass medium. Although this definition has merit and in some ways is correct, its traditional approach has been altered by new applications of mass media, such as the use of radio and television for internal corporate communication. Television now finds itself connecting the boards of directors of two corporations, located on different sides of the continent or even oceans away, for executive conferences. Meetings with participants scattered hundreds of miles apart take place regularly in this way. Similarly, television is used to disseminate messages to rather small audiences but ones which cannot communicate face-to-face or on an interpersonal basis. A state police commander may give a training lecture at his desk in front of a television camera. The videotape of his lecture is then played back at regional command centers throughout the state at which a group of ten or twelve troopers view the lecture. In each case the audience is relatively small, far from what we would normally consider a mass audience. It is in this way the definition of mass is changing. Technology of "mass" communication is being used to reach smaller, more specialized audiences.

the gatekeeper and delayed feedback

Besides the presence of a mass medium, two other factors have traditionally differentiated mass communication from intrapersonal and interpersonal communication—the presence of a gatekeeper and delayed feedback. With these additional concepts, our basic model of communication has been altered to represent the process of mass communication (Figure 1–5).

The term gatekeeper was first applied to the study of communication by Austrian psychologist Kurt Lewin, who defined gatekeeper as "a person or groups of persons governing the travels of news items in the communication channel."[11] Today, the term gatekeeper applies to not only groups of persons but to entire institutions. Within these institutions are both people and technology, all interacting to "govern the travels" of information between senders and receivers. That information is much more than news, as Lewin suggested. It may be strictly informative, such as an evening television news program compiled and produced by hundreds of reporters, camerapersons, editors, engineers, specialists in audio and video recording, researchers, writers, and many others. Or the message may be entertaining and involve producers, directors, costumers, scene designers, musicians, and countless more. The gatekeeper now becomes not only a person or group of persons, but people and technology through which the message must pass and be acted upon, sometimes altering it, before it reaches the consuming public.

Functions of the Gatekeeper

The function of the gatekeeper in altering what we receive from the mass media is one of *limiting* and *expanding* our information. Assume one morning that a television assignment editor dispatches a news crew to cover a music festival. When they arrive, the news crew finds the festival

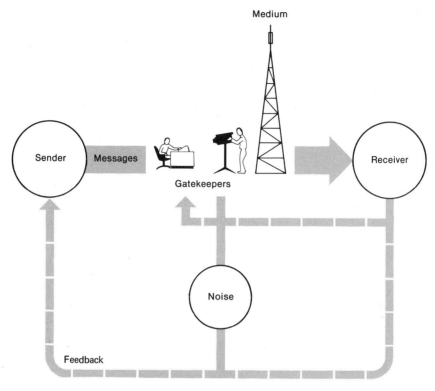

Figure 1–5 Basic model of mass communication.

spread out over a city block with every type of musician participating. Violinists, pianists, guitar players, and groups of musicians are playing everything from bagpipes to kazoos.

Upon seeing the television crew arrive, all of the musicians begin to play, each trying to gain the attention of the news crew. The reporter in charge of the story decides to focus on the bagpipe players. She bases this decision on a number of things. For one, the colorful costumes of the musicians will look good on color television. The bagpipes are also something the average viewer does not have the opportunity to see very often. In addition, the leader of the group is from Scotland and has a distinct Scottish accent. His voice alone will help hold the viewers' attention. That night our bagpipe players appear on the evening news.

Now let's examine how the news crew (gatekeepers) affected the information we received. First, they *expanded* our informational environment by offering us information we otherwise would not have received. The music festival may have been in an outlying community, and we either did not have the time or did not want to go to the trouble of driving all that way to attend it in person. On the other hand, the crew also *limited* the information we received. For instance, many more per-

11

formers were at the music festival than just those who played bagpipes. However, because the news crew chose to focus upon that one group, we were not exposed to any of the other performers. Had we been present at the music festival, we probably would have seen everyone perform. But because we watched a report of it on the evening news, we were seriously restricted in the amount of information we received.

In summary, gatekeepers actually serve three functions: (1) they can alter the information to which we are exposed; (2) they can expand our information by making us privy to facts of which we would not normally have been aware; and (3) they can limit the information we receive by making us aware of only a small amount of information compared to the total amount we would have been exposed to if we had been present at an event.

Delayed Feedback

Another distinction between mass communication and other types of communication is delayed feedback. Remember when you were sitting in the classroom listening to the mathematics lecture? In that situation, you could give instant feedback to the instructor. You could raise your hand, ask a question, and probably have your question immediately answered. However, when you were watching the mathematics lecture on television, this immediacy vanished. If you did not understand something and wanted to ask a question, you only could telephone the station if the program were live, or perhaps write a letter to the professor. Either of these alternatives is feedback, but this time, it is *delayed* feedback.

New Technology: Altering Delayed Feedback

New developments in broadcast technology have in some cases altered the "delayed feedback" of mass communication. New two-way, interactive media do permit instant feedback under some circumstances. For instance, the instructor teaching the mathematics course via television may have two different television monitors in front of the lecturn, each permitting her to view students in two different classrooms hundreds of miles away. In turn, all of the students can see and hear the instructor on their television monitors located in each classroom. A two-way voice connection permits the instructor to hear any questions the students may ask and to answer them immediately. Although messages are being directed toward a large number of people through a mass medium, instant feedback is possible in this case. New developments in cable technology promise to bring similar response possibilities to the home viewer on a broad scale.

Communicative Noise

Noise can exist in mass communication just as it can in intrapersonal and interpersonal communication, as Figure 1–3 illustrates. First, noise can appear in the processing of information through the gatekeeper. Keep in mind the gatekeeper can be many different people or groups of people, all part of the processing of information. When information is passed from one gatekeeper to another, it can become distorted.

One example of noise in the communicative process occurred when a group of reporters covered an incident along an interstate highway in the Midwest. A truck loaded with two canisters of phosgene gas stopped at a truck stop. The driver of the truck smelled a peculiar odor and decided that one of the canisters was leaking. He became sick and was taken to a local hospital. When state police learned from the invoice what the truck was carrying, they notified authorities at a local army depot. The state police then blocked off an exit on the interstate highway almost twenty miles away. It was the logical place to divert traffic since it was next to a main feeder highway, which made an excellent detour in case the highway immediately adjacent to the truck stop had to be blocked off.

When all of this information was processed into the news media, all under the pressure of deadlines and semicrisis conditions, it was distorted considerably. First, news reports left the impression that the entire truck was loaded with phosgene gas, not merely two canisters. Obviously, a leak in a tank of gas the size of a gasoline tanker was much more serious than a single canister about five feet high and less than two feet in diameter, strapped to the back of a flatbed truck. Second, because phosgene gas had been used in World War II, the wire services began to refer to the canisters as containing "war gas." Added to this was the news of the roadblock twenty miles away, which left the impression that everyone in a twenty-mile radius of the truck stop was in danger of inhaling war gas!

The network of gatekeepers that covered the story included a group of reporters from three radio stations, at least two newspapers, two wire services, two television stations, and the local and military authorities who also were dispensing information. The "institution" of gatekeepers was substantial, and much information was processed and eventually distorted.

Reducing Communicative Noise

Just as new technology has altered the concept of delayed feedback, it also has altered noise, primarily by reducing it. Thirty years ago, it would have been almost impossible to carry live pictures and sound from one continent to another. Back then, the speech of a European leader would have been reported first by a correspondent and then fed to a wire service editor in the United States. The wire service editor would have then rewritten the correspondent's report before sending it over the teletype to subscribers. This entire process was subject to much distortion and noise, because of the number of gatekeepers involved.

Today, although that process still takes place, it is now possible for a videotape of a speech to be sent by satellite into the homes of viewers thousands of miles away. Thus, the viewer watches the picture and listens to the voice of the political leader on the evening news in place of the correspondent's interpretations, thereby reducing the possibility of noise. Even the newspaper reporter can carry a small recorder, almost as inconspicuous as a note pad, and reduce the chance of misquoting a source. Still, few systems of processing information are perfect. Remember that although broadcast technology can reduce noise, the human factor is always present to return some noise back into the system.

the social context of mass communication

Our discussion thus far has been concerned with messages being sent, processed, and received. Although we have seen how gatekeepers act upon those messages, we should also realize that social forces act upon senders, gatekeepers, and receivers, and influence how they react to and process messages.

Consider the analogy of the computer. Data is fed (sent) into the computer in which it is processed and then presented, usually in the form of a printout. You might feed the computer a series of numbers, of which the computer will add and print out the answer. If you fed the same set of numbers into the computer each time, the computer's answer would be the same each time. Such is not the case with messages sent, processed, and received by mass communication. People are not computers, and we do not live in a vacuum. Messages causing one reaction at one time may cause an entirely different reaction another time. A politician's speech that attracted the attention of one gatekeeper may not attract another gatekeeper's attention. Let's examine this in more detail.

Social Context of Senders

Assume that you have decided to run for a political office, and it is time to begin the long arduous trail to election day. In writing your speech to kick off your campaign, you want to convey to the crowd those qualities you feel will truthfully express your character, your position on the issues, your background, and your intentions. As you approach the podium in a small rural community, you think about the times you have seen scenes like this before. The serenity of your childhood, the familiar faces of people you do not know but really do know, the soft mellow breeze—everything is there—including a reporter from each of the two local radio stations.

You begin your speech. You talk about things and people that have influenced your life. You talk about farm prices, having grown up on a farm, and know what you are talking about. You relate your personal experiences of meeting expenses during the harvest season and borrowing money to buy tractors. You also talk about the plight of small business people, for after the farm failed, your family opened a clothing store. All of these social forces had a direct affect on your campaign speech. Now how did your speech affect the two radio reporters (gatekeepers)?

Social Context of the Gatekeeper

When you listen to the newscasts of the two radio stations later that afternoon, you are surprised to find they each covered different parts of your speech. One reported your comments on farm prices and only briefly mentioned statements about small businesses. The other station detailed your statements about the small businesses but skimmed your comments about farm prices. Although you considered both reports objective, you wondered why they focused upon different subjects. You later discover that the reporter who reported your comments on farm prices not only grew up on a farm but also owned one. The other reporter grew up in the suburbs, his father had a small business, and he had no love whatsoever for farming. Each reporter had interpreted your speech in this instance, in accordance with his own particular background. Unlike a computer programmed to select and process certain information, the two reporters were as different as were the forces influencing them.

Research has called these phenomena *selective perception* and *selective retention*. Selective perception means you perceive only certain things, such as those which are most familiar to you or which agree with your preconceived ideas. The reporters' backgrounds and resulting selective perception created two different interpretations of the speech. Selective retention means we tend to remember things which are familiar to us or which we perceive as corresponding to our preconceived ideas. Research implies that these can become even more prominent when covering controversial issues.[12]

Another influence on the story might be the reporters' peers. The reporter may belong to a professional association and adhere to a code of ethics. The code of ethics in turn could have a direct effect on the stories processed by the gatekeepers and consequently received by the public. What if the music festival we discussed earlier had charged a $10.00 admission fee? And what if the assignment editor, as part of his or her professional ethics, had prohibited any of the staff from accepting free tickets to any event while assigned to cover that event? Admission to the festival, therefore, for the entire news crew would have come to $30.00. But what if the manager of the station had refused to pay the $30.00 admission fee for something "as unimportant as a music festival." The editor might have decided finally not to assign a news crew to the festival. Do you agree with that decision?

Social Context of Receivers

Just as gatekeepers do not operate in a vacuum, neither do receivers of mass communication. Our family, co-workers, peer groups, and organizations all affect how we receive and how we react to messages from the mass media. In this realm, interpersonal communication also is very important. For instance, upon hearing the report of your campaign speech over one of the radio stations, one local listener thinks your speech has some strong, positive merits. Yet her friend has an entirely different opinion. Since the listener respects her friend's opinion, she, in turn, changes her opinion of your speech. In this case, the friend acted as an *opinion leader, a person who is relied upon to provide us with an interpretation of messages originally disseminated from the mass media.* [13]

Let's use another example. Suppose you are watching television and see a commercial about a new headache remedy. The remedy claims to be better than aspirin, to cause fewer side effects, and to work much faster. You have been having trouble with headaches, but instead of running out to buy the new remedy you call your friend, a nurse whose opinion you respect. The nurse recommends the new remedy, and the following day you purchase it and take two pills. It works. Notice, however, that it was not the commercial that convinced you to purchase the medicine. Although the commercial helped, your friend ultimately convinced you. In this case she served as an opinion leader. Had she not recommended the remedy, the chances are you might not have bought it then.

Interrelationships of Senders, Gatekeepers, and Receivers

In reviewing examples of what occurs when information is processed through the mass media, you should begin to see many interrelationships among senders, gatekeepers, and receivers. In dealing with the concept of homophily again, it was this perceived "sharing" or "overlap" of experiences between you and the two radio reporters that caused each to report different parts of your campaign speech to listeners. Similarly, the radio listeners interpreted your speech in certain ways, also because of this sharing or perceived sharing of experiences, attitudes, and other factors. In fact, listeners may even have selected one radio station over another because of the perceived similarities between them and the reporter. Similar in some ways to selective perception and selective retention, selecting one radio station over another in this case would represent *selective exposure*, exposing yourself to information which you perceive to support your beliefs or ideas. We shall examine more of these interrelationships in greater detail when we talk about broadcast audiences.

broadcasting as mass communication

By now you should begin to see how broadcasting enters into the process of mass communication. Looking closer, you will notice that between the senders and receivers of broadcast communication are the broadcasting stations. These, along with supporting and allied organizations, have a direct effect on the messages sent through this medium of mass communication. The stations represent both standard broadcast radio and television stations as well as cable television, commonly called community antenna television or (CATV), and closed circuit television (CCTV).

Many secondary organizations affect the operation of these broadcasting stations (Figure 1–6).[14]

Program Suppliers. Program suppliers provide stations with programming ranging from Hollywood game shows to major spectaculars. Many of these are already familiar to us. They include such major television networks as the CBC in Canada; the BBC in Great Britain; NHK in Japan; ABC, CBS, NBC, and PBS in the United States; and numerous others worldwide. Television production houses, such as MTM Enterprises, creators of "Lou Grant" and "The Mary Tyler Moore Show," now in syndication, are other program sources. Their programs either are sold directly to the networks or are distributed through major distribution companies, such as Viacom. Not all program sources deal with entertainment. News program sources have become increasingly important as communication links with satellites continue to shrink the world and whet our interest in international events. Two widely used radio news program sources are United Press International Audio and Associated Press Radio.

Supporting Industries. In addition to program suppliers are advertising agencies, which place commercials on stations, and station representatives, who act as national salespersons for a station or group of stations.

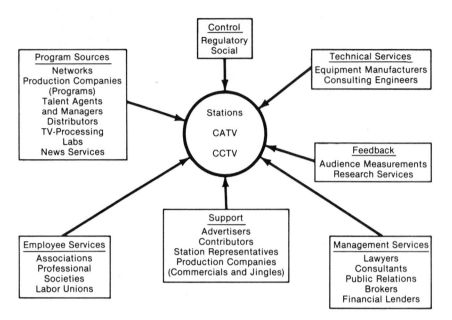

Figure 1–6 The institutions of broadcasting. Many allied organizations support broadcasting. A production company producing a network show, a communications lawyer, a technician, an advertising executive, all are part of these support organizations. (Adapted from Wilbur Schramm and Janet Alexander, "Broadcasting," in *Handbook of Communication,* ed. Ithiel de Sola Pool and others, p. 586, © 1973 Rand McNally College Publishing Company. Reprinted by permission.)

Professional Organizations. Within any industry or profession are employee services linking employees together for a variety of reasons, professional to purely social. Broadcasting's version of these include the National Association of Broadcasters (NAB), the National Cable Television Association (NCTA), and the National Association of Radio Broadcasters (NARB). Among the more narrowly defined organizations are the Radio Television News Directors Association (RTNDA) and American Women in Radio and Television (AWRT). There are over 100 other broadcast employee services in the United States alone. Labor unions also comprise a large share of the broadcast employee membership, especially at metropolitan stations and networks. Major unions with a foothold in broadcasting include the International Brotherhood of Electrical Workers (IBEW) and the Communication Workers of America (CWA).

Control. Control of broadcasting ranges from governmental to social. At the national level, governmental control is represented by the Federal Communications Commission (FCC) and the National Telecommunications and Information Administration (NTIA). In the former, control is in the form of specific laws and regulations. In the latter, it is oriented more toward policy issues. State and local governments may also control broadcasting, cable in particular.

In the social control arena, public interest groups, such as Action for Children's Television (ACT), lobby both legislators and the stations themselves. The National Congress of Parents and Teachers (PTA) became active in holding hearings on television violence in 1976 and 1977. Those hearings culminated in a report to the industry and pressure to reduce violence on television.

Advertisers and stockholders also exercise control over broadcasting. In fact, a small market radio station may be just as fearful of losing its biggest advertiser as it is of a visit from an FCC inspector. Why? Because advertisers, especially in smaller communities, can often "influence" the content of broadcast programming. If the local car dealer spends a huge sum of advertising money on a station, his drunk driving charge may conveniently be absent from the morning news, all on the strong suggestion of the station manager. Or sponsors may refuse to air their ads during violent programming.

Technical Services.   The hardware components of broadcasting have spawned a giant industry of people in everything from producing television and radio receivers to engineering consulting. Names like General Electric, Zenith, SONY, Panasonic, RCA, Motorola, and others all vie for this lucrative broadcasting market. In addition, companies and governments are active in producing and servicing satellite and microwave systems which span the globe. As in any industry, broadcasting fosters its own technical service—the consulting engineer. When an antenna on a 2,000-foot tower needs fixing, (Figure 1–7) it is hardly the job for the local TV repairperson.

Audience Measurement Services.   An audience is the lifeblood of any mass medium, broadcasting being no exception. Measuring this audience uses the talent of a host of survey companies. Others specialize in customized surveys, such as measuring the effectiveness of a station promotion, undertaking a station image survey, or initiating a personality recognition survey among the viewers.

Management Services.   With the increasing complexity of broadcasting, few broadcast managers have the skills necessary to handle all the functions of the station, so they must rely on management consultants. Among the most important of these are consulting attorneys hired by broadcasters to help them process the mountain of governmental forms they now must file, and to give complicated legal advice. Most of the major communication law firms are in Washington, D.C., close to the

Figure 1–7 Technical services supporting broadcasting include highly trained technicians. Even the smallest stations use consulting engineers and tower construction companies. Pictured is a tower climber servicing the receiving antenna for a major cable system. (NCTA)

FCC. They include Cohn & Marks; Pittman, Lovett, Ford, and Hennessey; and many others.

Broadcast promotion services help with specialized campaigns, such as a station contest or station-sponsored fairs or concerts. Many stations now employ people full-time to handle these tasks.

When it is time to buy or sell a broadcasting property, the broadcast broker becomes important. The "realtor" of broadcasting, the names of George Moore & Associates; Richard A. Shaheen, Inc.; and R. C. Crisler & Co., Inc. are household words to broadcast managers and investors wanting to acquire interests in broadcast properties.

Although we have discussed each of these allied organizations and services as being separate, keep in mind that they interrelate to each other. The production company is just as concerned about the FCC's stand on obscenity as the broadcaster is. The attorney's advice is just as valuable to the advertising agency producing a broadcast commercial as it is to the station manager. These organizations and interrelationships are the interactive "process" of broadcasting in our society.

where this book will take us

Our first look will be at the early years of broadcasting, the years of new discoveries, which in their time rivaled the discoveries of our space age. It was a time of anticipation and little hope lost at any failure. We shall examine the development of broadcasting, its programming content, and its criticism. We also shall look at its future.

From there, we shall explore the systems that carry broadcast messages not only around the world but also around the corner. We shall learn about national networks and state networks, formal and informal networks, and wired and wireless networks. We also shall examine international broadcasting from London to Leningrad and from Switzerland to Southeast Asia. And we shall go inside the corporate and educational institutions to see how broadcasting is transforming them.

To understand how specific controls affect the broadcast media and consequently the messages we receive, we shall examine such organizations as the FCC and the laws that govern broadcasting. We shall look at social controls, as exercised by such formal groups as ACT and NAB.

The broadcast audience and how research has evaluated the medium will comprise the fourth section of this book. Terms like "demographics," "psychographics," "ratings," and "rating points," will become important.

To view the world of decisions and decision makers, we shall explore the inside of a broadcasting station, meeting the managers and the managed. We shall plan a rate card, make an advertising buy, and understand the "bottom line." To help us use the library more effectively in our research about broadcasting, we shall conclude with a library guide.

summary

In order to understand broadcasting, you first must understand communication. Communication includes intrapersonal communication, or communication within ourselves, interpersonal communication, or communication between people, and mass communication, which is messages directed toward people via a mass medium. Regardless of the type of communication, the potential for noise is always present to disrupt the flow of messages between sender or receiver. Traditionally, delayed feedback and the presence of gatekeepers have distinguished mass communication from intrapersonal and interpersonal communication. However, new technology, the increase in the use of closed-circuit television, and broadcasting in business and education, have all altered the definition of "mass" as it applies to broadcasting. Today mass media may be directed to a small and specialized audience.

The primary components of broadcasting are the stations themselves; cable television, commonly called community antenna television or CATV; and closed-circuit broadcasting or CCTV. All secondary organizations affect these institutions and include program sources, employee services, controls, support industries, technical services, audience measurement services, and management services.

spotlight on further learning

AUSTIN-LETT, GENELLE, and JANET SPRAGUE, *Talk to Yourself: Experiencing Intrapersonal Communication.* Boston: Houghton Mifflin Company, 1976.

BITTNER, JOHN R., *Mass Communication: An Introduction* (2nd ed.).Englewood Cliffs, N.J.: Prentice-Hall, Inc., 1980.

CHANEY, DAVID, *Processes of Mass Communication.* London: Macmillan Press Ltd., 1972.

DeFLEUR, MELVIN L., and SANDRA BALL-ROKEACH, *Theories of Mass Communication* (3rd ed.). New York: David McKay Company, Inc., 1975.

DeLOZIER, M. WAYNE, *The Marketing Communications Process.* New York: McGraw-Hill Book Company, 1976.

DEL POLITO, CAROLYN M., *Intrapersonal Communication.* Menlo Park, Calif.: Cummings Publishing Company, 1977.

SCHRAMM, WILBUR, *Men, Messages, and Media: A Look at Human Communication.* New York: Harper & Row, Publishers, Inc. 1975.

SCHRAMM, WILBUR, and DONALD F. ROBERTS, eds., *The Process and Effects of Mass Communication* (2nd ed.). Urbana: University of Illinois Press, 1971.

SMITH, ALFRED G., ed., *Communication and Culture.* New York: Holt, Rinehart & Winston, 1966.

TUBBS, STEWART, and SYLVIA MOSS, *Human Communication.* New York: Random House, Inc., 1974.

WRIGHT, CHARLES R., *Mass Communication: A Sociological Perspective* (2nd ed.). New York: Random House, Inc., 1975.

part

1

History and Development

2

the beginning of wireless

FOCUS   After completing this chapter we should be able to

Explain the work of Henrich Hertz and the discovery of electromagnetic waves.

Discuss Marconi's early experiments in Italy, England, and North America.

Trace the development of the Marconi companies.

List the contributions of Thomas Edison, J. Ambrose Fleming, and Lee de Forest to the early wireless.

Identify the contribution of Nathan B. Stubblefield to the early history of voice broadcasting.

Explain how Reginald Fessenden and Ernst Alexanderson contributed to long-distance voice broadcasting.

Understand the relationship between the crystal detector and amateur radio.

88680

LIBRARY
College of St. Francis
JOLIET, ILL.

We are about to undertake a journey, one that will begin in the mid-nineteenth century and lead us into the twentieth century.[1] It is a journey through history, but do not be afraid of the word "history." History is exciting if you relive it and meet the people who were part of it. The historian Alan Nevins compares history to a sextant on a ship. History tells us where we have been so we can better understand where we are going. The television program showing us sporting contests at an Olympic village, the view from outer space, or the sounds of contemporary music from our hand-held transistor radio, all were conceptualized long ago. These receivers of communication that today we take for granted were developed in the laboratories of scientists with the same elation and excitement that scientists have today when discovering new knowledge. Let's share some of this excitement in learning how broadcast communication came to make such an important impression on our lives.

Keep in mind that not all of the players in this epoch of scientific achievement can be recognized, for three reasons: (1) many of them conducted research, but never took the time to publish their results so that other scientists could share in their discoveries; (2) others conducted research and, although in some cases published it, never patented their inventions. Thus, someone else was given that recorded place in history; and (3) still others were interested in pure science as opposed to the practical application of their discoveries to society. The result, again, was recognition of those who made practical application of their research, sometimes overshadowing their academic peers in general publicity. The latter also tended to reap the greatest economic rewards, not only from patents but also from the development of full-scale corporate endeavors, sometimes extending worldwide.

Henrich Hertz and electromagnetic waves

In 1857 Henrich Rudolph Hertz (Figure 2–1) was born to a middle-class family in Hamburg, Germany. Taught an hour a day by tutors and obtaining the rest of his knowledge in his spare time, young Hertz developed a keen interest in science and outfitted himself with his own home laboratory. Engineering first whetted his appetite, but after a year of study at the University of Munich, he moved to Berlin to study pure science under the well-known German scientist, Hermann Ludwig Ferdinand von Helmholtz. It was there under the lure of the Berlin prize of

26

LIBRARY
College of St. Francis
JOLIET, IL.

Figure 2–1 Henrich Hertz. His early experiments began in a home laboratory and eventually reached Berlin. Using an experiment in static electricity, he stumbled upon proof of the presence of electromagnetic waves. (der Universität Karlsruhe)

1879 that von Helmholtz encouraged his twenty-two-year-old apprentice to examine further the inquiry into electromagnetic forces. The early experiments were not very fruitful, and for a while, Hertz occupied his time with other experiments. He never, however, ceased to be fascinated by the potential proof of the travels of electromagnetic energy through space.

One day while lecturing, Hertz was using two pieces of spiraled wire in discharging static electricity. He noticed that when a spark gap was introduced into one coil, it produced a current in another coil. What Hertz had stumbled across was the presence of very high-frequency electromagnetic waves which were generated by the spark. From there the investigations proceeded systematically, and from 1886 through 1889, using both transmitting and receiving spark gaps at high frequencies, he was able to provide further proof of the presence of electromagnetic waves. Today we call these radio waves, which catapult radio, television, and other communication around the world and into outer space.

Both von Helmholtz and Hertz died in 1894. Some scientists picked up Hertz's pure science. Others, more concerned with the practical application of these principles, concentrated on improving long-distance communication between people. It is these applied efforts to which we will now turn our attention.

wireless is born: Marconi the inventor

The telegraph had captivated America and Europe. Samuel Morse, its inventor, died two years before the second son of Giuseppe and Anna Marconi was born, on April 25, 1874. Guglielmo Marconi (Figure 2–2) was a child of parents who, by late nineteenth-century standards, were quite well to do.[2] But the young, restless Marconi was not like the rest of his family, comfortable with gracious Italian living. Often he irritated his father when he interrupted the quiet conversations at an evening meal with persistent, unrelated questions. This did not improve when, after reading a scientific magazine, Guglielmo developed a keen interest in the work of Henrich Hertz. Finally, from the top floor of their home, with his father's rancor and his mother's reinforcement, Guglielmo Marconi began to experiment. With crude tables, boards, hanging wires, and other paraphernalia, he began to duplicate the experiments of Hertz.

Early Experiments in Italy

To the family, the work of the young son in his upstairs laboratory was intriguing but of questionable value. The boy's father felt that he was

Figure 2–2 Guglielmo Marconi. As much a businessman as an inventor, Marconi created a worldwide wireless empire with companies in England, the United States, Canada, and other countries. (RCA)

wasting the best years of his life, but became more interested when Guglielmo asked him for money to advance his work beyond the experimental stage.[3] Being a stern and practical businessman, his father first wanted a demonstration. This was followed by a long discussion as to how his father would get a return on his money. The whole scenario is rather humorous, considering in retrospect that the boy's corporate empire would eventually gross billions. Finally the two agreed to an initial investment, and Guglielmo went to work building his first transmitting device. Using a reflector sheet strung between two poles (Figure 2–3), he first managed to receive a signal across the room. The receiver used a device called a *coherer*—a small glass tube with wires in each end and filled with metal filings which would collect between the two wires whenever electricity was applied.

Marconi, already familiar with the work of Samuel Morse, immediately realized the potential for long-distance communication.[4] He also had a sense of urgency, because to him the principle was extremely simple. Why had someone not thought of it before, or more importantly, applied it? His experiments became more and more frequent and the range of his signals, more and more distant. The distances became farther and farther, until on top of a hillside twenty minutes from his home, the experiments reached a threshold. Could the signal go beyond the hilltop?

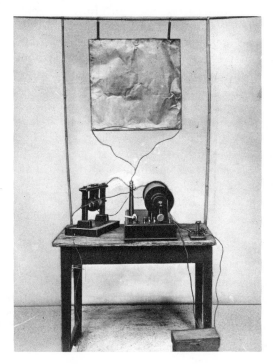

Figure 2–3 Marconi's first transmitter used in his early experiments in 1895. The large piece of tin suspended above the table served as the antenna. (The Marconi Company Limited, Marconi House, Chelmsford, Essex)

If the invention were to be a success, it would have to be able to leap over hills, mountains, buildings, and oceans. The crucial test sent his brother and two helpers carrying the receiver and antenna apparatus over the hilltop out of sight of the villa. Guglielmo's brother also carried a gun with instructions to shoot to confirm the signal. No sooner had Guglielmo fed current to the transmitter than the shot rang out. Now the capital that his father had provided had to be increased before the experiments could reach the next level. In attempting to obtain government backing for Guglielmo, a letter was sent to the Italian Post Office Department. The reply was negative. But if Italy were to say no, perhaps the great naval power of the day would say yes. With his mother's encouragement and accompaniment, Marconi was off to England.

Experiments in England

The first stop was customs. Here the journey hit one of its low points as ignorant customs inspectors ripped at the equipment until it was all but destroyed. Marconi managed to reconstruct the broken pieces which had been so carefully crated in Italy. The next step was to be sure no one else captured the idea. For four months, Marconi and his mother slaved over the papers which were to be presented to the London Patent Office.[5] The first specifications were filed on June 2, 1896. The complete diagrams and detailed specifications were filed on March 2, 1897 under the title, "Improvements in Transmitting Electrical Impulses and Signals, and an Apparatus Therefor." On July 2, 1897, patent number 12,039 was granted to the twenty-three-year-old Italian inventor. The experiments now continued, but it still was necessary to get support from the government for capital to develop the invention to its full potential.

The help and support Marconi needed came first from the chief engineer of the British Post Office, William Henry Preece. Preece liked young Marconi, and with Preece's support, he began his experiments in England, first between two buildings and then a major demonstration across the Bristol Channel, a distance of about three miles. The press noticed Marconi's wireless and published the news to the world. Attention was being bestowed on the device beyond any measure ever dreamed of by the young inventor (Figure 2–4). Along with offers to buy the rights to his invention came the more humorous offers of marriage from women who said Marconi's waves made their feet tickle.[6] The distance of his experiments increased from three to thirty-four miles. Publicity abounded again when Marconi was commissioned to install a wireless on a tugboat to report the sailing races at the Kingston regatta. He secured other patents. One of the most important, patent 7,777 for a selective tuning device, was granted in 1901.

Figure 2–4 Three officials of the British Post Office Department examine the equipment Marconi used to test the first successful wireless across the Bristol Channel in 1897. The British Post Office Department provided both encouragement and financial support of Marconi's early work. (The Marconi Company Limited, Marconi House, Chelmsford, Essex)

Wireless across the Atlantic

The year 1901 also was the year of the most convincing experiment of the power of wireless communication. Still to be hurdled was the vast expanse of the Atlantic Ocean. Marconi left England for America in February, 1901 and headed for Cape Cod, the point he felt best suited to the test of his wireless. But like any stretch of New England coastline, harsh winter winds can play havoc with any structure not built for permanency. Likewise, the Atlantic can do the same on the English coast. For Marconi, 1901 had double disasters. News arrived that storms had toppled the antenna at his installation at Poldhu, England (Figure 2–5). Within weeks, the same fate befell the Cape Cod station. Marconi now decided to transfer operations to Newfoundland, then a British colony. Using a bit of intrigue, he told local officials he was attempting to communicate with ships at sea, with no mention of the real purpose, transatlantic communication. Instead of antenna towers, he planned to use balloons, and packed six kites as a back-up.[7]

Figure 2–5 The first circular antenna arrangement constructed at Poldhu, England and designed for Marconi's transatlantic test. There were 20 wooden towers, each 200 feet high. The structures collapsed under gale-force winds which hit England in September, 1901. (The Marconi Company Limited, Marconi House, Chelmsford, Essex)

The experiments in Newfoundland started on December 9, 1901. First, the balloons were tested, but a line broke and the balloon headed for open sea. The next decision was to try one of the large kites. Marconi's assistants, George Kemp and P. W. Paget worked with the kite until it soared hundreds of feet, stringing behind it the antenna wire connected to the essential receiving equipment on top of Signal Hill, Newfoundland (Figure 2–6). Serious monitoring started on December 12, 1901. There were no results in the morning; nothing was heard from Poldhu. Spirits were low as the men continued to listen for the tapping signal that would indicate that England was calling. At 12:30 P.M., Guglielmo Marconi listened intently as the tapping sound of three dots, signaling the letter *S*, crackled through the earphone. Handing the earphone to Kemp, the assistant verified the signal.

Reaction to Transatlantic Wireless

The world would spend the rest of December reading about it.[8] *The New York Times* called it "the most wonderful scientific development of recent times" and headlined the story *WIRELESS SIGNALS ACROSS THE*

Figure 2–6 Signal Hill, Newfoundland where Marconi
made his first successful transatlantic wireless contact.
(Newfoundland Department of Tourism)

ATLANTIC. Across the ocean, the *Times* of London headlined *WIRELESS TELEGRAPHY ACROSS THE ATLANTIC.* The London paper described how Marconi had authorized Sir Cavendish Boyle, the Governor of Newfoundland, to "apprise the British Cabinet of the discovery, the importance of which is impossible to overvalue." Not forgetting his beloved Italy, Marconi informed the Italian government himself. Magazines were equally enthusiastic about the feat.[9] *Century Magazine* called Newfoundland "the theatre of this unequaled scientific development." *The World's Work* labeled the transatlantic transmission "a red letter day in electrical history." *McClure's Magazine* demanded: "Think for a moment of sitting here on the edge of North America and listening to communications sent *through space* across nearly 2,000 miles of ocean from the edge of Europe!"

Not all, however, was as happy as in Newfoundland. The apparent threat of competition between wireless and the cable telegraph surfaced immediately. Cable stocks declined shortly after the announcement of the Atlantic broadcast.[10] The Anglo-American Telegraph Company, which had a monopoly on telegraph communication in Newfoundland, was quick to threaten reprisals if Marconi did not stop the experiments. A few days later, the inventor received a letter from the company stating:

Unless we receive an intimation from you during the day that you will not proceed any further with the work you are engaged in and remove the appliances erected for the purpose of telegraphic communication, legal proceedings will be instituted to restrain you from further prosecution of your work and for any damages which our clients may sustain or have sustained; and we further give you notice that our clients will hold you responsible for any loss of damage sustained by reason of your trespass upon their rights.[11]

The Canadian government, obviously seeing the chance to emulate its neighbor, immediately contacted Marconi and offered full cooperation with the inventor. Public sentiment for the action taken by the telegraph company was unfavorable on both sides of the Atlantic. *The New York Times* criticized the action, and letters to the editor of the London *Times* expressed similar sentiments. The threats, comments, and letters to the editor soon became history as the world began to use the results of the December experiments of 1901.

wireless expands: the Marconi companies

Despite the respect Marconi had for pursuers of pure science, he was much more interested in applying results and harvesting financial rewards.[12] Thus, it was only a short time after his patent had been issued in England that he began formulating a world corporate empire that would stretch over the seven continents and involve millions of dollars in capital. The company that had the most direct effect on wireless development was the Marconi Wireless Telegraph Company, Limited, first formed on July 20, 1897 as the Wireless Telegraph and Signal Company, Limited. It was Marconi's father who insisted that the family name should be attached to the venture.[13] The beginning capital amounted to 100,000 English pounds, of which 15,000 went to Marconi for his patents. It was from this 15,000 pounds that he paid the cost of organizing the company. He also received 60,000 of the 100,000 initial shares valued at 1 pound each. The remaining 40,000 went on the open market.

England: Marconi's Wireless Telegraph Company, Ltd.

A year after its formation, the operating capital increased by another 100,000 pounds, and although wireless had captured the imagination of the British, there were warnings for unwary investors. *Investors World,* a financial publication, advised in 1898 that, "from all we can gather, the public will be well advised to keep clear of this concern. . . . Marconi's ingenious ideas do not seem to have made much headway, and it would be interesting to learn what the government officials reported about them."[14] The warning had little effect, and although the investment for years to come did not show much success in paying dividends, the public was always ready to buy up new shares whenever they were placed on the market.

By March of 1912, however, rumors of a contract between the company and the government had a taint of corruption. One rumor suggested that Marconi was treated favorably because of his close friends in Parliament. Some government officials had made a huge profit by selling their Marconi stock when it peaked after the news of the contract being signed. The second set of charges was of manipulation of stock by the American Marconi Company. A committee was appointed by Parliament to investigate the matter. After due deliberation, they came out strongly in favor of Marconi, but the matter was not over. Another committee investigated the role of middlemen and stockbrokers, and still another committee investigated the role of the House of Lords. Libel actions were taken and the stock tumbled. The publicity from the scandal meant that the company only briefly enjoyed the prosperity for which Marconi had long hoped. The future development of Marconi in England would have to wait until after the end of World War I. In North America, the story was much the same.

Marconi's Interests in Canada

Marconi's corporate interest in Canada came after the Newfoundland experience with the cable authorities. He erected a station at Glace Bay, Canada and began major experiments in achieving reliable transatlantic wireless. The first transatlantic service opened on the night of December 15, 1902 when the London *Times* correspondent at Glace Bay cabled his newspaper report across the Atlantic. Two nights later, it was arranged that the American station at Cape Cod would send a message from the United States to the King of England. The signal would be relayed to Glace Bay and from there to Poldhu. As it turned out, the atmospheric conditions were so good that the station in England picked up the signal directly from America.

The London *Times* was infatuated enough by the prospects of transatlantic service that it convinced Marconi to open the station again to send news flashes to England. That lasted for little more than a week until an ice storm sent the Glace Bay antennas crashing to the ground. The station was later reconstructed using a large umbrella antenna.

The American Marconi Wireless Telegraph Company

When Marconi came to the United States in 1899 to report the American Cup races by wireless, he also began an American subsidiary of his English company. To utilize Marconi patents in America, the American

Marconi Wireless Telegraph Company was incorporated in the state of New Jersey in the fall of 1899. The first equipment was used to warn ships of bad weather and coastal conditions, and was installed on the Nantucket Light Ship with a shore station on the eastern shore of Nantucket Island.

That same year, the American company ran into trouble over a proposed United States Navy contract for the installation of Marconi wireless on Navy ships. After a series of tests, the Navy first recommended buying the Marconi equipment. Then after asking Marconi what the cost would be for wireless, the Navy received word that the company would not sell the equipment, but that the Navy would have to rent it. At that point the Navy backed out of the deal. Captain L. S. Howeth, later writing about the negotiations, said: "In light of future events, the Marconi leases and stipulations have proved a blessing in disguise. The foresight of the authorities in not permitting themselves to be shackled with its restrictions, which would have persisted for more than a decade, allowed the Navy a free hand in guiding and assisting in the development of radio in this country."[15]

Despite the loss of the Navy contract, the American company received a boost in assets in 1912 when it won a patent suit against the United Wireless Company. Using a case in England as a legal precedent, the American company charged United Wireless with patent infringement of patent 7,777, the Marconi tuner that could select different signals from a single aerial. United Wireless pleaded no defense, and Marconi assumed control of the company and all of its assets and contracts. It was an unusual way to obtain a corporate merger, but it was a positive boost for the American assets. The U.S. Navy ended up using Marconi equipment after all when World War I began, since the Navy either had taken over or had closed all commercial and amateur wireless stations, many using Marconi equipment.

A young boy named David Sarnoff had been hired by the American company in September of 1906. Shortly after World War I ended, the American company was purchased by the newly formed Radio Corporation of America (RCA). Sarnoff became part of RCA management, later heading the company. We shall learn the reasons for the sale and RCA's early development in chapter 3.

working to improve wireless reception

Although Marconi had successfully transmitted signals across the Atlantic and had developed a world corporate empire, his success had been greatly aided by subsequent developments in wireless communication.

One of the most important areas of scientific investigation was the need for a device that would more efficiently detect and receive electromagnetic waves. The receiving and sending apparatus had to be so big that antennas were the size of football fields, both for transmitting and receiving. Yet the current that entered a radio antenna and received an electromagnetic wave was minute. The great challenge was how better to detect these tiny, almost indistinguishable currents of energy hitting the antenna of a radio receiver. For radio to become a household appliance, the football-sized receiving antennas had to be eliminated.

Edison's Contributions

Some of the first experiments leading to an improved detector came not during the study of radio, but during the study of electric light.[16] Thomas Edison, while in the process of inventing the light bulb, had experimented with a two-element bulb, but had found it impractical. The bulb consisted of two metallic elements in a vacuum, called the *plate* and the *filament*. If a battery were attached to the bulb with the positive connector of the battery attached to the plate and the negative connector of the battery attached to the filament, current would flow through the bulb. If the connectors were reversed, the current would stop. What Edison had invented but discarded was an early two-element vacuum tube, or what was later to be called a *valve*, since it could "shut off" current running in one direction, much like a valve controls steam or water.

The Fleming Valve

J. Ambrose Fleming was an employee of Marconi when one of the big secrets to unlocking future development in wireless was to find some way to measure electromagnetic waves in order to understand better their behavior and their frequencies. Fleming determined that the best way to do this would be to invent a means of measuring the waves as they flowed in only one direction. The secret was in Edison's two-element light bulb. Fleming went to work perfecting the device, patenting the "valve" in England in 1904 and, through the American Marconi Company, in the United States in 1905. The device worked by attaching the plate to the antenna, attaching a wire from the filament to the ground, and then hooking a telephone receiver into this completed circuit. With the "valve" in position, the telephone receiver was able to hear the presence of the waves. His invention was called the oscillation valve or Fleming valve. It was not long, however, before Fleming's device was greatly improved by the inventive hand of an American, Lee de Forest (Figure 2–7).[17]

Figure 2-7 Lee de Forest, inventor of the "Audion," the three-element vacuum tube. Lee de Forest based his work on the foundation laid by Hertz. De Forest received a Ph.D. from Yale in 1899 with a dissertation titled "Reflection of Hertzian Wave from the Ends of Parallel Wires." Much of his career was marked by legal feuds with J. Ambrose Fleming. He is seen at the 1939 World's Fair with a display tracing the evolution of the vacuum tube. (AT & T Co.)

de Forest and the audion

The work of de Forest (Figure 2–7) ranks close to that of Marconi in the development of radio. Born in Council Bluffs, Iowa in 1873, de Forest was the son of a Congregational minister who was later to become president of Talladega College in Alabama. After attending Mt. Hermon School in Massachusetts, de Forest entered a mechanical engineering program at the Sheffield Scientific School at Yale University. With a dissertation entitled "Reflection of Hertzian Waves from the Ends of Parallel Wires," he was granted a Ph.D. in 1899. The research done and knowledge gained at Yale and his desire to apply pure science, first to inventions, then to patents, and then to profits, led him on a remarkable career that spanned much of his more than eighty years. He died in 1961 in Hollywood, California where he was closest to one of his most beloved works, talking motion pictures. Our emphasis here, however, is on the invention of the audion, a three-element vacuum tube which revolutionized radio.

Adding the Grid to the Vacuum Tube

Lee de Forest discovered that a third element could be added to Fleming's two-element vacuum tube valve. De Forest inserted a tiny grid of iron wires. The result was characterized in an early radio publication:

> *This may not seem much to the uninitiated, but that miniature gadget was the truest "little giant" in all history. . . . that the brain of man ever created. It set unbelievable powerful currents in motion, magnifications of those which flicked up and down the antenna wire, and thus produced voice amplification which made radio telephony a finished product. By adding another tube and another, the amplification was enormously increased.*[18]

With the third element, the vacuum tube now had a filament, plate, and grid (Figure 2–8). Named the audion by de Forest's assistant, C. D. Babcock, de Forest first announced the tube in a paper presented to the October 26, 1906 meeting of the American Institute of Electrical Engineers in New York. After the paper was reproduced in the November 3,

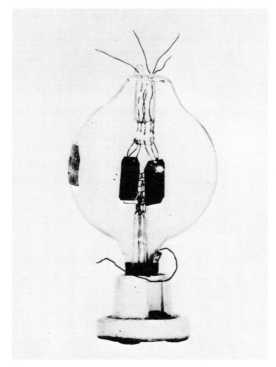

Figure 2–8   Lee de Forest's "Audion" tube. (AT & T Co.)

1906 issue of *Electrical World,* it was not surprising that one of the first reactions to the de Forest discovery came from Fleming.[19] In a letter to the editor of *Electrical World* on December 8, 1906, Fleming attempted to diminish some of the importance of the de Forest invention. Fleming wrote in part:

> *There is a remarkable similarity between the appliance now christ ened by de Forest as an "audion" and a wireless telegraphic receiver I called an oscillation valve.... Dr. de Forest's method of using this appliance as an electric wave detector appears, so far as I can judge from published accounts, to be a little different from mine, but nevertheless the actual construction of the apparatus is the same.... Even if Dr. de Forest has discovered some other way of employing the same device as a receiver, I venture to think that my introduction and use of it should not be ignored, as I believe I was the first to apply this device... as a means of detecting electric oscillations and electric waves.*[20]

De Forest did not sit back and let Fleming's suggestions go unchallenged. He replied to Fleming's letter with a letter to the editor of the same magazine, published in the December 22, 1906 issue. De Forest credited German scientists Johann Elster and F. K. Geitel, not Fleming, as deserving credit:

> *Prof. Fleming has done me the injustice of expressing an opinion based on an extract only of my paper regarding the "audion." In a more complete abstract of that paper published in the* Electrician *of London, it is seen that I mention not only the device described by Prof. Fleming in 1904, but point out the real genesis of this device by Elster and Geitel in 1882, or eight years prior to its rediscovery of Prof. Fleming in 1890.... The difference which Prof. Fleming questions may be tersely stated as that between a few yards and a few hundreds of miles; between a laboratory curiosity and an astonishingly efficient wireless receiver employing the same medium but operating on a principle different in kind.*[21]

The Feud with Fleming

The rift between Fleming and de Forest did not end in the pages of *Electrical World.* Lee de Forest went on to patent his audion, but the Fleming valve also had been patented in both England and the United States under the American Marconi Company. It was the United States patent that provoked a lawsuit brought by the American Marconi Company. The case went in favor of the company, which contended that Lee de Forest had read the paper presented by Fleming to the Royal Society of England in 1905 in which Fleming described the oscillation valve, and that de Forest then had used this knowledge to begin experimenting, with the end result being the audion.[22] The case was appealed, and again the court ruled in favor of the Marconi Company and the Fleming patent. Two years passed between the lower court's decision and the appeal. Both de Forest and the Marconi Company continued to manufacture the tubes. To make matters more complex, the court held that, although de Forest had infringed on the Fleming valve, the Marconi Company had infringed on the audion. The result was that neither company could manufacture the devices without the other's consent.[23] The situation was chaotic until the Fleming patent with Marconi expired in 1922. Even more incredibly, the United States Supreme Court in 1943 ruled that the Fleming patent had never been valid in the first place!

Although we might consider Lee de Forest's account of all this as rather biased, his summation of the conclusion of the Fleming-de Forest dispute is worth reading partly because of its humor, and most importantly because it captures the intense rivalry between two men, typical of feuds between companies and inventors in the early development of radio. In his book, de Forest comments on the state of affairs between him and Fleming in 1943 shortly after the Supreme Court decision. Lee de Forest wrote:

> *About this time Sir John Fleming, still unregenerate at ninety-two, published an amazing article in which he ignored all the earlier work . . . claiming even the discovery of the so-called "Edison effect," but never mentioning Edison's name! For this omission I wrote him in righteous reproach, incidentally calling to his attention the recent Supreme Court decision. Fleming's reply evinced profound disdain for what a mere Yankee court might think of his best-loved child. Having married a young opera singer at 84, he lived to the ripe old age of 95, dying in 1945. He never yielded in his firm conviction that he was radio's true inventor![24]*

Lee de Forest's own modesty is not excused when we remember that he entitled his autobiography, *The Father of Radio*. He also had some words of his own about what radio had become when he told a group of radio executives: "The radio was conceived as a potent instrumentality for culture, fine music, the uplifting of America's mass intelligence. You have debased this child, you have sent him out in the streets in rags of ragtime, tatters of jive and boogie-woogie, to collect money from all and sundry."[25]

breaking the voice barrier: radio telephony

The second challenge to be hurdled in early radio was how to go beyond the "dit-dahs" of the Morse code and the "What Hath God Wrought" of the telegraph to the "O Holy Night" of Reginald Fessenden's Christmas Eve radio broadcast in 1906. The story of voice transmission starts long before 1906, back when early experimenters examined the phenomenon of using the ground and water to act as a conductor for transmitting and receiving "wireless telephone" conversations.

The system had been used by wireless operators in 1838.[26] It applied a process known as *conduction,* in which the ground or water would be the conductor, providing the "second wire" in a telegraph hookup. Samuel Morse, inventor of the telegraph, used it in his New York experiments. It was not long before inventors discovered that they did not need any wire at all to communicate between the transmitter and receiver over short distances. Because a current in one antenna would produce a current in another one nearby, a process called *induction,* two antennas in close proximity to one another would make the system work. This was a different principle from that of electromagnetic waves traveling through space that Marconi and others used. Induction created an electrical disturbance in the atmosphere that was detectible only in the immediate vicinity of the transmitter.

Nathan B. Stubblefield and His Wireless Telephone

Before Marconi mastered the Atlantic and while de Forest was studying at Yale, a Kentucky farmer and experimenter developed a means by which voice could be transmitted as much as three miles by using induction. His name was Nathan B. Stubblefield (Figure 2–9).[27] Near his

Figure 2–9   Nathan B. Stubblefield (left) and his wireless telephone. As early as 1892, Subblefield is reported to have sent voice by wireless over a short distance at his farm in Murray, Kentucky. His son, Bernard (right) later became an employee of Westinghouse. (Murray, Kentucky Chamber of Commerce)

home in Murray, Kentucky, and later on the Potomac River in Washington, D.C., he successfully transmitted voice without wires. It was in Murray, Kentucky, however, that he received his first publicity. Dr. Rainey T. Wells witnessed Stubblefield's experiments and recollected:

> He [Stubblefield] had a shack about four feet square near his house from which he took an ordinary telephone receiver, but entirely without wires. Handing me these, he asked me to walk some distance away and listen. I had hardly reached my post, which happened to be an apple orchard, when I heard "Hello, Rainey" come booming out of the receiver. I jumped a foot and said to myself, "This fellow is fooling me. He has wires somewhere." So I moved to the side some 20 feet but all the while he kept talking to me. I talked back and he answered me as plainly as you please. I asked him to patent the thing but he refused, saying he wanted to continue his research and perfect it.[28]

The demonstration was reported to have taken place in 1892. A modified, Bell-type transmitting device provided the signal which emanated from a large, circular, metal antenna. Other residents of the small town of Murray witnessed a similar demonstration six years later in 1898. Claims of his accomplishments were published in the *St. Louis Post Dispatch* and generated enough interest to bring him to Washington, D.C. for a public demonstration on March 20, 1902. Following the demonstration, Stubblefield said, "as to the practicality of my invention—all I can claim for it now is that it is capable of sending simultaneous messages from a central distributing station over a wide territory. . . . Eventually, it will be used for the general transmission of news of every description."[29]

Commercial exploitation of the invention was not far behind, and in 1903, Stubblefield became director of the Wireless Telephone Company of America. Demonstrations in Philadelphia and Washington, D.C. created more interest in the device; yet it is here that the rest of Stubblefield's life becomes somewhat obscure. There are various reports of what happened to him. One suggests that Stubblefield became disillusioned with how the stock for the company was being handled and on one occasion even charged it with fraud.[30] Stubblefield returned to Kentucky and with the help of local citizens, received a patent for the device on May 12, 1908. Obviously disenchanted with the business developments of his wireless telephone, Stubblefield went into seclusion and continued to research in his workshop-shack near Murray.

If you travel through the Kentucky countryside near Murray today, you may pass the place where Stubblefield was found dead on March 30, 1928, the cause of death listed as starvation. Or you might drive by Murray State College where students are acting in the play of Nathan B. Stubblefield's life, *The Stubblefield Story*. Then as you go downtown, you can tune your car radio to 1340 KHZ and hear a blend of rock, easy listening, and country, and the news "centrally distributed" from WNBS radio. At a certain time the announcer will tell you: "You are tuned to WNBS, 1340 on your radio dial in Murray, Kentucky, the birthplace of radio."

Fessenden at Brant Rock

Although Nathan Stubblefield's wireless telephone worked, someone had yet to master the ability to transmit voice beyond very short distances. Some of the most productive experiments toward this goal were carried out in Pittsburgh, Pennsylvania, in 1901. Reginald A. Fessenden (Figure 2–10), a Canadian by birth and a professor at the University of Pittsburgh, worked to improve both the detection of electromagnetic waves and a means by which a human voice could be placed "piggy back" on electrical oscillations and sent into the atmosphere.[31] He first developed an improved detector, which would later be called the *hetrodyne circuit.* Fessenden applied for patent papers for the improved circuit in 1905. Simultaneously, he continued to improve the transmitting and antenna systems for wireless. The excitement of a breakthrough into reliable voice broadcasting was enough to encourage two Pittsburgh financiers, Thomas H. Givens and Hay Walker, Jr., to put $2 million into Fessenden's work and form, with Fessenden, the National Electric Signalling Company.

Figure 2–10 Reginald Fessenden. With financial support from two Pittsburgh investors, Fessenden applied his earlier work at the University of Pittsburgh to a successful test of long-distance voice transmission in 1906. With a new antenna design, he transmitted signals from Brant Rock, Massachusetts which were received as far away as Scotland and Puerto Rico. (Archives of the University of Pittsburgh)

Fessenden, besides $300 a month in salary, also received stock in the new venture.

The early experiments were conducted on the Chesapeake Bay, then were moved in 1905 to Brant Rock, Massachusetts. Here, the next chapter in wireless history would be written. Continually trying to improve Marconi's original invention, Fessenden constructed a high-power station at Brant Rock and radically altered the antenna design. Instead of the series of umbrellalike wires used in Marconi's experiments, Fessenden constructed an "antenna tower." It stood 420 feet high and was a series of telescopic metal tubes three feet in diameter at the bottom, held in place by supporting guy wires, and insulated at all points from the ground. The result was penetrating signals which beamed through the atmosphere which Marconi's station could not reach. Signals were received in Puerto Rico and at a station in Scotland even during the summer months when static normally interferes with transatlantic broadcasts. These first achievements at Brant Rock were shadowed by later excitement as voice broadcasting moved out of the laboratory.

Alexanderson's Alternator

The problem that continually plagued Fessenden was how to increase the number of transmitted oscillations so that the human voice would be audible. The inclusion of a telephone-type receiving apparatus already had been successful. Marconi had used it to hear the signals from England in his famous Newfoundland experiment, and wireless operators on ships used headphones to listen to messages in Morse code. The secret was to generate enough cycles so that the voice would travel with the signal and not be drowned out by the sound of the current passing through the headphones. To accomplish this, Fessenden enlisted the help of the General Electric Company in Schenectady, New York. There in the GE laboratories, a young Swedish scientist named Ernst Alexanderson (Figure 2–11) was placed in charge of the engineering team assigned to produce the Fessenden alternator. Both trial and error and difficulties in meeting Fessenden's wishes slowed down the project.

First, Alexanderson developed an alternator using a revolving iron core called an *armature*. Fessenden, however, demanded a wooden core, and the work started over. Fortunately for Fessenden, his two Pittsburgh financial backers continued to pour money into the project. Fessenden tried another company, the Rivett Lathe Manufacturing Company in Boston, but their device failed when the bearings burned up at the high speeds necessary to produce 50,000 cycles, the amount Fessenden felt would be needed for voice transmittals. Finally, Alexanderson and the GE team delivered the wooden armature alternator in September, 1906.

Ships within a few hundred miles of Brant Rock, Massachusetts were

Figure 2–11 Dr. Ernst F. W. Alexander-son, a General Electric engineer who developed the high-frequency alterna-tor that gave America a big edge in early long-distance voice broadcasting. This alternator, one of several devel-oped by Alexanderson between 1905 and 1920, was used to send transatlan-tic broadcasts from the RCA station at Rocky Point, Long Island. (General Electric Research and Development Center)

filled with crews celebrating the mixed merriment and loneliness of Christmas at sea on that December night in 1906. In the wireless rooms, the operators were on scheduled duty exchanging messages and receiving the food and good cheer of fellow officers when the splitting sound of CQ, CQ came through their headphones. The universal call alerted them that a message immediately would follow. Then, instead of the dit-dah of Morse code came the sound of a human voice. Officers were called to the wireless station to witness the phenomenon. The voice was that of Reginald Fessenden. "O Holy Night" rang out through the cabin followed by the words, "Glory to God in the highest, and on earth, peace to men of good will." Voice broadcasting had reached as far away as Norfolk, Virginia and the West Indies, shouting the world of wireless into a new era.

The Canadian Controversy: Fessenden Is Bankrupt

Many of the early wireless experimenters managed to amass considerable financial gain from the new medium, and even those who at first had lost money later reaped a profit. For Reginald Fessenden, fate had the opposite in store.[32] With the Brant Rock experiments a success, Fessen-den's backers wanted to develop some profit potential for the company, which up until now had been devoted to pure research. But Fessenden was at odds with Givens and Walker over a proposal to open a Canadian subsidiary. The Canadian company had evolved from a design on the part of all three men to give Marconi competition in transatlantic broadcasts. Fessenden went to England and made an agreement with the British Post Office in which if his station at Brant Rock could communicate with a station in New Orleans, a distance of about eighteen hundred miles, the British Post Office Department would approve a fifteen-year license for Fessenden's company to establish a reliable communication link between Canada and England.

Fessenden successfully completed the Brant Rock-New Orleans experiments, and then the trouble started. Fessenden, a Canadian by birth and the chief negotiator in the British contract, felt the Canadian subsidiary company should be controlled mainly by himself, the Canadians, and the British. Despite providing the capital for the new venture, to say nothing of the millions already invested, Walker and Givens were not to serve in any position of authority. Naturally, both men objected strongly, and Fessenden resigned and sued, collecting $460,000. The National Electric Signalling Company declared bankruptcy in 1912, and Marconi and his companies were once again the undisputed leaders in wireless communication.

de Forest Gains Publicity

After inventing the audion, Lee de Forest began to experiment with voice communication at the same time Fessenden was developing his hetrodyne circuit and conducting the Brant Rock experiments. Using a high frequency arc to modulate the signal, de Forest succeeded in transmitting a voice across the length of a room during the same year in which Fessenden gained recognition for his Brant Rock experiments with ocean vessels. De Forest was quick to see the potential of voice broadcasting and felt that good publicity would bring investors to his own company. Although voice broadcasts were well known in the United States, they were unknown in Europe. So in the summer of 1908, de Forest traveled to France and conducted demonstrations of radiotelephony from atop the Eiffel Tower to stations about twenty-five miles away.

The European experience whetted de Forest's appetite for more publicity at home. Always an opera buff, the inventor contacted the Metropolitan Opera in New York. He arranged to place a transmitter in the attic and connect it to the microphones on stage in the music hall. Although not very clear by modern standards, the microphones were the new Acousticon models manufactured by the National Dictograph Company.[33] On January 13, 1910, Enrico Caruso's and Ricardo Martin's booming voices bellowed *Cavalleria Rusticana* and *Il Pagliacci* to a small audience listening to receiving sets in New York. A master at gaining publicity, de Forest could rival the likes of Buffalo Bill Cody at obtaining press coverage for a show. The opera broadcasts were no exception. "The newspapers had been tipped off in advance and reporters were listening in at the Terminal Building, 103 Park Row, the Metropolitan Tower station, at the Hotel Breslin, on one of the ships downstream, and at our factory in Newark."[34] Although World War I and patent squabbles would slow the growth of modern radio until the late teens, de Forest's publicity helped to set the stage and arouse the public's enthusiasm for what would occur in the decades ahead.

wireless gains popularity: crystals and hams

Up to this time, the wireless had remained in the hands of the large companies, such as Marconi, and the major users—the Navy in the United States and the Post Office Department in England. That all changed in 1906 with the invention of the crystal radio receiver by General Henry C. Dunwoody. That same year, Greenleaf W. Pickard perfected a silicone crystal detector. These two devices contributed two important words to the wireless vocabulary: "availability" and "inexpensive." Remember, the audion was still being perfected, and vacuum tubes were expensive. Early ads for radio-receiving equipment as late as 1915 ran anywhere from $20.00 to $125.00. For a young experimenter attracted to the lure of wireless, those prices were beyond his or her reach. But by using the silicone crystal and a long, outside antenna, the general public could listen in on everything from opera to Navy broadcasts.

These early experimenters were called *amateur radio operators*, better known today as *hams*.[35] They were primarily of two types, (1) those who were interested in radio to experiment with and test new equipment, and (2) those who wanted to use the new medium to communicate with other people. In each type the spirit of the other was fostered. It was these early, home-town inventors who did much to see radio mature. Although the inventors and the big companies provided capital backing for international expansion, the ham operators were responsible for many of the early developments and experiments aimed at improving radio. In 1909, the first known amateur radio club was formed in New York City. The group started with five youngsters, and their advisor was Reginald Fessenden.

A second organization also acquired members, the Wireless Association of America. Started by Hugo Gernsback, the publisher of *Modern Electrics*, the membership roster jumped from 3,200 in 1909 to 10,000 by November, 1910. The association published the first *Wireless Blue Book,* which listed ninety amateur stations as members. A second *Blue Book* followed a short time later, and *Modern Electrics'* circulation soared to 52,000 by 1911. Sensing a lucrative market, the D. Van Nostrand book publishing company put *Wireless Telegraph Construction for Amateurs* on the bookstore shelves. By now, other radio clubs were rapidly forming, including the Radio Club of Salt Lake City, the Wireless Association of Central California, and the Radio Club of Hartford.

Ham radio was also gaining stature because of its ability to communicate when other systems failed. In March of 1913, a major storm hit the Midwest, knocking out power lines and telephone communication. Ham radio operators, including those at Ohio State University in Columbus and the University of Michigan in Ann Arbor, carried on communication and relayed emergency messages for seven days following the storm. It was this idea of relaying messages that sparked Hiram Percy Maxim, famous as an inventor of an automobile and engine silencer, to form the American Radio Relay League in 1914, an outgrowth of the Hartford Radio Club of which Maxim was a member.

Ham radio has continued to thrive as a hobby and has developed throughout the entire field of wireless communication. When radiotelephony replaced wireless, hams began to chat "in person," but the Morse code remains even today a cherished language of those amateur experimenters. They communicate worldwide using teletype, teleprinters, video display terminals, and television. In cooperation with NASA, relay satellites have been launched for use by hams in international communication.

summary

In chapter 2, we traced the beginnings of wireless from the early work of Henrich Hertz to the successful transmission of voice broadcasting by Reginald Fessenden and Lee de Forest. Hertz's major experiments on the presence of electromagnetic waves took place between 1886 and 1889. Working under the German scientist von Helmholtz, Hertz made the historical discovery while discharging static electricity through two metal coils. Based on Hertz's discoveries, Guglielmo Marconi first transmitted wireless signals over a short distance near his home in Italy. After the Italian government showed little interest in his invention, he traveled to Great Britain where he received financial support from the British Post Office Department. After tests near the Bristol Channel, he successfully received wireless signals from across the Atlantic in 1901.

Marconi continued with his experiments while expanding his corporate interests. Companies began to spring up in many countries, including the Marconi Wireless Telegraph Company Ltd. in England and the American Marconi Wireless Telegraph Company in the United States.

Improvements in the wireless were also made by J. Ambrose Fleming with the Fleming valve, and Lee de Forest with the audion.

It was not long before people started to transmit voice over the airwaves. With the devices that had been developed for telephone communication, scientists came closer and closer to quality voice transmission. A Kentucky farmer named Nathan B. Stubblefield performed short-distance wireless voice transmission. Then Reginald Fessenden developed the hetrodyne circuit. With this improved detector of electromagnetic waves and with the help of a large alternator developed by GE and Ernst Alexanderson, Fessenden transmitted voice in December 1906. After disagreements with his financial backers, Fessenden was overtaken in his quest for what was called radiotelephony by de Forest and others. At the same time, radiotelephony became practical as inexpensive receiving sets using silicone crystal detectors were manufactured and sold. The general public was becoming interested in what was now being called radio, and amateur ham operators talked across city blocks and eventually across continents.

spotlight on further learning

AITKEN, G. J., *Syntony and Spark—The Origins of Radio*. New York: John Wiley & Sons, Inc., 1976.

BAKER, W. J., *A History of the Marconi Company*. New York: St. Martin's Press, Inc., 1971.

BLAKE, G. G., *History of Radio Telegraphy and Telephony*. London: Chapman & Hall Ltd., 1928 (reprinted by the Arno Press, 1974).

DUNLAP, ORRIN E., *Marconi: The Man and His Wireless*. New York: The Macmillan Company, 1937 (reprinted by the Arno Press, 1971).

FAHIE, JOHN J., *A History of Wireless Telegraphy*. New York: Dodd, Mead & Company, 1901 (reprinted by the Arno Press, 1971).

HANCOCK, HARRY E., *Wireless at Sea*. Chelmsford, England: Marconi International Marine Communication Company, Ltd., 1950 (reprinted by the Arno Press, 1974).

JOLLY, W. P., *Marconi*. New York: Stein & Day Publishers, 1972.

MARCONI, DEGNA, *My Father Marconi*. New York: McGraw-Hill Book Company, 1962.

STERLING, CHRISTOPHER, and JOHN M. KITTROSS, *Stay Tuned: A Concise History of American Broadcasting*. Belmont, Calif: Wadsworth Publishing Co., Inc., 1978.

VYVYAN, R. N., *Marconi and Wireless*. Yorkshire, England: E P Publishing Limited, 1974.

3

the development of modern radio and television

FOCUS After completing this chapter we should be able to

Trace the development of the pioneer radio stations.
Tell what led to the formation of RCA.
Understand how cross-licensing agreements affected radio's early development.
Describe how commercial or "toll" broadcasting started.
Explain the events that shaped the beginning of early radio networks.
Describe the background and impact of FM broadcasting.
Describe early attempts by Nipkow to reproduce an image using mechanical television.
Explain the later contributions of Zworykin and Farnsworth.
Discuss the FCC imposed "freeze," the battle over color, and the struggle of UHF.
Trace the development and application of modern television technology.
Tell how home VTRs, videodiscs, and games could fragment the television audience.

The excitement of the first wireless signals, the thrill of the first voice broadcasts, and the world of the radio amateur, all came from an era of pioneer spirit and experimental technology. Radio was magic, and people welcomed it with open arms. They could set a black box on their kitchen table, stretch a wire into the evening sky, and pick voices and music right out of the air. There was no need to have it delivered by the paper carrier, no need to walk to the country store to get it. The sounds of presidents, operas, big bands, and sporting events were live and immediate. Needless to say, people wanted all the radio they could get, and the stations that gave it to them grew in stature and power. Some of the earliest stations are still household words, and by learning about them, we can catch some of the spirit of early radio.

the pioneer stations

Much like trying to identify the inventor of radio, it is hard to put a label on the town, the place, or the person responsible for the first broadcasting station.

Basic Criteria of a Broadcasting Station

R. Franklin Smith has established basic criteria for broadcasting stations of modern standards.[1] These are: (1) *A broadcasting station transmits by wireless.* The signals must travel through space using electromagnetic waves. Smith does not consider ETV a form of broadcasting nor closed-circuit, "wired" college stations; (2) *A broadcasting station transmits by telephony.* The sounds of the station should be intelligible to the general listener; (3) *A broadcasting station transmits to the public.* It is distinct from other types of communication such as telephone or telegraph, and such special services as safety, aviation, and marine use; (4) *A broadcasting station transmits a continuous program service.* Smith characterizes this as programming which is interconnected and occurs as a pattern recognizable as a program service; and (5) *A broadcasting station is licensed by government.* In the United States, this would be the Federal Communications Commission.

Although these criteria are too limited for our purposes, they are helpful in outlining the history of broadcasting. Even finding the station which first fit the five criteria is difficult since definitions of broadcasting were changing even in the early 1920s. Service, licenses, call letters, and ownership were often short-lived and sporadic.

Still, four stations rank as being important to understanding the historical development of broadcasting. These are KCBS in San Francisco, which evolved over the years from an experimental station located in San Jose, California in 1909; noncommercial WHA at the University of Wisconsin in Madison; WWJ in Detroit; and KDKA in Pittsburgh, Pennsylvania.

Charles David Herrold Begins in San Jose

Professor Charles David Herrold is credited with operating one of the first broadcasting stations in America (Figure 3–1). Others broke the airwave silence before him, but in 1909, residents of San Jose, California could spend a Wednesday evening with their crystal sets tuned to news

Figure 3–1 Charles David Herrold is seen in the doorway of one of his early radio stations in San Jose, California, this one constructed in 1913. He first signed on the air in 1909 and is credited with starting one of America's pioneer radio stations. Others in the picture include (left to right) Kenneth Saunders, E. A. Portal, and Frank G. Schmidt. (KCBS radio, Gordon B. Greb, and the Sourisseau Academy of San Jose State University)

and music broadcast by this owner of the School of Radio in San Jose.[2] For the school, the radio station was its medium of advertising, advertising which was aired more than ten years before KDKA in Pittsburgh and WWJ in Detroit began regular programming. A classmate of Herbert Hoover at Stanford, Herrold constructed a huge umbrellalike antenna in downtown San Jose. From the Garden City Bank Building, the wire structure hung out in all directions for a city block. Although it was a far cry from the eastern giants that could carry football games and political speeches, the little San Jose wireless station became one of the famous firsts in the broadcasting industry.

After 1910, the station handled regularly scheduled programs with operators on regular shifts. Even Herrold's first wife, Sybil M. True, had an air shift and was one of the earliest female disc jockeys.[3] She would borrow records from a local store and play them as a form of advertising. When listeners went to the store to purchase the recordings, they would register their name and address, giving the station an indication of its extent and influence. The California station gained national recognition at the Panama Pacific Exposition in 1915. When Lee de Forest spoke in San Francisco in 1940, he credited Herrold's station as being "the oldest broadcasting station in the entire world."[4]

WHA, Madison, Wisconsin

WHA traces its inception all the way back to 1904 in the physics laboratory at the University of Wisconsin where Earle M. Terry (Figure 3–2) worked as a graduate student pursuing a Ph.D.[5] Graduating in 1910, he stayed on as an assistant professor, and in 1917, with the help of colleagues and assistants, he began experimental broadcasting of voice and music. The equipment was makeshift, and the three element tubes were not the sturdy successors of the 1920s. Instead, they were a mixture of creative craftmanship, hand-blown glass, and immense frustration, especially when they burned out.

By 1922, station 9XM had been legitimized by the Department of Commerce with a license and the new call letters WHA. The same year, Professor William H. Lighty (Figure 3–2) became WHA's program director, developing the station into one of the first "extension" stations responsible for bringing universities to the public with everything from news to college courses. Major programming also made great strides by featuring the University of Wisconsin Glee Club, regular weather and road reports, farm and market reports, symphony broadcasts, and the famous Wisconsin School of the Air.

To aid listeners, Professor Terry taught them how to build their own radio sets. He even distributed some of the raw materials free of charge. The radio rage of the early twenties caught many of the large equipment

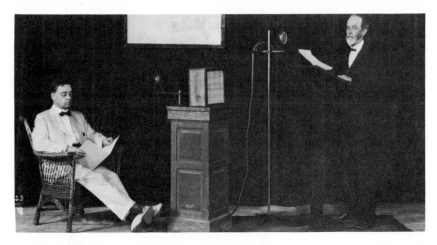

Figure 3–2 Professors Earl M. Terry and William H.
Lighty in an early studio of WHA radio at the University
of Wisconsin, Madison. (University of Wisconsin)

manufacturers unprepared. Loudspeakers had not yet replaced the
earphone, and Professor Terry's demonstrations of amplified radio
reception made their first appearance in the Wisconsin exposition hall.

Meanwhile, WHA's farm and market reports and weather broadcasts
were being picked up by the newspapers, and weather forecasting stations
were using WHA data to aid predictions as far away as Chicago. Letters
continued to pour in from listeners as far away as Texas and Canada.
WHA has since been joined by WHA-FM and WHA-TV. At the University of Wisconsin in Madison, a historical marker reads: "The Oldest
Station in the Nation . . . the University of Wisconsin station under the
calls 9XM and WHA has been in existence longer than any other."

WWJ and the Detroit News

After leaving the historical marker at the University of Wisconsin, you
can travel east around Chicago and the tip of Lake Michigan to another
pioneer station still operating—WWJ in Detroit.[6] When broadcasting was
still in its infancy, some forward-thinking newspaper publishers realized
that it would be better to reap some of its profits rather than always to
compete against it. William E. Scripps of the *Detroit News* had such a
vision. Presenting the idea to his colleagues, they appropriated money to
construct a makeshift radiotelephone room on the second floor of the
Detroit News Building.

At 8:15 P.M. on the night of August 20, 1920, an Edison phonograph
played two records into the mouthpiece of the de Forest transmitter, and
probably no more than one hundred amateur operators heard the signal.

No advance warning gave a hint of the trial broadcast; no publicity draped the pages of the *Detroit News*. Everything worked perfectly, and the staff began preparations for the next day's broadcast of a Michigan election. When the election returns began to trickle in, it was the radio, not the newspaper, that first brought them to the public. Like a proud parent doting on a child's accomplishments, the September 1 issue of the *News* reported: "The sending of the election returns by the Detroit *News* Radiotelephone Thursday night was fraught with romance, and must go down in the history of man's conquest of the elements as a gigantic step in his progress."

The early programming of WWJ, originally licensed under the call letters 8MK, reflects much of the same programming that other early stations tried, experimented with, and sometimes nurtured into long-running popular fare. The election returns were supplemented the following day with a sportscast, a preview of the World Series on October 5, and WWJ's reports of the Brooklyn-Cleveland match-up. Returns of the Harding-Cox election were heard on November 20, 1920, the same election returns that later became KDKA's claim to the "first station" honor.

So important was music, that WWJ organized the sixteen-piece *Detroit News* Orchestra expressly for broadcast use. It also expanded its studios to auditorium proportions which were described as "magnificent" with perfect acoustics, two-toned blue walls, and a white ceiling with a silver border.

The latest equipment took WWJ's news microphone onto the road and into the air. A single engine prop aircraft with NEWS painted in big letters on one wing was equipped for direct broadcast, and a news and photographic unit thus became one of the first "mobile units" (Figure 3–3), now a common element of radio stations in even small communities.

With all of these early credits to its name, it is not surprising that on the front of an antique microphone illustrating the promotional literature of WWJ reads the inscription: "WWJ RADIO ONE, WHERE IT ALL BEGAN, AUGUST 20, 1920."

KDKA at East Pittsburgh

Station KDKA also established its place in broadcasting's history in 1920.[7] The story of KDKA's begins with Dr. Frank Conrad (Figure 3–4), the assistant chief engineer at the Westinghouse Electric Plant in East Pittsburgh, Pennsylvania, who constructed a transmitter licensed in 1916 as 8XK. After the World War I ban on nonmilitary uses of radio was lifted, Conrad began his experimental programming. Through an arrangement with a local record store in the nearby community of Wilkinsburg, Pennsylvania, he received records in exchange for mentioning the name

Figure 3–3　A broadcasting station, photographic dark room on wheels, and a newspaper office is how the *Detroit News* described this early mobile news unit. (WWJ and the *Detroit News*)

Figure 3–4　Dr. Frank Conrad, who conducted the experimental work that led to the establishment of KDKA radio. The station began broadcasting on November 2, 1920. (KDKA)

of the store. Its popularity grew so rapidly that Horne's Department Store in Pittsburgh ran an ad for inexpensive receiving sets.

To H. P. Davis, a Westinghouse vice president, the ad was the inspiration for a license application with the call letters KDKA, granted on October 27, 1920. A month later, the new call letters identified the station as it sent Harding-Cox election returns to listeners with amateur receiving sets and to a crowd gathered around a set at a local club. The crowd called for more news and less music, and KDKA's mail reported receiving the signals even at sea. The success of the broadcast gained widespread publicity, overshadowing WWJ's similar efforts, because KDKA had official call letters. Moreover, the combination of a publicized event coupled with a major effort to get receivers into the hands of the public made the KDKA broadcast an historical milestone.

For Westinghouse, manufacturing receivers was its definition of "commercial" broadcasting. Addressing an audience at the Harvard Business School, H. P. Davis remarked: "A broadcasting station is a rather useless enterprise unless there is someone to listen to it. . . . To meet this situation we had a number of simple receiving outfits manufactured. These we distributed among friends and to several of the officers of the company."[8]

As the popularity of the station grew, so did the staff, and when a Westinghouse engineer walked into the transmitting shack, he became the first full-time announcer in radio. Harold W. Arlin's (Figure 3–5) experiences were quite a change from his duties as an electrical engineer. During his career he introduced to KDKA's listeners such famous names as William Jennings Bryan, Will Rogers, Herbert Hoover, and Babe Ruth.

Today, clear-channel, 50,000-watt KDKA can be heard over a wide area of the northern hemisphere during late night periods and with good atmospheric conditions. If you are traveling in the Pittsburgh area, you might even pass the former home of Dr. Frank Conrad in the suburb of Wilkinsburg where a plaque reads: "Here radio broadcasting was born. . . ."

RCA is formed

Although it may seem as though the pioneer stations and their owners were to become the corporate giants of broadcasting, by 1920, a new worldwide corporation was already operating and had the blessing of the United States government. It would soon become a giant not only in broadcasting but in other communications interests as well. The company was the Radio Corporation of America (RCA). Its beginning is full of international intrigue, skilled corporate maneuvering, presidential politics, and even the United States Navy.

Figure 3–5 Harold W. Arlin, who became KDKA's first full-time announcer. Besides introducing many well known personalities to KDKA listeners, he is also credited with broadcasting the station's first play-by-play football game. (KDKA)

The play begins at the close of World War I when the United States still had control of all wireless communication. The thought of turning a major share of the American wireless interests back to Marconi was more than President Woodrow Wilson wanted to do. After all, the Marconi Company was still substantially British in influence if not in stock ownership. Communication and transportation were now realized as important keys to world international power. Great Britain had a network of cable systems in Europe and the United States, and its shipping industry and strategic location had an edge on transportation. Although not necessarily a threat, the British were at least to be treated with caution. President Wilson was also a fan of radio in his own right, having seen the benefits of his famous Fourteen Points spread throughout Europe by an American station using the huge General Electric alternator of Ernst Alexanderson.[9]

Government Attempts at Keeping the Alternator

In 1918, two bills were introduced in Congress, both designed indirectly to bring wireless under control. Although seemingly harmless at the time, they suggested the use of technical school radio stations for experimental use but failed to mention anything about the amateur or "ham" stations.

Although the legislation had the support of President Wilson and the Department of the Navy, neither counted on the lobbying efforts of the amateurs. In chapter 2, we learned of the mushrooming popularity of radio and the growth of the amateur organizations. When World War I began, and the government took control of broadcasting, it was silence for the hams, and their equipment did little more than collect dust. Now, with all of their pent-up enthusiasm of going back on the air, their exciting tales of radio's war escapades, and a flood of war-trained operators wanting to continue their experiences as hobbyists, the legislation did not have a chance. Scathing attacks on the bills even claimed that they would prohibit the youth of the country from participating in investigation and invention.[10] Finally, the bills were tabled, permanently.

Bullard, Young, and Sarnoff

The next scene cast the General Electric Company, President Wilson, and Admiral William H. G. Bullard in leading roles. For GE, the war was a period of considerable government support, especially for its Alexanderson alternator. When the war ended, the company faced serious layoffs because of the lack of government contracts. Although trading patriotism continued during the hostilities, an end to the conflict meant GE was free to trade with any company it chose. By coincidence, that trading was about to begin with the British Marconi Company. But President Wilson wanted the new technology of radio to remain in American hands. Although the details of the conversation are unclear, we know President Wilson at least spoke to Admiral W. H. G. Bullard (Figure 3–6), chief of Naval Operations Service, about keeping the Alexanderson alternator on home ground.[11] Bullard then took it upon himself to speak to General Electric's general counsel, Owen T. Young (Figure 3–6). Whatever persuasive arguments Bullard used, he managed to convince Young and GE to take the giant leap to form a new, all-American company in the wireless business.

In the tradition of people known for significant corporate maneuvers, Owen T. Young managed to coordinate an international negotiation that formed not only the Radio Corporation of America, but also facilitated the purchase by RCA of the American Marconi Company. GE also bought the stock of the American Marconi Company that was owned by the British Marconi Company. The new corporation had American directors and stipulated that no more than 20 percent of its stock could be held by foreign nationals. For American Marconi, the decision to become part of RCA was a necessity in order to overcome its "British" image in the face of American patriotism. It also needed the alternator to succeed just as much as GE needed customers. As it turned out, the merger maintained the jobs of American Marconi employees and directors.

Figure 3–6 Two men important to the early formation of the Radio Corporation of America. Admiral William Hannum Grubb Bullard (left) became the chief intermediary between the U.S. Navy, the White House, and General Electric. Owen D. Young, serving as General Electric's general counsel, was persuaded by Bullard's recommendations and began the maneuvers culminating in the formation of RCA. (Department of the Navy and RCA)

One of the more famous directors was David Sarnoff (Figure 3–7). As a wireless operator, he had "worked" the messages from the ships rescuing the survivors of the *Titanic*. Later in 1916, he wrote his boss, Edward J. Nally, at the American Marconi Company suggesting the modern application of radio. In the now famous letter, Sarnoff said:

I have in mind a plan of development which would make radio a "household utility" in the same sense as the piano or phonograph. The idea is to bring music into the house by wireless. . . . The receiver can be designed in the form of a simple "Radio Music Box"; . . . supplied with amplifying tubes and a loudspeaking telephone, all of which can be neatly mounted in one box.

Aside from the profit to be derived from this proposition the possibilities for advertising for the company are tremendous, for its name would ultimately be brought into the household, and wireless would receive national and universal attention.[12]

David Sarnoff, named commercial manager of RCA when the merger took place, later headed the corporation.[13]

Figure 3–7 David Sarnoff taught himself Morse Code and landed a job with the American Marconi Company as a wireless operator for the station at Siasconset on Nantucket Island. (RCA)

patents, cross-licensing, and competition

KDKA's experiments and their accompanying publicity put the major corporations in wireless communication into a small turmoil. The giants of RCA, GE, and American Telephone & Telegraph had entrusted their futures in a joint block of power that would effectively, if not completely, control the development of radio. But to the triumvirate, the vision had been marine communication and radiotelephony, not the communication KDKA created with its November, 1920 demonstration. Now the memo of David Sarnoff that originally went politely unheeded took on new significance. Perhaps there was money to be made from using broadcasting for mass appeal. The empire that Owen T. Young had built already had acquired allies in GE and AT&T—each had previously acquired important broadcasting patents which the three now shared by agreement.

Sharing the Discoveries

Some of the earliest patents belonged to Lee de Forest. The audion, which was the forerunner of a whole series of improved vacuum tubes, was the most important link to the future of communication, at least in the way AT&T saw it. In 1913, AT&T began buying de Forest's patents to the vacuum tube and then worked with their own engineers to improve the device. By 1915, using the latest equipment, including German-manufactured vacuum pumps to suck the air out of the tubes, the company perfected the first commercially successful device (Figure 3–8).[14] AT&T used it for the first transatlantic telephone call.

As we learned in chapter 2, the courts ruled that the audion infringed on the vacuum tube invented by Ambrose Fleming and that Fleming's patents belonged to the American Marconi Company. Yet war has its peculiar benefits, and breaking this AT&T-American Marconi conflict was one of them. The United States government stepped in and called for all companies to forge ahead as part of the war effort; thus all became immune from patent infringement suits.

The demand for vacuum tubes also involved GE and Westinghouse. Each had the capacity to manufacture light bulbs. The equipment that could suck air from a light bulb also could perform the same task in manufacturing vacuum tubes.[15] General Electric, as we learned, also had the Alexanderson alternator.

So for the duration of World War I, everyone worked in harmony, each cooperating with the other, but each with an important part of the pie that could be reheated after the war ended. When it did, each had

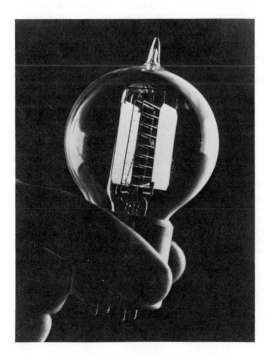

Figure 3–8 The first commercially successful vacuum tube developed in 1915. (AT & T Co.)

something the other needed. Thus for the future of radio, it was advantageous for RCA, GE, and AT&T to enter into a complex arrangement of cross-licensing agreements, permitting each to share in the developments of the others but clearly dividing the way in which radio would be marketed to the public.

Armstrong's Superhetrodyne: Westinghouse Asset

Westinghouse, meanwhile, had been scrambling to compete with the RCA-GE-AT&T alliance. It acquired the patents to a new type of circuitry invented by a graduate student at Columbia University, Edwin H. Armstrong (Figure 3–9). While Armstrong was serving in France in World War I, he became interested in finding a way for antiaircraft guidance systems to home in on the radio waves emitted by aircraft engines.[16] Although his invention never aided the war effort, it did spark the development of the *superhetrodyne circuit,* an improvement on Fessenden's hetrodyne circuit. The superhetrodyne changed the frequency of incoming radio waves, amplified them, then changed them to an audible signal. Westinghouse, just a month before the November 1920 broadcast, shrewdly bought the Armstrong patents and also some held by Michael Pupin, a Columbia professor who had worked with Armstrong, permitting Armstrong to use his laboratory and financing some of Armstrong's work.[17]

When KDKA showed its stuff, Westinghouse was invited to become the fourth member of the RCA-GE-AT&T alliance. Still another com-

Figure 3–9 Edwin H. Armstrong, who developed the superhetrodyne circuit and later, a successful frequency modulated (FM) transmission system. (Columbia University)

pany, United Fruit, joined because of its patents on crystal detectors. Thus under agreements among the big four: (1) GE and Westinghouse would manufacture radio parts and receivers; (2) RCA would market and sell them; and (3) AT&T would make, lease, and sell radio transmitters.[18] All of them were free to start their own broadcasting stations, and they did. But the agreements were mostly concerned with wireless telephony and telegraphy.[19]

When the stations did get underway, they signed on fast and furiously. KDKA was only the beginning. More and more amateurs with numbered-prefixed call signs applied and were granted licenses to operate broadcasting stations in the same fashion as KDKA. Westinghouse did not stop with that Pittsburgh station; it was not long before it signed on WBZ in Springfield, Massachusetts; WJZ in Newark, New Jersey; and KYW in Chicago, later assigned to Philadelphia. WJZ was sold to RCA in 1923.

RCA started its own station in 1921, WDY in New York. Although it stayed on the air only three months, it did try some innovative programming, including a remote broadcast from the New York Electrical Show featuring Metropolitan Opera star Ann Case (Figure 3–10).

Figure 3–10 An early remote broadcast from WDY's improvised studio at the New York Electrical Show of 1921. Standing (left) is Major J. Andrew White, WDY program manager, and Ann Case of the Metropolitan Opera. Opera singers often were used in early experimental broadcasts, not only because of their popularity, but also because their powerful voices could compensate for the less sensitive microphones. (RCA)

General Electric entered broadcasting by signing on WGY in Schenectady, New York. But of all the stations on the air in the early 1920s, the one to stir the attention of the public and the industry alike was AT&T's WEAF in New York.

toll broadcasting: WEAF

The idea of commercial broadcasting was developed at AT&T on June 1, 1922 with the licensing of WEAF. WEAF initiated the concept of "toll broadcasting." This meant that anyone wishing to use the station could do so by paying a toll. Sponsoring a program meant buying the entire time segment and using it for whatever purpose desired. At first, the idea had few takers. So the station used AT&T personnel as announcers to fill the programming void, one of the earliest being Helen Hann, a member of AT&T's Long Lines Department (Figure 3–11). The first sponsor to try the new toll concept was the Queensboro Corporation of New York which used a set of five short programs over five days to sell real estate.[20] On August 28, 1922, at a cost of $50.00, the Queensboro Corporation's first program began the era of modern commercial broadcasting.

Figure 3–11   WEAF's early studio with Helen Hann, the announcer. (AT & T Co.)

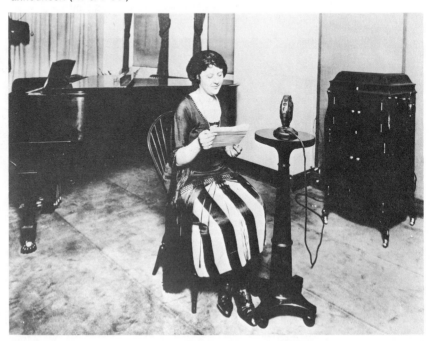

Criticism of Toll Broadcasting

Not everyone thought the idea was good. Arguments against commercial radio started surfacing in the trade press. The *American Radio Journal* suggested three alternatives: (1) have municipalities undertake programs on a civic entertainment basis; (2) charge the public and collect revenues from a large number of "radio subscribers"; or (3) tax the manufacturers of radio equipment, the people who distribute it, and the people who sell it.[21] *Printer's Ink,* the trade journal of early advertising, concluded: "Any attempt to make the radio an advertising medium, in the accepted sense of the term, would, we think, prove positively offensive to great numbers of people. The family circle is not a public place, and advertising has no business intruding there unless it is invited. . . . The man who does not want to read a paint ad in the newspaper, can turn the page and read something else. But the man on the end of the radio must listen, or shut off entirely. That is a big distinction that ought not be overlooked."[22] Despite questionable reviews, advertising revenue gradually dribbled in to WEAF.

Through some political strategy with the Department of Commerce, the station managed to secure a more favorable frequency and extended hours. That in itself was important since stations did not have the protection from interference that they have today. In fact, sometimes three or more stations had to split up the broadcast day on the same frequency with each vying for the audience when the other signed off.

As WEAF attracted more advertisers, AT&T began pouring money into the station, building new studios and the finest equipment Western Electric would manufacture. That fine equipment also became the envy of the broadcasting industry, and when other stations started to request it, AT&T was reluctant to fill their orders. The short-term profit of a transmitter sale was less important to AT&T than the potential of a national advertising medium under the telephone company's control. When AT&T increased WEAF's remote broadcasts, the audience clamored to listen, and when WEAF's competition did make remote broadcasts, AT&T responded financially.

Finally, AT&T concluded that it would be in its best interests to block remote hookups on AT&T lines from its old allies, RCA, GE, and Westinghouse. Resentment, fueled by profits, lit a spark that inflamed the industry. While the other three competitors were scrambling to use Western Union lines for broadcasts, AT&T was arguing that it alone should be permitted to engage in toll broadcasting, based on nothing less than the 1920 cross-licensing agreements which spelled out manufacturing and distribution rights to radio equipment.

As time went on, the stakes grew higher. WEAF's income from its toll venture continued to climb, while hundreds of smaller companies ate away at the profits of RCA, GE, and Westinghouse by manufacturing radio receivers in defiance and sometimes in ignorance of patent rights. It was clear that the future belonged to commercial broadcasting to a mass audience. AT&T even went so far as to collect license fees from some stations before permitting hookups into AT&T long lines. The company also gathered together a group of stations on which an advertiser could buy time separately or as a group. The stations, or "chain" as early network broadcasting was to become known, presented a sizeable example of how toll broadcasting could work.[23] Although AT&T was receiving some severe criticism in the press, it continued its toll concept.

Finally, the accusations of infringement on the 1920 agreements elevated into open confrontation, and an arbiter was called in to mediate. The parties agreed to adhere to a final verdict issued by Boston lawyer Roland W. Boyden.[24] Simultaneously, the Federal Trade Commission, apparently completely unaware of the arbitration action, issued a report claiming a monopoly in the radio industry and placed the blame on none other than AT&T, RCA, GE, Westinghouse, and the United Fruit Company.

The Antitrust Issue

The Federal Trade Commission's report was sobering, but the radio allies already had agreed to abide by Boyden's decision. Taking his time in this delicate matter, Boyden finally presented his position to the litigants.[25] His draft opinion which effectively ended AT&T's claim to exclusivity in toll broadcasting caused the telephone company to try an end-run. The AT&T attorneys first leveled a reaction to the draft opinion:

> We believe that the referee's unavoidably incomplete knowledge of the extremely intricate art involved in this arbitration with his effort to cooperate in the attempt of the parties to work out this situation, have misled him into a radical departure from the contract which the parties actually made, and into conclusions which amount to an attempt to make a new contract for them.[26]

They then got an opinion from none other than John W. Davis, who had helped draft the Clayton Act, that major piece of antitrust legislation passed the same year that the Federal Trade Commission was formed. Davis argued that if Boyden were correct, then the original cross-licensing agreements of 1920 were illegal and an infringement of antitrust laws. It was a crafty move on AT&T's part, effectively suggesting that it did not have to agree to the arbitration because the agreement was illegal in the first place.

With all the turmoil, AT&T was still very conscious of public opinion. An image of waging open warfare to gain control of broadcasting was something it did not want to acquire. The next scene would see the power structure of American broadcasting change dramatically.

Whether RCA either realized AT&T did not want to begin battle, or decided it was time it went into toll broadcasting is open to speculation. Undoubtedly, both thoughts crossed the mind of David Sarnoff as he and other RCA officers watched AT&T organize its broadcasting interests into a separate corporation in May, 1926 and call the new subsidiary the Broadcasting Company of America. At RCA, a similar move was afoot as the RCA broadcasting interests were consolidated in September, 1926 into a company called the National Broadcasting Company. Shortly thereafter, for the sum of $1 million, WEAF became the property of RCA, which was eventually consolidated into WJZ, which RCA had previously purchased from Westinghouse.

NBC had been successfully launched, and for AT&T, the future forecast a healthy income from fees paid by broadcasters to use long lines for remote and network broadcasting. It also lifted the weight of negative public opinion from AT&T's shoulders. Although it might have won the court battles and the arbitration, and even survived the wrath of the Federal Trade Commission, AT&T felt comfortable with its network of "wires," letting NBC shoulder public opinion on the new "national network." In a major display of public pronouncement, NBC advertised its new venture in newspapers with a promise of "better programs permanently assured by this important action of the Radio Corporation of America."

NBC's Red and Blue

NBC operated two basic networks as part of its nationwide coverage plan. The Blue network served some stations exclusively as did the Red network, and a number of the stations had the option of drawing programming from either. Although still consolidated under NBC, the flagship station of the Red network was none other than WEAF. The Blue network chose its old rival WJZ. It is not surprising that the rivalries continued with each attempting to outdo the other. In 1932, NBC executives began to consider a separate status for the Blue network, having it operate even more competitively with the Red. One of the executives to conceive the idea was Mark Woods (Figure 3–12) later to play a key role in ABC's development. Things remained the same at the Blue until 1939 when a separate Blue sales department was established, followed by other departments separate from the Red network. Undoubtedly, an impetus for the changes came with the Federal Communications Commission's announcement in 1938 that it was planning a full-scale inquiry of network broadcasting.

The FCC's Report on Chain Broadcasting

Out of the inquiry came the FCC's 1941 *Report on Chain Broadcasting*. Among other things, it was critical of NBC's interest in talent management. This interest developed early in 1931 when, because of its need for talent, NBC acquired a 50 percent share of the Civic Concert Service, Inc. to complement an artists' management division of the company. Increasing its share in the Civic Concert Service until it owned it, the network became the target of conflict-of-interest charges by the FCC. The *Report* stated: "As an agent for artists, NBC is under a fiduciary (hold in trust) duty to procure the best terms possible for the artists. As employer of artists, NBC is interested in securing the best terms possible from the artists. NBC's dual role necessarily prevents arm's-length bargaining and constitutes a serious conflict of interest."[27] Criticism of the artists' service was only part of the scrutiny. The *Report* also examined NBC's growing interest in its transcription business, which included recordings for libraries and other services.

When the FCC concluded its investigation, it reported that stations could not be bound by exclusive network contracts prohibiting them from airing programming from other networks; network contracts were to be for periods of one year; and stations were to be the sole determiner of programming, that right not to be delegated to the networks. The most important statement, however, hit at the very heart of NBC's dual network concept: "No license shall be granted to a standard broadcast station affiliated with a network organization which maintains more than one network."[28] Seasoned veteran David Sarnoff, now president of RCA, set the wheels in motion to protect RCA's investment. He immediately organized the Blue network as a separate corporation. The action was an attempt: (1) at least temporarily to pacify the FCC, and (2) to get an accurate reading of exactly how much the Blue network was worth by creating a separate accounting system. The handwriting was on the wall—Blue had to be sold.

Edward J. Nobel Launches ABC

When it became clear that NBC's disposal of the Blue network was inevitable, major industrialists began to consider the jump into broadcasting. They included the Mellons in Pittsburgh, Marshall Field, Paramount Pictures, and Edward J. Nobel (Figure 3–13), who had developed a sizeable fortune making and selling Lifesavers candy. A former undersecretary of commerce, Nobel in the summer of 1943 posted $1 million of Blue's purchase price and made arrangements to pay RCA the remainder from his own pocket and with loans from three New York

Figure 3–12 Mark Woods, who helped to establish the NBC Blue network which became ABC with Mark Woods as its president. (ABC)

Figure 3–13 Edward J. Nobel, who used the profits of Lifesaver candies to purchase the NBC Blue network. (ABC)

banks.[29] The FCC, meanwhile, had delayed enforcing the 1941 *Report* to permit the sale of the Blue network in a calm atmosphere that wouldn't depress the price.

On October 12, 1943, the commission announced it was approving the sale of the Blue network to Edward J. Nobel.[30] Mark Woods was retained as president. In approving the sale, the FCC stated the transaction "should aid in the fuller use of the radio as a mechanism of free speech. The mechanism of free speech can operate freely only when controls of public access to the means of a dissemination of news and issues are in as many responsible ownerships as possible and each exercises its own independent judgment."[31]

For Edward J. Nobel, the challenge to develop the Blue network was sizeable. World War II was raging, and American business, although geared up for war production, was in a state of uncertainty. A total of 168 stations and 715 employees were now Nobel's responsibility. Already on the climb, however, were Blue's credits as an independent organization.

While still part of NBC, the Blue network showed promising oppor-

71

tunities as an investment. It instituted a special daytime-rate package, permitting advertisers to buy at a discount over a series of daytime hours. Another discount package provided savings for advertisers who steadily bought time on more stations to carry their programs. Institutional advertising permitted companies to sponsor one-time programs publicizing important accomplishments. Typical were the famous "Victory Broadcasts" calling attention to the war effort. Nobel also inherited the "strip" broadcasts, which permitted companies to sponsor programming over a strip of four to seven evenings per week. Some of the early takers included Metro-Goldwyn-Mayer which sponsored the comic antics of Colonel Lemuel Q. Stoopnagle, heard five nights a week for five minutes a night and carried by 54 stations.[32]

Despite all its recent accomplishments, the Blue network still had not made a profit, and Nobel pulled together his own team of experts. He named a New York advertising executive to be vice president of programs. Adrian Samish, in his mid thirties, had had previous stage work and realized the Blue network did not have the big-name talent that was pulling audiences to the other networks. He was also faced with a die-hard group of female followers on other networks who lived for the soap operas of sound with their tensions, intrigues, and love affairs. To compete with these, he instituted a series of game shows, and although not setting the world on fire, did provide the Blue with alternative programming.

Working with Samish was Robert Kinter, a former Washington correspondent. As vice president in charge of special events, Kinter at first seemed like a public relations trouble-shooter until he began showing everyone he had a head for management decisions. By the turn of the decade, he was serving as executive vice president, later to be named president of the network.

Nobel had formed a separate corporation, the American Broadcasting System, Inc., to purchase the Blue network. On June 15, 1945, affiliate stations heard announcer James Gibbons say, "This is the American Broadcasting Company." The influence of the war effort and the patriotic mood of the country were reflected in Mark Wood's comments about the new name. The name was chosen, he said, "because 'American' so completely typifies all that we hope, and believe, this company will be and will represent to the people of the world. The tradition of independence and of free enterprise, liberality in social philosophy, belief in free education for all and in public service—all of this and much more is inherent in the name."[33]

To some, the words today might be criticized as overstated, but for a nation headed toward victory in global conflict, Woods was appealing to the heart of the hinterland. Patriotism was also present later that same evening, when ABC officially retired the label of "Blue" with an hour-long program titled "Weapon for Tomorrow" discussing the "importance to a democracy of a freely-informed people."[34]

CBS Is Born

When ABC began network broadcasting, it had three formidable competitors—the Red network which later became NBC, and two other networks, CBS and Mutual. CBS can trace its beginnings to January 27, 1927 when a company called United Independent Broadcasters, Inc. was formed to serve the dual purpose of selling time to advertisers and furnishing programs for stations. Acting as the sales arm of United was another company and stockholder, the Columbia Phonograph Broadcasting System, Inc. Sixteen stations were included in the original United network, and it was not long before financial difficulties made the venture less than profitable.

United had devised a plan by which it would pay the sixteen stations $500 per week to furnish it with ten specified hours of broadcasting. But the cost was simply too high, and in the fall of 1927, the Columbia people withdrew from the venture, and United bought the stock. United also changed the name of the organization to the Columbia Broadcasting System, Inc. The network fortunately revised its rate agreement with the affiliate stations, having suffered losses of over $220,000 in its first nine months of operation.[35] The new agreement cut the losses, but it was not until a man named William S. Paley (Figure 3–14) arrived on the scene that things began to look up.

Paley's father owned the Congress Cigar Company which had been one of the sponsors on the old United network. When cigar sales jumped from 400,000 to 1 million per day in six months, radio got the credit. Congress's advertising manager, the owner's son, went to New York with an eye on buying the faltering 16-station network. Taking control of 50.3 percent of the stock, the Paley family entered the broadcasting business.

The growth of CBS is somewhat legendary in broadcasting history. The very next year after Paley assumed control, the network jumped into the black and continues to operate at a profit as one of the largest advertising media in the world. Ten years after Paley arrived, the network grew from 16 to 113 affiliates. In its first year of operation, it sank more than $1 million into programming and moved its facilities to new quarters.[36] Paley himself took an active interest in network programming, personally supervising CBS's coverage of the 1928 election returns. By 1930, CBS was holding its own against NBC and was actively participating in the era of "experimental" broadcasting that characterized early radio. We'll learn more about this later in the chapter.

Figure 3–14 William S. Paley, who guided the corporate development of CBS after his father's cigar company purchased the floundering network in 1928. (CBS Inc.)

The Mutual Broadcasting System

The Mutual Broadcasting System started much in the same way as the United network did except for two major differences.[37] First, Mutual did not enter into agreements to pay unmanageable sums of money to affiliate stations, and second, it started small. Mutual began with four stations—WOR in Newark, New Jersey; WXYZ in Detroit; WGN in Chicago; and WLW in Cincinnati. The four stations agreed that Mutual would become the "time broker" and pay the stations their regular advertising rate, deducting first 5 percent sales commission and other expenses such as advertising agency fees or line charges. It expanded in 1936 when Mutual added 13 stations in California and 10 in New England. In 1938, a regional network in Texas added 23 more stations to the chain. By 1940, Mutual had 160 outlets. Yet the network operated more like a co-op than a profit-making network like NBC and CBS. A special stock arrangement even gave some stations a greater voice in the network's operation as well as in special sales commissions. But unlike United, which had to scramble to stay in business, Mutual grew slowly, making it the third competitor in early network broadcasting.

A system of noncommercial radio stations also developed and we shall learn more about them in the chapter on educational-public broadcasting.

fm broadcasting

Many stepping stones dot radio's path of development. Some of these are milestones, such as KDKA's first broadcast. Others mark decisions made in corporate board rooms, decisions that charted the medium's course. Still others represent developments from inside the laboratory.

Armstrong Applies the Principle

Frequency modulation, changing the frequency of a wave to modulate a signal, was not new to Edwin Armstrong. He had studied it and did not believe the words of his predecessors that FM had no real application to broadcasting. One day while talking with David Sarnoff, Armstrong agreed that a device was needed that would clear the static from radio transmission.[38] To Armstrong, that challenge meant years of research at Columbia, culminating in 1933 when RCA engineers accepted an invitation to witness his latest efforts. Although the equipment worked, RCA was not enthusiastic. Still, it gave him permission to continue the experiments at the Empire State Building. There, with successful tests ranging up to 65 miles, Armstrong was sure FM held the key to revolutionizing radio.

But the vision in RCA's eye was television. Tests had already been successful, and the company was undoubtedly thinking a few years ahead to the public relations splash a television demonstration would make at the World's Fair. Armstrong grew increasingly suspicious of RCA's and Sarnoff's intentions. Deciding not to wait, he launched a lecture tour of his own and demonstrated FM to dozens of audiences across the United States.[39] Selling his RCA stock and receiving encouragement from the Yankee and Colonial networks in New England, Armstrong built his own FM station in Alpine, New Jersey. There, after battling the FCC for a license, he continued his experiments and managed successful broadcasts of up to 300 miles while spending a personal fortune of between $700,000 and $800,000.[40] The World's Fair came and went, and Armstrong was left with his fledgling experiments. But even with the thrill of television, FM was beginning to catch on, so much so that on January 1, 1941, the commission authorized commercial FM broadcasting. Although it might have seemed that Armstrong could look his old friend David Sarnoff in the eye with an "I told you so," that was not the case. In 1945, RCA won a victory when the FCC moved FM to its higher frequency, making room on the spectrum for television.[41]

Finally in 1948, after seeing RCA get away without paying royalties on FM sound transmission for TV, Armstrong brought suit. For five years the legal battle went on until Armstrong finally agreed to a settlement.

75

He died shortly thereafter, reportedly committing suicide. But FM, even with setbacks, corporate lobbying, government tampering, and changed frequencies, continues to develop and win audiences.

Factors Affecting FM Growth

The growth of FM broadcasting can be attributed to ten reasons. *First,* even though it was set back by World War II, the FCC gave FM permission in 1941 for full-scale development which prospered in the brief prewar period. *Second,* the perfection of sound recording gave the public a new appreciation for quality reproduced music, quality that FM could provide better than AM could. *Third,* FM was boosted by the development of stereo sound recording and the corresponding public demand for stereo FM. *Fourth,* the June 1, 1961 decision by the FCC permitting FM to broadcast stereo signals gave FM the ability to supply that demand. *Fifth,* the crowding on the AM frequencies prompted new broadcasters to enter the industry on the FM spectrum. *Sixth,* FCC requirements which gradually eliminated the once common practice of simulcasting the same program on combination AM/FM stations under the same licensee forced licensees to develop the FM stations. *Seventh,* more radio receivers are capable of receiving FM signals.

Eighth of the ten reasons for growth is that FM stations are now presenting a diversity of programming which appeals to a wide range of tastes (Figure 3–15). A research project conducted by Cox Broadcasting Corporation examined broadcast diversity.[42] The research found 21 percent of FM stations programmed contemporary music, more than what AM stations programmed. So the image of FM as devoted mostly to classical music is already outdated. Cox found that "beautiful" music occupied only 6 percent of FM programming. Twenty-four percent of the stations programmed "beautiful" music, 13 percent light music, 9 percent country, 8 percent contemporary, 5 percent soul, and 14 percent "other" formats. *Ninth,* although it is changing, FM has fewer commercials than AM does. *Tenth,* automated programming equipment permits licensees with an AM/FM combination to program the FM station without increased staffing.

FM's growth has been substantial. Cox researchers predict that in 1981 there will be 3,500 FM stations in operation as compared to 4,600 AM stations (Figure 3–16). The same research sees FM as equaling AM in popularity, capturing 57 percent of the total American listening audience (Figure 3–17) in 1981.[43] So the work of Edwin Armstrong, despite his frustrations, did open a new era in radio which has had not only a profound effect on the industry, but also on the radio programming which we receive.

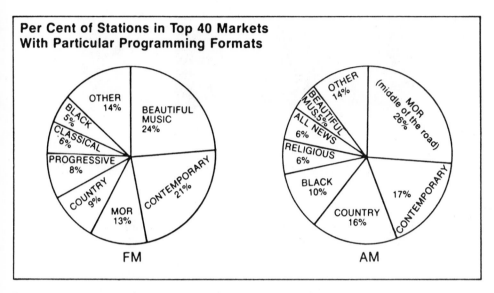

Figure 3–15   (Cox Broadcasting Corporation)

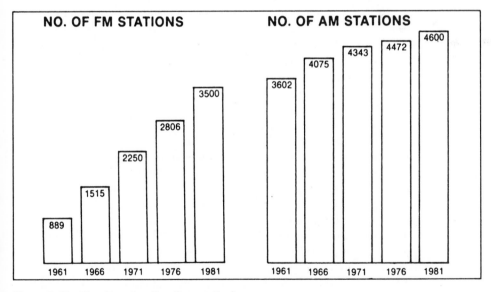

Figure 3–16   (Cox Broadcasting Corporation)

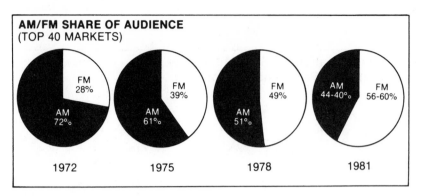

Figure 3–17   (Cox Broadcasting Corporation)

the transistor

The story of the transistor begins in the Bell Laboratories in 1947 when Dr. William Shockley sent a memo to colleagues inviting them to observe an experiment he had perfected using crystals, much like those used in early radio receivers. What Dr. Shockley was experimenting with was the "transistor effect." Using a small silicon crystal, scientists at Bell Labs discovered that the crystal could be made to react to electrical currents in much the same way that the vacuum tube reacted. Dr. Shockley, working with colleagues Walter H. Brattain and John Bardeen, perfected the transistor, early models of which were not much bigger than a grain of sand. Today, scientists have perfected the transistor to the point at which thousands of them can fit onto a tiny chip smaller than the end of your finger (Figure 3–18).

Transistors function in a process similar to a switch controlling an electrical current. The transistor in your portable radio consists of a wafer-thin crystal in three layers. A wire is attached to each of the three layers. One wire is used to detect the radio signal being sent through space. When the signal is detected, it allows current to flow through the transistor in sequence with the incoming signal. By attaching a battery to the transistor, the radio signal can trigger a circuit, releasing current from the battery. Because the current is released in exact sequence to the incoming signal, the transistor permits the signal, through the battery's current, to be amplified tens of thousands of times.

The small size of the transistor revolutionized radio. When its practical applications were realized, the radio receiver could be taken outside the home with no more power than a tiny battery. Radio receivers were suddenly everywhere—on the beach, at the ball game, on picnics. There was a new gift-giving spree as transistor radios became *the thing* to own. We now take for granted the tiny pocket device that can put us in instant

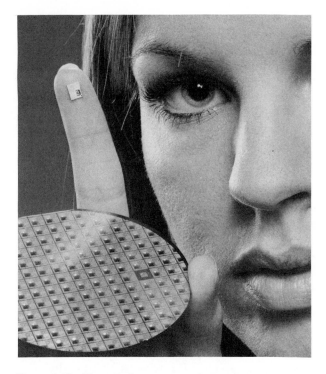

Figure 3–18 The transistor chip on the woman's finger has a circuit capacity of thousands of single transistors. (Rockwell International)

touch with dozens of AM and FM radio stations. For William Shockley, Walter H. Brattain, and John Bardeen, their discovery won them the Nobel Prize in Physics.

reproducing an image

Even before Henrich Hertz had proved the existence of electromagnetic waves, scientists were working to find a way to reproduce images and send them from a transmitter to a receiver. The person credited with the first breakthrough was German scientist Paul Nipkow. As Marconi continued to build his worldwide corporate ties, as the giant American corporations fought over the control of wireless, as the pioneer stations signed on the air, and as the transistor was being developed, television continued through its experimental era to become a mass medium that by the 1950s equaled radio as a mass medium.

Nipkow's Scanning Disc

Paul Nipkow's scanning disc proved that images could be transmitted electrically and mechanically, using a series of wires between the transmitter and receiver. By punching holes in the disc and arranging the holes in a spiral, the revolving disc would enable the holes to scan a picture placed behind the disc. By rapidly changing the picture, the illusion of a moving image could be transmitted over wires. The system worked even in 1884; yet it lacked many components needed to make television a reality. First, the system was mechanical, not electronic. Compared to today's television, it was slow and cumbersome. Second, the wires limited the distance that the image could be transmitted, because stringing wires to many different locations was impractical. Third, the image was unclear because coordinating the scanning disc with the changing pictures still had not been perfected. Experimentation on the scanning disc continued. Ernst F. Alexanderson, inventor of the Alexanderson alternator, continued to work on mechanical television, experimenting with both small and large screen systems. Television technology continued to advance to the point at which it could transmit pictures of crude clarity using high intensity lights. The most famous demonstration of this technique was with the cartoon character, Felix the Cat (Figure 3–19). Perched on a converted phonograph turntable in New York, experimental receivers as far away as Kansas City relayed the revolving Felix, who looked very much like a venetian blind. Two men in the 1920s were to make those experiments obsolete.

Zworykin and Farnsworth

A scientist at RCA and an Idaho schoolboy gave the world electronic television. Working far apart, the two transformed television from a clumsy mechanical device into a sophisticated electronic eye. Vladimir K. Zworykin, who immigrated to the United States from Russia, went to work for Westinghouse in its Wilkensburg, Pennsylvania plant a year before KDKA signed on the air. He asked and received permission from Westinghouse to work on a device that used electrons instead of scanning discs to detect and transmit pictures. In 1923, Zworykin patented an all-electronic television pickup tube called the iconoscope Figure 3–20) which signaled the end of the scanning process. By 1929, he and his research team demonstrated an all-electronic television receiving tube called the kinescope.

But 60-line television was still a far cry from public acceptance. It was this very thought that crossed the mind of a teenager from Rigby, Idaho. Philo Farnsworth, later to become known as the "father of television,"

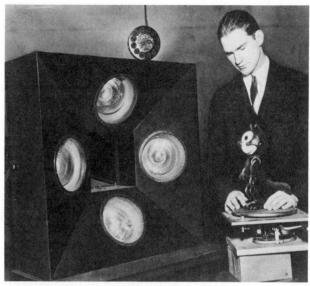

Figure 3–19 In 1928, television was progressing through its experimental era which showed the clarity of a 60-line picture in black and white (left). The television picture was produced by putting Felix the Cat in front of a television camera flanked by high intensity spotlights (right). (RCA)

Figure 3–20 Vladimir K. Zworykin holding his invention, the iconoscope, an all-electronic television pickup tube. (RCA)

first came up with the idea of electronic pickup and scanning systems in 1921 at the age of fifteen.[44] To his science teacher, Justin Tolman, Philo explained the solution to the electronic scanning and pickup problem. Tolman encouraged the boy, as much as could be expected in deference to a fifteen-year-old with an inquisitive mind. By the late 1920s, Farnsworth had formed the Crocker Laboratories in San Francisco, later to become the Capehart-Farnsworth Corporation. In 1930, Farnsworth was asking for permission to experiment on the air with a system which could get a 300-line picture quality.

By 1931, he had provoked Zworykin and RCA's interest enough to entice them to travel to California to view his system. Upon seeing it, Zworykin claimed that Farnsworth had nothing he needed. But RCA quickly realized it was going to have to reckon with the boy from Idaho. Farnsworth, with gall that probably made the corporate giant a bit uneasy, told RCA that it could forget about buying his system. The company would have to rent it in a royalty contract. Finally, after much negotiation, RCA did just that. Later in 1941, Justin Tolman testified in a court suit over patent rights. In the courtroom, he drew the same diagram Farnsworth had placed on his blackboard that day in 1921 in Rigby, Idaho, winning an important patent decision for Farnsworth who would later hold over 150 patents, some common to all television receivers.[45]

The Experimental Era

With both electronic pickup tubes and an improved scanning system, television was ready to make its mark. The first United States station to sign on was W2XBS in 1930, owned by NBC in New York. The following year, an experimental RCA-NBC transmitter and antenna were in operation atop the Empire State Building. At RCA, a committment of $1 million was earmarked for field tests. From these tests came the forerunner of big-screen television as RCA's electron "projection" gun made history by producing television pictures on an $8' \times 10'$ screen. RCA-NBC mobile television arrived in 1937. The following year, scenes from the play *Susan and God* were telecast from NBC studios in New York, and RCA president David Sarnoff announced that television sets would go on sale at the World's Fair in 1939.

President Roosevelt opened that World's Fair and became the first president ever seen on television by the general public. The fair-going public flocked to look inside the special television receiver prototype displayed by RCA (Figure 3–21). An $8'' \times 10''$ screen reflected on the lid kept fair-goers asking questions about how it worked and how they could buy one. The same year, AT&T lines linked an NBC camera at a Madison Square bicycle race to a broadcast transmitter, proving that both wires and airwaves could complement each other in aiding television's growth.

Figure 3–21 A crowd gathers in front of a clear glass model of an early television receiver on display at the 1939 New York World's Fair. The set had an 8″ × 10″ screen reflected on a mirrored cabinet top. (RCA)

By 1941, the FCC realized both the potential and the demands of television, and authorized commercial licensing of television. But the glory was short-lived. As war raged in Europe, the United States needed skilled technicians to work in electronic plants and laboratories at home. There was little use for television. In fact, when the Japanese attacked Pearl Harbor, pushing the United States officially into World War II, it was radio, not television, that brought the sounds of bombs and gunfire to American living rooms. For that dubious distinction, television would have to wait for Vietnam.

The Freeze, UHF, Color

Between 1948 and 1964, three events helped mold television's future. Although not directly related, they occurred somewhat simultaneously and represent an era which is best described as an era of decision and indecision. After World War II was over, the industry once again began to gear up for television. At the FCC, concern was beginning to mount over signal interference from those stations which wanted to begin broadcasting. The bombardment of requests was overwhelming, and the commission instituted the famous television "freeze" of 1948, placing a hold on all new licenses. In 1952 the freeze was lifted. The FCC assigned 12 channels in the very-high frequency (VHF) area (channels 2 through 13) of the electromagnetic spectrum, and 70 channels in the ultra-high frequency (UHF) area of the spectrum (through 83).

In theory, UHF was on a par with VHF. In practice, they were far apart. One big reason was the lack of receiving sets with UHF tuners. UHF simply could not compete in the marketplace. If people did not watch UHF, the UHF stations found it difficult to attract advertising dollars. Finally, in 1964 the FCC began requiring manufacturers to install both VHF and UHF tuners on all television sets. In 1976, the electronics firm of Sarkes Tarzian developed a device called a Uni-tuner which tunes both UHF and VHF channels with the same "click knob."

Although UHF still has a long way to go toward reaching its full potential, its future is beginning to brighten significantly. Many UHF stations are not network affiliated, and expensive network advertising rates are sending many national advertisers to these independent stations. Moreover, a wider assortment of syndicated programming is permitting independent stations to capture a larger viewing audience once reserved for network affiliates.

At the same time as the freeze was taking place and the FCC was deciding how to allocate frequencies, two giants were battling it out over the future of color television. RCA and CBS went to battle over what type of color television system should become the national standard. CBS won the first round when the FCC approved a noncompatible color system for commercial broadcasting.[46] Noncompatible meant the color signals could not be received on sets built for black and white reception. Meanwhile, RCA, which had been developing a compatible system, slapped CBS with a law suit. The appeals went all the way to the Supreme Court which upheld the FCC's approval of the CBS system. CBS, elated over the victory, bought a company called Hytron Electronics and its subsidiary Air King. The new company manufactured receivers capable of picking up the CBS color telecasts. Unfortunately the joy was short-lived. Realizing the importance of compatible color, the FCC in 1952 reversed its decision, and CBS's venture into color television came to an abrupt halt.

television technology

The FCC eventually approved a 525-line resolution system for American television, meaning the picture would be scanned 525 times in rapid succession. It was a giant improvement over Zworykin's initial 60-line system and a considerable improvement over the 441-line system used in Europe before a later 625-line system was adopted.

Iconoscopes to Plumbicons

Some of the most important improvements in television technology were in the sensitivity of camera tubes. We already have learned about Zworykin's iconoscope tube and the work of Philo Farnsworth. Although helping picture clarity, the necessity for high intensity lights made working with the iconoscope tube at best uncomfortable. The *orthicon* tube followed, providing improved clarity, and its successor, the *image orthicon,* reduced considerably the amount of light necessary to obtain a clear picture.

After the image orthicon tube came the *vidicon* tube. Even more sensitive, the vidicon paved the way for even greater picture quality. The vidicon is best described as a very sensitive light detector which turns light into electricity. All television since the scanning disc has done this, but the key is sensitivity combined with a high-resolution electronic scanning process. The vidicon is like an electron gun with a heater in the bullet chamber and a lens over the gun barrel. The electrons are fired at the lens and are controlled by a photoconductive layer just behind the lens. When the lens sees a light image, such as a snow scene, more electrons pass through the photoconductive layer to the signal electrode which generates the picture (Figure 3–22).

Improvements on the vidicon tube, especially as applied to color television, continue to be made. A major advancement came with the Amperex Plumbicon® tube which has the ability to capture color images with the sensitivity of the human eye. Experiments in holography, using

Figure 3–22 *(Educational & Industrial Television)*

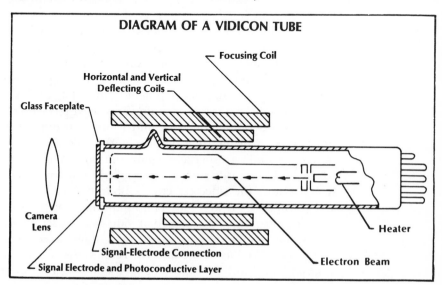

refracted laser beams, further indicate that three-dimensional pictures are not far away.

Magnetic Recording

Capturing the live image was only part of television's progress. Recording that image for future playback would give the medium a new, flexible, dimension. Thus, film and magnetic recording, later to be called video-tape, developed side by side, one using the technology of the other.

Recording television fascinated John L. Baird as early as 1927 when he conducted successful experiments using a magnetic disc.[47] The quality, however, was too unsatisfactory even to try using it for future television recording purposes. Building on Baird's work, researchers spent the next twenty years trying to perfect a video-recording device by using such modes as a combination-television camera and standard 16 mm film, and even large-screen television using 35 mm film, the unsuccessful brainchild of Lee de Forest.

Early color television recording concentrated on combining color clarity with picture clarity. Although most early attempts were marginal, in 1948, Eastman Kodak introduced a 16 mm system developed in cooperation with NBC and the Allen B. DuMont studios. Using a Navy camera, Kodak film, and a CBS receiver, the first "completely successful" recording of color television was made in February, 1950.

The birth of videotape came a year later. The Electronic Division of Bing Crosby Enterprises demonstrated a videotape recorder in 1951, improving the quality a year later. In 1953, RCA demonstrated its version of a videotape recorder. The big videotape breakthrough and publicity, however, came in April 1956 when Ampex engineers demonstrated their videotape recording to a CBS-TV affiliates' meeting. RCA meanwhile, demonstrated a color videotape in 1957, but Ampex was to carry the banner for some years to come. Ampex engineer Charles P. Ginsburg is credited with much of Ampex's success, although a team of engineers worked on videotape development.[48] In 1964, SONY Corporation of Japan introduced a system claiming improved recording head design and simplified operation for black and white recording. Portable videotape units were proving their worth in the mid-1960s with schools and businesses finding especially useful the one-inch, reel-to-reel videotape which could be easily stored and applied to instructional purposes. Next to arrive were the video cassettes. CBS introduced the first video-cassette system called EVR in 1968. The following year, SONY Corporation of America introduced the first color videotape cassette recorder.

Further refinements in videotape storage were developed by CBS under the direction of the late Dr. Peter C. Goldmark. The CBS Rapid Transmission and Storage (RTS) system became operable in 1976, permitting up to 30 hours of programming to be stored on one video cassette. Different programs can be played back from the tape simultaneously over different transmission systems, such as different cable channels.

John L. Baird's research efforts in 1927 ushered in a new era in video recording for everything from full-length movie features to electronic news gathering (ENG).

The Role of Film in Video Recording

Although videotape is currently the center of attention because of its quick playback and reusable tape, film continues to be important. An intermediate film transmitter which "scanned" film was introduced at the Berlin Radio Exhibition in 1932. In 1933, an intermediate film receiver was demonstrated at the same exhibition. Kinescope recording, a quick-developing, film-recording process, was used widely in the early 1950s. In fact, when a nationwide microwave link was completed in 1951, kinescope recording became popular for network transmissions until the conversion to videotape. Even after the conversion, the 16 mm camera continued to be essential to the television news production process and still remains a favorite of many television newsrooms.

Super 8 film also has become a favorite of some broadcasters. Less expensive than 16 mm, super 8 uses one-third more area on the film and an improved camera, thus making it adaptable to many broadcast uses. Professor Ron Whittaker has examined the uses of super 8 film in broadcasting, noting that it can serve broadcasting because: (1) electronic image enhancers can increase image sharpness and provide clarity on television comparable with 16 mm film; (2) advances in the film emulsion process have reduced graininess; (3) super 8 works well in low light conditions; (4) super 8 equipment is still more portable and lighter than ENG equipment; (5) in low light levels, a picture "lag" or "smear" can occur with many electronic cameras, whereas film can handle the greater brightness range; (6) film can be processed in as little as fifteen minutes; (7) the super 8 camera is small and inconspicuous compared to most ENG equipment, which can be especially important when covering such things as civil unrest in which the presence of television cameras can trigger crowd reaction; and (8) stringers can use super 8 cameras easily without much training and at less cost than an ENG setup.[49]

Electronic News Gathering

As refinements continued in videotape recording, bringing higher quality, lighter weight cameras and smaller microwave transmitting and receiving equipment, the stage was set for electronic news gathering (ENG). By the early 1970s, stations were beginning to jump on the ENG bandwagon, some disregarding film altogether. It changed much of the news and public affairs programming as local anchorpersons switched to live coverage of events in the midst of their local television news as easily as they switched to a commercial. Today, we see live pictures from helicopters, boats, lettuce fields, and courthouse steps. We also have become accustomed to live aerial shots of a football stadium from one of the Goodyear blimps (Figure 3–23). The new technology has moved live television far beyond the confines of the television studio.

Changes in Receiver Design

Changes in television receiver design have been as dramatic as the rest of television's facelift. A comparison of the receiver displayed by RCA at the 1939 World's Fair with today's average home set illustrates the considerable difference in both size and design. The transistor's application to television permitted a vast reduction in size, and miniature,

Figure 3–23 Developments in microwave technology have made remote broadcasting commonplace. Pictured is the television camera inside the Goodyear blimp, "Mayflower," focusing on a test picture. Pictures from the blimp are sent by microwave to a receiving dish on the ground and in turn are sent over network distribution systems to local television stations which broadcast the picture to home receivers. (The Goodyear Tire & Rubber Company)

computerlike processing devices called microprocessors further aided its development. Already, pocket televisions are rapidly becoming common (Figure 3–24). Scientists are experimenting with television screens the thickness of a standard picture frame, and using holography, predictions of three-dimensional television receivers are more than mere science fiction. Dick Tracy's two-way wrist TV of comic strip fame may someday be commonplace.

While some manufacturers are working to reduce the size of receivers, others are working to increase the size of the screen. Big-screen television, nothing new, is now becoming popular as a home medium and is especially attractive to restaurants as an inexpensive form of entertainment. Using three projection lenses to display the three primary colors of light in much the same way the set receives a picture, the primary colors are projected onto a large screen which can be viewed from a distance with picture quality equal to that of a standard television receiver.

fragmenting the audience: home vtrs, discs, and games

Videodiscs are also available for home use (Figure 3–25). Looking much like a large record, the discs are played on a recorder that resembles a standard record player. Complete programs lasting an hour and longer can be contained on a disc.

Figure 3–24 Small circuity has made possible portable television receivers the size of portable radios. The set shown measures 4″ × 6 3/4″ × 1 3/4″. A 2″ diagonal screen produces black and white images with the detail of larger sets. (Sinclair Radionics, Inc.)

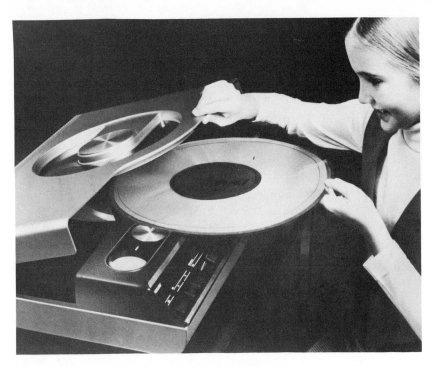

Figure 3–25 Videodiscs could make television programs and other material as easily accessible as phonograph records are now. There still are questions as to what effect this new technology might have on the television viewing audience and on traditional program distribution systems. (North American Philips Corporation)

In addition to these discs, video-cassette recorders like the SONY Betamax permit you to set a special timer and record your favorite program while watching another channel or doing something completely away from the set. Preselected timers automatically turn the recorder on and off, permitting you to spend an evening out and later to return to watch the program you missed.

Television games are becoming as popular as the recorders. Financially within reach of many families and becoming less expensive all the time, video games can turn any home television receiver into a competitive contest. Hockey, football, tennis, racing cars, and airplane dogfights all can be controlled in a penny-arcade atmosphere but with a one-time expense.

All of these new devices worry the television industry. The new technology can play havoc with advertisers, for example. It is difficult to obtain a true reading on how many people watch a television program if some of them watch when the program is first broadcast while others watch on their home video-cassette recorders. Even though a television set may be turned on, no program garners rating points when a video game is under way. The problem has yet to reach serious proportions. However, if these new novelties continue to gain popularity, networks and stations may face competition from more than just each other. Moreover, with inexpensive discs, program distribution systems can completely bypass the local stations. With these thoughts in mind, some companies have entered into litigation to try to stop the growth of home video-cassette recorders and playback systems, claiming infringement on copyrights. The future of these new devices will be determined by technology, law, and public demand.

news applications: subscription tv and teletext

As we approach the end of the twentieth century, television's future may very well be shaped by the adaptation of the medium to new uses for program distribution and information processing.

Subscription-Pay TV

While we think of traditional over-the-air television transmission systems distributing programming to anyone who has a home receiver, growing interest in what is called *subscription TV* or *pay TV* means new potential for the medium. With pay TV, the signal is scrambled as it leaves the station transmitter and descrambled with a special attachment on the home receiving set. The individual pays a monthly fee for the descrambler attachment and can receive special programming such as first-run movies, sports programming, and other programming generally not available to nonsubscribers.

More and more cities are becoming competitive pay-TV markets, with subscribers willing to pay fees substantially above the costs for standard cable services. Chicago, Los Angeles, and Detroit are three examples of pay-TV markets, and fees can run approximately $25.00 per month, with an installation and deposit fee of $100. Major companies such as Time, Inc., see pay TV as having a substantial future and have made substantial investments in pay-TV systems.

Teletext

Also gaining attention is the use of *teletext,* a system where a computer is interfaced with a television transmission system and can send data to home receivers equipped with a special decoder and keyboard. The home viewer, using the keyboard, can select either standard television programming, the teletext signal, or both superimposed. Teletext can send "digital pages" of information using scanned lines of the television picture not visible without an encoder (Figure 3–25). The system pi-

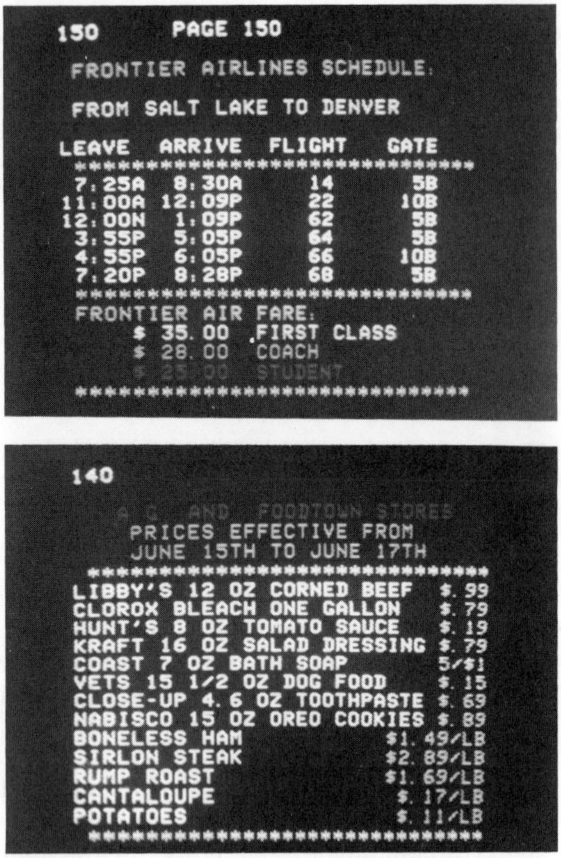

```
150      PAGE 150

FRONTIER AIRLINES SCHEDULE:

FROM SALT LAKE TO DENVER

LEAVE   ARRIVE  FLIGHT   GATE
*****************************
 7:25A   8:30A    14      5B
11:00A  12:09P    22     10B
12:00N   1:09P    62      5B
 3:55P   5:05P    64      5B
 4:55P   6:05P    66     10B
 7:20P   8:28P    68      5B
*****************************
FRONTIER AIR FARE:
   $ 35.00 .FIRST CLASS
   $ 28.00  COACH
   $ 25.00  STUDENT

*****************************
```

```
140

A C AND FOODTOWN STORES
  PRICES EFFECTIVE FROM
  JUNE 15TH TO JUNE 17TH
*****************************
LIBBY'S 12 OZ CORNED BEEF  $.99
CLOROX BLEACH ONE GALLON   $.79
HUNT'S 8 OZ TOMATO SAUCE   $.19
KRAFT 16 OZ SALAD DRESSING $.79
COAST 7 OZ BATH SOAP       5/$1
VETS 15 1/2 OZ DOG FOOD    $.15
CLOSE-UP 4.6 OZ TOOTHPASTE $.69
NABISCO 15 OZ OREO COOKIES $.89
BONELESS HAM            $1.49/LB
SIRLON STEAK           $2.89/LB
RUMP ROAST             $1.69/LB
CANTALOUPE              $.17/LB
POTATOES               $.11/LB
*****************************
```

Figure 3–26 "Pages" from the experimental teletext system at KSL-TV in Salt Lake City. (Courtesy Bill Loveless and KSL-TV)

oneered in Britain with the operation of the BBC's CEEFAX operation. In the United States the system gained attention after experimental progress was made with a prototype unit installed at Bonneville International Corporation's KSL-TV in Salt Lake City. Much of the work there is under the direction of Bill Loveless, director of engineering for Bonneville.

The information potential for teletext is almost limitless. From grocery lists to newspaper pages, the system promises still new frontiers in communication as television becomes more and more a part of our information society.

While in this chapter we have touched the early history as well as the new technology of broadcasting, the chapters that follow will give us a broader understanding not only of how broadcasting operates but of the social, political, and economic aspects of broadcasting, both domestically and internationally.

summary

When Charles David Herrold started his station in San Jose, California in 1909, it became one of the earliest of a string of pioneer stations. Such names as WHA at the University of Wisconsin, WWJ in Detroit, and KDKA in Pittsburgh were added to the list. The stations expanded in power and in audience and were joined by thousands of others as radio matured.

One of the major developers was the Radio Corporation of America. Formed in 1919 as part of a scheme to keep the Alexanderson alternator in the United States, RCA's direction was charted by former Marconi employee David Sarnoff.

The 1920s were marked by agreements and disagreements among the major radio powers. Although Westinghouse, GE, AT&T, and some smaller concerns joined together to share inventive efforts, they competed in developing commercial broadcasting. AT&T's WEAF attempted "toll broadcasting," and when it tried to monopolize stations' use of the long lines, matters went to court. A corporate agreement resulted, and AT&T went back to the telephone business while the others forged ahead with broadcasting.

NBC, CBS, and Mutual emerged as the major networks. When NBC was required to dispose of half of its dual network concept, the American Broadcasting Company was born. In 1967, the Public Broadcasting Act became the foundation for still another major network, National Public Radio.

New technology has also been important to radio's development. Through the work of Edwin Armstrong, radio gained a sizeable "sound" advantage with FM. The FCC's support of FM, requiring separate programming from AM, and the development of stereo FM broadcasting opened up new possibilities for this area of the spectrum. And just when it was needed, the invention of the transistor by three Bell Lab scientists made radio a portable medium.

Early attempts to reproduce an image, however, used a scanning disc developed by Paul Nipkow. This mechanical process was soon replaced by electronic reproduction. The work of Vladimir Zworykin and Philo Farnsworth gave us the iconoscope television tube and electronic scanning of high resolution, permitting improved picture quality. The early experimental era of television saw station W2XBS sign on the air in New York, followed by a series of breakthroughs in improved television technology.

The first true introduction of television to the American public took place at the World's Fair in 1939. In 1941, the FCC approved commercial television. Then during the television freeze which began in 1948, the commission spent five years deciding frequency assignments and standards for color.

93

Television cameras, meanwhile, improved from the iconoscope tube to the orthicon, the image orthicon, the vidicon, and the Plumbicon® tube, which is based on the vidicon concept but provides color clarity equivalent to that which can be detected by the naked eye. Magnetic recording of television programs progressed from the early kinescope methods to videotape. Film advanced to super 8 technology. But at the same time, industry people became concerned about the fragmenting television audience caused by such developments as video cassettes, home recorders, videodiscs, and video games. Subscription TV and teletext systems promise new applications for television program distribution and information processing.

spotlight on further learning

BARNOW, ERIK, *The Image Empire.* New York: Oxford University Press, Inc., 1970.

BARNOW, ERIK, *Tube of Plenty.* New York: Oxford University Press, Inc., 1975.

BROWN, LES, *Television: The Business behind the Box.* New York: Harcourt Brace Jovanovich, Inc., 1971.

CATER, DOUGLASS and others, *Television as a Social Force: New Approaches to TV Criticism.* New York: Praeger Publishers, Inc., 1975.

COWEN, GEOFFREY, *See No Evil: The Backstage Battle over Sex and Violence on Television.* New York: Simon and Schuster, Inc., 1979.

FOLEY, SUE K., *The Political Blacklist in the Broadcast Industry: The Decade of the 1950s.* New York: Arno Press, 1979.

KALBA, KAS, *The Video Implosion: Models for Reinventing Television.* Palo Alto, Calif.: Aspen Institute, 1974.

LARKA, ROBERT, *Television's Private Eye: An Examination of Twenty Years Programming to a Particular Genre, 1949 to 1969.* New York: Arno Press, 1979.

MACFARLAND, DAVID T., *The Development of the Top 40 Radio Format.* New York: Arno Press, 1979.

PALEY, WILLIAM S., *As It Happened.* New York: Doubleday & Co., Inc., 1979.

PRIMEAU, RONALD, *The Rhetoric of Television.* New York: Longman, Inc., 1979.

SMITH, RALPH LEWIS, *A Study of the Professional Criticism of Broadcasting in the United States.* New York: Arno Press, 1979.

SMITH, ROBERT R., *Beyond the Wasteland: The Criticism of Broadcasting.* Falls Church, Va.: Speech Communication Association, 1976.

SMYTHE, TED C., and GEORGE A. MASTROIANNI, eds., *Issues in Broadcasting: Radio, Television, and Cable.* Palo Alto, Calif.: Mayfield Publishing Co., 1975.

STEIN, ALETHA HUSTON, and LYNETTE KOHN FREIDRICK, *Impact of Television on Children and Youth.* Chicago: University of Chicago Press, 1975.

part

2

Basic Technology

4

radio waves and
the spectrum

FOCUS After completing this chapter we should be able to

Understand the theory of the electromagnetic spectrum and electro-magnetic waves.

Describe how AM and FM broadcasting works.

Explain the difference among FM stereo, quadraphonic, and AM stereo.

Describe how television signals are transmitted.

Describe the paths electromagnetic waves follow.

Understand the allocation of broadcast station operating frequencies.

Explain the differences between directional and nondirectional stations.

Discuss international spectrum management.

It is doubtful if Henrich Hertz could imagine that almost 100 years after his discovery of electromagnetic waves, those same waves would carry messages around the world and even to planets millions of miles away. In this chapter, we shall discover how radio waves travel and how radio and television programs get from the station transmitter to our home receivers.

the electromagnetic spectrum

To understand how broadcast signals are carried between transmitter and receiver, it is necessary to understand the *electromagnetic spectrum*. Consider the spectrum as a "measuring stick" for electromagnetic energy (Figure 4–1). At the lower end of the measuring stick are radio waves. At the upper end of the spectrum we find visible light and X-rays. For our purposes, we shall concentrate on radio waves.

Defining Frequency

What differentiates radio waves from light waves or X-rays? The answer is their *frequency*. You have heard the term used when referring to the dial on your standard radio or perhaps the channels on your CB radio. Two radio stations in the same community operate on different "frequencies" so they do not interfere with each other. When current is applied to the transmitter of a radio station, the antenna emits electromagnetic radiation. This radiation is actually a series of electromagnetic waves, coming one after another. The next time you throw a rock into a pool of water, watch the series of waves which ripple one after the other in all directions from the point at which the rock entered the water. This is what happens when electromagnetic energy travels through the at-

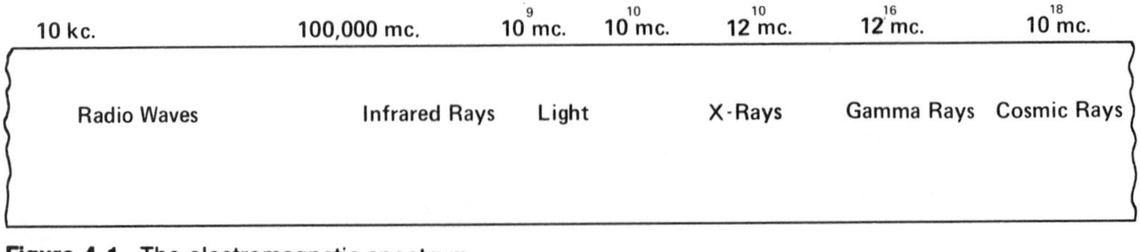

10 kc.		100,000 mc.	10^9 mc.	10^{10} mc.	12^{10} mc.	12^{16} mc.	10^{18} mc.
	Radio Waves		Infrared Rays	Light	X-Rays	Gamma Rays	Cosmic Rays

Figure 4–1 The electromagnetic spectrum.

98

mosphere or the vacuum of outer space. *The number of waves passing a certain point in a given interval of time* is the *frequency*. In broadcasting the waves are *electromagnetic waves*, but for our purposes in this book we will sometimes call them *waves*.

Defining Wavelength, Cycle, and Cycles per Second

The actual *distance between two waves* is called the *wavelength*. For instance, if we take a stop-action picture of the ripples (waves) in our pond and then figured the distance between two ripples, that would be the *wavelength*. Cycle is closely related to wavelength (Figure 4–2). When one complete wave passes our counting point, we have observed one cycle. The "point" can be any geographical location. For instance, if we were watching ripples in a pond we might stand at a certain point on the edge of the pond. The same applies to counting electromagnetic waves. However, electromagnetic waves travel much too fast to count and cannot be seen.

We know that all electromagnetic waves travel at the speed of light, or at 186,000 miles/second. Thus, since we measure the speed of light in miles per second or, using the metric system, in meters per second, the second becomes the commonly used time interval. The number of waves passing a certain point in one second is called *cycles per second*. One thousand cycles per second is called a *kilocycle*, one million a *megacycle*. With this in mind, we can determine the wavelength of radio waves by

Figure 4–2 When one complete wave passes a given point it is called a "cycle." The term "kilocycle" (also called "kilohertz") is used to denote 1,000 cycles, and "megacycle" (also called "megahertz") to denote 1,000,000 cycles. (FCC *Broadcast Operators Handbook*)

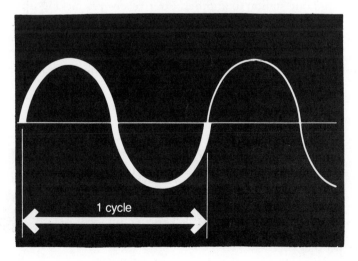

1 cycle

simple division. For example, knowing the speed of light is 186,000 miles per second, if 10,000 cycles (10,000 complete waves) pass a given point in one second, the wavelength of each wave would be 18.6 miles (186,000 ÷ 10,000). Now let's figure the wavelength of a higher frequency, 535 kilocycles (535,000 cycles per second). We divide 186,000 by 535,000. The answer is .3477 miles. Since there are 5,280 feet in one mile, we can convert our wavelength to feet by multiplying 5,280 by .3477. The answer is a wavelength of 1,836 feet.

Computing Frequency

Now that we understand cycles per second, we can easily compute frequency. If, for instance, 1,000 waves pass a given point in one second (1,000 cycles per second or one kilocycle), the frequency or location on the electromagnetic spectrum is 1 kilocycle. Similarly, 535,000 cycles per second is represented as 535 kilocycles. On your AM radio dial, that particular frequency would be at the lower end of the dial (Figure 4–3).

Do not be alarmed if you purchase an AM radio reading 540–1600. Each radio station is assigned 10 kilocycle ranges on the electromagnetic spectrum; thus the station assigned the lowest frequency is assigned 540 kilocycles which permits it to operate between 535 and 545 kilocycles. Some radios, even though capable of receiving 540 kilocycles, begin numbering their dials at 550 kilocycles or abbreviate it as the number 55.

Tuning to a Wavelength

In simplified terms, a radio receiver "counts" the waves or cycles per second to determine a frequency. Your radio does this when you tune from one station to another. Your receiver is picking up only those waves which are being transmitted on the same frequency to which you tune. Thus, different frequencies on your radio dial correspond to different positions on the electromagnetic spectrum.

Figure 4–3 illustrates this concept. At one position on the spectrum is

Figure 4–3 The typical AM radio dial reads from 540 kilohertz to 1600 kilohertz. (FCC *Broadcast Operators Handbook*)

the frequency allocated to AM broadcasting. This represents that portion of the spectrum between 535 and 1605 kilocycles. Progressing higher on the electromagnetic spectrum are citizens' band radio, television, and FM broadcasting. Later in this chapter, we shall examine even higher frequencies of microwaves.

Common Terms: Meters and Hertz

Before going on to our discussion of AM and FM broadcasting, we need to mention two terms: *meters* and *hertz*. In our discussion of wavelengths, we computed the wavelength of 535 kilocycles using miles and feet. You should be aware that international wavelength is based on the metric system; thus, wavelength is figured in *meters*. Although you may find feet and miles easier to use, meters are the common international measure used in broadcasting. The speed of light is 300,000,000 meters per second (186,000 miles per second). A meter is equal to 2.8 feet. Also, in our discussion we used the term cycles to denote a specific frequency on the electromagnetic spectrum. In recent years, the term cycles has been replaced by the word *hertz* (Hz), in honor of Henrich Hertz who discovered electromagnetic waves. Thus, 1,000 cycles per second becomes one kilohertz (1 kHz), and 1,000,000 cycles per second becomes 1 megahertz (1 mHz). Frequencies of 535,000 cycles per second become 535 kilohertz (535 kHz), referred to simply as a "frequency" of 535 kilohertz.

how *am* broadcasting works

Now that we understand the electromagnetic spectrum and how waves are radiated into space, we'll learn how voice and music use those waves to reach radio listeners. We'll begin with AM broadcasting, or that portion of the electromagnetic spectrum which falls between 535 and 1605 kHz.

AM stands for *amplitude modulation*. Amplitude is defined as breadth of range.[1] Modulation means to adjust or adapt to a certain proportion.[2] Now let's apply both of these words to radio waves. Figure 4–4 illustrates an unmodulated radio wave. Now examine Figure 4–5, a radio wave which has been altered by adjusting the amplitude or "breadth of range" of the wave. The wave characteristics of music and the human voice are transformed into the wave, which in turn "carries" them between the transmitter and receiver. Notice that the wavelength, or frequency, remains constant. The change takes place in amplitude, not in frequency. When the wave is adjusted to carry changes in sound, it is said to be *modulated*.

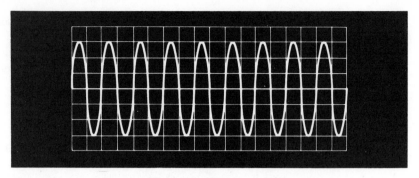

Figure 4–4 Unmodulated wave. (FCC *Broadcast Operators Handbook*)

how *fm* broadcasting works

From our discussion of AM broadcasting, you should have already started to figure out how FM, or frequency modulated, broadcasting works. Instead of changing the amplitude of the wave, we change the frequency, or wavelength. Figure 4–6 illustrates a frequency modulated wave. Notice there is no change in the amplitude of the wave. Instead, the frequency, or wavelength, varies. Different sounds indicate different wavelengths (cycles per second). FM broadcasting to the general public operates between 88 and 108 mHz (Figure 4–7). Each FM station is allocated a width of 200 kHz.[3]

Whether the 200 kHz protection between FM stations will remain is questionable. The FCC has considered narrowing the space allocated to

Figure 4–5 Amplitude modulated wave. Notice the frequency (width of the wave or distance between waves) remains the same, but the amplitude varies. (FCC *Broadcast Operators Handbook*)

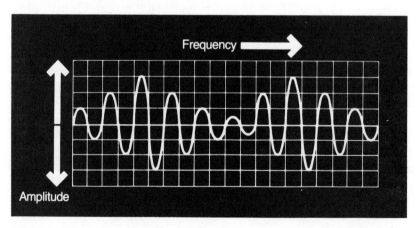

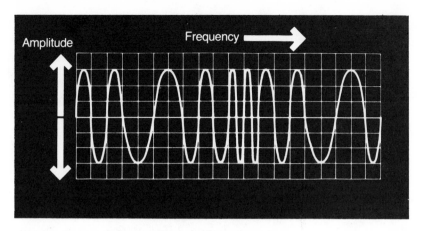

Figure 4–6 Frequency modulated wave. Notice the frequency varies, but the amplitude remains constant. (FCC *Broadcast Operators Handbook*)

each station to 100 kHz to allow room for more stations. Major opposition to the proposal has been voiced by proponents of quadraphonic FM broadcasting, which uses four separate channels instead of the two channels common to FM stereo. Characteristic of this opposition is the editorial opinion of Lou Dorren, whose company, Quadracast Systems Incorporated, developed one of the first quadraphonic systems. Dorren stated: "From the very beginning of FM's history, it has been a medium with special accent on quality. That is how FM grew, and it is really how FM maintains its present market position; it would be folly to relegate it to a lower quantity level." Dorren also contended that "FM has always maintained an advantage over AM in that it could provide multichannel performance. Now, as possibilities for AM stereo are being considered, FM's performance edge for discrete quadraphony should be maintained."[4]

Dorren's arguments are countered by others who want to see as much "room" as possible for the development of future FM stations, expected to outnumber AM stations by the 1980s.

Figure 4–7 The FM radio band goes from 88.1 to 107.9 megahertz. Most FM radio dials are shown in whole numbers without the decimals and read from 88 to 107. (FCC *Broadcast Operators Handbook*)

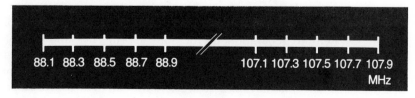

how *fm* stereo and quadraphonic work

Radio broadcasting has taken three major strides to improve the quality of sound reproduction from the studio to the living room. One was the development of FM broadcasting. The second was the development of stereo FM broadcasting. In the 1970s there has been an interest in quadraphonic, or four-channel FM broadcasting. Let's examine stereo and quadraphonic broadcasting in more detail.

Stereo FM

Within the 200 kHz space allocated to FM broadcasting is ample room for the separation of broadcast signals, permitting the same station to broadcast on *two* slightly different frequencies. In the dual-frequency concept there is room for a special tone sent at the same time, which automatically triggers specially equipped radios to receive the stereo signal. Radios equipped to receive stereo actually have *two* separate receiving systems which, when triggered by the tone, separately receive each of the two frequencies being broadcast by the stereo station. When the tone is not transmitted, the radio still receives the monophonic signal. Many of us have seen a small light flip on in a stereo FM radio receiver when we tune the receiver to a station broadcasting in stereo. This signals us that our radio is tuned to a stereo station, that the station is broadcasting in stereo, and that our receiver is receiving both channels of the stereo system.

Stereo broadcasting has steadily grown in popularity. The reproduction of quality music with the added dimensions of space and depth similar to having the orchestra in your living room has been the main distinction between stereo FM and standard FM and AM broadcasting.

Quadraphonic FM

What added dimensions that stereo broadcasting brought to FM, quadraphonic doubled. With quadraphonic systems, four instead of two frequencies are employed. Quadraphonic broadcasting is still in its experimental stages, and the FCC is currently evaluating various quadraphonic systems. Serious evaluation began in 1972 with the formation of the National Quadraphonic Radio Committee (NQRC) of the Electronic Industries Association. The problem in transmitting four-channel sound via broadcasting is the development of systems which will provide definite separation of the four frequencies and still allow radios not equipped for quadraphonic reception to receive stereo FM and monophonic FM. Without widespread standardization of equipment to receive quadraphonic broadcasting, it is difficult to measure the potential demand for this system in the first place.

Experiments in Quadraphonic Broadcasting

Some attempts have been made, however, to educate the public in the potential benefits of quadraphonic broadcasting. There was one promotional experiment on July 24, 1976. Two San Francisco radio stations, both stereo, joined in a broadcast permitting listeners with two radios, both capable of receiving stereo, to hear what quadraphonic broadcasting of high quality sounded like.[5] Two Sacramento, California stations repeated the experiment by simultaneously rebroadcasting the signals of the two San Francisco stations.[6] Whether quadraphonic broadcasting becomes a reality on a mass scale will depend on public demand, support from quadraphonic equipment companies, and the FCC.

Dimensions of Quadraphonic FM

When there is true separation of the four quadraphonic channels, the listening experience is literally being surrounded by sound. For example, assume you are listening to a live performance of a chamber orchestra. On stage is a violinist, a pianist, a trumpeter, and a flutist. As you listen to the music, all of the instruments are in front of you. If we placed microphones in front of the orchestra, one to the left and one to the right, and broadcast the music in stereo, you could hear the sound simultaneously from the same two directions in your living room.

Now assume we use four microphones. One is directed toward the violinist, the other toward the pianist, the third toward the trumpeter, and the fourth toward the flutist. We now broadcast the music on a quadraphonic four-channel system, and you listen to the music on a radio capable of receiving all four channels. A separate speaker is connected to each channel. If you place one speaker to the left front of you, a second to the right front, and the third and fourth speakers to your rear, the sound will be like sitting in the *middle* of the orchestra instead of in front of it.[7] Now consider a radio commercial in quadraphonic sound. Close your eyes and imagine sitting in a new automobile as the salesperson walks around the car telling you about its features.

Pros and Cons of Quad

Along with the "total" sound experience, proponents of quadraphonic broadcasting use the coming of stereo AM as an argument to develop quadraphonic broadcasting. They claim that with AM stereo, FM broadcasters will need this added quadraphonic dimension to retain FM as a special attraction. More conservative watchers say there is a need to determine first from where the software (records and tapes) will come before leaping ahead with full-scale quad. Moreover, sales figures indicate that quadraphonic systems have not been overly popular for in-home use.[8] Somewhere among all of these arguments lies the future of quad.

am stereo

Like quadraphonic systems, AM stereo is still in the experimental stage. For many years, AM broadcasting has not seriously considered stereo beyond laboratory ventures, primarily for two reasons. The first was the narrow channel width of AM stations, 10 kHz compared to 200 kHz for FM. Second, since FM has not really been a serious competitive threat to AM, there has not been widespread interest in the system. However, as FM gradually has cut a wide swath through the AM listenership, AM broadcasters have begun to search for something which would regain their competitive edge. The answer seems to be AM stereo.

As with quadraphonic, AM stereo is available through many different systems, and the FCC is examining each to determine a compatible standard for the industry. Active evaluation of all systems began in the mid 1970s by the National AM Stereo Committee (NASC) of the Electronics Industries Association. The collection of field performance data on AM stereo was completed in 1977.[9] One of the possible systems proposed by RCA uses both amplitude and frequency modulation. In this system, one part of the signal is transmitted by frequency modulation, the other by amplitude modulation. A special stereo AM radio receives the amplitude modulated signal separately from the frequency modulated signal. As with FM stereo and quadraphonic systems, it is necessary for AM stereo receivers to be able to pick up both stereo and monophonic signals.

By 1980 the FCC had authorized stations KFRC in San Francisco, KHJ in Los Angeles, KCMO in Kansas City, and WOW in Omaha to test AM stereo on the air.

The future widespread use of AM stereo looks bright for two reasons. First, there is considerable backing for AM stereo from the many AM broadcasters who already have felt the competitive sting of FM.[10] Second, the technology is ready. The future will depend on how quickly equipment can be standardized, and how the public accepts it, as demonstrated by the sale of AM stereo receivers.

transmitting tv signals

Our discussion would not be complete without mentioning television transmission. Television stations broadcast on frequencies which are located both above and below the standard FM radio frequencies of 88–108 mHz. Radio waves are used to carry the television picture. The width of the spectrum allocated for television transmission is established by the FCC at 6 mHz. Part of the frequency is used to transmit the video portion of the signal and part of it, the audio portion.[11]

Processing the TV Picture

Earlier in this book we learned about Paul G. Nipkow and his "mechanical" television, which consisted of a scanning disc with holes punched in the pattern of a spiral. When the disc turned, the holes would pass over a small opening through which could be seen a picture. In one complete revolution of the disc, the entire picture would be scanned and transmitted to another receiving unit, which would then reproduce the image. If the image were replaced with another in rapid succession, it would give the illusion of motion.

Further refinements changed the process and with the invention of Vladimir Zworykin's iconoscope tube, electronics picked up where mechanics left off. Using electrons instead of a spiraling disc, the image could be scanned with increasing clarity and speed. The result was the electron-scanning process as we know it today. A simplified example is to imagine a flag with a series of red and white stripes. The scanning process first scans the white stripes and then the red stripes. Now imagine this process taking place 525 times per second with the picture (flag) rapidly changing. The result is a series of rapidly scanned and broadcast pictures which appear on our television set as an illusion of motion.

Processing Color Television

A similar process is used with color television except that the television camera separates the three primary colors of light: red, green, and blue. All other colors are made up of a combination of these three primary colors.[12] When the television camera scans an image, it separates the red, blue, and green hues. These are transmitted individually and then appear as tiny microdots on our television screen. The microdots are too small for us to see with our naked eye, tending to "run together," creating the color picture. This, plus the rapid scanning process, creates a picture in both color and motion.

the path of electromagnetic waves

Melinda drives in and out of the rush-hour traffic looking anxiously for the exit which will send her east on the interstate, over the Allegheny Mountains and on to the Atlantic coast. The beltway seems a little less harrassing with the company of her favorite AM radio station. She decides to keep the station on as long as she receives clear reception. Here comes the exit sign; she is on her way. The station continues to provide clear signals, and Melinda listens attentively to weather reports. A major storm is ahead, but she will not reach it for at least an hour.

About 90 minutes from home, the radio station begins to fade. Finally, it is necessary to push another button on the car radio. This time, she also switches to the FM frequencies. As she turns the dial, she hears another station from her home town. The signal is clear, and once again she settles back to enjoy the drive as she heads up the west side of the mountain. It is beginning to rain when she reaches the summit, and lightning flashes as she heads down the other side. Less than ten minutes after crossing the summit, she loses her home town FM signal and switches back to the AM frequencies to pick up a nearby station. Heavy static garbles the receiver, but she finally finds a nearby signal and learns that the rain should stop in another hour. Sure enough, 50 miles farther, the moon breaks through the clouds and casts a soft glow on the open countryside.

Now she is beginning to feel the strain of the drive. The bright lights from the diner up ahead look inviting. She decides to stop for a sandwich. Walking through the door of the diner, she hears the familiar sound of her home town AM station, the one she listened to when she started her trip more than three hours ago. Melinda tells the waitress not to hurry. She just wants to unwind and listen to the music.

Ground Waves

To understand how radio waves travel, let's retrace Melinda's route. When she was on the beltway and on the interstate approaching the climb over the mountain, she had little difficulty listening to her home town AM station. That station was on the lower end of the electromagnetic spectrum, and the signal from the transmitter was carried partially by *ground waves*, which are *electromagnetic waves which adhere to the contour of the earth* (Figure 4–8). As a result, Melinda was able to listen uninterrupted until the waves finally died out and she had to change stations. In radio terminology, the area covered by the ground wave is the *primary service area*, or "the area in which the ground wave is not subject to objectionable interference or objectionable fading."[13] It is also that portion of the station's signal which is most protected by the FCC when licensing other stations which could interfere.

Sky Waves

When Melinda arrived at the diner, she again heard her home town AM station. The reason the radio at the diner was able to receive the station was because of *sky wave propagation* (Figure 4–8). With sky wave propagation, the *radio waves actually travel into the sky instead of along the earth's contour*. However, they all do not remain in the sky, some bounce off

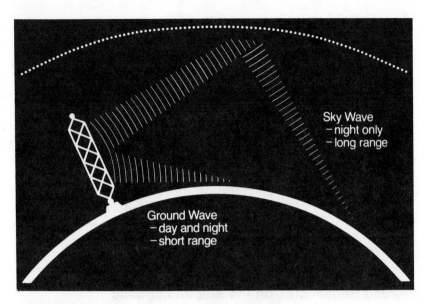

Figure 4–8 Radio waves travel in different patterns. Ground waves stay close to the earth's surface, and sky waves bounce off the upper layers of the atmosphere. Direct waves travel in straight lines, usually adhering to line of sight. (FCC *Broadcast Operators Handbook*)

different layers of the ionosphere and head back to earth. In other words, they are reflected back to the earth's surface. As a result, there is a section on the earth's surface which neither ground nor sky waves reach (Figure 4–8). In Melinda's drive to the coast, this occurred just before she started up the mountain. The diner, however, was in the path of the sky waves. This phenomenon is also referred to as *skip*, and the distance from where a sky wave touches the earth to the transmitter is the *skip distance*.

That area in which a station's signal is heard clearly because of sky wave propagation is referred to as its *secondary service area*. The secondary service area is defined as "the area served by the sky-wave and not subject to objectionable interference."[14]

Do you remember when the signal from Melinda's home town AM station began to fade, and she finally changed stations? At that point, Melinda had reached the *intermittent service area* of the station, or that "area receiving service from the ground wave but beyond the primary service area and subject to some interference and fading."[15]

Melinda reached the diner at night. If she had reached it at noon, the AM station may not have been audible. The ionosphere has different reflective qualities at different times during the day. The sun warms it and decreases its ability to reflect sky waves.

Direct Waves

When Melinda reached the mountain and the ground waves died out, she changed frequencies on the car radio. She also switched to the FM band. Here she was again able to hear clearly a station from her home town, an FM station. At the higher frequency of FM (88 to 108 mHz), are radio waves called *direct waves*, which *travel in a straight line*. As Melinda began to climb the mountain, the antenna on her car was in a direct line of sight position with the transmitter of the FM station. When she crossed the summit, the mountain blocked the waves, and she could no longer hear the station.

Keep in mind that because the FM station was on a higher frequency, the signal traveled in a direct, "line-of-sight" path between the transmitter and the receiver, not because the station was FM. As a general rule, the higher the frequency, the more "direct" the wave propagation will be.

assigning frequencies on the electromagnetic spectrum

Through a long series of policy decisions, based partly on definite planning and implementation and partly on "squatters' rights," the FCC has developed a systematic allocation of the available frequencies on the electromagnetic spectrum.

Allocating TV Channels

For television, the allocations are based on eliminating interference between channels. For channels 2 through 13, this means at least one channel separation between stations serving the same community. However, the development of more sophisticated transmitting and receiving equipment coupled with the demand for more channel utilization may someday change this allocation pattern.

Allocation of AM

Between 535 and 1605 kHz are 107 channels, or "frequency locations," in which AM stations can operate. Each occupies 10 kHz of space; thus, one is located at 540 kHz and every kHz thereafter until 1600 kHz. The characteristics of the frequency determine the ability of a station to reach a geographic region.

To provide maximum opportunity for developing AM staions, to foster free enterprise competition, and to protect those pioneer stations which staked early claims to the airways, the FCC uses three major classifications and numerous subclassifications in assigning frequencies. The major classifications are clear channels, regional channels, and local channels. Channels are specific frequencies (540, 550, 560, 570 kHz and so forth) on the broadcast band.

Clear Channels. A clear channel is one "on which the dominant station or stations render service over wide areas and which are cleared of objectionable interference within their primary service areas and over all or a substantial portion of their secondary service areas."[16] From this definition, it is easy to see that clear channels operate at a high power and have priority over a given frequency. Within the clear-channel allocations are Class I and Class II stations. Class I stations are protected from interference in both their primary and secondary service areas. Class II stations can operate on clear-channel frequencies but must protect Class I stations by either using directional antennas, operating on reduced power, signing off at sunset, or combinations of all three. The maximum operating power of a clear-channel station is 50,000 watts, and the minimum operating power varies from 250 watts to 10,000 watts. Some of the more famous clear-channel stations include KDKA in Pittsburgh, Pennsylvania and WSM in Nashville, Tennessee.

Regional Channels. Regional channels are assigned where several stations operate, none having a power of more than 5,000 watts.[17] There are many more regional channels than clear channels. The primary area serviced by stations operating on regional channels is the city or town in which the station is located and its adjacent areas. Regional channels are found over the entire range of the AM band, yet their specific channel assignments do not duplicate those of clear channel stations. Their secondary coverage areas and sometimes portions of their primary coverage areas are not protected from interference as are those of clear channel stations.

Local Channels. Local channels usually are located at the upper end of the AM band and operate with power no greater than 250 watts at night and 1,000 watts during the day. Some are required to sign off at sunset to protect other stations, even those which also operate on local channels. Local channel stations are the backbone of medium- and small-town radio. Many operate with a maximum power of 250 watts at all times, and their programming runs the gamut from automated country-western to want-ad radio.

Allocation of FM

The allocation of FM frequencies is similar to that of AM. However, the maximum range of an FM station, regardless of its power, is usually to the horizon or to a distance of about 70 miles. Generally, the lower end of the FM band between 88.1 and 91.9 MHz is allocated to noncommercial broadcasters.[18] The remaining portion of the band is allocated to commercially operated FM stations. Commercial FM stations also fall into station classes. As a rule of thumb, Class C FM stations serve the widest area with the highest power, regulated to a maximum of 100,000 watts. Class B and Class B-C FM stations serve smaller communities. Noncommercial FM stations fall into the three classes assigned to commercial FM stations but have an additional Class D category which is designated for noncommercial stations not exceeding 10 watts.

directional and nondirectional stations

The importance of any station's frequency allocation is its ability to serve a given population base without interference from other stations. The FCC is careful to require special antenna systems to prevent interference wherever possible. Although the problem is not yet critical in the FM band, spectrum space is at a premium. To avoid interfering with other stations, many AM stations operate as either directional or daytimer stations or as a combination of both. Daytimer stations are those which are authorized to operate only between sunrise and sunset, local time. If the station were allowed to operate past the sunset hour, its signal would travel great distances because of the reflective qualities of the ionosphere.

Directional antenna systems are designed also to protect against interference. The next time you are out for a drive, scan the horizon for a cluster of antenna towers. That cluster has a purpose. The strategic location of the towers in the cluster permits the signal from one primary tower to be radiated among the other towers, creating a specially shaped, broadcast-coverage contour that does not interfere with the other stations' contours (Figure 4–9). Broadcast-coverage contours come in all shapes and sizes, all designated by the FCC to maintain relatively interference-free airwaves.

international spectrum management

The need for international agreements sharing the available space on the electromagnetic spectrum is not new. When Marconi set up shop on Newfoundland's coast for a transatlantic broadcast, the Anglo-American Telegraph Company of Newfoundland promptly told him if he did not "remove his apparatus forthwith," he would face an injunction. Such rigidity has been softened over time by numerous agreements between nations on not only experiments but on allocation of frequencies.

Some of these are coordinated by the International Telecommunica-

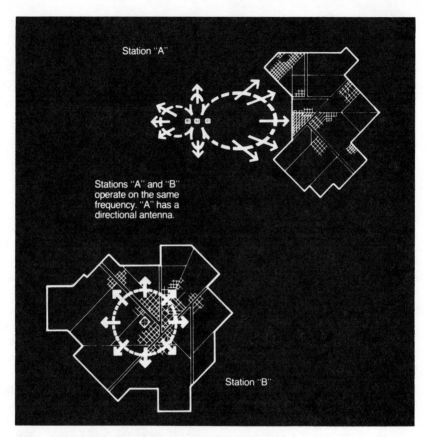

Figure 4–9 Directional antennas permit stations to operate on the same frequency without interfering with each other. (FCC *Broadcast Operators Handbook*)

tion Union, a United Nations organization responsible for worldwide coordination of frequency use.[19] The ITU's origin dates back to 1865 with the formation of the International Telegraph Union in Paris. It was not until 1906, however, that the organization considered the issue of wireless communication. Then the medium captured the world's enthusiasm as well as provoking its ire. As we learned in chapter 2, Germany accused the Marconi Company of monopolizing the wireless. From these charges came the Berlin Conference of 1906 and the first major international agreement on the use of the airwaves. In this spirit, eighty-five countries met in Geneva, Switzerland in 1959 and signed an agreement which became the governing body of the ITU. The ITU's International Frequency Register keeps track of the use of every available space on the electromagnetic spectrum. Periodic World Administrative Radio Conferences (WARC) are held to review decisions and formulate future use of the spectrum, the most recent being in 1979. We shall learn more about the ITU in chapter 16.

The United States entered into the North American Regional Broadcasting Agreement (NARBA) in 1937 at the Inter-American Radio Convention in Havanna, Cuba. Although the NARBA contemplated working relationships for such broadcasting policies as standards for engineering practices, these relationships exist only between the United States and Canada.[20]

summary

Radio waves can be compared to a point on the electromagnetic spectrum, a yardstick of electromagnetic energy which includes such forces as microwaves, light waves, and X-rays. Radio waves travel at the speed of light at a frequency dependent on the length of each wave. They vary in amplitude and frequency. Variations in amplitude are used to modulate radio waves in the AM broadcast band, and variations in frequency are used to modulate radio waves in the FM broadcast band.

Because of the greater width allocated to FM stations, both stereo, or dual-channel, and quadraphonic, or four-channel, broadcasting are becoming popular. In an effort to retain a competitive edge, AM proponents are developing AM stereo.

Radio waves are of three types: ground, sky, and direct waves. The Federal Communications Commission, through its assignment of frequencies, guards against signal interference of these waves between radio and television stations. The FCC in cooperation with other regulatory agencies, through the State Department, and with other countries, participates in an international effort to use responsibly the spectrum throughout the world.

spotlight on further learning

BECK, A. H. W., *Words and Waves: An Introduction to Electrical Communication.* New York: McGraw-Hill Book Company, 1967.

KITTROSS, JOHN MICHAEL, *Television Frequency Allocation Policy in the United States.* New York: Arno Press, 1979.

LEVIN, HARVEY J., *The Invisible Resource: Use and Regulation of the Radio Spectrum.* Baltimore: The Johns Hopkins University Press, 1971.

PIERCE, JOHN, *Electrons and Waves: An Introduction to the Science of Electronics and Communication.* New York: Doubleday & Co., Inc., 1964.

5

broadcasting's use of microwaves and satellites

FOCUS   After completing this chapter we should be able to

Explain how microwaves are used for such things as television programming, satellite relay, cable and pay-subscription TV.

Explain how microwaves are used in electronic news gathering, and in educational, and industrial television.

Trace the role of satellites in early television programming.

Describe the theory of the synchronous orbit satellite.

Describe the uses of NASA's Application Technology Satellite and Canada's Communications Technology Satellite.

Discuss the international implications of direct-broadcast satellites.

The atmosphere and standard over-the-air broadcasting are necessary to deliver broadcast messages to the public. Other important parts of the worldwide system of broadcasting are applications of microwave technology, satellites, and telephone systems. In any broadcasting day, the chances are excellent that the broadcast messages we receive traveled on one or all of these three carriers. In this chapter, we shall examine how each is applied to broadcasting and how they join to carry the broadcast message.

using microwaves

Not all broadcast transmission and relay are at standard AM, FM, and TV frequencies. The electromagnetic spectrum also is used at much higher frequencies, in the thousands of megahertz range. It is in this area that *microwave* transmission is found. Microwaves are *very short waves*. We have learned that the higher the frequency is, the farther the electromagnetic waves will travel in a direct line-of-sight path between transmitter and receiver. Thus, microwaves always travel by line-of-sight transmission.

Microwaves also allow many more channels of communication to operate because of its shorter wavelengths. Because the waves are shorter, many more will fit into the same space on the electromagnetic spectrum. Many thousands of channels are possible, and the frequency width is in the billions of cycles per second. When we realize that an AM radio station is allocated a width of only 10,000 cycles, it is easy to see how much more information can be transmitted at higher frequencies. Microwaves are necessary to bring us our evening television programs or to relay satellite pictures across oceans. The current state-of-the-art in microwave technology has barely scratched the potential of this multifaceted carrier of information, much of it affecting the broadcasting industry. We'll now examine the uses of this technology in more detail.

Relaying Television Programming

A network television program may travel many thousands of miles before it reaches your local television station. The path it follows may very well use microwave relay systems. These dishlike antennas dot almost every landscape from the roofs of skyscrapers to the peaks of snowcapped mountains. Using high frequency line-of-sight transmission, these systems can carry crystal clear signals over long distances through a series of relay antennas which are approximately thirty miles apart. Their advantage is the lower cost and increased efficiency of transmission over traditional land-line systems.

Consider, for example, a television station in the rugged Colorado mountains which receives its network signal from Denver. To string a cable over the Rocky Mountains would be far too costly. Instead, microwave towers on mountain tops (Figure 5–1), all within sight of each other, become the path over which the signal travels.

But mountain country is not the microwaves' only domain. Because

Figure 5–1 Microwave antennas, such as these atop a mountain near Boulder, Colorado, provide long-distance communication links without wires. (AT & T Co.)

flat areas are free from natural obstructions, microwave systems are scattered over the plains of the farm belt and the deserts of the southwest for an efficient transmission system.

Keep in mind that the program you receive in your home does not arrive there directly by microwave. The local television station receives the signal via microwave then retransmits it to your home receiver at a frequency regularly assigned to television transmission. You also should be aware that the television station probably does not own the microwave system, but rather rents its frequency, just as it would rent a line from the telephone company. Many private companies including major telephone companies, own microwave systems. These systems currently comprise about 30,000 transmission miles in the United States alone.

Satellite Relay Systems

Ground-based microwave systems are not the only route used to carry television programming. Widespread use of microwave satellite transmission is evident in any week's television fare, with the familiar "live via satellite" message across the bottom of your television screen. Even if you do not see a live satellite-relay picture on the evening news, the chances are that foreign correspondents' reports were first fed by satellite to the network's headquarters, videotaped, and then played back as inserts in the evening news. Because space is a vacuum, microwaves travel over long distances unimpeded by the earth's heavy atmosphere near ground level. This permits a transmitting station in London, for example, to transmit a television picture by microwave to a satellite thousands of miles in space, which relays it back to an earth-receiving station in the United States. Satellites are also used to bring television signals to many outlying regions in which even microwave links would be too costly (Figure 5–2). We shall learn more about satellites later in this chapter.

Cable and Pay-Subscription TV

Although you may receive television programs through your local community antenna television system, often called *CATV* or *cable*, the original signal probably reached the local cable company through a microwave-relay link. A cable company often leases a microwave channel, receives the direct-broadcast signal from a television station, then retransmits by microwave many hundreds of miles to a cable system in a community far removed from the original television station (Figure 5–3). In fact, this use of microwave is one reason cable systems have come under FCC jurisdiction.

A newer use of microwave in certain metropolitan areas sends special programming to pay-television subscribers. An example of this is the

Figure 5–2 The Valdez earth terminal of RCA Alaska Communications is typical of the remote communication links provided by microwave and satellite communication. (Scientific-Atlanta)

Figure 5–3 Microwave connections permit cable systems to receive signals from distances far greater than those stations can broadcast. (NCTA)

Microwave Interconnection
Of Cable Television Systems

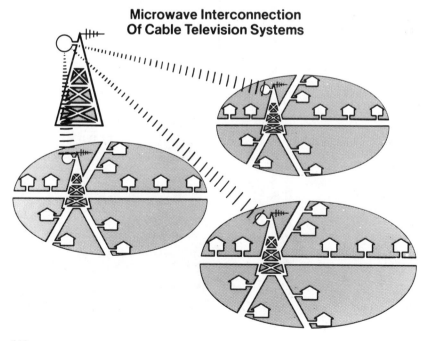

Chicago-based, pay-TV channel.[1] From a studio control center, a signal is sent via microwave to a receiving dish antenna on top of a Chicago skyscraper. From there, it is retransmitted by microwave to other microwave-receiving antennas on top of high-rise apartment complexes and hotels. Hotel and apartment television sets are then connected by cable directly to the microwave-receiving antenna on top of the building. These systems not only can receive but also can initiate live programming from a studio in the way a local television station can. The studio programs are then sent over the system to subscribers. Also called subscription television, the system is gaining popularity as an alternative to over-the-air broadcasting.

Electronic News Gathering

The application of microwave to electronic news gathering has given television the flexibility that only radio once enjoyed. A mobile van and portable camera (Figure 5–4) can provide live programming from a

Figure 5–4 Through well-placed microwave links, a television station has the ability to broadcast live news from virtually anywhere in its coverage area. News crews theoretically can go on the air live from almost anywhere in the world. (KCMO-TV)

community arts fair or live aerial scenes of a football stadium, and the unrehearsed moments of a newly elected politician's acceptance speech can give television news a dynamic dimension. New developments in microwave ENG systems now bounce the microwave beam off the side of a building to a relay antenna, permitting the eyes of a live television camera to peer into almost every nook and cranny of the largest metropolitan areas.

Educational and Industrial Television

In Salt Lake City, Utah, a group of students watches an instructional television program sent from a classroom forty miles away at Brigham Young University in Provo.[2] The audio and video transmission between the two schools is made possible by a microwave link. Students in Salt Lake City can even answer the professor in Provo by microwave. In Indiana, all of the state-supported universities are connected by a microwave link, and they thus can exchange their instructional television programs.[3]

Businesses using television use similar microwave hookups. A plant with two locations in the same city can use microwave for an intracorporate television link. With this, special sales-training seminars can be broadcast from the main corporate television studio to a special seminar room on the other side of town. Or using the leased facilities of a national microwave system, a special videotaped management-training program can be distributed to plants throughout the nation. Businesses and television stations alike also can use microwaves to transfer computer data, providing connections between systems hundreds and even thousands of miles apart.

When microwave systems were first developed, they opened up whole new frontiers for communication. The wide channel width permitted much more information to be sent than was possible with lower frequency systems. Microwave systems were also relatively free from interference. In many ways, they were superior to cable systems, especially for long-distance communication.

The existing, well developed microwave systems will undoubtedly remain in service. Yet two words stand squarely in the way of microwave: fiber optics. These microscopic glasslike fibers through which light passes use laser beams to carry information equivalent to what can be transmitted on thousands of cable channels. In fact, fiber optics now permit telephone systems to compete with cable companies and microwave systems as major carriers of information.

The future interrelation between fiber-optic technology and our need and desire for its potential will determine the future of worldwide broadcasting systems. It will also determine how much regulatory control we exercise over its use and whether such multibillion-dollar corporations as AT&T will be the true "gatekeepers" of the future.

satellites and broadcasting

The wedding between satellites and broadcasting took place on a warm New England evening in Andover, Maine.[4] From this outpost the first television pictures were relayed by satellite across the Atlantic to Europe on July 10, 1962. The pictures were of then Vice President Lyndon B. Johnson and several AT&T officials gathered in the Carnegie Institute Auditorium in Washington, D.C. The event covered many columns in major newspapers in the United States and Europe as well as news bulletins on the television networks.

Telstar

The hero of the evening was a 170-pound payload named Telstar (Figure 5–5). It had been launched into space in a cooperative effort by NASA and AT&T. This beach-ball sized satellite sent signals to television stations in France and England by first receiving them from United States's earth-

Figure 5–5 Telstar, launched in 1962, ushered in the era of live satellite-relayed television pictures. (AT & T Co.)

based antennas, amplifying them ten million times, and retransmitting them to European receiving stations. Telstar was powered by solar energy. On any given pass over the United States or one of the European receiving stations, the satellite was in viewing range for only about 45 minutes. The highest point in Telstar's orbit, the apogee, was 3,502 miles away. The closest point to earth during its orbit, the perigee, was 593 miles away. Directing the project for AT&T was 44-year-old Eugene Frank O'Neill, a Columbia University engineering graduate.

In return, the United States received signals from Europe the following evening. From France came a seven-minute taped program with an appearance by Jacques M. Marette, the French Minister of Postal Services and Telecommunications, and musical entertainment by French performers. The British signal followed shortly thereafter, consisting of a test pattern and a live broadcast by Britain's Deputy Chief Engineer, Captain Charles Booth.

The Politics of Telstar

It is not surprising that the breakthrough of transoceanic broadcasting brought with it a series of political issues, some based on age-old rivalries, others on contemporary concerns. The most intense rivalries were between England and France, long-time sparring partners in economic and political warfare. Telstar merely set the scene for their combat. On the night of the first transoceanic broadcast, British pride was hurt when its receiving station was not able to monitor clear audio and video signals from the United States. The French, meanwhile, using a station that was not supposed to be ready for tests, monitored clear signals which were described as though they came "from about twenty-five miles away." The British, however, did achieve a victory that night when they relayed a live television program to the United States in contrast to France's taped program.

In the United States, the press was quick to report that AT&T had paid NASA to launch Telstar. This agreement also called for the free availability of any inventions arising from the Telstar project to any company that wanted them. President John F. Kennedy formally called for a national corporation to oversee all satellite communication development in the United States. The Kennedy administration also emphasized the need for commercial broadcasters to participate in examining the potential of satellite communication. That bit of political rhetoric tried to pull down the fence that FCC Commissioner Newton Minow had put up a year earlier between the administration and commercial broadcasters with his famous "vast wasteland" speech before the National Association of Broadcasters' convention.

Economic and Social Implications of Telstar

Telstar created much more than international television communication. The morning after Telstar's broadcast, AT&T became the most active stock on the New York Stock Exchange with 105,800 shares traded. Opening at 109⅞, it closed at 113¼, the day's high and a gain of 3½ points. Less than a month later, David Sarnoff, chairman of the board of RCA, proposed the concept of a single company to deal with international communication matters.[5] Western Union quickly supported Sarnoff's suggestions, saying it had proposed similar concepts before.[6] The concept, along with the Kennedy administration's support, was accepted with the passage of the Communications Satellite Act of 1962 and the formation of the Communications Satellite Corporation called COMSAT.

The prospect of domestic television programming crossing national boundaries opened up a new arena for discussion and heated debate. The vast wasteland was one thing at home but something entirely different when it reached France and England. The initial Telstar broadcast itself caused some concern. The program was produced by AT&T, which provoked CBS to break away from the initial network-pool coverage and not to carry the remarks of AT&T board chairman Frederick R. Kappel. Jack Gould, television critic of *The New York Times*, said of the event, "The sight of Government dignitaries serving as a passive gallery for private corporation executives was not very good staging, particularly for consumption in foreign countries." Gould went on to predict, "The crucial decision that will determine the lasting value of international television—a willingness of countries and broadcasters to clear the necessary time on their own screens to see and hear other peoples of the world—cannot be made in laboratories in the sky but in offices on the ground."[7]

Programming beyond Telstar

The early 1960s brought many international satellite experiments. Perhaps the most vivid occurred eighteen months after Telstar when the funeral procession of President John F. Kennedy in Washington, D.C., was seen in the halls of the Kremlin. By 1965, the Ecumenical Conference in Rome had been seen on both sides of the Atlantic.[8] European viewers saw and heard Washington dignitaries react to the unveiling of the Mona Lisa in the National Gallery of Art. Special programming from the 1964 Olympic Games in Tokyo traveled far beyond Japan.

stopped in space: the synchronous orbit satellites

A little more than a year after Telstar, there was another breakthrough in satellite technology. At Lakehurst, New Jersey, in 1963, a crew waited for a satellite called Syncom II to "lock" into position for a transmission that would be heard half-way around the world on a ship stationed at Lagos, Nigeria.[9]

Out of a static-born receiver aboard the U.S. Navy's *Kingsport* came the clearly heard words, "*Kingsport*, this is Lakehurst. *Kingsport*, this is Lakehurst. How do you hear me?" The words came from space, relayed back to earth from Syncom II, no ordinary satellite. The technical crews did not have to adjust their receiving equipment this time as the satellite passed within viewing range, because Syncom II was technically "stopped in space," the first successful synchronous orbit satellite. Its baby sister, Syncom I, had failed six months earlier. So, for Hughes Aircraft Company engineers Harold A. Rosen, Donald D. Williams, and Thomas Hudspeth, the team chiefly responsible for the satellite's development, Syncom II's success was especially welcome. This is how it worked.

Before Syncom II, scientists could utilize the communication capabilities of a satellite only when it passed over a given region of the earth. This meant it could be used for only about 45 minutes at a time. However, scientists felt if they could (1) position a satellite over the equator, and (2) place it at the right height (22,300 miles) at which it would travel at a speed similar to the earth's rotation, it would appear stationary above the earth. The launch of Syncom II achieved the desired position by placing it in *synchronous* orbit, also called *geo-stationary* orbit, over the equator. Since Syncom II, major communication satellite systems have used the synchronous orbit positions to create so-called microwave towers in space, permitting worldwide transmission of television signals, computer data, and telephone service.

tv and INTELSAT

Satellite communication mushroomed after the success of the synchronous orbit satellites. Entire plays from Europe could reach a living room in North America. The Olympics could travel from one side of the earth to the other. Understandably, international development and cooperation became paramount. COMSAT, formed by the Communication Satellite Act of 1962, quickly became the manager of the International Telecommunications Satellite Organization called INTELSAT, a group of countries which jointly owned a group of satellites.

On April 6, 1965, the first step in this international system was taken with the launch of an Early Bird satellite called INTELSAT I. It was positioned over the Atlantic and permitted live commercial television broadcasts on a regular basis for the first time. Early Bird was just one in a series of INTELSAT satellites circumnavigating the earth, providing a global system of communication. Construction of an INTELSAT V (Figure 5–6) system began in 1976 for a 1979 launch.

Figure 5–6 The Intelsat V spacecraft measures approximately 50 feet from the end of one solar panel "wing" to the other and has twice the capacity of previous Intelsat spacecraft. (Aeronutronic Ford)

etv and application technology satellites

Commercial television is not the only benefactor of satellite systems. In the United States, NASA developed the Application Technology Satellite program (ATS) for educational purposes.[10] It launched six such satellites in all, with the sixth sending satellite signals to small earth stations. This was the key to important applications of satellite technology, from log cabin schools in the mountains of Idaho to mud huts across the world in India.

In the United States, such towns as Gila Bend, Three Forks, Battle Mountain, Wagon Mound, Sundance, and Arapahoe, all participated in the ATS experiments. The local school yard had a new visitor in the form of a microwave dish and its strange antenna with cork-screw wires around a long metal tube. With the help of their visitor, students could sit in a classroom in West Yellowstone and talk via satellite to a classroom in Denver. Where no land-line or microwave system had been developed, the ATS-6 satellite (Figure 5–7) would beam a signal simultaneously across half a continent.

The program generated a cooperative, although somewhat reluctant,

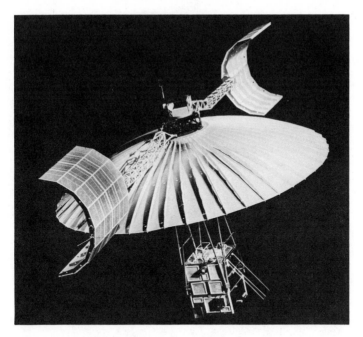

Figure 5–7 ATS-6 began a new series of educational television experiments first bringing direct in-school programming to outlying areas of the Rocky Mountains. (NASA)

effort between state governments. For example, issues affecting the local autonomy of schools usually formed a political thicket. But the Federation of Rocky Mountain States, Inc., composed of Arizona, Colorado, Idaho, Montana, New Mexico, Utah, and Wyoming, joined together to bring two-way educational television to the outlying communities in the eight-state region. In some areas, the satellite-receiving systems were hooked directly into the local cable systems or microwave translator systems, permitting the signals to be received at home. The ATS project also provided similar educational television programming for the Eastern United States, especially Appalachia.

After the American experiments, special earth-controlled rockets on the satellite shifted the ATS-6 from the Galapagos Islands to a new orbit over Kenya in Africa. From that position in 1976, it conducted educational television experiments in India where some of the population had never even seen television or motion pictures. Programs on modern agriculture, health, and family planning were part of the television fare. After the Indian experiments, the satellite traveled back into the western hemisphere, stopping along the way for demonstrations sponsored by the United States Agency for International Development (USAID).

domestic satellite systems

Not all satellite systems belong to or serve the purposes of an international audience. Although not in synchronous orbit, the first domestic satellite system named Molniya was developed by the Soviet Union and launched in 1965. Canada captured the first synchronous orbit satellite honors for its domestic communication satellite named Anik, launched in 1972. Western Union claimed the American satellite "first" with the launching of its two Westar satellites in the 1970s (Figure 5–8). The Westar system made Western Union the first United States corporation to use its own spacecraft. The company constructed earth stations at key points in the United States and hooked them into the regular land-line Western Union system. Television pictures can be carried by Westar as can telephone calls and computer data.

RCA also plunged into satellite communication and inaugurated America's first domestic satellite communication system in December, 1973 using leased channels first on Canada's Anik II and later on Westar II. RCA put its own satellites into orbit, Satcom I and Satcom II (Figure

Figure 5–8 Western Union's Westar satellite system showing its integration with the Western Union microwave network. (Western Union and *Communications* News)

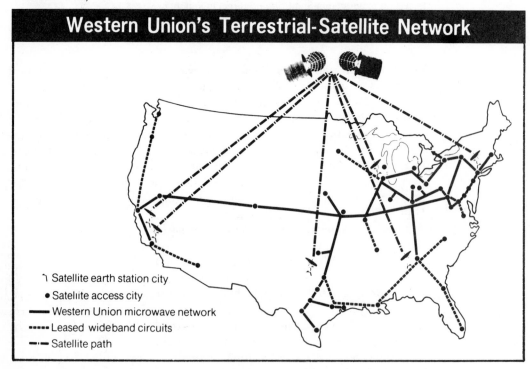

Figure 5–9 RCA's Satcom satellite, which is part of RCA's domestic satellite system. Satcom II has 75 square feet of solar cells mounted on two panels continuously pointed at the sun. The cells produce 740 watts of power sufficient to charge its batteries and drive the operating functions of the 1,000-pound spacecraft. Satcom III, operational in 1980, is devoted primarily to serving cable television systems.

5–9). A subsidiary company, RCA American Communications, Inc., owns and operates the domestic satellite system.

In 1976, NASA launched a new COMSTAR satellite system which is leased by AT&T from COMSAT. Each satellite has a 14,400-circuit system, and four satellites are planned for orbit by the 1980s. Each satellite is expected to remain operational for about seven years.

tv station in the sky: Canada's CTS satellite

The domestic and international politics of satellite communication are mind-boggling. There, positioned 22,300 miles away in space, is a television station that can beam a signal to any home in the world with the proper receiving antenna. This is the ultimate in direct broadcast satellites. The programs are received not through a cable or translator microwave system, but rather right onto the rooftops where an antenna, called a "dish," the size of an automobile tire, assimilates signals from space. That is exactly how Canada's Communications Technology Satellite (CTS) works, and the world is watching its progress.

The CTS idea is not new. But think of the implications. Theoretically, CTS enables any local television station serving a small community to beam its signal into the sky and retransmit it into every home on an entire continent. Launched on January 16, 1976 in cooperation with NASA, the CTS system is the basis for many ongoing experiments, including some in the United States. An Atlanta, Georgia, station is already beaming satellite-relayed programs to cable systems.

This is just the beginning. The world of 1980 sees Japan with a similar system beaming two television channels down to home receivers.[11] The receiving dish antenna rising above Tokyo will cost a relatively inexpensive $500, the key to bringing the general public into direct-broadcast satellite programming. For Germany, it is a satellite system three-and-a-half times the size of Canada's CTS. Other countries are sure to follow.

international implications

Satellite technology is only one part of the whole picture. Major issues of culture, politics, economic development, and similar topics are also at the heart of satellite communication development.

Intercultural Considerations

What will happen when the world communication systems are developed to the point at which the dish antenna on the roof is as common as the television set in the living room or the radio in a car? American programs in foreign countries have become popular overnight. Conversely, foreign programs broadcast in America have had similar effects. The evening news reports a president's visit to the Peoples' Republic of China, and major department stores immediately feature Chinese fashions. Stop and contemplate the cultural fusion of evening television fare of Russian lessons from Moscow, sports events from Germany and China, and a cooking demonstrations from Egypt. What will happen when societies are bombarded with dozens of cultural stimuli every day?

Communications attorney Leonard H. Marks asked similar questions in 1965:

Will the man in the street in New Delhi be asked about the Hindu-Moslem problem so that the factory worker in Detroit will have a first-hand report? Will programs of this type be designed to encourage a common language and break down the barriers which currently exist for communications between peoples of different

> *cultures? Is there any reason why we shouldn't feature international "town meetings of the air" with participants from Berlin, Rome, Cairo, or other distant points with their counterparts in Des Moines, Seattle and San Francisco?*[12]

Such questions are at the heart of the direct-broadcast satellite programming issue.

Political and Economic Implications

Direct-broadcast satellites are an example of technology advancing faster than legislation can deal with it. Today satellites provide a multitude of services (Figure 5–10). The world currently does not have a governing body, commission, or regulatory structure to keep up with this rapidly changing technology. Back in 1960, Dallas Smythe, former head of the FCC's economics division, predicted this dilemma: "The danger inherent in the development of space-satellite communications lies in the additional strain it will place upon international relations in the absence of international agreements on policy and organizations to control its use."

Figure 5–10   The multiple uses of satellite technology. (Scientific-Atlanta)

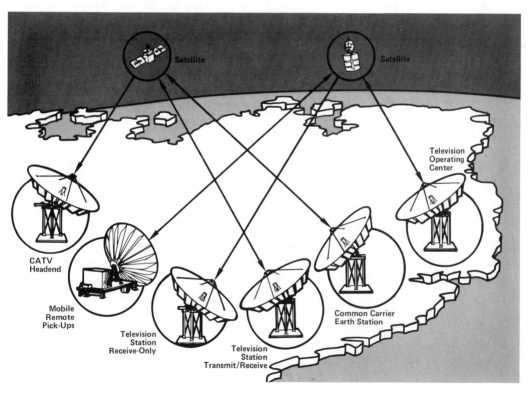

Satellite

Satellite

Television Operating Center

CATV Headend

Mobile Remote Pick-Ups

Television Station Receive-Only

Television Station Transmit/Receive

Common Carrier Earth Station

Predicting cold-war rivalry, he said, "The first power to begin extensive use of this new means of communication will initiate the deadly cycle. The second power will then try to outdo the first with a rival space-satellite communications system, and so on until international agreements are almost impossible to achieve."[13]

Problems are already developing, especially over the issue of what rights countries have in sending or receiving broadcasts across international boundaries. Mexico has already banned certain American television programs which it felt were too violent. Canada has taken economic steps to curtail American commercials. But these are only two countries. What happens when a Baptist church service in Alabama reaches a sacred culture in Thailand? How can a dictatorship retain power with massive amounts of televised propaganda from a democracy? Conversely, can a democracy succeed in the wake of a dictator's propaganda? What happens when the FCC rules that certain films are too sexually explicit for American television fare, yet every American living room has instant access to foreign, X-rated television films?

What will direct-satellite broadcasting do to the world economy? The local marketplace can literally become worldwide. A major department store with international branches can choose a popular world television program on which to advertise. A media buyer for an international ad agency may have to choose between purchasing time on a London channel or on one in Micronesia. Theoretically, a popular world program produced by a small, less-developed country could attract enough world advertising to affect significantly that country's balance of payments.

What happens when a television network moves its entire headquarters to a small country on the other side of the world where labor costs are a fraction of what it was previously paying? With direct-broadcast satellites, the network can reach the viewers back home with scarcely more effort or cost than it took to operate the domestic system. With this in mind, the research firm of Arthur D. Little, Inc., predicts that "the potential for competitive advantage afforded by the use of satellites might ultimately lead to the elimination of local TV broadcast stations."[14]

The AT&T—COMSAT Interface

Perhaps nowhere do economic issues surface faster than in the arena in which two corporations fight for the rights to operate future communication technology. Two companies with some of the biggest stakes are the American Telephone and Telegraph (AT&T) Corporation and the Communications Satellite Corporation (COMSAT). The former is a private industry, the latter quasi-governmental, formed by the Communications Satellite Act of 1962.

COMSAT is responsible for overseeing satellite development in the

United States. It even supplies AT&T with satellite circuits, and AT&T in turn, leases them to its users. Satellites are COMSAT's only business, but AT&T not only operates satellites but also ocean cables (Figure 5–11). Herein lies the crux of the problem. Will messages be sent via satellite or cable?

If satellites are used exclusively, then COMSAT benefits. If the government continues to permit a percentage of all messages to be sent by cable, then, theoretically, COMSAT could suffer. Of the two, satellites have the greater number of circuits. AT&T's COMSTAR satellites have over 14,000 circuits each, but its transatlantic cable handles only about 4,000 circuits. The complexity of which one is more economical is compounded by the fact that the life of a cable is much longer than the life of a satellite; yet cables are much more costly to construct and operate.

Some writers have argued that AT&T is taking advantage of a regulation that permits them to set user rates according to the investment necessary to develop the system.[15] Similarly, COMSAT's position is succinctly stated in its own report to stockholders: "Any policy of the FCC

Figure 5–11 Ocean dredges lay thousands of miles of cable which serve many of the functions of satellites, aggravating the discussion of the regulatory framework and cost-effectiveness of the two communication systems. (AT & T Co.)

which would permit COMSAT's carrier customers to bypass satellites in favor of new cables would have an adverse impact on the growth of COMSAT's international traffic."[16] The future of both carriers will depend on the future direction of telecommunication policy.

summary

At very high frequencies on the electromagnetic spectrum are microwaves which have many applications to broadcasting. They relay television programs between stations and networks. They also are used by satellites to beam television signals around the globe, and by cable systems to import distant television signals for redistribution to cable subscribers. Microwave technology is also used in electronic news gathering and in educational and industrial broadcasting.

Two other systems that carry broadcast messages are satellites and telephones. The Telstar satellite, launched in 1962, was the first to broadcast international television signals. Since then, many satellites have been applied to broadcasting. Synchronous orbit satellites provide twenty-four-hour communication among almost any points on earth, making continuous, live television coverage of international events a reality. Application Technology Satellites (ATS) beam educational television programs to outlying areas of the world, and domestic satellites provide a new national communication system for many countries. Telephone systems are also used extensively in broadcast communication. The future of both satellite and telephone systems for broadcasting depends on the developments in telecommunication policy regulating the two carriers.

spotlight on further learning

PELTON, JOSEPH N., *Global Communications Satellite Policy*. Mt. Airy, Md.: Lomond Books, 1974.

PELTON, JOSEPH N., and MARCELLUS S. SNOW, eds., *Economic and Policy Problems in Satellite Communications*. New York: Praeger Publishers, Inc., 1977.

SNOW, MARCELLUS, *International Commercial Satellite Communications*. New York: Praeger Publishers, Inc., 1976.

part

3

Systems

6

cable

FOCUS After completing this chapter we should be able to

Trace the development of cable television and to identify the components of a cable system.

Understand how two-way cable works.

Compare the arguments for and against pay cable.

Explain cable radio.

Identify the many types of locally produced cable programming.

Understand the concept of local access.

Discuss the differing opinions of cable as expressed by broadcasters and cable operators.

Discuss the future of cable.

Understand the economics of cable systems.

Understand the human implications of our cable environment.

It is 1940, and television is in its infancy. Large, bulky home receiving sets strain to tune into the preciously few television stations broadcasting the magic of pictures over the airwaves. And strain they do. In fact, if you live in a remote area, it takes a large, well directed rooftop antenna even to focus on a picture. Rooftops are a maze of aluminum and steel.

Still, the excitement of this medium does not dampen your spirits. Instead you purchase numerous antennas. Newspapers are filled with ads claiming this or that make or model of antenna will give you clear reception. Stores even sell indoor "rabbit ears," which are two telescopic rods about three feet long connected to a base that sits atop your television set. For the most part, though, "rabbit ears" cannot compete with outdoor antennas.

the concept and its beginnings

Finally, someone realized there must be a better way. That way is with *community antenna television,* commonly called *CATV* or cable. The concept is simple: Erect a single tower and antenna on a high elevation and then run cables from that tower to individual homes. The result is clear reception without the housetop clutter of antennas. Five hundred companies which owned the antennas and charged a fee to people who want a hookup were created. A new medium was born.

The idea of cable was especially important to people in hilly or mountainous country. Because both television video and audio signals are broadcast at a relatively high frequency, they travel in an almost straight line from the transmitter. When there is a mountain between the television station's transmitting antenna and a home antenna, it blocks these signals (Figure 6–1).

Today, cable is advantageous to people other than those living in mountainous regions. Philadelphia cable subscribers, for example, can receive New York City's television channels. Although over-the-air reception necessitates one-channel spacing between stations to avoid spillover interference, cable does not. Also, building structures such as high-rise apartment houses, tightly spaced row houses, or clustered office buildings can obstruct even local television station signals. Consequently, cable has become popular for urban reception as well as for distant signals.

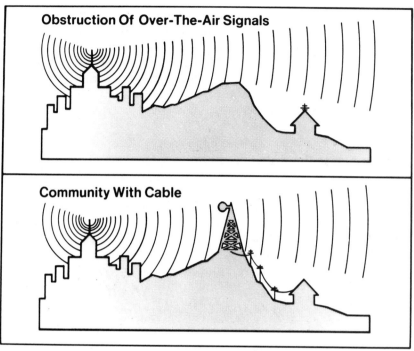

Figure 6-1 (NCTA)

Starts in Oregon and Pennsylvania

There was a bit of friendly rivalry in the origin of cable television. Two
people claim that famous first, one in Oregon and one in Pennsylvania.
L. E. Parsons is credited with a working cable system in Astoria, Oregon
in 1948. John Walson is believed to have had a cable system operating in
Mahoney City, Pennsylvania that same year. Some of the discrepency
results from the definition of what is, or was, a true cable system. Parsons
apparently did build a reception system but sold it to small cooperative
groups.[1] The system did not operate on a monthly fee basis until
sometime after August, 1950. In 1952, fire destroyed Walson's records of
operations in Mahoney City. Research efforts thus far have failed to turn
up bills, newspaper accounts, or other documentation. But Walson and
others are unequivocal about the 1948 operation in Mahoney City.

The Size of the Industry

Since those first beginnings in 1948, cable has grown considerably.
Although it is still a long way from being connected to all of the television
sets in use, it has developed to the point at which there are enough

139

subscribers to make it profitable. Recent figures indicate approximately 3,500 cable companies operate in the United States today.[2] Those companies reach about 10 million subscribers, with revenues fast approaching $800 million annually. Pennsylvania has the largest number of cable systems, about 300, and California has the most subscribers, about 1.3 million. The cable company with the most subscribers is Tele-PrompTer with over 1 million.

components of the cable system

To understand how a cable system operates, let's examine its parts (Figure 6–2). The center of any cable system is the *head end*. The head end is a combination of human beings and technology. The human side includes the personnel who actually operate the cable system. The head end's technical components include the *receiving antenna* which receives the signals from a distant television station. The receiving antenna system usually is a tall tower on which are attached a number of smaller antennas specially positioned for receiving the distant signals. The tower can be located anywhere from a hillside outside of town to the top of a mountain

Figure 6–2   (NCTA)

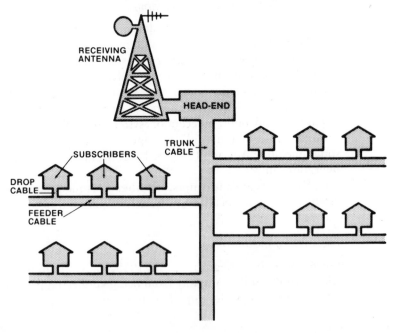

Basic Cable Television System

far from a residential area. Installing the tower and antennas entails major construction using everything from lumber cutting crews to giant helicopters (Figure 6–3).

The head end may also be television production facilities such as cameras, lights, and other studio hardware, depending on the size of the cable system and how much locally originated programming there is in the studio. The facilities can range from a small, black and white camera to full-scale color production equipment. With all of this in mind, we will define the head end as *the human and hardware combination responsible for originating, controlling, and processing signals over the cable system.*

Another important cable system component is the *distribution system,* which disperses the programming. The main part of the distribution system is the cable itself. A coaxial cable (Figure 6–4), which is used in most cable systems, consists of an inner metal conductor shielded by a

Figure 6–3   Critical to the effective operation of the cable system is the master antenna. Often located on mountain tops and in rough terrain, constructing the antenna can be dangerous and expensive. (NCTA)

Coaxial Cable

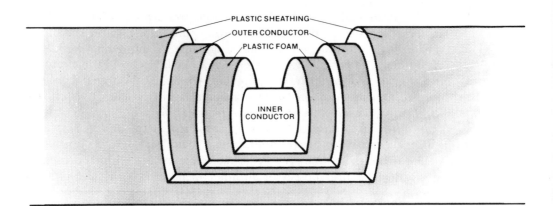

Figure 6–4 (NCTA)

plastic foam. The foam is then covered with another metal conductor, and that, in turn, is covered by plastic sheathing. This protected cable may either be strung on utility poles or buried underground. The primary cable, or main transmission line, is called the *trunk cable*. It usually follows the main traffic arteries of a city, branching off into a series of smaller feeder cables, also called sub-trunks. The feeder cables usually travel into side streets or into apartment complexes.

The actual connection to the home is made with a *drop cable*. This coaxial cable goes directly into the house where it connects with a *home terminal*. The home terminal, in turn, connects directly to the back of the television set. In most cable systems, the home terminal is simply a splicing connector that adapts the drop cable to a two-wire connector which fits onto the two screws on the back of every television set. In two-way cable, which we shall learn about next, the home terminal is more complex and may even include a small keyboard (Figure 6–5). Some cable systems install these more sophisticated home terminals even though two-way cable may not be operative yet on the system. When it does become operative, the system and the subscriber will be ready.

two-way cable television

Two-way cable systems, sometimes called two-way interactive television, permit the subscriber to feed back information to the head end. They are quickly becoming popular, and two-way is able to bring a wide variety of services into the home (Figure 6–6).

142

Figure 6–5 Warner Cable's 30-channel home terminal, which enables participation from the home in a variety of television programs such as interactive games, educational tests, and public opinion surveys. The terminal, connected to the subscriber's television set, is activated by pressing the program channel choices. (Warner Cable of Columbus, Inc.)

Application to Instructional Television

Two-way cable can also be applied to home instruction programs. For example, a handicapped child who cannot attend regular school can receive lessons at home via two-way cable television. The program reaches the two-way subscriber in the same manner as it would on one-way cable systems. However, the student can communicate with the teacher using a feedback loop. Feedback loops are generally of three types (Figure 6–7). One is a single cable used for both transmitting to and receiving information from the subscriber. Another uses two separate cables. Incoming signals reach the subscriber through one cable, and outgoing signals return from the subscriber to the head end using the second cable. A third is called a round robin cable loop, an adaption of single cable but with separate drop cables.

In our example of instructional television, special questions appear on the screen at regular intervals in the program. The student answers the questions on his or her home terminal, which can be as complex as a typewriter keyboard or as simple as a touch-tone telephone. The home terminal is hooked into a central computer, which aids the instructional

Two-Way Cable Transmission Techniques

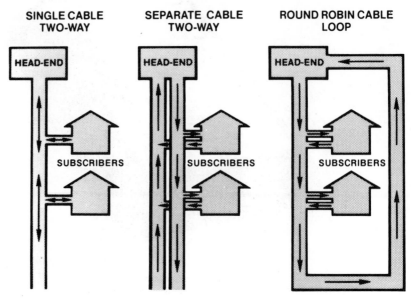

SINGLE CABLE TWO-WAY   SEPARATE CABLE TWO-WAY   ROUND ROBIN CABLE LOOP

HEAD-END   HEAD-END   HEAD-END

SUBSCRIBERS   SUBSCRIBERS   SUBSCRIBERS

Figure 6–6 (NCTA)

process in two ways. First, it notifies the student if he or she has selected the right answer. Secondly, it tells the instructor how many students have selected the right answer. If the proportion of correct answers is low, the instructor can repeat the lesson. We shall learn more about this process in the chapters on educational and industrial television and feedback.

Industry and Consumer Resistance to Two-Way Cable

Although there are many advantages of two-way cable, there is still considerable resistance to the concept. Some is financial. The cost of two-way cable systems is much greater than that of one-way systems. The advanced computer technology, sophisticated home terminals, and other necessary hardware add up to more expensive subscriber fees. Consequently, businesses and institutions have become two-way cable's heaviest users, and even they use it only to transmit data.

There also is fear of the unknown. Two-way cable conjures up images of privacy invasion and "Big Brother." The thought of your living room being hooked up to a central computer has given some people doubts about the whole process. The future of the system will depend on being

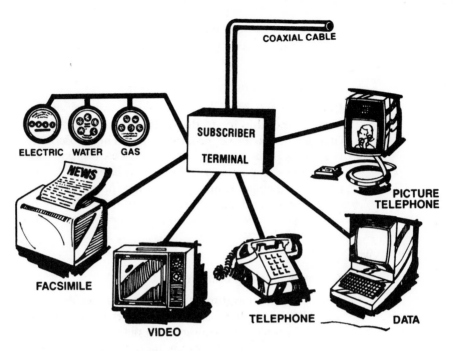

Figure 6–7 Cable's multiple uses beyond television programming can include such services as reading utility meters, in-home electronic newspapers, telephone, data, and picture telephone. (Communications News)·

able to alleviate this apprehension and making two-way cable economically feasible for widespread use by the general public.

pay cable

Pay cable is defined as *the delivery of information and/or services to cable subscribers by assessing fees beyond the regular rental fee.* Pay cable should not be confused with "subscription TV," or "pay TV," terms defining a special over-the-air TV distribution system where the signal is "scrambled" as it leaves the transmitter and is descrambled using a special device on the home receiving set. The advantage of pay cable is the opportunity to see first-run movies, exclusive viewings of major sports events, and other special entertainment programs. Overall, pay cable has not been tremendously successful. Only within the hotel-motel industry is it gaining wide acceptance. Part of the problem is in the necessity of charging additional fees for programs, fees which the public has not been willing to pay, especially when there is a rich diet of alternative programming. This does not, however, mean that pay cable has no future. Alternative services such as data processing may revitalize the pay cable concept.

Pay Cable Connection Arrangements

People can subscribe to pay cable in a number of ways. The simplest is to pay a monthly fee to the cable company in order to receive a special channel of exclusive programming. The subscriber pays this set fee regardless of how much he or she views the pay cable channel. Or the subscriber can pay on a per-program basis. In this system, some type of two-way capability is necessary so that the subscriber can signal the cable company to send the desired program into his or her home. A credit card inserted into a special home terminal to signal the cable company is one method.

Arguments for and against Pay Cable

One of the main arguments in favor of pay cable is its ability to broadcast to small select audiences, something over-the-air broadcast stations cannot do. The problem is economic. To obtain profitable advertising revenue, over-the-air station programming must attract audiences in the millions, especially network programming. Programs which cannot achieve this type of saturation are simply kept off the air. Fine arts programming, educational programs, and similar fare are on what the pay-cable proponents base their appeals. By charging for the programs, the need for advertising revenue on pay cable is eliminated, or at least substantially reduced.

There also are arguments against pay cable. Commercial broadcasters charge that pay cable destroys the free system of broadcasting upon which the nation's broadcast communication systems were built. Other groups claim that pay cable discriminates against the poor who cannot afford the additional rates. The opposing arguments intensify with suggestions that pay cable could make exclusive contracts with program producers, placing programs on pay cable which would normally have attracted an audience large enough to warrant airing them on over-the-air broadcast stations. That suggestion makes commercial broadcasters bristle.

Theater owners showing first-run movies have still another argument against pay cable. They contend that when first-run movies are made available through pay cable, people will stay at home and watch them instead of going to the neighborhood movie theater. Examples supporting and denying these contentions are used by both proponents and opponents of pay cable.

cable radio

Television is not the only medium channeled by cable systems. This wired concept also applies to radio. The principle for cable radio is the same as it is for television. Distant station programming is cabled into a local community. As expected, commercial radio broadcasters vehemently oppose cable radio. Radio is a very local medium. When a small community's cable system imports one or more stations from outside the local market, the radio broadcaster feels the economic pinch. The importation usually translates into a reduced audience and consequently, reduced advertising dollars. The problem is not as serious for television since many communities do not have local television. But few do not have local radio stations. To compensate for this, the FCC has strongly suggested that cable operators not import radio signals from more than 75 miles away into markets of less than 50,000 persons. But so far, it is only a suggestion.

the potential for programming

The most obvious advantage of cable is the increased selection of programming channels that it offers. Even though a community may have a wide selection of local channels, channels from neighboring communities can be "cabled" in, enabling residents to see independent stations' top-rated syndicated programming. The fine arts programming of an area's public television station can also be "cabled" in. Major sports events not carried on local stations can also be made available through cable.

Instructional Television through Cable

Cable companies often contract with a local university to fill one of its cable channels with instructional programming. Under these agreements, the university can offer a complete curriculum with a wide variety of courses that can be taken in the living room instead of in the classroom. The continuing education and outreach functions of colleges thus gain a whole new perspective through cable. Students can enroll in one college while taking courses by cable television from another. Theoretically, as we creep toward the new wired-nation concept, colleges could even become specialized in one type of instruction which is "syndicated" beyond the local campus. For instance, a school in New Mexico could use cable and national satellite hookups to teach Indian culture. Or a school in Wyoming could offer courses in mining economics.

Colleges and universities are not the only educational institutions in cable television.[3] High schools have produced programs explaining school activities, which they then air over the local cable system for the parents' benefit. Special programs for classroom use are also shown over cable systems, permitting taxpayers to see what is being done with their tax money. Schools even use the local cable system for such highly specialized programming as drug education. High-school sports programs are also broadcast on cable.

Cable is also becoming a means of communication among individuals and institutions which in the past have been distant, making communication impossible. The public that looks inside a school building, watches teacher-training programs, and hears school officials discuss issues learns more about and participates more in the institution. This social and political responsiveness, in turn, permits the school officials to be more accountable to those forces affecting policy.

Programming the Arts

Perhaps one of the greatest potentials of cable is in performing arts programming. In the past, commercial broadcasting has not been able to program the arts successfully for two reasons. The first was the lack of viewers for such specialized programming. The second was caused by the first, an inadequate profit foundation upon which to produce and program these shows. Although public broadcasting has inched toward this type of programming, it still must produce programs which appeal to a mass audience. Cable provides the alternative. Now local fine arts programming can be produced and funded on a local level.

Fine arts programming provides another benefit for cable. If a city has a good local symphony orchestra, for instance, a group of local sponsors and friends of the arts often contribute to its development with considerable zeal. This can spill over into the cable system. The local symphony can play to its audience over cable even during prime-time hours. On a commercial broadcasting station, the only profitable time to air such programming might be in the wee hours of the morning. Exceptionally good fine arts programming even can be duplicated and syndicated to other cable systems.

cable television and the political process

Along with education and the fine arts, cable has provided new opportunities for citizen interaction with the political process.

Broadcasting Public Meetings

School officials becoming accountable to community forces is just one example of this political involvement. Cable gives the public access even to the smallest governmental body. Television cameras and microphones can be in the audience of a school board meeting, a city council meeting, or a zoning commission meeting. Experience has also shown that live cablecasting of such meetings has increased attendance. Those with an active interest attend the meetings in person, and those with a borderline interest watch from home.[4]

Candidate Access

Commercial broadcasting, and to some extent public broadcasting, is limited in the amount of advertising time which can be given to candidates for political office. It is just not economically feasible to turn a station's complete advertising availabilities over to the politicians. The reason for this is that the FCC has decreed that political advertising be sold at the lowest rate the station charges to an advertiser. Although this helps politicians with scanty campaign coffers to obtain valuable television exposure, the procedure is not the favorite rule of commercial broadcasters. Some stations simply appropriate an amount of free time to candidates and dispense with selling political advertising altogether. In addition, election laws now restrict campaign financing. These have substantially trimmed candidates' budgets, budgets which used to produce lucrative television campaigns.

Cable television provides a number of alternatives to these dilemmas. First, in small communities without a television station, cable television permits candidates to reach the voters through a visual medium. Second, most local access rates permit even the candidate running for dog catcher of Possum Hollow to campaign on the local cable channel. Third, when commercial availabilities are not open, cable availabilities may be. Fourth, cable permits the candidates to reach highly specialized audiences not normally reached by commercial television. In short, cable is helping our society recapture the old-fashioned, town-meeting approach common to democracy yet difficult to attain in our modern age.

cable's local access concept

Most cable systems allow any member of the public access to a cable channel. We shall learn why some of them don't and the regulations accompanying cable in chapter 15. Some larger cable systems also provide, at a nominal cost, equipment for people to produce local programming. Let's look at a local example.

An Example of Applied Local Access

Local access programming is not the glittering lights of Hollywood. Nor is it the elaborate production studios of a major network. It more than likely materializes as this one did, on a October day in a small community of 8,000 people and 1,500 local cable subscribers. It is evening, and on a drive in the country, a local resident spots a poster tacked to a utility pole. "Halloween Parade—Everyone Welcome—6:30 P.M., October 31, The Fire House Parking Lot." The perfect opportunity for local-access programming. The next morning, a call to the local cable company produces the name of a student at the nearby college in charge of cable equipment available for public use. The cost is minimal.

A second telephone reveals that two small, portable, black and white cameras, a videotape recorder, and a switching unit are available.

"We can set them up in the alley behind the firehouse. That's where the parade starts."

"What about lights and electricity?"

"Whoops, I never thought of that."

Time to check with the fire department. Another telephone call . . .

At 6:30 P.M., the parking lot is filled with children. Costumes are everywhere. A hay wagon is in position with the parade's director and her megaphone astride. The children line up to march around the block and back into the parking lot for an awards ceremony. Wait a minute. The light from the fire department isn't enough. Time to move some cars into position. Park some at the head of the alley and turn the headlights on high beam. It works; we're ready to go. Lights, cameras, action, the parade is on! A few hours later, the 90-minute production of the parade and the awards ceremony is seen over the local cable channel. A full-scale television production? No, but to the parade participants it was just as exciting. For the parents and children who participated, they'll remember that Halloween for a long time to come.

The above example is what local access television really is. It is the grassroots side of mass media not possible with standard broadcasting stations. Creative opportunities on many cable systems await not the seasoned professional, but *the amateur citizen*.

Local Access and Specialized Audiences

One advantage of cable is its ability to reach specialized audiences. In Philadelphia, for example, as part of the cable company hardware capability, a series of "mini-hubs" brings new concepts in cable programming.[5] The mini-hub is a series of local origination points along the cable route. With this, cable programming can be limited to a few city blocks. Then, perhaps three blocks away, there is another program from another mini-hub. This highly localized access concept permits programming to small neighborhoods which might have similar ethnic or religious backgrounds or other common ties.

Another example of local-access programming to specialized audiences is programming for the elderly. Cable is connected to living complexes for the elderly such as nursing homes. It is popular with the elderly because of the added leisure time retirement permits. Local access cable programming permits the elderly to "communicate" with each other, alleviate loneliness, and feel more a part of the community. Special programs about social security benefits, Medicare information, transportation, shopping bargains for senior citizens, and similar offerings all are possible through local-access cable programming.

Issues Facing Local Access

Up to this point, our discussion about cable television has been positive. Instructional television (ITV) programming, the fine arts, cable in the political process, and reaching specialized audiences all are beneficial. But not everything is this positive, and the local-access concept is perhaps the least so.

The major problem of local-access television is that its viewing audience has been difficult to measure. Where major commercial rating services are employed, the local-access channel is not usually included in their survey measurements. When audience measurements have been taken, they have shown very few viewers. A research study by the Institute for Communication Research at Indiana University sampled the television viewing habits of the community of Columbus, Indiana.[6] Results showed that "the total public-access viewing time, for all persons in the sample, was two tenths of 1 percent (.2 percent) of the total of all television viewing for the week."[7] Explanations of the small audience were a lack of promotion by the cable company, "causal" scheduling which included intermittent programming, blacked-out periods, and perhaps sampling error. However, a promotional campaign did slightly increase subsequent viewer levels.[8]

Another problem area for local-access television is the programming itself. Pamela Doty, a researcher for the Center for Policy Research in New York, spent two months viewing local-access channels in New York.[9] Because local-access programming did not expect a mass audience nor did it need advertising revenue, she felt she would find a "higher percentage of hard-hitting social criticism and controversy" on local-access channels than on typical commercial television. This was not the case. She found it was "rare to see a debate or dialogue between two persons who even mildly disagree, let alone have major differences."[10] Moreover, the rather bland format of people sitting around a coffee table talking to each other, the "talking heads" format, on local-access channels, was rather boring.

Doty's recommendations for improving local-access programming include true debating of local issues, "explorational tours" of communities with "behind-the-scenes" glimpses, and programs showing the audience how to edit videotape. She also states that the public access users need to develop a clear-cut sense of what they want their programs to accomplish and how to go about it in such a way that interests the audience they want to reach.[11] When done properly, local-access programming has great potential. We have already learned about programming the local symphony. Other special events such as Little League baseball, children's parades, community fairs, picnics, church services, and others can fill a cable system's programming schedule.

cable versus the broadcaster

With cable's ability to carry broadcast messages beyond the coverage area of the over-the-air station, it would seem that broadcasters would support cable's efforts more. But the two have seldom coexisted harmoniously, and at times their opposition has been bitter. We shall examine both the broadcasters' and the cable operators' arguments. Each has valid points. It is up to the regulatory bodies to judge which will prevail.

The Broadcaster's Arguments

Place yourself in the position of a commercial broadcaster in a medium-sized community. You are against cable. The main lobbying group for your position is the National Association of Broadcasters. You, yourself, give your arguments everywhere from the FCC to your local city council. What are these arguments?

The strongest argument is economic. The importation of broadcast signals slashes your audience. You used to be able to offer a substantial audience to advertisers for a healthy profit. Second, you are charged copyright fees for certain programs, yet a cable operator can rebroadcast those same programs without paying them. "That's fair?" you exclaim. Third, you, as a broadcaster, are serving the public's interest, convenience, and necessity, and are providing that service free of charge to the viewer. On the other hand, a large interconnection of cable systems can successfully negotiate exclusive programming with a college football team, for example, and charge viewers to see the games. You, in turn, because of the cable systems' exclusive contract, would not be permitted to carry the game. "Outrageous!" you fume. Fourth, economics usually dictates that cable be installed only in densely populated areas with the most subscribers, especially pay TV. Yet you, while competing with cable, are also serving the rural public, regardless of the population density.

Your fifth, but more general argument, is that cable has developed as a parasite industry of broadcasting and now is trying to compete with it. Sixth, you argue that cable has been favored by the FCC with a general relaxation of rules, thus permitting it to compete better with you. You compare this to fighting with one hand tied behind your back while your opponent's hands are free. Seventh, since cable companies have the ability to interconnect among systems, you claim that the theory of local accountability and service has been destroyed. Eighth, while you operate on a limited spectrum space, cable can carry large numbers of channels, many of them programmed by the local cable systems themselves.

Some of these arguments are good; others are not. Nevertheless, the broadcasting industry is quite concerned about cable.

Cable's Rebuttal

Now put yourself in the cable operator's position. Your major lobbying organization is the National Cable Television Association. You, too, take your case before any interested body. How would you argue?

First, you contend that over-the-air broadcasters are severely restricted in serving their viewing audiences, since even in the largest markets, only a few stations can operate within a limited spectrum space. You feel that cable serves its audiences far better with its variety of channels. Second, you point out that precisely because of their limited spectrum space, broadcasters have made giant profits. You state that those profits are sometimes at the expense of viewers, who long for more innovative, perhaps more costly, programming. Third, restricting legislation, you argue, means that cable has not been able to build a similarly firm financial base. Fourth, while still on the subject of regulation, you argue the cable industry is regulated by legislation designed to protect the broadcaster, the Communications Act of 1934 being the principal example. This envisions a system of over-the-air broadcasting. Even amendments to this act can control cable. Fifth, when broadcasters criticize cable's exclusive contracts with program distributors, you remind them of their exclusive contract advantage with the major networks. Sixth, although commercial broadcasters answer to only one master, the FCC, you face regulatory control by three levels of government—local, state, and national. "Is all of this fair?" you ask.

All of these arguments, in varying degrees of detail and intensity, are used throughout the broadcasting and cable industries. They have been presented in cloakrooms to members of Congress, at special legislative hearings, and in public relations literature. Meanwhile, mass consumers are living in both worlds, unaware that the future is bound to bring some dramatic changes in how we receive our daily television fare.

Healing the Split

Despite all the rivalry and rhetoric, broadcasters and cable operators must begin to consider how each can complement the other in a common goal. Increasingly, new developments in telecommunication are drawing the two out of their warring camps. Television stations broadcasting by satellite to distant cable systems when both are operated by the same company find it awkward to cut the other's throat. Still, with powerful lobbying groups like the NAB and the NCTA, the chasm will not be bridged overnight.

A call for unity has come from Clifford M. Kirtland, Jr., president of Cox Broadcasting Corporation, which has holdings in both cable and broadcasting stations. Addressing a meeting of the Institute of Financial Management, Kirtland states he feels there needs to be "a recognition by all broadcasters, cable operators, and producers of programming that the viewer and listener in the home are *not* served by high-toned rhetoric lambasting the opposition and feeding the critics. What is needed is a recognition that—even after all the compromises, rule changes, techno-logical changes, and criticism from all sides— the audience today is better served then ever before." He concludes: "Perhaps a greater spirit of cooperation among all parties working toward balanced regulations in a positive way . . . a greater acceptance of technological change . . . and a greater reliance on the free enterprise system to work its wonders . . . will, by 1985, further enhance the total communication service of our country."[12]

It is difficult to predict what future scenario will bring cable operators and broadcasters together. Perhaps a regulatory issue demanding com-mon lobbying efforts, or perhaps a foreign competitor beaming signals into North American living rooms may turn the tide. For now, both are working feverishly to protect their individual economic domains.

cable's future: the major issues

Despite the respectable growth and impact of cable, it is in its infancy as a technological and social force. Standard broadcasting, both in size and influence, makes cable minute in comparison. Still, it is a force with which to be reckoned. Let's briefly examine some of the issues pertaining to its future. A more detailed discussion of these issues is found in other chapters of this text.

Regulatory Considerations

The cable industry looked forward to the rewriting of the Communications Act of 1934. Cable is the stepchild of that piece of legislation. Although legislation regulating cable was added to the act, those regulations have been greatly tempered by the strong lobbying efforts of commercial broadcasting organizations. Many cable operators feel that wiping the slate clean and starting over would give cable a better footing.

Besides the federal regulatory stance is the influence of state and local governmental bodies. Here, cable suffers a decisive disadvantage over commercial broadcasters. A cable system can find itself regulated by three systems—local, state and federal. Standard broadcasters answer only to the FCC. To make matters worse, some regulations actually conflict with each other, creating a maze of court-cases ranging anywhere from rate structures to local-access programming. Directly related to these problems is the futility of trying to regulate, on the basis of state and local boundaries, a communications system that transcends boundaries. For example, communications attorney Anne W. Branscomb, using New York as an example, feels that the New York metropolitan area should coordinate its telecommunications planning and development with New Jersey and Connecticut, rather than with New York State.[13]

Cable's Interference with Legal Precedent

Another problem cable faces is its relationship with laws indirectly affecting its operation. Consider the case of local-access. A local community group decides it wants to use the local-access channel to broadcast the school board meeting live and in its entirety. The state's open-meetings law permits public access to all public meetings. But the school board says no. The school board's attorney contends that cable television cameras are not people and therefore can be barred. The community group reminds the school board that it permits the local television station to film and videotape portions of its meetings. In fact, when major issues are being discussed, the board even allows the station to broadcast live, mini-cam reports. But the school board replies that cable is not considered a bona fide news gathering organization and does not come under the protection of a free press.

This is just one of many "gray" areas cable faces. Many laws, such as open meetings statutes and reporters' shield laws, have yet to define their applicability to these situations. Until they do, cable has an uphill climb for its legal identity.

fiber optics

Whatever policy decisions are made on cable, their full social implications will be felt only when there is widespread use of fiber optics. Fiber optics, as we have already learned, are thin strands of glass fiber through which light passes. This light, which travels at a very high frequency on the electromagnetic spectrum, carries the broadcast signals (Figure 6–8). The use of fiber optics will dramatically increase the amount of information that can be carried on any single cable system. The thought of 1,000 cable channels on a single glass fiber boggles the mind.

Telephone Companies: A Competitive Edge?

Control of something usually gives power to the controller. It is no different with fiber optics. This time, the telephone companies could be in the driver's seat. Before development of fiber optics, only the cable companies had a channel capacity large enough to transmit and receive computer data directly into the home. The telephone company's home hookups just were not satisfactory for transferring those large amounts of information. Fiber optics has changed all of that. In its report on the future of telecommunications policy, the research firm of Arthur D. Little, Inc. stated that, "current research into the use of fiber optics would lead to telephone companies becoming the most logical providers of home television access in the long term."[14] Remember that the telephone companies already have an established system of installation much larger than that of the entire cable industry. Their personnel, equipment, and

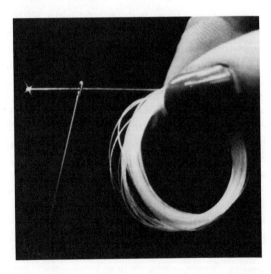

Figure 6–8 Tiny strands of glass can carry thousands of channels in the space coaxial cable fills. (Western Electric)

even utility poles are ready now to install fiber optics in any home, anywhere.

A future scenario even envisions telephone companies with their own television stations and networks feeding a "wired nation" as today's cable operators do, only on an extremely small scale. The social implications of such a phenomenon are far-reaching. How would the computer of the future adapt to such a system? Behavioral scientists are currently studying how children are affected by constant association with television from birth. What will be the future of fiber optics and its ability to adapt in-the-home computer technology to television? Children may grow up not only watching television but also "communicating" with a central computer which continually processes and adapts to his or her programming likes and dislikes. It is all possible with the increased channel capacity that fiber optics offers. A future educational program may project a math instructor to a national viewing audience with individual computer terminals, enabling them to work the problems at home. At the same time, the math instructor could look at the computerized results on his or her computer terminal and determine thousands of miles away how well the lesson is being understood.

The Future Transition

The future influence of fiber optics is not clear. It definitely has renewed interest in CATV, especially by the telephone companies which were at a disadvantage because of the inability of home telephone wires to carry multiple television channels. Both the cable industry and the telephone companies are interested in the transmission of data communications, such as connecting two distant computers, to which fiber optics will be essential.

Economics will partially dictate the future. Although AT&T has fiber optics at its disposal, it has already invested close to $60 billion in its current hookup system. Converting to the new fiber optics hardware, depreciating the $60 billion loss, and amortizing the debt for new development could put the cost of this new technology to the consumer out of reach. In short, for the telephone company, a short-range shift to fiber optics could be prohibitively expensive.

the economics of cable

The future of cable and how we interact with and use it is directly tied to its economic issues. It is important to understand them. You may find yourself voting in a local referendum to raise the rates for your local cable system. Or your community may determine whether the local cable company should install its cable underground or, more economically, attach it to telephone poles. It may even decide what supplementary services, such as electronic funds transfer or bank-from-home services, should be added to the local television fare. To make intelligent decisions, you will need to understand the economic forces affecting cable.

The Capital-Intensive Factor

Cable is a capital-intensive business. By capital intensive, we mean that maximum costs occur immediately. A standard radio or television station can go on the air with a minimum amount of equipment, some of marginal quality, and a skeleton staff. Cable does not enjoy this luxury. Cable systems are designed for permanency, and the system must be taken to the total potential audience before it can even begin operation. Therefore, hiring skilled technicians, installing miles of cable, requiring elaborate antenna systems, purchasing head end equipment, and incurring similar expenses must all be done before the first subscriber is hooked on.

Construction Costs

Starting a cable system requires planning construction of the head end and production facilities, the distribution plant, subscriber equipment, and preoperating expenses.[15] Costs for underground construction of the distribution system can run as high as $60,000 per mile in crowded metropolitan centers. Some radio stations have gone on the air for less. Even above-ground pole attachment systems are conservatively estimated by operators at $50,000 per mile. The location and type of antenna can also raise the cost. An antenna that must be constructed on top of a mountain is going to cost much more than a tower built in a level field outside of town. A head end with production facilities for locally originated programming is going to cost more than one without local production capabilities. All of this determines how much the subscriber must be charged, how long it will be before the cable system makes a profit, where financing can be obtained, how high the interest rate will be, and whether it is economically feasible to construct the system at all.

Operating Costs

Construction costs are followed by the costs of operating the cable system. The main cost is system maintenance. Although cable operators usually install the best possible equipment for long life and maintenance-free service, nothing is infallible. Breaks due to storms and equipment repair at the head end are just part of the regular maintenance schedule. Second, a subscriber cannot simply turn on a television set and tune to the cable channel without first having the set connected to the cable. That requires a service call, and service calls are responsible for much of the cable company's personnel time.

A third expense is vehicle operation. Unlike a radio or television station which may have its entire operation under one roof, the cable company literally can be spread all over town. Servicing this territory in larger markets can require an entire fleet of trucks, many with aerial ladders. Because of the high price of gasoline, this is a spiraling cost for the cable system. Future developments in technology, however, will permit more and more switching and connecting functions to be done at the cable head end.

Fourth, utility pole and underground duct rentals are a large expense for the cable system. For example, if pole attachments are used, the cable company must rent them from the telephone company, and this can cause strained relations between the two. Pole attachments cost an average of $4.00 to $5.00 per year.[17] Although that may not sound like much, it is almost equivalent to a monthly cable subscription fee. When a major cable company like TelePrompTer rents more than 800,000 poles, the cost is considerable.

Fifth, operating costs continue. Local municipalities may charge franchise fees, money the cable company pays the local government for the privilege of operating. Sixth, cable companies may be charged copyright fees by program distributors. Seventh, although construction of the antenna and other head end facilities is usually figured into the construction costs, the cost to bring in distant signals may require separate lease agreements with telephone companies or private microwave carriers. Eighth, local origination costs also can be substantial. Here a local cable company can incur some of the same studio expenses as a small television station does. Although it can boadcast with black and white equipment, color capabilities help to develop programming which can successfully compete for viewers of other channels. This does not mean that high quality black and white programming with special local appeal cannot be successful. With local meetings, special seminars, and similar "individualized" programming, the interest will be high no matter what the quality of production is.

Income from the Cable System

For the cable company to make a profit, it must receive income in the form of subscriber fees. The number of subscribers and the amount of the fees are the key components. Most subscribers pay approximately $12 per month for all cable services. That may include additional fees for pay TV programs, such special services as electronic funds transfer, or even two-way, interactive instructional television programming. As the variety of these services increases, the subscriber fee increases. Research by Cox Broadcasting has predicted that by 1985, the amount paid by a subscriber for all types of cable services could be $17.00 per month (Figure 6–9).[18]

Subscriber fees are not the only source of cable's revenue. Another is advertising. Cable companies have been successful in selling advertising in the same way that standard broadcasting stations do. Moreover, the interconnection between cable systems through satellite and microwave makes the concept of a cable network a working reality. In a network, a group of cable companies joins together to carry the same programming, deriving income from sponsors who buy advertising seen throughout the entire cable network. As these cable networks develop, a larger percentage of cable's income will come from advertising (Figure 6–10). Predictions by researchers for Cox Broadcasting place advertising revenue at 10 percent of cable's total revenue by 1985, an increase of 100 percent over 1980 figures.[19] Pay cable revenue is based on a figure of 20 percent of total cable revenue in 1980 to 30 percent in 1985.

In summary, income from the cable system is classified into four broad categories: (1) subscribers' monthly rental fees for standard television

CABLE REVENUE PER SUBSCRIBER
(ALL SERVICES)

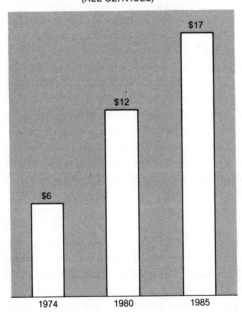

Figure 6–9 (Cox Broadcasting Corporation)

DISTRIBUTION OF CABLE REVENUE

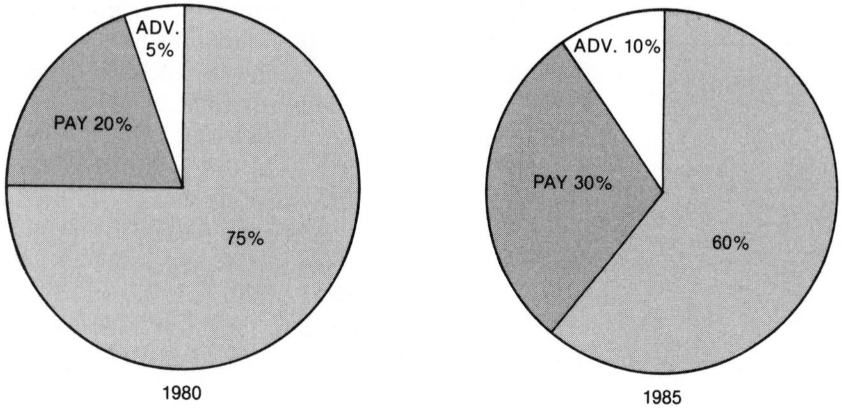

1980   1985

Figure 6–10   (Cox Broadcasting Corporation)

services; (2) pay cable fees for special programming, much of it exclusive, which is usually either a set charge above the regular monthly rental fee, or an individual, per-program assessment; (3) revenue from such special services as at-home banking; and (4) revenue from advertising.

Approaching the Profit Margin

A cable system has no set formula for success. But the enterprising operator does follow some basic guidelines. Among them is the delicate balance between the amount of money that can be charged to a subscriber and the number of subscribers needed to make the system profitable. For example, if a cable system has 1,000 subscribers, each of whom pays a $10 monthly subscription fee, the total income would be $10,000 per month (if there is no income from other revenue sources). Now we will assume that the cable operator decides to increase the subscription rate to $12 per month. Theoretically, this raise would net the company $12,000 per month. But what if the rate increase drove away 200 current subscribers? The income to the cable operator would then drop to $9,600 per month (12 x 800), a loss of $400 per month.

Another economic balance for the cable operator is between the amount of the original construction cost and the number of subscribers necessary to equalize that cost. The key here is subscriber cost, not to be confused with the subscription fee. Subscriber cost is what the "cable operator must obtain in revenues in order for the system to operate at a profit."[20] The more subscribers there are, the less the subscriber cost needs to be.

To understand this principle, consider the following example of

Rolland C. Johnson and Robert T. Blau in their report on a consulting project with an Indiana cable company. Imagine that a city of 40,000 people with 10,000 homes (100 homes per mile) grants a cable television franchise. A cable operator is able to construct the entire system for $1 million, including 100 miles of cable, head end, and miscellaneous costs. At the end of one year, let us assume that 1,000 homes (10 percent of the total) are being served. At this point, the cost per subscriber is $1,000 ($1 million ÷ 1,000)—drop costs are assumed to be covered by installation fees. During the second year, an additional 1,500 homes subscribe, and the cable operator now has 2,500 subscribers, or one-fourth of the total potential. Per subscriber cost is now only $400. If, at the end of ten years, 80 percent (or 8,000 homes) of the community is served, per-subscriber costs drop to $125 (assuming the critical equipment is still in working order).[21]

The above example is hypothetical. *Actual situations include a number of variables.* New independent television stations may be constructed, which beam clear signals into the areas already served by the cable. Or the potential for large subscriber blocks, such as apartment houses, may vanish when a landlord decides to prohibit the cable company from hooking up to his or her complex. Even a competing cable company may appear on the scene. Such competition would have been unlikely a few years ago. Installing expensive lines along an existing cable route simply was not practical. But new technology has changed all that. A small satellite dish on the top of an apartment complex and its accompanying roof-to-residence cable hookups can turn the complex into an instant cable market. This comparatively inexpensive competitive system can upset the most solid projections for success.

Nevertheless, with the right marketing techniques, cable systems can do a sizeable business. A report by R. E. Park identifies the influences of subscriber penetration:

The more television stations of various types a cable system carries, the higher its saturation will be.

The fewer of each type of station receivable locally over the air, the higher the system's saturation will be.

The farther from the television transmitters the system is, the higher its saturation will be.

The more stations that broadcast on UHF rather than on VHF channels, the higher the system's saturation will be, because of a variety of reception and tuning problems in UHF stations.

The less the system charges for its services, the higher its saturation will be.

The higher the average income of households in the community served by the system, the higher the system's saturation will be.

The older a system is, the higher its saturation will be.[22]

Park bases his conclusions on data from cable systems already in operation. Constantly changing technology and our own changing media-use habits require cautious optimism when using such data to predict success for all new cable systems.

Cable will certainly continue to grow as a viable medium. However, new technology may greatly change its current definition. The ability to beam broadcast signals directly to small satellite dish antennas positioned on rooftops may eliminate the actual "wires" used in current cable transmission. If the telephone companies gain, through regulatory protection, primary development and use of fiber optics technology, then they may become the cable companies of the future.

The research firm of Arthur D. Little, Inc. conducted an impact study on cable's outlook through 1985. The study predicts continued but slower growth in the early 1980s "as markets with limited over-the-air broadcast TV reception become saturated. New service options will be needed for further penetration of existing markets."[23] Predictions are "that in 1985, pay cable subscribers will number five to seven million and basic service subscribers 20–26 million, roughly 30 percent of the projected $4.8 million TV households." Arthur D. Little, Inc.'s projections are similar to those of Cox Broadcasting Corporation, which predicts cable penetration of the total United States population to reach 29 percent in 1985 and 17 percent in the top 25 markets (Figure 6–11).[24]

Figure 6–11   (Cox Broadcasting Corporation)

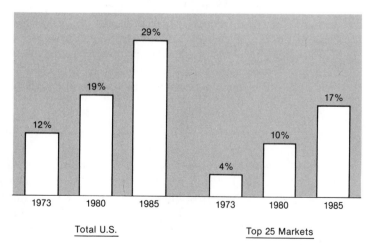

CABLE PENETRATION

human implications of our cable environment

Before concluding our discussion of cable, it is appropriate to stop and ponder what this technology means to our human environment.

Civilization has developed and prospered partly because of its ability to originate and maintain systems of communication, not only between people but between cultures. Research has shown that we spend approximately six hours per day watching television. Although we may not be silent during television viewing periods, we are not communicating with others as much as we would during periods of conversation. We do have other opportunities during the day to participate in interpersonal relations. We may go to the grocery store, to the bank, or simply window-shop around our community or campus. In each instance, we are around other people. Walking between classes is a perfect time to greet those we meet on the sidewalks, and we love to converse with our friends in snackbars and coffee shops. In short, although many of us spend considerable time with television, we still spend a great deal of time communicating with other people.

Now project yourself into the future. The lure of television still attracts you, but it is joined by many more. Instead of taking a break between classes at the local coffee shop or walking across campus to attend another class, you spend the majority of the day in your room taking courses via cable television, from not one but four or five colleges around the country. Chemistry from the University of Washington, physics from the University of Notre Dame, English from the University of Texas, and geology from the University of North Carolina all would be part of your daily academic routine. The time you spent in front of a television screen theoretically could increase to twelve or more hours per day. Then there is shopping by cable, banking by cable, and endless other services by cable. What would be the psychological effect on people with this concentrated media interaction and human isolation?

As you learn about the economic, political, and technological development of broadcast communication, it is important to remember its effect upon the interrelationship with people and society.

summary

Cable television, also called CATV or community antenna television, was developed in the 1940s as a way of bringing distant television signals to outlying communities through use of a common or "community" antenna. The system consists of the head end and the distribution system, which is broken up into trunk cables, drop cables, and home terminals. The development of two-way cable has made possible such services as interactive instructional television. An offspring of cable is pay TV, which charges the subscriber an additional fee to see exclusive programming. Although it has had very strong resistance from standard broadcasters, even cable radio has developed on a limited scale.

Because of its revenue system, cable has the economic ability to reach much smaller, specialized audiences than over-the-air television, which must reach a large, mass audience in order to attract advertising dollars. Thus it can specialize in such programming as instructional, fine arts, and political. Although not fully utilized, local-access programming on cable permits even the average citizen to create television programming.

Cable has received strong opposition from organized groups of over-the-air broadcasters. They argue that cable has fragmented broadcast audiences and has developed as a competitive, parasite industry. The cable industry counters that it is overregulated and because of this, has not been able to build a strong financial base. Cable also faces major economic hurdles because of its capital-intensive nature. Although its future looks stable, we need to consider the human implications of our cable environment.

spotlight on further learning

ADLER, RICHARD, and WALTER S. BAER, eds. *Cable and Continuing Education.* New York: Praeger Publishers, Inc., 1973.

ADLER, RICHARD, and WALTER S. BAER, eds. *The Electronic Box Office: Humanities and Arts on the Cable.* New York: Praeger Publishers, Inc., 1974.

BABE, ROBERT E., *Cable Television and Telecommunications in Canada.* East Lansing: Michigan State University Graduate School of Business Administration, 1975.

BAER, WALTER S., *Cable Television: A Handbook for Decision Making.* Santa Monica, Calif.: The Rand Corporation, 1973.

GILLESPIE, GILBERT, *Public Access Cable Television in the United States and Canada.* New York: Praeger Publishers, Inc., 1975.

KLETTER, RICHARD C., *Cable Television: Making Public Access Effective.* Santa Monica, Calif.: The Rand Corporation, 1973.

LEDUC, DON R., *Cable Television and the FCC.* Philadelphia: Temple University Press, 1973.

LEMELSHTRICH, NOAM, *Two-Way Communication: Political and Design Analysis of a Home Terminal.* Beverly Hills, Calif.: Sage Publications, Inc., 1974.

MUTH, THOMAS A., *State Interest in Cable Communications.* New York: Arno Press, 1979.

PILNICK, CARL and WALTER S. BAER, *Cable Television: A Guide to the Technology.* Santa Monica, Calif.: The Rand Corporation, 1973.

POOL, ITHIEL DE SOLA, ed., *Talking Back: Citizen Feedback and Cable Technology.* Cambridge, Mass.: The MIT Press, 1973.

PRICE, MONROE, and JOHN WICKLEIN, *Cable Television: A Guide for Citizen Action.* Philadelphia: Pilgrim Press, 1972.

Ross, Leonard, *Economic and Legal Foundations of Cable Television.* Beverly Hills, Calif.: Sage Publications, Inc., 1974.

Seiden, Martin H., *Cable Television U.S.A.—An Analysis of Government Policy.* New York: Praeger Publishers, Inc., 1972.

Smith, Ralph Lee, *The Wired Nation.* New York: Harper & Row, Publishers, Inc., 1972.

Veith, Richard, *Talk Back TV: Two-Way Cable Television.* Blue Ridge, Pa.: TV Books, 1976.

Yin, Robert K., *Cable Television: Citizen Participation in Planning.* Santa Monica, Calif.: The Rand Corporation, 1973.

7

networks and syndication

FOCUS   After completing this chapter, we should be able to

Discuss the concept of network programming, affiliate relations, and criticism of the network.

Explain the corporate development of ABC, CBS, and NBC.

List the businesses that comprise the networks' corporate structure.

Discuss contemporary radio networks.

Give examples of ethnic, educational, cable, and religious networks.

Explain how broadcast syndication functions as a network alternative.

Disucss the many facets of radio syndication.

Understand format control in syndicated programming.

List the reasons why the concept of a fourth network may someday be a reality.

ABC, CBS, NBC, and PBS—those magical initials of living room entertainment. We all have grown up with them, and they have shaped much of our lives. They still do. These are just the television networks. Radio has its own versions of national programming, followed by the syndicated networks. Major broadcasting networks are in most countries, the largest and most dominant systems of broadcasting. In this chapter, we shall view them as disseminators of information and as carriers of news and entertainment programming.[1]

the network concept

Networks not only provide the public with broadcast programming, but they also serve the advertiser and the broadcaster. For the advertiser, the networks provide a medium through which they can reach large numbers of people at economical rates. Even though the cost of a minute-long commercial on a major network can be staggering, the cost of reaching the same number of people through another medium, such as direct mail, would be far greater. Although direct mail can be used to reach a highly specialized audience, regional networks, a wide variety of programs, and selective scheduling of commercials help broadcasting remain competitive in reaching specialized audiences.

For the broadcaster, airing network programming is much more economical than the cost of producing and airing comparable local programming. In addition, affiliates usually receive compensation from the network for airing the programs.

Acquiring Programs

Within the networks, the news and public affairs units are responsible for informational programming. Supported by major staffs of reporters, producers, directors, and technicians, the units produce daily newscasts seen and heard by millions. They also provide special coverage of such important events as elections, inaugurations, press conferences, and space flights. The units offer special features and documentaries, which can range from a superficial look at a national fad to an investigative report of political corruption. Joint cooperative documentaries and reports with the networks' owned and operated (O & O) stations are still another source of network news programming, as are reports contributed by local affiliates.

168

Entertainment programming has another story. Here, various independent companies called *production houses* work closely with the networks to supply everything from detective thrillers to situation comedies. The ideas for programs come from the network, the production house, writers, or anyone else connected with the creative process. If a network is interested in an idea, both the network and the production house make a major investment to produce a pilot program. Audience reaction to the pilot program is then evaluated. If the ratings prove that it will be a hit, the production house moves into full-scale production. A complete season of shows is usually produced only after testing the program during a new television season to see whether its popularity will last more than a few weeks. Conversely, if the pilot is disappointing, the program may never get on the air. Other factors determining an air date may include affiliate reaction to a program, scheduling availabilities, and competition from other shows.

Affiliate Relations

How much attention do the networks really pay to their affiliates' wishes? On such major issues as clearing programming (agreeing to air a network program), the networks carefully heed them. This is a form of direct feedback, which is measured in terms of how many affiliates agree to carry the program, called *clearance ratios*.

Clearance ratios are critical to networks. If a group of stations decides not to carry a program, then the network's audience suffers, the ratings go down, and the advertising dollar follows. The clearance aspect of the network-affiliate relationship has taken on considerably more importance with the concern over violent and sexually explicit television programming. In many cases, local affiliate stations simply have refused to carry the network programs. In other cases, they have rescheduled the programming to a later hour when children will not be watching. Although that idea may sound reasonable, to a network, it can spell disaster since the late-night viewing audience is only a fraction of that in prime time. As a result of these continuing concerns, the *affiliate organizations* (also called advisory boards), groups of affiliate stations that elect a representative to carry their concerns to the network, are becoming increasingly important.

Affiliate Organizations

The organizations, some of which have grown out of advisory boards, have a set of bylaws and charge their members dues. There is no particular advantage of one over the other. Both function as feedback mechanisms to the network. Affiliates which do not wish to become part of the organization or advisory board still have a voice in the network.

No affiliate is overlooked. Each provides an audience for network programming and commercials. In some cases, the large-market affiliates purposely avoid belonging to the boards or organizations because they feel they have more impact as individual stations. In the same sense, some very small affiliates do not belong because they feel their voice is too small to be heard. These are the exceptions, not the rule. In theory, the affiliate organizations and advisory boards represent all of the affiliates.

The affiliates' organizations and advisory boards do not control the network directly, nor do they have any legal relationship beyond the contract between the network and the affiliate. The affiliates do, however, influence network decision making. Typically, the organizations and boards elect representatives from their membership, although some are appointed by the network. It then becomes the responsibility of the elected representatives to be the "voice" of the affiliates.

Voicing Affiliate Concerns

For example, imagine yourself an elected representative of a network affiliates' organization serving a group of states in the West. About two months before the network-affiliate representatives' meeting, you circulate a questionnaire to all of the affiliates asking them what information they want communicated to the network executives. You ask if they are satisfied with the quality of network programming, the scheduling of programming, whether they want more news and less sports or vice versa, and other pertinent information about the operation of their station, which is affected by their network affiliation. You also ask them to respond in a letter to any points not covered in the questionnaire. About a month before the meeting, you forward to the network a summary of the results.

At the meeting, the discussion centers on your summary and those provided by other representatives from around the country. One radio representative points out that many of the affiliates have a contemporary music format, and ten minutes of network news on the hour is hurting their listenership. On the other hand, the news and information affiliates love the ten minutes. The network must make a decision. It decides to continue to air ten minutes of network news every hour but to permit the stations with contemporary formats to cut out of the news at five minutes past the hour. The network will, however, need to cut the *local availability*, that *commercial slot in the network feed that is sold locally by the station*. The contemporary affiliates do not mind. They are willing to sacrifice the commercial availability to return to musical programming and please their audience.

Or perhaps you are at a television affiliates' meeting when a representative, speaking for affiliates in a conservative part of the country, feels the sex and violence on network programming has become too explicit for the audience. Complaints continue to pour in to their stations from viewers and advertisers. The affiliates want something done to "clean up" the programming. Another representative from a large city is not quite as upset over the programming. She says the affiliates she represents are aware of the sex and violence but also like the high ratings the network shows are receiving. Obviously the network must make some tough decisions based on this affiliate feedback. Should it cancel the shows with sex and violence, irritating the affiliates from the big cities, or should it keep the programs on the air and risk losing affiliate support in some conservative markets? Whatever decision the network makes, even in our hypothetical examples, they will want to know what the affiliates are thinking and how they will react to any changes in network programming.

Criticisms of the Network Concept

Any time an organization becomes a powerful conglomerate, it attracts its share of criticism. Networks are no exception. Some of the criticism has been directed at the very concept of hugeness and the control it fosters. More than a decade ago, Vice President Spiro Agnew lashed out at network control in a famous speech to a Republican political gathering, claiming Eastern (liberal) bias in network news coverage. His speech caused a stir among journalists and gave the public a target for wide-spread criticism.

More recent criticism of network news came when CBS paid Watergate offender H. R. Haldeman for an exclusive interview. The incident launched a wave of discussion about "checkbook journalism," but CBS survived relatively unscathed. Similar criticism came when a one-time cooperative venture between the CIA and CBS became public knowledge. Criticism has also been satirized, in the movie *Network* starring Faye Dunaway. The movie went to the ludicrous point of depicting an anchorperson murdered on the air just to gain top billing in the ratings. The movie was a farce, but not everyone saw it that way. More than one theater owner reported people leaving the lobby shaking their heads about how terrible it actually was inside the network decision rooms.

Let us turn our attention away from criticism now as we take a closer look at major network operations. We shall examine the commercial networks, leaving public broadcasting for the next chapter. In chapter 3, on the development of modern radio and television, we discussed the formation of the early networks. The youngest major commercial network is ABC, formed when NBC was forced to dispose of its dual-network operation.

When ABC acquired its own identity in 1943, it began a concentrated effort to compete with its two closest rivals, CBS and NBC. Because of some advantageous breaks, the late 1940s was a profitable time for ABC radio. One of its biggest breaks was singing star Bing Crosby. When Crosby wanted to produce a prerecorded show instead of meeting the demands of the weekly radio appearances, ABC gave the innovative idea a try. It proved that even a prerecorded show could be a hit, and other stars followed Crosby's example. No longer plagued by the image of an unsteady toddler, ABC moved forward into television.

Gaining Corporate Maturity: Television and Mergers

In a special program originating from Broadway's famous Palace Theater, ABC launched its television ventures on Tuesday, August 10, 1948. ABC's flagship station, Channel 7, with the call letters WJZ-TV, showed a documentary on the progress of New York City, narrated by Milton Cross. Cameras caught the live action of a parade outside the Palace, street dancing, music from Times Square, an eighty-piece combined police and fire department band, majorettes, and horse-drawn fire engines and streetcars. Later that evening, viewers "watched" "Candid Microphone" with Allen Funt, a radio version of the future "Candid Camera." Shows in the coming weeks included "The Fitzgeralds," "Hollywood Screen Test," "Ethel and Albert," "You're Invited," "The Singing Lady," and "Quizzing the News."

A month later, ABC regional network programming began in the Midwest with hookups between stations in Chicago, Milwaukee, Cleveland, and Toledo. The Chicago Cardinals and Pittsburgh Steelers football teams launched the regional network.

Not all of ABC's corporate maturity came from television. Needing capital to make inroads against their competition, ABC announced plans for a merger with United Paramount Theaters, Inc. (UPT) in 1951. That same year, both the ABC and UPT boards of directors approved the merger, and the FCC held hearings on it in 1952. By early 1953, the merger was complete, and cash reserves of $30 million were added to ABC's bank account. In a shrewd personnel move, Robert T. Weitman, a vice president at UPT, was placed in charge of ABC talent. Weitman had previously been instrumental in advancing the careers of Frank Sinatra, Danny Kaye, Red Skelton, Betty Hutton, and Perry Como.

The mid-1950s signaled changes in both station operations and corporate structure at ABC. New call letters were assigned to the network's

O & O New York station, and WJZ became WABC. Within the company, five new divisions were formed: the ABC Radio Network, the ABC Television Network, ABC Owned Radio Stations, ABC Owned Television Stations, and ABC Film Syndication. By the end of the decade, ABC had become a formidable opponent.

Edging the Competition

Unfortunately, the "formidable opponent" status was where ABC television remained. Although profitable and popular, moving out from under the dominance of NBC and CBS was no easy task. It took the combination of a greater number of television stations on the air and popular programming to bring ABC out of the cellar.

The magic formula started to work in the mid-1970s. The network had already managed to excel in one important area, sports programming, with its popular "ABC Wide World of Sports." From Johnny Carson's puns on NBC's "Tonight Show" to major national programming awards, "Wide World of Sports" was clearly a favorite. Then, when weekends seemed saturated with football, ABC introduced "Monday Night Football." With stars Howard Cosell, Frank Gifford, and Don Meredith, the show became the "in" pastime for people and parties huddled in front of the television set. Excellent Olympic coverage added more gold to ABC's pot.

But sports could not manage everything. Entertainment had to put in its share. ABC launched a talent raid that plucked comedian Redd Foxx and "Today" host Barbara Walters from NBC and programming executive Fred Silverman from CBS. Walters gained publicity for her reported one-million-dollar salary, and Silverman for making ABC stock jump upward the day he announced his resignation from CBS. Perhaps ABC's biggest push into dominant prime-time television occurred with its presentation of *Roots*. First aired in January, 1977, the story of black struggle traced through the "roots" of author Alex Haley's family set new records in television viewing. The twelve-hour production included stars such as John Amos, Madge Sinclair (Figure 7–1), LeVar Burton, Lorne Greene, Ed Asner, Cicely Tyson, and others. A second showing of the same series again brought ABC the lead over its competitors, and the second series titled *Roots: The Next Generations* topped another week of prime-time for ABC in early 1979.

Meanwhile, Fred Silverman (Figure 7–2) had exited ABC in 1978 for a new job at NBC. And while he was still attempting to get NBC out of the ratings cellar, ABC continued to introduce new shows in its situation-comedy genre, one of the most popular becoming *Mork and Mindy*, the story of a Martian who lives with his girlfriend in Boulder, Colorado.

Figure 7-1 ABC's "Roots" achieved a rating success for the network and at the same time strengthened its position in prime-time competition with NBC and CBS. A host of stars appeared in the program, including Madge Sinclair and John Amos (pictured). The first series ran in the fall of 1977 and was repeated in 1978. "Roots, The Next Generations," first appeared in February, 1979. (Courtesy ABC, Wolper Productions, Inc.; Warner Bros.; Phil Gersh Agency; Bresler, Wolff, Cota and Livingston)

CBS

CBS had both the resources and personnel to be a formidable opponent of ABC. By the late 1940s William Paley had already established a track record as an excellent administrator and builder. He was joined at CBS by an exceptional management team.

Klauber, Kesten, and Stanton

Speculating on how CBS might have developed with different executives is difficult, but for its record of corporate growth, the report card of Paley and his management team shows high marks. With net sales of $1.3 million in 1928, the network, exclusive of allied businesses, had an

Figure 7–2 After bringing ABC to the top of the ratings, Fred Silverman left the network in 1978 to become NBC's president and chief executive officer. (Courtesy NBC)

annual revenue of $1 billion by the late 1970s, making it the largest single advertising medium in the world.

In addition to Paley, three people greatly contributed to CBS's early growth—Ed Klauber, Paul Kesten, and Frank Stanton. Klauber was a newspaper reporter by profession until Bill Paley coaxed him off the night city-editor's desk of *The New York Times* in 1930. Klauber, hired as Paley's assistant, is credited with shaping the character of early journalism at CBS. Paul Kesten came from advertising. Recruited as head of sales promotion in 1930, Kesten earned his stripes with the New York ad agency of Lennen & Mitchell. One of his first decisions was to hire the accounting firm of Price, Waterhouse to conduct an "audit" of NBC's claim to having the highest radio listenership. In the audit, CBS came out on top.

Dr. Frank Stanton joined CBS five years later in 1935. A psychology professor at Ohio State University, Stanton had an interest in measuring radio listenership. When CBS learned of his work, it brought him to New York at a $55 weekly salary and gave him the number-three position in a three-person research department. Frank Stanton continued to measure radio listenership and developed an electronic device to measure immediate responses to radio programs. He later left the research department and by 1942 was an administrative vice president. As much a statesman for the entire broadcasting industry as a CBS executive, Stanton rose to the top of CBS in 1946 giving the company what many said was a sense of character and responsibility.

Trial and Error in Corporate Expansion

CBS did not remain strictly in the broadcasting business. As with ABC, CBS began to apply its profits to acquisitions which in some ways directly supported, yet were different from network operations. As we learned earlier, a lucrative artist-management business gave the network ready access to top talent. However, the FCC questioned the propriety of the network's control of talent, so CBS sold its interests to the Music Corporation of America. Similar to those discovered by the other networks, the more profitable acquisitions turned out to be radio and television stations in large markets, not only giving the network an affiliate station, but also a share of the affiliate's profit.

CBS ventured into sports, publishing, and the recording industry. The sports venture was not successful. After CBS bought the New York Yankees in 1964, the team promptly sank into the doldrums. Attendance dropped, and CBS sold the baseball team in 1973.

In 1951, Frank Stanton asked CBS's creative director of advertising and sales promotion, William Golden, to design a corporate trademark for all these business ventures. Golden came up with the famous CBS "eye." Early depictions had it superimposed on a clouded sky to lessen the fuzzy edges of the sharp lines which showed up poorly on some of the early black and white receiving sets. Although dropping the background, the eye today remains one of the few corporate trademarks that has not undergone radical change over the years and is consequently one of the most recognized corporate logos in the world.

Programming

Until ABC began inching its way up the programming ratings ladder in the 1970s, CBS for many years dominated television programming. Some of the early favorites were "I Remember Mama," an evening family drama about an immigrant family with three growing children. "I Love Lucy" belonged to CBS and set the stage for a generation of situation comedies. The network opened CBS Television City in Hollywood which produced such shows as "Playhouse 90." Children's programming made a hit with the long-running "Captain Kangaroo." Such adult variety programs as the "Ed Sullivan Show" continuously capped the ratings. CBS News also had its share of successes with Edward R. Murrow (Figure 7–3) and "See it Now" becoming a classic in broadcast journalism. "Douglas Edwards and the News" (Figure 7–4) and exclusive interviews with such notables as President John Kennedy and Soviet leader Nikita Khrushchev kept CBS news on top and anchorperson Walter Cronkite topped polls as the United States's most credible person.

Figure 7–3 Edward R. Murrow. (WNET/13 & CBS Inc.)

Figure 7–4 Douglas Edwards.

NBC

The oldest of the three commercial networks is NBC. With its Red and Blue dual-network concept, it gained momentum early and was well entrenched when CBS arrived on the scene.

Sarnoff and Goodman

If we look at NBC's past and present, two people stand out as sharing the largest responsibilities—David Sarnoff (Figure 7–5) and Julian Goodman. For Sarnoff, the impetus came from RCA, NBC's parent company.

Figure 7–5 David Sarnoff, seen here holding magnetic tape which RCA demonstrated in 1953, ushering in the era of "electronic photography" for television. (RCA)

The shore-bound radio operator during the *Titanic* disaster, Sarnoff came to RCA from the American Marconi Company. His energies at RCA were directed to two areas—developing new broadcast technology and promoting television stars as a means of winning audiences. In technology, he encouraged the development of FM and committed millions of dollars to Zworykin to continue his experiments on an electronic television camera. With Sarnoff as its leader, RCA pioneered in color television, devloping the system finally approved by the FCC for full-scale color production (Figure 7–6).

Julian Goodman is a product of NBC more than of RCA. His lifetime career started with the network in 1945. Goodman was to NBC what Ed Klauber was to CBS. His stints for NBC News were as an editor in Washington and as head of news and special events for the NBC radio network. He attracted the attention of top NBC echelons after holding key news directorships during both the 1952 and 1956 national political conventions. Both years were during television's golden age when its merging dominance was watched closely by the public and industry alike. By the end of the 1950s, Goodman was head of news and public affairs. Six years later in 1966, he held the NBC presidency. Subsequent positions moved him to the chairman of the board of NBC.

Figure 7–6 The first commercial RCA color television receivers came off the production line on March 25, 1954, at the company's plant in Bloomington, Indiana. (RCA)

During Goodman's tenure, the network continued to grow, based on a three-pronged attack that Sarnoff had started and perpetuated: stars, color, and innovation.

Radio had been a medium of programs, but Sarnoff realized that television made people bigger than life. So people would be in what NBC invested its efforts. Major moves were made to attract and sign top talent. Names like Milton Berle (Figure 7–7) and the "Texaco Star Theater" gave America a new night at home with the television set. Sid Ceasar with "Your Show of Shows" and Eddie Cantor's "Comedy Hour" all contributed to the "people" orientation.

Two other factors helped even more—color television and innovative programming. With the FCC's approval of RCA's color system, it was only natural that NBC moved ahead full-speed to air as many color programs as possible. Regardless of whether you were watching in color or in black and white, the famous NBC peacock spreading its tail feathers spelled credibility. The first network colorcast was the 1954 Tournament of Roses Parade. Ten years later, NBC was producing almost all of its programs in color.

NBC initiated a series of firsts in programming, many later copied by other networks. With host Dave Garroway, the "Today Show" dawned in

Figure 7–7 Milton Berle.

1952 followed two years later by the "Tonight Show" with host Jack Parr, then Steve Allen, and finally Johnny Carson. By the 1970s, insomniacs were watching the postmidnight "Tomorrow" program. Westerns were also NBC's glory. The Ponderosa swept the imagination of millions with "Bonanza" (Figure 7–8). Then came the NBC "specials" like "Satin and Spurs" with Betty Hutton and "Peter Pan" with Mary Martin. In movies, the network pioneered with "Saturday Night at the Movies" and later "World Premiere Movies." News and public affairs programming included in-depth coverage of the political conventions, and NBC capitalized on the popularity of two anchormen named Chet Huntley and David Brinkley.

Today, with the network race even more hotly contested than in past years, NBC has worked hard to match CBS in news and ABC in sports. By acquiring programming whiz Fred Silverman from ABC and signing the rights to air the 1980 Winter Olympics, NBC has placed itself in a position again to rise to the top slot in prime-time ratings.

Figure 7–8   "Bonanza" became one of the longest running Westerns on television. Dan Blocker (left) has since died, Michael Landon (center) went on to produce and star in "Little House on the Prairie," and Lorne Greene appeared in the fall-1978 series, "Battlestar Galactica."

networks and allied businesses

Although broadcasting contributes a major share of the networks' income, it is interrelated with a wide range of other businesses, including publishing, recordings, amusement parks, and electronic supplies. For example, ABC not only owns the network as well as the ABC owned and operated radio and television stations, but it also owns ABC Theaters. Located in eleven southern states, the theater division constructs multiscreen theaters in new urban areas. With this large chain of movie houses, the network is able to negotiate profitable rental rates for first-run movies for its theaters. Publications such as *Prairie Farmer, Wisconsin Agriculturist,* and *Modern Photography* are owned by ABC as is the tourist attraction, Historic Towne of Smithville, located near Atlantic City, New Jersey.

At CBS, the organization chart (Figure 7–9) shows a Broadcast Group, Records Group, Columbia Group, and Publishing Group. We have already learned about the Broadcast Group which consists of the network, the owned-and-operated stations, and CBS News. The CBS Records Group is the world's largest producer, manufacturer, and marketer of recorded music. Bob Dylan, Chicago, Barbra Streisand, and Lou Rawls are just some of the stars recording on CBS labels. Record and tape clubs, retail stores, Steinway pianos, and Creative Playthings toys are part of a company division called the Columbia Group. The publishing house of Holt, Rinehart & Winston and the proprietary schools of Brown Institute and the Kansas City Business College in Missouri are part of the CBS Publishing Group.

For NBC, the umbrella of allied corporate interests is RCA. Electronic parts and equipment continue as big businesses. Brand names such as the XL-100 TV line are part of RCA's division of Electronics-Consumer Products and Services. Commercial electronic products are manufactured and sold as part of RCA's Electronics-Commercial Products and Services. If you buy a Banquet brand frozen dinner, rent a Hertz Rent-A-Car, read a book published by Random House, or talk to Alaska through the Globcom satellite networks, you are contributing to RCA's income.

contemporary radio networks

When television reached its golden age, some predicted the demise of radio, especially the radio networks. This simply has not happened. Although they do not provide the same amount of programming they once did, radio networks are still a vital part of radio broadcasting. In fact, with the soaring costs of television advertising, radio networks have been increasingly attractive to advertisers.

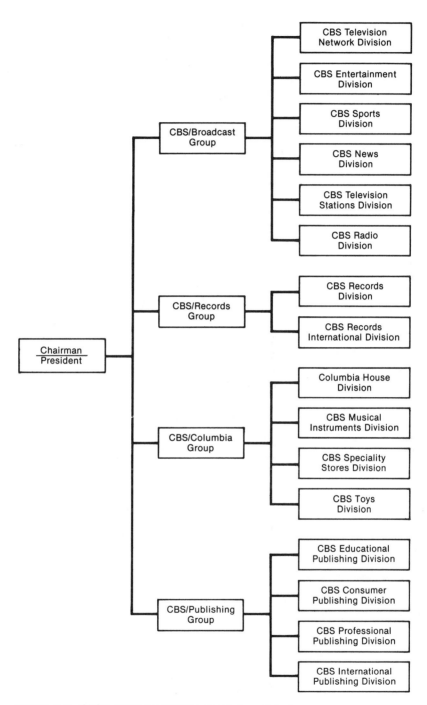

Figure 7–9 CBS's allied businesses range from television, to recordings, to retail sales, to publishing.

183

ABC's Demographic Networks

One of the more innovative ideas in modern radio was the decision by ABC in 1968 to break off into four separate radio networks. The split created the American Information, American Contemporary, American Entertainment, and American FM Radio Networks. The four networks are designed to serve affiliates reaching audiences with different demographic characteristics. For example, stations programming rock music and reaching a younger audience might affiliate with the American Contemporary Network which has short, fast-paced newcasts. A station with news and information programming might affiliate with the ABC Information Network. These same demographics are equally attractive to advertisers wishing to reach those audiences.

The content of the news on the different networks also varies. Designed for younger audiences, the Contemporary Network has less emphasis on foreign news and more on consumer-oriented, "pocketbook" stories. The Information Network emphasizes foreign and domestic politics and political aspects of the economy. Many affiliates with the American FM Network program progressive rock music, reaching a young, college-age audience. Thus, stories on the FM Network emphasize such things as careers, young politicians, and information of interest to an audience ABC characterizes as "thoughtful and involved." American Entertainment Radio News serves affiliates that offer easy listening, middle-of-the-road music, reaching an older audience. Along with regularly scheduled newscasts, the Entertainment Network features "Paul Harvey News."

Using the same feed lines, each of the networks broadcasts its separate newscast at different times during the hour. Thus, ABC gains maximum efficiency and minimum line charges per network. On the hour, Information Radio programs five minutes of news. FM then airs news from 15 to 19 minutes after the hour. Entertainment takes over at the half-hour for five minutes, and Contemporary Radio comes on with a minute-long "News in Brief" at 50 minutes and 30 seconds after the hour, and a 4 1/2-minute newscast at 54 minutes and 30 seconds after the hour. What started out with 400 affiliates in 1968 when the four networks began has grown today to over 1,500 affiliates.

Mutual, NBC, CBS

Of the major networks, only the Mutual Broadcasting System, MBS, remains exclusively radio. Its identity has been shaped by regularly scheduled newscasts, including such familiar names as Fulton Lewis, Jr. and his son Fulton Lewis, III, airing regular commentaries on the network. Mutual affiliates also receive such major sports events as Notre

Dame football, championship boxing, NFL football, PGA Golf, and the Sugar Bowl. Colorful commentaries are heard from regulars Jack Anderson and oddsmaker Jimmy the Greek.

In 1975, trying to capture some of the still unclaimed affiliates market, NBC launched its News and Information Service (NIS) which fed approximately forty-five minutes of news per hour to all-news affiliate stations. The idea was good. All-news stations, although reaching mainly an educated, affluent audience, incur major expenses. Costs for broadcast journalists can be high, especially in smaller markets. To affiliates, NIS was costly but less expensive than hiring local personnel to cover national news. Unfortunately, the number of affiliates needed to make NIS profitable did not materialize, and in 1977, NBC dropped the service. Some NIS affiliates scrambled to develop an alternative network, others switched to wire-service audio or other networks, and still others went back to programming music. For some affiliates, the dissolution of NIS was costly. Major promotional campaigns had publicized their all-news operations, and they suddenly either had to change formats and consider their promotional campaigns as lost revenue, or make major expenditures to hire local reporters to cover a full news schedule. Today, the NBC Radio Network still serves affiliates nationwide, but NIS is a part of history.

CBS, perhaps watching the NBC experience, has yet to specialize its radio operations. CBS radio affiliates still receive a regular schedule of network newscasts, many of which are anchored by nationally recognized newscasters from CBS television.

UPI Audio and AP Radio

Although not networks in the traditional sense, many radio stations affiliate with wire-service audio. Local stations pay a fee to subscribe to the wire-service audio. The stations then can sell time within the audio newscasts to local advertisers, which many station managers feel is a more profitable arrangement.

United Press International first launched an audio service in 1956 with state news telephone feeds in North Carolina and California. By 1960, the wire service established a New York audio headquarters and three years later added a Washington, D.C. audio bureau. The next year an audio bureau in London was opened, and by 1965, coast-to-coast hookups were operable. The 1970s brought hourly newscasts, expanded sports coverage, and experimental satellite transmission. In 1977, UPI initiated a regional audio service which operates out of its Chicago and Los Angeles bureaus. In addition, UPI Audio feeds affiliates a daily set of audio actualities which stations can incorporate into locally produced newscasts.

Associated Press began an audio service in 1974 called AP Radio. Based on the same principle as UPI Audio's national news service, affiliates pay a fee to receive the service, then sell local advertising time within the regularly scheduled newscasts.

Wire Service and Audio Advisory Boards

Along with the networks, the wire services also use affiliate advisory boards. These can operate at either the state or national level. Typical of such organizations is the UPI Broadcast Advisory Board. The history of the organization can be traced to 1976 when it was conceived at a steering committee meeting of UPI executives and broadcasters.

The committee formulated plans for the UPI Broadcast Advisory Board and constructed a set of initial objectives.

1. Evaluation and improvement of UPI Services.
2. Better understanding of UPI technology by subscribers and public.
3. Improvement of general image of broadcast news.
4. Encourage input, discussion, and recommendations at all levels.
5. Forum for discussion of major complaints affecting the general performance of the service.

These objectives were later used in writing the board's bylaws.

The UPI Broadcast Advisory Board met for the first time in December 1976 in Chicago. To assure representative membership and orderly replacement of board members, the bylaws provide for a board "generally representative of geographical areas, large and small radio and television markets, independent and group ownership, management and working news people." Membership on the board consists of "a minimum of 13 and a maximum of 16 members, and the Board is empowered to expand its membership to include international members." The bylaws also provide for the election of officers, terms, and special committees. Evidence of the rapidly changing developments in broadcast journalism comes under the heading of "technology," which charges the board with encouraging "research and development of new methods and systems in broadcast news."[2]

ethnic, educational, cable, and religious networks

Whereas ABC divides its listeners into such demographic categories as age, education, and income, two other networks have been successful in directing programming to audiences based on race and national origin. In radio, the National Black Network formed in 1973 serves more than

80 affiliates covering 90 black markets and reaching approximately 70 percent of the United States' black population. With headquarters in New York, news on the network emphasizes events and issues of importance to black Americans. Owned and operated by blacks, the network currently produces 120 hourly newscasts each week. Other programs heard on the network include the 30-minute news forum, "Black Issues and the Black Press," patterned after NBC's "Meet the Press." Commentators and entertainment programming supplement the regular newscasts.

What the National Black Network offers to the black audience, the Spanish International Network (SIN) provides for the Hispanic-American audience. Formed in 1961 to reach Spanish-speaking households, SIN is a television network which airs Spanish-language programming and programming pertinent to Hispanic-American culture. Affiliated with television stations reaching a large percentage of Hispanic-Americans, there are SIN affiliates in Chicago, New York, Miami, and similar markets stretching from Texas across the Southwest and up the California coast to San Francisco.

Although we shall learn later in this text about public broadcasting networks, we should be aware that many noncommercial radio and television stations not only provide instructional programs for in-school use, but also are a transmitter and source of programming for many state and regional networks. Often these networks are part of a state system of higher education. One of the pioneers is the Alabama ETV network. In Indiana, the Indiana Higher Education Telecommunication System (IHETS) links both state and private colleges and has two-way television capabilities. In this way, a professor can lecture from one campus to students on another. The students immediately can ask questions of the professor through a two-way talk-back system. The talk-back system is another example of the changing definition of mass communication which we discussed in chapter 1—delayed feedback being replaced by immediate feedback.

Cable television's growth has spurred interest in networks linking cable systems. With satellite interconnection systems, the technology is already operable to offer satellite-distributed programs to cable systems. Other cable networks operate more as an exchange, sharing programs or joining to purchase such special programming as sports events or first-run movies.

Multipoint distribution service—using microwave to distribute television signals over a small regional area—has enabled such organizations as schools and churches to enter the world of network television. One example of this is operating at the Catholic Television Network of Chicago (CTNC). With its own television studios and distribution system, the network broadcasts educational television programs to Chicago's Catholic Schools.

syndication: the network alternative

Syndication is the distribution of a self-contained program directly from a production house to the station, bypassing the networks. The station schedules the program at its convenience, paying a fee for the program or bartering it through trade-out arrangements, such as agreeing to air advertising already contained in the program. Shows like "Lawrence Welk," "The Mary Tyler Moore Show," and "Hee Haw" all are syndicated.

Breaking the Ice: "Mary Hartman, Mary Hartman"

In order to go into syndication, most programs first started as network material, proved they could attract a loyal following, then made the break with the network and wooed the individual stations. The first program to go directly into syndication on a mass national scale and achieve popularity was Norman Lear's somewhat controversial "Mary Hartman, Mary Hartman" (Figure 7–10).

Having been turned down by the television networks, Lear resorted to direct syndication for the show, and its popularity took off, drawing

Figure 7–10 Louise Lasser starred in "Mary Hartman Mary Hartman," one of the first series to go directly into national syndication and be successful. (T.A.T. Communications Company)

large audiences. Some bold, metropolitan stations even backed the program up against their competitors' late-evening news programs, scoring rating points that gave news consultants a headache.

But "Mary Hartman, Mary Hartman" started a chain of events that may do more than merely attract television viewers. Local affiliates have become increasingly dissatisfied with the type of programming coming out of network hoppers. With "Mary Hartman, Mary Hartman," station managers are realizing that there may be other alternatives to the network distribution system. What is popular in one city may not be in another, and syndication gives managers a freedom of choice that network television does not.

Advantage of Network Exposure

"Mary Hartman, Mary Hartman" was an example of direct syndication. But airing a show on the network and then placing it in syndication is still the preferred route. When the network takes the program and promotes it in a major prime-time lineup, it reaps the benefit of a coordinated, national promotional effort that few syndicators could afford to match. Plugs in prime time, newspaper advertising, and other publicity give the program the best possible chance of success. Remember, the network wants the show to succeed just as much as the company that produces it does.

With direct syndication, individual stations are responsible for promoting the program within their own schedules. That may mean a huge advertising campaign or hardly a mention. Although promotional literature and photographs are made available to the subscribing stations, the station is under no obligation to use the material. Besides, local news often is favored in the allocation of the local advertising budget.

If a program does become a network hit, then syndication is almost assured. The profits to be made in syndication will probably far exceed the money earned from the network. In fact, some production companies sell a program to the network at a loss, hoping to regain the loss when the program reaches syndication.

Although a big gamble, the money syndication can bring makes it worth the risk. The money can continue to trickle in for years to come. The cartoon program "The Flintstones" was aired on network television almost twenty years ago. It is still in syndication today with its lifetime earnings climbing over $30 million in 1980. A program of equal success currently on the network will bring many times that much syndicated revenue, based on current syndicated rates, which can easily bring in tens of thousands of dollars for a single half-hour program in a major market like New York or Los Angeles.

Direct Syndication

Despite the publicity a network can give a program, direct syndication's popularity is growing. We have already learned that direct syndication's customer list is expanding due to the dissatisfaction of some station managers and viewers with network programs. The success of syndicated programs has lured more syndicators into the marketplace, increasing the programming variety and keeping costs competitive.

Direct syndication eliminates the need to "fit" the network audience. For example, if you own a production company that produces a weekly series on skiing techniques, the network probably will not be interested. Although its affiliates in New England and the Northwest would consider the show, stations in the deep South would very likely preempt the program with something more popular in their local areas. Yet by going into direct syndication, you can create your own network of stations. In fact, you might find enough stations in New England and the Northwest interested in airing your series to make your venture profitable.

The same principle applies to the expanding enterprise of regional syndication. Perhaps your area hosts a salt-water fishing tournament. A documentary of the tournament would be of interest to other stations located near the ocean. Conversely, generating wide acceptance for the show in Nebraska and Iowa would be difficult.

Major events are becoming attractive to syndicators. Tying up rights to special productions can make a long-term syndication package very lucrative. For example, although the Miss America Pageant is a network production seen internationally, you may be surprised to learn there is also a Miss Rodeo America Pageant. Held in the Southwest in conjunction with the National Finals Rodeo, contestants from the 50 states and Canada compete for the title of Miss Rodeo America. Although the networks shied away from the pageant, syndicators saw major audience appeal, especially with pretty girls, colorful Western outfits, and such hard-fought competition as barrel racing and showing horses. These all spell color, action, and entertainment—three appealing ingredients to a television syndicator. Public broadcasting saw the potential first and gave the pageant regional television syndication. The event is just one more example of the growing variety and potential of direct syndication.

Many syndicated programs find their way out of the broadcasting station and into an educational institution, corporation, library, or church. A syndicated religious documentary that first airs on television can find additional audiences at a Sunday morning Bible class or at a convention of lay preachers. As you can see, the market possibilities for syndicated material are turning it into a booming business.

syndicating radio

So far we have been talking about television syndication. Syndication also is alive and well in radio, especially since the networks take up so little of the staion's programming schedule. With the increased number of automated staions, syndication has found a ready market. With automation, the local station still programs commercials and local newscasts, but much of its remaining programming comes from syndication. It can choose from a wide variety of syndicated radio programming, anything from music to interviews and talk programs. In addition, both nonautomated and automated stations use syndicated production aids called "jingles" to insert between records, introduce newscasts and weather reports, and adopt as a musical background for commercial and public service messages.

Why Syndicate?

Syndication is an economical way to operate a station while at the same time providing a "competitive sound." But before deciding to use syndicated programming, you first should evaluate the competition. Is there a programming need not being met or an audience not being reached by the competing stations? Second, you must decide that when your station does air locally produced programming, it will be of the highest quality. Commercials, local newscasts, weather reports, and supplemental, local disc jockey entertainment must be able to blend professionally with the syndicated music. Poor quality in local production is only accentuated by high quality syndicated programming.

Formats

A wide variety of syndicated formats are available, but not every syndicator defines the same format in the same way. Thus, managers must preview the syndicated "sounds" to make sure they fit the need of their market. Most syndicators offer a limited number of formats, keeping to a reasonable minimum the size and range of their own musical libraries. Syndicators specializing in contemporary music might syndicate progressive-rock, country-rock, and soft-rock formats. Each of the formats tends to overlap, and hit songs which are on more than one chart, such as *Billboard*, can be inserted in more than one syndicated format.

Other syndicated packages can be even more specialized, offering "middle-of-the-road string orchestras" or "up-beat string orchestras." Still another may contain "middle-of-the-road orchestras and bands." To offer these specialties, the syndicator purchases mostly albums and interchanges the different cuts on the albums to fit the different formats.

191

Talk radio is another specialized format in which a talk-show host interviews listeners who call in to the station. The original program can be taped in one market and aired in another. The host is careful not to make any reference to the city in which the show originates, and the topics do not have a highly localized emphasis. If the show does become localized, that part can be deleted before the program is syndicated.

The Consultant as Syndicator

In television syndication, the syndicator has little involvement with the station except to sell the program. In radio, however, the association is much closer. Radio formats are a more "finely tuned" type of programming than television formats. First, they last longer, in some cases for the entire 24-hour broadcast schedule. Second, the competition may be much greater. Instead of a market of three or four television stations, it may have twenty or more radio stations. Third, the local radio station directly participates in the programming because of its locally produced and inserted commercials and newscasts.

As a result, the radio syndicator often doubles as a broadcast consultant, recommending how the programs should be utilized. Judging whether the local commercial production is up to par or if the musical background of the commercials matches the syndicated format becomes the consultant's major concern. If the station is going to continue to use syndication, it must see an increase in audience and, ultimately, in profits. Moreover, the reputation of the syndicator is at stake. A station at the bottom of the ratings is not good publicity for the syndicator. But one which is on top can be a valuable asset in selling other stations. Thus, along with buying the syndicated programming, management's willingness to work closely with the consultant can determine the difference between success or failure.

Anthologies and Concerts

Along with regular radio programming, syndicators have found profits and popularity in special anthologies of music. With heavy promotional efforts, the anthologies are programmed as a continuing special over whatever schedule fits the station. Some of the most successful include "The History of Rock and Roll," "The Golden Year of Rock," "Top 100 of the 60s," and "The Golden Years of Country."

Rock concerts are another popular syndication feature, and "The King Biscuit Flower Hour" has one of the biggest followings. The unusual name, drawn from old-time radio, is the brain child of two former network executives, Bob Meyrowitz of NBC and Peter Kauff of ABC. The "Hour," which sometimes runs 90 minutes, is a recording of an actual rock concert. The recording is then syndicated by D.I.R. Broadcasting Corporation. Meyrowitz and Kauff are the president and executive vice president, respectively, of D.I.R. Recording very popular rock groups, the "Hour" is syndicated to more than 200 radio stations and has become a Sunday night ritual for more than 6 million listeners. In many markets, it effectively competes with prime-time television.

Spinoffs of "The King Biscuit Flower Hour" have also been successful for D.I.R. Repeating some of their best concerts, a show aptly entitled "The Best of Biscuit" airs the first Sunday of each month. D.I.R.'s show "Conversation" contains interviews with rock stars interspersed with their music. "Conversation" is hosted by Dave Herman, a radio personality from WNEW-FM in New York. Other D.I.R. programs have aired in England under the title of "The British Biscuit."

format control in syndicated programming

With the increase in both the number of automated stations and the reliance on syndicated musical programming, the issue of format control also has become important to the field of syndicated programming. Keep in mind that the FCC still feels the licensee should be responsible for its format in order to serve the public interest. As a result of some rather binding contracts offered to broadcasters by syndication companies, the FCC has adopted guidelines for broadcasters to follow when agreeing to carry syndicated programming. The commission suggests stations should not enter into contracts which:

1. Fix the number of broadcast hours;
2. Prohibit AM-FM duplication;
3. Prohibit sub-carrier authorizations;
4. Require the exclusive use of any music format service, or prohibits other sources;
5. Fix the amount of format service company music broadcast; [*sic*]
6. Prohibit any announcement by the station;
7. Fix the number of commercials broadcast;
8. Limit the content or source of any nonmusical programming;
9. Fix the amount of air time for news, music, or other programming;
10. Prohibit automatic gain control of company-supplied material; or
11. Allow termination in the event of program format changes by a licensee exercising his responsibility for the public interest.[3]

The key to the contracts is to retain flexibility. This is especially important in long-term contracts. The FCC does not want the licensee obligated to the degree that programming "in the public interest" might not air because of restrictions placed upon the station by the syndicator.

Syndication contracts have also been scrutinized because of recent changes in rules affecting simulcasting. Before 1977, an FM station owned with an AM station could not duplicate the programming of the AM station more than 50 percent of the time in markets having a population of over 100,000. In 1977, the rule was changed to 25 percent duplication in markets of over 100,000, and in 1979, the 25 percent duplication rule becomes effective for stations in markets over 25,000. Additional syndicated programming thus is often used to meet these nonduplication requirements.

the fourth network concept

The success of syndication has provoked discussion of the possibilities of a fourth network. Part of the discussion is on semantics. What exactly is a fourth network? A major network of the size and scope of one of the commercial networks or the public broadcasting system is not yet foreseen. But an alternative network with quality programming, although with fewer stations and fewer broadcasting hours, is conceivable.

Four reasons give this new network its potential. First, with the revenue that syndication is generating, industry planners are beginning to feel that a network would have a chance of financial success. Profits of independent stations, those most likely to carry the new network, are up, and they can thus afford to pay higher prices for quality programming. Second, even network affiliates are receptive to the idea, especially those who are unhappy with the programming of the three commercial networks. Third, the technology is ready. With satellite systems capable of transmitting programs to cable companies with earth stations, a "wired" fourth network is already functional in some areas. A fourth reason is the current advertising rates on the three commercial networks. Rates, although still economical considering the audience reached, have reached a level at which only large companies can afford major advertising campaigns. Thus, dollars are being channeled into such other media as direct mail and magazines, dollars which theoretically could be used for advertising on a fourth network.

A fourth network could be successful if advertisers were willing to make a commitment beyond the "experimental" stage. If there were a commitment the network would need quality programming which can compete with that of the commercial networks. It would require stations willing to air programs simultaneously in order that national ratings could be made for specific day-parts and programs. If these prerequisites were met, then the fourth network concept could become a reality.

summary

Chapter 7 discussed networks and syndication, mostly the commercial networks with public broadcasting to be discussed in another chapter. Networks give advertisers a means of reaching a mass national audience at economical rates. Their entertainment programs are usually bought from production companies, but their news and public affairs programming is self-produced. At the heart of the networks are the affiliate stations which, with the public, have strongly criticized network operation.

The three major commercial networks, ABC, CBS, and NBC, remain viable and profitable. ABC, formed when NBC sold its Blue Network, quickly acquired an identity and became a formidable competitor of NBC and CBS. CBS gained an early reputation as a leader in broadcast journalism under Ed Klauber and in sound management under William Paley and Dr. Frank Stanton. With little more than a million dollars in revenue in 1928, the network had its first billion-dollar year in 1976. NBC, meanwhile, developed under the RCA umbrella and the guidance of David Sarnoff. Using big-name stars, color, and innovative programming to make its mark, NBC continues to make profits. Although the networks themselves are big business, all are interconnected with other business ventures. These include such wide-ranging enterprises as publishing, amusement parks, rent-a-cars, and frozen foods.

All three of the major commercial networks are in radio, as is the Mutual Broadcasting System. ABC, in attempting to satisfy the specialized audience of radio, split into four demographic networks in 1968. Both UPI Audio and AP Radio provide alternatives for radio stations not affiliated with the commercial networks or for those which want to supplement their network programming. Other examples of radio and television networks include ethnic, educational, cable, and religious networks.

A growing alternative to network programming is syndication, which received a positive push from the success of "Mary Hartman, Mary Hartman." Radio syndication is also booming, as is the number of automated broadcast consultants. With the high cost of advertising on the big three commercial networks, the concept of a fourth network could very well become a reality.

spotlight on further learning

BAILEY, ROBERT LEE, *An Examination of Prime Time Network Television Special Programs, 1948 to 1966.* New York: Arno Press, 1979.

CAMPBELL, ROBERT, *The Golden Years of Broadcasting: A Celebration of the First 50 Years of Radio and TV on NBC.* New York: Simon and Schuster, Inc., 1976.

DREHER, CARL, *Sarnoff: An American Success*. New York: Quadrangle/New York Times Book Co., 1977.

FRIENDLY, FRED W., *Due to Circumstances beyond Our Control* . . . New York: Random House, Inc., 1967.

GERANI, GARY, and PAUL H. SCHULMAN, *Fantastic Television*. New York: Harmony Books/Crown Publishers, 1977.

HESS, GARY NEWTON, *An Historical Study of the Dumont Television Network*. New York: Arno Press, 1979.

KIRKLEY, DONALD H., JR., *A Descriptive Study of the Network Television Western During the Seasons 1955–1956 to 1962–1963*. New York: Arno Press, 1979.

LONG, STEWARD L., *The Development of the Television Network Oligopoly*. New York: Arno Press, 1979.

METZ, ROBERT, *The Today Show*. New York: Playboy Press, 1977.

MORRIS, JOE A., *Deadline Every Minute: The Story of United Press*. Westport, Conn. Greenwood Press, Inc., 1969.

NBC, *Broadcasting the Next Ten Years*. New York: NBC, 1977.

PALEY, WILLIAM, *As It Happened*. New York: Doubleday, 1979.

PEARCE, ALAN, *NBC News Division: A Study of the Costs, the Revenues, and the Benefits of Broadcast News*. New York: Arno Press, 1979 (includes in the same volume *The Economics of Prime Time Access*).

SHANK, BOB, *The Cool Fire: How to Make It in Television*. New York: W. W. Norton & Co., Inc., 1976.

8

educational and public broadcasting

FOCUS After completing this chapter we should be able to

Trace the beginnings of educational and public broadcasting.

List some of the earliest uses of educational television.

Tell how the MPATI experiments affected the development of educational television.

Explain how educational broadcasting fits into the learning experience.

Discuss the transition from educational to public broadcasting.

Explain the functions of PBS and NPR.

Describe how the Station Program Cooperative operates.

Understand the scope of programming on public stations.

List questions posed to the second Carnegie Commission studying public broadcasting.

At 6:55 A.M., radio and television stations are keying up for another broadcasting day. For some, sign-on came in the predawn darkness. For others 7:00 A.M. will arrive with the "Today Show," "CBS Morning News," or "Good Morning, America." In foreign countries, 7:00 A.M. ushers in news, special-feature programming, entertainment, and public affairs.

Sign-on preparations are also bustling at this hour in many school systems. A closed-circuit television system warms up to "broadcast" the daily calendar of events, students begin to produce a morning news program, and teachers preview instructional television lessons that they will incorporate into afternoon lectures. At a nearby college, a professor is preparing a lecture that will be aired over a statewide educational television network. In a famous medical school, a television camera focuses on the operating table (Figure 8–1), broadcasting a color picture to interns in an observation room across campus. It is all part of the world of educational broadcasting.

Figure 8–1 (Division of Audio/Visual Services, Johns Hopkins Medical Institutions)

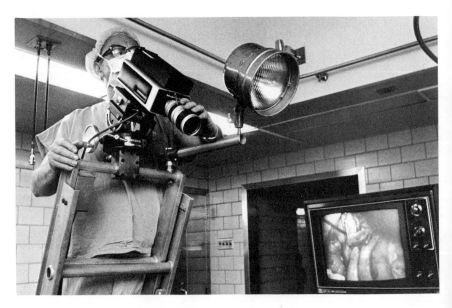

ETV—the beginnings

Although a closed-circuit television system was in use at the State University of Iowa as early as 1932, it was in 1938 that the first over-the-air experimental broadcast for educational purposes took place in cooperation with NBC and the School of Commerce of New York University.[1] The broadcast was arranged by Dr. James Rowland Angell, who was then educational director of NBC. From a studio on the third floor of the RCA building, the program consisted of "an explanation and demonstration" of television, broadcast from the transmitting tower on top of the Empire State Building. A class of 250 students seated in a large auditorium on the sixty-second floor of the RCA building viewed the program. A two-way radio hookup connected the studio with the auditorium. Capturing the flavor of the event, program instructor Professor C. C. Clark recalled that he asked one student in the auditorium to come to the studio to have a question answered. About ten minutes later when the student arrived at the studio and appeared on the screen, the group in the auditorium broke into applause.

The Experimental Era

The early 1940s were years of continued experiments with the new medium. The Metropolitan Museum of Art in New York arranged with CBS to televise its painting collection. Francis Henry Taylor, director of the museum, predicted television would be "just as revolutionary for visual education as radio was for the symphony and the opera."[2] The following year saw such mass-oriented educational television programs as New York's WCBW broadcast of a series of first aid programs in cooperation with the American Red Cross. Programs of a more informative and educational nature during the height of World War II, included one from Schenectady, New York, on blood plasma.

Early enthusiasm for educational television was sidelined by World War II, but when the war ended the industry began to concentrate again on educational television. NBC announced the first "permanent" series in educational broadcasting, "Your World Tomorrow."[3] Some of the early program titles in the series included "The Mighty Atom," "Jet Propulsion," and "Huff-Duff, the Radio Detective." The network secured the cooperation of the New York City Board of Education to have students watch the programs in special "viewing rooms." The students then evaluated the programs. On a very limited scale, this was an early example of the systematic evaluation of educational programming.

Despite the encouragement from the networks and the willingness of certain school officials, educational television was a long way from widespread acceptance. As late as 1947, the *Journal* of the National

199

Education Association reported the efforts of the state of Virginia to make the transition to what was termed "visual education."[4] Although Virginia was known for its pioneering efforts in the field, the report never even mentioned educational television. The medium had not been able to rise above all the movie projectors, slide projectors, charts, models, and posters of the typical classroom. In the same year, a survey of elementary teachers studied the skills and knowledge necessary to use audio-visual aids.[5] Out of 42 survey items, which included mechanics, utilization, production, and facilities, none referred to television.

ETV Gains Acceptance

Finally at the turn of the decade, it began to change. In 1949, Crosley Broadcasting awarded a fellowship to a Kentucky high school principal, Russel Helmick, to "carry on research of how education by television can best serve the needs of the general public." The broad descriptions assigned to Helmick's charge illustrate the early survey approach to researching educational television:

1. Careful sifting of the literature of radio education to discover mistakes to be avoided and lessons helpful in investigating the educational possibilities of television.
2. Analysis of the television programs available for possible correlation with adult-education programs and the curriculum at university, high-school, and elementary-school levels.
3. Canvassing of school and home facilities and equipment for utilizing video programs.
4. Study of teacher interests and attitudes toward correlation of their school offerings in the curriculum with cultural and educational programs from television stations.
5. Investigation of pupil attitudes and interests in such cultural and educational areas as history, geography, English, science, and physical education and sports in relation to utilization of appropriate television programs.
6. Investigation of the educational levels at which television can be made most effective—adult education, colleges and universities, high schools, and elementary schools.[6]

Thirty years later, more narrowly defined and highly sophisticated research was still conducted within the very broad parameters of Helmick's goals.

Then in 1950, public awareness of the importance and potential of educational television rose when Dr. Earl J. McGrath, U.S. Commissioner of Education, appeared at hearings before the FCC and called for at least one channel in every broadcasting area to be reserved for educational purposes. McGrath suggested "that it is vital to the continuous improvement of public education that every school system and college competent to produce educational television programs and financially able to construct and operate a station be assured that, when the time comes that it is ready to start construction of a television broadcast station, a suitable locally usable transmitting frequency will be available."[7] The FCC responded favorably.

Organized Support for ETV

Organized support for educational television came in the early fifties when the American Council on Education coordinated the formation of the Joint Committee on Educational Television (JCET). The committee brought together seven supporters of ETV, all organizations which had originally called upon the FCC to hold hearings on the subject. Financial commitment was also provided by a $90,000 grant from the Ford Foundation. One of the committee's main goals was to assist educational institutions in establishing stations. The first chairman of the JCET was Dr. Edgar Fuller, then executive secretary of the National Council of Chief State School Officers, one of the seven member-organizations of JCET.

Other ETV financial support went directly to colleges and universities. Syracuse University received a $150,000 gift to offer graduate programs in radio and television. In the fall of 1950, Syracuse announced the new degree of Master of Science in Radio and Television.[8]

The medium even gained some artistic legitimacy when *Variety*, the show business weekly, reviewed a "University Tele-course" program aired on the Cleveland, Ohio station, WEWS. *Variety* called the program a "preciously packaged mine of informational nuggets," and "a fast-moving, easily digested, highly accredited performance."[9] Few professors in the classroom could have been so enthusiastic.

The FCC, as part of lifting its freeze on new licenses, allocated 242 channels for educational use. The first to take advantage of the newly allocated frequencies was the University of Houston. Station KUHT went on the air in 1953 (Figure 8–2). But the expensive facilities and television equipment necessary for transmission prohibited the rush to the marketplace that the FCC had anticipated. Nevertheless, educational television had delivered its firstborn.

As the decade progressed, forecasters had educational television helping to teach the huge enrollments predicted in the 1960s and 1970s. Educational television was used more widely, and state systems of educational television were created. But by the end of the decade, the honeymoon was beginning to pall, and educators were beginning to take a more critical look at teaching by television.

Figure 8–2 KUHT-TV's early studio. (Office of Information of the University of Houston)

Critical Evaluation of ETV

Arguments both for and against ETV surfaced rapidly.[10] In some ways, the controversy was a compliment to the influence of the new teaching tool. Both sides did agree that more research was needed to determine the true assets of ETV. Teachers were beginning to object to administrators' requests for teaching time on television on top of their already crowded classroom schedules. They also wanted to participate more in planning instructional television from the very beginning of program formulation. Such practice is now routine for developing good instructional programming. The thought of television as a substitute teacher, understandably, received considerable resistance. Classroom teachers were fully aware that the interaction between student and teacher contributed to much of the learning process. They appreciated television's particular abilities but were protective of their own role in the student-teacher interaction.

The public's focus on the economics of educational television also concerned educators. The politicians' and taxpayers' misconceptions of a medium that could eliminate personnel and save taxpayers' money was worrisome. They realized that they had to reemphasize the importance of the teacher in the classroom. They faced still another economic concern—the proportionate share of financing that educational television would receive. Television studios could easily incur bills in the hundreds of thousands of dollars. Money like that not only could pay several salaries but also could provide many of the "traditional" classroom teaching aids. One of the most ambitious projects undertaken was the Midwest Program for Airborne Television Instruction (MPATI).

airborne ETV—the MPATI experiments

The lumbering DC-6 (Figure 8–3) taxied down the runway, its motors roaring in unison. Inside, highly trained technicians sat with their sophisticated electronics gear and waited until the big bird was airborne. It was early morning. Continuing its climb to 20,000 feet, the pilot leveled off. Reaching its destination over northeastern Indiana, the plane was in position for its target—thousands of school children waiting patiently below. The cargo was television programs, broadcasting such subjects as Spanish, French, history, and science to the outlying rural areas of Indiana, Illinois, Kentucky, Michigan, Wisconsin, and Ohio. Called MPATI, for Midwest Program for Airborne Television Instruction, it was an idea conceived seven years earlier in 1944 by a Westinghouse

Figure 8–3 Circling above the Midwest, the specially equipped plane for the Midwest Program for Airborne Television Instruction brought instructional television to a wide geographical area. In concept, it was the forerunner of the ATS satellite systems discussed in chapter 5. (Purdue University Archives)

Electric Corporation radar expert named Charles E. Nobles and with the persistence of a colleague, Westinghouse engineer Ruben Lee.[11]

The Theory behind MPATI

The idea of taking the television station into the sky had been fixed in Nobles's mind. Theoretically, it would raise the television tower to 23,000 feet. The distance then covered by this signal would expand to include a potential half-million pupils. Nobles reasoned that the plane must have a special gyroscope-controlled antenna that would always remain vertical, regardless of the position of the aircraft. Upon reaching its position over Indiana, the DC-6 would then fly in a tight pattern of twenty miles, beaming its signals to the states below.

Funding and Software

The project succeeded with support from a number of organizations. MPATI contracted with Purdue University to furnish maintenance and other support personnel for the aircraft. Due to its proximity to the target area, the plane was based at the Purdue airport. Local school systems using MPATI programs contributed a nominal amount of money to buy equipment to receive the programs. Dr. J. A. Hutcheson, then Westinghouse vice president in charge of engineering, aided the project by applying to the Ford Foundation for funds. Ford responded with an initial $6 million. Major national education organizations such as the National Education Association, the Parent-Teachers Association, and the U.S. Office of Education threw their support behind the experiment.

The MPATI program, despite its popularity, did not last. Satellite communication soon overshadowed its usefulness with antennas 23,000 miles high and signals that could span large sections of entire continents. The MPATI plane that made its first broadcast in 1961 made its last in 1968. The organization stayed in limited operation until 1971. Yet the novel effort definitely pioneered research for future satellite experiments when small communities again would receive educational television programs, this time from outer space.

educational versus instructional broadcasting

In its formative years, the range of television's experiments from MPATI to televising art masterpieces for the public all were considered examples of educational television or ETV. The programs were truly designed to educate. For our own purposes, we shall define ETV as *all noncommercial television programming and commercial programming produced especially for educational purposes, whether or not the program is used for direct classroom instruction.*

As educational television matured and more and more uses of in-school programming were developed, the word *instructional television* evolved. Although both terms still tend to be used interchangeably, instructional television refers to programming specifically designed for use in the classroom or in a direct teaching role. Notice we did not use the words "in-school" in our definition. Both ETV and ITV are employed beyond the confines of the classroom. Differentiating between the two terms also is important for economic and political reasons.[12]

Remember that criticism of ETV reached full force in the 1960s. Part of this criticism went beyond the classroom to the policy issues. Because much of the programming of ETV stations was designed for in-school use, and because major federal funding was emerging, some saw the potential threat of a national school system under federal control. Congressional advocates of ETV were charged with maintaining the ETV movement as "a sinister conspiracy directed from the U.S. Office of Education to homogenize the nation's moppets by a standardized curriculum spread from sea to shining sea."[13] This furor caused ETV stations to stop and seriously consider from where their future programming dollars would come.

During this same time, smaller portable videotape equipment arrived in the marketplace. Such equipment permitted many educators still to tinker with television but this time without the threat of outside controls. For school administrations, the portable equipment was particularly satisfying, since they could buy inexpensive television equipment with local money and avoid the public criticism caused by large federal expenditures. Although new, the idea of airplanes and satellites beaming federally funded programming to local school systems was sold most easily to both school officials and taxpayers as "strictly experimental" in nature. The portable equipment satisfied those who demanded new technology in the classroom. Instead of educational television for large numbers of students, something new appeared, instructional television (ITV) used strictly for in-classroom use with programs often produced by the teachers themselves.

integrating educational broadcasting into the learning experience

To understand fully educational broadcasting's relationship to the teaching and learning process, it is important to learn its relationship to other forms of educational presentations. To illustrate these interrelationships, we shall use a model called the Cone of Experience (Figure 8–4), developed by audiovisual specialist Edgar Dale.

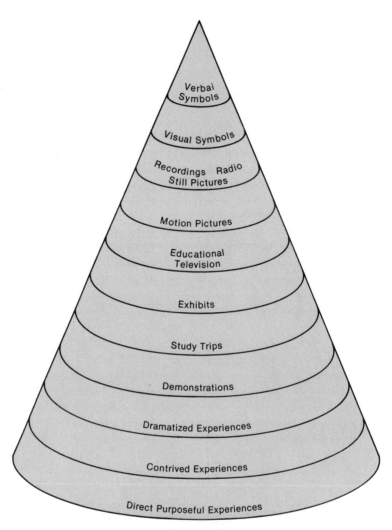

Verbal
Symbols

Visual Symbols

Recordings Radio
Still Pictures

Motion Pictures

Educational
Television

Exhibits

Study Trips

Demonstrations

Dramatized Experiences

Contrived Experiences

Direct Purposeful Experiences

Figure 8–4 Edgar Dale's Cone of Experience. (From *Audiovisual Methods in Teaching*, Third Edition, by Edgar Dale. Copyright © 1946, 1954, 1969 by Holt, Rinehart & Winston. Reprinted by permission of Holt, Rinehart & Winston.)

Dale's Cone of Experience

Viewing the cone from the base up, we can see that it represents different levels of abstraction, each part of the learning experience. Dale uses Jerome S. Bruner's concepts as analogies to his Cone of Experience. Bruner analyzes three major modes of learning: "the *enactive* (direct experience), the *iconic* (pictorial experience), and the *symbolic* (highly abstract experience)."[14] At the base of Dale's cone are the more direct experiences, or enactive ones. The mid-section represents the more abstract experiences, and at the top area of the cone, the symbolic experiences. It is important to understand that the cone represents different levels of abstraction, not of difficulty. The bands of the cone also are interrelated. They are "fluid, extensive, and continually interacting."[15] For example, although the very tip of the cone represents verbal symbols, an educational television program would also contain those same verbal symbols, as would other experiences lower on the cone.

Let us apply the Cone of Experience to a game of tennis. We can make the analogy that at the very bottom of the cone, the direct purposeful experiences refer to our actually playing the game. Farther up on the cone we might experience a demonstration by watching a game of tennis, either in person, or still higher on the cone, on educational television. At the very top of the cone, in our own minds we might conjure up the "image" of tennis upon hearing the word, or verbal symbol "tennis."

Choosing the Correct Presentation

By now you are probably beginning to see how we choose which presentations to use to teach tennis. If, for example, you had a champion tennis instructor, superb courts and equipment, a select group of students, and plenty of time for individual instruction, the best teaching method would not be to make a program about tennis. The best instructional choice would be "direct, purposeful experiences" on the tennis court with actual equipment playing with the instructor as your partner.

Choosing the level of abstraction to teach our game of tennis now begins to include other factors. For example, perhaps the students have already had experience with the game and merely want to review the basics. The television program may be just the help they need. Perhaps the class is too large and there is not enough time or personnel to teach each student individually, but before working with them on the court, it is more efficient to teach them first the basics through the television. The program thus frees the instructors to concentrate on teaching skill and strategy.

We could produce this program and place it in a learning resource center from where it can be checked out and viewed in special study carrels at the students' convenience. If they do not understand the program, they can play it back as many times as necessary until they feel they have mastered the material.[16]

Now that we understand how educational broadcasting fits into the learning process, let us learn more about the actual process of producing an accountable program.

the quest for accountability:
developing quality ITV programming

Almost anything significant that occurs in a local school system becomes the concern of the community. Students, parents, teachers, school-board members, and even the legislature all are involved. The educational process, in its simplest terms, attempts to provide the most for the taxpayers' dollars. This quest can reach extreme levels, as in Oregon where taxpayers have refused to approve school budgets, forcing some schools to close. It also prompts hard questions asked at PTA or school board meetings. Thus, it is only natural that people are concerned about ITV. The public, parents, and policy makers want to know if ITV is worth the money. Are the expensive facilities paying for themselves in student benefits? Does the type of programming really teach something or is it just background entertainment?

The result of this probing is the trend toward increased accountability in developing ITV programming. In a statement before a congressional hearing, learning authority P. Kenneth Komoski noted that less than one percent of the tapes now on the market are learner verified.[17] Komoski meant that although programs are produced to teach something, it does not mean learning will be achieved by simply viewing the program. Most programs simply have not been tested, so program developers are beginning to ask questions. What do we really want to teach with this program? What are its objectives? What is the best way to present this material? What techniques can we use to keep the student interested while learning? How much entertainment can we include without digressing from the basic subject matter? In a very real sense, such accountability affects people as well as programs. The person in charge of ITV program development is responsible for the overall effectiveness of ITV in the school setting. The director of the instructional television studio also must justify the increased budget to the school board.

transition to public broadcasting

In the mid 1960s as ITV programming switched to local production and distribution, educational television stations, their programming supplemented by locally produced videotape material, began to move toward a greater variety of programs beyond in-school programming. At about the same time, foundations and government agencies were supporting not only in-school production but also the development of state and national systems of noncommercial broadcasting. In 1962 the Educational TV Facilities Act provided over $30 million over a five-year period to develop state systems of educational broadcasting. Many of these state systems operate today as an integral part of a larger national system of noncommercial radio and television stations.

The Carnegie Commission and CPB

Full-scale planning for what would eventually become a national system of noncommercial radio and television stations began in 1965 with the Carnegie Commission for Educational Television. The Commission, whose members were a broad range of industry leaders, was charged with conducting a "broadly conceived study of noncommercial television and to focus attention principally, although not exclusively, on community owned channels and their service to the general public." The commission's report concluded that a "well financed, well directed educational television system, substantially larger and far more persuasive and effective than that which now exists in the United States, must be brought into being if the full needs of the American public are to be served."

The Public Broadcasting Act of 1967

The commission's recommendations were the impetus for the passage of the Public Broadcasting Act of 1967, which allocated $38 million dollars to the improvement and construction of new facilities for noncommercial radio and television in the United States. Also formed by the act was the Corporation for Public Broadcasting, a quasi-government company established to administer the public broadcasting funds appropriated by Congress.

The act authorized the CPB to:

Facilitate the full development of educational broadcasting in which programs of high quality, obtained from diverse sources, will be made available to noncommercial education television or radio broadcast stations, with strict adherence to objectivity and balance in all programs or series of programs of a controversial nature;

Assist in the establishment and development of one or more systems of interconnection to be used for the distribution of educational television or radio programs so that all non-commercial educational television or radio broadcast stations that wish to may broadcast the programs at times chosen by the stations;

Assist in the establishment and development of one or more systems of noncommercial educational television or radio broadcast stations throughout the United States;

Carry out its purposes and functions and engage in its activities in ways that will most effectively assure the maximum freedom of the non-commercial educational television or radio broadcast systems and local stations from interference with or control of program content or other activities.

With passage of the act, noncommercial radio and television stations became known as "public" broadcasting stations, signifying their ability to secure income from the public as well as from corporations, foundations, and government agencies.

distributing programs: PBS and NPR

Public broadcasting stations which meet certain operating standards are eligible for financial support from the CPB. Such stations also belong to National Public Radio (NPR), the radio network of public broadcasting, or the Public Broadcasting Service (PBS), the television network of public radio.

The Public Broadcasting Service—PBS

To help meet the goals of the 1967 act, CPB joined with many of the licensees of noncommercial television stations in the United States and in 1970 formed the Public Broadcasting Service (PBS), which today is the primary distribution system for programs aired on public broadcasting stations. In 1979 the PBS board of directors approved a three network satellite-fed distribution system permitting stations to choose from (1) general entertainment mass appeal programs; (2) regional programs of special interest; and (3) instructional programs.

In some ways PBS is similar to commercial television networks, in that it is a distributor of programs. At that point, however, most of the similarity stops. Whereas the commercial affiliates are part of affiliate organizations which advise networks, PBS is in many ways more sensitive to its affiliates. In a sense the affiliates represent the public who, at least in theory, "owns" the public broadcasting stations through public contributions and tax dollars.

Part of this sensitivity is generated by the PBS board of directors responsible for the governance of PBS. The board represents the general public and station managers and gains insights into the success and failure of its system. It is a good, broad-based indicator on which to base decisions.

In addition to the board meetings, PBS initiates a series of semiannual regional meetings called *round robins*. In round robins, PBS officials travel around the country and talk to public television station management. Later, all managers get a briefing paper, a synopsis of what occurred in the meetings. Another means of feedback are vote tabs, which are electronic questionnaires sent to affiliate stations via the nationwide teletype system. The system connects all of the PBS stations with each other and with the network offices in Washington, D.C. (Figure 8–5). If PBS wants to poll the affiliates about something, it sends a question by teletype to each of the PBS stations. Each station can then answer the question by teletype. The responses from all of the PBS affiliates then are tabulated at the network headquarters.

National Public Radio—NPR

Radio also benefited from the Public Broadcasting Act, and in 1971 many noncommercial radio stations became members of National Public Radio, the radio equivalent of PBS. The two differ, however, in that NPR also produces programs, whereas PBS's chief responsibility is distribution. NPR affiliates also produce programs which often are syndicated by NPR and made available to member stations. One of the most famous programs aired on NPR is the daily news-magazine program "All Things Considered," acclaimed by both educators and the public for its informative, in-depth coverage of news and public affairs. NPR's growth has been closely aligned to the overall growth of noncommercial radio.

Like PBS, NPR also is sensitive to its affiliates. It holds semiannual round robins, and the NPR Board is sensitive to feedback from its members. NPR also conducts four workshops per year, which are more "how-to-do-it" meetings than discussions of issues, but nevertheless en-

Figure 8–5   Connecting point for the PBS communication teletype link with affiliate stations. (Stan Cahill & PBS)

courage interaction between NPR officials and local public radio broadcasters. An annual conference of station managers is another means by which the network can learn about issues affecting its affiliates.

selecting programs for public television: the Station Program Cooperative

One of the biggest differences between public and commercial television is the process by which PBS selects the programs that will become part of the regular schedule fed to affiliate stations. Each year, Public Broadcasting Service affiliates participate in a series of "bidding" rounds to determine which television programs you will see on public television. These bidding rounds use direct feedback in the form of financial commitments from affiliates. In commercial broadcasting, programs are picked by network planners and executives, but in public broadcasting, the individual stations make the decisions. The process is called the *Station Program Cooperative (SPC)*. It began in 1974 as an experiment to see if local public broadcasting affiliates were willing to commit money to purchase programs for network distribution. The process has since been revised considerably. Currently, the "bidding" in the cooperative is held three different times during the year.

To understand how the SPC works, imagine you are the manager of a public broadcasting station. You have a certain amount of money to spend on programming and, along with other public broadcasting stations nationwide, you will help determine which programs will reach network distribution. You will receive a catalogue of the shows being considered for the late winter bidding round and are even invited to preview them. You decide to show them to your audience and give them a chance to vote on which show they would like to see. You publish a ballot for your viewers in the local newspaper (Figure 8–6). You tally the returned ballots and decide to commit a portion of your programming budget to the most popular shows. You send your financial commitment to PBS, which tallies yours with the rest of the affiliates to determine which shows received the most support, both popular and financial. These shows then will appear as part of the fall lineup.

You still have two other bidding rounds in which to choose programs and to participate in the network program selection process. Also, if there were a special program offering at some time other than the regularly scheduled SPC rounds, you could bid on this as well. The entire process is one of feedback from your station and the other public television stations to determine network programming. It directly involves both the audience and the station before the programs even are aired.

WHAT DO YOU WANT TO SEE ON PUBLIC TV 9

On Monday, February 14 from 8 p.m.-11:30p.m., Channel 9 will broadcast pilots for 24 proposed series. Which ones do you think we should buy? Which ones would you like to see on TV this coming fall and spring? Fill out this ballot and let us know. Place a check next to those programs that you liked. Place two checks next to those shows you really liked. Then return your ballot AS SOON AS POSSIBLE to:

VIEWERS' CHOICE, KCTS/9
4045 Brooklyn Ave., N.E.
Seattle, Wash. 98105

●

Only ballots received by FEBRUARY 23 can be counted. The 24 pilots will be rebroadcast Saturday, February 19 from 3 p.m.-6 p.m. Watch either or both times— and VOTE.

●

Optional—to help us in our scheduling: Age____Sex: M____F____ Number in Household____ I watch KCTS/9____hours per week. City/Town____

Clip & Mail Immediately

_____ FRENCH CHEF WITH JULIA CHILD
_____ IN SEARCH OF REAL AMERICA
_____ MOTHER'S LITTLE NETWORK
_____ DRUM CORPS CHAMPIONSHIPS
_____ ZOOM
_____ FOLK GUITAR
_____ SPORT OF THE WEEK
_____ CITY KID
_____ WOMANTIME
_____ YOUR RETIREMENT DOLLAR
_____ JUST PLAIN COUNTRY
_____ MUSIC FROM SUMMERFEST
_____ FERTILE CRESCENT
_____ MARK RUSSELL COMEDY SPEC.
_____ OPERA THEATRE
_____ SOUNDSTAGE
_____ ANTIQUES
_____ FIRING LINE
_____ LOWELL THOMAS
_____ SPECIAL EVENTS
_____ MACNEIL/LEHRER REPORT
_____ STUDIO SEE
_____ THE ADVOCATES
_____ EVENING AT DOCK STREET
_____

PUBLIC TV 9
PBS

Figure 8–6

the scope of public television programming

Today, public television has become a system of broadcasting with a wide variety of programs ranging from the traditional "instructional" programs used for in-school purposes to programming with a wide national and even international appeal.

"Sesame Street" (Figure 8–7) and "The Electric Company," for instance, both have wide audience appeal beyond the preschool and elementary school audiences they are designed to attract. Another example with similar appeal is "The Big Blue Marble" which seeks to assure children that other children around the world have experiences similar to theirs. Still another program with national distribution and appeal is WNET-TV's "The Adams Chronicles," depicting the history of America's

Figure 8-7 Big Bird of "Sesame Street." The television program has been the recipient of many awards for creative programming. (Courtesy of the Children's Television Workshop)

famous Adams family of presidential fame and its role in history from 1750 to 1900. Some programs, such as "The Adams Chronicles," have accompanying teacher and curriculum guides, which are especially helpful when the program is used in the classroom.

Regional programs also have reached a broad-based audience (Figure 8–8). "Mr. Rogers," which started out as a regional program on WQED-TV in Pittsburgh, has become popular nationally. This and other regional programs are making inroads into the typical cartoon fare so common to children's television.

Many colleges are taking advantage of the "outreach" programs, employing ITV to reach adults who may not want to take the trouble of coming to campus, or because of the distance and time, simply cannot come. As a result, television colleges are prospering. The Public Broadcasting Service has offered such "nationwide" courses as "Classic Theater" and "The Ascent of Man." The programs were telecast over PBS stations, and many colleges offered credit for viewing them and passing examinations on them. Colleges list the television courses as part of their regular schedules, enabling students to enroll for them as they would for any other course.

Figure 8–8 Fred Rogers, star of "Mr. Rogers' Neighborhood," is one of the pioneers of public broadcasting. (Copyright, Family Communications Inc.)

In addition to the TV college, public television also finds its way into classrooms as a supplement to lectures. A perusal of one ETV supplier's catalogue, the Great Plains National Instructional Television Library of Lincoln, Nebraska, reveals how broad this scope is. For example, under the heading of language arts, you can choose from "Language Corner" for grade 1, "Ride the Reading Rocket" for grade 1, "Word Magic" for grade 2, and eleven other language arts possibilities. "Language Corner" has thirty, 15-minute programs with such titles as "Listening," "Fairy Tales," "Story by the Teacher," "Letter Writing," and "Speech and Telephone." These are just some of approximately 150 different programming series from which to choose. Programs in a given series may have from two or three lessons to more than sixty lessons. The program "Mathemagic," produced by Channel 33 in Huntington, West Virginia, features sixty-four lessons for second-grade youngsters (Figure 8–9). The goals of the program include improving computational skills through a better understanding of the numbering system and developing problem-solving ability through an interchange of mathematical and verbal language.

Some of the programs are specially designed for summer viewing to help students retain the skills achieved during the academic year. For example, a language arts series entitled "Catch a Bubble" was produced by WNIN-TV of Evansville, Indiana, for a local school corporation. The program is now nationally syndicated and has four thirty-minute lessons for students who have completed the second grade. The star of the show

Figure 8–9 A scene from "Mathemagic." The program features 64 lessons for second-grade youngsters. (Great Plains National Instructional Television Library)

is a seahorse named Salty. The children share Salty's adventures while reinforcing learning skills and maintaining their interest in reading during the summer vacation. Along with viewing the program, the student also works in the "Catch a Bubble" activity book.

Among programs offered by other suppliers is "Time-Life Multimedia," which explores the social sciences, language arts, humanities, sciences, business, and recreation. If you are interested in archeology, you might want to view "How Old is Old?," which discusses the age of the Grand Canyon, explores the history of man, or probes the longevity of the ice cover of Antarctica. Or if you are interested in dinosaurs, you might want to watch "The Dinosaur Hunters." This program studies how the dinosaurs reproduced. Then there is "Digging up the Past," which examines relics of ancient civilizations. Other archeological programs include "Cracking the Stone-Age Code," "Lost World of Maya," and "Ancient Egypt." Many of the programs are in both English and Spanish and are on film or videotape.

With the interest in metric conversion, mathematics programs which teach the metric system are becoming popular. Typical of these is "Metrify or Petrify," a series of eight, 30-minute programs produced by KLCS-TV in Los Angeles and distributed nationally. The programs include an introduction to the metric system, linear measurement, volume measurement, mass versus weight, temperature, times, and an overview of the metric units of measurement.

public radio as instructional radio

Because of television's dominance, radio tends to be overshadowed as a medium with both educational and direct instructional value. It should not be, because in many places, radio is as much and sometimes more a part of the educational scene as its visual counterpart is. It certainly has been in use longer.

As we learned in chapter 3, one of the first radio stations in the United States crackled on the air at the University of Wisconsin in 1919. Through the years, WHA radio served a wide spectrum of listeners with direct instructional programming. In fact, it was the flagship station of an entire state instructional radio network. What was WHA's effect? By the late 1950s, schools in Wisconsin had been listening to instructional radio for almost thirty years. The director of the Wisconsin State Broadcasting Service and former WHA director, Harold B. McCarty, told a Washington, D.C. conference on educational television that educational radio was alive and well.[18] He told television enthusiasts of some tough goals to match. At that time the Wisconsin system had 100,000 pupils enrolled in creative art classes by radio, 70,000 in music, and 43,000 in a social studies class.

Today, educational and instructional radio programs are found at many colleges. WBAA at Purdue University has long had a viable instructional program and has not only been responsible for major research in the field, but also has had a regularly scheduled offering of college courses taught exclusively by radio. The university charges a nominal fee for the course, and the student enrolls just as he or she would for any other course. Lectures are broadcast mostly in the evenings and are repeated on Sundays. Tapes of the courses are available in Purdue's audio-visual center and can be checked out for review. The student receives credit by taking an examination at the end of the semester. Although credit is given, no final grade is assigned. Students must, however, maintain at least a *C* to obtain credit. The university maintains no record of those who fail the test and consequently receive no course credit.

Some states also operate major instructional radio networks. South Carolina, for example, simultaneously links four noncommercial FM stations, WEPR at Greenville, WLTR at Columbia, WMPR at Sumter, and WSCI at Charleston, into a special instructional network. The stations are located at strategic points in the state, so schools over a wide region are within earshot of the broadcasts (Figure 8–10). The broadcasts are directed toward elementary grades through high school as supplements to regular classes, and teachers have found them effective in generating student interest. Each broadcast lesson is self-contained so if one is missed, it does not interrupt the regular classroom schedule. To facilitate reception, the state offers special radio receivers pretuned to the educational stations, although a standard radio receiver also works (Figure 8–11). Each participating station designates a coordinator to help teachers gain the maximum benefit from the broadcasts. Actual airings are scattered throughout the school day, and supplementary teaching materials are available to help teachers plan lessons around the broadcasts.

South Carolina is just one example of states with many instructional radio networks functioning at city and state levels. The medium is

Figure 8–10 (Public Telecommunication Review)

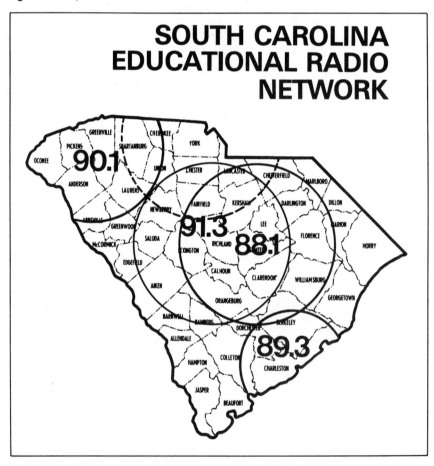

Figure 8–11 (South Carolina Educational Television and Radio Network and the South Carolina State Department of Education)

especially effective in teaching the great works of literature, music, foreign languages, and other subjects which particularly lend themselves to "audio" concepts.

funding and the future

The public broadcasting system in the United States has managed to survive criticism and erratic congressional support to mature into a viable and even competitive chain of stations serving most of the United States.

Yet many issues need to be resolved to ensure public broadcasting's future. Perhaps the most essential one is funding. In theory, if not in practice, the support for long-range funding of the system came in 1975

219

with the passage of the Public Broadcasting Financing Act of 1975. Signed into law by President Gerald Ford, it called for authorization of a five-year appropriation, but found the actual money tied to a separate appropriation measure, a measure which ran into trouble in the House Appropriations Committee. Currently, CPB-qualified stations receive $1.00 of federal funds for every $2.50 raised locally. Although clearly the burden of operating a local station rests on the public, few stations could operate without federal money.

In 1977 a task force successfully obtained funds for a second Carnegie Commission inquiry into public broadcasting. The questions asked were:

What is the mission of public broadcasting in American society, and how can Americans best be served by this important national resource?

How are quality programming and creativity to be fostered?

What should be the nature of citizen involvement in public broadcasting?

How can the local station cooperate with other community organizations— schools, churches, libraries, community groups, museums, and other institutions?

How adequately can public broadcasting meet the needs of the special audiences such as minorities, women, children, adults interested in life-long learning, the disabled, or groups with particular interests?

How will public broadcasting develop as a multichannel system and interact with growing technologies such as satellite, cable, and videodiscs?

Over the next ten to fifteen years, how should public broadcasting be funded and at what levels?

What should be the central and regional organizations and institutions in the system, and what should be their functions?[19]

General answers to the charge of the second Carnegie Commission came with a report released in January, 1979. Along with some strong criticism of commercial broadcasting, the report called for better ways to insulate the public broadcasting system against government influence. Funding increases were also called for which included raising the budget to $1.16 billion by 1985, $590 million to come partially from a spectrum fee charged commercial broadcasters. The report recommended a Public Telecommunications Trust, which would be autonomous and administrate the system and, under the Trust, a Program Services Endowment to deal with programming.

The report received wide play in the media and trade publications. As expected, commercial broadcasters were in some cases highly critical of the report. What effect the report will have and what the future holds for public broadcasting will be dependent on specific legislation. Overall the system is regarded as an important source for quality programming, much of which is not possible on commercial networks because of the competitive nature of the system and the need to garner audience ratings. Although public broadcasting is also concerned about getting and holding an audience, it has latitude in creative programming that commercial broadcasting simply has not been able to match because of competitive commercial restraints.

summary

The beginnings of educational and public broadcasting can be traced back to WHA at the University of Wisconsin. Signing on the air in 1919, WHA is considered the pioneer public broadcasting station in the United States. Educational television was used at Iowa State University as early as 1932. Support for educational television came in the early 1950s with the formation of the Joint Committee on Educational Television (JCET). With a grant from the Ford Foundation, the JCET became the early representative of educational television and assisted educational institutions both in getting stations on the air and in using educational television in the classroom.

One of the more novel early experiments was the Midwest Program for Airborne Television Instruction (MPATI). Using a converted airplane, television programs were sent over a wide area of the Midwest via a special airborne transmitter. Although the system eventually became obsolete with the introduction of satellites, it did prove that it was possible to integrate television into the learning experience in many schools simultaneously. With the increased use of portable equipment and educational television in the classroom, television became important to direct classroom instruction. At the same time, many stations which had been used mainly for instructional purposes began broadcasting to a more general audience.

After a series of foundation-supported evaluations of noncommercial broadcasting, Congress enacted the Public Broadcasting Act of 1967 which designated noncommercial and educational stations as public broadcasting stations. The act created the Corporation for Public Broadcasting, a quasi-governmental corporation funded by Congress to assist in developing a national system of public broadcasting. To aid in distributing programs to public stations, the Public Broadcasting Service and National Public Radio were established. Although PBS's first function is distribution of television programs to CPB-qualified stations, NPR both distributes and produces programs. The future of public broadcasting is tied to a number of important issues, such as how the system will continue to be funded, what relationship the public will have to the operation of the system, how minorities and handicapped people will benefit from the system, and what influence institutions and organizations should have on the system.

spotlight on further learning

AVERY, ROBERT K. AND ROBERT M. PEPPER, *The Politics of Interconnection: A History of Public Television at the National Level.* Washington, D.C.: National Association of Educational Broadcaster, 1979.

BURKE, JOHN E., *An Historical-Analytical Study of the Legislative and Political Origins of the Public Broadcasting Act of 1967.* New York: Arno Press, 1979.

DALE, EDGAR, *Audio Visual Methods in Teaching.* Hillsdale, Ill.: The Dryden Press, 1969.

DIAMOND, ROBERT M. ed., *A Guide to Instructional Television.* New York: McGraw-Hill Book Company, 1964.

GERLACH, VERNON S., and DONALD P. ELY, *Teaching and Media* (2nd ed.) Englewood Cliffs, N.J.: Prentice-Hall, Inc., 1980.

GORDON, GEORGE N., *Classroom Television.* New York: Hastings House, 1970.

KIBLER, ROBERT J., DONALD J. CEGALA, LARRY L. BARKER, and DAVID T. MILES, *Objectives for Instruction and Evaluation.* Boston: Allyn & Bacon, Inc., 1974.

MOIR, GUTHRIE, ed., *Teaching and Television.* London: Pergamon Press, Ltd., 1967.

PEPPER, ROBERT M., *The Formation of the Public Broadcasting Service.* New York: Arno Press, 1979.

SCHRAMM, WILBUR, ed., *Quality in Instructional Television.* Honolulu: University of Hawaii Press, 1972.

9

television in business and industry

FOCUS After completing this chapter we should be able to

Discuss the growth of television in business and industry.

Identify how television aids both internal and external corporate communication.

Discuss examples of in-house corporate news and information programming.

Explain how television is used for marketing and sales.

Describe how television is used for in-service training.

Understand the management-development uses of corporate television.

Explain the application of television to in-house advertising agencies.

Discuss the future of corporate television.

Across town at the manufacturing plant, the 9 o'clock whistle is about to blow. Inside, a television crew has just received an urgent message. "Contact the air mobile units! Alert the camera crews! We have a breakdown at Arctic Station One. They want us there by tonight." Telephone calls, checks with management, notification to the crew of the company jet—it all sets in motion a chain of events that by nightfall will lead to a complete television crew thousands of miles away at a power station in northern Canada. A generator manufactured by the company has broken down. When the repair crews start their job, television cameras will be there to record it. Later, back at the manufacturing plant, the videotape will be edited into an instructional television program to train future repair crews. It all is just one example of industrial television in action, a growing area of mass media far removed from the typical commercial television station.

growth and impact

The purpose of this chapter is to acquaint you with television in business and industry. It is an expanding field with many applications. How big is it? Based on 1977 data, more companies will use television in the 1980s than there are commercial television stations in the United States.[1] In addition, a variety of corporate television networks are developing. Communication consultant Judith M. Brush of D/J Brush Associates in New York City classifies a network as "an organization which distributes programming at least once a month to six or more locations away from the point of origin."[2] She also notes that "more then 40 of these networks have more than 50 viewing locations with at least half that number distributing programs to 100-plus locations. For example, Pepsico has more than 300 locations in its network, IBM has some 1,300 locations in 400 countries, and Bank of America has 1,100 locations."[3] What kinds of firms use television? One survey of the top 500 companies listed in *Forbes* magazine indicated "just over half were engaged in manufacturing, followed in order by financial institutions, utilities, retailers, natural resources, and transportation companies."[4]

Let's learn more about how the medium of television is used in the corporate setting.

224

in-house corporate news and information programming

Most of us think of television as the local network prime-time program or as the local television station's daily newscasts. For these programs, there are decisions on what stories to use, how to edit them, what the audience wants and what it should have, what graphics to use, which audio cuts to include, and many more. Those same decisions also are made everyday in places far from the network and newsrooms. They're made at corporations, where television production crews and "corporate newscasters" are preparing the daily newscast to be sent to employees at the downstate plant or through "corporate networks" to international offices.

Applications of Corporate Newscasts

Dow Chemical Corporation is one company which has daily television newscasts for its employees. When Dow president Paul F. Oreffice wanted better communications with Dow's 10,000 employees, he thought that television would be the medium to do the job.[5] Produced at Dow headquarters in Midland, Michigan, the Dow corporate newscast is five to seven minutes long and is broadcast through closed-circuit television systems to lunch-hour viewers. Topics such as company news, stock market reports, and safety procedures are featured. Other companies follow a similar process. Some even produce commentaries. Union Carbide Corporation produced one 29-minute tape reporting on its gases, metals, and carbon divisions.[6]

To diversified companies which are spread out over wide areas, corporate news programming is especially valuable. An oil company may consist of oil exploration, refineries, and gas stations, as well as the corporate office. How is it possible to link the people and activities of these varied enterprises? The main instrument for many companies has been, and still is, the corporate magazine or newsletter. Filled with pictures and articles about the corporation's activities, it is sent to all employees. Different parts of the company have their own "reporters" or "stringers" who contribute to the magazine. Now, although continuing the corporate magazine, corporations are turning to television. A gas station owner wins a community award; an oil rig worker is promoted; a pipe line crew starts a new project (Figure 9–1); a secretary is married—they all appear on the lunch-hour news program. The corporate news cameras catch it all and in living color.

225

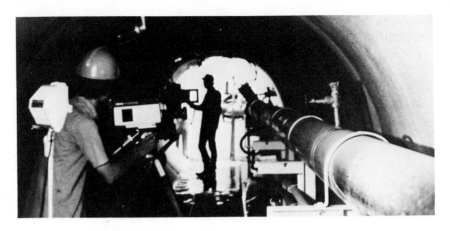

Figure 9–1   (Phillips Petroleum Company)

Content of Corporate News Programming

The scope of corporate news programing can be seen in one company, First National City Bank in New York. Consultant Eugene Marlow describes a daily news program called "Channel 6" seen by approximately 10,000 bank employees scattered throughout three buildings in New York City.[7] The show runs about 15 minutes and has three segments— news, a feature on some aspect of the bank, and one on entertainment. Content ranges from an interview with a bank employee to a report on the opening of a new headquarters a continent away. Features might be on computer programming, an interview with a professional magician, or one with a nationally famous artist.

Unique approaches are often used to communicate somewhat dry topics. First National City Bank had to inform employees of company benefits but wanted to do so in an interesting manner. "Channel 6's" solution was to use a puppet called Professor Wienerschnitzel (Figure 9–2). The professor finds himself in various settings all designed around a company benefit. On one occasion, he runs for mayor of New York, his platform being the scholarship benefit's program for the bank's employees. On another occasion, he practices his "voice and diction" for a speech he is to deliver on employees' insurance benefits.

Not all of what appears on "Channel 6" is limited to internal-bank programming. Cameras for "Channel 6" also venture into New York City to capture the after-hours activities of the bank's employees. For example, when a group of employees purchased group tickets to attend a hockey game, "Channel 6" interviewed some of the employee spectators on their reactions to the game. When the Ringling Brothers Circus came to town,

Figure 9–2 A discussion of employee benefits between the puppet "Professor Wienerschnitzel" and host Terese Kreuzer of Citibank's employee television news program "Channel 6." (Citibank)

"Channel 6" cameras had "ringside" seats. The bank's involvement in community affairs is also part of corporate news programming. This corporate programming combination has two effects: it is entertaining and informative—not much different from the evening's prime-time network news program.

Not all corporate news programming is as elaborate as that of First National City Bank. Dana Corporation's Reading Frame Division, for example, places news and information on a motor-driven wheel which revolves slowly in front of a fixed-position camera.[8] Each message or news story remains in view for fifteen seconds. The messages are broadcast around the clock, and television monitors are scattered throughout the plant's facilities. For Dana Corporation, the problem is reaching workers while they circulate through the plant (Figure 9–3), not at a centralized lunch location. Consequently, the monitors have become a substitute for many of the bulletin boards. At Dana, the concept of more elaborate corporate news is secondary to that of sending brief messages to employees in a short period of time via television.

Corporate newscasts are not the only way to communicate with employees. Many companies produce special television programs to inform employees of the corporation's issues and concerns. For instance, a company's public relations efforts may be just as interesting to the employees as they are to the general public. How do the employees learn of these activities? The company produces special television programs on such topics as how the company is volunteering executives to help with the local United Way drive, how special volunteers are canvassing the city for the March of Dimes, or how children are being taught to swim and play sports at the corporation's summer camp in the mountains. Such programs may also fill the slots of public service programming on many commercial television stations.

Figure 9–3 On-line television monitors informing employees at Dana Corporation. (Dana Corporation)

Corporate policy can be disseminated effectively through television. New employees in large companies need to know such things as how to file insurance claims, where claims offices are located, and how they are staffed. The company can produce a television program spelling this out in detail, and the employee can view it at his or her own convenience. At first it might seem that a booklet could be just as effective. It would contain the same information. But the sound and motion of a courteous claims officer can set a positive example for the new employee, reducing tension and facilitating the process when he or she comes to file a claim. In addition, a welcoming statement from the president of a large corporation might mean much more when the president can be "seen" and "heard."

Another programming concept is the "state-of-the-corporation" address. Although we are familiar with the "state of the State" and the "state of the Union" addresses, we do not usually hear chief executives speak on the "health" of their corporations. Such information is usually given in the annual report accompanied by statistics, charts, and accounting jargon. However, many chief executives are learning that speaking to employees in an understandable language via television about how well the corporation is doing and its prognosis for the future can be an extremely effective way to communicate.

With the increasing number of federal regulations, many companies have had to scramble just to keep up with them. This has been especially true with safety regulations. A new governmental safety regulation may require educating thousands of employees and gaining their compliance. To help solve this problem, companies have produced television programs on safety procedures in their own particular plant surroundings. A manufacturing firm produced an "on-location" program detailing its fast moving equipment and their dangerous areas. An oil company took a television crew to a drilling rig to portray the protection necessary to prevent on-the-job accidents. Then, before new employees are assigned to one of the many drilling operations, they watch this series of safety programs. Such a series is much more efficient and effective than a foreman taking time and energy necessary to educate each new employee individually.

using television for marketing and sales

Television video cassettes have opened up a whole new world for many companies in the area of marketing and sales. Assume you are one of the nation's major automobile manufacturers. Your showrooms are full of new models, styles, colors, special performance features, and sales personnel who want to communicate all of this wonderful information to all of their potential car buyers. How do you secure the edge in this vastly competitive business? First, you amass the best sales tips and techniques from all of the dealers. Then using these, you have the central office prepare a special video-cassette tape showing a test driver effortlessly maneuvering the cars through their paces, while the accompanying monologue uses a narrator who would rival a network radio newscaster. It is simple electronic persuasion. By placing the video-cassette tape in the showroom playback unit, the customer is entertained with a professional television presentation (Figure 9–4). The local salesperson is still there to answer questions, add the personal touch, and sign the contracts.

Training Sales Personnel

One example of a company using television to train sales personnel and dealers is Deere and Company, manufacturers of the famous John Deere tractors and farm equipment. The company has a full-scale television

Figure 9–4 Videotape recorders are becoming part of the "tools" of the sales presentation at larger auto dealers in the United States. (Provided courtesy of Lincoln-Mercury Division)

production facility at Moline, Illinois, its company headquarters.[9] Replete with sophisticated computer editing equipment, the facility produces programs for all corporate activites, including videotapes which can explain to a dealer the advantages of using John Deere equipment (Figure 9–5). The company started to use television in 1968. At that time it was part of the audio-visual department and consisted of black and white equipment and two people to operate it. In 1973, the financial commitment was significant enough to equip a mobile van for color television production. A year later, television became a separate department in the company, outgrew its space in the corporate headquarters building, and moved to an office building in Moline. The facility now includes a 40′ × 40′ studio and 20′ × 32′ control room besides the mobile van and storage and office space.

The programs for dealers are only a few of the more than 75 programs produced each year by the facility. The Industrial Equipment Division alone uses two or three new programs every month as sales training aids. Programs are distributed to about 250 John Deere dealers in the United States and Canada, each equipped with playback units. They also are distributed to each of Deere's factories which also are equipped with playback units. The company has even enlarged its television production capabilities. Its Dubuque, Iowa Industrial Training Center now has television facilities and produces an average of 18 programs each year,

Figure 9–5   An example of major corporate commitment to television is John Deere's production facilities. (Deere & Company)

helping to inform both salespersons and customers about John Deere products.

Reliance Electric also uses television to help train sales personnel. The company produced its first videotape in 1975 and since then has been expanding the use of television throughout the company. Jerry Wilson, employee supervisor for the Electrical Group of Reliance Electric points out: "We have all the normal channels of communication between our main offices and the far plants and offices but, unfortunately, we have had no way to show motion. Motion is important to Reliance because that's what we sell . . . motors, drives, power transmission equipment and so forth. We build hard and softwear that cause things to move."[10]

Television can also benefit a typical sales meeting. Television solves the problem and communicates this "motion." Capturing the emotion and content of these brainstorming, pep-talking sessions on television has many advantages. First, it provides a record of what happens. The videotape takes notes, eliminating many procedural and secretarial burdens. Second, the tape can be given to sales personnel unable to attend. Perhaps the meeting is regional. The meeting can be videotaped and distributed to other regional offices for other sales personnel. In this way, the meeting also is not slowed down by too many participants. With professional editing, the highlights of the meeting can be produced into a half-hour program of high-intensity sales training, to be viewed by sales personnel throughout the company.

Consider another use. A company is having difficulty selling one of its products. Perhaps it is a special attachment for garden tractors. The problem is not company-wide, however; five sales outlets have had very high sales. Management decides to fly in the five sales representatives who have tallied the highest sales. For over an hour, they discuss how they sell the attachment, why customers find it useful, and the special techniques they use to convince customers that it is worth the money. The session is videotaped and edited into a 15-minute training program on that one product. The company then distributes the program to all of its retail sales outlets for the sales representatives. Immediately after viewing the program, sales begin to climb, and the program is hailed as a success.

A related use of the medium is keeping customers informed. When Owen-Corning Fiberglass® found certain raw materials in short supply, it needed to tell its customers of this problem, clearly and openly. So the company produced a television program with its purchasing managers discussing the problem. Ben Coe, Architectural and Home Building Products branch manager in Los Angeles, said the customers "could see what we were doing in our own mind to minimize the adverse effects. And since the program showed our own purchasing people in frank, candid discussion, the information came across very believably."[11]

Customers need to know how to use the products they buy. This may seem unnecessary for items like clothespins and detergent, but consider the computer. Many businesses are integrating computers into their overall operations. Mini-computers permit even the smallest companies to use this new technology. However, just because the computer is small does not mean that it is simple to operate. Even some small electronic calculators are difficult for the uninitiated to use. Although some manual training is normally included with many major computer purchases, it is time-consuming and does not solve the problem of training the employee hired after the computer is installed.

To help solve this problem, many computer manufacturing companies are developing their own training videotapes. These permit instruction in the new equipment without tying up the time of company personnel. For example, Honeywell has produced videotapes that instruct not only its own customers but also anyone else who needs to learn the basis of computer language. Honeywell's curriculum of video programs includes topics on FORTRAN, BASIC, Decision Tables, PERT, and DATA Base.[12] Its Video-Assisted Learning (VAL) program uses a multimedia approach combining video lecture material with readings and examinations. The courses are designed by educators, writers, and computer experts who develop, test, and review the courses.

in-service training

Besides being used to help sales and train customers, television is also used to teach employees new skills. Company X has just converted its order and shipping departments to computer technology. Before the conversion, an order was received, checked, and verified, and separate slips were made out for every item on the order. The slips were then distributed to the various warehouses at which the items were assembled. Finally, the shipping department received the individual slips and the master order from which to package the goods and prepare the mailing label. But the conversion ended all of this. Even typewriters have been replaced by visual display terminals. Now comes the task of training two entire divisions of the company to use the new equipment as well as to train the personnel of the company's branch plants located throughout the world. Company X accomplishes this through specially produced television training programs. It could have just as easily trained its personnel in the uses of a new telephone system or any other device. Television is especially useful in being able to go places where people may not. Hazardous locations, off-limits to many, become readily accessible to the television camera. The inside of a factory, the welder in a steel plant (Figure 9–6), the equipment operator in a coal mine—all can be captured on videotape using portable equipment.

One of the most difficult training assignments any company faces is educating equipment maintenance personnel. Every year new models are produced, new parts are required, and refinements are made. What about the person who must repair this equipment when it breaks down? How can he or she keep abreast of the latest developments? Again, television comes to the rescue. When a company introduces a new piece of equipment, it automatically produces a new television training program along with it. Company service representatives thus can learn how to fix the new equipment at their home locations scattered throughout the

Figure 9–6 (Inland Steel Company)

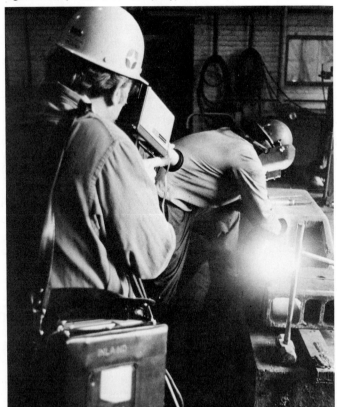

world. It is much less expensive than bringing together all the repair personnel each year for a new training program.

Along with management and employee training services, corporations are also using television to offer college-credit classes to employees. One example is in the Chicago area where employees at such companies as Illinois Bell, Western Electric, Motorola, and Standard Oil receive college credit from the Illinois Institute of Technology. The college programs are made in the ITT classrooms, but are sent via a special microwave frequency to the plants where they are received on standard television sets. Without the corporate link with ITT, students either would have to miss work or attend ITT night classes.

management development

High- and middle-level management people are constantly learning. Let's examine how corporate television helps.

The Executive Communicator

One area in which television is used effectively and frequently is in teaching management personal communication skills. Public speaking is a regular part of executive life. Speaking effectively is an important asset. With television, special speaker-training sessions can be used to videotape an executive's speech and then to play it back for criticism. In some cases, the speech is delivered before an executive panel that also criticizes the playback. Portable equipment also enables the executive to use television after hours in the privacy of his or her office without worrying about special technicians. Many companies also offer videotaped, short courses to help executives learn essential writing skills. In addition, principles of organizational communication can be taught using videotape. New executives can learn the structure of the organization as well as its rules and procedures in this way.

More and more often, television settings are being used to train executives in press relations. The business community frequently finds itself the subject of inquiry from the press, and few public relations directors can substitute for the chief executive when news such as a gas shortage surfaces. Many corporate executives have not been effective or convincing in such encounters. As a result, they now are receiving training in how to act and what to say in front of live television cameras while being pumped by reporters. This training is conducted under the lights of a corporate television studio while being interrogated by corporate personnel. The executives have dress rehearsals to gain skills in handling public encounters with the press.

Training Management Decision Makers

Television is also used to train management in everyday decision making, and especially in dealing with personnel. Videotaped models simulate such situations as firing employees, reprimanding them, or counseling them about a personal problem. After studying the tapes, a group of executives usually act out the different roles in these situations. Their role playing is also videotaped, then played back and compared to the videotaped models. This type of training often includes the advice of trained professionals who criticize the executives' performances.

Communicating with middle-level management can be aided by television, especially in highly technical industries in which the company is spread out over a large geographic area. For instance, suppose a supervisor in Mexico must train a group of assembly line workers in constructing a new product. This construction will mean changing many workers' jobs. Since such changes can cause serious personnel problems, the manager in Mexico watches a videotaped lecture by the supervisor at another plant at which the product already has been introduced. The supervisor tells what problems to watch for, how to solve them, and the effect of their solutions on other workers. The entire program is in Spanish. In other parts of the world, other supervisors can view the same lecture given in French, German, or Italian. Produced in the corporate headquarters, the program is an alternative to having executives fly all over the world with interpreters in order to start a new product down the assembly line.

An applied example of communicating with management by television is that of Holiday Inns.[13] With more than 1,600 motels around the world, the company has many managers to reach. Keeping them abreast of new corporate developments is accomplished by equipping each Holiday Inn with a videotape playback system. The company offers over 24 hours of video-based training programs from which to choose.

Another example is AT&T's Picturephone systems. These systems permit management at one location to communicate with management at another, across town or across the country, but this time they all view one another via television. This is done with the help of individual microphones in the Picturephone meeting rooms which voice-activate the cameras so that the cameras automatically switch to the person speaking. With the push of a button, the camera will select a composite picture of everyone at the meeting table.

We already learned about Dana Corporation's Reading Frame Division in the section on corporate newscasts. However, that is just one phase of the system. Much of the rest of it is used to help management make quick decisions with a minimum of paperwork.[14] At the desk of key executives is a television monitor equipped for multichannel reception. The executive can "tune in" almost limitless amounts of information, including charts and graphs of the day's production. In addition, at a given time each day, the company broadcasts on closed-circuit television investment and stock reports on one channel for the convenience of any Dana manager who wants to watch. This also can be seen on the monitor in the office.

in-house advertising

Although many companies hire advertising agencies to produce their television and radio commercials, some companies produce their own. After hiring skilled advertising professionals away from the agencies, the companies furnish them with creative facilities and establish in-house agencies, with most of the responsibilities and rights of a regular agency. One of these rights is the lucrative media commission paid to agencies by the stations for providing them with advertising business. Since this commission usually is 15 percent of the advertising budget for that station, in-house agencies even can make a profit for the company. At least they can make a return on their investment. One argument against such a system is that in-house agencies cannot view the company objectively and therefore often overlook the most creative approach. Another is that a company should stick to what it knows best—manufacturing the product—and leave its broadcast advertising up to the ad agency professionals. Still another argument is that not all media recognize in-house agencies and allow them the media commission. Among all of these pros and cons is the best arrangement for each company.

future perspectives of corporate television

How will technology and the relationship between employees and corporate media affect the future of television in business and industry? Already there has been much research on organization communication, but the study of television's position in this organizational setting is still not complete. We need to know what happens when a television screen replaces human interaction on a face-to-face basis. To what extent can a corporation employ television without decreasing the company's sensitivity to people, or can television help increase this sensitivity?

Despite these questions, corporate television is expanding. With the development of inexpensive and compact satellite receiving equipment, the 1980s will see a large company like IBM or AT&T produce a coporate newscast and send it simultaneously into offices and plants throughout the world. The public relations coordinator once trained in print journalism and skilled in editing the company magazine will be producing a video magazine and a television newscast. All of this will necessitate broader perspectives of understanding broadcasting and how it can be used effectively to communicate in a corporate atmosphere. We must not forget that the current generation moving up the corporate ladder is the "television generation." When these younger executives reach top management positions, what decisions will they make about corporate television?

Smaller components and more sophisticated delivery systems also promise to change television. For example, the night watchman of the future will approach a troubled area of the plant with a tiny television camera strapped to his belt, constantly monitoring and videotaping the path before him. Confronting a thief will mean simultaneously taking his picture.

A perspective of the future of corporate television was offered by Will Lewis, vice president of communication at International Paper, a major user of television for corporate communication. Lewis states: "When the contribution of employees is not principally physical strength or physical speed but, in fact, the ability to think and make rational decisions, the Company has an obligation to extend the vision of employees so that they become better informed." Lewis notes that, "as the need for more timely information increases, we must seek new and better ways to provide access to information."[15] Lewis's predictions will be based on more companies realizing the potential for television as a medium for both internal and external corporate communication.

summary

The use of television in business and industry is growing. In fact, it has been predicted that in the 1980s, more companies will be using television than there will be commercial television stations in the United States. Some of these uses currently include producing and disseminating corporate newscasts, developing information programming for employees, training sales personnel, and informing customers. In-service training for new skills is also a function of corporate television. Another is helping managers acquire effective communication skills, deal with employees, and communicate with other managers. Televised surveillance can be used not only for security but also for production control as well. Some companies are trying even to cut costs by using corporate television in their in-house advertising agencies to produce their own commercials.

Producing accountable corporate ITV programming, on the other hand, requires many of the same steps used in producing ITV for education. The future of corporate television will be guided by new television technology and its human implications.

spotlight on further learning

BUNYAN, JOHN A. AND OTHERS, *Practical Video: The Manager's Guide to Applications.* White Plains, N.Y.: Knowledge Industry Publications, Inc., 1979.

BUNYAN, JOHN A., and JAMES C. CRIMMINS, *Television and Management.* White Plains, N.Y.: Knowledge Industry Publications, Inc., 1977.

MARLOW, EUGENE, "Programming for a Company Television News Show," *Educational & Industrial Television,* 6 (April 1974), 30, 33–37.

WILLE, WARREN R., "The Dana Approach—A Management Information System," *Educational & Industrial Television,* 6 (January 1974), 10–12.

10

international broadcasting

FOCUS After completing this chapter we should be able to

Compare and contrast broadcasting systems in different parts of the world.

Understand how Canadian broadcasting operates.

Explain how both the government and private sector operate broadcasting in Mexico.

List and explain the two systems of broadcasting in the United Kingdom.

Describe broadcasting on the European continent.

Discuss broadcasting in the USSR.

Explain Japan's broadcasting system and the country's use of new fiber optics technology.

Explain broadcasting in Australia.

Describe broadcasting in Ghana, Rhodesia, and South Africa.

Discuss educational broadcasting in developing nations.

To confine the study of broadcasting today to one country is to have a very narrow view of both the world and the broadcast media. The international flow of broadcast programming, direct broadcast satellites beaming signals across national boundaries, the growing importance of World Administrative Radio Conferences (WARC), and the expanding role of the International Telecommunication Union (ITU), all demand a universal perspective of these electronic media.

We learned in chapter 2 that broadcasting developed simultaneously in many different parts of the world, especially after World War I. Some countries progressed more rapidly than others, with some introducing broadcasting as late as the 1970s. The different political, economic, and social conditions in which broadcasting operates are as varied as the countries themselves. Not every country permits commercial advertising, has a free press, nor allows private ownership of broadcasting. Each country has its own system serving both domestic and international audiences.

This chapter will discuss broadcasting systems around the world so that we can gain a broader perspective of our own system.[1] We will not examine every country, nor will we view each country from the same perspective. What we will do on our world tour is acquaint ourselves with many different elements of international broadcasting.

the scope of international broadcasting

As we visit different countries, remember that the United States is only one model of broadcasting. Our commercial radio and television stations are supported by advertising, and our public broadcasting is supported by government funds, the public, corporations, and institutions. In other countries, advertising also supports some systems of broadcasting, as do public contributions. We also can find systems totally supported by the government. We can find systems for which a special tax is paid on radio and television receivers to help support the system. We can find systems for which the public obtains a license to listen to radio or to watch television, and the license fees support the system. In many countries, different methods of financial support are found side by side.

Also keep in mind that not every country has the freedom of expression found in the United States; some countries are even freer. And the content of radio and television programming, like that in the

United States, is determined by many things, such as the ratings, the government, and advertising. Some countries operate very large systems and some, very small ones. In some parts of the world you might be able to watch television only a few hours per day. In other parts of the world you can watch television beamed in from many different countries. Do not view any country from a narrow perspective. Compare and contrast the different systems and ask yourself questions about the advantages and disadvantages of each.

Canada

Canada's place in the history of international broadcasting is well defined. Marconi ensured that when he selected Newfoundland to test his trans-atlantic wireless. Those dots and dashes started a chain of events that eventually gave birth to the Canadian Marconi Company. Today, the country is the home of a broadcasting system that stretches from Eskimo villages in the north to the United States in the south. It is a country in which the evening news can be delivered by satellite; yet in which television can be a strange phenomenon to an inhabitant of the northern tundra. Canada also is a country whose fierce pride and loyalty are reflected in everything from its broadcast regulation to its television programming.

Regulatory Framework

Canadian broadcasting is regulated by the Canadian Radio-Television Commission (CRTC) created by the Broadcasting Act of 1968. Comparable to the United States' FCC, the CRTC is composed of fifteen members. Five are full-time, forming the CRTC's Executive Committee, and ten are part-time. The part-time appointees are the key to CRTC's operation. With appointments of up to five years, they are drawn from throughout the country and must be consulted before any major decisions can be made, including issuing, renewing, revoking, or amending the license of a radio, television, or cable company. The full-time members are appointed for seven-year terms and can be reappointed.

Canada's broadcasting policy is stated in the 1968 act:

(a) broadcasting undertakings in Canada make use of radio frequencies that are public property, and such undertakings constitute a single system, herein referred to as the Canadian broadcasting system, comprising public and private elements;

(b) the Canadian broadcasting system should be effectively owned and controlled by Canadians so as to safeguard, enrich and strengthen the cultural, political, social and economic fabric of Canada;

(c)   all persons licensed to carry on broadcasting undertakings have a responsibility for programs they broadcast, but the right to freedom of expression and
the right of persons to receive programs, subject only to generally applicable
statutes and regulations, is unquestioned;

(d)   the programming provided by the Canadian broadcasting system should be
varied and comprehensive and should provide reasonable balanced opportunity for the expression of differing views on matters of public concern, and
the programming provided by each broadcaster should be of high standard,
using predominantly Canadian creative and other resources.

The "balanced opportunity for expression" is equivalent in intent to
the Fairness Doctrine in the United States. Enforcement is not taken
lightly. For example, station CFCF-AM was called to task after broadcasting what the CRTC felt was one-sided coverage of the Official Language
Act of the Province of Quebec, with the CRTC charging that the station
had "failed to provide adequately in its own programming for a reasoned
and responsible discussion of the subject."[2] That prompted the CRTC in
1976 to establish a task force on freedom of broadcast information.
Pending any sweeping changes resulting from the task force, the CRTC
renewed CFCF's license.

The policy to "safeguard, enrich, and strengthen the cultural, political,
social and economic fabric of Canada," is also of great concern to
Canadians and has opened up a series of electronic confrontations
between the United States and Canada, some involving the State Departments of the two countries. Two examples of these are clamping
down not only on American advertising which reaches Canadian audiences but also on Canadian firms buying advertising on American media.
When Canada decided to stop permitting Canadian businesses from
claiming tax deductions for advertising expenditures on American media,
the U.S. National Association of Broadcasters protested vigorously, and
even Secretary of State Henry Kissinger became involved. Both Canadian
and American cable companies import the other country's signals, and
Canada has moved further to delete American commercials from Canadian cable. Although these actions may seem arbitrary, the problems lie
in preserving national interests. Nevertheless, Canadians have no control
over what American media send into Canada and vice versa, since
broadcasting signals simply do not honor national boundaries.

Violent programs imported from the United States are a major worry.[3]
Some Canadians have even claimed that American program suppliers
dump those violent cartoons on Canada which cannot be aired in the
United States. Citing French, Mexican, and Swedish limitations on
imported violent programming, some Canadian politicians suggest that
Canada should follow the same course. As a start, the American program
Cannon was struck from the government-owned Canadian Broadcasting
Corporation's (CBC) television lineup. Violent television is a popular
topic of the politically and socially aware in Canada. Much of the attention

has been focused through the Ontario Royal Commission on Violence in the communications industry which amassed approximately 2,500 studies from throughout the world which purport to show a relationship between crime and media violence. Whatever the results of the Canadian debate, violence in the media will remain as important an issue in Canada as it is in the United States.

Programming Systems

Like the United States, Canada has both public and private broadcasting sectors. The private sector has a well-established system of commercial broadcasting stations serving the entire country. The main commercial television network is the CTV Television Network Ltd., which is owned by broadcasters. Inaugurated in 1961, CTV reaches about 95 percent of Canadian television households and transmits over 66 hours of programming per week. One popular program is "A.M. Canada" (Figure 10-1), the early morning information show available in Canada. There also are numerous news documentaries. CTV entertainment fare features such

Figure 10–1 "AM Canada," the Canadian version of America's early morning television programming, broadcasts from 7:00 A.M. until 9:00 A.M. from a Toronto skyscraper. (CTV Television Network Ltd.)

shows as "Stars on Ice," indicative of the popularity of ice skating in Canada and its viewing appetite for everything from hockey to ice ballet.

Comparable in many ways to the U.S. Public Broadcasting Service, the Canadian Broadcasting Corporation (CBC) is the government-owned public broadcasting service. Although reporting to Parliament through a designated cabinet minister, the responsibility for CBC's programs and policies lies with the CBC's own directors and officers. CBC is financed by public funds and by advertising. Several national services are operated by the CBC, including a French-language television network, an English-language television network, and English- and French-language AM radio networks, FM stereo networks broadcast in both English and French, and northern radio services serving the Indian and Inuit peoples. Programs represent a wide range of tastes and can be received by 99 percent of the Canadian population. Typical of the cultural presentations on CBC is the all-Canadian production of "Madame Butterfly," which was seen on the French-language network. Radio Canada International is CBC's overseas service. With headquarters in Montreal, it broadcasts in eleven languages and distributes programs throughout the world.

In addition to the CTV and CBC networks, the Canadian Association of Broadcasters has a program exchange service for its members. Despite its huge land mass, Canada has managed a coordinated policy of broadcast development using the latest technology to reach its diverse population.

Mexico

The government takes an active part in Mexican broadcasting, including owning and programming its own stations and requiring privately owned stations to provide 12.5 percent of their air time for government use. The government also owns educational television production studios and produces and distributes programs through a nationwide microwave link. Three government agencies participate in Mexican radio and television: the Ministry of Transport and Communications, dealing directly with station regulations; the Ministry of Internal Affairs, which oversees government supported programming; and the Ministry of Education, which coordinates educational programming.

Along with the government stations and government-supported television networks, privately owned radio and television networks crisscross the country. Channels 2, 4, 5, and 8 comprise a federation of four coordinated television channels called Televisa which serves most of Mexico. The channels to some degree represent an attempt to reach more specialized audiences in Mexico. Channel 2's signals, for example, cover most of Mexico with a predominantly middle-class viewing audience. Its programming includes state-produced programs for very young children with a Mexican version of "Sesame Street" entitled "Plaza Sesamo," soap operas, the Mexican version of "Today," weekend sports events, and family entertainment programs. Channel 4 covers metropolitan Mexico City and the immediate surrounding area. Aimed at the mass urban public, programs include specials on Mexico City's neighborhoods and regularly televised block parties (Figure 10-2). Evening programs feature films, amateur hours, and variety artists.

Channels 5 and 8 are geared to the younger, better educated Mexican population. Channel 5's signals reach about half the population of the country via a series of repeater stations which receive and retransmit the signal. Its target audience is the youthful middle class, including university students. Programs on current issues are featured as are American, British, and Japanese programs. Approximately thirty hours of national productions are also seen on Channel 5. Reaching about ten million viewers in the urban valley of Mexico, channel 8 produces many cultural programs. Academic groups and people with differing political beliefs comprise its target audience. Much of Mexico's programming is derived from recognized artists, award-winning international firms, and international productions.

Figure 10–2 A remote broadcast for Mexican television. "Block party" telecasts are a popular format. (TELEVISA, S.A.)

Like Canada, Mexico is very concerned about protecting its national spirit and has excluded what it considers to be offensive programming imported from the United States and other countries.

the United Kingdom

While North America enjoyed the fruits of Marconi's labor, Europe saw the seeds of his communication germinate much earlier. Perhaps nowhere was the new medium's impact felt more than in England where its naval vessels, merchant fleet, Post Office Department, high-powered stations linking continents, and British Marconi Company all used it from its inception. Today, the British Broadcasting Corporation (BBC) and the Independent Broadcasting Authority (IBA) comprise the two systems of broadcasting which have grown out of the wireless.

The British Broadcasting Corporation—BBC

The BBC started in 1922 as the British Broadcasting Company, becoming a corporation in 1927. Today, it acts as an independent broadcasting organization, although it receives its budget for overseas broadcasting from Parliament. The fees it collects from licenses are also determined by Parliament. Directed by a twelve-person board of governors appointed by the Queen, the corporation operates under the advice of a series of advisory boards. These include the General Advisory Council, National Broadcasting Councils for Scotland and Wales, and advisory bodies in such areas as religion, education, and local radio. Originally founded as a nonprofit, public corporation by royal charter in 1927, the BBC does not receive income from advertising.

Radio programming is disseminated throughout the United Kingdom by four domestic radio networks. BBC-1 and BBC-2 are the popular formats with BBC-1 more progressive and BBC-2 attracting a more general audience. Together they capture about 80 percent of Britain's listening audience. They provide news and information, and BBC-2 presents shipping forecasts. BBC-3, on the other hand, programs more classical music as well as dramatic and cultural programs. Live concerts both in Britain and in other countries are emphasized. Masterpieces of world theater and discussions of scientific and philosophical subjects round out BBC-3's programs. BBC-4 is devoted to speeches, news and information programming, dramatic entertainment, and current events. Such programs as "Today," "The World at One," "PM Reports," "The World in Focus," and "The World Tonight" are typical of its extended magazine-type news programs. Phone-in programs, panel games, plays, and readings are heard on BBC-4. Together, these four national networks complement the local radio stations which serve small geographic areas.

Although radio has and continues to be the foundation of the BBC, television does not take a back seat. Experimental television was launched in 1936. Suspended during the Second World War, it went back on the air in 1946. Today, two BBC television networks serve nearly the entire United Kingdom. Over 80 percent of the programs are produced by the BBC, with the remaining 20 percent from independent producers and other countries. Like the domestic radio system, television is financed by license fees collected from owners of receiving sets. Although the licensing fee system brings in income, recent economic difficulties in England have caused serious inflation and some cutbacks in the overall operation of BBC television.

Despite this, the BBC continues to export programming and has received international acclaim for quality. Over 100 countries use BBC productions, and on the average 500 programs are seen per week in various parts of the world. The BBC has a policy of not relinquishing editorial control over any of its programs and does not tailor its programs to any specific region.

Radio is also "exported," not only through direct broadcasting but also through a transcription service that permits selected BBC radio programs to be played back in other countries. At the heart of both transcription and direct broadcasting is the BBC External Service. The BBC receives respect and attention from a global audience. "This is London" uses 39 different languages to broadcast news, information, cultural, and entertainment programming. As the BBC has stated, the External Service transmits values "of a society governed by laws voted democratically, yet willing to listen to dissidents, both within its own frontiers and outside them. It mirrors a national community retooling itself economically and ideologically for the 21st century."[4] Three major services comprise the External Service: the European Service and the World Service (Figure 10-3), which broadcast 24 hours a day in English, and the Overseas Service. Complementing the External Service are BBC radio and television regional services transmitting to Northern Ireland, Scotland, Wales, and the English regions served by television.

Independent Broadcasting Authority

Along with the BBC, the Independent Broadcasting Authority (IBA) operates its Independent Television (ITV) and Independent Local Radio (ILR). A commercial broadcasting system, IBA was created by Parliament in 1954 to broadcast side by side with the BBC. Its sole income is from advertising with commercials airing between programs, not in the middle of them. IBA's structure was amended by the Independent Broadcasting Authority Act of 1973. IBA supervises a developing system of local radio outlets and 15 television production companies which operate much like

Figure 10–3   The BBC World Service. (BBC)

American television stations, serving different regions of the United Kingdom. IBA also monitors the quality of programs carried by its radio and television stations.

The production companies, although supporting themselves by regional advertising, receive their assignments from the IBA, which also operates the transmitters and receives a percentage of the production companies' income as remuneration. Like the BBC, the IBA is free to sell its programs to other countries. At home, it has the authority to assign time to such specific programs as education, news, religious broadcasts, and documentaries. Similarly, the nature and amount of advertising is controlled by the IBA in keeping with the mandate of the 1973 legislation. Other guidelines are provided by the IBA's Code of Advertising Standards and Practice. Television advertising is limited to an average of six minutes an hour, and radio advertising is limited to approximately nine minutes each hour. Like the BBC, the IBA is advised by a group of quasi-citizen-government advisory committees on advertising, medicine, religion, education, local radio, and other operations.

European broadcasting systems

Broadcasting's development and prosperity continue to affect other countries on the European continent.[5]

Scandinavia

In northwest Europe lies the land of the midnight sun—Scandinavia, which includes the countries of Iceland, Sweden, Denmark, Finland, and Norway.

Icelandic broadcasting is controlled by the Icelandic State Broadcasting Service which derives its income from receiver license fees and advertising. Television finally was introduced into the country in 1966. The U.S. Armed Services Radio and Television Service also provides programming for Iceland.

Sweden's broadcasting is under the exclusive control of the Swedish Broadcasting Corporation, Sveriges Radio. Although the government does have ultimate control over the size of the radio and television budget assigned to the corporation, the government does not have control over programming. The Radio Act of 1967 forbids public authorities and agencies to examine programs before they are aired. But the government has indirect control, as it appoints the chairman of the board of governors, the board having final authority over the operation of Sveriges Radio.

There are three radio networks in Sweden, continuing a radio broadcasting service that started in 1925. The second and third networks began service in 1955 and 1962. All three share a variety of programming: Program 1 (P1) airs mostly talk and informational programs; Program 2 (P2) broadcasts classical music; and Program 3 (P3) broadcasts mostly light music interspersed with regional and national news. All are financed by receiver-license fees, as is Swedish television.

Television and overseas broadcasting also come under the jurisdiction of Sveriges Radio, the Swedish Broadcasting Corporation. Introduced in 1957, two television networks currently serve the country—TV1 and TV2. In a somewhat unusual arrangement, the two, even though part of the same system of broadcasting, compete with each other. Overseas broadcasts are conducted by Radio Sweden and reach four continents via short wave transmissions. Programs are broadcast in English, German, French, Spanish, Portuguese, Russian, and Swedish.

The chief regulatory body in Sweden, comparable in some ways to the FCC, is the Swedish Radio Council. Having seven members, the council has the authority to examine programs that already have been broadcast and resolve complaints made by organizations and individuals. At the same time, a group of chief program editors have final authority over all programs and can be held personally responsible for any libel action that might be taken. The editors, not the Swedish Broadcasting Corporation, can be prosecuted for libel.

Denmark's broadcasting system is somewhat similar to Sweden's. Danmarks Radio is a public corporation with receiver-license fees constituting its entire income. Advertising is prohibited. Through a series of transmitters located across the country, the corporation operates three radio program services. Danmarks Radio also has a television service. A radio council with 27 members represents the listeners and viewers. Reporting to the Minister of Cultural Affairs, the council is responsible for carrying out the provisions of the Radio and Television Broadcasting Act of 1973, the most recent major broadcast legislation. The Voice of Denmark has a regularly scheduled overseas broadcast.

Experimental broadcasting began in Finland as early as 1923. In 1926, regular service began with the formation of a public radio company. A year later, legislation placed broadcasting under the Ministry of Communications. Today, all radio and television is controlled by a state monopoly called Yleisradio (YLE). Two networks cover the country's heartland, and a third covers the coastal regions. Broadcasting in Swedish and Finnish, radio is financed entirely by receiver license fees. The External Service broadcasts beyond Finland's borders with programs aimed at three groups: English-language listeners, Finns living in North America, and those at sea.

Finnish television, starting in 1955 at the Technical University, began full-fledged service on January 1, 1958 in a dedication broadcast by then President Urho Kekkonen (Figure 10-4). A second channel was added in 1964. Along with license fees, advertising is permitted on television and is sold by Mainos-TV (MTV), a private company which rents time from the YLE. Founded in 1957, MTV transmits its own programs on both

Figure 10–4 President Urho Kekkonen's New Year's speech on January 1, 1958 initiated YLE's regular TV broadcasts.

television channels. Commercials are transmitted in groups of two to three minutes at the end of and at natural breaks in the programs. The agreement between YLE and MTV has certain restrictions. These restrictions include the prohibition of political programs and the unnecessary use of children in commercials.

In Norway, radio and television is operated by an independent government-owned company, the Norwegian Broadcasting Corporation. Advertising is prohibited and the system is financed by license fees paid by registered listeners. Started in 1923, the broadcasting system of Norway includes a single radio and television network, the programming of which consists largely of educational and cultural fare.

France and West and East Germany

One of the most diversified broadcasting systems is in France. The country has a seven-part broadcasting system ranging from government-controlled to citizen-access channels. There are four independent program societies and three support societies, all formed by the reorganization of French broadcasting in 1974. The seven-part system consists of Télévision Francais 1 (TF1), which has been operating since 1948 and most closely resembles pre-1974 French broadcasting when the Office de Radiodiffusion-Television Francaise (ORTF) operated it. Second is Antenna 2 (A2) created in 1964 and which has a UHF color channel. Third is the French regional radio and television service, France-Regions 3 (FR3), allowing the public considerable access and fostering the widest latitude of public opinion. Fourth is Radio France, under whose auspices all of French broadcasting is organized. The three support societies are the Société Francaise de Production et de Creation Audiovisuelles (SFP), responsible for program production; Télédiffusion de France (TDG), responsible for transmission services; and the Institute Nationale de l'Audiovisuel (INA), responsible for auxiliary services such as research, technical training, and archives. The system is supported mainly by listener fees and, on TF1 and Antenna 2, by advertising.

France's seven-part system still has transition and adjustment problems. Because the change was greatly influenced by French politics, it is only natural that there are varying opinions of the system's effectiveness. Critics generally agree, however, that there is less government control of the system now than before and that there are more opportunities to present diverse programming and political views.

West German broadcasting includes radio, television, and a highly developed cable system. Radio broadcasting is coordinated by the ARD, a federal organization made up of nine members of the *Land* public broadcasting corporations. Two radio organizations, DLF and DW, are operated by the government and do mainly international broadcasting.

The *Land* broadcasting stations all exchange programs. Private commercial broadcasting, Radio Free Europe, Radio Liberty, the United States's Voice of America (VOA), a BBC station, and military stations all are part of broadcasting in Germany. The *Land* corporations also engage in television broadcasting as does the ZDF, which broadcasts both separately and jointly with ARD's *Land* stations. Receiver license fees, government grants, and advertising provide income for broadcasting in West Germany.

East German radio is controlled by the State Broadcasting Committee of the Council of Ministers. Four radio networks are in operation. Television is controlled by the State Television Committee of the Council of Ministers. Broadcast income is obtained from receiver license fees. Interestingly, although East Germany has banned most West German printed media, it cannot ban West German television. Thus, instead of watching East German television, which many consider rather dull, East Germans often watch West German signals, which are easily received across the border.

Switzerland: SBC Short Wave Service

Along with the BBC and other broadcasting systems in international broadcasting, the Swiss Short Wave Service operates an extensive European and overseas service with high-powered transmitters beaming news, information and cultural programs to all continents of the world (Figure 10-5). The Swiss, because of their reputation for neutrality in international affairs, have substantial credibility among many nations, whereas other countries may be perceived as having vested interests in the content of their programming. The Swiss service is well accepted, especially in third-world countries and developing nations.

The Swiss short wave service has two goals—providing timely information to Swiss citizens living in other countries and spreading Swiss culture. Special care is taken in news programming with no report of world significance being broadcast until it is verified by at least two other sources. Entertainment programs are also part of the worldwide service, and recorded programs are mailed to over three hundred stations throughout the world.

Switzerland also operates a domestic broadcasting system. Within this system, Swiss radio is financed by receiver-license fees, and Swiss television is financed by fees, advertising, and government loans.

The Netherlands's Open Door System

The Netherlands's four domestic radio services and two television services stand out as somewhat unique in the world of broadcasting. The radio

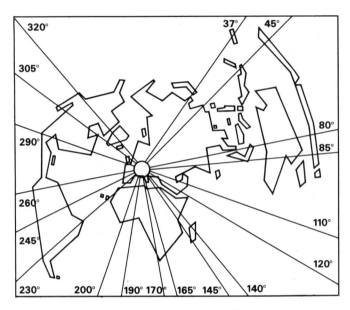

Figure 10–5 Coverage map of the Swiss Short Wave Service.

services—Hilversum 1, 2, 3, and 4—program everything from news to radio drama, with most programs in stereo. The television services, Netherland 1 and 2, broadcast about 80 hours a week, mostly in color and mostly at night. What is unique about these services is that they will give citizens and other organizations air time, if they qualify for it by providing evidence of significant membership and support. Approximately 30 organizations have qualified for broadcasting time on the radio and television services. Some 70 percent of the programming is used by these "qualified" organizations, with the remaining time devoted to jointly produced programs and educational services.

To be granted air time, Dutch citizens must fulfill certain requirements as defined in the Broadcasting Act. An organization must claim at least 40,000 Dutch citizens as members or as supporting the aims of the organization. At this point, the organization is given the status of Candidate Broadcasting Organization by the Ministry of Culture, Recreation, and Social Welfare, the supreme governing body over broadcasting. The candidacy period is used mostly to recruit new members, since the organization's membership must reach 100,000 within two years, or it loses permission to broadcast. There are three categories of qualified organizations—A, B, and C—each permitted different amounts of air time. Which category an organization falls into is determined by the size of its membership. Along with air time comes the right to publish program guides. Despite what may seem like arbitrary controls, the

Dutch government cannot censor any of the organization's programming, whose content is determined solely by the qualified organization.

The system is financed both by advertising and receiver license fees. Every Dutch citizen who owns a radio or television must pay an annual fee. Appropriate joint programming is permitted and even encouraged between different organizations. The Netherlands Broadcasting Foundation (NOS) coordinates the activities of the approximately 30 qualified organizations.

USSR

Soviet broadcasting is controlled by the state under the Union Republic State Committee of the USSR Council of Ministers. Income is derived from the state budget and from the sale of programs, announcements, and public concerts. Approximately 60 million television sets are in use or about 98 for every 100 families. The Molina satellites help relay programming to the Pacific coast and Central Asia. Central operations are housed in the Moscow TV Center Tower Building, which is higher than the Empire State Building and has full production facilities for everything from small studio to large auditorium productions (Figure

Figure 10–6 U.S.S.R. Broadcasting complexes. (Courtesy Radio Moscow)

10-6). Color television is seen in many Russian areas using the Soviet-French SECAM system.

The government operates four television networks. Channel I is a network serving all of the Soviet Union. Moscow and its immediate surrounding area are served by Channel II. Channel III is an educational channel, and Channel IV carries drama, music, film and literary programs. Typical programs seen on Soviet television are "Time," a half-hour news program; "The 13 Chairs Tavern," a musical satire; and "Come on Boys," an audience participation show for boxers, wrestlers, weightlifters, and motorcycle racers. Broadcasts are in the 66 different languages of the various Soviet nationalities.

Russian radio broadcasts are heard in 60 different languages and are received by approximately 65 million radio sets. The Home Service of Radio Moscow has four networks. Channel I, like its television counterpart, is heard throughout the country and includes news, commentary, drama, music, and children's programming. News and music are heard on Channel II, with news every half hour. Literary programs and drama are presented on Channel III, and Channel IV concentrates on FM musical broadcasting. FM stereo currently can be heard in 26 major Soviet cities. The USSR also has cable radio. In U.S. dollars, the cost for cable is about 61¢ per month. Cable radio, with three different channels, reaches about 400 Soviet communities.

The USSR's external and overseas service is the responsibility of Radio Moscow. Broadcasting in 64 languages, Radio Moscow programs are mostly on life in the USSR, the Soviet view of international issues, and Russian drama and entertainment. Programs are distributed free of charge to stations in other parts of the world. Weekly programs include "Soviet Press Review," an editorial and commentary program, and "Moscow Meridian," a short commentary. Other biweekly and monthly programs are available on such topics as politics, science, and art.

The Radio and Television Committee also cooperates with other countries wishing to produce programs about Soviet life. One example is "Pravda," a documentary produced by Finnish broadcasters showing the inside workings of the Soviet Union and *Pravda,* a newspaper with one of the world's largest circulations.

Despite its desire to broaden its influence through these broadcasts, the USSR finds television a difficult vehicle through which to spread propaganda. Interestingly, the popularity can do the same to a Soviet propaganda broadcast that poor ratings can do to American television. A dull show is simply not watched, neither in America nor in the Soviet Union. Since there is more than one channel to choose from, a Soviet audience will more than likely choose sports or entertainment over political broadcasts. In fact, if given the choice of watching political broadcasts or not watching, many times they will choose not to watch.[6]

Australia

Both a private-commercial and a government-sponsored system of broadcasting operate side by side in Australia.

Radio Services

The system, from radio's perspective, is a competitive battle for audience listenership. The Australian Broadcasting Commission operates the government radio system, called the National System. Responsible to Parliament, the A. B. C. controls fifty transmitters scattered throughout the continent and is programmed by two major networks. License fees help finance the government system with one network servicing local and regional areas and the other crossing state boundaries as a country-wide system. Government subsidies supplement the license fees. Programming has emphasized cultural entertainment and classical music presentations. Specially held concerts by world-recognized musicians have been a popular broadcast feature on A.B.C. stations. Since the concerts are often held in large Australian cities, an admission charge helps defray the cost of the concerts.

There is also a comprehensive commercial broadcasting system with more limited service areas but with the ability to sell advertising, which also covers the Commonwealth. Commercial stations, partly because they must depend on local advertisers, orient their programming to their own locales. As in the United States, commercial stations compete directly with each other as well as with the government stations. A voluntary code of ethics helps control the content of Australian broadcasting and acts as a buffer to increased government control, similar to the NAB Code in the United States. Australia also has public stations, similar to those in the United States, which are supported by subscription, universities, and organizations.

Radio Australia

Overseas broadcasting is the responsibility of Radio Australia. With 24-hour service concentrating on news and information interspersed with some music and entertainment programs, Radio Australia broadcasts in eight languages and some dialects. Current language breakdowns include approximately 54 percent English, 17 percent Indonesian, 7 percent Chinese, 7 percent French, 4 percent Vietnamese, 4 percent Japanese, 2 percent Thai, and 6 percent simple English and Neo-Melanesian.[7] The fate of Chinese refugees has been an important part of its Chinese broadcasts. Reports to relatives, messages from students studying in

Australia, and songs and interviews with children of Chinese refugees are typical broadcast fare.

Television

As with radio, there are two systems of television in Australia: the government-supported system and the private-commercial system. More than 50 government stations are in operation compared to just under 50 commercial stations. Government stations come under the A.B.C's jurisdiction, and commercial stations are part of the Federation of Commercial Television Stations (FACTS). Drama, public interest programs, and sports make up the majority of government television programs. About 60 percent of the government programs are produced by the A.B.C., 12 percent by the BBC, 7 percent by other United Kingdom and Commonwealth countries, and 21 percent by the United States and other overseas countries. The most popular programs on commercial television include "Number 96," an adult serial drama; news and weather; drama; "Disneyland"; and "Matlock Police," a police drama series.

Japan

With so much attention focused on Japanese imports of television sets and related electronic gear, we tend to think of broadcasting in Japan in such terms as SONY and Panasonic. However, beyond the manufacturers is a broadcasting system geared to new technology.

The Wired City: Optical Visual Information System

Japan has been one of the world leaders in making practical application of fiber optics. Indicative of this is the experimental prototype system for Higashi Ikoma New Town, selected for its location and population base. The system contains a central computer terminal with fiber optics transmission lines permitting not only television signals, but also computer processing and all of its related services to be available on a residential basis. Although experimental uses will vary, the system will enable shopping, banking, and other purchasing decisions to be made from the living room. Television program guides, newspaper delivery, and still-picture transmission are just a few of the many uses.

This experimental fiber optics system has nine components:[8] (1) a *computer* which will automatically select programs and other services in response to home terminal commands; (2) a *subcenter system* which supplies the programs and information to the home terminals; (3) a *TV request system* which provides motion pictures to home screens through video-cassette tape recorders; (4) a *still-picture system* for such activities as shopping and news; (5) a *studio system* permitting full-scale production at the center; (6) a *retransmission system* for standard cable television transmission; (7) a *mobile center* for live coverage outside the studio; (8) an *optical transmission and reception system* which converts electrical signals to light signals compatible with fiber optics transmission; and (9) *home terminals* which permit viewers to control the system. The home terminal system also includes a camera and microphone for *two-way audio and visual* communication (Figure 10-7).

Using fiber optics, the amount of information that the system can carry becomes almost infinite. For the Japanese, they view such a system as increasing, not decreasing, personal communication between individuals that high technology often discourages.

Nippon Hōsō Kyōkai (NHK)

Correspondent, critic, and network executive Sander Vanocur has called Japan's governmental television service, NHK, the best he has ever seen and ranks it above the BBC. To many Japanese, NHK means television. Financed by license fees, the system provides a blend of information and both Japanese and western cultural programming (Figure 10-8). Oper-

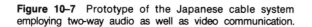

Figure 10–7 Prototype of the Japanese cable system employing two-way audio as well as video communication.

Figure 10–8 A scene from the NHK program "Kashin"
(God of Flower), a serial drama airing on Sunday evening.
(NHK)

ating two channels, one for education and the other for information and
entertainment, NHK television reaches every corner of the country. It
also operates three radio networks (Figure 10-9).

Along with the domestic service, NHK's Radio Japan broadcasts
worldwide in 21 different languages. The Regional Service of NHK is
designed for specific areas, and the General Service airs worldwide.
NHK's posture in domestic and international broadcasting is best described in the 1950 Broadcast Law:

*(1) To broadcast well-balanced, high-quality programs in the fields of news
reporting, education, culture, and entertainment in order to meet the various needs
of the people;*

*(2) To undertake construction of broadcasting stations even in remote mountainous
areas and isolated islands to bring broadcasting to every corner of the country;*

*(3) To conduct research and investigation necessary for the progress and devel-
opment of broadcasting, and to make public the results thereof;*

*(4) To foster correct understanding of Japan by introducing this country's culture,
industries, and other aspects through the overseas broadcasting service; also, to
provide international cooperation, such as program exchange and technical aid,
for overseas broadcasting organizations.*

Figure 10–9 Transmission control room at the NHK Broadcasting Center. (NHK)

Commercial Service

With the end of World War II, the democratization of Japan, and the introduction of television technology, the important 1950 Broadcast Law recognized private commercial broadcasting as a competitor of NHK. Together, NHK and commercial companies operate over 6,000 radio and television stations covering even Japan's mountainous terrain. The public has a wide choice of programming. Tokyo, for example, has the two television services of NHK, five commercial television stations, four commercial radio stations, and three NHK radio networks.[9] There are approximately 50 commercial companies in radio and 85 in television programming. With Japan's interest in high technology and a well supported government broadcasting system competing with a commercial system, the future for broadcasting in this nation is particularly bright.

Africa

The wide ranges of technology, the varied cultures, the great expanse of land, and the political and economic issues of the time all blend together to make Africa a fascinating place from which to view international

261

broadcasting. Within the continent, we can both find the most modern facilities in the world and the first introduction of a primitive tribe to the "voice in the box."

Ghana

Called the Gold Coast when broadcasting was first introduced, Ghana's association with radio goes back to 1935.[10] Introduced to the country by Governor Sir Arnold Hudson, radio at that time had only about 1,000 listeners. Station ZOY began in a bungalow as a "wired-relay station" and then, bolstered by government funds, moved into Broadcasting House in Accra in 1939. Eight technician-announcers manned the station in pre-World War II days, then it increased its staff and expanded its facilities to other regions before the war actually started. ZOY served as a news and information source during the war, and signals were retransmitted via sixteen relay stations. In 1952, a government-appointed commission of broadcasters and academicians reviewed broadcasting and upon the commission's recommendations, the Gold Coast Broadcasting System was formed in 1954. When independence was gained three years later, the name was changed to the Ghana Broadcasting Corporation (GBC).

Today, the GBC operates two services, National and External. Two networks operate out of the National Service, GBC-1 and GBC-2. The External Service is described by the GBC as "a true voice of Africa helping us forward in the fight for emancipation, helping us struggle for total emancipation and political Union of African States, a voice raised forever in the cause of peace and understanding between men and between nations of the world."[11]

Television in Ghana dates back to 1959 when a Ghana-Canada Technical Assistance Program promoted the installation of facilities of the Accra Broadcasting House. Service began in 1965, mainly to provide Ghana with an understanding of domestic and world affairs. Drama presentations have been especially popular with illiterate segments of the population. Overall, the viewing audience is estimated at 660,000 out of a population of approximately 8.5 million.

Ghana participates in a broadcasting exchange program with other African states. Its broadcasting systems are financed by government subsidies, advertising, and subscriptions to the broadcast relay service, which extends the availability of radio and television to outlying areas.

Rhodesia

The exact structure and organization of Rhodesian broadcasting is somewhat uncertain, depending on the changes that may occur with black majority rule. But through 1978, broadcasting remained under control of the Rhodesian Broadcasting Corporation.[12] Three Post Office Department engineers began public broadcasting back in 1941 with a converted transmitter originally used for aircraft guidance. For years, the Federal Broadcasting Corporation (FBC) in Rhodesia operated radio broadcasting. Television, meanwhile, was produced by R.T.V. Ltd., a production company affiliated with the FBC, in much the same way production companies produce the programs under the jurisdiction of the Independent Broadcasting Authority in the United Kingdom. R.T.V. Ltd. was owned by a consortium, including the local newspaper which owned 51 percent of the stock. The remaining 49 percent was held by the FBC which had the authority to prohibit any change in stock ownership. After the country became independent in 1964, the Rhodesian government changed this shareholding arrangement. The Rhodesian Broadcasting Corporation now owns 51 percent of the stock with the remaining 49 percent owned by the public and listed on the stock exchange.

In 1976, the television services were changed from their status as production companies to an actual part of the RBC. Under this system, about 20 to 30 percent of RBC's programming is locally produced with the remainder being imported from such sources in the United States.

Rhodesian independence has been a time of tension and strife for not only the country but for broadcasting as well. Previously contributing news and entertainment programming to the FBC, the BCC has cut all ties with the RBC. The loss was a severe blow, and the new RBC had to scramble to plug the holes left by the vacant programs. The transition was roughest for the African Service broadcasting to Black Rhodesia. Loyalists threatened the studies and harassed the personnel. Special escape routes even were devised to transport the staff in and out of the broadcasting buildings without incident.

Today, the African Service continues to exist and is responsible for educating and maintaining the culture of Rhodesia's African tradition. Although modernization of the entertainment programming is evident, the sounds of a tribal drummer and tales of cultural heritage are still a part of African Service radio programming.

Republic of South Africa

Among those countries involved in international broadcasting, the Republic of South Africa contains a fully developed broadcasting system. Controlled by the South African Broadcasting Corporation (SABC), the country currently has both radio and color television services which are programmed in two languages for the white population and in seven

African languages for the Bantu peoples. The South African Railways launched the very first broadcast transmissions in July 1924, signaling the beginning of three separate broadcasting services. Three years later in 1927, the three services were consolidated into the African Broadcasting Company (ABC).

Financial problems continued to beset the consolidated service, and the ABC limped along until 1936 when legislation responding to a study group's recommendations formed the SABC. There was equal parity in the control of the SABC until World War II when the English-speaking, British descendants and the Afrikaans-speaking Dutch descendants, once known as the Boers, split along political and philosophical lines. The Afrikaners had been a rural minority with very little representation in early broadcasting, save for an-hour-per-week service from the heavily British ABC. Resentment continued until the legislation creating the SABC called for an equal number of programs in both English and Afrikaans. However, equal numbers did not always mean equal political content, and the Afrikaners began to complain about the political bias of programs, about being forced to report statements of the ruling English, and about the required loyalty oath to the SABC.

World War II brought two important changes. First, it sent abroad many of the English-speaking staff of the SABC volunteering to fight for the Allied cause, leaving behind a majority of Afrikaners. Second, the Afrikaner Nationalist party won the 1948 elections and was in a position to regain a strong voice in the affairs of the SABC. Today, the black-white struggle is publicized more from the white standpoint and is influenced heavily by the political philosophy of those earlier struggles between the English and Afrikaners.

The SABC retains control of South Africa's broadcasting, although private radio has represented both white factions. With headquarters in new broadcasting facilities in Johannesburg, SABC has one of the most lavish broadcasting complexes in the world (Figure 10-10). A huge office complex and production facilities overlook a giant broadcasting tower nearby, which houses the antennas of the SABC services. Radio is heard throughout the country by the English service, the Afrikaans service, regional commercial services, Radio Bantu, and the Voice of South Africa which is beamed to 23 different areas of the world.

Because of the mountainous terrain and sparse population, television arrived late in South Africa. Approved by the government in 1971, a one-channel service devoting equal time to broadcasting in both English and Afrikaans, began in 1976. Four mobile units in Cape Town, Durban, and two in Johannesburg provide supplementary programs. Television service to the Bantu people is already planned and expected to be operational in a couple of years.

Figure 10–10 Headquarters of the South African Broadcasting Corporation. The tall building is the administration building and the tower in the background houses the transmitter facilities.

educational broadcasting in developing nations

Many developing nations are reassessing the importance of education to their systematic growth. To understand a common language, participate in political processes, and provide local government leadership all require a literate populace. For most countries, however, this renewed emphasis on education has been a painful and deliberate process. Many nations lack the funds to train a national teacher corps. Others find it difficult to

265

find qualified personnel who will both be sensitive to cultural heritage and acceptable to local communities. As a result, these nations have turned to the media for help, often using foreign-aid dollars to defray the cost.

To understand how broadcasting has helped the educational growth of these developing nations, we first shall examine radio. Emile G. McAnany has categorized radio's role into five strategies of use.[14]

Five Strategies for Using Radio

McAnany's first strategy is open broadcasting: the unorganized audience. This category describes the broadcasts to the general population. For example, natives in African Zaire hear the voice of "Dr. Massikita" bringing them information on feeding newborn children or on nutrition. Other African countries have instituted similar health and agricultural programs which are usually repeated in different languages. Latin American programming strategies also follow this broadcasting pattern.

A second strategy is that of instructional radio: the organized learning group. Here radio is used as a substitute teacher. Australian children often receive these broadcast lessons in remote rural settings, far from a traditional classroom. In Tanzania, instructional radio teaches basic skills by correspondence. Around the globe in Thailand, radio beams music, social studies, and English courses to the populace. In Sudan, instructional radio teaches Arabic, tribal history, and etiquette.

A third strategy is radio rural-forums: the decision group. These programs offer information on rural news, answers to listeners' questions, or a discussion or lecture topic. Many listeners get together in groups, listen to the broadcasts, then a discussion leader guides them through a dialogue on the subject. Countries using this approach, among others, are Ghana and India. McAnany illustrates this strategy using Benin, a small country in western Africa. There, ten to thirty villagers gather to listen and discuss the programs, which are broadcast in ten native languages at different times during the week. A group discussion leader called an *animateur* is chosen from the village to lead the dialogue. This person then sends a monthly report back to the production center, which may be used as feedback in producing future programs. A second person called an *encadreur* serves as a technical resource person for the village. The *encadreur* provides guidance for any projects which result from the discussions.

Another strategy is radio schools: the nonformal learning group. This audience is mostly illiterate, rural adults, so the programming is basic training in reading, writing and "some figuring." Nations often use multimedia approaches in this strategy combining radio, newspapers, film strips, booklets, and charts.

McAnany's fifth strategy is radio animation: the participating group.

Although at first participation may seem like rural forums or radio schools, it is not. Radio's chief contribution here is to define and identify community problems rather than to provide solutions. Discussion leaders then promote dialogue among community members to solve these local problems.

The future of educational radio in developing nations is brighter than that of educational television. A television receiver in some ways can be more intrusive than a radio receiver, especially in areas unaccustomed to new technology. Moreover, many mountainous areas can not receive television signals, and radio is the only alternative. Radio also lets local discussion leaders maintain their identity, unchallenged by a visual commentator.

Applications of ITV

Television also is influencing major educational reforms. Wilbur Schramm has documented one example in the small, Latin American country of El Salvador.[15] ITV was introduced into El Salvador on a pilot basis in 1967.

The government selected work in the seventh, eighth, and ninth grades as the major test curricula. El Salvador's schools at that time were faced with a serious attrition rate and inadequately trained secondary school teachers. Schramm's report noted that although scheduling problems and a teachers' strike did cause some difficulties, the ITV pilot program did show positive results. He also observed "tentative evidence" that classes using ITV contributed to greater equalization of learning gains.[16] At the same time, however, Schramm noted that ITV was but one portion of the educational reform in El Salvador, although it was the most publicized. Other reforms were extensive teacher training and curriculum revision.

El Salvador's students seemed to like television in the classrooms as did teachers and parents.[17] Teachers did, however, recognize the medium's limitations, such as the inability of students to "ask" television questions. In fact, after the first year of the evaluation program, the initially positive reaction to ITV diminished. Schramm's report contributed this waning popularity partially to a "spillover effect" of general dissatisfaction resulting from the teachers' strike. Parents were receptive to ITV especially after having it explained to them. Although less expensive than placing teachers in the classroom, the future of ITV in El Salvador will be linked to its cost-effectiveness. Can ITV replace teachers in some areas? If it can, will the educational results harm or help the students? The answers are crucial both to El Salvador and to ITV.

Another example of educational broadcasting at work in educational reform is in American Samoa. Here, the government introduced television into the classroom in 1964 as a way of rapidly trying to upgrade the educational system and teach English to elementary school students. Samoa's governor, H. Rex Lee, even proposed a television network to teach some basic material in elementary classes.[18] As in El Salvador, ITV programs are used as a supplement to teachers, not as replacements. The format of each lesson has four phases: preparation, television lesson, follow-up and feedback.[19] The advantages of the format are uniform content and, through the lesson plan, as much control as possible over the discussions that precede and follow the broadcast.

Despite the widespread use of ETV and strict controls over the use of the programs, English proficiency in Samoa is still not up to U.S. mainland levels. One analysis blames it on the lack of sensitivity to the local culture.[20] Although English is important, Samoan culture, the analysis states, is more important. A group from the United Nations Educational, Scientific, and Cultural Organization (UNESCO) commented that in one of the broadcasts "the stateside teacher used a Samoan adult as a prop, asking him to touch his nose, go to the door, and so on. All of the talking was done by the stateside teacher. This may have implied an ethnic superior-subordinate relationship."[21] Efforts are still made to bridge the cultural gap. Stories are written in English using native names for the characters, and mathematical concepts are taught using such objects as shells, stones, and leaves.

ETV in Samoa, although widely used, illustrates the problems of intercultural communication which can develop whenever new technology and "foreign" teachers, even used only as advisors, are sent to help a developing nation improve its educational processes. A country's internal characteristics, such as those of Malaysia where two different native languages are spoken, can also create problems.[22]

As in El Salvador and Samoa, Malaysia uses television as a supplement to classroom instruction. Again, acceptance of it was slow. The country's British instructional system emphasizes rote learning and examinations; the people felt more comfortable with lecture recitation than with television. But these problems are slowly being resolved by training local educational leaders who, in turn, are introducing television to others. In 1973, UNESCO sponsored a crash program to train nine school officials to conduct workshops for 5,000 primary school teachers. This was supplemented by a computer-assisted evaluation system which analyzed the effects of teacher feedback on the new programming.

The future of educational broadcasting in developing nations will depend on how successfully cultural barriers are bridged and how well instructional broadcasting can represent and blend into local culture.

Although our world tour has stopped at only a limited number of countries, we have been able to sample some of the different broadcasting systems and the many issues common to other parts of the world. New technology will undoubtedly bring us all closer together, and it remains to be seen how much influence different countries will have on each other's communication systems. Even though countries are interested in adopting the technology of different parts of the world, each wants to retain its own system of broadcast control and cultural integrity of programming.

summary

In our tour of world broadcasting systems, we stopped first in Canada where broadcasting is controlled by the CRTC, with jurisdiction over both public and private broadcasting. The Mexican government is active in broadcasting in that country where it owns and operates its own stations and requires other stations to provide certain amounts of program time for government use. In the United Kingdom, there are two systems of broadcasting: the BBC and the IBA. A series of advisory committees also influences the British system of broadcasting.

On the European continent, the Scandanavian countries have somewhat similar broadcasting systems. Broadcasting in France is controlled by a quasi-government organization. West Germany has radio, television, and a highly developed cable system. Many West German programs are also popular in East Germany where broadcasting is controlled by the government. Switzerland, while providing domestic radio and television, also provides short-wave broadcasts popular in many parts of the world because of Switzerland's neutral political stands. In the Netherlands, any organization which can show support of a large following of people within the country can obtain air time. Soviet broadcasting is a government-controlled system which distributes programs free of charge to many other parts of the world.

In Australia, much as in Britain, both government and private broadcasting stations compete for the broadcast audience. In Japan, NHK is looked upon as one of the world's most highly regarded broadcasting systems. We also examined broadcasting in Ghana, Rhodesia and the Republic of South Africa. Broadcasting has found especially valuable applications in developing nations to continuing education as well as to direct in-school learning.

spotlight on further learning

ABSHIRE, DAVID M., *International Broadcasting: A New Dimension of Western Diplomacy.* Beverly Hills, Calif.: Sage Publications, Inc., 1976.

CASMIR, FRED L., ED., *Intercultural and International Communication.* Washington, D.C.: University Press of America, 1978.

The CBC—A Perspective. Ottawa: Canadian Broadcasting Corporation, 1978.

EDELSTEIN, ALEX S., ED., *Information Societies: Comparing the Japanese and American Experiences*. Seattle: International Communication Center, School of Communication, University of Washington, 1979.

HEAD, SYDNEY, ed., *Broadcasting in Africa*. Philadelphia: Temple University Press, 1974.

HINDLEY, PATRICIA, GAIL MCMARTIN, and JEAN MCNULTY, *The Tangled Net: Basic Issues in Canadian Communications*. Vancouver, B.C.: J. J. Douglas, Ltd., 1977.

KATZ, ELIHU, and GEORGE WEDELL, *Broadcasting in the Third World: Promise and Performance*. Cambridge, Mass.: Harvard University Press, 1977.

LENT, JOHN A., ED., *Broadcasting in Asia and the Pacific*. Philadelphia: Temple University Press, 1975.

MCANANY, EMILE G., *Radio's Role in Development: Five Strategies of Use*. Washington, D.C.: Academy for Educational Development, 1973.

PAULU, BURTON, *Radio and Television Broadcasting in Eastern Europe*. Minneapolis: University of Minnesota Press, 1974.

PIRSEIN, ROBERT W., *The Voice of America: A History of the International Broadcasting Activities of the United States Government, 1940–1962*. New York: Arno Press, 1979.

SANFORD, JOHN, *The Mass Media of the German Speaking Countries*. Ames: Iowa State University Press, 1976.

SCHRAMM, WILBUR, *Instructional Television in the Educational Reform of El Salvado*. Washington, D.C.: Academy for Educational Development, 1973.

SMITH, ANTHONY, *The Politics of Information: Problems of Policy in Modern Media*. London: Macmillan Press Ltd,. 1978.

Television and Radio 1978 (Annual Report). London: Independent Broadcasting Authority, 1978.

THOMAS, RUTH, *Broadcasting and Democracy in France*. Philadelphia: Temple University Press, 1976.

TUNSTALL, JEREMY, *The Media Are American: Anglo-American Media in the World*. London: Constable, 1977.

UNESCO, *World Communications*. New York: Unipub, 1975.

WILLIAMS, ARTHUR, *Broadcasting and Democracy in Western Germany*. Philadelphia: Temple University Press, 1976.

part

4

Regulatory Control

11

early attempts at government control

FOCUS   After completing this chapter we should be able to

Explain the provisions of the Wireless Ship Act of 1910.
Explain the provisions of the Radio Act of 1912.
Discuss the National Radio Conferences.
Trace the collapse of the 1912 legislation.
Explain the provisions of the Radio Act of 1927.
Describe what led to the passage of the Communications Act of 1934.

By the second decade of the twentieth century, the British Marconi Company had a well developed, worldwide corporate empire. Ships were comfortable with wireless and had demonstrated its effectiveness in numerous cruises. Experimental radio stations were popping up everywhere, and ham radio operators were toying with a hobby which would soon become a major social influence. As we learned in chapter 3, in the 1920s some of the historical giants of broadcasting took to the air—WHA at the University of Wisconsin, KDKA in Pittsburgh, WBZ in Springfield/Boston, Chicago's WGN, and WWJ in Detroit. The public was being entertained with Big-Ten football, live orchestras, election night fervor, and presidential speeches. They also were being harassed by higher powered stations, jumbled interference, and rampant competition. There simply was not enough room on the electromagnetic spectrum for everyone to jump on without someone being pushed off.

The government first became concerned about radio's impact when Marconi started prohibiting ships and shore stations from communicating with each other unless they were equipped with Marconi equipment. The idea may seem incredible by today's standards, but Marconi managed to get away with this for some time. Germany was especially affected by Marconi's antics since it housed the competing Slaby-Arco system. The Germans finally took the initiative and called a conference in Berlin in 1903, at which a protocol agreement was reached on international cooperation in wireless communication. Three years later, Berlin hosted the first International Radiotelegraph Convention, from which an agreement was signed by twenty-seven nations. In the United States, the stage was now set for domestic legislation which would embody the spirit of the Berlin agreement and foster safety and cooperation among American shipping interests. It is this 1910 legislation that we'll study as part of the development of government involvement in broadcasting.

the Wireless Ship Act of 1910

In 1910 there were few visions of commercial broadcasting stations as we know them today. Transatlantic experiments were less than a decade old, and Congress was thinking only now about safety applications of the new medium, especially to ships at sea. Some ships, although by no means all, had installed wireless apparatus (Figure 11–1). It was in this atmosphere

274

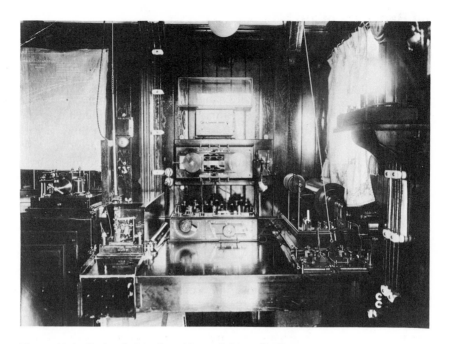

Figure 11-1 Early wireless found immediate application in ship-to-shore communication. Typical of the elaborate Marconi-equipped facilities was this wireless room aboard the *Lusitania*. (The Marconi Company Limited, Marconi House, Chelmsford, Essex)

that the Wireless Ship Act of 1910 was passed.[1] Containing only four paragraphs, it set the stage for maritime communication. Among other provisions, the act made it illegal for a ship carrying more than 50 people not to be equipped with radio communication. The equipment had to be in good working order and under the direction of a skilled operator. The range of the radio had to be at least 100 miles, day or night. Exempted from the provisions were steamers traveling between ports less than 200 miles apart.

The act also specified that the "master of the vessel" should see that the apparatus could communicate with both shore stations and other ships. Violations meant a $5,000 fine, and a vessel could be fined for every infraction of the law and cited in the district court with jurisdiction over the port where the ship arrived or departed. Enforcement of the law was spelled out: "That the Secretary of Commerce and Labor shall make such regulations as may be necessary to secure the proper execution of this Act by collectors of customs and other officers of government."

the Radio Act of 1912

By 1912, wireless had achieved international recognition and cooperation. Yet the United States had been lax in these agreements, partially because wireless was not under total government control as it was in some other countries. That all changed on an April night in 1912 when an iceberg took the American ship *Titanic* to the bottom of the North Atlantic. The following days and months were filled with news of the sinking and the role of wireless in the event. Reports centered on everything from the way in which wireless shipboard operators might have prevented the sinking to the brilliant performance of the medium in relaying news of survivors. Ironically, four months before the tragedy, the provisions of the 1906 Berlin treaty had been taken out of congressional mothballs for discussion in Senate committees. Those discussions, hastened by the sinking of the *Titanic*, led to the August passage of the Radio Act of 1912.

The 1912 act was much more comprehensive than the 1910 legislation. It defined authority between federal and state governments and established call letters for government stations. The law read: " . . . but nothing in this Act shall be construed to apply to the transmission and exchange of radiograms or signals between points situated in the same State. . . ." Along with providing clauses for revocation of licenses and fines for violators, it also established the assignment of frequencies, stating the license of the station would "state the wave length or the wave lengths authorized for use by the station for the prevention of interference and the hours for which the station is licensed to work. . . ." In addition to these specified wave lengths, stations could still use "other sending wave lengths." Licenses were to be granted by the Secretary of Commerce "upon application thereof." The president of the United States was given the power to control stations during wartime with compensation to the station's owners. The 1912 act also established the famous SOS distress signal allowing it to be broadcast with a maximum of interference "and 9" maximum of radiation. It was required to reach at least 100 miles. For the first time, the act defined *radio communication* as, "any system of electrical communication by telegraphy or telephony without the aid of any wire connecting the points from and at which the radiograms, signals, or other communications are sent or received." Other provisions of the act covered secrecy-of-messages restrictions to protect government stations' signals, rules for ship-to-shore communication, and a ban on stations refusing to receive messages from those not equipped with apparatus manufactured by a certain company.[2]

Despite its particular reaction to the sinking of the *Titanic*, the 1912 act was a valiant effort to control wireless communication. But few legislators could have foreseen the huge popularity of wireless, and even if they could have, legislative processes could not have begun to keep up with the new technology. It was not long before the regulatory framework began to crumble.

The National Radio Conferences: The 1912 Law in Trouble

By the mid-teens, both the United States and radio were in World War I. For the U.S. Navy, it meant hurriedly constructed wireless towers on board warships (Figure 11–2). Having taken over the country's radio stations, the government stopped all radio development except for wartime service. But when the war ended it was like uncapping a bottle. All the pent-up enthusiasm was released, and new experimenters eagerly flocked to their equipment. Although the Radio Act of 1912 had survived the World War, it was headed for rough sailing in a radio industry exploding with popularity and technology. By the 1920s, the chaos had mushroomed out of proportion. In 1922 alone, receiving-set sales had climbed 1,200 percent. The airwaves were flooded with everything from marine-military operations to thousands of amateur radio experimenters. Added to this was the advent of commercial radio and its powerful stations booming onto the air.

Figure 11-2 Ships using early wireless equipment had large masts constructed to support antenna wires. (Department of the Navy and *Sea Power*)

On February 27, 1922, groups of government officials, amateur radio operators, and commercial radio representatives met for the First National Radio Conference in Washington, D.C.[3] The conference was addressed by representatives of all opposing factions. Amateur radio operators were afraid their privileges were going to be curtailed under the influence of such large commercial firms as General Electric and Westinghouse. The large commercial firms were afraid their privileges were going to be relegated to the military. After the rhetoric subsided, the conference split into three committees: amateur, technical, and legislative. Since interference was still the biggest problem, it was not surprising that the technical committee's recommendations received the most attention. Based on that report, legislation was introduced in Congress in 1923 but never emerged from a Senate committee.

The second conference began on March 20, 1923. This one reaffirmed the problems of interference and recommended discretion in frequency allocations. Taking into account the commercial interests of the new medium, the conference suggested that allowing more stations on the air would only aggravate an already shaky financial condition. By today's standards of competition among almost 8,000 stations, the proposal seems inappropriate. Realizing that different geographical areas had different problems, the second conference suggested splitting up the country into zones, with each tackling its own problems locally. As he had after the first conference, Congressman Wallace White of Maine introduced legislation which again did not budge from the congressional committee.

The deafening interference continued straight into the convening of the Third National Radio Conference on October 6, 1924. Two major developments captured the attention of these delegates. Network broadcasting had become a reality. AT&T's wire system and Westinghouse's short-wave system were proving interstation connection was not only possible but also might be successful. Almost simultaneously, David Sarnoff announced RCA was going to experiment with the concept of super-power stations crisscrossing the country. It is little wonder that the third conference recommended resolutions opposing monopoly and even encouraged government intervention. Nevertheless, the conference supported the development of network broadcasting but, although agreeing to let the super-power experiments proceed, warned that they "should only be permitted under strict government scrutiny."[4] On the request of Secretary of Commerce Herbert Hoover (Figure 11–3), Congressman White refrained from introducing legislation. A third defeat would have been bad politically, and the decision was made to wait until still another conference was called.

Convening on November 11, 1925, the Fourth National Radio Conference ended with proposals which later became the foundation of the

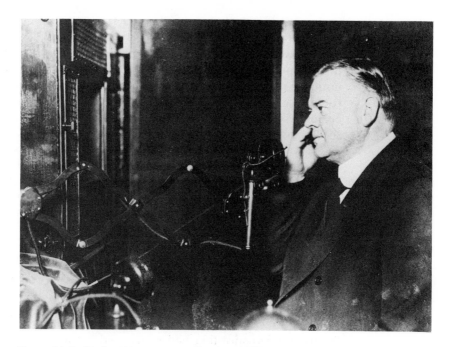

Figure 11–3 Herbert Hoover appearing in a 1927 experimental television demonstration. (AT & T Co.)

Radio Act of 1927. This conference suggested a system of station classifications and admonished Congress to pass some workable broadcasting legislation. The delegates recommended preventing monopoly, installing five-year terms for licenses, requiring stations to operate in the public interest, providing for licenses to be revoked, and giving the secretary of commerce the power to enforce regulations. They also wanted to guard against government censorship of programming, provide for due process of law, give the president control of stations in wartime, and suggest that broadcasting not be considered a public utility. But the good intentions were too late.

Judicial Setbacks for the Radio Act of 1912

Despite the radio conferences' valiant efforts to make it workable, two law suits and an opinion from the United States Attorney General made it clear that the 1912 law was in serious trouble. Highlighting the problem in 1923 was *Hoover v. Intercity Radio Co., Inc.* Intercity had been engaged in telegraph communication between New York and other points under a license issued by the secretary of commerce and labor. Upon expiration, Intercity applied for and was denied a renewal because there was not space available on the spectrum for a frequency assignment that would not interfere with government and private stations.

The issue went to court, and the judges ruled that the secretary had overstepped his bounds in refusing to renew Intercity's license. Cited as justification was a statement made by the chairman of the committee on commerce when the bill was passed, that: "it is compulsory with the Secretary of Commerce and Labor that upon application, these licenses shall be issued." The interpretation meant that the secretary of commerce and labor, although having the power to place restrictions on licenses and to prevent interference, could not refuse to issue a license as a means of reducing that interference. The court stated: "In the present case, the duty of naming a wavelength is mandatory upon the Secretary. The only discretionary act is in selecting a wavelength within the limitations prescribed in the statute, which, in his judgment, will result in the least possible interference." The court went on to define the relationship between the restrictions and license, stating: "The issuing of a license is not dependent upon the fixing of a wavelength. It is a restriction entering into the license. The wavelength named by the Secretary merely measures the extent of the privilege granted to the licensee."[5]

For the secretary of commerce and labor, the ruling was extremely frustrating. Broadcasting had progressed beyond the experimental and military stages. The secretary was faced with regulating a limited resource, and the court was telling him that he had to give some to everyone who wanted it. The act had given the secretary broad responsibilities, but the provisions of the act did not give him the power to implement them.

This was only the first of the secretary's setbacks. Three years later came the case of the *United States* v. *Zenith Radio Corporation et al.* Zenith had received a license authorizing it to operate on a "wavelength of 332.4 meters on Thursday night from 10 to 12 P.M. when the use of this period is not desired by the General Electric Company's Denver station." Zenith clashed with the secretary when it operated at other times and on another, unauthorized frequency. Yet the court ruled in favor of Zenith. The legal catch was a section of the 1912 law reading: "In addition to the normal sending wavelength, all stations may use other sending wavelengths. . . ."[6]

The crowning blow came when Acting Secretary of Commerce Stephen Davis (Figure 11–4) answered a request from the Chicago Federation of Labor.[7] Although the federation had planned to apply for a license, the application did not even reach Washington before Davis wrote the federation telling them that all the wavelengths were in use, and if the federation constructed a station, there would be no license forthcoming. Davis put the blame on the Fourth National Radio Conference on which it did not belong since the conference did not have the power to dictate

Figure 11-4 Stephen Davis. (AT & T Co.)

policy. Some politicians began to be concerned, and as the situation grew worse and the stations continued to interfere with each other, the office of the secretary of commerce sought an opinion from the attorney general.

In a letter dated June 4, 1926, the secretary asked the attorney general for a definition of power. The questions in the letter, as interpreted by the attorney general, were:

1. Does the 1912 Act require broadcasting stations to obtain licenses, and is the operation of such a station without a license an offense under that Act?
2. Has the Secretary of Commerce authority under the 1912 Act to assign wavelengths and times of operation and limit the power of stations?
3. Has a station, whose license stipulates a wavelength for its use, the right to use any other wavelength, and if it does operate on a different wavelength,is it in violation of the law and does it become subject to the penalties of the Act?
4. If a station, whose license stipulates a period during which only the station may operate and limits its power, transmits at different times, or with excessive power, is it in violation of the Act and does it become subject to the penalties of the Act?
5. Has the Secretary of Commerce power to fix the duration of the licenses which he issues or should they be indeterminate, continuing in effect until revoked or until Congress otherwise provides?[8]

The attorney general's answers made it clear that the problems were going to get worse, not better. The answer to the first question was affirmative. The act definitely provided for stations to be licensed, and stations operating without a license were clearly in violation. To the second question, the attorney general replied that the secretary had the right to assign a wavelength to each station under one provision of the act, but for the most part, the stations could use whatever other frequency they so desired, whenever they wanted. With the exception of two minor provisions, the attorney general also stated that the secretary had no power to designate hours of operation. Also lost was the contention over limiting power. The act stated that stations should use "the minimum amount of energy necessary to carry out any communication desired." The attorney general said: "It does not appear that the Secretary is given power to determine in advance what this minimum amount shall be for every case; and I therefore conclude that you have no authority to insert such a determination as a part of any license."

The third answer was obvious. Stations could use any other wavelength they desired. The act and the courts had confirmed that. That also answered question four. Since the secretary could not limit power or operating times beyond the actual license, stations were free to use other wavelengths with different power outputs and at different times from those stated in the license. The attorney general said in answer to the fifth question that he could, "find no authority in the Act for the issuance of licenses of limited duration."

Clearly a law which only a decade earlier had seemed firmly in control of the new medium now was almost worthless. Four months later on December 7, 1926, President Coolidge sent a message to Congress. He called for legislation to remedy a chaotic situation that threatened to destroy radio broadcasting.[9] The next day, he signed a joint resolution of Congress placing a freeze on broadcasting until specific legislation could be passed.

the Radio Act of 1927

Congress had been working on the Radio Act of 1927 before Coolidge's message. The act passed in both houses of Congress and received the president's signature on February 23, 1927. The Radio Act of 1927 was administered by the secretary of commerce and provided for the formation of a Federal Radio Commission (FRC) to oversee broadcasting. The act was intended to remain in force for only a year but was subsequently extended until 1934. With court decisions to guide them, Congress did an admirable job of plugging the holes left by the 1912 law.

The most important provision of the 1927 act was the formation of a

Federal Radio Commission, "composed of five commissioners appointed by the President, by and with the advice and consent of the Senate, and one of whom the President shall designate as chairman. . . ."[10] The law specified that each commissioner must be a citizen of the United States and would receive compensation of $10,000 for the first year of service. The commissioner system, as well as many other provisions of the 1927 legislation, became part of the Communications Act of 1934.

Other provisions in the 1927 act divided the United States into zones represented by individual commissioners. No more than one commissioner could be appointed from any one zone. One zone covered New England and the upper tip of the middle Atlantic states and included the District of Columbia, Puerto Rico, and the Virgin Islands. The second zone included the upper middle Atlantic states west to Michigan and Kentucky. The third zone covered the South and the fourth and fifth zones, the Great Plains and the West, respectively.

The act provided for the licensing of stations, but only for a specified time, and gave the government considerable control over the electromagnetic spectrum. The act also set out to define states' rights. Keep in mind that federal regulation over intrastate commerce, for which wireless was used, was not popular. So it was not surprising that the Radio Act of 1927 tried to avoid direct control of intrastate communication while at the same time retaining control of communication crossing state borders. The act stated that the law's jurisdiction would extend "within any State when the effects of such use extend beyond the borders of said State. . . ." The most quoted provision came from Section 4 in its statement that stations should operate "as public convenience, interest, or necessity requires. . . ."

Section 4 also prescribed "the nature of the service to be rendered by each class of licensed station and each station within any class." Control over frequency, power, and times of operation were covered by the act, giving the FRC power to "assign bands or frequencies or wavelengths to the various classes of stations, and assign frequencies or wavelengths for each individual station and determine the power which each station shall use and the time during which it may operate." Coverage areas for stations were to be fixed by the FRC, and the commission was to have power over "chain" or network broadcasting. Stations also were required to keep operating logs.

In addition to regulating the industry, the 1927 act gave the commission quasi-judicial powers with "the authority to hold hearings, summon witnesses, administer oaths, compel the production of books, documents, and papers and to make such investigations as may be necessary in the performance of its duties." The secretary of commerce was empowered "to prescribe the qualifications of station operators, to classify them according to the duties to be performed, to fix the forms of such licenses, and to issue them to such persons as he finds qualified." The secretary also was empowered to issue call letters to all stations and to "publish" the call letters. But before issuing a license, the government made certain that the prospective licensee gave up all rights to the frequency. The applicant had to sign "a waiver of any claim to the use of any particular frequency or wavelength. . . ." Once granted, station licenses were limited to three years.

To close the wavelength loophole of the 1912 legislation, the 1927 law stated: "The station license shall not vest in the licensee any right to operate the station nor any right in the use of the frequencies or wavelength designated in the license beyond the term thereof nor in any other manner than authorized therein." The act also discouraged monopolies and prohibited transferring licenses without the commission's approval. It also gave the commission power to revoke licenses for "issuing false statements or failing to operate substantially as set forth in the license."

Wording for the famous Section 315 of the Communications Act of 1934 came from the 1927 legislation: "If any licensee shall permit any person who is a legally qualified candidate for any public office to use a broadcasting station, he shall afford equal opportunities to all other such candidates for that office. . . ." Commercial broadcasting gained instant recognition and regulation with the requirement that paid commercials were to be announced as paid or furnished by the sponsor.

Putting a station on the air was another important provision of the act. As we shall see later in this chapter, this issue also arose in the appeals process. Specifically, the act stated: "No license shall be issued under the authority of this Act for the operation of any station, the construction of which is begun or is continued after this Act takes effect, unless a permit for its construction has been granted by the licensing authority upon written application thereof." The law acknowledged that construction permits for stations would specify "the earliest and latest dates between which the actual operation of such station is expected to begin, and shall provide that said permit will be automatically forfeited if the station is not ready for operation within the time specified. . . ."

The anticensorship provision, later to become incorporated into Section 326 of the Communications Act of 1934, was also included. Ironically, that provision was immediately followed with "no person within the jurisdiction of the United States shall utter any obscene, indecent, or profane language by means of radio communication."

We can see immediately the conflicts which could develop between not only these two provisions but also in the "convenience, interest, and necessity" clause. It was not long before the broadcasters and the government were indeed arguing. Yet keep in mind that the 1927 law is the very foundation of contemporary broadcast regulation. It was simple and straightforward, and the courts gave it strong support.

From 1927 to 1934, the Radio Act of 1927 withstood challenges from all sides. It achieved the ability to regulate effectively the expanding medium of "wireless," which now blanketed the nation with entertainment and news programming envisioned by few of the 1910 pioneer regulators. It is little surprise that the 1927 law was liberally quoted in the Communications Act of 1934, the law governing contemporary broadcasting. This law took broadcasting entirely out of the Department of Commerce and gave it separate status as an independent agency of government.

passing the Communications Act of 1934

It was becoming clear that broadcasting needed a new, more comprehensive regulatory agency. The FRC was still limited in its scope, having to share responsibilities with the U.S. Department of Commerce. Although the Commerce Department had at one time been an appropriate home, the prevailing trend was toward the public consumption of radio, overshadowing its commercial uses. Although commercial stations would still far outnumber those directing their signals to the public, guarding the public's convenience, interest, and necessity was no small task. After a number of proposals to coordinate regulation had been examined, President Franklin D. Roosevelt sent to Congress on February 26, 1934 a proposal for a separate agency known as the Federal Communications Commission. Roosevelt's message said that the FCC should have the authority "now lying in the Federal Radio Commission and with such authority over communications as now lies with the Interstate Commerce Commission—the services affected to be all of those which rely on wires, cables, or radio as a medium of transmission."[11]

Congress responded to Roosevelt's proposal by passing the Communications Act of 1934. With it came the Federal Communications Commission, which in the next 45 years was to see its domain come to reign over everything from citizens' band radios to satellite communication, from intrastate to international communication. Although it took only five months for Roosevelt's message to become law, the scope of the FCC already had been hammered out in court challenges to the 1927 law. In fact, much of the 1927 law was left intact, including the guiding phrase, "public convenience, interest, or necessity," which was retained as a nebulous but very powerful componenet of the 1934 legislation.[12] There were a few minor changes in the actual wording of the law. "Wavelength" was changed to "frequency," and whereas the 1927 law was concerned with "wireless communication," the FCC was to govern both wire and wireless.

As with most laws, the 1934 legislation has been amended many times. Although it would take volumes to discuss the decisions and cases which have molded today's version, part 3 of this text will examine some of the specific provisions and amendments of the 1934 act which directly affect current broadcasting.

summary

Chapter 11 traces the government's role in early broadcasting. An outgrowth of the Berlin meetings of 1903 and 1906, the Wireless Ship Act of 1910 provided an early safeguard for ships at sea. It required them to be equipped with radio apparatus which could communicate with other ships and shore stations. Violations meant possible fines and court proceedings. Two years later, the Radio Act of 1912 expanded the 1910 legislation but could not even begin to deal with radio's exploding growth in the 1920s. Four National Radio Conferences convened and discussed how to bring the new medium under government control in a way which was acceptable to the industry yet permitted the orderly use of the spectrum. The combination of these conferences and two landmark court cases which threatened the legality of the 1912 legislation generated enough support in Congress to pass the Radio Act of 1927. This act created the Federal Radio Commission which was renewed on a year-to-year basis while it fought a series of court battles to affirm its control over radio. Seven years later, Congress passed the Communications Act of 1934 and established the Federal Communications Commission, a separate, independent government agency.

spotlight on further learning

BENSMAN, MARVIN R., "Regulation of Broadcasting by the Department of Commerce, 1921-1927," in *American Broadcasting: A Source Book on the History of Radio and Television*, ed. Lawrence W. Lichty and Malachi C. Topping, pp. 544-55. New York: Hastings House, Publishers, 1975.

KAHN, FRANK J., ed., *Documents of American Broadcasting* (3rd ed.). Englewood Cliffs, N.J.: Prentice-Hall, Inc., 1979.

JAMESON, KAY C., *The Influence of the United States Court of Appeals for the District of Columbia on Federal Policy in Broadcast Regulation, 1929-1971.* New York: Arno Press, 1979.

McMAHON, ROBERT S., *Federal Regulation of the Radio and Television Broadcast Industry in the United States, 1927-1959.* New York: Arno Press, 1979.

NOLL, ROGER G., MERTON J. PECK, and JOHN J. McGOWAN, *Economic Aspects of Television Regulation.* Washington, D.C.: The Brookings Institution, 1973.

12

the federal communications commission

FOCUS   After completing this chapter we should be able to

List the primary responsibilities of the Federal Communications Commission.

List the areas over which the FCC does not have jurisdiction.

List the items on a typical agenda of an FCC meeting.

Understand the influence of individual commissioners on regulatory policy.

Explain the organization of the FCC, including its offices and bureaus.

Identify the different enforcement powers of the commission.

Discuss the criticism of the commission.

Few government agencies have had such a direct effect on the public as the Federal Communications Commission. Nearly everything we watch on television and hear on radio is in some way touched by the FCC's control over broadcasting stations, cable, satellites, even the telephone systems. Growing out of the Federal Radio Commission, the FCC is an independent agency accountable directly to Congress. In this chapter, we'll learn about the jurisdiction of the FCC, how it conducts business, its organization, its enforcement powers, and current criticism of its actions.

primary responsibilities

The FCC's thirteen areas of responsibility are:

1. The orderly development and operation of broadcast services and the providing of rapid, efficient nationwide and worldwide telephone and telegraph service at reasonable rates.
2. The promoting of safety of life and property through radio, and the use of radio and television facilities to strengthen national defense.
3. Consultation with other Government agencies and departments on national and international matters involving wire and radio communications, and with State regulatory commissions on telephone and telegraph matters.
4. Regulation of all broadcast services—commercial and educational AM, FM, and TV. This includes approval of all applications for construction permits and licenses for these services, assignment of frequencies, establishment of operating power, designation of call signs, and inspection and regulation of the use of transmitting equipment.
5. Review of station performance to assure that promises made when a license is issued have been carried out.
6. Evaluation of stations' performance in meeting the requirement that they operate in the public interest, convenience, and necessity.
7. Approval of changes in ownership and major technical alterations.
8. Regulation of cable television ...
9. Action on requests for mergers and on applications for construction of facilities and changes in service.
10. The prescribing and reviewing of accounting practices.
11. Issuance of licenses to, and regulation of, all forms of two-way radio, including ship and aviation communications, a wide range of public safety and business services, and amateur and citizens radio services.
12. Responsibility for domestic administration of the telecommunications provisions of treaties and international agreements. Under the auspices of the State Department, the Commission takes part in international communications conferences.

13. Supervision of the Emergency Broadcast System (EBS), which is designed to alert and instruct the public in matters of national and civil defense.[1]

As we can see from the preceding list, the commission's functions cover much more than just radio and television in the familiar sense. Telephone, telegraph, and cable all are within the FCC's jurisdiction as are applications of communication to public safety, transportation, industrial, amateur, and citizen service. In some cases, the regulation of these services is shared with other government agencies, such as local municipalities in the case of cable. A television station in New York City or a CB radio in Wyoming are both in the FCC's domain. This domain stretches beyond the 50 states into Guam, Puerto Rico, and the Virgin Islands.

what the FCC does not control

It is equally important to understand over what the FCC *does not* have jurisdiction.[2] Many people perceive the FCC as having broad powers of regulation. This is especially true when consumers are unhappy about something they have seen or heard on local radio or television. We already have learned that the commission has very little control over the content of broadcast messages. With the exception of obscene and indecent programming, and even that is vaguely defined, lotteries and deceptive advertising are the only areas the FCC can regulate without infringing on First Amendment rights. Even when it does act in these areas, a court battle is bound to arise over the First Amendment.

In the same sense, the FCC cannot tell a station when to air a program. Nor can it tell a station when to run commercials or public service announcements. The FCC will not substitute its judgment for that of the local broadcaster in those areas. Although some network contracts prohibit editing of certain programs, that is solely between the network and the station, not the FCC. Despite the nonediting clauses, the licensee retains control over local programming with the right to delete the network's entire offering, if it feels that it would not be in the local public interest to air it.

Although lotteries are forbidden, the FCC has little jurisdiction over the conduct of legitimate contests, especially over the awarding of prizes. If a station has a contest, and you win a prize which, for some reason, does not satisfy you, the best recourse would be to deal directly with the station or the manufacturer of the prize. The FCC would not have the authority to tell the manufacturer to give you a different prize or to help you obtain repairs for a defective item. Similarly, although stations broadcast a variety of sporting events, the FCC has no jurisdiction over the promoters or organizers of those sporting events. If your favorite boxer fails to appear on the local televised "golden gloves" championship, you can write the boxing commission, but the FCC will not be able to help you.

Similarly, the commission does not have any jurisdiction over countries whose radio or television signals cross into the United States. Although there are reciprocal international agreements on the use of the electromagnetic spectrum, the consumer in Michigan who complains to the FCC about a Canadian radio station would receive little satisfaction. A listener in southern California complaining to the FCC about a station in Mexico would have similar frustration. An exception would be if the Canadian or Mexican station were operating off its frequency and interfering with American stations, although even in these cases, the FCC would go through the regulatory agencies in Canada and Mexico to solve the problem.

The FCC also has no jurisdiction over news-gathering organizations, either local or national.[3] Press associations, such as United Press International, Associated Press, and Reuters are independent of the broadcast stations they serve and are not regulated by the FCC. To the extent that such organizations use radio frequencies or satellites to transmit information, the FCC does have jurisdiction, but only in a technical sense. Similarly, the commission does not directly control the networks but does control network-owned broadcasting stations. Musical-rights organizations, although directly serving stations and collecting royalties from them for airing performers' works, also are not under the jurisdiction of the FCC. Firms such as ASCAP, BMI, and SESAC are independent organizations not involved in activities which the commission controls. In addition, audience-measurement firms such as Nielsen and Arbitron are independent of the FCC, although a station's fraudulent use of audience ratings reflects on the licensee's commitment to serve the public interest.

The commission has instituted rules affecting the duplication (simulcasting) of programs on commonly owned AM and FM stations, but the FCC has no authority to tell a radio station to broadcast in stereo or quadraphonic sound nor to tell a television station to broadcast a program in color. Although the FCC can act in the public interest to question overcommercialization of radio and television, it does not have direct authority to tell a station to air so many commercials per hour. Likewise, the commission views the amount of public service programming as a condition for license renewal but has no authority to tell a station what public service programming to air. If the local licensee chooses to air public service announcements for the Red Cross instead of for the American Cancer Society, that is the station's prerogative. The exception to this would be if the announcement were on a controversial issue. Then, because of the Fairness Doctrine, the commission would want to ensure that the station aired a balanced presentation of the issue through whatever type of programming the station chose.

Libel and slander during radio and television broadcasts is another area over which the FCC has no jurisdiction. If you feel you have been libeled or slandered (both terms sometimes apply to "broadcast" speech which is "published"), your best recourse would be to consult an attorney, not the FCC. The FCC even shies away from these in matters of license renewal. In fact, when defamation did become an issue in a license renewal, the FCC stated in part:

> *It is the judgment of the Commission, as it has been the judgment of those who drafted our Constitution and of the overwhelming majority of our legislators and judges over the years, that the public interest is best served by permitting the expression of any views that do not involve, quoting from Supreme Court decisions, "a clear and present danger of serious substantive evil that rises far above public inconvenience, annoyance or unrest." . . . This principle insures that the most diverse and opposing opinions will be expressed, many of which may be even highly offensive to those officials who thus protect the rights of others to free speech. If there is to be free speech, it must be free for speech that we abhor and hate as well as for speech that we find tolerable or congenial.*[4]

Once again we see the First Amendment rising to protect free speech, even when that free speech is unpopular. In the same vein, program content which contains derogatory comments about sex, race, or religious beliefs also enjoys the protection of the First Amendment. Ethnic humor on such shows as "Sanford and Son" and "All in the Family" may offend some people, but any attempt to control this area of programming would be clearly outside the FCC's jurisdiction.

making decisions at the FCC

The commissioners hold weekly meetings and executive sessions to oversee commission activities. Their meetings are open to the public, a procedure started in 1977 under a congressional mandate. Closed meetings can be called by a majority vote of the entire commission. These meetings usually are on matters of national defense, manufacturer's trade secrets, criminal matters, or when the parties concerned with the FCC's decision specifically request that the meeting be closed.[5]

Meeting Agenda

A typical FCC agenda is classified so as to reflect the organization.[6] In order of business the categories include: Hearing, General, Safety and Special, Common Carrier, Personnel, Classified, CATV, Assignment and Transfer, Renewals, Aural, Television, Broadcast, and Complaints and

Compliance. The commission deals with these agenda items usually after it has heard a series of briefings by the appropriate FCC bureaus and offices. In a *Hearing,* the FCC acts as the final tribunal in an appellate process entailing decisions previously made by the FCC Administrative Law Judge and, in some cases, by the FCC Review Board. The *General* covers items not included in the other categories. Perhaps a representative from another federal agency will discuss the FCC's compliance with that agency's rules. *Safety and Special* deals with the application of broadcast communication to such areas as fire department, taxicab, and police department radios. Other industrial applications would be business use of mobile radios, citizens-band radio, and amateur (hams) services. The next item on the agenda, *Common Carrier,* is on the FCC's regulation of telephone and telegraph systems. Here, the commission acts as a quasi-public utility on issues concerning microwave and satellite systems, among others. Next comes *Personnel.* FCC staffing matters and promotion and appointments come under this agenda category. Promotions generally are routine, since three other FCC officers—the Bureau Chief, Personnel Chief, and Executive Director—have usually approved the promotions before they reach the seven commissioners.[7]

If national security, manufacturer's trade secrets, or other classified matters need to be discussed, they will be in the *Classified* category. *Cable* is next on the agenda. Approval of a new link-up between two cable systems, mergers of cable companies, approval for new cable systems to begin operation (issuing a certificate of compliance), and matters concerning a public-access channel can be resolved here. If you buy or sell a radio station, the transaction will be approved or rejected during the next order of business, *Assignment and Transfer.* The commission's deliberations may dwell upon previous inquiries about the transactions such as a Hearing or recommendations by the administrative law judge. If a group of stations is seeking to acquire more broadcasting properties, the discussion might center on the possibly powerful influence of a single owner of multiple broadcast properties and whether the public interest would be served by approving such a sale.[8] Transfers of licenses would be first approved during this order of business.[9] If you already own a station and it comes up for license renewal, the renewal will be acted upon during the next agenda category, *Renewal.* Most of the renewals reaching the commissioner level are contested renewals. Uncontested renewals are usually approved at the staff level.

If you are applying for permission to start a new radio or television station, a decision will be made on your application during either the *Aural* or *Television* agenda categories. Altering the service your station is already licensed to provide will also be acted upon at this time. If your station is on the air and for some reason wishes to seek a waiver of FCC rules, your request will be considered during the next item on the commission's agenda, *Broadcast*. For example, a network may request a waiver of the prime-time access rule to offer a special sports program. Or a station operating in an area in which there already is one network affiliate may request permission to affiliate with the same network.[10] Rule violations are considered during *Complaints and Compliance*. A station which has seriously violated FCC rules, complaints about the Fairness Doctrine, and fraudulent operating practices all would be considered at this time. It goes without saying that not every single violation is discussed by the entire commission. However, when a violator feels that there has been a legitimate injustice, then the case could reach this level.

Commissioner Influence on Regulatory Policy

Individual commissioners can help shape regulatory policy. Researchers Lawrence Lichty[10] and Wenmouth Williams, Jr.[11] studied the impact of commissioners' influence on FCC decisions. It is not surprising that during the early years of the Federal Radio Commission, the commissioners, four of whom were trained in law, were comfortable in the atmosphere of the frequent court challenges which surrounded these early decisions. The fact that the FRC added a legal division one year after it was formed demonstrates the importance that the commissioners placed on not only fighting but also winning those court challenges.

The FCC carried on this tradition when it began its six-year "trust-busting" campaign in 1939, breaking up networks and setting up rules for chain broadcasting. Two FCC chairmen, Frank R. McNich and James L. Fly, led the fight and weathered appeals which tried to claim that the regulations were unconstitutional. McNich had served on the Federal Power Commission and was a lawyer; Fly had headed the legal department of the Tennessee Valley Authority and had been in charge of judicial proceedings defending TVA's constitutionality.

The growth of television was also influenced partially by the attitudes of the FCC during the 1950s as Williams found the Kennedy administration to be characterized by a commission advocating strict regulation. Newton Minow set the pace with his "vast wasteland" speech and was joined by liberal FCC Democrats E. William Henry and Kenneth A. Cox. During the Kennedy administration, the FCC passed nonduplication rules governing simulcasting on AM and FM, and also brought cable under the regulatory umbrella. President Nixon's appointment of Benjamin Hooks emphasized the importance of minorities to broadcasting. The chairmanship of Richard Wiley under Presidents Nixon, Ford, and Carter was characterized by attempts, many successful, to streamline FCC decision making.

organization

Now that we have a basic understanding of how the FCC functions, let's examine its organization.

The Commissioners

At the top of the commission hierarchy are seven FCC commissioners headed by a chairperson. Appointed by the president of the United States and confirmed by the Senate, commissioners are prohibited from having a financial interest in any of the industries they regulate. This includes industries which are only partially in FCC-regulated businesses. No more than four commissioners can be from the same political party, and their seven-year terms are staggered so that one position opens up each year. Appointees who fill the unexpired term of a commissioner may or may not be reappointed when that term expires.

FCC Offices

Directly under the commissioners are the Office of Plans and Policy, Office of Opinions and Review, Office of Administrative Law Judges, Review Board, Office of General Counsel, Office of Chief Engineer, and Office of the Executive Director.[12]

Office of Plans and Policy. The Office of Plans and Policy is responsible for developing long-range policy decisions for industries coming under FCC jurisdiction. It also is responsible for assessing the policy implications of FCC decisions, providing policy analyses and recommendations to the commission staff, and coordinating policy research. The chief of the Office of Plans and Policy recommends budgets and priorities to the commission policy research program, and functions as the central account manager for all contractual research studies funded by the FCC.

Office of Opinions and Review. When the FCC makes a major decision, the document outlining that decision is written in consultation with and with the assistance of the Office of Opinions and Reviews. This office serves as the commission's legal staff, advising it on procedural matters, researching judicial precedent, and overseeing hearings ordered by the commission. The office then recommends action to the commission based on the evidence presented by the parties involved.

Office of Administrative Law Judges. This office is the first ladder in the appeals process. The administrative law judges preside over hearings and make initial decisions. It is not unusual for their decisions to be

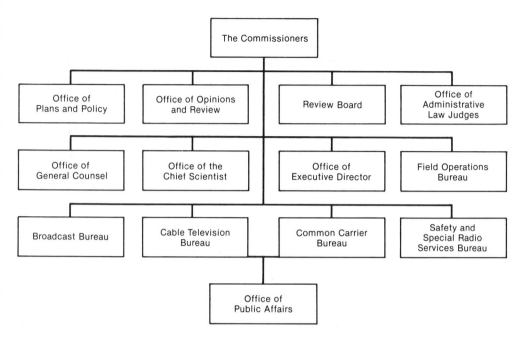

Figure 12–1 The Federal Communications Commission

appealed. When two applicants for a broadcast license appear at a hearing, both have a major investment at stake, and a ruling in favor of one party will prompt the other to continue the appeals process.

Review Board. The Review Board is the second step in the FCC appellate process, between the administrative law judges and the commissioners. In some cases, the decisions of the administrative law judges are reversed by the Review Board and then reversed again by the seven commissioners. This is not so much a reflection on the ability of the judges to adhere to judicial procedure as it is on the desire of the offended to exhaust every administrative possibility. In special cases, initial decisions even can be reviewed directly by the commissioners. If, for example, a renewal decision goes against a licensee, the licensee can appeal to the Review Board, which is made up of senior-level employees of the commission. Individual FCC bureaus also can appeal to the Review Board. If, for instance, a bureau rules against a licensee, and an administrative law judge rules in the licensee's favor, then the bureau can appeal to the Review Board. If the ruling still goes against the bureau, or the licensee for that matter, the party can appeal that ruling to the commissioners, who, as a body, choose which cases to accept for review. The seven commissioners are the last appeals step before the matter goes to the Federal Court of Appeals.

Office of the General Counsel. The Office of the General Counsel is the commission's attorney, representing it before the courts. The office also aids in preparing legislative programs supported by the commission and works closely with the attorney general and the Justice Department in cases which entail prosecution or jurisdiction across agency boundaries. An example of this cooperation would be in prosecuting violations of the Criminal Code or such other violations associated with a wrongdoing beyond those under the jurisdiction of the FCC. If a person steals radio equipment and then uses it to broadcast illegally, both the Justice Department and the FCC would become involved. The Office of the General Counsel also works closely with the Office of Opinions and Review, since the decisions which the latter writes may be the basis for the former's defense of the commission in court.

Office of the Chief Scientist. This office is the top "technical" office at the commission. The responsibilities of administering the electromagnetic spectrum and all of the policies associated with its implementation are developed by the Office of the Chief Scientist. About half the staff are engineers, and they consider such matters as determining the number of stations in a given market, equipment testing and certification, frequency allocations and modifications, and requests for increases in power output. The Office of the Chief Scientist also operates a Laboratory Division near Laurel, Maryland. Here, new equipment is tested to see if it meets FCC specifications. For example, manufacturers of radio and television transmitters must first receive authorization before they can sell them for broadcast use. The commission usually uses the technical data submitted by the manufacturer as a basis for its authorization, but on occasion, it spot-checks equipment to verify the test data. Citizens-band radios, for example, are tested at the Laurel, Maryland facility.[13] With the help of this testing, the FCC issues approximately 1,000 authorizations per year over a wide range of equipment. The office also works with other organizations testing new equipment and its applicability to broadcasting.

Office of Executive Director. Although the commissioners are the highest ranking officers of the FCC, the FCC Executive Director coordinates the overall operation of the commission. The position is somewhat analogous to a city manager running a municipality, even though the city council is the highest level in the administrative hierarchy. The executive director coordinates the activities of the different staff units, including the personnel division, the internal review and security division, the financial management division, and the public information officer.[14]

If it can be said that decisions are made by FCC offices, then they are implemented by FCC bureaus. Here, the day-to-day administrative services are performed which control the thousands of broadcast stations and licensees. The commission is divided into five bureaus: Broadcast, Safety and Special Radio Services, Cable Television, Field Operations, and Common Carrier. Those concerned most directly with broadcasting are the Broadcast Bureau, Cable Television Bureau, and Field Operations Bureau.

The **Broadcast Bureau** handles matters concerning commercial and noncommercial broadcasting stations. License renewals, for example, are handled by the bureau's Renewal and Transfer Division. Other divisions within the bureau include the Office of Network Study, the Broadcast Facilities Division, the Complaints and Compliance Division, the Hearing Division, the Policy and Rules Division, and the License Division.[15]

The **Cable Television Bureau,** as the name implies, is responsible for overseeing the daily operations of the cable television industry. Within this bureau are five divisions: the Policy Review and Development Division, the Compliance Division, the Research Division, the Special Relief and Microwave Division, and the Records and Systems Mangement Division.[16]

The prime enforcement arm of the FCC is the **Field Operations Bureau.** The Field Operations Bureau maintains a number of field offices in the larger cities across the United States, as well as mobile monitoring stations in specially equipped vans. Special investigative teams are assigned to make on-location inspections of stations, and a separate unit concentrates solely on CB radio violators. The field offices are also contact points for the public at which people can get information about the FCC and the communications industry. In addition, this bureau is responsible for administering FCC license examinations.

The Field Operations Bureau maintains sophisticated equipment which can trace a signal and pinpoint its location. It can thus catch illegal CB transmitters, violating amateur stations, and even "pirate" broadcasting stations operating on frequencies assigned to commercial AM and FM radio stations. Someone caught operating an illegal station will be raided by the FCC and U.S. Marshals, with equipment being seized as evidence. The Field Operations Bureau has four divisions:

(1) The Field Enforcement Division directs field enforcement programs, including the monitoring and inspection of stations. The Division also conducts investigations. (2) The Regional Services Division directs the public service programs including, among other things, radio operator

licensing. (3) Responsible for receiving and processing enforcement reports, such as violation files and investigations, is the Violations Division. (4) The Engineering Division of the Field Operations Bureau is responsible for providing engineering support and equipment specifications and construction for the field facilities.

In addition to the three previously mentioned bureaus, the **Common Carrier Bureau** oversees such areas as telephone and telegraph, and the **Safety and Special Radio Services Bureau** oversees such areas as aviation and marine communication. An **Office of Public Affairs** directs liaison between the Commission and the public.

enforcement power

The Communications Act specified that violators of its provisions would be penalized, and the commission has at its disposal a number of enforcement measures for the industry. Depending on the type of violation, the Commission may impose penalties ranging from a simple letter, a cease-and-desist order, a forfeiture (fine), short-term license renewal, license revocation, or denial of renewal.

Letters

Letters are usually used in less serious matters or ones in which the FCC accepts amends instead of imposing a forfeiture. Letters can be used to reprimand stations for incomplete community needs and ascertainment surveys, failure of programming to meet Fairness Doctrine requirements, or improper submission or failure to submit required FCC documents, such as employment reports or exhibits for a license renewal application. The letters are not always a reprimand, but in cases of license renewal, for example, they state that the license renewal is being withheld pending receipt of the required exhibit, and that after a certain date the license will be forfeited.

Cease and Desist Orders

Cease and desist orders are rare, partly because of the effectiveness of the commission's forfeitures and other sanctions. Gillmor and Barron cite one case of a minister asking the FCC to issue a cease and desist order prohibiting a religious program from being dropped by a station. The FCC declined to issue the order under the anticensorship provision of the Communications Act, although it confirmed that it did have the authority to issue it.[18] The cease and desist order was issued, on the other hand, by the commission to an AM station for broadcasting off-color remarks.[19]

Forfeitures

The most common sanction imposed on a station is a forfeiture, usually for a technical-rule violation or the more serious offense of fraudulent billing, although the latter can set the stage for a license revocation. The forfeitures vary and are based not only on the violation but also on the ability of the station to pay. They can cost up to $10,000 for serious violations by major market stations. Typical forfeiture notices for alleged violations are on the following partial list of apparent liabilities announced during a single week of commission activity:

—Broadcast Bureau ordered licensee to forfeit $250 for failing to calibrate remote ammeters to indicate within 2% of regular meter.
—Broadcast Bureau ordered licensee to forfeit $1,000 for failing to maintain actual antenna input power as near as practical to authorized power.
—Broadcast Bureau ordered licensee to forfeit $500 for failing to keep proper log as required.
—Broadcast Bureau ordered licensee to forfeit $500 for operating with antenna input power greater than 105% of authorized power during daytime operation.
—Broadcast Bureau notified licensee that it had incurred apparent liability for $1,300 for failing to maintain receiver capable of receiving Emergency Broadcast System tests or emergency action notifications and terminations at nighttime control point.
—Broadcast Bureau ordered licensee to forfeit $2,000 for operating with modes of power other than those specified in basic instrument of authorization.[20]

Notice that, with the exception of logging violations and failure to have equipment to monitor the Emergency Broadcast System, these alleged violations are infractions of technical rules. Now consider the following list of more sizeable, apparent liabilities:

—$10,000 for logging violations and for fraudulent billing practices.
—$5,000 for failure to make time available to political candidates at the lowest unit charge, charging different rates for political announcements of the same class and duration to legally qualified candidates for the same office, and failure to comply with logging requirements.
—$8,000 for failure to comply with logging requirements (program-length commercial.)
—$10,000 for falsification of operating logs.
—$10,000 for fraudulent billing practices.
—$8,000 for broadcasting information concerning a lottery.[21]

Notice the increased importance the commission assigns to alleged commercial violations. This is one area in which a maximum fine is not uncommon, and even stations in smaller communities can incur substantial liabilities from these violations. Remember, these listings **do not** necessarily imply that the stations are guilty, only that forfeiture notices were served.

When the FCC conducts an investigation, the station does have certain rights in seeking refuge from the penalties. First, the FCC cannot simply impose a fine on a station. Procedures outlined in the Communications Act state that a written notice of the apparent liability must first be sent by certified mail to the "last known address" of the permittee. The licensee or permittee then has thirty days in which to pay the fine or submit in writing the reason why it should not be held liable. The notice sent by the commission also must include the date, facts, and nature of the act or omission and must identify the "particular provision or provisions of the law, rule, or regulation or the license, permit, or cease and desist order involved." The fine is payable to the United States Treasury and can be collected in a civil suit if the violator refuses to pay. Of course, the station can appeal the commission's action through the usual administrative processes. In many cases, however, logs are powerful evidence as documents, and the excuse that an unsupervised or unqualified employee is to blame is no defense.

The commission issued its first letter of apparent liability in March 1961, one month after it outlined its policy and procedures regarding forfeitures. Authority to issue forfeitures had been granted in September 1960.[22] Researchers Charles Clift, Fredric Weiss, and John D. Abel studied the pattern of FCC forfeitures over the decade immediately after the law was enacted and found the highest percentage (87.1%) of forfeitures occurred because of failure to observe a provision of the act or a rule or regulation of the commission.[23] Included in this category were such infractions as logging violations, fraudulent billing, unlicensed or underlicensed operators, improper station identifications, and failure to conduct equipment performance measurements. The second highest category (8.0%) of forfeiture notices was failure to operate the station as set forth in the license. Violations of broadcasting hours, power, and presunrise authorization accounted for 3.4% of all forfeiture notices, including violations of sponsorship identifications and "rigged" contests. The fourth category—violations of lottery, fraud, or obscene-language sections of Title 18 of the United States Code—accounted for 1.4% of the forfeiture notices. The researchers found no forfeiture notices resulting from failure to observe a commission cease and desist order.

Short-Term Renewals

Next to forfeitures, the most severe sanction that can be imposed on a station is a short-term license renewal. These short-term renewals can range anywhere from six months to up to two years. Their purpose is to give the commission an early opportunity to review alleged past deficiencies.[24] Typical of short-term license renewals are those issued for the following infractions:

1. Station's equal employment.
2. Utilization of broadcast facility to gain competitive advantage in nonbroadcast business activities; fraudulent billing.
3. Fraudulent billing; inadvertent misrepresentations to the commission, falsification of logs; violation of logging rules; nonfulfillment of prior proposals concerning public service announcements; lack of supervision and control over station operations.
4. Broadcast of false, misleading, or deceptive advertising in connection with the promotion of a contest.
5. Predetermining the outcome of a contest.
6. Fraudulent billing.
7. Conducting contests during audience survey periods (hypoing)[25]

Notice again that alleged violations centering on commercial matters were responsible for most of these short-term renewals, indicating the seriousness with which the FCC views such actions.

One study investigated 156 short-term license renewals granted by the commission in the decade immediately following the passage of the statute.[26] It showed that 113 (72%) received one-year renewals, 29 stations (19%) received renewals for more than one year (but less than three), and 14 (9%) were licensed for less than a year. Three reasons accounted for the majority of short-term renewals: (1) improper control over station operation, which generally means that an owner was not adequately supervising employees; (2) repeated rule violations, both technical and programming; and (3) performance versus promise, or in other words, the licensee was not living up to the promises made in the previous license renewal.[27] Research has yet to tell us if any of these trends have changed in the second decade of their issuance, but a perusal of current short-term renewals at least finds the same reasons justifying FCC action.

Renewal Denials and Revocation

The most serious penalty that the FCC can impose on a licensee is to deny it the right to operate, either through revoking its license or denying renewal of its license. In a sweeping action, the FCC revoked the licenses of the entire Alabama Educational Television Commission in 1975. It was a precedent demonstrating that the commission was not going to tolerate what it considered lack of service to an audience, in this case, the black audience. The action came before public broadcasting stations were required even to conduct community needs and ascertainment surveys. Nevertheless, the FCC acted on the premise that the licensee still has the responsiblilty to determine the needs of its audience and to program in accordance with those needs.

Two years later, an administrative law judge denied renewal of a noncommercial station licensed to the board of trustees of the University of Pennsylvania. The FCC upheld the decision, and among the criticisms leveled at the station was that the licensee had delegated and subdelegated authority to students. Although the FCC accepted the station's application for a new license, the renewal denial woke many boards of trustees up to the fact that even they had the responsibility to see that a broadcasting station is operated in the public interest. If it is not, the university can be held responsible.

criticism of the commission

Perhaps because it regulates a very "visible" industry, and perhaps because that industry directly affects all of us every day, the FCC has received criticism from the public, the Congress, and even commissioners within its ranks.

Conflict with Judicial Precedent

One criticism is that the FCC has issued rulings that conflict with judicial precedent. Nicholas Johnson and John Dystel cite a case in which the FCC issued permission for AT&T to build a 350-foot tower near a residential area of Finksburg, Maryland.[28] Despite opposition from citizens' groups, the commission granted the request, partly because AT&T had already conducted an environmental impact study and found that the tower would not harm the environment. Johnson and Dystel note that the tower was approved although the courts had ruled that federal agencies cannot rely on interested parties' environmental-impact statements.[29]

Staff Organization

The relationship of the FCC's middle staff to the commissioners is another bone of contention among critics. Erwin Krasnow and Lawrence Longley in their book, *The Politics of Broadcast Regulation*, point out that the middle staff exerts influence over the commissioners by controlling the channels of communication at the FCC.[30] Thus, when the commissioners need to choose among alternative policies, they must rely on information which their staff feels is relevant. The authors also state that since hundreds of decisions are made every day, implementation of policy must be delegated to middle-staff personnel.[31]

Frequency Allocation Matters

Furthermore, not everyone feels that the way in which the FCC allocates frequencies on the electromagnetic spectrum is in the public's best interest. For example, the designation of certain frequencies for marine use means that there are wide areas of the country in which these frequencies go unused, simply because there is no demand for marine communication.[32] Moreover, because this policy has been perpetuated for years, trying to change it now would entail major capital expenditures for the industries affected. The commission's local station concept, allocating certain frequencies to lower-powered stations serving small communities, has drawbacks in that it ties up a sizeable portion of the spectrum for local station use, especially since one way of reducing crowding on the spectrum is to switch to regional allocations. The result of this would be fewer but higher powered stations serving large regions. The idea, although technologically sound, seems somewhat impractical when we think of the "local" service that would be lost.[33] A regional station in Chicago serving a small town in Illinois would be hard pressed to include that Illinois community's local news in its regional programming.

EEO Policies

Two other criticisms have landed squarely on the FCC's Equal Employment Opportunity (EEO) policies and the effect of citizens' groups on FCC decision making. A report by the Citizen's Communications Center claims the criteria for stations' compliance with FCC-EEO requirements are vague and can be met by broadcasters who still discriminate.[34] The report also asserts that the commission requires an unrealistically high standard of proof of discrimination practices before designating a hearing in a renewal case. Another report critical of the Commission's EEO policies was issued by the U.S. Commission on Civil Rights.[35] This report suggests that the FCC should improve the image of women and minorities in television programming, an area many would argue is clearly outside the commission's jurisdiction and would violate the First Amendment.

Citizen Participation

A report by the Rand Corporation suggests that the commission should do more to encourage citizen participation, one effort being to support legislation which would provide financial assistance to citizens' groups participating in commission proceedings.[36] Giving citizens' groups access to evidence which might support their cause also is high on the list of recommendations. In current judicial processes, a person or group can gain access to information, called the right of discovery, only after proceedings have begun in the courts or, in the case of the FCC, after a hearing has been designated. The FCC has started a new Actions Alert program designed to solicit advice on FCC rule making. Written in plain English, the Actions Alerts are issued to citizens' groups and interested parties who can then give written opinions to the FCC (Figure 12–2).

Decision-Making Processes

One of the most serious criticisms of the FCC is its sluggishness in making important decisions. Krasnow and Longley state that the FCC is "incapable of policy planning, of disposing within a reasonable period of time the business before it, of fashioning procedures that are effective to deal with its problems."[37] A classic case is the proceedings affecting the assignment of WHDH-TV in Boston. The case started in 1947 when WHDH filed an application for a license to operate Channel 5. This channel allocation was the subject of competing applications and FCC decrees for 25 years. It was one of the longest proceedings ever to come before the FCC. Professors Robert Smith and Paul Prince reviewed the chronology of the WHDH-TV case and concluded that if there were no clear winner in the proceedings, "one party was a significant loser: the public."[38] Sterling Quinlan, writing a lighter book about the case, quotes a former commissioner as saying, "Let's face it. This was the 'Whorehouse Era' of the commission. When matters were arranged, not adjudicated."[39]

Figure 12–2

December 13, 1976

⑨≡ FCC Actions Alert

Special Feedback Edition

FCC Seeks Comments
On Cable Franchise Standards

The Federal Communications Commission invites your comments on questions raised regarding the current standards governing contracts between local authorities and cable companies. As the enclosed news

The commission often has been called to task for potential conflict of interest because of its staff-owned stocks of corporations regulated by the FCC. A staff report of the House Oversight and Investigations Subcommittee has criticized members for transferring shares of stock in communications-related industries to immediate members of their families (although the law, as it now stands, does not prohibit that practice.) Some of the stock ownership reported by the subcommittee included shares of General Electric owned by the spouse of a staff member in the Office of the Chief Engineer, shares of AT&T owned by the spouse of a staff member in the Common Carrier Bureau, and shares of AT&T owned by the spouse of an engineer in charge of an FCC field office.[40]

Johnson and Dystel divided their own criticism of the FCC into seven areas. They contend that (1) the FCC delves into areas beyond its expertise and issues beyond its ken; (2) it takes years to resolve important cases; (3) the FCC is manipulated by its own staff and the industries it is supposed to regulate. The results of this are precedents which return to "haunt" the commission; (4) principled decision making does not exist because the FCC no longer approves of its own rules and precedents, and instead ignores them—either by waiving them or evading them; (5) the commission ignores its own administrative principles and those established by the judiciary; (6) the commissioners decide cases they do not understand; and (7) the FCC has yet to develop "rational" policies for governing its day-to-day decisions.[41]

Criticism of the commission will undoubtedly continue, regardless of future changes. However, it is time for an in-depth evaluation of the entire commission. It is operating under 1934 procedures, a time when cable, satellites, microwaves, and fiber optics were only a dream. As it stands today, the prospects of the communications industry simply becoming unmanageable are real. The commission has established bureaus responsible for specific areas of the industry, but because so much is at stake when two competing corporations seek allocations or permission to develop technology, a ruling against one sends the matter through an appeals process which eventually reaches the seven commissioners. Those seven may very well be forced into a decision they are not qualified to make. As a result, the numerous reversals between the administrative law judges and the courts play havoc with anything that even resembles judicial precedent.

summary

Chapter 12 discussed the operation of the Federal Communications Commission. We learned that the FCC has thirteen areas of responsibility. Some of these include the orderly development and operation of broadcast services; control over AM, FM, TV, telephone, common carrier, cable and satellite communications; starts of new stations and transfer of ownership of those already operating; responsibility of domestic administration of the telecommunications provisions of treaties and international agreements; and supervision of the Emergency Broadcast System. The FCC does not have jurisdiction over such things as program scheduling, awarding of prizes in contests, broadcasting outside the United States and its possessions, news-gathering organizations, and libel and slander.

The typical agenda of an FCC meeting includes the following items, corresponding in many ways to the functions and organization of the commission: hearing, general, safety and special, common carrier, personnel, classified, cable television, assignment and transfer, renewal, aural, television, broadcast, and complaints and compliance. Research has taught us that, although selected as a bipartisan group, individual commissioners can have considerable influence over the policy directions the FCC takes. Organization of the FCC includes, along with the commissioners, the Office of Plans and Policy, Office of Opinions and Review, Office of Administrative Law Judges, the Review Board, Office of General Counsel, Office of Chief Engineer, and Office of Executive Director. The bureaus of the Commission include the Broadcast Bureau, Cable Television Bureau, and Field Operations Bureau.

No government agency can function effectively as a regulator without enforcement powers. The FCC is no exception. At its disposal are such measures as letters, cease and desist orders, forfeitures, short-term renewals, renewal denials, and revocations.

Criticism of the FCC has surfaced in recent years and has included such issues as decisions conflicting with judicial precedent, the influence on and separation of the middle staff from the commissioners, the FCC's allocation of frequencies, affirmative action policies, citizen participation in commission decisions, and the decision-making processes of the FCC.

spotlight on further learning

COLE, BARRY, and MAL OETTINGER, *Reluctant Regulators: The FCC and the Broadcast Audience.* Reading, Mass.: Addison-Wesley Publishing Company, 1978.

The FCC and Broadcasting. Washington, D.C.: Federal Communications Commission, 1977 (Broadcast Bureau Publication # 8310-100) (looseleaf).

The FCC in Brief. Washington, D.C.: Federal Communications Commission, 1977 (looseleaf).

KRASNOW, ERWIN G., and LAWRENCE D. LONGLEY, *The Politics of Broadcast Regulation.* New York: St. Martin's Press, Inc., 1973.

QUINLAN, STERLING, *The Hundred Million Dollar Lunch.* Chicago: J. Philip O'Hara, Inc., 1974.

ULLOTH, DANA ROYAL, *The Supreme Court: A Judicial Review of the Federal Communications Commission.* New York: Arno Press, 1979.

13

control of broadcast programming

FOCUS   After completing this chapter we should be able to

Tell how the concepts of a limited spectrum and mass influence affect broadcast regulation.

Discuss the application of Section 315 to political broadcasting.

Trace the development of the Fairness Doctrine.

Explain how the FCC regulates obscene, indecent, and profane material.

Define *prime-time access.*

Describe how the broadcast press is attempting to bring television cameras into judicial proceedings.

Understand broadcasters' efforts at self-regulation.

To understand why there is control of broadcast programming, we first must understand the legal forces which have helped mold both domestic and international broadcast communication. They include both basic constitutional documents and obscure local ordinances, ranging from major international treaties affecting satellite communication and multinational networks to agreements guarding against "electronic" border disputes.

limited spectrum and mass influence

The control of broadcasting centers on supply and demand. We know that if there is a great demand for a product and a shortage of that product, certain rules will be written to avoid chaos. Imagine a group of children all wanting a piece of candy, but there are only half as many pieces of candy available as there are children. Who gets the candy? Perhaps the children who have perfect behavior records will get the candy. Perhaps only those who agree to share their candy with others will get it. Perhaps those who eat responsibly and do not gobble will be rewarded. Or perhaps only those who can afford to buy the candy will get some. Our example illustrates the need for controls both to regulate the allocation of the product and to maintain order.

Now transpose our example to the allocation of frequencies on the electromagnetic spectrum. The spectrum has only so much space upon which radio and television stations can operate. Consequently, certain rules to govern the allocation and operation of stations are necessary. This limited-resource concept is the reason behind much broadcast regulation.

The second important concept is the influence of broadcasting on a great number of people. The citizens-band radio that sends out a five-watt signal to a passing motorist has little impact on a "mass" audience. If the operator decides to sing songs into the microphone, tell a joke, or provide "smokey" reports, the chances are the FCC will not be overly concerned. On the other hand, if a 50,000-watt clear-channel radio station decides to forego all its regular programming for a steady diet of songs sung by announcers, jokes, and traffic reports, then the station will have a difficult time justifying its privilege to operate. The fact that broadcasting sends messages to the public has a considerable effect on society. Thus, to assure that society is protected from abuse, there are certain rules.

At this point you may say, "Fine, we set up certain rules, people follow the rules, and the system functions." Unfortunately it is not that simple, and everyone from FCC commissioners, to citizens' groups, to broadcasters argue the legitimacy of the regulatory process. Part of the discussion centers on the legal philosophy upon which our society operates. America is considered a free country. Nelson and Teeter write: "Seventeenth and Eighteenth Century thought in much of Western Europe and America turned to faith in man's reason as the safest basis for government."[1] Lee Loevinger describes the practical application of this philosophy as negative or proscriptive rather than positive or prescriptive. That is, law in America, for example, forbids behavior which might harm society, but it does not require behavior which society has determined to be beneficial.[2] Nor does it require the best behavior of which one is capable, or even behavior that is socially desirable. At first, this attitude may seem as though it would undermine the good of society. Not so, Loevinger assures us, "when the law prohibits antisocial conduct, it leaves an extemely wide area of personal choice and individual liberty to the citizen."[3]

From the standpoint of broadcasting, we can see the head of regulatory conflict beginning to protrude. Although we must control the allocation of frequencies on the electromagnetic spectrum, to control programming on those frequencies goes against traditional American legal philosophy.

The arguments run between two extremes. One point of view suggests a total lack of control; supporters point out that the First Amendment assures free press and free speech. Some legal scholars even suggest that one freedom embodies the other.[4] The other point of view supports total control of broadcasting. Those arguing for total control argue their case using four assumptions: (1) there is a reliable and authoritative basis for determining program quality; (2) the public interest can be determined in one broadcast without reference to all other broadcasts; (3) there are programs which meet the assumed authoritative government standards; and (4) if the government commands it, then quality programs will be produced.[5]

Where does all this lead us? It has led to a regulatory system that greatly affects what radio and television stations will program. We now will examine in detail regulatory control and some of the areas in which federal regulation directly influences broadcast programming. We'll begin with Section 315 of the Communications Act, which concerns political broadcasting, examine the Fairness Doctrine, examine such areas as obscenity and profanity, prime-time access, and coverage of judicial proceedings.

Section 315 of the Communications Act

Of all of the provisions of the Communications Act of 1934, few have received as much attention or notoriety as Section 315, which regulates political broadcasting. The provision instructs the broadcaster and candidate in how the electronic media are to be used as part of our political system. It, along with the Fairness Doctrine, has an effect on how we, the consumers of broadcast communication, are informed of our electoral process.

The most crucial wording of Section 315 is its "equal-time" provision which states: "If any licensee shall permit any person who is a legally qualified candidate for public office to use a broadcasting station, he shall afford equal opportunities to all other such candidates for that office in the use of such broadcasting station."

Definitions Guiding Equal-Time Provisions

The Communications Act defines a legally qualified candidate as:

> any person who has publicly announced that he is a candidate for nomination by a convention of a political party or for nomination or election in a primary, special, or general election, municipal, county, state or national, and who meets the qualifications prescribed by the applicable laws to hold the office for which he is a candidate, so that he may be voted for by the electorate directly or by means of delegates or electors, and who:
> (1) has qualified for a place on the ballot or
> (2) is eligible under the applicable law to be voted for by sticker, by writing in his name on the ballot, or by other method, and
>   (i) has been duly nominated by a political party which is commonly known and regarded as such, or
>   (ii) makes a substantial showing that he is a bona fide candidate for nomination or office, as the case may be.[6]

In addition to this definition are hundreds of state and local statutes further clarifying political eligibility. Yet broadcasters are prohibited from deciding themselves who may be considered legally qualified. It makes little difference if the candidate has a chance of winning. If the law says the candidate is qualified and the candidate has publicly announced his or her candidacy, then the equal-time provisions will apply. Those provisions also apply to cable television systems.

Anticensorship Provisions

As a further safeguard against unfair treatment to political candidates, Section 315 expressly prohibits the broadcaster from censoring the content of any political message. The law succinctly states the licensee

310

TV Station Cancels Program
To Avoid Political Hassling

Rep. Patricia McDermott, D-Pocatello, threw a monkey wrench into plans for live television coverage Friday night of the Republican version of the State of the State message by demanding equal time for a Democratic reply.

KTVB (Channel 7) in Boise originally had planned a live broadcast of the message by legislative leaders Sen. Phil Batt, R-Wilder, and Rep. Allan Larsen, R-Blackfoot, with comments by leading Democrats to balance the program, said Sal Celeski, KTVB director of public affairs.

He initially proposed that the Republican message be broadcast live on his weekly news program "Viewpoint" during its usual 6:30 p.m. time slot.

"Then Patty called me and started saying that wasn't going to be fair. She said the Republicans would be there in that room (in the Statehouse) with a partisan crowd, and we'll be in your cold studio, it won't be equal. She thought we should give them another time slot.

"I said 'Hey, I don't have time slots to give,' " Celeski said. "Then I got to thinking about it, and it's just not worth the hassle." So he canceled plans for the live broadcast.

Figure 13–1   (*The Idaho Statesman*)

"shall have no power of censorship over the material broadcast under provisions of this section."

Until 1959, broadcasters were confused by the noncensorship rule, fearing it was only a matter of time until some candidate blatantly libeled an opponent, and the station was sued for damages. The dreaded event occurred in 1959 in North Dakota when U.S. senatorial candidate A. C. Townley charged on the air that the North Dakota Farmers' Union was Communist-controlled. The Farmers' Union sued the station and Townley for $100,000. But the North Dakota Supreme Court ruled that the station was not liable and that the suit should have been brought against Townley alone. Undoubtedly, the Farmers' Union thought about that, but since Townley made only $98.50 a month, the prospect for recovering damages was not bright.[7]

The Farmers' Union appealed to the Supreme Court. Justice Hugo Black, in delivering the opinion of the Court, stated: "Quite possible, if a station were held responsible for the broadcast of libelous material, all remarks even faintly objectionable would be excluded out of an excess of caution . . . if any censorship were permissible, a station so inclined could intentionally inhibit a candidate's legitimate presentation under the guise of lawful censorship of libelous matter."[8]

Exemptions to the Equal-Time Provisions

Exempt from the equal-time provisions are appearances by candidates on these types of news programming:

1. bona fide newscast,
2. bona fide news interview
3. bona fide news documentary (if the appearance of the candidate is incidental to the presentation of the subject or subjects covered by the news documentary), or
4. on-the-spot coverage of bona fide news events (including but not limited to political conventions and activities incidental thereto), shall not be deemed to be use of a broadcasting station within the meaning of this subsection.

In the fall of 1975, the FCC added to the exemption list political debates and news conferences, as long as they were broadcast in their entirety, and if the broadcaster made a good-faith judgment that they constituted a bona fide news event. In the spring of 1976, a three-judge panel of the U.S. Court of Appeals in Washington, D.C. upheld the FCC's right to include the added exemption. The court split in a two-to-one decision and in offering the verdict, noted that it took comfort in the fact that Congress could correct the FCC if it had overstepped its authority in the added exemption.[9]

The exemption itself is a hot political issue, since party loyalty as well as congressional autonomy tends to surface during an election year. For example, the FCC's exemption permitted Gerald Ford and Jimmy Carter to participate in nationally televised debates in 1976. For John F. Kennedy and Richard Nixon to debate in 1960, Congress had to suspend Section 315.[10] Without the suspensions and exemptions, networks and local stations would have been faced with a plethora of minority-party candidates demanding equal time.

Selling Time: The Lowest Unit Charge

Besides granting equal time to candidates, Section 315 also spells out how much they are to be charged for the use of broadcast facilities:

(b) The charges made for the use of any broadcasting station by any person who is a legally qualified candidate for any public office in connection with his campaign for nomination for election, or election, to such office shall not exceed-

   (1) during the forty-five days preceding the date of a primary or primary runoff election and during the sixty days preceding the date of a general or special election in which such person is a candidate, the lowest unit charge of the station for the same class and amount of time for the same period, and

   (2) at any time, the charges made for comparable use of such station by other users thereof.

The above is known as the "lowest unit charge" rule. To understand it more clearly, assume that you are the sales manager for a television station. The station's rate card charges an advertiser $1,000 to buy a single, one-minute commercial in prime time. An advertiser purchasing two commercials receives a discount and is charged only $850 per commercial. We'll assume that the rate card permits an advertiser purchasing 25 commercials to receive an even bigger discount, when each commercial costing $500. Along comes candidate John Doe who is running for municipal judge. Doe wants to buy just one commercial to remind his friends that he is running for office. He wants it to run in prime time. What will you charge him for the cost of his one commercial? You will charge him $500. Even though he is buying only one commercial, the law states that you must charge him the "lowest unit charge." If he wanted to purchase a commercial in a fringe-time period during which the rates are lower, then you would charge him the "lowest unit charge" for that time period.

Access: The Relationship of Section 312 to Section 315

Our discussion of Section 315 would not be complete without mentioning another section of the Communication Act of 1934, Section 312, and how it relates to Section 315. Section 312 is actually a prerequisite to 315, since 312 succinctly states that the station must not deny access to any candidate for federal office, regardless of what form that access takes. Section 312 cautions the broadcaster that a station license may be revoked, "for willful or repeated failure to allow reasonable access to or to permit purchase of reasonable amounts of time for the use of a broadcasting station by a legally qualified candidate for Federal elective office on behalf of his candidacy."

Notice that the law reads *"federal elective office."* This clause has been a bone of some contention and confusion in interpreting Section 315, especially when candidates on other than the federal level are involved. Some stations have used Section 312 as grounds for refusing to sell commercial time to candidates other than those running for federal offices. The advantage to such a policy is mainly economic. First, there are fewer federal candidates than local candidates, translating into fewer political commercials. You may ask whether the station is not in business to sell commercials. Yes, but remember the lowest unit charge. If a department store pays a nondiscounted rate for commercials but cannot get on the air because of the many political commercials sold at the lowest unit charge, then the station is losing money. Second, federal candidates often place their advertising through advertising agencies. Although the station still must give a discount to the agency, the number of commercials purchased is usually more than what candidates would purchase on their own. Thus, the total amount spent is closer to the actual profit made from typical business advertising. Third, the commercials from the agency are usually prerecorded, eliminating the need for the local station to tie up its staff and facilities helping a local candidate produce a commercial which may run only one time at the lowest unit charge.

By inserting the term *federal*, Section 312 left no definition of "reasonable access" for candidates running for state and local offices. Historically, the station has been flexible in such cases. In its *Guidelines* to political candidates, the commission says: "The licensee in its own good-faith judgment in serving the public interest may determine which political races are of greatest interest and significance to its service area, and therefore may refuse to sell time to candidates for less important offices, provided it treats all candidates for such offices equally."[11]

the Fairness Doctrine

The Fairness Doctrine was first issued in 1949 as an FCC report to broadcasters on handling controversial issues with "fairness" to all sides.[12] The FCC reexamined the doctrine in policy statements issued in 1964, 1974, and 1976.

Mayflower Decision

The Federal Radio Commission, in discussing the limited spectrum space, noted that if issues "are of sufficient importance to the listening public, the microphone will undoubtedly be available. If not, a well-founded complaint will receive the careful consideration of the commission."[13] Attention to this fairness issue crystalized in 1941 with the "Mayflower" decision, involving station WAAB in Boston.[14] The Mayflower Broadcasting Corporation petitioned the FCC to give Mayflower the facilities of WAAB, which were up for renewal. Although the FCC ruled in favor of WAAB in a review of WAAB's past performance, the station was strongly criticized by the commission for its practice of "editorializing." The FCC stated that it was "clear that with the limitations in frequencies inherent in the nature of radio, the public interest can never be served by a dedication of any broadcast facility to the support of his own partisan ends."[15] The FCC offered the opinion that "a truly free radio cannot be used to advocate the causes of the licensee. . . . In brief, the broadcaster cannot be an advocate."[16] The Mayflower decision successfully discouraged other stations from jumping on the editorial bandwagon.

While the Mayflower decision was stifling editorials, the Code of the National Association of Broadcasters was stifling discussion of controversial issues by prohibiting the purchase of commercials to air those issues. It was not long before one station was caught in the triangle between the FCC, the NAB Code, and the First Amendment. Station WHKC in Columbus, Ohio, believing it was operating in the public interest, adhered to the NAB Code and promptly found itself in a dispute with a labor union.[17] The union claimed that the station had refused to sell it time and had censored its submitted scripts.

The union filed a petition against WHKC's license renewal. The FCC held a hearing on the matter between August 16 and 24, 1944, and heard the argument about the NAB Code. By October, the union and the station had agreed to a compromise. The agreement broke with the code and prohibited any further censorship of scripts, dropping the station's policy of banning selling time for controversial issues. The FCC stated the station must be

> ... sensitive to the problems of public concern in the community and ... make sufficient time available on a nondiscriminatory basis, for full discussion thereof, without any type of censorship which would undertake to impose the views of the licensee upon the material to be broadcast.

Further support for airing controversial issues came in 1946 when Robert Harold Scott of Palo Alto, California, filed a petition asking the FCC to revoke the licenses of radio stations KQW, KPO, and KFRC. Scott claimed that he wanted time to expound his views on atheism, balancing the station's "direct statements and arguments against atheism as well as ... indirect arguments, such as church services, prayers, Bible reading and other kinds of religious programs."[18] Scott did not get the station's license revocation, but in its decision, the FCC stated:

> The fact that a licensee's duty to make time available for the presentation of opposing views on current controversial issues of public importance may not extend to all possible differences of opinion within the ambit of human contemplation cannot serve as the basis for any rigid policy that time shall be denied for the presentation of views which may have a high degree of unpopularity.

Issuing the Doctrine

The commission began to tackle the editorial issue in March and April of 1948. In eight days of hearings on the subject, it heard from 49 witnesses; 21 other persons filed written motions. From the hearings came a statement issued by the FCC on June 1, 1949 under the heading,

In the Matter of Editorializing by Broadcast Licensees. It was to become known as the "Fairness Doctrine." In the doctrine, the commission reasserted its commitment to free expression of controversial issues of public importance as stated in the WHKC and Scott decisions. It also reversed the Mayflower decision by supporting broadcast editorials. The commission came "to the conclusion that overt licensee editorialization, within reasonable limits and subject to the general requirements of fairness . . . is not contrary to the public interest."[19]

The Fairness Primer

As expected, there was a series of court cases and complaints about abuse of the Fairness Doctrine. Finally, it became necessary in 1964 for the FCC to issue some clarifying guidelines. The 1964 document, commonly called the "Fairness Primer" was a compilation of representative FCC rulings from over those years.[20] It gave people an opportunity to study the FCC's decisions, shedding light on other stations' practices and policies, seeing when complaints might be warranted, and guiding stations on how to meet Fairness Doctrine requirements.

Still waiting, however, was a major legal test of the constitutionality of the Fairness Doctrine. It came in an appeals court case in 1967, reaching the Supreme Court in 1969. Called the "Red Lion decision," it affirmed the constitutionality of the Fairness Doctrine. We'll examine this landmark case in more detail.

The Red Lion Decision

The Red Lion decision involved the Red Lion Broadcasting Company of Red Lion, Pennsylvania. In November, 1964, the Reverend Billy James Hargis lashed out on Red Lion's radio station against the author of a book about Barry Goldwater. The author, Fred J. Cook, was held in low esteem by Hargis, who spelled out what he felt to be the less favorable aspects of Cook's career as a writer. Cook contacted the station for a chance to reply to Hargis. But the station claimed that it did not have to offer free time to Cook unless he could prove that there was no commercial sponsorship available to present his views. Cook went to the FCC, which ruled in his favor, citing the Fairness Doctrine. In the case of *Red Lion Broadcasting Co.* v. *Federal Communications Commission,* the appeals court upheld the FCC's decision.[21]

At that point, the Radio-Television News Directors Association entered the picture and appealed the case once more, this time to the United States Court of Appeals for the Seventh Circuit in Chicago. In the case of *Radio-Television News Directors Association* v. *United States,* the court ruled that the Fairness Doctrine's personal attack and editorial rules would "contravene the first amendment."[22] But RTNDA's victory was short-lived. The FCC then took the case to the Supreme Court which reviewed both the circuit and appeals courts' decisions. The Supreme Court ruled: "In view of the prevalence of scarcity of broadcast frequencies, the Government's role in allocating those frequencies, and the legitimate claims of those unable without governmental assistance to gain access to those frequencies for expression of their views, we hold the regulations and ruling at issue here are both authorized by statute and constitutional."[23] With this, the Supreme Court upheld the FCC and reversed the decision in the RTNDA case. The Fairness Doctrine now was not only a broadcast regulation, but one reaffirmed by judicial precedent by the highest court in the land.

Personal Attack Rule

One area of the Fairness Doctrine which remained somewhat nebulous was the broadcast of direct personal attacks on individuals or organizations. When the Red Lion issue came to the FCC's attention, it decided that this was the time for a ruling. Becoming effective on August 14, 1967, the FCC's rules regarding personal attack read:

> *(a) When, during the presentation of views on a controversial issue of public importance, an attack is made upon the honesty, character, integrity or like personal qualities of an identified person or group, the licensee shall, within a reasonable time and in no event later than one week after the attack, transmit to the person or group attacked (1) notification of the date, time and identification of the broadcast; (2) a script or tape (or accurate summary if a script or tape is not available) of the attack and (3) an offer of a responsible opportunity to respond over licensee's facilities.*

The rules exempt foreign groups or foreign public figures, certain types of attacks made by political candidates during campaigns, and with the same provisions as in Section 315, various bona fide news events.

At the same time that it spelled out the new personal attack policy, the FCC also spelled out new rules covering editorials:

> *(c) Where a licensee in an editorial, (i) endorses or (ii) opposes a legally qualified candidate or candidates, the licensee shall, within 24 hours after the editorial, transmit to respectively (i) the other qualified candidate or candidates for the same office or (ii) the candidate opposed in the editorial (1) notification of the date and the time of the editorial; (2) a script or tape of the editorial; and (3) an offer of a reasonable opportunity for the candidate or a spokesman of the candidate to respond over the licensee's facilities:* **Provided, however,** *that where such editorials are broadcast within 72 hours prior to the day of the election, the licensee shall comply with the provisions of this paragraph sufficiently far in advance of the broadcast to enable the candidate or candidates to have a reasonable opportunity to prepare a response and to present it in a timely fashion.*

With these 1967 rules, broadcasters now know exactly what is expected of them when such incidents occur on their stations. They do, however, have the discretion to determine what constitutes a personal attack. Here, although not completely free, the FCC has permitted broadcast management to remain in charge of its local programming, somewhat unimpeded by a federal agency.

Broadcast Advertising

The FCC's position on the fairness issue is that the overall programming of a station should reflect its commitment to fairness, not just a single program. The position of advertising in this programming became a contested issue when WCBS-TV was approached by New York lawyer John W. Banzhaf, who requested equal time to reply to cigarette commercials. WCBS-TV refused to grant him time, but the FCC agreed with Banzhaf. The FCC's decision was upheld by the appeals court which tried to confine the decision to cigarette advertising. But that was too much to hope for, and over the years, the Fairness Doctrine has applied to many factions of advertising. (Cigarette advertising meanwhile was banned on radio and television after 1971 by the Public Health Cigarette Smoking Act of 1969.)

In a sweeping FCC order, eight California stations were caught in a Fairness-Doctrine controversy over programming on nuclear power plants. At a time when people in California were being asked to sign a petition for a referendum on nuclear power plants, the stations aired commercials sponsored by the Pacific Gas and Electric Company. The commercials promoted nuclear power and power plants. Citizen-action groups brought the matter to the FCC's attention and in 1974, filed an action against thirteen stations. The commission found that five stations had presented the issue fairly with programming advocating the anti-nuclear stand. Eight others were required to show the FCC how they intended to comply with the Fairness Doctrine. The commission felt that the issue was controversial and of public importance and investigated "to the minute" the amount of time the stations had devoted to different sides of the issue.[24]

The 1974 Report

In 1974 the FCC reopened hearings on the Fairness Doctrine and concluded the hearings by issuing an updated report on the applicability of the Fairness Doctrine. More importantly, the 1974 Report also attempted to create an atmosphere of flexibility in interpreting the doctrine. What the FCC, the broadcasters, and the public had been worrying about was the absence of guidelines defining such sensitive issues as "contro-versial issue" or "reasonable opportunities for contrasting viewpoints." The commission summed up its feelings on these matters as follows:

> *The Fairness Doctrine will not ensure perfect balance and debate, and each station is not required to provide an "equal" opportunity for opposing views. Furthermore, since the Fairness Doctrine does not require balance in individual programs or a series of programs, but only in a station's overall programming, there is no assurance that a listener who hears an initial presentation will also hear a rebuttal. However, if all stations presenting programming relating to a controversial issue of public importance make an effort to round out their coverage with contrasting viewpoints, these various points of view will receive a much wider public dissem-ination.*

The 1974 Report did not diminish the debate over the Fairness Doctrine.

Reconsidering the Fairness Doctrine: 1976

The commission decided to reconsider the Fairness Doctrine in 1976 after citizens' groups wanted more access to broadcasting. The FCC generally reaffirmed its decisions in the 1974 report. It felt that the doctrine should continue to be applied to advertisements of public issues, not of specific products. It agreed that broadcast editorials should come under the aegis of the doctrine and reaffirmed the right of the broad-caster to decide how the doctrine should be applied locally. If the FCC did have to intervene, it was felt the probable action would be simply to require that the station provide time for opposing viewpoints.

regulating obscene, indecent, and profane material

One of the most complex areas of broadcast regulation is obscene and indecent programming. The statutes governing such programming have evolved from both the Radio Act of 1927 and the Communications Act of 1934. The Radio Act of 1927 provided for penalties of up to $5,000 and imprisonment for five years for anyone convicted of violating the act, including its obscenity provisions. The Communications Act of 1934 changed this to $10,000 and two years in jail, stating that the violator's license could be suspended for up to two years. In 1937, the penal provisions covering obscenity were amended to include license suspension for those transmitting communications containing profane or obscene "words, language, or meaning." The license suspension was no longer limited to two years, and the word "meaning' became even more appro-priate as television became more popular.[25]

The U.S. Criminal Code

In 1948, Congress took the obscenity provisions from the Communications Act of 1934 and put them into the United States Code. Thus, the U.S. Criminal Code, Section 1464 states: "Whoever utters any obscene, indecent, or profane language by means of radio communication shall be fined not more than $10,000 or imprisoned not more than two years or both."[26] "Radio communication" includes television. Both the Department of Justice and the FCC have the power to enforce Section 1464. Penalties include forfeiture of a license or construction permit and fines of $1,000 for each day the offense occurs, not to exceed a total of $10,000. The Justice Department also can prosecute under Section 1464 and send a licensee to jail.

Although there are other cases in which the FCC has acted against stations which have broadcast obscene, indecent, or profane material, two stand out. One was an Illinois station's "topless" format and the other a New York station's broadcast of comedian George Carlin's monologue on words that cannot be said on radio or television.

Topless Radio and Seven Dirty Words

In 1973, the FCC found itself in the obscenity arena with a case involving a station in Oak Park, Illinois. The topic for the call-in program on February 23 was oral sex, and female listeners called moderator Morgan Moore with graphic descriptions of their experiences. The format, also employed at other stations, was known as "topless radio." Female listeners were not the only ones to contact the station. The FCC notified them of their apparent liability of $2,000 for violating *both* the indecency and obscenity clauses of the Criminal Code.[27]

Two groups, the Illinois Citizens Committee for Broadcasting and the Illinois Division of the American Civil Liberties Union, asked the FCC to reconsider the ruling. When the commission declined, the Illinois Citizens Committee for Broadcasting appealed in *Illinois Citizens Committee for Broadcasting* v. *Federal Communications Commission*. On November 20, 1974, the court upheld the FCC's action and in effect ruled that the commission was acting constitutionally.

On the afternoon of October 30, 1973, WBAI-FM warned its listeners that the following broadcast included sensitive language which might be offensive. What followed was a recording by comedian George Carlin from his album "George Carlin: Occupation Foole." Carlin's monologue was a satire on seven four-letter words which could not be used on radio or television because they depicted sexual or excretory organs and activities. A month later, the FCC received a complaint from a man who said that he had heard the broadcast while he was driving with his son. It was the only complaint received about the broadcast, which had been aired as part of a discussion on contemporary societies' attitudes toward language.

The FCC issued a declatory ruling against WBAI-FM and stated that such language " . . . describes, in terms patently offensive as measured by contemporary community standards for the broadcast medium, sexual or excretory activities and organs, at times of the day where there is a reasonable risk that children may be in the audience."[28]

The commission also rationalized that broadcast media should be treated differently from print media in regulating indecent material, because broadcast media are instrusive, based on four considerations:

(1) children have access to radio and in some cases are unsupervised by parents; (2) radio receivers are in the home, a place where people's privacy interest is entitled to extra deference; (3) unconsenting adults may tune in a station without any warning that offensive language is being or will be broadcast; and (4) there is a scarcity of spectrum space, the use of which the government must therefore license in the public interest.[29]

The commission reiterated that it was not in the business of censorship but did have a statutory obligation to enforce those provisions of the criminal code that regulated obscene, indecent, or profane language.

Whatever good intentions the commission had in issuing its declatory order, the U.S. Court of Appeals for the District of Columbia did not agree and did little to uphold it. Striking down most of the commission's major arguments, the court gave the FCC a judicial setback bordering on embarrassment. It first found that the commission's order was in direct violation of Section 326 of the Communications Act of 1934, which prohibits the FCC from censoring programming. Although the FCC clearly stated that it was not censoring, the appeals court felt it was doing just that simply by issuing the order. The issue did not stop there, however. The case went to the United States Supreme Court where the FCC found itself back in favor. The single complaint from a father about what his son heard on radio had set a strong precedent for future FCC action against questionable material on the air and made it clear that there were at least seven words that could cause broadcasters much trouble if they decided to use them on the air.

prime-time access

Another area of programming which involves the FCC is the amount of time local stations devote to network television programming.

Concern over the dominance of network programming prompted the FCC to take measures assuring that alternative programming would also be aired during the evening hours. From these measures came the prime-time access rule. The latest Prime Time Access Rule III (PTAR III), charges stations in the top fifty markets, which are either network-affiliated or network-owned, to clear an hour from network prime-time programming, which is 7:00 P.M. to 11:00 P.M. in Eastern and Pacific time zones and 6:00 to 10:00 P.M. in Central and Mountain time zones.[30] PTAR III was "refined" by order of the United States Second Circuit Court of Appeals in the case *National Association of Independent Television Producers and Distributors et al.* v. *FCC.*[31] The rule is designed to (1) give independent producers and syndicators a market for their programming and (2) encourage local stations to develop creative programming. By applying the rule to the top fifty markets, the FCC has successfully covered the nation. Yet the rule has been more successful in providing time for syndicated programming than in stimulating local creativity. The result has been a plethora of quiz and game shows in the 6:00 to 8:00 P.M. time periods across the country.

PTAR III still allows a series of exemptions. Stations can broadcast network or off-network documentaries, public affairs, and children's programming. Public affairs programming is defined the same as it is in the FCC logging rules, as " . . . talks, commentaries, discussions, speeches, editorials, political programs, documentaries, forums, panels, roundtables, and similar programs primarily concerning local, national, and international public affairs." Feature films also can be broadcast as can network news programming of special interest to the viewing audience. In other words, if a network provides affiliates with coverage of a major developing news event, such as an assassination or natural disaster, the local affiliates can carry the program and have it count as prime-time access. If a television station produces an hour of local news which immediately precedes the prime-time access hour, for example, the local news from 6:00 to 7:00 P.M., then the station can carry network news programming up to one-half hour into the access period, or until 7:30 P.M.

Sports programming also is exempted. If a sports event is scheduled to end at the beginning of prime-time access but lasts longer, stations are permitted to continue their sports coverage. Major sports events in which all prime time is devoted to their coverage, such as New Year's Day football games or coverage of the Olympic games, receive the same exemption. Under continued scrutiny is the antiblackout law (Public Law 93-107), which permits the telecast of a home football game during prime time, but only if the game is sold out 72 hours before kickoff.

coverage of judicial proceedings

Although not an issue of direct concern to the FCC, the broadcast press has been deeply concerned over the Court's reluctance to allow broadcast coverage of judicial proceedings. At the heart of the issue is permission to have television cameras record the proceedings and for stations to broadcast live and video-taped courtroom activity, including testimony. The idea itself of cameras in the courtroom was conceived before television became part of the American scene. When Bruno Hauptmann was tried for the kidnapping of Charles Lindbergh's son, the courtroom resembled more a county fair than a judicial proceeding. Reporters were falling over reporters, vendors were selling souvenirs, and when the judge barred cameras from the courtroom, an enterprising chap managed to sneak a camera into the balcony and snap a picture of the courtroom that appeared in papers across the country.

Since the Lindbergh trial, everyone from the Supreme Court to bar associations has grappled with the difficult issue of how much publicity is too much and how or if television cameras and recording equipment interfere with a fair trial. The American Bar Association approved its famous Cannon 35 two years after the Lindbergh trial. Amended in 1963 to include television, Cannon 35 forbade either the taking of photographs or the broadcasting of court proceedings. Individual states were quick to affirm Cannon 35's principles and place it in statutes affecting court proceedings. The Federal Rules of Criminal Procedures, specifically Rule 53, carries the prohibition of cameras to federal courts. A special committee of the Judicial Conference of the United States reaffirmed Cannon 35 in 1968, calling for prohibition of " . . . radio or television broadcasting from the courtroom or its environs, during the progress of or in connection with judicial proceedings." Clearly, from the standpoint of the courts and of many lawyers, there is popular support for the Sixth Amendment's position.

Such claims for constitutional priority are not founded merely in supposition or conjecture. The annals of case law are filled with overturned verdicts, appeals, and charges of biased juries because the news media have been less than restrained in their coverage. Cases which stand out include *Rideau* v. *Louisiana*.[32] In this case, the suspect was interviewed by a country sheriff, and the interview was filmed and played on local television. The suspect's confessions made during the interview and the subsequent televising of those confessions prompted the defense attorney to request a change of venue. A denial and subsequent guilty verdict were all that was needed for the United States Supreme Court to reverse the conviction and state that the jury should have been drawn from a community whose residents had not seen the televised interview.

The case of Texas businessman Billie Sol Estes added fuel to this constitutional fire. Estes was tried and convicted of swindling. An appeals court affirmed the conviction, but when the case reached the United States Supreme Court in 1965 in *Estes* v. *State of Texas*, the conviction was reversed.[33] Massive national publicity surrounded the trial, and when it first went to court, the trial judge permitted television coverage of portions of the trial. In fact, the initial hearings were carried live. The scene was described by Justice Clark, who delivered the opinion in the case: "Indeed, at least 12 cameramen were engaged in the courtroom throughout the hearing taking motion and still pictures and televising the proceedings. Cables and wires were snaked across the courtroom floor, three microphones were on the judge's bench, and others were beamed at the jury box and the counsel table. It is conceded that the activities of the television crews and news photographers led to considerable disruption of the hearings."

Justice Clark summarized four areas in which television could potentially interfere with a trial: (1) Television can have an impact on the jury. The mere announcement of a televised trial can alert the community to "all the morbid details surrounding" the trial. "Every juror carries with him into the jury box those solemn facts and thus increases the chance of prejudice that is present in every criminal case." (2) Television can impair the quality of testimony. "The impact upon a witness of the knowledge that he is being viewed by a vast audience is simply incalculable. Some may be demoralized and frightened, some cocky and given to overstatement; memories may falter . . . ." (3) Television places additional responsibilities on the trial judge. Along with other supervisory duties, the judge also must supervise television. The job of the judge, "is to make certain that the accused receives a fair trial. This most difficult task requires his undivided attention." (4) For the defendent, television "is a form of mental if not physical harassment, resembling a police lineup or the third degree. The inevitable closeups of his gestures and expressions during the ordeal of his trial might well transgress his personal sensibilities, his dignity, and his ability to concentrate . . . ."

Despite the Supreme Court's decision, broadcasters continued their fight for courtroom access for the omnipresent television camera (Figure 13–2). There were breakthroughs in 1972 when the American Bar Association's House of Delegates approved a Code of Professional Responsibility, permitting the use of television in the courtroom for such activities as presenting evidence. Another breakthrough came in 1974 when the Washington State Supreme Court instructed a county superior court to select a trial and experiment with televising it, for "educational" purposes. The experiment was generally successful. In Las Vegas, Nevada, in the fall of 1976, KLAS-TV televised in color a criminal court trial. Sixty hours of courtroom activity, including interviews with the defendant, jury, and attorneys, were videotaped and edited for a three-part, prime-time special. One of the most publicized trials was of a teenager accused of murder, which took place in Florida in 1977. The trial was televised, and segments appeared regularly on network television, calling national attention to the camera-courtroom issue (Figure 13–3). A few weeks later, when the verdict was read in an Indiana kidnapping case, cameras again were present, and the courtroom once again appeared on national television.

Figure 13–2 Smaller cameras and progressive judges have made television cameras in courtrooms more and more common. (Dick Wetmore and WPBT)

Figure 13–3 A national television audience witnessed courtroom scenes during the trial of a Florida teenager accused of murder. (George Chase and WPBT)

The future of broadcast coverage of judicial proceedings rests on two factors: (1) the willingness of the courts to recognize the public's right of access to trials by permitting in the courtroom the apparatus necessary to capture the actual sounds and sights of the court in session, and (2) the willingness of the broadcast press to use restraint and the highest professional attitude and activity while covering a trial. Certainly, not all of the courts across the country are going to open their doors overnight to broadcasting. The process will be slow and gradual, and many trials will remain closed at the request of the parties involved. Even with television technology, the familiar and talented courtroom artists employed by many of the networks and larger television stations will continue their craft of capturing on the sketch pad the activity judges bar from the eyes of the television camera (Figure 13–4).

self-regulation

Although the government has been active in controlling the content of broadcast programming, the broadcasters themselves, through their individual professional associations, also have contributed. The amount of self-regulation that broadcasters employ seems in many ways to be in direct proportion to the amount of government regulation. Specifically, self-regulation, when effective, can actually displace government regulation before it begins. Within the broadcasting industry and related fields there are numerous guidelines and codes of ethics which are at least part of the by-laws of many organizations, even if they are not always followed by their members.

Stations which belong to the National Association of Broadcasters also can belong to the NAB Radio Code and the NAB Television Code. The codes themselves are not new. Organized in 1922, the NAB was originally founded

> *to foster and promote the development of the arts or aural and visual broadcasting in all forms; to protect its members in every lawful and proper manner from injustices and unjust exactions; to do all things necessary and proper to encourage and promote customs and practices which will strengthen and maintain the broadcasting industry to the end that it may best serve the public.*

By 1937 the broadcasting industry, through the NAB, was adopting its first set of self-regulatory guidelines. There was a major revision in 1945, and since that time the codes have been regularly updated to reflect both the pressure of government regulation on one side and the pressure of NAB members on the other. The Code Authority of the NAB administers the codes with the approval of the NAB board of directors. The executive

Figure 13–4 Defense attorney F. Lee Bailey rises to his feet as the scene from the Patty Hearst trial is captured by NBC courtroom artist Walt Stewart of KRON-TV in San Francisco. (Walt Stewart)

staff of the Code Authority enforces the codes with regular monitoring of NAB member stations. Broadcast advertising and programming are the two principal areas of both the Radio and Television Codes, and membership in the code is by subscription and open to "any individual, firm or corporation which is engaged in the operation of . . . a broadcasting station or network . . . or which holds a construction permit . . . for a radio or television station." Membership in the code requires an agreement by the station to abide by NAB advertising standards and also permits the station to use the code seals (Figure 13–5) on on-air promotions and advertising literature used by the station.

Along with the NAB Codes are codes of other organizations such as the Radio-Television News Directors Association, the Society of Professional Journalists-Sigma Delta Chi, American Women in Radio and Television, and advertising associations. Press councils operate nationally and in many states and communities in which other than legal complaints of the print as well as broadcast press can be aired.

Figure 13–5

summary

Government control of programming is a major concern of the broadcasting industry. Its basis is in the fact that the electromagnetic spectrum on which radio and television waves travel is a limited resource and must have safeguards to assure its responsible use. Coupled with this is the tremendous influence that radio and television have, which in the case of satellites crosses international boundaries.

Two areas of great concern to both broadcasters and the public are regulations resulting from Section 315 of the Communications Act and regulations from the Fairness Doctrine. Section 315 is mainly concerned with political broadcasting and assures that candidates for public offices will have the same opportunities to gain access to the broadcast media as other candidates for the same offices do. Key parts of Section 315 include its definitions of equal time, its anticensorship provisions, and the lowest unit charge. The Fairness Doctrine traces its roots back to the 1940s when the FCC prohibited editorializing and then reversed itself in 1949. Since then the doctrine has been revised considerably, mostly through FCC policies and court decisions. It now covers all areas of radio and television broadcasting, including advertising, and as far as overall programming is concerned, broadcast news programming.

An area in which the FCC has found one of its strongest footholds is in control of obscene, indecent, and profane material. Supported by the U.S. Criminal Code, the FCC has levied sanctions against numerous stations. Two of the most famous cases were the frank sexual discussions found in "topless radio" formats and the use of comedian George Carlin's monologue on words prohibited on radio and television.

Broadcasters' relationships with their networks also have come under the FCC's jurisdiction. The prime-time access rule requires stations to permit a certain amount of local programming to air in the early evening hours preceding prime-time network programming. Although some stations have developed a creative "magazine" format and other television fare, others have chosen to rely on syndicated programs, such as game shows.

Although not actually under FCC jurisdiction, the broadcast press has been waging a continuing battle to permit television cameras in courtrooms and to bring entire judicial proceedings to the television screen. Although the courtroom artists will remain an important part of courtroom coverage, more and more states are experimenting with television cameras in judicial chambers.

Self-regulation has been practiced by many broadcasters. Through such organizations as the National Association of Broadcasters, broadcasters have managed to avoid some federal regulations and have attempted to improve the quality of radio and television programming.

spotlight on further learning

Applicability of the Fairness Doctrine in the Handling of Controversial Issues of Public Importance. Washington, D.C.: Federal Communications Commission, 1964 (FCC 64-611).

BOSMAJIAN, HAIG A., ed., *Obscenity and Freedom of Expression.* New York: Burt Franklin & Co., 1976.

DEVOL, KENNETH S., ed., *Mass Media and the Supreme Court: The Legacy of the Warren Years.* New York: Hastings House, 1976.

Fairness Doctrine and Public Standards. Washington, D.C.: Federal Communications Commission, 1974 (Docket 19260).

FRANKLIN, MARC A., *The First Amendment and the Fourth Estate.* Mineola, N.Y.: The Foundation Press, Inc., 1977.

FRIENDLY, FRED W., *The Good Guys, the Bad Guys and the First Amendment: Free Speech vs. Fairness in Broadcasting.* New York: Random House, Inc., 1976.

KURLAND, PHILIP B., ed., *Free Speech and Association: The Supreme Court and the First Amendment.* Chicago: University of Chicago Press, 1976.

LEWIS, FELICE FLANERY, *Literature, Obscenity, and Law.* Carbondale: Southern Illinois University Press, 1976.

National Association of Broadcasters, *Political Broadcast Catechism.* Washington, D.C.: NAB Legal Department, 1976.

National Association of Broadcasters. *Radio and Television Codes.* Washington, D.C.: National Association of Broadcasters, 1979 (updated periodically).

Recommendations of the Communications Law Committee, Section on Science and Technology, American Bar Association, "Electronic Journalism and First Amendment Problems," *Federal Communications Bar Journal,* 29, no. 1 (1976), 1–61.

RUCKELHAUS, WILLIAM, and ELIE ABEL, eds., *Freedom of the Press.* Washington, D.C.: American Enterprise Institute for Public Policy Research, 1976.

SADOWSKI, ROBERT P., *An Analysis of Statutory Laws Governing Commercial and Educational Broadcasting in the Fifty States.* New York: Arno Press, 1979.

14

regulating station
operations

FOCUS After completing this chapter we should be able to

Identify the sections of a station's programming log.
Identify the sections of a station's operating log.
Tell what is in a station's public inspection file.
Explain the basic procedures in conducting a community needs and ascertainment survey.
List the steps in a typical license renewal.
Discuss cross-ownership rules.
List the steps in starting a new station.

Most businesses, whether regulated or not, must keep certain records. Accounting ledgers, tax returns, and profit-loss statements are essential to any enterprise. Businesses which are highly regulated, such as broadcasting, are burdened by even more "paperwork." Although broadcasters complain that the paperwork drains their time and is a considerable expense, the reason the FCC requires such detailed information is to assure operation in the public interest. Station program logs, for example, tell what programs aired at what time, what commercials were broadcast, when news programming aired, what time the station signed on and off the air, whether a program was live or recorded, and whether that program was news, entertainment, or something else.[1] Such records help prevent an unscrupulous broadcaster from interfering with other stations, refusing to meet the promises stated in the license renewal, overcharging advertisers, and other violations of FCC rules.

In chapter 14, along with examining program logs we also shall discuss operating logs, the station's public inspection file, conducting community needs and ascertainment surveys, license renewal, cross-ownership rules, and steps to starting a new broadcasting station.

program logs

FCC regulations require every radio and television station to keep logs detailing what they broadcast to the public. Since our discussion of logs will be in general terms, you should consult the specific FCC regulation before actually making entries in a log or examining it to see if it meets precise FCC standards. Logging requirements, although fairly standard, are fulfilled in different ways by different stations.

FCC rules require program logs to be kept by someone who is familiar with FCC standards, in most cases, the person operating the station's control board. A secretary not in programming is not permitted to make log entries. Neither is a general manager sitting in a back office, although the general manager, program director, or representative of the licensee can make corrections on a program log when the person who made the original entry is not available.

Regulations also require that logs be legible and organized in an easily accessible location at the station. An FCC inspector would have little sympathy for logs containing illegible entries, logs scattered around the station at different locations, logs not filed in chronological order, or worse yet, missing logs. Some stations make duplicate copies of logs just to protect against loss or damage. More and more stations are using automatic logging systems (Figure 14–1) which take many of the logging tasks, although not the responsibility, away from the person operating the control board.

To understand program logs, let's examine the one in Figure 14–2. It illustrates one type of program log which meets basic FCC requirements. Program logs also must have a cover sheet which explains all of the categories, sources, and types of programs which are abbreviated on the individual program entries.

At the top of the program log is the station's call letters (in this case WXXX), the name of its parent corporation, and the town in which it is licensed. The log is clearly identified ("program" log) to differentiate it from other station logs which we shall learn about later. The log also tells the page number, (page 2), the day of the week, (Wednesday), the date, (7/16/80), and the time zone, (Eastern Standard Time [EST]). Logs always must be kept in *local* time. To the left of the first column is a number for each line. This is not required by the FCC but can be helpful in giving directions to the operator. For example, if a mistake is made on the log, the program director can leave a note for the board operator referring to a specific line and page number on which the correction is to be made.

Figure 14–1 (Harris Corporation, Broadcast Products Division, P.O. Box 4290, Quincy, Illinois 62301)

```
 TIME       DUR   VIDEO  AUD   MID#              COMMENTS         FCC
--------   -----  -----  ---  --------    ----------------------  --
 6:05:23AM 15:18  TST                     TEST PATTERN             LO

 6:20:41*   0:44  VT1 MS        S0IJ7890  SIGN ON V918 CUT 1        I
 6:21:25    5:00  BRS    C1               COLOR BARS               LO
 6:26:25    1:55  VT2 /M         UGGH2345 NATIONAL ANTHEM           S
 6:28:20    0:30  VC1            CWRT9876 HALLELUAH                 P
 6:28:50    0:30  VC1            CWRT3324 UNITED WAY CRUSADE #1     P
 6:29:20    0:30  VC1            INHS5149 CHANNEL 97 MOVIE          P
 6:29:50    0:10  F3SX   C2      INHS5381 ID BUMPER                 I
 6:30:00*  -MAN-  NET MM                  MORNING COFFEE SHOW 1    NE
[6:42:58]   0:30  FIA FF         IERD0198 ZONK COMM'L               C
[6:43:28]   0:30  FC1            FISH2645 TANKORISHNESS COMM'L      C
[6:43:58]   1:00  VC1            ARGH1480 LARGO NUT CO. NO. 1       C
[6:44:58    0:02] F3SX   #    #  NUMB0045 ID SAFETY                 I
 6:45:00*  13:00  NET MM                  MORNING COFFEE SHOW 2    NE
 6:58:00    1:00  VC1 FF         DLIX1694 US ARMY                   S
 6:59:00    0:45  VC1            DSNU5777 THREE TERMITES            C
 6:59:45    0:05  FIS KI    #    IXOU6789 INSERT I SLIDE-MATTE
 6:59:50    0:05  VC2 KS  ---    MNUV6790 INSERT 2 SELF KEY
 6:59:55    0:05  VC1                     THREE TERMITES CONT.
 7:00:00*  -MAN-  NET MM                  MORNING COFFEE SHOW 3    NE
```

STATION WXXX DAILY PROGRAM LOG

Chevrolac Broadcasting Co., Inc.

Littletown, Plainstate

page 2

day Wednesday

date 7/16/80

time EDT

6. Commercial Matter or Announcement Type: Commercial Matter (CM); Public Service Announcement (PSA); Mechanical Reproduction Announcement (MRA); Announced as Sponsored (√).
7. Program Source: Local (L); Network (Identify); Recorded (REC).
8. Program Type: Agricultural (A); Entertainment (E); News (N); Public Affairs (PA); Religious (R); Instructional (I); Sports (S); Other (O); Editorials (EDIT); Political (POL); Educational (ED).

	Station Identification Time 1	PROGRAM TIME Begin 2	PROGRAM TIME End 3	PROGRAM TITLE — SPONSOR 4	Commercial Matter or Announcement Duration 5	Commercial Matter or Announcement Type 6	PROGRAM Source 7	PROGRAM Type 8
1–	8:00	8:00	9:00	RHYTHM MELODIES			REC	E
2–				James Brothers	60	CM		
3–				XYZ Laundry	60	CM		
4–				Alan Tires	60	CM		
5–				ABC Ice Cream	30	CM		
6–				Red Cross		PSA		
7–				Sureway Food	60	CM		
8–				Stop-Start Driver Training School	60	CM		
9–				Shady Hill Summer Theatre	60	CM		
10–		8:30	8:35	NEWS HEADLINES - Country Journal	1:30	CM	L	N
11–				John's Donut Shop	60	CM		
12–				Blackacre Real Estate	60	CM		
13–				Wright Insurance	60	CM		
14–				Rong Shoe Store	60	CM		
15–	9:00	9:00	9:14	JOE SMITH DEM. County Democratic Com.			L	PA-POL
16–				Cosmo Drugs	30	CM		
17–		9:15	9:28	FARM REPORT Coles' Tractor Co.	3:00	CM	L	A
18–		9:29		Announ. per Sec. 73.1202				
19–		9:30	9:59	LITTLE ORPHAN PUNJAB			MBS	
20–				Frank's Carpet Center		CM		
21–	10:00	10:00	10:29	LITTLETOWN LIBRARY TOPICS			L	I-ED
22–				Petite Clothes	60	CM		
23–		10:30	10:44	HEAVENLY MOMENTS - Coun. of Churches			L	R
24–				Lehi Beverage Co.	60	CM		
25–		10:45	10:59	MAN ON THE STREET Ford's Used Cars	3:00	CM	L	PA
26–				John's Garage	60	CM		
27–	11:00	11:00	11:24	Coca Cola Melodies	3:30	CM	REC	E
28–				Tony's Pizzeria	60	CM		
29–		11:25	11:29	MORNING HEADLINES			L	N
30–				Sta. Promo - Sports Windup (Schmaltz Beer)	10	CM		
31–		11:30	11:59	JOHN'S OTHER LIFE			MBS	
32–				Ray Hay Rep. Back Hay Com.	20	CM		
33–				Weekday Religious Education		PSA		
34–	12:00	12:00	12:14	MID-DAY NEWS			MBS	
35–		12:15	12:30	AIR FORCE TUNE TIME			REC	E
36–				Air Force Recruiting		PSA		
37–				Air Force Recruiting		PSA		

On	9. Operator or Announcer *Wm Schultzing*	Off 9:00	On 9:00	9. Operator or Announcer *Gerald McBuck*	Off
On	9. Operator or Announcer	Off	On	9. Operator or Announcer	Off

Comments: ABC Ice Cream spot was not run during Rhythm Melodies and log-keeper forgot to delete entry. *Bob West, Program Manager* WXXX 7/17/80

Figure 14–2 (National Association of Broadcasters)

Times, Program, and Station Identification

The first column on the log indicates the time of station identifications (ID). Station identifications are required when the station begins and ends its programming day. Sometimes stations are forced off the air unexpectedly, such as when lightning strikes the antenna. The minute they return to the air, they must broadcast an ID and make a notation in the "Remarks" section of the interruption. Station identifications must also be broadcast "hourly, as close to the hour as feasible, at a natural break in the program."[2] Some exceptions are permitted. If a station is broadcasting a symphony orchestra concert, interrupting with a station ID would not exactly thrill the serious listener. In these cases, IDs can be given as close to the hour as possible but at a natural break in the programming, such as at intermission or at the end of a particular selection.

A station ID must consist of the call letters and the name of the community served by the station. Television stations also must announce their channel. The station may insert the name of its licensee between the call letters and the community, such as "This is WXXX, Tower Broadcasting, in Anytown." But that is the *only* fact that can be inserted. Additional information can be added before and after the mention of the call letters and the community, such as "At 1410 on your dial, this is WXXX in Anytown, the voice of the Wabash Valley."

The FCC also prohibits stations from "inflating" their primary service area by substituting a town in the ID other than the one to which the licensee is assigned. For example, a Newark, New Jersey station could not announce "This is WXXX serving greater New York City." Even though Newark is across the river from New York and might be heard in New York, it is still *licensed* to Newark. Thus, the word *Newark* must be used with the call letters.

This does not mean that all announcements must avoid reference to their larger metropolitan area. For example, if our hypothetical station in Newark could show that its broadcast contour actually covered New York City, then it could identify itself as "This is WXXX, Newark, serving the greater New York City area." Notice, however, that the city of license was mentioned in conjunction with the call letters. In addition, when the FCC has licensed a station for dual-city designation, then the name of both cities can be announced in the order that they are listed on the station's license. This occurs in such twin cities as "This is WXXX, Minneapolis-St. Paul."

Classifying Programming: Type and Source

The next two columns on the log are the begin and end times of a program. Examine line 10 and column 4 of our sample log. The program title is NEWS HEADLINES. We can tell from column 2 that the program started at 8:30 and ended at 8:35. Column 4 also tells us that the sponsor for "News Headlines" is *Country Journal.* Logs require identification of the program and its sponsor(s). To distinguish the program from the sponsor, we have typed the program in capital letters and the sponsor in lower case letters.

Column 6 lists the actual duration of the commercials. For example, we can see by examining line 10 and column 6 that *Country Journal* commercials ran for 1 minute and 30 seconds during "News Headlines." This might have occurred as one 60-second and one 30-second commercial. Most commercial announcements run in ten-second multiples. The short, 10-second ones are often promotional announcements, reminding listeners, for example, that a grand opening is taking place or a special sale will start next week. Although 30- and 60-second announcements have been popular with most sponsors, the recent high costs of advertising time on radio and television have made the 10- and 20-second commercials a hit.

Another important part of the log is a listing of the commercial type, found in column 6. Most of the announcements in our sample log are listed as CM, which, by consulting the top of the log, we see stands for commercial matter. Another common symbol is PSA, standing for Public Service Announcement. In examining line 36 and column 6, we see that a PSA is logged for Air Force Recruiting. Stations are required to make available a certain amount of time for airing PSAs. These are free to whatever nonprofit organization is fortunate enough to receive the time. Although our example was for a national organization, local, nonprofit organizations in the station's home community also qualify for public service time.

Column 7 on our sample log lists the program's source. Three main categories of source are used by the FCC—local, network, and recorded. Letters are used to abbreviate the source with a network abbreviation often identifying the actual network. In our sample log, line 34 and column 7 tell us that the program originated from the Mutual Broadcasting System (MBS). The FCC defines these sources as:

1. *Local:* Any program originated or produced by the station, or for the production of which the station is primarily responsible, employing live talent more than 50% of the time. Such a program, taped or recorded for later broadcast, shall be classified as local. A local program fed to a network shall be classified by the originating station as local. All non-network news programs may be classified as local. Programs primarily featuring records or transcriptions shall be classified as recorded (REC) even though a station announcer appears in connection with such material. Identifiable units of such programs which are live and separately logged as such may be classified as local (e.g., if during the course of a program featuring records or transcriptions, a non-network two-minute news program is given and logged as a local news program, the report may be classified as local).
2. *Network* (NET): Any program furnished to the station by a network (national, regional or special). Delayed broadcast programs originated by networks are classified as network.

3. *Recorded* (REC): Any program not otherwise defined, above, including, without limitation, those using recordings, inscriptions or tapes.[3]

In our sample log, we can see on line 10, column 7 that "News Headlines" is logged as local (L) since it was produced by the station and used live talent. Line 1, column 7 shows "Rhythm Melodies" logged as recorded (REC). Even though the disc jockey might make announcements during the program, the program consists primarily of recorded material in the form of records.

Column 8 refers to program types. This is the most complex of all the logging requirements because of the wide variety of programming types available. Programming types fall into one of the following FCC categories:

Agricultural Programs (A) include market reports, farming or other information specifically addressed, or primarily of interest, to the agricultural population.

Entertainment Programs (E) include all programs intended primarily as entertainment, such as music, drama, variety, comedy, quiz, etc.

News Programs (N) include reports dealing with current local, national, and international events, including weather and stock market reports; and, when an integral part of a news program, commentary, analysis, and sports news.

Public Affairs Programs (PA) include talks, commentaries, discussions, speeches, editorials, political programs, documentaries, forums, panels, round tables, and similar programs primarily concerning local, national, and international public affairs.

Religious Programs (R) include sermons or devotionals; religious news; and music, drama, and other types of programs designed primarily for religious purposes.

Instructional Programs (I) include programs other than those classified under Agricultural, News, Public Affairs, Religious or Sports involving the discussion of, or primarily designed to further an appreciation or understanding of, literature, music, fine arts, history, geography, and the natural and social sciences; and programs devoted to occupational and vocational instruction, instruction with respect to hobbies, and similar programs intended primarily to instruct.

Sports Programs (S) include play-by-play and pre- or post-game related activities and separate programs of sports instruction, news or information (e.g., fishing opportunities, golfing instructions, etc.).

Other Programs (O) include all programs not falling within other definitions.

Editorials (EDIT) include programs presented for the purpose of stating opinions of the licensee.

Political Programs (POL) include those which present candidates for public office or which give expressions (other than in station editorials) to views on such candidates or on issues subject to public ballot.

Educational Institution Programs (ED) include any program prepared by, in behalf of, or in cooperation with, educational institutions, educational organizations, libraries, museums, PTA's, or similar organizations. Sports programs shall not be included.[4]

Program-type loggings become very important at license renewal time. Each station must submit logs for a composite week of broadcasting as part of its renewal application and must determine the percentages of programming on those logs devoted to each programming type. These then are compared by the FCC to see if the station did devote the time to programming types that it "promised" in its previous license renewal.

A program log also tells us who the board operators were and what time they signed on and off the log. Under the section marked *Comments,* we can see a program change for ABC Ice Cream that was written in by the program manager. On line 5, we can see a correction has been listed. In this case, the commercial for ABC Ice Cream did not air. If a correction is made while the operator is on duty, all that is necessary is to draw a single line through the entry. If the correction is made after the operator signs off the log, then a special notation must be made in the comments section, and be dated and signed by either the operator who made the correction, or a representative of the licensee, the general manager, or the program director. Some stations have started using a special statement at the bottom of each log adjacent to the operator's signature, similar to, "I hereby certify that this log is an accurate and true representation of that material broadcast during the period I was on duty as a station operator."[5]

operating logs

Besides logging broadcast programming, FCC regulations require that logs be kept for the operation of the station's transmitter.[6] This includes regularly scheduled monitorings of power output, voltage measurements, and such operating functions as monitoring tower lights. Our discussion will be confined to the very simplest operating logs, those for standard radio-broadcast stations. More complex logging is required for television stations and directional radio stations.

Figure 14–3 illustrates a basic operating log for an AM radio station. Like the programming log, it contains the station's call letters, its location, the date, and the city and state of its license. The time of each entry is logged in the left column. The person making the entry must be licensed by the FCC. We shall learn more about licenses later in this chapter, but for now keep in mind that whereas programming logs may be kept by people who do not have an FCC license, operating logs may not. As with programming logs, operating logs must be legible and must be available for inspection.

For an AM or FM station with no special logging requirements, an operating log is rather simple. Besides those for the time of the entries are columns for three "power" readings affecting the operation of the transmitter. Although it is not important at this point to understand the precise meaning of these terms, they include the plate current (column 2) the plate voltage (column 3), and the antenna current (column 4).

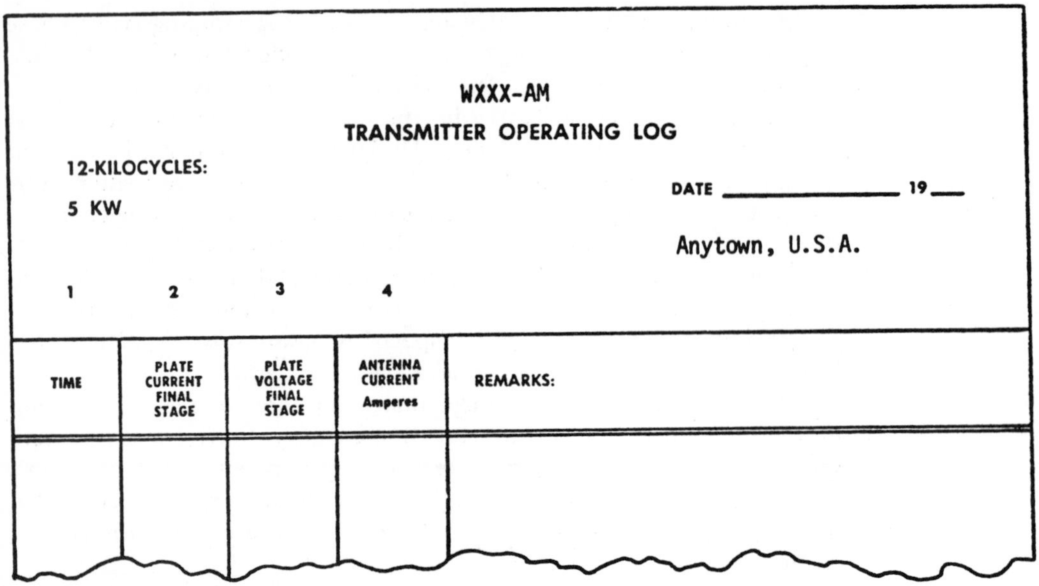

```
                          WXXX-AM
                  TRANSMITTER OPERATING LOG

12-KILOCYCLES:
                                            DATE _____ 19__
 5 KW
                                            Anytown, U.S.A.


  1          2          3          4

        PLATE      PLATE      ANTENNA
        CURRENT    VOLTAGE    CURRENT    REMARKS:
TIME    FINAL      FINAL
        STAGE      STAGE      Amperes
```

Figure 14–3 (National Association of Broadcasters)

Although not shown in our figure, the operating log also has spaces to note the times the carrier (power to the antenna) is turned on and off at the beginning and end of the broadcast day, and spaces for the licensed operators to sign on and off the log.[7] Additional information requested from most AM and FM stations includes the time the tower lights are turned on and off, the time they are checked daily for satisfactory operation, and if a light is extinguished for malfunctions.

Different modes of operation for certain stations also must be logged. A *mode* can best be defined as *a particular combination of transmitter, operating power and antenna pattern.* Consider the following example: "If an AM station is operating at 1,000 watts with a nondirectional antenna, that is its mode of operation. If it reduces its power to 250 watts and changes to a directional pattern at sunset, that is a different mode of operation. If it shifted from a main to an alternate transmitter at midnight, that is a third mode, and if at 6:00 A.M. . . . . it increased power to 500 watts, that is a fourth mode of operation."[8] Remember, any change in the mode of operation must be entered in the operating log.

the public inspection file

As the trustee of the public domain, broadcasting stations are required by law to keep certain documents and information open for public inspection. This means members of the general public are entitled to

338

documents from the file. A station may be visited any time during normal business hours, and although the station has the right to ask for personal identification, you should not be "interrogated" about your motives for wanting to see the file. If you would like certain documents copied, then the station can charge you a reasonable fee to have the material reproduced.

What is contained in the public file varies somewhat among different types of stations—AM, FM, and TV as well as noncommercial versus commercial stations. So be sure the station is required to keep a certain document before you request to see it. Your best bet in this instance is to consult the FCC rules. Our discussion of public files will be general, and remember, it varies from station to station.

A public file contains technical information directly related to the construction and daily operation of the station. Construction permits, major changes in frequency, output power, a change in station location or the transmitters are typical inclusions. Do not expect to find minor technical information, such as pointers on the new antenna support wires, information about a new control board, or information about the new record racks. If a new construction permit has been granted, and the FCC grants an extension of the permit, the extension is in the public file. The file also will include correspondence related to these changes. A copy of the station's coverage area (contour maps) should be there, as should reports listing the ownership of the station, and any FCC decisions arising from a hearing on the station's license renewal. A copy of the license renewal and the logs submitted as part of the renewal's "composite week" should also be available.

Politicians will be interested in examining the file's political documents. Most of what a station does in the way of political programming is an open book to the public, including the candidates and their opponents. Requests for political time by legally qualified candidates, a record of what was done with the request, and the rate charged for that time are kept for two years from the date of request. The spirit of the law behind the political file is to keep access to the airwaves open to any and every legally qualified candidate. It prevents an unscrupulous broadcaster or politician from claiming that a candidate has not talked with the station nor has bought any advertising, thus discouraging an opponent from buying time when in fact the candidate has purchased and aired a series of political commercials.

Other information in the file includes the FCC procedural manual, *The Public and Broadcasting,* and copies of letters from the public unless they are obscene or the sender specifically requests that they be confidential. Letters of little importance to the station, such as love letters to a movie star or fan mail for the local anchorperson, may be absent. What should not be missing is a copy of the latest listing of problems affecting the community as determined from the community needs and ascertainment survey.

Although not considered part of the public file, program logs or copies of them are open for public inspection, beginning 45 days after the date on the log. The requirements for viewing program logs are stricter. You will need to make an appointment, identify who you are and whom you represent, why you want to see the logs and, if you are part of a large group of people wanting to view the logs, the station may limit the number of people viewing them. You can obtain copies of the logs but probably will be asked to pay for reproduction. You will be given a reasonable time to inspect the logs, but if you want to come back again, the station may charge you for the time necessary for personnel to supervise your efforts.

community needs and ascertainment surveys

Before public or commercial broadcasting stations can be granted an operating license, have that license renewed, or even continue to operate while the license is in force, it must conduct regular surveys of the problems facing its community and direct its programming to meet the needs of those problems. These community needs and ascertainment surveys are another means of obtaining feedback from the broadcasters' communities. They can be quite detailed, depending on the size of the station's market. Their importance cannot be overlooked, considering that such surveys have been used as evidence in license challenges.[9]

The guidelines broadcasters follow in conducting the community needs and ascertainment surveys were first spelled out in the *Primer on Ascertainment of Community Problems* issued by the FCC in 1971.[10] Further clarifying those guidelines in 1975, the FCC added noncommercial broadcasting stations to the list of those required to conduct the surveys.[11]

The ascertainment process has three parts, the first of which is a *demographic profile.* After checking census data, the broadcaster determines the population of the community served by the station, the percentage of males and females in the population, the percentage of minorities, the percentage of older people (over 65), and the percentage of youths (under 17). This demographic profile shows the broadcaster what proportion of people will provide a good cross section of information about the community's problems. For example, if the demographic profile shows that 30 percent of the residents are over 65 years of age, yet only 5 percent of the station's general public survey consists of older people, the broadcaster will need to conduct additional interviews with this population. Although the FCC has avoided requiring broadcasters to match their survey precisely with the demographic profile, the profile does act as a guide.

The second part of the community needs and ascertainment survey is *community leader interviews.* Here the broadcaster interviews the leaders of different elements in the community. The FCC has established a general list of community elements from which the community leaders can be drawn:

1. Agriculture
2. Business
3. Charities
4. Civic, neighborhood, and fraternal organizations
5. Consumer services
6. Culture
7. Education
8. Environment
9. Government (local, county, State and Federal)
10. Labor
11. Military
12. Minority and ethnic groups
13. Organizations of and for the elderly
14. Organizations of and for women
15. Organizations of and for youth (including children) and students
16. Professions
17. Public Safety, health, and welfare
18. Recreation
19. Religion
20. Other

If you were the general manager of a radio or television station and were interviewing people who represented your community's educational elements (7), you might interview the local college president or administrators and perhaps some professors. You might also interview local school board members, the principal of the local high school, the principal of the local elementary schools, teachers, and officers in the local parent-teacher organizations. How many of these community leaders you interviewed would depend on the size of your community. A good rule-of-thumb would be the following combinations suggested by the FCC:

Population of City of License	Number of Consultations
10,001 to 25,000	60
25,001 to 50,000	100
50,001 to 200,000	140
200,001 to 500,000	180
Over 500,000	220

Although you would probably conduct more community leader interviews, the FCC expects stations to fulfill at least these minimum requirements.[12] The community leaders must be contacted by station management or personnel under direct management supervision. You would need to keep track of how many women and minority community leaders you contacted, the recommendation being that those interviews be conducted directly by management-level personnel.[13]

You also would need to place in the station's public file the (a) name and address of the community leader; (b) institution or element he or she represents; (c) date, time, and place of the interview; (d) problems, needs, and interests discussed (although the leader can request that this information be confidential); (e) name of the person who conducted the interview (if the interviewer was a supervised person, the name of the management-level person who reviewed the interview report); and (f) the date the report was reviewed.

The third phase of ascertainment would be the *general public survey.* Here you need to select a random sample of the community. You would interview each person in that sample either in person, by telephone, or by mail.[14] Whatever method you chose, you would want not only to find out their opinions about the community, but also obtain their demographic characteristics. Again, the information in your demographic profile would be your guide, and you would want to match this as closely as possible to be sure you had a representative cross section of the general public.

You would probably ask one of two types of questions—open-ended or close-ended (Figure 14–4). Open-ended questions permit the greatest flexibility in their answer.[15] Consider the following: *What do you feel is the most important problem facing our community?* The person answering this open-ended question has many choices. Now consider this close-ended question: *Is there a problem with public transportation in the community? Yes☐ No☐.* The only acceptable answer is either yes or no. The advantage of the open-ended question is its less restrictive nature. Its disadvantage is

Figure 14–4   (Corporation for Public Broadcasting)

11.	ARE THERE PARKING PROBLEMS?	YES	NO
12.	IS THERE A PROBLEM WITH URBAN DEVELOPMENT?	YES	NO
13.	IS THERE A PROBLEM WITH POLICE PROTECTION?	YES	NO
14.	IS THERE A HOUSING PROBLEM?	YES	NO
15.	IS CRIME A PROBLEM?	YES	NO
16.	IS POPULATION A PROBLEM?	YES	NO
17.	ARE THERE WELFARE PROBLEMS?	YES	NO
18.	IS THERE A PROBLEM WITH ELECTRIC UTILITIES?	YES	NO

the difficulty in tabulating the various answers. Whatever questions you decide to ask, you would then conduct the survey and organize your results, ranking in importance those problems affecting your community.

license renewal

Standard AM, FM, and TV stations are licensed by the FCC. Each license is valid for three years. The license, also called *station authorization* or *instrument of authorization,* specifies the authorized power of the station, its hours of operation, the brand name and model of the transmitter and the antenna, the location of the transmitter, the latitude and longitude of the antenna, and the name and address of the licensee. Television and AM licenses contain such additional information as directional antenna patterns or video transmission frequencies. The expiration date is coordinated to expire along with all other stations' licenses in the state.

A license is the single most important document the station possesses. Every three years, the station applies to renew that license and continue operating. The license renewal is essentially a forecast of what the station will do over the next three-year period in terms of entertainment and commercial programming, news and public affairs broadcasting, affirmative action, and in other specific terms required by the FCC. Once forecasted, the station must adhere to those predictions. Although the exact application procedures are slightly different for radio and television and for commercial and noncommercial broadcasting, we'll examine a typical radio-license renewal form, Form 303-R, commonly called the short form.

Parts I through III: General Information, Legal, Engineering

Part I of Form 303-R asks for general information, including the name of the applicant and an address to which communication regarding the the renewal application should be sent, because the person handling the renewal may not be at the station, as in the case of absentee ownership or central offices handling renewals for their group of stations. The second section of Part I requires the call letters, frequency and channel,[16] power, hours of operation, location of the station, and whether renewal is requested for subsidiary communication authority. SCA, as you remember, is the subcarrier frequency used, among other things, to pipe music into businesses or restaurants.

Part II is the "legal" part of the renewal. The commission wants to see a copy of the station's ownership report, and in item 5 of Part II asks if "the applicant is in compliance with the provisions of Section 310 of the

Communications Act ... relating to the interests of aliens and foreign governments." Section 310 prohibits licensing a broadcast station to:

> *Any corporation of which any officer or director is an alien or of which more than one-fifth of the capital stock is owned of record or voted by aliens or their representatives or by a foreign government or representative thereof or by any corporation organized under the laws of a foreign country.*

Guarding against the threat of monopoly, item 6 of Part II checks to see if the applicant owns other media outlets, and item 7 of Part II examines the applicant's qualifications. A licensee convicted of a felony or other unlawful activities can make a license renewal difficult, if not impossible. The commission wants full details of any convictions or charges including "identification of the court or administrative body" handling the proceedings. Part III of the renewal application is on engineering and is verified by the station's technical director, the chief operator, consulting engineer, or a registered professional engineer.

Parts IV and V: Programming and Equal Employment Opportunity

Part IV is one of the most important sections. Here the FCC examines a station's community ascertainment efforts, its community leader checklist, and the percentages of different types of programming, down to the minute.[17] The FCC carefully studies what was promised in the previous application, how well the station met those promises, and what is promised for the next three years. More than any other part of the license renewal, failure to perform as promised will place the application in serious trouble.

The FCC also wants to see if the applicant stayed within the acceptable 18 minutes of commercial time per hour[18] and the percentage of programming which may be duplicated with other stations, such as a simulcast by an AM and FM station licensed to the same applicant. The 18-minutes-per-hour guideline is just that, a guideline. License renewal applications containing more than 18 minutes of commercial time an hour for radio and more than 16 minutes for television receive close scrutiny to determine if they are operating in the public interest. But the FCC has no power to dictate the amount of commercial time, since that would violate the anticensorship provision of the Communications Act. This license-renewal process, however, generally "encourages" stations to stay within acceptable commercial limits. Part IV concludes with item 22, for additional information, permitting the applicant to call to the FCC's attention anything else that may reflect on its qualifications for renewal.

A separate section of Part V deals exclusively with the station's affirmative-action program. Item 23 of Part V notes the model, 10-part affirmative-action program, and item 24 informs the FCC of any affirmative-action complaints brought against the station as well as the status of those complaints. Serious, unresolved affirmative-action issues can provoke the issuance of a short-term license renewal, and although broadcasters cannot be threatened by unwarranted complaints, blatant disregard for affirmative-action policies can spell trouble.

The quickest way to have a license revoked is to lie on the renewal application. Not only will the station face license revocation, but the United States Criminal Code also provides for additional penalties. For most broadcasters, responsible operation of a broadcasting station on a daily basis turns a license renewal into almost a routine procedure.

crossownership

Although many broadcasting stations are profitable ventures, ownership is regulated by the FCC to avoid monopolies. The FCC rules include (1) the seven-station rule, (2) the duopoly rule, (3) the one-to-a-market rule, (4) the regional concentration-control rule, and (5) the newspaper-broadcast ownership rule.

Seven-Station, Duopoly, and One-to-a-Market Rules

One of the most inflexible rules is the seven-station rule. It prohibits an owner from having more than seven stations of any one type—AM, FM, or TV. Thus, a total of 21 stations is permitted to the same owner. Only five of the TV stations can be VHF. The duopoly rule prohibits crossownership when two stations of the same type (such as two AMs) have certain overlapping contours.[19] Directly related to the duopoly rule is the one-to-a-market rule which prohibits a radio-TV crossownership in which certain contours of the radio and television station overlap. An owner can operate an AM/FM combination in the same market.

Regional-Concentration Rule

Guarding against a monopoly of viewpoints over what the public receives in any given area, the FCC prohibits regional concentrations of ownership. To understand the regional-concentration rule, imagine an owner with three stations in three different markets. In drawing a triangle connecting the three markets, we discover that one side of the triangle is 100 miles long. If this is the distance, and the primary contours of any of the stations overlap on any side of the triangle, the crossownership component is illegal. Naturally, when the stations were purchased or started, the FCC would have prohibited the concentration in the first place. Thus, finding an illegal crossownership is highly unlikely. The regional-concentration rule also is applied frequently when an owner wants to increase the power of one of the stations. Even though the owner may be within legal limits operating the stations less than 100 miles apart, increasing the power of one of the stations may result in the contours overlapping on one side of the triangle; thus, the power increase would place the station in violation.

Newspaper-Broadcast Crossownership Rule

If the seven-station rule were considered the most inflexible, the newspaper-broadcast crossownership rule could be considered the most controversial. The controversy surfaced in 1977 when the U.S. Court of Appeals in Washington, D.C. ruled against an FCC policy of not requiring long-standing newspaper-broadcast crossownerships to be dissolved but prohibiting certain new ones from being formed.[20] Brought to the court by the National Citizens Committee for Broadcasting, the case placed hundreds of millions of dollars of crossownerships at stake, leaving an appeal to the U.S. Supreme Court the only alternative for over 150 newspaper-broadcast crossownerships. Except when there was a clear indication that the public interest would be harmed if the newspaper-broadcast crossownership continued to exist, the FCC permitted those crossownerships to stand. The appeals court took an opposite view to the long-standing FCC policy. Refuting the rationale of allowing existing crossownerships to continue, the appeals court said: "We believe precisely the opposite presumption is compelled, and that divesture is required except in those cases where the evidence clearly discloses that crossownership is in the public interest."

Although buying and selling stations will continue, both FCC and court rulings have made it clear that owners can have no more than their share of properties and that the public interest will be served even if it means prohibition and disruption in the broadcast industry.

starting a new station

For those who can find the market and frequency available, starting a station from scratch is still an attractive venture. Although the job is complex and the competition for a new station in a multistation market substantial, many still try it and succeed.

Preliminary Steps

The first step in starting a new station is to find an area in which a frequency is available. For an AM radio station, the search will mean not only consulting the engineering data of stations already in the market, but also having a qualified engineer conduct a frequency search. The frequency search entails checking the exact broadcast contours of stations currently serving the area and determining what type of signal will not interfere with those already operating. Thus, researching possible wattage, contour patterns, and available frequencies all must proceed the application process.

FM radio and TV starts are a bit different. An applicant for an FM radio license either must select an available frequency already assigned by the FCC to the area in which the applicant wants to operate or a place within a specified radius to which no FM frequency has been assigned. TV applicants must request a UHF or VHF channel assigned either to the community or to a place in which there is no channel assignment within 15 miles of a community.

Once the frequency search has been completed, the next step is a community needs and ascertainment survey. We already learned about these surveys earlier in this chapter.

Construction Permit to License

When the community needs and ascertainment survey is completed, the applicant applies to the FCC for a construction permit. The applicant also must have the financial capability to operate the station for at least one year after construction. Notice of the pending application must be made in the local newspaper, and a public inspection file must be kept in the locality in which the station will be built. Once filed with the FCC, others can comment on the application or, in the case of competing applicants, file against it. If necessary, the FCC will schedule a hearing on the application. Following the hearing, the FCC Administrative Law Judge will issue a decision which can be appealed.

If everything in the application is found satisfactory, and there are no objections, the FCC then issues the construction permit. Construction on the station must begin within sixty days after the date the construction permit is issued. Depending on the type of station, a period of up to eighteen months from the date the construction permit is issued is given to complete construction. If the applicant cannot build the station in the specified time allotted, then the applicant must apply for an extension in time.

After the station is constructed, the applicant then applies for the license. At this time, the applicant can also request authority to conduct program tests. These tests will usually be permitted if nothing has come to the attention of the FCC to indicate the operation of the station would be contrary to the public interest. When the license is issued, the station can go on the air and begin regular programming.

Although the procedure is somewhat systematic, putting the station on the air is anything but simple. The paperwork, dealing with engineers and communication attorneys, and securing the financing necessary not only to buy land and equipment but also to keep the station running for a year all can be difficult and time-consuming. If objections or competing applications become an issue, the court costs can discourage an applicant from completing the application process. Still, for those who do succeed, the rewards can be substantial, both in personal satisfaction and income.

summary

Program logs are an important part of every station operation. The FCC requires, among other information, that the station note the times of programs and station identifications and log all programming according to specific program categories. These include: agricultural programs, entertainment programs, public affairs programs, religious programs, instructional programs, sports programs, and other programs which do not fall into any other category. In addition, editorials, political programs, and educational institution programs also must be logged. Operating logs are also required and concern the technical operation of the station, such as the output of the transmitter and such things as antenna lighting and periodic inspections by qualified licensed engineers.

So the public can readily inspect the ability of the station to serve the public interest and additional information about station operations, stations also must maintain a public inspection file. Most stations must also conduct regular community needs and ascertainment surveys which become part of the station's public records as do license renewal applications. Such surveys require management to consult a cross section of the community, then to draw up a list of specific problems facing the community and how the station programming is directed to these problems. The FCC and the courts have established regulations on the number of stations that can be owned by the same firm or individual in a given market.

Crossownership has been an important issue in broadcast regulation, not only in the operation of stations already on the air but also of stations just starting. The courts have continually moved toward breaking up multiple ownership, and many owners have had to sell one or more of their broadcasting properties. The FCC has also established guidelines that stations must follow when constructing new or altering old facilities.

spotlight on further learning

Broadcasting and the Federal Lottery Laws. Washington, D.C.: National Association of Broadcasters, 1974 (periodically revised).
In the Matter of Nondiscrimination in the Employment Policies and Practices of Broadcast

Licensees. Washington, D.C.: Federal Communications Commission, 1976 (Docket No. 20550).

In the Matter of Revision of FCC Form 303. Washington, D.C.: Federal Communication Commission, 1976 (Docket 20419).

NAB Legal Guide. Washington, D.C.: National Association of Broadcasters, 1977 (updated periodically).

PALETZ, DAVID L., ROBERTA E. PEARSON, and DONALD L. WILLIS, *Politics in Public Service Advertising on Television*. New York: Praeger Publishers, Inc., 1977.

15

cable, satellites, and future perspectives

FOCUS   After completing this chapter we should be able to

Define the terms *cable system, cable television system community unit,* and *subscriber.*

Explain carriage exclusivity.

Understand the concept of state regulation of cable.

List the major responsibilities of the FCC as found in the Communications Satellite Act of 1962.

Grasp the human implications of satellite communication.

Trace the early support in Congress for rewriting the Communications Act of 1934.

Identify the major provisions which could occur in a Communications Act rewrite.

Although traditional broadcasting stations have always been under the control of federal agencies, new technology such as cable and satellites also have not been exempt from government regulation. With the beginning uses of microwaves to link cable systems, the FCC gained early court approval for control over cable systems. This control is shared with local communities and even states. It is the one area of broadcasting in which a single cable system can find itself faced with regulations from three different levels of regulatory authority: local, state, and federal. Satellite communication has also come under government scrutiny and has become directly involved in international issues such as cooperative and peaceful uses of space. With all of these changes in technology, including but not limited to satellites and cable, it is not surprising that Congress is having serious discussions on overall communications legislation. Efforts in both the U.S. House of Representatives and the Senate are directed toward rewriting the basis of the Communications Act of 1934 and replacing it with what some people feel would be a more up-to-date document. In this chapter we shall examine regulation of cable and satellite communication as well as proposals to rewrite the Communications Act.

regulating cable

In writing about cable television, Lee Loevinger noted, "Each new development of man's inventive genius is a threat to the timid and the indolent and an opportunity to the bold and enterprising. For the pessimist, the golden age is always in the past; for the optimist, it is in the future."[1] The regulatory background of cable television has alternated between the pessimist and the optimist, and the timid and the bold. From cable's beginnings in the late 1940s to its new technology of two-way communication in the 1970s and 1980s, many forces have been at work.

The FCC began exercising its authority over cable in 1962. In 1965, the commission established rules governing cable systems that received signals by microwave. A year later, the FCC added rules and regulations for cable systems not using microwave. Knowing a court case would soon test its authority to regulate cable, the FCC decided to prepare for the inevitable when it issued a decision limiting the ability of a San Diego, California cable system to import signals from Los Angeles. The test came in *United States* v. *Southwestern Cable Co.* in which the Supreme Court upheld the FCC's right to regulate cable as part of its mandate under the Communications Act to regulate "interstate commerce by wire or radio."[2] By 1968, the FCC had started an official rule-making proceeding to develop comprehensive regulations for cable, which it finally issued in 1972. In the midst of all of this, cable came under the aegis of not only the federal government but state and local governments as well. In examining cable regulations, we shall begin by exploring the FCC's definition of cable, local franchise requirements, and service to the public. Finally, we shall examine the people in the middle of cable regulation— the state authorities.

Definitions

To understand cable regulations, we first need to understand some cable definitions.[3] The FCC defines a *cable television system* as a "nonbroadcast facility consisting of a set of transmission paths and associated signal generation, reception, and control equipment that distributes or is designed to distribute to subscribers the signals of one or more television broadcast stations." The location of the cable television system is called the *cable television system community unit*, which is defined as "a cable television system, or portion of a cable television system operating within a separate or distinct community." The person receiving the cable system's services is the *subscriber*, defined by the FCC as a "member of the general public who receives broadcast programming distributed by a cable television system and does not further distribute it." The FCC points out that for regulatory purposes, these definitions do not include cable systems operating with less than 50 subscribers or multiple-unit dwellings.

Our discussion of cable regulations will be general rather than specific. The complexities of regulations affecting different communities and cable systems are written in difficult legal and technical terminology. However, if you are interested in the regulations will be general rather than specific. The complexities of regulations affecting a particular cable system, then you should obtain a copy of the latest cable regulations from the FCC. Always consult an attorney familiar with communication law should you desire precise interpretation of a rule.

Local Franchise Structure

The foundation of many cable regulatory concepts is found at the local level. Unlike over-the-air broadcasting, cable can be regulated by its local community, which has the authority to place certain service and operational requirements upon it, to levy fees, and to determine community-access channels. Types of local control vary considerably. Professor Vernone Sparkes studied these different types and classified them into five agency organizations. The first is an *administrative office*, in which the local government establishes a regulatory agency much like the FCC. It might be found in the mayor's office or the city planner's office. A second type is the *advisory committee* which can be appointed by the mayor or the city council and "advises" city government on cable regulation. Closely related to the advisory committee is an *advisory committee with administrative office,* which "combines an appointed advisory committee with a full-time salaried executive office." Sparkes points out that the executive usually works independently from the advisory committee, with the latter advising the city council on policy matters. A fourth organization calls for the creation of an *independent regulatory commission,* which administers and participates in rule making. A fifth plan provides for *elected boards* answering to the electorate on cable regulations rather than to another elected body.[4]

Local Franchise Registration Requirements

A license is not required to operate a cable system, but the FCC has established some registration procedures. Cable systems beginning operation must notify the commission of such facts as the legal name of the cable operator; the identification of the entity operating the system; whether the operator is an individual, private association, partnership, or corporation; and the legal name of the individual responsible for communicating with the FCC. Such data as the mailing address, date the system reached fifty subscribers, names of communities served, signals to be carried on the system, and certain employment reports must also be filed, including a statement of the proposed equal-employment opportunity program. When this information is complete, the commission announces the filing in a public notice. Smaller systems contemplating operation with between 50 and 1,000 subscribers have somewhat different registration requirements.

FCC Recommendations for Local Franchises

Although the FCC keeps a regulatory distance between itself and local communities engaged in awarding cable franchises, the commission has established *recommended* franchise standards for communities, as follows:

(1) The franchising authority should approve a franchisee's qualifications only after a full public proceeding affording due process.

(2) Neither the initial franchise period nor the renewal period should exceed fifteen years, and any renewal should be granted only after a public proceeding affording due process.

(3) The franchise should accomplish significant construction within one year after registering with the commission and make service available to a substantial portion of the franchise area each following year, as determined by the franchising authority.

(4)   A franchise policy requiring less than complete wiring of the franchise area should be adopted only after a full public proceeding, preceded by specific notice of such policy.

(5)   The franchise should specify that the franchisee and franchisor have adopted local procedures for investigating and resolving complaints.[5]

The FCC also advises the local franchisee to adopt a local complaint procedure, identify a local person to handle complaints, and specify how complaints can be reported and resolved.

Sensitivity to the Public

Meeting these franchise standards is not the end of a cable system's responsibilities. It still must be sensitive to its public. Complaints can always be filed with the FCC, the cable company, and local and state authorities. In addition, a special FCC Cable Complaint Service operates in Washington, D.C., as part of the Cable Television Bureau. Its functions include: "(1) attempting to clear up misunderstandings between subscribers (and/or state and local governments) and cable systems with regard to Commission rules; (2) dealing with complaints about a system's service; and (3) helping local governments to structure complaint procedures."[5] Local franchisees are encouraged by the FCC to adopt some form of local complaint procedure, identify a local person who will handle complaints, and notify new subscribers who this person is.[6]

Carriage Requirements

In servicing a community, a cable system faces certain requirements of what it can and cannot carry on its cable channels. Again, standards differ from community to community, and you should check the rules which apply to your area. The requirements are usually based on the size of the system, with systems of fewer than 3,500 subscribers free from some of the controls imposed on larger systems. For most cable systems, however, the service requirements are broken down into two areas: (1) "must carry" signals which are signals the cable system is required to carry at the request of a television station and; (2) "may carry" signals which are signals permitted on the system after the "must carry" signal requirements are met.[7] The new two-way cable services are being evaluated to see how they affect these current signal requirements.

The "must carry" generally entails carrying the signals of all television stations licensed to communities within thirty-five miles of the cable system's community. If a noncommercial television station's grade B contour (signal) covers the community, then the cable system is also required to carry that signal. By using the B contour as the criterion, all stations whose A contours touch the cable system are automatically covered, since the A contour falls inside the B. As we learned earlier, some stations boost their signal over wide geographic areas by using translator stations which receive and retransmit the programming from the main television transmitter. Thus, if either a commercial television translator station with at least 100 watts of power or a noncommercial translator station with at least 5 watts of power serves the community, then the cable system is required to carry those translator stations' signals.

Stations that are significantly viewed in the cable system's home community must also be carried. The term "significantly viewed" is defined by the commission as network stations with a minimum 3 percent share of viewing hours and 25 percent net weekly circulation. For independent stations, the criterion is a minimum 2 percent share and 5 percent net weekly circulation. In addition, cable systems in certain small markets must carry all stations from other small markets whose grade B contour covers the cable system's community.

Additional "must carry" requirements apply to those cable systems located in communities outside of a television market, such as a sparsely populated area of the country. In those communities, the cable system is required to carry all stations whose grade B contours cover the community in which the cable system is located, stations which are significantly viewed in the cable system's community, and all educational stations licensed to communities within thirty-five miles of the cable system's community. In addition, the system must meet the carriage requirements for translator stations mentioned earlier.

The "may carry" requirements are based on certain combinations of stations. Keep in mind that all television markets in the United States are ranked by the FCC according to their size. These market rankings are then the basis for "may carry" requirements. For example, cable systems in the *first fifty* major television markets, as defined by the FCC, may carry a compliment of three full network stations and three independent stations. In the *second fifty* major television markets, the compliment is three full network stations and two independent stations. *Smaller* television markets may carry three full network stations and one independent station, and systems *outside* all markets have no restrictions on number or type of television signals they may carry. Cable systems serving fewer than 1,000 subscribers per headend may carry any additional signals.

The FCC does permit exceptions to these four categories. For instance, in markets with an abundance of local stations, the cable system can add the local stations to its complement. Major markets may carry two "bonus" independent stations but must deduct this bonus from any distant independent stations that have been imported to fulfill the complement.[8] The FCC points out that their "must carry" and "may carry" rules do not capriciously give to one community the right to more television service than another. Instead, the differentiation is based on the fact that densely populated areas have a larger advertising base which can therefore support more television "stations" than sparsely populated areas can.[9] In communities with no television stations, the criterion becomes simply to bring television into these outlying areas.

Additional exceptions apply to late-night programming.[10] Cable systems are permitted to carry the late-night programming of any television station when local "must carry" stations sign off the air. Moreover, network news programs which are not usually carried by the cable system can be broadcast when no local station is broadcasting a news program.[11] The FCC also makes provisions for speciality television stations, such as those programming foreign-language, religious, or automated broadcasts.[12] In certain cases, these stations do not come under the definition of an independent station when meeting their required complement; thus cable systems can carry more signals if speciality stations are involved.

Program Exclusivity

In addition to these carriage requirements, the FCC has rules which protect a television station originating a signal. These program-exclusivity rules vary for different sized cable systems, but for the most part, a station can request its local cable system to carry its signal at the expense of a distant station's signal which may duplicate the local station's programming. To understand the rule, assume that you own television station WXXX, and your grade B signal covers Elmville, which is served by the Anytown Cable System. Anytown Cable also imports a distant station's signal which duplicates your programming. You have the right to request Anytown to carry your signal instead of the distant signal. The advantage for you is that your programming, and consequently your advertisers, reach a larger audience. This protection also applies for same-day broadcasting and is called same-day nonduplication protection. The rules vary somewhat for network programming, in which simultaneous nonduplication protection is provided in most time zones. Provisions also apply to syndicated programming, and the FCC states "that a cable system serving a community in the first 50 major markets, upon receipt of notification, cease carrying syndicated programming from a distant signal during a pre-clearance period of one year from the date that such programs are sold for the first time as syndicated programs in the United States."[13] An FCC Notice of Inquiry is currently reevaluating the syndicated exclusivity rules. Pay-cable systems charging a separate rate for a certain channel or movie received a big boost in 1977 when the U. S. Court of Appeals in Washington, D. C. struck down, in part, the FCC's pay-cable rules as an unwarranted intrusion on the First Amendment. Although some groups indicated an appeal was forthcoming, the case of *Home Box Office* v. *FCC* paved the way for pay-cable systems to show first-run movies and major sporting events previously prohibited in smaller markets.

Sports programs pose a special problem for cable systems. Consequently, the FCC has a separate set of guidelines which apply to nonduplication of sports programming. It also has a sports blackout rule which protects home teams not having sold all their tickets. Cable systems are prohibited from carrying local sporting events broadcast on distant television stations when the local station does not carry the events. To understand this rule, assume that you operate a cable system in Cedarville. The Cedarville Tigers football team is playing this Saturday, but the local station will not be carrying the game since it is not sold out. Yet you know your subscribers would jump at the chance to watch it on television. So you decide to carry the coverage of the game, not from the Cedarville television station but from a distant station not required to black-out the game. Would you be within the law? *No.* You would not be permitted to carry the game since the purpose of the FCC rule is to protect local television sports blackouts.

State Regulation of Cable

State government also controls cable.[14] However, state control is not widespread and varies in degree. State laws can be classified into three categories. First are preempt statutes. These are the strongest laws and take precedent over local regulations. With preempt statutes, cable falls under the jurisdiction of the Public Utility or Public Service Commission in some states. Based on many of the same criteria specified by the FCC and the local municipality, that commission may also issue a compliance or "license" to accompany the federal authorization to operate. In addition, preempt statutes give considerable clout to the state commission, permitting it to issue and enforce a separate set of state cable regulations. These rules can govern everything from the daily operation of the cable system, to collecting fees on gross revenue, even to demanding financial collateral before allowing construction. A second approach is for a state to pass appellate statutes. With these statutes, the local municipalities still retain some control over franchising, but the state has the power to review local agreements and be the final arbiter of disputes. Everything works well until the state and the municipality disagree, then the municipality has less than an even chance against the state. A less powerful state statute is the advisory statute. Advisory statutes are obviously more popular with cable systems and municipalities since they do not exercise either the clout or the enforcement power of a state commission. Some are best described as "general guidelines" which serve as a reference for local government.

It is still difficult to predict a trend in these state statutes. Proponents of state control argue for the need for consistency among cable systems within a state. Such arguments gain support when two municipalities cannot resolve their jurisdiction differences over a cable system or when significantly differing fee structures provoke public outcry. Control of cable can also be a political plum for legislators since it means control of a communication system, and communication influences public opinion. Since cable commissions can have a significant effect on cable growth within a state, appointment to the commission can be a sweet political reward for a party in power.

Arguments against state control are equally vociferous, asserting that it presents an unnecessary duplication of law. With duplication comes conflict. The state is caught between local and federal control, and meeting the requirements of one can violate the other. The only solution to this dilemma may be a long and expensive court battle. Other opponents claim that state control merely throws local interests into a political arena with representatives who are looking out for their own interests, not those of the local community. Still other arguments warn that if the state eventually becomes involved in direct programming, that the programming will be directed more toward propaganda than public interest.

State cable regulation seems less probable today than in the past. Many state statutes now in force were passed before cable's major growth and impact. Since then, the public has grown more aware of cable's importance, and lobbying groups such as the National Cable Television Association are more organized and effective in their opposition to state control. Yet lurking on the horizon is the proposed revision of the Communications Act, which may change the entire future of cable regulation.

control of direct broadcast satellites

In a special committee meeting at the United Nations, there are discussions on governing the use of satellites in transporting television signals to developing nations. In a small community in the Rocky Mountains, a school board discusses the ramifications of educational television signals beamed into the local high school via satellite communication. From hallways to hamlets, the new technology of satellite communication has altered many of our approaches to broadcast regulation. For example, what new laws and what effect on old ones will develop when television stations use satellites to beam their signals across the United States? What new regulatory issues will arise when a cable system in Nebraska provides live coverage of an Atlanta baseball game via satellite signals beamed from Georgia?

Domestic regulation of satellite communication finds its base in a number of laws. At the federal level is the Communications Act of 1934 which regulates common carrier communication, such as telephone and telegraph, as well as standard broadcasting. Satellites with their thousands of circuits, each circuit carrying a specific amount of information, find some of those circuits falling under the common carrier provisions and others falling under the broadcasting statutes of the act. When satellites join with cable systems to complete a broadcast, state and local regulations become involved.

The framework for satellite regulation in the United States was built in 1962 with the passage of the Communications Satellite Act. This act established the Communications Satellite Corporation called COMSAT. Referring to the eventual participation of the United States in a global system of satellite communication, the act stated:

> *In order to facilitate this development and to provide for the widest possible participation by private enterprise, United States participation in the global system shall be in the form of a private corporation, subject to appropriate governmental regulation. It is the intent of Congress that all authorized users shall have nondiscriminatory access to the system; that maximum competition be maintained in the provision of equipment and services utilized by the system; that the corporation created under this Act be so organized and operated as to maintain and strengthen competition in the provision of communication services to the public; and that the activities of the corporation created under this Act and of the persons or companies participating in the ownership of the corporation shall be consistent with the Federal Antitrust Laws.*

The act assigned three areas of authority to the president of the United States, the National Aeronautics and Space Administration, and the Federal Communications Commission. The FCC's duties covered eleven broad areas: (1) ensure competition in the procurement of apparatus, services, and equipment for the developing satellite system; (2) ensure nondiscriminatory access and use of the satellite system; (3) require the construction of the system to provide services to foreign countries when advised to do so by the State Department; (4) ensure compatibility for interconnecting with other systems; (5) prescribe accounting regulations which ensure that any savings are reflected in rates for system use; (6) approve technical characteristics of the system; (7) grant construction permits for the system; (8) authorize the corporation to issue stock; (9) ensure no substantial additions are made to the system unless they are in the public interest; (10) require additions to the system when they do meet the public interest; and (11) make rules and regulations to carry out the provisions of this act.

Satellite communication also gained international prominence in 1962. The General Assembly of the United Nations passed a resolution that year stating that it "believes that communication by satellite offers great benefits to mankind as it will permit the expansion of radio, telephone and television transmissions, including the broadcast of United Nations activities, thus facilitating contact among peoples of the world."[15] A more explicit statement on the use of satellite communication came indirectly in the *Treaty of Principles Governing the Activities of States in the Exploration and Use of Outer Space, Including the Moon and Other Celestial Bodies.* The treaty said: "Outer Space . . . shall be free for exploration and use by all states without discrimination of any kind, on a basis of equality . . . and there shall be free access."[16]

While studying the acts and treaties which govern satellite communication, we must also understand the human implications of this technology that warrant this regulation. In a study prepared under the auspices of the American Society of International Law, Paul L. Laskin and Abram Chayes reviewed the issues facing international satellite communication.[17] The first is the problem of signal spillover. Satellite signals do not conform to political boundaries, and when one country has the ability to beam signals to another country, threats to cultural integrity, national security, and national goals take on new meaning. For socialist countries, combatant political propaganda beamed via satellites by other countries becomes a major concern. Some nations fear that those countries with huge economic and technological resources, such as the United States, will dominate international programming. Others share the opinion that American programming, although visually attractive, is for the most part trivial, banal, and violent.[18]

Laskin and Chayes also point out another consequence of commercial television. Commercials from American programming may not only displace local merchandise with foreign-made goods, but also may create a demand for consumer goods which could frustrate plans for orderly social and economic development. Underlying these arguments is another position that television systems and their development belong to each country as a product of national sovereignty. In addition, the very conception of international satellite regulation becomes a problem. Laskin and Cheyes state that "the desire for some form of international control over direct satellite broadcasting may reflect a nation's legal philosophy. Where the Anglo-American countries, for example, proceed pragmatically, formulating the rules of legal behavior as they acquire experience, the civil law tradition tends to rely on the codification of rules in advance of action."[19]

All of these issues reflect the difficult path ahead for a consistent and mutually acceptable code of international law affecting satellite communication. Thus, the immediate future of international satellite control will probably remain in the abstract terms of international treaties.

future perspectives: issues in rewriting communications law

Swirling around the broadcast industry are issues which are an outgrowth of policy decisions, new technology, and changing emphases in the industry. These issues are on everything from changing the very structure of regulatory agencies to modifying rules and regulations controlling the daily operation of broadcasting stations. Of all these issues, none has greater interest or more potential effect on the broadcast industry than the rewriting of the Communications Act of 1934.

The impetus for a major rewriting of the Communications Act of 1934 came out of congressional hearings held in 1976 on the proposed Consumer Communications Reform Act, which would have overhauled telephone regulation. Commonly called the "Bell Bill," the proposed legislation went into hearings before the U.S. House of Representatives' Communications Subcommittee under the chairmanship of Lionel Van Deerlin (Figure 15–1), a California Democrat. Cable television also was scrutinized by Van Deerlin. In October of 1976, Van Deerlin announced

Figure 15–1 Congressman Lionel Van Deerlin, chief supporter of the rewriting of the Communications Act.

a full-scale inquiry into rewriting communications legislation. The Radio-Television News Directors Association's legal counsel, Larry Scharff, prepared a position paper saying, in essence, that the best way to revise the law would be to assure that broadcast journalists received the same First Amendment rights as print journalists do.

Rewriting became the topic of numerous symposia. Gaps began to grow between rewriting supporters and those who felt a radical change in the act would not be in their best interests. The first major forum was the February, 1977 meeting of the National Association of Television Program Executives (NATPE). All the members were present, including Congressman Van Deerlin. Donald H. McGannon, chairman and president of Westinghouse Broadcasting Company, supported the rewriting. He suggested a cabinet-level Department of Communication, bringing under one roof all the agencies now controlling use of the electromagnetic spectrum. Russel Karp, president of TelePrompTer, also supported the rewriting and cited the need for cable to be given more independent regulatory status. Bill Leonard, a vice president for CBS, Inc., was on the other side of the fence. He characterized the American system of communication as the finest on earth and in reference to the 1934 act dating before current technology, said that the U.S. Constitution did not mention anything about railroads, cars, telephones, and a number of other technologies.

The next rewriting forum was at the March 1977 meeting of the National Association of Broadcasters in Washington, D. C. There, Congressman Van Deerlin stated that "the broadcaster should be entitled to the same First Amendment protection afforded newspapers." He also came out in favor of repealing the Fairness Doctrine. For those favoring more competition, the legislation was insufficient to regulate the new technology which had been developed since its passage—technology which included satellites, microwave, cable, fiber optics, citizens-band radio, radar, land mobile communication, and light wave or laser beam communication. Van Deerlin later toned down the word "rewrite" and substituted "review." The House Commerce Committee stamped its approval by doubling Van Deerlin's committee's budget appropriations.

The subcommittee's review announcement prompted industry professionals, government bureaucrats, academicians, and citizens' groups to come out of the woodwork supporting or opposing the revision. Not wanting to be left behind, the U.S. Senate announced a special Senate hearing, conducted by Senator Ernest Hollings, a South Carolina Democrat, which would review the Communications Act and everything associated with telecommunication policy. Joining with Hollings, Senator Warren Magnuson, a Washington Democrat, directed the Office of Technological Assistance (OTA) to gear up for telecommunications capability. The OTA is a research organization serving both the House and the Senate.

political dilemmas of rewriting communications law

On June 7, 1978, House Bill No. 13015 was introduced in Congress. Based on months of congressional hearings and discussions in the industry, the draft of the new revision of the Communications Act of 1934 was now public. Called the Communications Act of 1978, it was only the first such bill to be introduced. Van Deerlin himself reintroduced changes in the original draft of the first bill and, by 1979, the Senate was getting into the rewrite business. In March, 1979, even before Van Deerlin could get his revision ready, Senator Ernest F. Hollings introduced what was called the "revision" of the Communications Act and suggested changes in the current law, some of which were different in thrust than the Van Deerlin measure. To make things even more complicated, Senator Barry Goldwater also introduced his version of a bill which would overhaul the Communications Act.

The introduction of Van Deerlin's original bill, the versions introduced by Goldwater and Hollings, and Van Deerlin's own revision, all brought speculators to their feet making predictions that, with so many divided legislative proposals, prospects for enacting any real legislation were remote. Lobby groups, already taking sides on issues brought about by the original Van Deerlin bill, had their issues even more diversified by the new proposals. If change does occur, it will certainly not happen overnight. The complexities are much greater now than those pertaining to the 1927 and 1934 laws. Moreover, the 1934 Act is now a much larger document with many amendments, rule changes, and judicial precedents. Change in the current legislation could well move into the 1980s, and there are those arguing that with rapid developments in technology, even new legislation could be obsolete by the end of the 1980s.

summary

Chapter 15 has examined control of cable and satellite communication as well as efforts to rewrite the Communications Act of 1934.

The FCC's control over cable systems was established in 1962. Rules for governing cable systems were announced in 1965 and upheld in a court case involving a San Diego cable system. Official rule making was started in 1968, and detailed regulations were announced in 1972. Key issues in the regulation of cable systems include the definition of a cable system which is a nonbroadcast facility consisting of a set of transmission paths and associated signal generation, reception, and control equipment that distributes or is designed to distribute to subscribers the signals of one or more television broadcast stations." For most cable systems, carriage requirements fall into the categories of "must carry" and "may carry" signals. States and local governments also govern cable systems.

Few technologies have a more direct effect on us than satellite communication does. There are concerns over the control of satellite communication in great international bodies as well as in the small communities in which ground systems bring educational television to local school systems. Most of the basis of satellite communication control is in the Communications Satellite Act of 1962 as well as in the Communications Act of 1934.

Chapter 15 concludes with a discussion of efforts to rewrite the Communications Act of 1934.

spotlight on further learning

SIGNITZER, BENNO, *Regulation of Direct Broadcasting from Satellites: The UN Involvement.* New York: Praeger Publishers, Inc., 1974.

Communications Law: 1976. New York: Practicing Law Institute, 1976 (continuing series).

FRANKLIN, MARC A., *Cases and Materials on Mass Media Law.* Mineola, N.Y.: The Foundation Press, Inc., 1977.

GILLMOR, DONALD M., and JEROME A. BARRON, *Mass Communication Law.* St. Paul: West Publishing Co., 1974.

GINSBURG, DOUGLAS H., *Regulation of Broadcasting.* St. Paul, Minn.: West Publishing Co., 1979.

JONES, WILLIAM K., *Cases and Materials on Electronic Mass Media: Radio, Television and Cable.* Mineola, N.Y.: The Foundation Press, Inc., 1976.

Media Law Reporter. Washington, D.C.: Bureau of National Affairs, 1977 (continuing compilation of media decisions).

NELSON, HAROLD L., and DWIGHT L. TEETER, JR., *Law of Mass Communications.* Mineola, N.Y. The Foundation Press, Inc., 1973.

RIVERS, WILLIAM L., and MICHAEL J. NYHAN, eds., *Aspen Notebook on Government and the Media.* New York: Praeger Publishers, Inc., 1975.

Use of Broadcast and Cablecast Facilities: Candidates for Public Office. Washington, D.C.: Federal Communications Commission, 1972 (FCC 72-231).

part

5

Economics and Evaluation

16

advertising and economics

FOCUS   After completing this chapter, we should be able to

Explain the different sources of station income.
Compare local and national rate cards.
Define *barter* and *co-op advertising*.
Understand the uses of a chart of accounts.
Explain broadcast financial statements.
Describe the role of the broadcast broker.
Describe the fourteen-point checklist for buying a station.
Identify a successful broadcast promotion campaign.

H. P. Davis was a vice president of Westinghouse when he delivered a lecture in 1928 to the Harvard University Graduate School of Business Administration, saying:

> *In seeking a revenue returning service, the thought occurred to broadcast a news service regularly from ship-to-shore stations to the ships. This thought was followed up, but nothing was accomplished because of the negative reaction obtained from those organizations whom we desired to furnish with this news material service. However, the thought of accomplishing something which would realize the service referred to, still persisted in our minds.*

H. P. Davis's persistent thoughts turned out to be the foundation of American commercial broadcasting. The purpose of this chapter is to examine more closely the different aspects of that foundation: building station revenues, station finances, buying and selling stations, starting new stations, and broadcast promotion.

building station revenue

In the United States, stations sell time to advertisers who, in turn, pay the station for having their advertising messages broadcast during that time to the listening or viewing public. That elusive commodity, time, is the product the station offers. Some "times" are better and more expensive than others. And buying a great deal of time is cheaper than buying a small amount of time.

The Local Rate Card

The rates the station charges for its time are listed on two types of rate cards— the local rate card listing advertising rates for businesses in the station's local community, and a national rate card for advertisers who buy large amounts of time nationally on many different stations. Figure 16–1 represents one station's local rates for two time classifications, AA and A. AA time runs from 6:00 A.M. to 10:00 A.M., Monday through Saturday and 3:00 P.M. to 7:00 P.M., Monday through Saturday. These are the station's most expensive times on the local rate card, representing the drive time, those heavy radio-listening times when, along with the home audience, radio captures the listeners driving to and from work. All other days and times are listed as A class time. Class A times are less expensive than drive time since they traditionally do not attract as large an audience.

National Rate Card: Reps and Ad Agencies

Now compare the rates listed in Figure 16–2 with those in Figure 16–1. You will notice a difference of approximately 17.65 percent. The higher rates in Figure 16–2 are from the same station's national rate card. National rates are more expensive than local rates and apply to advertisers who buy large amounts of advertising in many different stations. To eliminate all the paperwork and negotiations in placing advertising with every individual station, advertisers purchase their advertising through either station representatives, called "reps," or advertising agencies. Stations representatives, as the name implies, represent the station to large advertisers. The rep also represents more than one station and contracts to buy time on many different stations to reach the audience the advertiser requests. Advertising agencies, on the other hand, represent the advertiser but buy time in the same way the rep does.

The company using the rep or agency typically pays a rate 17.65 percent higher than the rate on the local rate card. Of this increase, the

Figure 16–1 Local rate card (left).

Figure 16–2 National rate card (right). (WAZY)

ANNOUNCEMENTS

(AA)   6:00 to 10:00 a.m., Mon./Sat.
3:00 to 7:00 p.m., Mon./Sat.

	60 Sec.	30 Sec.
1	19.00	15.20
52	16.00	12.80
156	13.00	10.40
312	11.00	8.80
624	9.50	7.60
1040	8.50	6.80

(A)   All other days and times

	60 Sec.	30 Sec.
1	16.00	12.80
52	13.00	10.40
156	11.00	8.80
312	9.00	7.20
624	7.50	6.00
1040	6.50	5.20

ANNOUNCEMENTS

(AA)   6:00 to 10:00 a.m., Mon./Sat.
3:00 to 7:00 p.m., Mon./Sat.

	60 Sec.	30 Sec.
1	22.40	17.90
52	18.90	15.10
156	15.30	12.20
312	13.00	10.40
624	11.20	9.00
1040	10.00	8.00

(A)   All other days and times

	60 Sec.	30 Sec.
1	18.80	15.10
52	15.30	12.20
156	13.00	10.40
312	10.60	8.50
624	8.80	7.00
1040	7.70	6.10

rep or agency takes a commission of 15 percent. The remainder goes to the station. This 2.65 percent increase is considered compensation for handling the long-distance account, promoting itself to and through the rep and ad agencies, the extra bookkeeping, and other related costs. The advertiser, in turn, saves the time and cost of placing each individual advertising order, a cost which on large buys would run much more than the 15 percent commission.

Despite the rate cards, stations in highly competitive markets still wheel and deal to entice advertisers to buy time on their station. This usually happens when an advertising agency tries to buy the most time for the least money and pits stations against each other to see which one can offer the best price. Sometimes they are successful. There is considerable price slashing; discounts of 35 percent below the rate card are not unusual, and some stations discount 50 percent below the rate card.

Announcements or commercials can be purchased in two lengths—60 seconds and 30 seconds. Some stations divide their times into even smaller categories of 20- or 10-second lengths. The shorter time periods are common in television for which rates are higher than those for radio. For smaller businesses, the shorter announcements make buying television time possible, whereas forcing them to buy longer commercials might price them out of the market.

Prices for commercial time become less expensive when bought in quantity. For example, on the local rate card purchasing only one 60-second announcement costs $19.00 per announcement in AA time (Figure 16–1). If you purchase 52 announcements, the cost per announcement drops to $16.00 per announcement. If you purchased 1,040 60-second announcements, the cost per announcement would drop to $8.50. Similarly graduated discounts are available for 30-second announcements, and these discounts apply to all announcements made in A time as well.

Stations periodically revise their rate cards just as supermarkets revise prices of meat and eggs. Successful stations revise theirs upwards to meet inflation and to represent increases in audience. The more viewers or listeners a station has, the more people an advertiser can reach, and the more the station can charge for its commercial time. The station's market can also reflect the rate change. If the station is the only broadcast outlet with no other competition, the rate may be higher to reach the same number of people than if other stations were offering competing rates. It is the old rule of supply and demand. The cost of local newspaper advertising can also influence the rate charged by the broadcast media. All media compete for those advertising dollars.

In addition to the ones we have discussed here, other time buys are available on most stations. These include special charges for remote broadcasts, such as live coverage of a store opening, or costs for larger time blocks in which to air entire programs.

Barter, or "trade-out," arrangements are another way in which stations receive income. Barter accounts mean the station provides advertising time in exchange for goods or services offered by the advertiser. These are anything from appliances to world cruises. The goods and services can also be used by the station at its discretion. Many stations give them away as prizes, and others may award them to top station account executives.

Most prizes awarded on television game shows are supplied by manufacturers paying a small fee to receive on-air announcements in return for their products' publicity. National television exposure is a relatively inexpensive way of obtaining advertising time, compared to the usual national television rate. For the game show, the prizes are, for all practical purposes, free merchandise, and there is almost no cost to the game show to announce the products on the air, beyond the salary of an announcer. The announcements are incorporated into the programming, adding the elements of excitement and dream fulfillment as the announcer describes the "fabulous prize" the contestant has a chance to win.

Many companies engage in barter advertising. Windjammer "Barefoot" Cruises Ltd. (Figure 16–3) is one example that trades on a dollar-for-dollar basis. The cruise can be a powerful incentive to the station's account executives to "beat the bushes" for advertising. Or a station contest with a Windjammer cruise as a prize can entice listeners to "stay tuned" for their chance to bask in the sun on the open sea. For the station, the cost of airing the commercials is minimal, since the announcements are usually scheduled in unsold time which generally would go unused without the trade-out advertising. Moreover, it is profitable for both the advertiser and the station to trade on a dollar-for-dollar basis, since each is getting something at a less expensive rate than if they had to buy it outright.

The easiest way to understand the trade-out, dollar-for-dollar advantage is to consider the example of a new car. We shall assume station WAAA furnishes its sales manager with a new car to use for business travel, such as calling on sponsors. In order to do this, the station enters into a trade-out agreement with the local car dealer. The car costs $8,000. In return for the car, the station will air $8,000 worth of commercial announcements for the car dealer. For the station, the cost of providing these announcements is less than $8,000, since the commercials are scheduled at unsold times when other commercials would not air anyway. Without the trade-out, the time would simply go unsold. Since the station must stay on the air in any case, the cost of operating the station remains the same. For the dealer, since the company bought the car wholesale, say for $6,000, that would be the total cost to the car dealer for $8,000 worth of advertising.

Figure 16–3 (Windjammer "Barefoot" Cruises)

Although trade-out advertising is common, stations try to avoid it, especially when the advertiser is willing to pay cash. Cash looks better than merchandise when the annual financial statement is prepared. Furthermore, when stations are sold, trade-out advertising is usually not considered part of the station's annual income, since the merchandise supplied by the advertiser usually cannot be used to pay the station's bills.

Co-op Advertising

Co-op advertising is an arrangement by which a local store splits part of the cost of advertising with the company whose products are mentioned in that advertising. For example, a radio announcement of a sale of Westinghouse appliances at the Ace Appliance Store costs $19.00. In a 10/90 co-op advertising arrangement between Westinghouse and the store, Westinghouse would pay 10 percent of the cost of the announcement, or $1.90, and the store would pay $17.10. The radio station still receives the rate card price for the announcement, Ace airs the commercial at a 10 percent discount, and Westinghouse receives local advertising exposure. Figure 16–4 shows the copy of a typical co-op radio commercial. This particular company manufactures Photo Charms, popular miniature reproductions of favorite photographs attached to charms, and jewelry products. The company pays 10 percent of the cost of radio commercials for which they supply the copy.

Not all companies use co-op advertising, and the amount paid by the manufacturer and the local store varies considerably. Newspapers continue to receive the majority of co-op advertising dollars, about 75 percent, since co-op advertising began with newspapers before broadcast advertising was widespread.[1] Some companies employ full-time co-op managers whose responsibility is to keep track of co-op advertising by retail outlets and to show retailers how to coordinate their advertising with the manufacturer's co-op program.

Combination Sales Agreements

Although not a common practice, some account executives are employed by more than one station. When this occurs, the FCC has some strict guidelines to guard against rate fixing or selling time on more than one

Figure 16–4   (Taylor Graphics Corp. Greencastle, IN)

```
ANNOUNCER:

LOOKING FOR THAT PERFECT GIFT FOR SOMEONE SPECIAL?
   (Dealer's Name)      AT    (Dealer's Address)
HAS THE ANSWER FOR YOU.....PHOTO CHARMS.  PHOTO CHARMS

ARE MINIATURE REPRODUCTIONS OF YOUR FAVORITE PHOTOS

BEAUTIFULLY MOUNTED ON CHARMS....HEIRLOOM LOCKETS....

EVEN ON MARBLE PAPERWEIGHTS OR HANDSOME ZIPPO LIGHTERS.

STOP IN AT   (Dealer's Name)      AT (Dealer's Address)
TODAY AND SEE THEIR LARGE SELECTION OF THESE TRULY

PERSONAL GIFTS.

SO WHEN YOU WANT TO GIVE SOMEONE SPECIAL A LASTING GIFT

THEY'LL CHERISH, CHARM THEM.....WITH PHOTO CHARMS FROM

   (Dealer's Name)      .
```

station for a single rate.[2] Although representing two stations is not illegal, selling time for two competing stations is. The definition of competing is any two stations whose signals overlap, regardless of which market they serve. Moreover, a radio and television station combining to offer a single rate is illegal, even if the two stations are jointly owned. Because other radio stations might not be able to team up with a television station, the FCC feels such arrangements are counter-competitive.

A single rate for an AM and an FM station engaged in simulcasting is permissible. But if combination rates are offered for two stations commonly owned but not engaged in simulcasting, management must be careful not to use the combination rate to "carry along" the weaker of the two stations. For stations not engaged in simulcasting, forcing an advertiser to buy a combination rate is illegal. If the advertiser wants to buy advertising on only one station, then that opportunity must be available to the advertiser.

station finances

Once it is earned, a station's advertising revenue is translated into a series of numbers that managers, bookkeepers, accountants, investors, and bankers spend many hours studying and using to answer some complex questions. How can we improve daytime sales? How can we cut expenses? How much money will we need to borrow? How much money should we invest? These queries and countless others plague station executives as they mull over their charts of accounts and financial statements.

Chart of Accounts

The basic ledger to record all station finances, both income and expenses (disbursements), is the chart of accounts.[3] Although charts of accounts vary, many in broadcasting use a system of numbers beginning with the prefix digits 10 and altering the prefix to represent a different account classification. For example, assume the numbers 100 to 199 represent assets. A specific number is assigned to each type of asset, using a two-digit prefix. The following represents cash accounts, which are considered assets, designated by the 10 prefix:

101 Cash in banks—regular
102 Petty cash
103 Cash in banks—payroll
104 Cash in banks—other

The prefix 14 might represent advances and prepayments:

141 Prepaid insurance
142 Prepaid rent
143 Prepaid taxes and payments of estimated taxes
144 Expense advance to officers and employees

Three-digit accounting systems provide about as much flexibility as the average broadcaster needs, but additional numbers can be added wherever appropriate. For instance, a group owner might want a prefix to identify each station in the group. So WAAA might have the prefix 5. Thus, if we wanted to know the petty cash at WAAA, we would look under account number 5–102, 5 representing the station and 102 representing petty cash. WBBB might be assigned the number 6, and so on. Some computer programs use additional numbers to facilitate more complex data analyses of the station's financial status.

Using the chart of accounts for identification, management can make periodic checks on the station's financial structure as well as to pinpoint and plot its activity over time. Such information is critical for management to ensure the station's profit structure. The information and accounting systems are equally as important to the broadcaster selling the station, the banker lending the station money, or the accountant preparing its tax returns.

Financial Statements

The chart of accounts is the basic accounting tool of the station, but the money recorded in using the chart of accounts is translated into many different financial statements. Figure 16–5 is a station balance sheet. Notice assets ($90,000) equal liabilities and equity ($90,000). Balance sheets must always balance. They represent a stop-action picture of the company on any given day, usually the last day of the year. Why must the balance sheet balance? Because all assets have a claim on ownership, equal to those assets. For example, suppose you drive a car worth $5,000 (a $5,000 asset). The title is in your name, and for all practical purposes it is your car. But what about claims of ownership of your car? We shall assume you still owe the bank $1,000 on your car. In accounting terms, the bank owns $1,000 worth of your car, and you own $4,000 worth of your car. The claims on ownership of your car represent $1,000 (the bank's claim) plus $4,000 (your claim), together totaling the exact value of the car, which is $5,000.

A comparative balance sheet compares assets and liabilities between two points in time. An income statement shows the amount of income

```
                          WBPE, Inc.
                        BALANCE SHEET
                       January 31, Year 1

        Cash ..........................................................      4 000
        Accounts receivable ..................................      2 000
        Depreciable assets .....................................     50 610
        Land ..........................................................     10 000
        Intangibles .................................................     23 390
     ASSETS ...........................................................     90 000
        Accounts payable ......................................      5 000
        Notes payable—sellers .............................     38 000
        Notes payable—stockholder .....................     50 000
     Liabilities ......................................................     93 000
        Contributed capital ...................................      1 000
        Loss since inception ................................     (4 000)
     Stockholders' Equity (Deficit) ..........................     (3 000)
     LIABILITIES AND EQUITY ..............................     90 000
```

Figure 16–5   (National Association of Broadcasters)

after the debits are subtracted from the credits, or stated another way, it tells how much money is left after all the bills are paid.

A comparative income statement compares income between two points in time. Figure 16–6 is a comparative income statement comparing two years of broadcast operation. Notice that the revenues are listed first, then the expenses (indented), followed by the total expense. In year 8, we see the station's income before taxes was $11,500, on which a federal income tax of $2,530 was paid, leaving an income of $8,970.

If you are the treasurer of a broadcasting station, an especially important financial statement would be the cash flow statement. The cash flow statement would tell you the amount of money necessary to keep

Figure 16–6   (National Association of Broadcasters)

	Year 8	Year 7
WBPE, INC. COMPARATIVE INCOME STATEMENT Year Ending December 31		
Revenues	168 900	162 200
Technical expense	12 300	12 100
Program expense	48 200	47 600
Selling expense	29 400	28 100
General & administrative	67 500	63 900
Expense	157 400	151 700
Income before tax	11 500	10 500
Federal Income Tax	2 530	
INCOME	8 970	8 190

the station running smoothly. It would also tell you if you needed to borrow money to pay the bills or to invest excess income. A cash flow statement is looked upon as the "state of health" of the station and can be used to help find where changes in operations may have to be made.

Figure 16–7 is a cash flow statement comparing one year of operation with the projected cash flow for the coming year. Notice year 9. If projections hold true, year 9 will see the station spend more money than it makes, resulting in a cash overdraft of $29,795. As treasurer, you will need to inform management of the necessity to borrow at least $34,795 to make ends meet and still have the $5,000 desired minimum cash balance. If management does not want to borrow the money, it must change that year's projected operating procedures. Management might decide to "make do" with the equipment on hand and save the $20,000 budgeted for "purchase of equipment." But as treasurer, you point out that never in the station's history has that ever been done, that the equipment is old, and that the station's transmitter cannot last out even this year. Can you suggest cutting other disbursements? What recommendations can you make, and what will be the consequences? These are the questions that financial statements both pose and help answer in running a broadcasting station.

Figure 16–7 (National Association of Broadcasters)

WBPE, INC.
CASH FLOW STATEMENTS
Years Ending December 31

	Projected Year 9	Year 8
CASH BALANCE AT JANUARY 1	8 505	9 113
Collections from advertisers	190 000	173 460
Cash Available	198 505	182 573
Technical disbursements	15 000	8 229
Program disbursements	50 000	45 861
Selling disbursements	52 000	39 400
Administrative disbursements	74 000	70 825
Agency commission payments	8 000	6 100
Repayment of stockholders loan		886
Dividend payments	200	200
Federal income tax payments	9 100	2 567
Purchase of equipment	20 000	
Total Disbursements	228 300	174 068
CASH BALANCE AT DECEMBER 31, YEAR 8		8 505
PROJECTED CASH AT DECEMBER 31, YEAR 9—overdraft	(29 795)	
DESIRED MINIMUM CASH BALANCE	5 000	
REQUIRED ADDITIONAL FUNDS	34 795	

buying and selling broadcast properties

Financial statements also are especially important when a station changes hands. Buying and selling broadcast properties, including CATV, is big business, involving everyone from small-town entrepreneurs to corporate conglomerates. Actually, buying a station requires knowledge of the broadcasting business and considerable money, not only to buy the station, but to operate it until it builds an income. The sale of a station must also be approved by the FCC, which scrutinizes everything from the buyer's and stockholders' characters to their financial worth, to their other media investments, to their history in managing other enterprises. A record of bankruptcy can quickly close the door to owning a broadcasting property.

The Broadcast Broker

At the heart of over 70 percent of broadcast property sales is the broadcast broker—the real estate person of the broadcasting business. Most are seasoned professionals with years of experience in station transactions. Many people wanting to sell or purchase a broadcast property begin by contacting the broker. Using personal contacts, referrals, advertisements, and in some cases direct solicitation, the broker knows what properties are for sale, at what price, and who the buyers are. Since the sale of a broadcasting property can run into millions of dollars, good brokers are highly paid, but most industry professionals would agree that they are well worth their commission.

Brokers can operate independently or as members of a large firm. One well-known independent broker headquartered in Chicago but handling properties nationwide is Richard A. Shaheen. His brokerage business, Richard A. Shaheen, Inc. (Figure 16–8), has handled a wide range of station transactions, and Shaheen himself, has been in the station brokerage business since 1955.[4] His job takes him wherever it's necessary to bring together a buyer and seller. In addition to this travel, correspondence, national and state conventions and the telephone are important elements of keeping in contact with people in the industry.

Who buys broadcasting properties? Richard Shaheen deals primarily with people already in the media business. "They understand the business. They have the know-how; they know the language; they understand the business," he says. Still, Shaheen receives inquiries from people outside the industry. "Many still see the pie in the sky, that you can walk in and pick it like you're picking apples in the fall." Of those prospective buyers who are currently active in the broadcasting industry, he said, "most come from management, sales or ownership rather than being program oriented. These people deal with dollars every day and they

MEDIA BROKERS • APPRAISERS

*AT YOUR SERVICE WITH OVER
20 YEARS EXPERIENCE*

RICHARD A.

SHAHEEN ,INC.

435 NORTH MICHIGAN AVE. • CHICAGO 60611

312/467•0040

Figure 16–8 This ad, used by Richard Shaheen, has become a trademark in key broadcasting publications.

understand the business side of broadcasting more thoroughly". He also says, "I find that newspaper owners are good buyers simply because they are accumulating considerable profits through their newspaper operations and encounter tax problems in the accumulation of this cash. Instead of paying high taxes on the profits, they prefer to reinvest it." Further, Shaheen says that broadcast group owners are finding similar problems today because of the high profitability of stations.

Who sells broadcasting properties and why? Shaheen lists retirement as one reason but certainly not the only one. "Some people just need a change. The station has matured, and the owner wants a new challenge. So he'll go out and buy a different one." Location is another reason. "Others want to change markets. They look for a new place to live, a new challenge." Still others, he says "have used the stations as a tax shelter and have depreciated the property as far as the tax benefits will allow." And then there is capital gain. Looking back to the 50s, Shaheen remarks, "Back then, some people might buy a station for $60,000, keep it for six months, and sell it for $100,000. They thought they made a fortune, and they did in those days." And although prices have changed and stations are held longer, Shaheen is quick to point out, "People still sell stations to get the capital gain."

How does the broker participate in the sale? "By spending a lot of time with the buyer and seller," Shaheen states. He prefers to be directly involved in the sale until the attorneys draw up the contracts between buyer and seller. Then, "I want to watch what happens in the contract," he declares. Although preferring to handle the sale exclusively because of the commission and keeping track of the proceedings, Shaheen does at times work with other brokers. Co-brokering is a very common practice among many brokers. It has benefits for buyers, sellers and the brokers, particularly if one broker is stymied in locating the right buyer for a property. He or she may find it expedient to contact other brokers. Shaheen feels that his first obligation is to the seller. Obtaining the price and terms is important, and sometimes it is necessary to co-broker to meet those ends.

Although commissions can vary, typical brokerage fees for handling a station sale are 5 percent of the first million dollars, 4 percent of the second million, 3 percent of the third million, and 2 percent of the balance.[5] Brokers also perform property appraisals for clients. For an appraisal service, $500/day and up are typical charges.

Broadcast Buyer's Checklist

Still, the primary business of the broker is handling the sale of broadcast properties. And for the buyer, this involves many factors to consider before the final transaction. A concise, fourteen-point buyer checklist has been prepared by the brokerage firm of Richard A. Shaheen, Inc. The prospective buyer of a broadcast property should carefully examine the following:

1. *Gross Sales*
   Review gross sales record for three previous years.
   Determine if any represent "trade" dollars.
   Check any undue use of promotions to inflate gross.
   Determine what share of the total market revenue the station enjoys.

2. *Operating Profit*
   Is it before or after all taxes?
   Does it come after depreciation?

3. *Cash Flow*
   The most accepted method of determining a true cash flow is that which calls for it being a total of the following ingredients:
   a. Net profit before taxes.
   b. Depreciation.
   c. Interest Paid.
   d. Officer's salaries.
   e. Director's fees.
   To this can be added any true non-recurring expenses. A buyer should also consider the effect a new, higher base of depreciation will have with respect to the net profit after taxes. An owner-operator's salary and others should also merit consideration as part of cash flow.

4. *Fixed Assets*
   Age and condition of major pieces of equipment?
   Do they meet FCC specifications?
   Does the sale include all fixed assets?
   Anything on a lease basis?

5. *Leases, Licenses and Contracts*
   Determine beginning and end date and all details of:
   a. Land and building rentals.
   b. Personnel contracts.

      c. News Service.
      d. Transcription or jingle services.
      e. Film contracts.
      f. Station-purchased advertising contracts.
      g. Music copyright services.
      h. Trade association memberships.
      i. Equipment purchased or leased.
      j. Time sale contracts.
      k. Representative firm arrangement.
      l. Union contracts.
      m. Network agreements.
      n. Music format contracts.

6. *Legal Actions*
   Are there any legal actions pending against the station?
   Are there any complaints filed with the FCC?

7. *Personnel*
   Salary and position of all employees.
   Salesmen commission arrangements.
   Bonus or percentage plans.
   Vacation policy.
   Status of station's "affirmative action" program.

8. *Advertising Rates*
   Obtain a copy of all current rate cards and determine both when these rates went into effect and how they compare with the competition.

9. *Station Coverage*
   Is it sufficient to cover the market?
   How does it compare with the competition?
   Can the signal be improved?

10. *Program Format*
    Determine:
    a. Rating and demographic position for past two years.
    b. Definitive explanation of present format.
    c. Type of programming on competitive stations.
    d. Promotions in the market—type and extent.

11. *FCC Record*
    Ascertain if the license is presently in effect and also whether there are any actions pending in the FCC that directly involve the facility. This could include engineering, programming, EEO policies, and minority hiring status.

12. *Market Frequencies Available*
    Check to see if there are any applications filed or construction permits granted for additional facilities servicing the market.

13. *Competition*
    Compare ratings, rates, and program formats of competitive stations. Also determine circulation and advertising strength of daily or weekly newspapers.

14. *Market*
    As much care should be spent in investigating the market as the physical facility of the station itself.
    Carefully review:
    a. Population growth–past and anticipated.
    b. Retail sales.
    c. Major sources of employment.
    d. Consumer spending power.

If a buyer finds positive information after reviewing the items in the checklist, then the buy is probably a good one, provided the price is right.

A fair price is a difficult thing to pinpoint, but for radio and television stations, it is usually based on about 2 1/2 times gross annual billing. Cable companies are appraised at about $500 per connection. Today, however, with the scarcity of space on the electromagnetic spectrum, inflation, the potential of AM stereo, the growth of FM, and the effects of other media, the formula for determining what a property is worth can vary considerably.[6]

the role of promotion in building station revenues

For both new stations and those with long records of operation, station promotion is an important part of the broadcasting business. Perhaps more so than with other businesses, broadcasting as an entertainment industry entails promotion. Successful station promotion is much more than a few announcements promoting a program or personality. Instead, it is a well planned and systematically executed campaign. More and more stations are realizing the importance of station promotion. Many have rested on the laurels of their programming without learning from the example of other businesses that competition requires promotion to convince the public that they should patronize your business or listen to your station. In fact, broadcast promotional campaigns are becoming a necessity in many markets.

Promoting Assets

For every station, the important commodity is reaching a listenership or viewership. Thus, how the station fares in the ratings becomes the central theme of many promotional campaigns. Although only one station can be first overall, many stations can find their own niche in the ratings data. A certain station, for example, may be first among women, another first among 18- to 34-year-olds, another first among men aged 34 to 49, another first during morning-drive time, and still another first in late-evening news viewership. Each placement in the audience surveys can provide opportunities to promote the station's accomplishments.

Another consideration is the cost-per-thousand (CPM) of reaching the station's audience, or stated another way, what it will cost an advertiser to reach each 1,000 viewers. Perhaps the station does not have a first place showing, but its CPM is lower than that of any of its competition. Pointing this out to cost-conscious advertisers can make ratings a moot issue in a sales presentation. Naturally, the station with both an audience and a low CPM is in an especially competitive position.

Other promotional campaigns can be centered on such elements as new call letters and format changes. If a station changes hands and consequently changes call letters, calling attention to the new "identity" is critical. Similarly, changing formats often results in reaching a new audience. Telling that new audience about the new sound, and telling advertisers about that new audience are equally important.

Everyone understands awards, and some of the most familiar belong to broadcast journalism. Although promoting only one department—the news department— publicizing a journalism award can boost other areas of the station as well. This is especially critical to television in which the local news is one, if not the single most important, item determining the audience's perception of the station. More than one television station has ended up first in the ratings with local news pulling the station to the top, while its network programming in other markets and nationwide was running a poor third.

Planning Successful Promotion

To be successful, a promotional campaign must have certain qualities. One of the most important is simplicity. Although ideas such as contests and giveaways can be effective audience builders, successful long-term promotions have a central, underlying theme. Such slogans as "Musicradio," "Candlelight and Gold," "Radio Indiana," and "Happy Radio" can be woven throughout the station's programming. If ratings are being publicized, a single, concise statement that the station is first is much better than many sentences about the station's share of the audience.

Consider the billboard in Figure 16–9. A play on two words is all that is necessary to explain the station's number one position in the overall ratings. The promotion is short, concise, and the word "one" takes on the dual meaning of being first and winning the ratings race. This same "we one" theme can then be repeated on an entire series of promotional ventures, from direct mail flyers to station letterheads. The slogan also can be presented in many different media formats without looking cluttered. Too many promotional campaigns try to tell too much and accomplish nothing. Going beyond a few words or a single symbol usually begets more confusion than clear identity.

Simplicity also belongs in station logos. Notice the two logos in Figure

we one.

WAZY
WAZY RADIO
TURNED ON MORE OFTEN BY MORE PEOPLE
THAN ANY OTHER STATION
IN THE GREATER LAFAYETTE METRO AREA.

Figure 16–9

KRON · SAN FRANCISCO   **KING TV5**

Figure 16–10 (KRON logo: courtesy KRON. King logo: reproduced with permission of King Broadcasting Company, Seattle, Washington; all rights reserved.)

16–10. King Broadcasting in Seattle, Washington uses a letter and a few dots to represent a king's crown. Channel 4 in San Francisco finds the channel number ideal for a logo illustrating the Golden Gate Bridge. As an important part of station image, many logos carry registered trademarks. Effective designs are uncluttered; come out well in black and white, which is essential to newspaper advertising; look good on letter heads; and can be easily recognized and understood.[7]

Besides visual identities, audio identities are basic, especially in radio. Most radio stations have a particular "sound" in addition to their musical format. These sounds are most often reproduced in a "jingle package"— a collection of sounds designed around a four- to eight-note sequence fitting the call letters. Variations in the jingle package are then incorporated into news introductions, bulletins, musical bridges between records, and backgrounds for commercial and public service announcements.

Equally important to simplicity is consistency. Too many stations make the mistake of constantly changing their promotional theme. Many mediocre promotional efforts have been successful simply because they have continued unchanged long enough for the audience to accept them as household words. The very best promotional effort lasting six months and then disappearing defeats the purpose of the campaign. Contests and giveaways will change from season to season, but there needs to be a consistent theme which represents the station over time, be it the logo, jingle package, or a combination of both. When done effectively, a promotional campaign can elevate the awareness of the station among its audience, and being aware of the station is the first step in building an audience.

summary

At the basis of building station income is the rate card. Local rate cards illustrate the rates charged to local advertisers. With rates approximately 17.65 percent higher than local rates, the national rate card is used for national advertisers placing orders through station representatives or

384

advertising agencies. Barter or tradeout advertising and co-op advertising are two other sources of station revenue. The FCC has certain guidelines guarding against monopolistic practices and prohibits certain revenue-building schemes which use combination advertising rates among competing stations.

In managing station finances, the chart of accounts becomes the numerical and descriptive list used to classify income and expenditures. Financial statements usually include the balance sheet, comparative balance statement, income statement, comparative income statement, and cash flow statement.

People investing in broadcasting are those who buy stations or start new ones. The broadcast broker is the real estate person of the broadcasting business and is responsible for arranging many of the transactions. A prospective buyer should examine fourteen key areas before buying a station. These are investigating gross sales, operating profit, cash flow, fixed assets, leases/licenses and contracts, legal actions, personnel, advertising rates, station coverage, program format, FCC record, market frequencies available, competition and market. Preliminary work in launching a new station includes a frequency search and obtaining a construction permit. When a license is granted, regular programming can begin.

Promoting a station is becoming essential as more stations sign on the air and competition increases. An effective promotional campaign contains two major qualities—simplicity and consistency.

spotlight on further learning

ABRAHAMS, HOWARD P., *Making TV Pay Off: A Retailers Guide to Television Advertising.* New York: Fairchild Publications, 1975.

Application for Authority to Construct a New Broadcast Station or Make Change in an Existing Broadcast Station. (FCC Form 301) Washington, D.C.: Federal Communications Commission, 1977.

BESEN, STANLEY M. *The Value of Television Time and the Prospects for New Stations.* Santa Barbara, Calif.: The Rand Corporation, 1973.

COUGHLAN, JOHN, *Accounting Manual for Radio Stations.* Washington, D.C.: National Association of Broadcasters, 1975.

HEIGHTON, ELIZABETH J., and DON R. CUNNINGHAM, *Advertising in the Broadcast Media.* Belmont, Calif.: Wadsworth Publishing Co., Inc., 1976.

How to Apply for a Broadcast Station. Washington, D.C.: Federal Communications Commission, 1977 (information Bulletin, periodically updated).

How to Read a Financial Report. New York: Merrill Lynch Pierce Fenner & Smith, Inc., 1975.

HOWARD, HERBERT H., *Multiple Ownership in Television Broadcasting: Historical Development and Selected Case Studies.* New York: Arno Press, 1979.

LARSON, TIMOTHY L., *Aspects of Market Structure in the Broadcast Brokerage Industry.* Salt Lake City: Media Research Center, Department of Communication, University of Utah, 1979.

OWEN, BRUCE M., JACK H. BEEBE, and WILLARD G. MANNING, JR., *Television Economics.* Lexington, Mass.: Lexington Books/D.C. Heath & Company, 1974.

ROBINSON, SOL, *Radio Advertising: How to Sell It and Write It.* Blue Ridge Summit, Pa.: TAB Books, 1974.

STERLING, CHRISTOPHER H. and TIMOTHY HAIGHT, eds., *The Mass Media: Aspen Guide to Communications Industries Trends.* New York: Praeger Publishers, Inc., 1977.

17

broadcast ratings

FOCUS   After completing Chapter 17 we should be able to

Describe the development and function of broadcast ratings.

Interpret broadcast ratings.

Give examples of how ratings are used in making management decisions in broadcasting.

Identify the issues in the criticism of broadcast ratings.

Understand the function of the Broadcast Rating Council and speculate on the future of broadcast ratings.

To be truly effective, any communication system must have an efficient means of gathering and evaluating data. We learned how intrapersonal communication sends electrochemical impulses from our senses through the nervous system, triggering our muscles to react. We also learned how interpersonal communication uses direct feedback to evaluate what people are saying and how other people react to this communication. Such feedback is also vital to mass communication and is found in many forms in broadcasting. A disgruntled viewer writes a letter to the television station or network complaining about programming. A sophisticated computer daily analyzes viewing habit data from meters attached to home television sets, from viewers' diaries, from an assortment of answered questions by interviewers, and from completed viewing questionnaires returned through the mails. A network pollster gathers data on election day. A radio program director examines data to prepare the weekly playlist. Management listens to its employees' opinions. All are examples of feedback necessary for the successful maintenance and operation of a broadcasting system. This chapter and the next are on this feedback. In chapter 17 we'll examine broadcast ratings, surveys, and public opinion polls. In chapter 18 we'll discuss specialized feedback, including letters from the broadcast audience, feedback to the FCC, responses to the networks, and the feedback process determining the music you hear on your favorite radio station. Let's begin by examining the broadcast ratings.

the background of broadcast ratings

Possibly the first broadcast was rated in 1929 when an interviewer for the Crosley Radio Company called a randomly selected number from the telephone directory. The person who answered was asked, "What radio stations did you listen to yesterday?"[1] Since that time, broadcast ratings have been criticized as being inaccurate, unreliable, and arbitrary. They have been accused of canceling quality programs, determining network policy, and influencing everything we see and hear on the broadcast media. They also have been perceived as being a subsidiary of the television networks and cooperating with sponsors. All of these misconceptions are just that, misconceptions. The broadcast ratings are those services which tell how many people are viewing or listening to what, when, and how often. Some of the most familiar include the A.C. Nielsen Television Index, Arbitron, and Pulse. The rating services have nothing to do with the networks and are not responsible for canceling programs. Rating services are separate corporations, not part of the network structure. However, stations and networks use the data from rating surveys to make decisions on what programs to cancel or keep on the air. As a result, the rating services get blamed for canceling programs when it is the stations and networks that make the decisions. Despite serious criticism, the ratings have also proved to be, for the most part, very reliable, producing some of the most accurate and sophisticated audience research data available. But they are by no means perfect. Along with the major ratings services, which are controlled and professionally responsible, are minor ratings services which do use some of the methodologies that legitimize criticism of the broadcast ratings. It is important from both a consumer's and a professional standpoint to learn how ratings work, what function they have, and how reliable they actually are.

The Function of Broadcast Ratings

Media subscribe to broadcast rating services in much the same way as they subscribe to syndicated material. But this product is feedback, feedback on the size and composition of the medium's audience. For instance, if you were operating a commercial television station in a major city, you would want to know how many people watched your station in comparison with other stations. You would also be interested in detailed demographic information on those viewers, such as age, sex, education, and income. This information is essential to your advertisers. Similarly, you would want to know when these audiences tuned to your station. You would also want to know whether a particular program commanded a larger share of the audience than some others did. What share of the audience did your newscast capture, compared with competing television stations? Your advertisers need all of this information in order to purchase air time wisely, your station's air time.

Advertisers also want to compare the cost of attracting these viewers. By combining the information found in these ratings with a listing of the cost of your station's commercials, commonly called a rate card, an advertiser might discover that the cost of reaching 1,000 people (cost-per-thousand) via your television station is less than the cost-per-thousand on another television station. If it can be proved that your television station reaches more potential customers than your competition does, the advertiser probably will realize that buying commercials on your station is a wise investment. Rating services provide the necessary proof.

With this proof in hand, you may find that certain programs need to be rearranged or even canceled because of limited viewership. It simply is not profitable to air them, or at least to air them in that particular time-slot. Notice the word you. As a media executive, you are making the decision to cancel or reschedule a program, not the rating service. Many viewers under this misconception have complained to the rating services about the cancellation of their favorite television program or the dismissal of their favorite radio personality.

The rating services do not make the decision to cancel programs. They do provide station management with information on how many viewers or listeners a station has, comparing this to other stations in the same community or, on a national scale, other networks. If a station is not attracting the number of listeners or viewers that management feels it should, and that same management feels a change in personnel or programming will improve the ratings, then the result may be a change in staff or programs. But keep in mind the many aspects of ratings. A station with few viewers may be perfectly comfortable with that fact since the specialized audience it is reaching may have high incomes and considerable buying power. Public broadcasting stations, knowing their ratings' position among commercial stations, direct quality programming to an audience with a higher educational level than average. Such a station may be perfectly satisfied with less than first place.

The entertainment industry also is interested in ratings because they specify the kinds of programs that will be a hit and those the public wants. If the public demands situation comedies, then producing science fiction for prime-time television might not be a wise decision. Similarly, the producer of a television series is interested in what share of the audience his program had in cities across the country or around the world.

Although broadcast management and the entertainment industry generally believe that the larger rating services are accurate and reliable, the public does not. A lack of understanding of the methodologies used creates considerable skepticism. Personal preference is also powerful, and all of the mathematical formulas in the world will not convince a devoted viewer that a favorite television program has been canceled because too few people watched. "That's impossible; all my friends watch that program every week!" This enthusiasm may be mistaken. Among your friends, perhaps everyone does watch the program. But perhaps because your friends watch the program and talk about it in the dormitory, their friends do not want to be left out, so they also watch. Neither "standard" is an accurate indication of how the rest of the viewing audience feels about the program. How is a broadcast rating determined?

Judging Accuracy: The Sampling Process

A typical ratings skeptic will claim that there is absolutely no way that a small group of people selected to tell what television programs they watch or what radio stations they listen to can possibly determine the viewing or listening habits of thousands or millions of other people. To some extent they are correct, but only if the group of people polled is extremely small.

At the heart of broadcast ratings is a process called *sampling*. Sampling means exactly that, examining a small portion of some larger portion to see what the larger portion is like. A chef in a large restaurant tastes a tiny teaspoon of soup from a five-gallon kettle to determine if the soup is ready for serving. A doctor can examine a small blood sample from your arm to ascertain the characteristics of the rest of the blood in your body. Obviously, sampling is a much better way of determining your blood characteristics than draining you dry!

The essence of the process is a type of sampling called *random sampling*. Random sampling means that when a sample is taken, *each unit of the larger portion has an equal chance of being selected*. If the population of Toronto, Canada is being sampled, each person in Toronto has an equal chance of being selected. If households are being sampled, *each household* in Toronto has an equal chance of being selected. There are even different types of random sampling. If you picked numbers out of a hat, you would be conducting a simple random sample. But let's assume you wanted to draw a sample of 100 names from a voting list of 1,000 names. So you chose every tenth name on the list to obtain your sample of 100. By doing this, you conducted a systematic random sample.

You may say, "OK, I'll get on the phone, randomly select ten people from my hometown telephone directory, ask them what radio station they're listening to, and find what the rest of the town is listening to." If you do this and your prediction turns out to be true, you simply would be lucky. Because the random sample you selected was so small that its sampling error was too *large* to make an accurate prediction. *Sampling error* is determined by the size of the sample. The larger the sample, the smaller the sampling error.

Mathematicians centuries ago proved that a truly random sample is all that is needed to tell the characteristics of a larger population. Moreover, once a certain number of people are chosen for the random sample, increasing that number will not significantly change the outcome. In fact, you may be surprised to find out how small that random sample needs to be. For example, a truly random sample of 600 people is sufficient to make a prediction about the entire city of New York with only a ±4 percent sampling error.[2] That means that if 75 percent of the 600 people you sampled were listening to a certain radio station, then you could predict that somewhere between 71 percent (75 percent − 4 percent) and 79 percent (75 percent + 4 percent) of the entire city of New York was listening to the same radio station. Increasing your sample size to 1,000 only would decrease the sampling error to ± 3.1 percent. Moreover, increasing the size of the population to include the entire United

States would not significantly change the sampling error. With these figures, it is easy to see that a rating service can predict the viewing or listening habits of an entire city or nation by sampling about 1,200 people and still be within a few percentage points of being accurate. Although there will always be skeptics, those who understand sampling and other factors in the ratings process rely on them to make programming decisions.

Gaining Cooperation for Data Collection

Some of the most critical steps in determining a broadcast rating are in the data gathering process. After the random sample has been selected, the rating service next must secure the cooperation of people willing to provide the information that it is trying to collect. People may be hesitant to cooperate with a rating service for a number of reasons. They may be apprehensive of the stranger at the door or on the telephone who is requesting their help. Other prospective candidates may not be at home or may have moved. To overcome these obstacles, field representatives employed by the major rating services are highly trained in everything from persistance to interpersonal relations, training that has paid off in a cooperation rate of approximately 80 percent. One interesting research finding has shown that the person who watches educational television is more inclined to cooperate with a rating service[3] than those who seldom watch educational television. Thus, a rating service that interviewed only "easy cooperators" could find its results leaning toward educational television programs. This leads us to conclude that, although a high cooperation rate is important to secure fairly accurate ratings, the rating service needs to persevere if it encounters resistant candidates in its random sampling procedure.

Interviews, Diaries, and Meters

The rating services use three different methods of gathering information: interviews, diaries, and meters. Interviews are either personal or by telephone. Different studies have produced differing results on which is more effective.[4] Some rating services use both. Regardless of which type is used, certain variables must be considered in determining their outcome. For example, differences in personality among interviewers is important. You know that you react differently to the same question asked by different people. The tone of their voices, their articulation, and their inflections all influence your interpretation of their question. Dress can be another variable. A neatly dressed interviewer can communicate a sense of importance to the interview. Showing up at the front door unshaven or with your hair set in rollers would communicate something entirely different. When many different people gather data, these same variables can distort the results of the survey. Even a slight rewording of the question can change the results. Rating services, therefore, conduct sophisticated training sessions to make sure their interviewers are asking the same questions in the same way.

Besides personal and telephone interviews, rating services frequently utilize the diary method of collecting data. With this method, the viewer or listener keeps a record of the programs and stations he or she tunes to at periodic intervals during one given week. This schedule is then mailed back to the company which tabulates the results of all diaries submitted. In some cases, a small monetary incentive is included with the diary, usually a half-dollar. Figure 17–1 is a page from an Arbitron diary

Figure 17–1   (Arbitron Radio)

WEDNESDAY					
TIME		**STATION**		**PLACE**	
(Indicate AM or PM)		WHEN LISTENING TO FM, CHECK HERE (✓)	FILL IN STATION "CALL LETTERS" (IF YOU DON'T KNOW THEM, FILL IN PROGRAM NAME OR DIAL SETTING)	CHECK ONE (✓)	
FROM —	TO —			AT HOME	AWAY-FROM-HOME (INCLUDING IN A CAR)

PLEASE CHECK HERE ◯ IF YOU DID NOT LISTEN TO A RADIO TODAY.

PLEASE MAIL TOMORROW

to measure radio listening. It specifies the day of the week which the particular page(s) covers, the time at which the radio station was listened to and the duration of listening, a special column to note FM listening, and a place to check if listening was done at home or away from home. Arbitron also leaves a space at the bottom of the page to check if you did not listen to radio that day.

The third method of rating service data collection is the meter method. The meter method is used by both A. C. Nielsen and Arbitron. Nielsen enlists the cooperation of a household to install on the television set a small, inconspicuous, monitoring device called a Storage Instantaneous Audimeter (Figure 17–2), which is connected through a telephone system to Nielsen's central computer. The computer automatically dials each monitoring device at specific intervals and records the readings. The monitoring not only reveals what channel is on at any time of the day but also tells if more than one set is in use in the household. Diaries in other homes also supplement and help verify the Nielsen meters. Although each of the three methods has its advantages and disadvantages, research has shown little difference in the reliability of data collected from any of the three.[5]

When the data reach the rating service headquarters, banks of computers (Figure 17–3) process the information and provide printouts in the form of ready-to-publish sheets which are bound in booklet form and made available to the local stations and networks. The computer data sometimes are also given to advertising agencies and station sales representatives to aid in making time buys. If a media buyer for an ad agency in New York wants to know on which stations to buy commercials

Figure 17–2 Nielsen Instantaneous Audimeter. (A. C. Nielsen Co.)

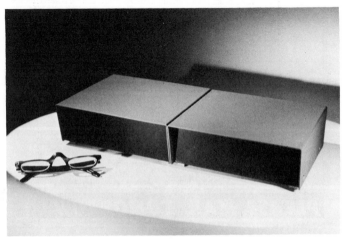

Figure 17–3 Arbitron computer center. (Arbitron)

to reach the female audience 18 to 34 years of age in Seattle, Washington, the computer will provide the information. Similar information can be supplied for national time buys when an advertiser wants to reach a certain type of audience.

interpreting the ratings

Up to this point we have talked about some of the methods used to collect broadcast ratings. Now we're going to learn how they are interpreted, some of the formulas used to interpret them, and the meaning of the terms commonly used in reporting ratings. This material is not difficult, and you do not need to know much math. But read carefully. You should remember that the terms we'll encounter sometimes have different meanings to different rating services.[6] Although we'll learn some widely used definitions, these can vary among sources. With this in mind, let's begin.

Formulas and Terminology

Rating. The first term we'll tackle is *rating*. You may say, "But, we've already been talking about ratings." That is correct. The term is used for all the processes employed to predict viewing and listening habits. But that is a more general definition of the term. More precisely, rating refers to the *percentage of people in a given population who are tuned to a radio or television station during a given time period.* For example, the formula to determine a rating for a radio station is the population (homes using television is used in TV) divided by the number of people who are listening:[7]

$$\frac{\text{Station's Listeners}}{\text{Population}} = \text{Rating}$$

Let's assume that the town of Elmsville has a population of 10,000 people and supports three radio stations. Using a random sample of the Elmsville population, we have projected that 1,000 people are listening to WAAA radio. Dividing 1,000 by 10,000 determines that WAAA radio has a rating of 10 percent.

$$\frac{1,000}{10,000} = 10\%$$

Share. Now let's consider another term—*share.* Share is also expressed as a percentage, which is the number of a station's listeners divided by the number of *all listeners during a given time period.* The formula for figuring share is:

$$\frac{\text{Station's Listeners}}{\text{Total Listeners}} = \text{Share}$$

Share does not determine the number of WAAA listeners by using the entire population of Elmsville, only those in Elmsville listening to radio. For example, when the survey was conducted, we found through sampling that between 3:00 P.M. and 7:00 P.M., only half, or 5,000 Elmsville residents, had their radios on. Thus, although we found that WAAA had a rating of 10 percent, its share was higher. In fact, the share for WAAA was 20 percent. Applying our formula, the 20 percent is reached as follows:

$$\frac{1,000}{5,000} = 20\%$$

Another way to understand share is to consider it as part of the listening pie.[8] For example, Figure 17–4 represents all of the Elmsville radio stations' listeners. Notice that WAAA occupies only 20 percent of the pie, and the other two stations have the remaining 80 percent.

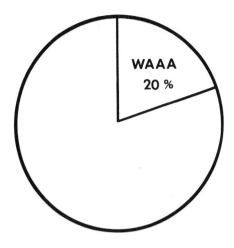

Figure 17–4 Of the total number of listeners, WAAA's share is 20 percent.

Average Quarter-Hour Persons. Now let's examine *average quarter-hour persons*, which is *an estimate of the number of persons listening to a station during any quarter-hour in a specified time period.* For example, many ratings are based on four-hour time blocks, such as 6:00 A.M. to 10:00 A.M. or 3:00 P.M. to 7:00 P.M. In radio, the morning time block covers the morning drive-time, and the evening block covers the evening drive-time. Time blocks can vary from market to market and from rating service to rating service. For television, the breakdown becomes more detailed, with time blocks as small as fifteen minutes. The smaller time block is necessary since television programming changes much more frequently than radio.

To define average quarter-hour persons, we'll use radio and the evening drive-time of 3:00 P.M. to 7:00 P.M. Remember, we are interested in average quarter-hour persons. Between 3:00 P.M. and 7:00 P.M. there are 16 quarter-hours (4 hours with 4 quarter-hours each). Let's assume that our random sample of Elmsville revealed that only 100 people from our sample were listening to station WAAA, and that those 100 people listened from 3:00 P.M. until 4:00 P.M. That means that they listened during four quarter-hours (four quarter-hours between 3:00 P.M. and 4:00 P.M.). We also know that each person in our sample represents 20 residents of Elmsville, since there are 10,000 people in Elmsville and our sample is 500. (10,000 ÷ 500 = 20.) Since our sample showed that 100 people were listening to WAAA, we can predict that between 3:00 P.M. and 4:00 P.M., 2,000 people were listening to WAAA (100 people from the sample multiplied by 20, the number represented by each person in the sample). Now to find WAAA's average quarter-hour persons between 3:00 P.M. and 7:00 P.M., we take 2,000 and divide it by 16 (the number of quarter-hours between 3:00 P.M. and 7:00 P.M.).

The answer is 125. Notice that our survey discovered that people were listening to WAAA only between the 3:00 P.M.-to-7:00 P.M. time block, and so projected those figures. We did this because when the ratings for Elmsville are published, all of the stations will compete in that same evening drive-time block.

"Cume" Persons. In determining the average-quarter hour persons, we also determined the number of *cume (cumulative audience) persons,* which was 2,000. Cume persons are *the number of different persons listening to a station at least once during a given time period.* It refers only to a specific time period and only to people listening at least once during that period of time. We never count a person more than once when we are figuring cume. For example, if of the 100 people who were listening to WAAA between 3:00 P.M. and 4:00 P.M., our sample discovered that ten of them turned their radio back on and listened again between 6:00 P.M. and 7:00 P.M., we would not count them a second time during that evening drive-time block.

The easiest way to understand cume is to compare it to a magazine subscription. When figuring the circulation of a magazine, the subscriber is counted only once, regardless of how many times the magazine is read. After picking up the magazine in the afternoon mail at 3:00 P.M., the subscriber might read it until 4:00 P.M., fix dinner, and finish the magazine between 6:00 P.M. and 7:00 P.M.. Although he or she reads the magazine twice before finishing it, he or she still is counted as only one subscriber. The same applies to cume persons, the number of different persons who listen at least once during a time period.

There are many more terms and formulas than we have covered here. If you are interested in learning more about ratings, consult the "Spotlight on Further Learning" section at the end of the chapter or books on advertising.

An actual rating also entails many more subdivisions of data than those we have encountered here. You might compare women between the ages of 25 and 49 who listen between 3:00 P.M. and 7:00 P.M., or you might compare men between the ages of 25 and 64 who listen between the hours of 3:00 P.M. and 7:00 P.M.. A portion of an actual rating from an Arbitron survey of radio listening in Dayton, Ohio shows how a listener survey is displayed for broadcast clients (Figure 17–5).

The survey represents radio listening during the 3:00 P.M. to 7:00 P.M. time block on Sunday. Notice how the age groups for women are actually broken down into categories of 25 to 49 and 25 to 64 years of age. Other age-category breakdowns for both men and women are included in the complete survey. Not all of the stations listed are actually in the Dayton area. Some are in outlying areas while others, such as WKRQ and WSAI, are in Cincinnati, southwest of Dayton (Figure 17–6).

STATION CALL LETTERS	WOMEN 25-49 TOTAL AREA		WOMEN 25-49 METRO SURVEY AREA				WOMEN 25-64 TOTAL AREA		WOMEN 25-64 METRO SURVEY AREA			
	AVG. PERS. (00)	CUME PERS. (00)	AVG. PERS. (00)	CUME PERS. (00)	AVG. PERS. RATING	AVG. PERS. SHARE	AVG. PERS. (00)	CUME PERS. (00)	AVG. PERS. (00)	CUME PERS. (00)	AVG. PERS. RATING	AVG. PERS. SHARE
*WAVI	4	9	4	9	.3	2.6	14	22	14	22	.7	6.1
WBZI	4	23	1	9	.1	.7	12	31	9	17	.4	3.9
WDAO	19	44	19	44	1.3	12.4	28	53	28	53	1.4	12.3
*WGIC	9	10	9	10	.6	5.9	19	20	9	10	.4	3.9
WHIO	13	75	12	64	.8	7.8	19	100	18	89	.9	7.9
WHIO FM	26	54	12	24	.8	7.8	72	139	41	89	2.0	18.0
WING	31	71	31	71	2.2	20.3	32	79	32	79	1.6	14.0
WONE	26	64	17	35	1.2	11.1	34	94	21	52	1.0	9.2
*WPTW WPTW FM	4	8	4	8	.3	2.6	9	15	9	15	.4	3.9
TOTAL	4	8	4	8	.3	2.6	9	15	9	15	.4	3.9
WTUE	7	21	7	21	.5	4.6	7	21	7	21	.3	3.1
WVUD	5	19	5	19	.4	3.3	5	19	5	19	.2	2.2
WKRC	2	18	2	18	.1	1.3	2	18	2	18	.1	.9
WKRQ	6	22					6	22				
WLW	11	54	3	19	.2	2.0	18	76	3	19	.1	1.3
WSAI	7	31					7	31				
WWEZ	11	19	11	19	.8	7.2	19	46	13	27	.6	5.7

Figure 17–5   (Arbitron Radio)

The survey was of radio listening habits of the Dayton area, and some residents in that area listen to radio broadcasts from other communities, thus these stations are listed in the Dayton survey data.

Understanding the Concept of Survey Area

Figure 17–5 lists categories for both the total survey area (TSA) and the metro survey area (MSA). These represent the geographic areas from which the random samples were drawn. The TSA and MSA are determined by a combination of data, including the signal contours of the stations, audience listening habits, and government census data. The TSA is larger than the MSA, encompassing outlying areas. The MSA corresponds more closely to the actual "metropolitan" district of a city. Figure 17–6 represents this. In Arbitron's map of Dayton, look closely at the areas of white and the horizontal lines. The area in white is the TSA. In the display of listenership data in Figure 17–5, the TSA is on the two

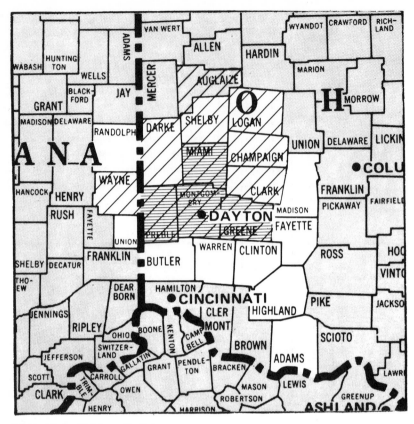

Figure 17–6 (Arbitron Radio)

columns on the left of each category. The area in which horizontal lines appear on the map is the MSA and is represented on the remaining four columns in the listenership data of Figure 17–5. The diagonal lines in the map represent the Area of Dominant Influence (ADI) which applies primarily to television viewing. It is an exclusively defined market area (no two ADIs overlap) of measurable viewing patterns.

Besides the terms we have already learned, you should know two more. The term *households using television* (HUT) refers to households in which one or more television sets are in use. *Television households* means households with at least one television set, but which are not necessarily in use at survey time.

applying ratings to management decisions

Ratings by themselves are not valuable. Their importance lies in how broadcast management uses them to make decisions.

If you were a television station manager and the ratings showed that your station was second in the local news ratings, you would have a number of alternatives. First you could do nothing, leaving everything as it is. Your station might be profitable, and you would see little need to change. Or you could replace your news department with different people. A third alternative would be to replace the news with another type of programming you feel would be more profitable. Whatever your decision, it could not be made without a careful analysis of the ratings.

You might find that although your station was second in the category of all adults 18 years and older, you had some other strength to offset this. You discover that among adults between 18 and 34 years of age, your station was first. The other station was reaching mainly adults 49 years and older. You are very satisfied with your station's showing in the younger group. It is an acquisitive group that still is forming buying habits and is making such major purchases as homes, home appliances, and automobiles.

Looking more closely at the ratings, you discover that for the time-slot immediately preceding local news, the ratings show the other station far ahead of yours. What does this mean? Perhaps the news department is not to blame for the second place showing as much as the program that preceeds the news is. Some viewers simply would rather stay in their comfy chairs than get up to change channels. So changing the program leading into the news may be all that is needed for your station to capture first place.

Ratings determine something else which your station's sponsors and advertising agencies will want to know—how expensive it is to reach an audience on your station compared to the competing station. The answer is in the cost-per-thousand (CPM). *Cost-per-thousand* is *the cost of reaching 1,000 people.* Let's assume that your station's rate card lists the cost of a one-minute commercial as $100. That same commercial on the competing station costs $150. Now let us assume that the Ace Garden Supply Co. wants to advertise topsoil to people 25 to 49 years of age. You examine your station's latest ratings and find that between 9:00 and 10:00 A.M., you are reaching about 12,000 viewers in the 25 to 49 year-old age bracket. But your competition reaches 14,000 viewers and has convinced the Ace people that they should buy advertising on their station. Now you compare the CPM of your station with that of the competition. First you figure your station's CPM by dividing the cost of your one-minute commercial by the number of people reached in thousands by that commercial.

$$\frac{\text{Cost}}{\text{Thousands of Viewers}} = \text{CPM}$$

Before reading further, substitute the figures from our sample rate card and ratings and figure the CPM. The correct answer is $8.33. You arrived at the $8.33 by dividing $100 by 12. If you did not get $8.33 you probably divided the $100 by 12,000. That would tell you how much it cost to reach just one person. But remember, CPM is the cost of reaching 1,000 people, and since there are 12 of those (12,000) we divide the $100 by 12.

Now you need to figure the CPM for your station's competition. To do this, divide $150 (the cost of the competition's one-minute commercial) by 14 (the "thousands" of people viewing). The answer is $10.71. Clearly the most economical way to reach the viewers is on your station! It is now up to you to convince Ace Garden Supply of that fact.

As a consumer of broadcast communication, you immediately will recognize that what you see on your television screen is the result of management's statistical scrutiny. A legitimate criticism of this decision making is that it creates bland programming to please the general public. That is true. Commercial broadcasting is business, and nowhere does this become more evident than in the use of broadcast ratings.

criticism of broadcast ratings and improvement of accuracy

Despite mathematical formulas and efforts toward accuracy, the broadcast ratings still receive considerable criticism. Some is warranted; some comes from ignorance.

Two Concerns: Sampling Error and Minority Audiences

One criticism of ratings is of the actual sample. Statistics tell us that although there may be a given sampling error for a random sample of an entire population, that sampling error increases as the population shrinks. For example, if you have an error of ± 4 percent for the total sample, that sampling error grows as you divide the sample into smaller units, such as those of sex or age. Although your total sample may have been 600 people, it may include 100 teenagers. Thus the sampling error for teenagers is based on 100, not 600.

There is much criticism of minority audience ratings. Since it usually is necessary for a station to reach a certain percentage of the audience before even being listed in the ratings books, those that reach small, specialized audiences can suffer. When the ratings do not show any listeners, an advertiser's reluctance to spend dollars is understandable. Some rating services use 10 percent listenership or viewership as the cutoff, below which a station is not reported. In a large city such as New York where there are approximately one-half million Spanish households, if each of these were tuned in to the Spanish station, it still would not be reported.[9]

Some criticism has made a difference. Arbitron withdrew a market report for McAllen-Brownsville, Texas after receiving complaints that the sample was not adequate.[10] The rating service attributed the problem to field staff members who helped viewers fill out diaries.[11] Rene Anselmo, president of the Spanish International Network, has charged that "both the Spanish and Black communities have been deprived of the variety and diversity of media to which they are entitled because rating practices have adversely affected the economics of minority-owned broadcasting."[12] There have been efforts to correct these alleged deficiencies, which include having diaries printed in Spanish and compensating more accurately for minority audiences.

Broadcasters have their own brand of rating service criticism. An Austin, Texas broadcaster complained about surveys of the Austin market. Examining the diaries used by one service, he was "particularly disturbed" that the people who were heads of households tended to be people over 50 years of age. He stated in his letter, "In one county, the average head-of-household age in the diaries was 60, in another county 65, and in still another county 62." He felt that "an accurate rating couldn't be taken from people in that age group when the over-all average is so much younger."[13] He went on to cite instances in which two different surveys of the Austin market differed substantially for the same time period.

The ratings received widespread attention in 1975 when Nielsen reported a significant drop in television use.[14] Then when the decline continued, finding out why a decline had taken place became more important than blaming the ratings for creating an artificial decline.[15]

The press has also leveled its share of criticism. The Associated Press reported the story of the man in Manhattan who became part of the Nielsen sample by not telling the rating service that his grandfather, the man they asked to complete the diary, had been dead for eleven years.[16] On another front, the Radio Advertising Bureau launched a task force to stimulate competition between services. The task force concentrated on improving the methods used in measuring radio audiences.

Working to Improve Broadcast Ratings

Because of the developments in cable television, people's changing lifestyles and the sophisticated needs of clients, there are continual efforts to improve the accuracy of broadcast ratings. Some of these efforts are credited to the services themselves, others to networks and broadcast associations. In the case of the latter, a state broadcasters' association complained to the Federal Trade Commission about a station which engaged in heavy promotion, called *hypoing*, during a rating period. Because of this complaint, the major rating services now include special notations within their reports that alert readers to any special promotions during the rating period which may have "artificially" increased the size of the viewing or listening audience (Figure 17–7).

In 1975, the three major commercial networks began developing monitoring systems to keep track of their affiliate stations' use of network programming. Remember that although a network program may be fed to affiliates, it is no guarantee that every affiliate is airing it. The affiliates may substitute something else. And although a meter attached to a viewer's television may record the channel being watched, it does not record the program being watched. Although the ratings showed homes were tuned to network-affiliate channels, the viewers may instead have been watching a local basketball game which preempted network programming. With the automatic monitoring system, the meter monitors the station and tells the rating service computer if the station is airing network programming.

Cable has been a particular problem for the ratings. Not only are channels from distant markets often preempted for local programming, but some cable operators also supply different stations at different times on the same channel. Such procedures keep the rating services hopping, and they continually attempt to distinguish the viewing habits of cable and noncable subscribers in their reports.

Figure 17–7

IMPORTANT STATION ACTIVITY
WAAA...

made direct reference to the current diary survey for this market. See page 5 for complete details.

ARBITRON RADIO

It has been reported to Arbitron that on Wednesday, January 12, 1977, direct reference to the current diary survey of Your City Market was made on WAAA's morning show from approximately 8:00 AM to 10:00 AM.

Ratings, the importance of ratings, diaries and Arbitron were discussed.

According to WAAA, this discussion was a direct result of an article titled, "John Doe Loses No. One Position", that appeared in Your City Press on Tuesday, January 11, 1977.

Rating services also are attempting to give their clients more demographic information. For example, Arbitron's Information on Demand Zip Codes gives clients such demographic breakdowns as income, education, occupation, and eighteen additional demographic characteristics (Figure 17–8). Such specialized target-audience measurement may revolutionize the station rate cards of the future. We may even see specialized rate cards for different sections of town, much like print-media circulation. If you own a neighborhood store in a high-income residential area, you may be able to purchase commercials based on your clientele's viewing habits. A television account executive may show you a ratings book which indicates that a special weekend sports program has a large viewing audience in your clientele's zip code area. You thus could receive a special advertising rate based on that particular program's "broadcast" circulation. It would be cheaper than the rate for reaching the entire viewing audience but perhaps more than you might pay for reaching a low-income audience. This specialized sales approach may become more common as cable television brings a multitude of specialized channels into our living rooms.

Some radio networks have arranged for rating services to base their radio surveys on the Areas of Dominant Influence (ADI) used in television ratings.[17] The problem in using a Metro Service Area (MSA) and a Total Survey Area (TSA) is accurately rating many of the smaller radio stations which are not in one of the larger metropolitan areas. Remember, for a station to be listed in a ratings report, it must have the minimum number of listeners for a metro survey area. As we learned previously, a station in an outlying community which does not have the minimum MSA audience may be left out of the rating (Figure 17–9). Even if an outlying station manages to reach one or more MSAs, most of its listeners would be outside the MSA, and adding the MSA audience would represent a much smaller audience than there actually is. Even if the TSAs are added together, it does not give a true audience measurement, since some of the audience would be counted twice and some three times (Figure 17–10). Using the ADI to measure the listenership solves all of these problems. Since ADIs do not overlap, the station that spills over into three different ADIs can add the listenership from each one to find out its total audience (Figure 17–11).

monitoring quality: the Broadcast Rating Council

Concern over the accuracy of the ratings and their influence on national programming prompted congressional hearings on the subject in 1963. The publicity of the hearings focused on the industry's need for a systematic, self-regulatory body to assure confidence in broadcast ratings.

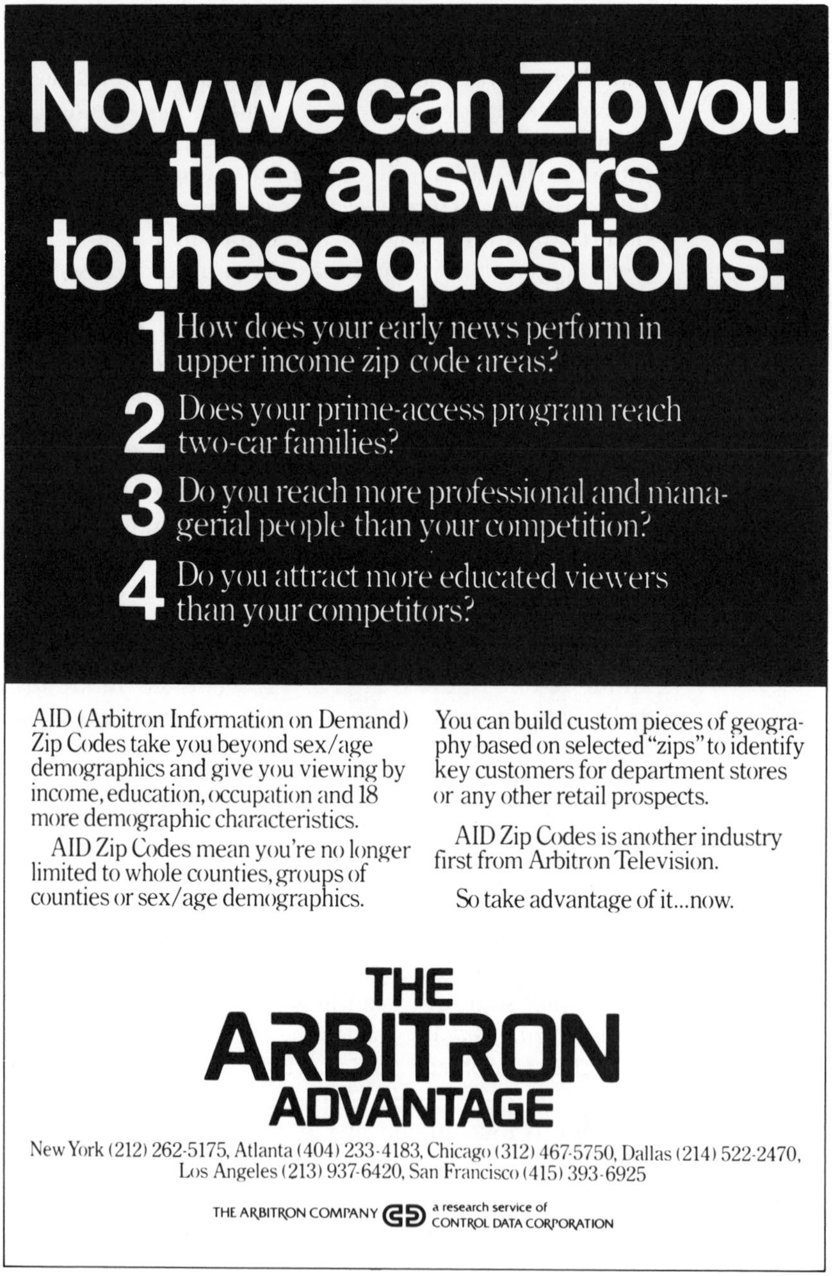

Now we can Zip you the answers to these questions:

1 How does your early news perform in upper income zip code areas?

2 Does your prime-access program reach two-car families?

3 Do you reach more professional and managerial people than your competition?

4 Do you attract more educated viewers than your competitors?

AID (Arbitron Information on Demand) Zip Codes take you beyond sex/age demographics and give you viewing by income, education, occupation and 18 more demographic characteristics.

AID Zip Codes mean you're no longer limited to whole counties, groups of counties or sex/age demographics.

You can build custom pieces of geography based on selected "zips" to identify key customers for department stores or any other retail prospects.

AID Zip Codes is another industry first from Arbitron Television.

So take advantage of it...now.

THE ARBITRON ADVANTAGE

New York (212) 262-5175, Atlanta (404) 233-4183, Chicago (312) 467-5750, Dallas (214) 522-2470, Los Angeles (213) 937-6420, San Francisco (415) 393-6925

THE ARBITRON COMPANY ⊖⊃ a research service of CONTROL DATA CORPORATION

Figure 17–8

406

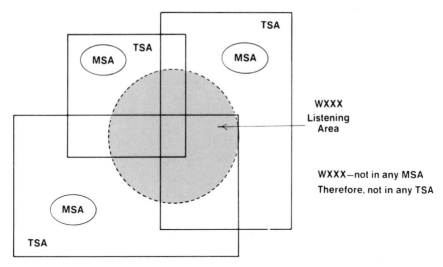

Figure 17-9 (Courtesy ABC Radio)

Thus, in 1963 the National Association of Broadcasters joined with ABC, CBS, NBC, and Mutual, with the blessings of the American Association of Advertising Agencies and the Association of National Advertisers, to form the Broadcast Rating Council (BRC). The council's main duty is to give the industry and the rating services credibility, assuring advertisers

Figure 17-10 (Courtesy ABC Radio)

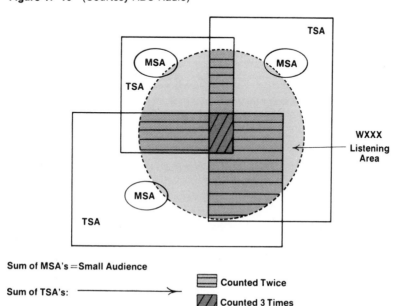

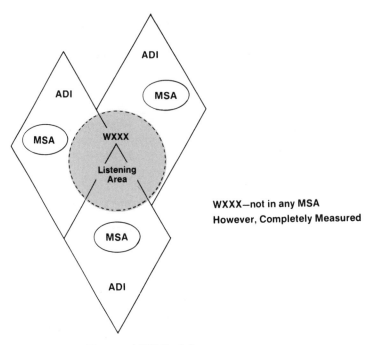

WXXX—not in any MSA
However, Completely Measured

Figure 17–11 (Courtesy ABC Radio)

that radio and television indeed are reaching the audiences they say they are reaching. The BRC's charges state:

> To secure for the broadcasting industry and related users audience measurement services that are valid, reliable, effective and viable; to evolve and determine minimum criteria and standards for broadcast audience measurement services; to establish and administer a system of accreditation for broadcast audience measurement services; to provide and administer an audit system designed to insure users that broadcast audience measurements are conducted in conformance with the criteria, standard, and procedures developed.

Since its inception, the council has policed the rating services, either granting or revoking its seal of approval on them and their surveys (Figure 17–12). The BRC coordinates a major auditing process of rating services practices, paid for by the rating services themselves. These audits cover all phases of the rating process, from the development of sample design, the gathering of data, data processing, to the published reports. The audits are unannounced and cover various market areas selected by the auditors. In return for the council's accreditation, the accredited services agree to: (1) provide information to the BRC; (2) operate under substantial compliance with BRC criteria; and (3) conduct their services as they represent them to their subscribers and the BRC.

Accredited by
THE BROADCAST
RATING COUNCIL
Inc.

Figure 17–12

Technological advancements will undoubtedly affect rating services' future operations. In the experimental stages are a special exterior monitoring device which records a household's television viewing habits from outside the home, and a device which monitors the radio-station listening habits of passing automobiles.

Although the criticism continues, rating services still are the only source of reliable, economical, broadcast-audience data. And rating is done quickly. A network executive can walk into the office on Tuesday morning and find Monday night's ratings on the desk. A local broadcaster can obtain data on local listeners without having personally to conduct the ratings or to hire, train, and supervise someone else to do it. Continual efforts are being made by broadcasters, networks, and rating services alike to improve the quality of this important area of feedback for the industry.[18]

the future of broadcast ratings

The entire realm of mass communication is undergoing rapid change. It will become even more pronounced as audiences become more specialized, as the number of media increases, and as new transmission techniques such as fiber optics transform the broadcasting spectrum from a limited to an almost unlimited resource. On the future of the ratings, Professor Robert E. Balon speculated on a scenario in the year 2016 predicting that as many as 26 networks would be in operation. Those networks would be trying to gain a mandatory 3 percent share and would be running reruns of "Bionic Grandson." Viewers would have 30 to 40 shows to watch every half-hour on multiscreen television, developed back in 1996. Prime time would be from 4:00 P.M. until 2:00 A.M. and from 6:00 A.M. until 10:00 A.M.. Minute populations would be the target audiences with advertisers trying to reach 18 to 21 year-olds. Balon predicts that almost "nothing short of a total, daily sensory monitoring of hundreds of selected sub-populations will be acceptable."[19]

Ask yourself how our current as well as our future society fits into the ratings scheme. Do you feel, after learning about random sampling, that the ratings are an accurate gauge of national preference in television and radio? Would there be a better way to analyze the audience and make decisions affecting the content of broadcast messages? If we actually do reach a point at which daily sensory monitoring becomes necessary, would you volunteer to walk around with a mini-computer strapped to your wrist, sending data about your viewing and listening habits to a central data processing unit? What would be the ethical considerations of such a system? Would there or should there be a need for government control over such a system? Discuss these questions with your friends and begin thinking about your own broadcasting philosophy for the future.

summary

Chapter 17 examined broadcast ratings. We learned that they are a controversial subject and that the public does not always believe them. Ratings are only as good as the sampling procedure employed. Random sampling, in which each person or household has an equal chance of being surveyed, is important to accurate ratings. The larger the sample, the smaller the margin of error (sampling error), although a sample of 600 people is sufficient to create a margin of error of about ± 4 percent. Once the sample is drawn, every attempt should be made to gain the cooperation of as many people in the sample as possible. Data for broadcast ratings are collected by three methods: interviews, diaries, and meters.

We also learned some of the ratings terminology, including "rating," "share," "average quarter-hour person," and "cume persons." We examined the different survey areas. We looked at the total survey area (TSA), metro survey area (MSA), and area of dominant influence (ADI). Some radio networks have used ADIs, a survey area usually reserved for television, to measure network radio audiences.

When accurate and reliable ratings are available, management can use the data to make programming, personnel, and marketing decisions. Nevertheless, criticism of the ratings, of such things as an adequate sample and neglect of minority audiences, continues. In most cases, the major rating services continually try to improve their procedures to provide more reliable and accurate data for their clients. The quality of these major services and their samples is monitored by the Broadcast Rating Council (BRC).

spotlight on further learning

Committee on Nationwide Television Audience Measurement (CONTAM), *How Good are Television Ratings?* New York: New York Television Information Office, 1969.

Committee on Nationwide Television Audience Measurement (CONTAM), and MARTIN MAYER, *How Good are Television Ratings?* New York: New York Television Information Office, 1966.

Committee on Nationwide Television Audience Measurement (CONTAM), *Television Ratings Revisited.* New York: New York Television Information Office, 1969, 1970.

Demonstration Report and User's Manual: Nielsen Station Index. Northbrook, Ill.: A. C. Nielsen Co., 1975.

Description of Methodology: Arbitron Television Market Reports. Beltsville, Md.: Arbitron (American Research Bureau), Inc., 1975.

How Arbitron Measures Radio. Beltsville, Md.: Arbitron (American Research Bureau), Inc., 1974.

NIELSEN, ARTHUR CHARLES, *Greater Prosperity through Marketing Research: The First 40 Years of A. C. Nielsen Company.* New York: Newcommen Society in North America, 1964.

Reference Supplement: NSI Methodology, Techniques and Data Interpretation. Northbrook, Ill.: A. C. Nielsen Company, 1974.

18

the research process

FOCUS   After completing this chapter we should be able to

Distinguish between historical, descriptive, experimental, and developmental research.

Grasp the scope of broadcasting research.

Discuss survey research common in local stations.

Describe how a network election poll is conducted.

Explain important procedures in reporting public opinion polls.

Describe new methods of data collection.

To the average undergraduate student, the word *research* is about as exciting as a flu epidemic. The word conjures up images of bloodshot eyes pouring over unsolved problems, stuffy laboratories with bubbling flasks, incomprehensible calculations, and all-night study sessions for impossible exams. Even related words can cause the research jitters— words like "theory," "numbers," "experiments," "statistics," or "computers."

The jitters would subside quickly if we stopped long enough to realize that we use research every day. We conduct research, we make decisions based on research, and we are affected by research. Turning to the radio-television section of the evening newspaper is a form of research. You survey the available programs and make a decision about which ones to watch or not to watch. When you decided to enroll in college you may have examined college catalogues, looked at college guides, written letters to different college admissions counselors, and even had on-campus interviews.

The professional researcher in a major television network does the same sort of thing everyday. He or she may look through back issues of annual reports to establish the image of the company, in much the same way you examined college catalogues. Letters might be sent to affiliate stations asking local managers to write back and describe network service. Just as you wrote to the admission counselors and studied their responses, the researcher will examine the managers' responses. If additional information is needed, the researcher may interview the manager over the telephone or even visit the station and talk to the manager in person.

Certainly broadcasting has much more complex and sophisticated research projects than these. However, with some training and the opportunity to conduct research projects of your own, you would be able to direct the same type of research common in many local broadcasting stations. So the next time the word research pops up, don't be afraid of it. By learning more about the broadcasting research process, you can open up a whole new frontier of exciting knowledge and experience.

types of research

We first should become acquainted with the different types of research, historical, descriptive, experimental, and developmental.

As the name implies, *historical research* is on the past. For example, you

413

might conduct a research study tracing the history of one of the local radio stations in your community. Perhaps a local broadcaster is nationally famous as an industry pioneer. A study of his professional career would be an historical study. Historical studies are not necessarily on subjects which are "ancient" or "old." A research study of the first five years of WAAA radio still would be considered historical even though WAAA signed on the air in 1975.

Descriptive research describes a current condition. The most common descriptive research is surveys. A survey of local listenership or a survey of television sets in use are examples of descriptive research. In fact, descriptive research is one of the most common types in the broadcasting industry.

Experimental research is sophisticated research which entails a "controlled" experiment. For example, we might compare children's aggressive behavior before and after they watched violent television programs. We would select two groups of children. One group would watch a violent television program while another group might watch the same program but with the violent episodes deleted. We then would compare the actions of the two groups. Two terms used frequently in experimental research are *independent variable* and *dependent variable*. An independent variable is the factor which is manipulated in the research study. In our example, it is the violent episodes in the television program. The dependent variable refers to the phenomenon of change, which in our example is the children's behavior.

A fourth type of research is developmental. *Developmental research* is applied mostly to instructional television. The example of continually perfecting an ITV program on teaching tennis is an example of developmental research. In this type, certain objectives first are established. Then an instructional radio or television program is "developed" by producing, testing, revising, and retesting the program until it meets the stated objectives.

the scope of broadcasting research

Broadcasting research reaches nearly every facet of the industry. Aside from government research studies and those conducted by such research corporations as A. C. Nielsen, broadcasting research extends from colleges and universities to local stations.

Research in Colleges and Universities

Some of the most sophisticated research on every aspect of broadcasting is done in an academic setting. Many colleges have major research centers which concentrate on radio and television. Some of these are the

Communication Research Center, Broadcast Research Center, Division of Communication Research, Center for Media Research, and Institute for Communication Research. They are anything from depositories for current studies and scholarly research papers to centers which administer major research contracts examining everything from the social effects of broadcasting to technical problems associated with fiber optics transmission. Doctoral dissertations provide another opportunity for research as do research projects by students and faculty. Many individual studies appear in scholarly journals.

Research in academia is funded by a multitude of sources. Major foundations are active in funding broadcasting research, including the Ford Foundation, the Rockefeller Foundation, the Lilly Endowment, and the John and Mary Markle Foundation. The networks, corporations, and the federal government contract with institutions of higher learning to do research for special projects of interest to the funding organization.

The Networks and CPB

Although we are more attuned to their prime-time programming, the networks' research arms are an integral part of their total operation. Perhaps the most famous project was that started in the early days of CBS when the network hired an Ohio State University professor named Dr. Frank Stanton to conduct research on radio listenership. Stanton went on to become a CBS executive, and his research efforts remain firmly entrenched.

Even in the networks' early years, research was clearly defined. Duties assigned to NBC's research department in 1948 included five areas which were to be developed under a "Master Plan for Television Research":

1. The audience: its size, characteristics, and viewing habits.
2. TV stations and their coverage.
3. Programs: the contribution of research to their better selection and presentation.
4. Advertising: measuring TV's advertising effectiveness.
5. The social impact of TV: its effect on the family and children, the psychological effects of TV, public attitudes toward TV, and TV as an educational medium.[1]

Thomas E. Coffin, who currently directs NBC's research, feels that some of the recent changes in NBC research have included the ability of research to communicate more effectively with creative people—news and program executives, personnel in program development, producers, and NBC's advertisers.[2] In turn, creative people are finding new ways in which research can help them, the result being that more and more research is devoted to "reality problems."

Noncommercial broadcasting also does much research. The Research Office of the Public Broadcasting Service conducts research in the areas of system information, system finance, and audience evaluation.[3] The research office offers six services: (1) to be a source of information for stations and other PBS departments using federal monies distributed by CPB to public stations; (2) to coordinate information on program funding and to calculate such information as PBS broadcast hours, hours of programming produced by independents, programming acquired from foreign producers, programming sold to foreign countries, and programming hours intended for or on minorities and women; (3) to answer information requests from such places as other PBS departments, Congress, the White House, the public, the press, and book publishers; (4) to measure the extent of usable signals transmitted by public television stations and the demographic breakdown of the population in each signal area; (5) to provide prompt accounting of each affiliate station's use of PBS programs; and (6) to publish reports on PBS activities.

Although it is not a network, the Office of Communication Research at the Corporation for Public Broadcasting both conducts research itself and contracts for research studies which benefit noncommercial public stations. The office keeps abreast of current research, both in and out of public broadcasting, and disseminates and interprets this research for public broadcasting stations (Figure 18–1).

Local Stations

Although we usually do not think of local stations conducting research, virtually all do in some way or another, from community needs and ascertainment surveys to advertising effectiveness studies. In its publica-

Figure 18–1

cpb **News Briefs**
Office of Communication Research

Vol. 1 - NO. 3   **Date:** OCTOBER 14, 1977
Subject: TELEVISION AUDIENCE 1977

We have just received A. C. Nielsen's Television Audience 1977, their annual report on the state of the medium. We want to share with you some of the highlights on the general trends in TV ownership, population growth and station growth.

OWNERSHIP

● As of September 1, 1977, the total number of U.S. households was estimated at 74.7 million. 98% of these households (72.9 million) have at least one TV set.

More th~ ~o out ~ ~v household~ ~ a col~

tion, *A Broadcast Research Primer,* the National Association of Broadcasters gives some examples of applied research which can be conducted by local broadcasters with some research background. They are: (1) how many radio or television sets are in the market served by the station; (2) the size of the station's audience; (3) what share of the total listening or viewing audience the station has; (4) the demographic characteristics of the audience; (5) the buying habits of the audience; (6) the use of products and services, such as how many people who use a certain product also listen to or watch the station; (7) how the station's coverage compares to the coverage provided by competing media; (8) researching audience reaction to station programming; (9) evaluating the effectiveness of air personalities; (10) determining the image of the station, which in many ways also includes #8 above, what the public thinks of the programming; and (11) conducting public opinion polls.[4]

survey research common to local stations

The research areas listed above are mostly examples of descriptive "survey" research. Partly because survey research is more easily interpreted and applied than other types, it is the type most often used by local stations. Let's now look more closely at the methodology used to conduct two of the most common types of broadcast surveys: the FCC required community needs and ascertainment survey and the station image survey.

Pitfalls in Community Needs and Ascertainment Surveys

In chapter 14 we learned how to conduct a community needs and ascertainment survey as required by the FCC. Now let's examine that survey procedure more closely to understand why research processes are important to station management. As a manager of a station you could select representatives from the nineteen community groups and interview them about the community. However, you might very well miss some of your most important listeners, groups which are not part of the main-stream of society. Let's look at one example, the "voiceless" community.

Researchers Orville G. Walker, Jr., and William Rudelius of the University of Minnesota examined the procedure for reaching this "voiceless" community. They defined these "voiceless" groups as, "people with a common problem who were not formally organized and who had no widely recognized leaders or spokesperson in the community."[5] The two researchers classified these groups into three categories.

The first category is the *Past-in Future-out* groups. These people were once in the mainstream of society but now watch from the sidelines. They include such people as the elderly, mental patients, and the deaf. The

FCC ascertainment guidelines make provisions for the elderly. But Walker and Rudelius found that others were equally concerned about medical facilities for the mentally ill or special captioned subtitles on television programs for the hard of hearing, yet did not voice these concerns. No one had asked them.

The second category is the *Past-in, Future-in* groups. Here would be found such people as "runaway teenagers, unwed mothers, VD victims, and prisoners." This group was in the mainstream in the past and intends to return in the future, once physical or personal problems are overcome. Although the Minnesota researchers found these people did not need communication from broadcasters, the unwed mothers wished for information on special parental care, and the prisoners wanted educational programs.

The third group is the *Past-out, Future-out* people, those who are minorities because of race or disabilities. They felt "more or less permanently removed from the mainstream of American life because of a lack of understanding or outright discrimination." The FCC ascertainment provides for reaching racial minorities and women, but ends there. The two researchers discovered that the principal desires for these groups were to have their story told, and for broadcasters to communicate the negative misconceptions and stereotypes that had been attached to them in the past.

Walker and Rudelius also pointed out that, consistent with the Past-out, Future-out groups' "desire for a more realistic and truthful portrayal of their cultures and lifestyles, most of these groups expressed a very strong desire for greater influence over the creation and execution of television programs about themselves." In other words, these groups were not satisfied with the type of messages being directed toward them. To these groups, media access was important. "Consequently, they see creative control and active participation—both in front of and behind the cameras—as the only guarantee that a television program or series would accurately reflect their viewpoint."

Station Image Surveys

Besides discovering what people think of their community, broadcasters are also interested in finding out what people think of broadcasters. One way of determining this is to conduct station image surveys. These surveys can be combined with ratings for an in-depth look at how the audience perceives the programming.

Imagine that you are managing a television station, and that you discover that certain programs are not getting a satisfactory share of the viewing audience. But you do not know why. Ratings tell only who is viewing what and when. They do not tell why a person likes or dislikes a certain program. A station image survey seeks to find out why people do or do not watch or listen to a station. It explores such nebulous areas as attitudes toward and opinions about programming and personalities, and it helps management make decisions. For example, if you discover that a program is being watched by a large share of your audience but find in a station image survey that the audience actually rates the program very low, then you could have a problem. The audience may be watching that program only because the competition is airing even more dismal programming. If the competition were to change their offerings and insert a popular program instead, you could have your audience swept away overnight.

Consider an example from radio. Assume that you are the program director of a radio station in a rather large market. You discover that although you seem to be capturing a large share of the audience for most of the broadcasting day, that share tends to dip during the local news. Why? You have already conducted research comparing the number of stories your news department with those in the rest of the stations in your market. You know your station has been consistently ahead of the others. So you decide to design a station image survey (Figure 18–2).

First you decide to determine if the audience even recognizes the name of your station's news director. Then you include a series of open-ended questions on why or why not people listen to your station's local news. From these questions, you will try to isolate those factors which are hurting the ratings.

Figure 18–2   (Arbitron)

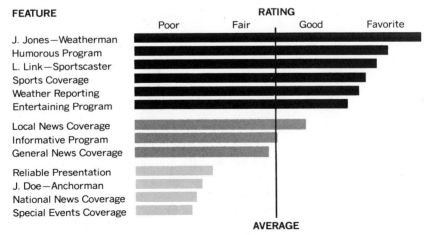

STRENGTHS AND WEAKNESSES
HOW WAAA NEWS VIEWERS RATE WAAA EARLY NEWS REPORT
MONDAY-FRIDAY, 6:00-6:30 PM   TOTAL ADULTS 18 +

FEATURE	RATING
	Poor Fair Good Favorite
J. Jones—Weatherman	
Humorous Program	
L. Link—Sportscaster	
Sports Coverage	
Weather Reporting	
Entertaining Program	
Local News Coverage	
Informative Program	
General News Coverage	
Reliable Presentation	
J. Doe—Anchorman	
National News Coverage	
Special Events Coverage	

AVERAGE

You discover that the public perceives your news director as being the only local news person employed by your station. Although you have a full staff of reporters, the audience is not aware of it. Second, you discover that the public thinks the competing stations cover national news much better than your station does. Finally, you find out that when there is fast-breaking news, the audience consistently turns to other stations.

A bit shocked by the results, you swallow your ego and go to work. Fortunately, you are too smart to fire the news director. If you did, you might lose the audience you already have. Besides, your budget does not have enough funds for a new publicity campaign necessary to announce the hiring of another director. You decide to develop the talent that is there.

Actually, your problem is not serious. You already have a fine staff of reporters. You just have not been using their voices on the air. When they cover a story, they have been telephoning the facts back to the news director, who then writes and airs the story. That now will change. You will have the field reporters phone in and record their reports, which then will be played back in the local news.

The national news programming is another problem. You cannot afford to hire additional reporters to cover national news. But your station does subscribe to one of the wire service audio networks. Each day the network feeds to its subscribers a series of correspondents' reports from around the nation over its telephone hookup. You have not been paying much attention to these reports since you air the regularly scheduled network newscast at the top of each hour. That also will change. You now will have the news director record those feeds and insert one or two of them in each local newscast.

Finally, to overcome the fast-breaking news problem, you will try to include a live, on-the-scene report from one of your reporters in every local newscast. For example, in the middle of the local news, the news director will switch live to the reporter at city hall, or one covering a fire. You also will have your news staff call the station and interrupt the regularly scheduled programming whenever important news breaks in the city.

Six months later, you begin to see the fruits of your labor. Your share of the local news audience now equals that of the rest of your programming. Although it is too early to tell, you see signs of an actual increase in that figure for your early evening newscast, meaning that people are tuning in especially to hear the news on your station. Without the station image survey your station's news might still be in the cellar.

Methods of collecting data for image surveys are as varied as the surveys themselves. However, remember that most station image surveys are designed to dig deeper than the ratings do. Figure 18–3 is a portion of a questionnaire used by WFIU in Bloomington, Indiana. Notice that

In order to assist us in evaluating how effectively we are serving our audience's needs, would you please take a few minutes to rate our programs. Any comments you wish to make will also be welcome.

PROGRAM TITLE   **Ranking**

	Low				High
Afterglow	1	2	3	4	5
Afternoon Concert	1	2	3	4	5
Alec Wilder & Friends	1	2	3	4	5
All Things Considered	1	2	3	4	5
Boston Symphony	1	2	3	4	5
Chamber Music	1	2	3	4	5
Chicago Symphony	1	2	3	4	5
Cleveland Orchestra	1	2	3	4	5
Duke Ellington	1	2	3	4	5

Figure 18–3 (WFIU)

the respondent has five different choices for each program. This is much more detailed than merely asking if that person does or does not listen to the program. Sophisticated computer analysis of this type of questionnaire can tell which programs are popular and how they compare with other programs.[6] Similar questionnaires can reveal how an audience feels about the station's hours of operation, the overall selection of programs, news coverage, and scheduling.

The semantic differential scale is another sophisticated instrument often used in station image surveys.[7] These scales use a series of bipolar adjectives, such as bad-good or like-dislike. Each bipolar adjective is separated by an odd number of spaces, usually between five and eleven, permitting the respondent to mark an X in the "semantic space" between the two adjectives. Numerical values then are assigned to each space. For example:

Qualified _____ : X : _____ : _____ : _____ : _____ : _____ Unqualified
Inexpert _____ : _____ : _____ : _____ : X : _____ : _____ Expert

If this scale measured an audience's reaction to a television newscaster, the newscaster's score would be 11. We got this score by adding the numbers corresponding to the appropriate spaces between the bipolar adjectives:

Qualified 7 : 6 : 5 : 4 : 3 : 2 : 1 Unqualifed
Inexpert 1 : 2 : 3 : 4 : 5 : 6 : 7 Expert

The highest number, in this case 7, corresponds to the positive adjectives of "qualified" and "expert." In our example, we reversed the adjectives, placing the positive ones at opposite ends of the scale, "qualified" on the left and "expert" on the right. By randomly alternating the position of the adjectives in longer questionnaires, we prevent someone from carelessly marking *X*s down one side of all the scales.

Semantic-differential scales are used most effectively with a statistical procedure called factor-analysis, which "factors out" and "clusters" the scales which correlate with each other. These groups of scales are called *dimensions* and can tell us a great deal about people's perceptions of everything from station personalities to musical programming. For example, if we are programming a hard-rock format on radio and find in a station image survey that our listeners prefer programming that falls into those "dimensions" appropriate to classical music, then we might revise the format to soft rock. We may find that a competing newscaster has certain "dimensions" based on our analysis of a semantic scale. The newscaster also receives the highest ratings in town. In analyzing the personnel in our own news department, we find that one of our street reporters is perceived by the audience as having very similar dimensions to the popular newscaster. We move the street reporter to the anchor desk, and the ratings immediately improve.

Many more data-gathering instruments are available than the two mentioned here. All are designed to provide an in-depth look at how the audience perceives the station in everything from its standing in the community to its programming.

Sales and Marketing Surveys

The sales manager had worked for months on the account. But the drugstore owner was convinced the newspaper was the only place in which to advertise and could see no reason whatsoever for trying the broadcast media. Finally after repeated calls, explanations, gentle persuasion, and a few lunches, the druggist decided to sponsor the local news. He demanded, however, that he be allowed to write the commercials and have an announcer at the station record them. The copy that reached the station told how long the drugstore had been in the community, how loyal its customers were, and the excellent service it provided. Now it was six months later and time to renew the contract. What the sales manager suspected happened. The drugstore canceled the account. The druggist claimed he just did not see how the commercials could increase sales. When he ran ads in the newspaper, the item he listed always sold out. The sales manager persuaded the druggist to try one more week of commercials, only this time to let the station write the commercials with the druggist's approval. It was June, and the sales manager suggested advertising a special suntan oil. He also convinced the druggist not to advertise the suntan oil in the newspaper. The commercial started on Monday, and by Wednesday afternoon the drugstore was out of suntan oil. The druggist extended the contract for another six months and took a new look at advertising on the broadcast media.

Our example shows the importance of another type of research—sales and marketing surveys.[8] The sales manager surveyed the marketing mix used by the drugstore and discovered that the tactic of advertising a single product generated noticeable sales. The manager used this information, and the result convinced the client to renew his contract. Monitoring sales is just as important to broadcasters as station image surveys and community needs and ascertainment surveys are. Some broadcasters would argue that it is even more important.

behind the scenes of a network election poll

"They told me at the doughnut shop we had a reporter from NBC here, and by golly we do!" The pollster looked up from her interviewing sheets to see a weathered farmer in bib-overalls smiling as if he had just reaped the biggest corn harvest in three counties . The woman's quick explanation that she was just freelancing for NBC as a precinct pollster for this election day did not seem to bother him in the least. It was obvious that his stories of meeting a real live "network person" in the flesh would be heard in front of many a crackling fire. ·

Planning and Pollster Selection

Public opinion polls in radio and television are mainly the domain of the broadcast journalist and are usually concentrated around election time. We take for granted the election night predictions of state and national winners only a few hours after the polls close, predictions made before all the votes are counted. We no longer sit in amazement as computers, network commentators, and sample precincts determine the fate of democracy. We expect it.

What is behind these public opinion polls? It all starts in the network planning rooms, where statisticians and polling experts pour over mounds of computer data, analyzing voting trends in every state. How many people voted in which county for what candidate? How do these figures correlate with state and national voting trends? Do they correlate often enough to be a sample precinct, those tiny subdivisions of each community used to predict winners long before the local election officials finish their work? Sample precincts may be in the heart of Los Angeles or a tiny Maine fishing village. They are chosen by the same random sampling techniques we learned about in broadcast ratings. Together, they become part of the state-by-state predictions that call the election of senators, congress-persons, governors, and presidents.

Equally critical to the election poll are the precinct pollsters selected by the networks to interview voters. Some of the pollsters are associated with a broadcasting station in that area and are "invited" by the network to represent it at the polling place. At the network's office in New York, other people are in charge of coordinating this national group of "volunteers," who will receive a modest fee for their day's work. Coordinating a national core of pollsters and providing "training" by mail are a big job and one which must be carefully planned and executed.

Imagine yourself as an interviewer participating in a network election-day poll. What will your job be like? It will actually start about a month before the election with a call from a network executive asking you to represent the network as a precinct pollster. About two weeks later, you will receive a large packet of instructions. You will be asked to read the instructions carefully and to call a special telephone number if you need advice or if something unexpected happens. The number belongs to a special state supervisor, also hired by the network. The instructions will list where you will conduct the polling, giving you the exact precinct and voting location, and the name and telephone number of the local precinct official, suggesting that you contact that person before election day. Since you are just one part of a national systematic effort you will be asked to follow the directions very carefully. You have two responsibilities (1) to conduct the poll and (2) to report the results to the network.

Each pollster is assigned certain time periods in which to conduct the precinct poll. For example, you might be assigned to interview people coming out of the polling place between 6:00 A.M. and 9:00 A.M., calling in your first set of results to the network at 9:15 A.M. The second interview time might be in the afternoon between 12:30 P.M. and 1:30 P.M., and you would call in your second report to the network at 1:45 P.M.

You also will be told which sequence of voters to interview, such as every fifth or every third voter. The intervals are different for different precincts, depending upon their size. Your instructions would also tell you your quota of completed interviews, such as 30 completed interviews within those two time periods.

Conducting the Poll

Finally election day arrives. In the predawn darkness, you start your drive to the polling place. You leave early, in plenty of time. We shall assume that you have been assigned to a precinct in a small, outlying rural area. The morning sun is just barely coming up when you roll into the sleepy little town. You pass an old general store, a gas station, and a cafe where a few people are downing their first morning coffee. The precinct voting place is in the town library. You have seen them before, one of the old Carnegie structures with an American flag outside. Inside, the precinct workers all are ready for the first voter. You are greeted by the local precinct official you talked to earlier on the phone. She eyes your official NBC badge and asks:

"How about a cup of coffee and some of Mabel's cookies?" You accept and begin to wonder what the day will bring.

"Who are you going to interview?" she asks.

"Well I'm supposed to interview every third person, starting with the first one to vote after 6:00 A. M."

Before the precinct official answers, Mabel interrupts, "I'll tell you who that's gonna be." She has a big grin on a complexion revealing years of hard toil and farm life. "You can betcha Harvey Clodfelter will be the first through that door. Twenty years and he ain't missed being first yet."

You sense that election day is one of those special days ranking with the church suppers, the day of the school picnic, and the beginning of the county fair. It seems light years away from the network control center in New York where computers and statisticians are adding the finishing touches for the marathon telecast to follow later that night.

But here is where it all begins. You begin, sure enough, with Harvey Clodfelter. Your day will be filled with Harveys. They will ask you who you are, where you live, why you are there, and when you will be back. They will offer you their own brand of election philosophy, tell you why this candidate or that candidate will win, why the electoral college should be abolished, and complain how the big city folks are going to sway the election. To each of them, you will give a special ballot, similar in some ways to the one they have just completed (Figure 18–4). The ballot will be labeled SECRET BALLOT, and when they have finished filling it out, they will put it in a big envelope. You later will open that envelope and tally the results.

The ballot requires much more detailed information from the respondent than merely for whom the person voted. It asks the voters whether the TV debates affected their choice, who they voted for in the last presidential election, their ethnic background, the occupation of the head of their household, and such basic demographic information as age and sex. The ballot also asks what issues influenced their decisions, what the candidates' positions on the economy and foreign policy are, and whether those positions affected their decision.

If Harvey Clodfelter, as your chosen first person to finish voting after 6:00 A. M., refuses to be interviewed, then you continue counting until the next interval is reached. Remember, your goal is to interview the

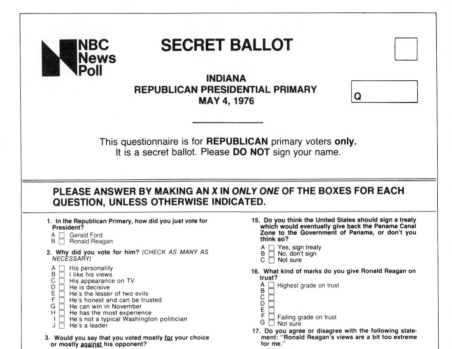

Figure 18–4

first, third, sixth, ninth, and twelfth voter, on down the line. It is a hard day. Some local people do not trust "outsiders," and you get completed interviews from only three of those chosen people. Nevertheless, you keep working toward your quota. You make certain you offer a questionnaire only to those people who have actually voted. That is one advantage the election day polls have over preelection polls. You can control your sample to include only people who have voted. You are also careful not to interfere in any way with the actual election process: voting and checking voter registrations.

So as not to influence the results of your poll, you try to remain polite while at the same time refraining from talking about candidates, issues, political beliefs, or other things which might bias the results. The only information you will supply is on the mechanics of completing the secret ballot; you will not want to answer any requests for political information. If a person does not understand a question or the directions, a polite suggestion to reread the ballot, not your rephrasing of the question or directions, is necessary. In other words, you will not want to influence, either by word or action, the response of the person being polled.

At precisely 9:15 A.M. you walk up the stairs to the library's office and call New York with the results of your first poll. You are careful to make sure every item is understood by the data collector.

To help ensure this, you will use word identifications for each of the possible answers: A = Alpha, B = Bravo, C = Charlie, D = Delta, and so on. A typical conversation, as explained in NBC's guide to its precinct pollsters, might go like this:

NBC Operator:	"NBC. Your location code, please."
Interviewer:	"This is Ohio, location 400." The NBC Operator then will repeat your code, and give you the corresponding name of your polling location, to be sure we have the correct information.
NBC Operator:	"I'm ready for the answers."
Interviewer:	"Question One, Alpha."
NBC Operator:	"One, Alpha, O. K. Go ahead."
Interviewer:	"Two, Charlie."
NBC Operator:	"Two, Charlie, right."
Interviewer:	"Three, Delta, etc."

You then walk back down the stairs, nibble another cookie, and wait until you can repeat the procedure at your specified 12:30 P. M.

In New York, the data from your poll and those of other pollsters scattered throughout the United States are fed into a computer which tabulates them and predicts the winners. So why does the network not start predicting at noon on election day? The answer to that is a matter of geography. Since the West, because of the time difference, votes later than the East does, such predictions could influence western voters. A news report stating that a certain candidate has won by a landslide in New York may convince people in California that they also should vote for the probable winner. On the other hand, that news might make the candidate's California supporters complacent to the point at which they stay home and "give" the opponent the election. Let's examine this reporting process more closely.

reporting public opinion polls

As either a consumer of broadcast communication or someone who will someday report polls, you should be aware that there is more to an opinion poll than just the results. Many times, a gullible public and press are the object of persuasion campaigns based on public opinion polls. Broadcast journalists, partly because of the time restrictions placed on them, find reporting all of the pertinent data of an opinion poll to be difficult. So the following could very likely happen. A wire service story reports that a senatorial candidate is leading his opponent by twenty percent. The radio station reports the story, only to find out later that the poll consisted of 25 people who stopped by for coffee at the candidate's office. The sampling error alone would make the results questionable, to say nothing of the probable bias. We can assume that anyone who takes the time to have coffee at a candidate's office is not vehemently opposed to the person.

Reporting public opinion polls can be just as important, perhaps more so, than conducting the poll. The American Association of Public Opinion Research (AAPOR) and the National Council on Public Opinion Polls (NCPP), have adopted guidelines for reporting public opinion polls. An emphasis on "precision journalism" also has fostered more responsible and intelligent interpretation of polls.[9] Other guidelines have been offered by both journalists and educators (Figure 18–5). The aim of this attention is to educate the public and the media in the polling process. Even if the press is unable, because of space or time limitations, to report all of the factors in a poll, it should be able to decide intelligently whether

Figure 18–5   (*The Quill*, published by the Society of Professional Journalists, Sigma Delta Chi.)

CHECKLIST—
WHAT TO LOOK FOR IN A POLL STORY:
- ✔ *Was a probability sample used?*
- ✔ *Who sponsored, and who conducted, the polling?*
- ✔ *Is the population sampled adequately defined?*
- ✔ *How were those polled contacted—by mail, by telephone or in person?*
- ✔ *Is exact wording of questions specified?*
- ✔ *How many people were surveyed?*
- ✔ *Is it clear when analysis and interpretation concern only part of the full sample?*
- ✔ *Does the story specify an error margin for results, and does interpretation outrun that error margin?*
- ✔ *Does the story provide data to compare the sample with demographic characteristics of the population from which the sample was drawn?*
- ✔ *Is the headline or teaser accurate?*

even to use the poll, and if used, whether it warrants detailed interpretation.

What are the important considerations in reporting polls?[10] First is the sample. If it is not a random probability sample, then it may be questionable. In other words, for the poll to be credible, the probability of one person being selected is equal to that of all other people who might be selected. If the sample covers only your state, then it is not good sense to apply these predictions to the entire country. A second important consideration is who sponsored and conducted the poll. A poll favoring a democratic governor which is sponsored by the state's democratic committee and conducted by democratic precinct workers legitimately would be suspect. Third, it is important to define the sample population. If a poll taken among Maine voters states that the senator from Maine has a 60 percent chance of winning his party's nomination for president, the chances are that the estimate is inflated. Knowing who was sampled permits a better judgment of the poll's results.

A fourth consideration in reporting polls is how the people were contacted. A telephone survey of poor families would automatically eliminate all those who could not afford to have a telephone. Fifth, it is important to know what questions were asked. "Will you be voting for our fine Senator Claghorn?" will elicit answers favoring Senator Claghorn. Sixth, to determine the margin of error, it is important to know how many people were surveyed. Seventh, the poll should state if a subsample is used to interpret the findings. For instance, a poll may be reported to have a sampling error of ± 4 percent, based on a random sample of 600 people. However, if 50 of those people were Indians, it would not be correct to predict how the Indian population would vote and still claim the sampling error was ± 4 percent. Because only 50 Indians were sampled, the margin of error would be closer to ± 14 percent.

Eighth, it is important to report the error margin. Merely knowing what it is but not telling the listening or viewing audience has little purpose. Error margins are one of the easiest and best understood factors in an opinion poll. If you claim that a candidate has a 5 percent edge over his opponent and do not tell the public that the margin for error is 10 percent, then you have misled them. Moreover, if you predict a winner when the margin of error makes it actually too close to call, then you also have misled the public. A ninth important item to include in reporting polls is how the sample compares with the total population. If the population is 50 percent Hispanic, yet only 33 percent of your sample is Hispanic, then you would be asking for trouble by predicting the election without compensating for that variation. Tenth, it is important to consider the all-important headline. People can be persuaded by headlines. The radio or television reporter who announces, "Opinion poll predicts Senator Doolittle will sweep the state," and then comes back after the commercial to report a poll of the senator's precinct workers, has been irresponsible.

Keep in mind that public opinion polls are not foolproof and can be used incorrectly. The key is to understand them and know their pitfalls. The broadcast manager who editorializes in favor of a candidate may be in hot water if he bases that support on a sloppy opinion poll. A television news director who sends her staff out to conduct an opinion poll on election eve and violates all the rules of random sampling, is asking for trouble if she reports the results of the poll as being representative of the total population. Take the opportunity to read more about opinion polling. Design a sample poll. Ask yourself if it meets all of these criteria. Would you be able to draw general conclusions from your results?

new forms of data collection and approaches to research

New technology is accelerating the polling process even beyond the computer analyses so common in today's surveys. It will give broadcasters the ability to conduct opinion polls faster and with fewer personnel than in the past. One system uses a series of tape recorders and a single operator. The operator dials a telephone number. When the phone is answered, the tape recorder takes over and plays a recorded message to the party answering the phone. It then records the person's response. Meanwhile, the operator is busy dialing another number for one of the other tape recorders. This system permits a single operator to make as many as 1,000 calls per day. Of course, the results of such a poll are only as accurate as the information being asked. If the questions are poorly phrased or if there are errors in the sampling procedure, then it is difficult for the poll to reflect an accurate representation of the population.

There is another electronic device called an oculometer. This tracks eye movements and has been used to examine how people "watch" television commercials. A special sensor monitors the eye movements of respondents and has shown, for example, that people are attracted immediately to people in television commercials, that attention spans are rather short, and that certain visual elements in a television commercial can actually interfere with the commercial message by distracting the viewer away from the product being advertised.

In special research centers, colleges and universities, advertising research firms, and almost anywhere there is broadcast-related research, the paper and pencil questionnaires often are being replaced by still more electronic data-gathering instruments.

Electronic Response Systems

Electronic data gathering, in which a member of an audience punches a button or has his or her pulse read electronically, is not new to broadcasting.[11] In 1937, Dr. Frank Stanton of CBS and Dr. Paul Lazarsfeld, a well known social scientist, developed "Big Annie." It was perhaps the first electronic response indicator (ERI) used to measure how an audience reacted to radio shows. The listener could push one of three buttons labeled "like," "dislike," and "neutral."

Big Annie had a short life, but she did spark interest in electronic data gathering. One instrument designed to do this was invented by a professor of telecommunication at the University of Oregon, Dr. Elwood Kretsinger. He called it a "Chi Square Meter." It simultaneously records an individual's responses to whatever he or she happens to be listening to or watching. Comparisons of responses among different individuals or groups are made using a statistical test called the chi square, thus the name Chi Square Meter. The Chi Square Meter was not Kretsinger's first venture into response systems. In the late 1940s, he invented a device called the "wiggle meter" which measured how much people fidget. The device had three components. A wire was strung on the back of a chair, which was fed into an amplifier, which in turn measured the amount of "wiggling" on a long roll of graph paper.[12]

The increased interest in ERIs has resulted in a number of commercial firms manufacturing the systems. Usually installed in a large auditorium, an ERI is attached to every seat or desk. At each desk the unit has five buttons that can electronically record anything from an answer to a five-part, multiple-choice question, to expressing one of five attitude responses such as "strongly agree," "agree," "neutral," "disagree," and "strongly disagree." The entire auditorium of ERIs can be monitored by a computer and can register continuously the response of the entire room, sections of the room, or calculate statistical comparisons between various groups of people.

Assume you have produced an instructional television program and now want to test its effectiveness. You arrange for a physical education class to use a room equipped with ERIs and, at periodic intervals in the program, ask the class to respond to a series of questions prepared to evaluate your program's effectiveness. When the program is over, a computer printout tells you how the class scored on every question. Those questions the students missed may indicate that you need to revise one or more sections of the program. If you had not been using ERIs, you could have waited until the program was over and given the students a paper-pencil test to evaluate what they had learned. However, you then would have had to score the tests and calculate the statistical computations yourself or wait until they could be done by computer.

Galvanic Skin Response

Not all of the methods to test audience reaction necessitate pushing a button or wiggling. There also are systems which measure the changes in the conductivity of the skin, called galvanic skin response (GSR). One firm uses electrodes attached to the palm of a hand to test audience reaction to such things as commercials, records, and radio station formats.[13] About 100 persons are selected for each study, and the system's developers have claimed that it has as high as 91 percent accuracy in predicting whether records will appear in the top 20 of the national charts. Another system uses a computer and analyzes the data from about 80 subjects, also wired for GSR readings. The developers claim that the device is successful in figuring out how to "recycle" listeners, making them return to listening to a radio station once they have sampled it.[14] Still other systems have used brain waves to measure audience response. Researchers at the Princeton Medical Center, using an electroencephalograph, examined the brain waves of people watching television.[15]

Gathering data from either ERIs or GSRs has two advantages over older methods of data collection. First, it is fast, especially when monitored by a computer. What formerly took hours of work and computation can now be calculated almost instantaneously. Second, it can measure detailed aspects of audience reaction, thus providing much more sophisticated data bases. When decision makers can monitor audience reaction every few seconds, they have a more useful account of how an audience is reacting to a message.

Such systems can be installed even in mobile vans which can stop outside a grocery store and ask shoppers to come in. Inside, customers watch a television commercial and have their reactions to the commercial tested. Then the operators watch the shoppers enter the grocery store and directly monitor sales of the product.

Disadvantages of the systems are in the limited number of responses that can be monitored at any one time. Although a major rating service can obtain feedback from an audience large enough to predict national trends, most electronic response systems are designed for a group no larger than can be accommodated in a small auditorium. This limitation is not, however, a technological weakness. Electronic response systems that connect the home to a computer, much in the same way as a two-way cable system does, already enable measurement of large audiences.

The future may see national populations watch television with their reactions to the program being monitored on a second-to-second basis. It is not beyond the reach of current technology.

Role Observation: Producing an Ethnography

Although such measures as ERI and GSR are useful for highly controlled laboratory conditions, the very artificiality of those conditions can sometimes limit the applicability of research results. Recently, researchers have been turning to less "laboratory-oriented" methods of data collection and instead to more real-life settings. Actually living with families and observing how they use radio and television is becoming a more frequently used research approach. The results of such approaches can produce an ethnography, an accounting of how individuals have interacted with their environment, with each other, and in the case of broadcasting, the role radio and television have played in this scenario.

Research is telling us of the increased importance media play in structuring our daily activities and interpersonal exchanges.[16] Further ethnographic approaches to the study of radio and television in our lives will provide new perspectives on how we relate and interact with our media-filled society.

summary

This chapter has examined the broadcasting research process. There are four kinds of research: historical, descriptive, experimental, and developmental. Of these, descriptive research is the one most often associated with research in the broadcasting industry. Experimental research manipulates an independent variable, then observes any phenomenon resulting from that manipulation. The phenomenon to be observed is the dependent variable. Research extends from that done at academic research centers to individual studies by students and faculty, to doctoral dissertations. Besides that done by the government and rating services, additional research is conducted by local stations and networks.

Survey research common in local stations includes community needs and ascertainment surveys, station image surveys, and sales and marketing surveys. Required by the FCC, community needs and ascertainment surveys include a demographic profile of the community, a community leader survey, a general public survey, and a survey of the "voiceless" community. The FCC suggests 19 different categories of community leaders and determines the approximate number of consultations by the size of the population. The general public survey should be a random sample of the total community serviced by the station. Station image surveys often use more sophisticated methodology than community needs and ascertainment surveys do. Ranking of programs, questions on station operating procedures, and semantic-differential scales are just some of the many methods used to discover an audience's perceptions of a station and its programming. Sales and marketing surveys determine if systematic broadcast advertising campaigns have had satisfactory results.

Public opinion polls are usually the domain of the broadcast journalist and are frequently used around election time. But merely conducting a good poll is not enough. Reporters should consider information about sample, sponsor, population, contact method, questions, sampling error, and headlines to communicate responsibly the results of a poll to a broadcast audience.

New technology is significantly altering the way in which research data are collected. Such methods as electronic response indicators and galvanic skin responses are being used more often to hasten the collection and processing of raw data. Role observation which produces an ethnography is also gaining increased attention.

spotlight on further learning

An Ascertainment Handbook for Public Broadcasting Facilities. Washington, D.C.: Corporation for Public Broadcasting, 1976. (The original research report upon which much of the *Handbook* is based was written by Bradley S. Greenberg, Thomas F. Baldwin, Byron Reeves, Lee Thornton, and Jack Wakshlag. CPB Contract 73-6229, December, 1975.)

A Broadcast Research Primer. Washington, D.C.: National Association of Broadcasters, 1974.

SHAW, DAVID,"Political Polls: Use and Abuse," chapter 3 in *Journalism Today.* New York: Harper & Row, Publishers, 1977.

Six Experiments in Ascertainment Methodology. Washington, D.C.: Corporation for Public Broadcasting, 1977. (Contains research studies by Robert K. Avery, Paige Birdsall, Antonio Rey, Alfred G. Smith, Patrick A. Nester, D. Lynn Pulford, James A. Anderson, Thomas A. McCain, C. Richard Hofstetter, Navita Cummings James, and James E. Hawkins.)

WILHOIT, G. CLEVELAND, and MAXWELL MCCOMBS "Reporting Surveys and Polls," in *Handbook of Reporting Methods,* ed. Maxwell McCombs, Donald Lewis Shaw, and David Grey. Boston: Houghton Mifflin Company, 1976.

19

the broadcast audience: approaches to studying effects

FOCUS After completing this chapter, we should be able to

Explain early approaches to understanding the mass audience.

Describe the Bullet Theory.

Define the *individual differences approach, categories approach,* and *social relations approach* to studying the effects of broadcast communication.

Define *opinion leaders* and the *two-step-flow.*

Understand the concepts of selective exposure, perception, and retention.

Define *demographics* and *psychographics.*

Explain what is meant by the functional uses of broadcasting.

Understand how broadcasting affects our socialization.

Discuss the issues and effects of televised violence.

Imagine that you are an advertising agency media director. Selecting a magazine in which to advertise your client's products requires you to examine the magazine's audience. You naturally need to know the circulation, but you also want to know something about how the readers think, what their interests are, how old they are, what their income is, and how much education they have had. If the magazine is well established, this information will be readily available to you in a readership survey.

Now assume that you need to buy time on a major television network. Here, identifying the audience becomes much more difficult. Although you could choose certain types of programming to reach certain audiences, such as Saturday morning programs to reach children, you are dealing mainly with a large, unidentified audience.

To say only that a broadcast audience is too big and complex to study would be to admit defeat much too soon. Instead, this chapter will examine what we *do* know about the broadcast audience, realizing all the time that researchers are continually working to learn more about it. We have been able only to scratch the surface of this inquiry, partly because human behavior is a highly complex phenomenon. Our studies of television violence and the effects of broadcast programming overlap with research in many other disciplines. Moreover, cooperative research across international boundaries is still another possibility which can be invaluable if political conditions permit.

understanding the audience

In chapter 1 we discussed broadcasting as mass communication. Now we shall discuss the broadcast audience as a mass audience.

Early Perceptions of the Audience: The Bullet Theory

How researchers view the mass audience has changed dramatically in the last fifty years. Theorists used to look at it as a disconnected mass of individuals who received communication in much the same way that sitting ducks receive buckshot. This approach to media effects was labeled the *bullet theory*, sometimes called the hypodermic theory. Part of the misconception developed during World War I with the scare tactics of the propaganda campaigns. As researchers began to realize that human behavior was more complex than that of sitting ducks, the idea of the mass audience began to change.

The first change was in the concept of mass. Instead of being viewed as a huge body of isolated people with similar reactions to media messages, the audience gradually was studied as a group of individuals held together by different social systems and reacting to messages based partly on other people within these social systems. The political commercial may affect different people in different ways. Those same people may interact with and talk to other people about the commercial. These other people may determine, just as much and many times more so than the media does, how the person reacts to the commercial.

Individual Differences, Categories, and Social Relations Approaches

In revising the bullet theory, three approaches to studying audience reaction were used. One was the *individual differences approach,* proposing that each of us has individual qualities which result in our reacting differently to media messages. Professors Melvin De Fleur and Sandara Ball-Rokeach in their book, *Theories of Mass Communication,* describe the individual differences approach: "Individual differences perspective implies that media messages contain particular stimulus attributes that have differential interaction with personality characteristics of audience members. Since there are individual differences in personality characteristics among such members, it is natural to assume that there will be variations in effect which correspond to these individual differences."[1] Variables in these differing effects are partially caused by the audience's exposure, perception, and *retention* of media content which we shall examine later in this chapter.

Another approach to the mass audience and the effects of media content is the *category approach.* Its origin was in the needs of advertisers to reach more specialized audiences. Although the simplest way to group an audience into categories is by demographics—sex, age, etc.—researchers are looking more and more at psychographics—values, beliefs, attitudes, and lifestyles. Looking at the audience using categories can be much more complex than the old bullet theory. Notice that we said *can be.* For the ad buyer wanting to reach 18- to 21-year-old females, the application of the theory becomes mechanical. But for social scientists wanting to know how categories of people think and how they interact with other categories of people, the approach becomes much more complex. Moreover, if we want to use these interrelationships to understand how people react to broadcast messages, the process becomes even more sophisticated. Buying an ad to reach the homemaker is one thing, but buying an ad to reach the homemaker who interacts with another homemaker viewing a competing commercial is something else.[2]

Concentrating on this interaction and the people taking part in it would describe the *social relationships approach* to studying the audience and the media effects on it. Interpersonal communication is important to the social relationship approach, as is the realization that although the media can help disseminate the initial message, how it is retransmitted, discussed, and rediscussed among audience members will in great part determine the effect of the message.

After considering the different approaches to studying the audience and the effects of broadcast communication on an audience, we readily can see that not only do the three approaches overlap, but that in some ways, all are part of the communicative process. It is much like viewing that process through different colored glasses. A psychologist concerned with an individual's behavior might feel more comfortable with an individual differences approach, although that same psychologist would be foolish to ignore the other approaches. Similarly, an advertiser wanting to reach a specific type of audience might be concerned with categories but cannot ignore the interrelationships among people that demand the attention of the social relations approach.

flow and processing of broadcast messages

Interaction among members of the broadcast audience is part of the natural process of information dissemination and processing.

Opinion Leaders and the Two-Step Flow

Imagine for a moment that you are discussing a new television series with your friend. Your friend tells you how great the series is, that you would enjoy the fast action, and suggests that you watch it. Since you value your friend's opinion, you decide you will take a break from studying to watch the show. Why did you make that decision? Because you undoubtedly valued your friend's opinion enough to be influenced to take action—to watch the program. In doing so, you demonstrated how other people, people we shall call *opinion leaders*, influence opinions of media content.[3]

Since you had not seen the program, your friend's description was all you knew about it. By first hearing about the program from your friend, you demonstrated how messages flow to and through the mass audience. We call this process the *two-step flow*, even though many more than two steps may be taken before the information reaches all the people who eventually learn about it. You should also remember that the two-step flow can apply to both acquiring information and being persuaded by it.

Selective Exposure, Perception, and Retention

In discussing the individual differences approach, we talked about exposure, perception, and retention. Each influences our interaction with the media and how it affects us. For instance, research has taught us that we selectively expose ourselves to certain types of programming, the process being called *selective exposure*. If a politician is delivering a televised address, you might tune in the program because you agree with the politician's views. On the other hand, you might tune in the program because you disagree with the politican. For either reason, you selectively exposed yourself to the program.

Second, the perceptions you hold before watching the televised address will also affect how you react to it. If you are extremely loyal to the politician, you might agree with everything she says regardless of *what* she says, so much so that if her opponent said the very same thing, you might totally disagree with him. You would be guilty of *selective perception*. It is not a serious crime, but one which can distort how you react to messages.

Third, because of your selective perception, you may retain only those portions of the address with which you agree. If you perceive the entire address as favorable, you may remember all of it. If you perceive it as unfavorable, you may wipe it entirely from your mind. If parts of the address affect you positively, you remember those parts while forgetting the negative ones. Or the negative ones may be the very ones you remember. Either way, how you originally perceived the address determines what you retain, a process called *selective retention*.

An interesting study examining selective perception among viewers of the television program "All in the Family" showed that selective perception caused differing opinions of two of the show's main characters—Archie Bunker, a very conservative factory worker, and his son-in-law Mike, a liberal college student. Professors Neil Vidmar and Milton Rokeach found that highly prejudiced people identified more with Archie and perceived him as making better sense than Mike. They also perceived Archie as winning between the two more often. Highly prejudiced persons indicated that they disliked Mike's personality much more than they disliked Archie's. Vidmar and Rokeach found that "persons who like Archie reported he is down-to-earth, honest, hard-working, predictable, and kind enough to allow his son-in-law and daughter to live with him." Conversely, less prejudiced persons disliked Archie's personality traits more than they disliked Mike's. People who liked Mike felt that he was tolerant and stood up for his beliefs. Those disliking him saw him as stupid, narrow-minded, prejudiced against the older generation, rebellious, lazy and a "banner waver."[4]

Source Credibility and Media Credibility

Two other factors which can affect how we perceive broadcast messages are source credibility and media credibility. Source credibility is the credibility of the original source of the communication. In the case of the politician's televised address, source credibility would be the credibility of the politician. If you perceive the politican as highly credible, then the chances are that your reaction to her would be favorable. Research has assigned many subordinate factors to source credibility, among them, dynamism, trustworthiness, and competence.

We know that source credibility lies partially in the source and partially in how the source is perceived by the audience. In other words, how broadcast messages affect us is related to how we perceive the source of those messages. Source credibility and both interpersonal communication and mass communication all are part of the communication process.

Media credibility refers to how you perceive the overall credibility of a medium, such as a local radio station. Media credibility can also determine the effect of a broadcast message. Two types of media credibility are important, intermedia credibility and intramedia credibility. Intermedia credibility is the relative credibility of various media, such as determining that television is a more credible medium than radio or newspapers. Intramedia credibility is the relative credibility within the same medium, such as the credibility between two radio stations. Over a number of years, studies on media credibility have asked such questions as, "In the face of conflicting news reports, which medium would you be most likely to believe: television, newspapers, radio, magazines, or other people?" Much of the research has found television to be the most frequent response. Notice that we said "most frequent response," not necessarily the "most credible medium." Such studies fail to allow for the many possible intervening variables which truly reflect the characteristics of media credibility.

Based on such research, we might justifiably ask whether, if television is listed first, it is because it is the most credible or because we spend more time with it than with other media? Is there a wide variance among different media in different communities? In some communities, would radio or newspapers come out on top because of the credibility of the local press? How many times do we really hear or even recognize conflicting news reports? Is television really the most credible medium or do we just think so because two of our senses, sight and hearing, can consume the information instead of one and do so in color and in motion?

Some research which had listed television as most credible, listed radio as least credible. If we were to interpret those results literally, we might have some difficulty explaining some recent effects of radio programming. Back in 1938, Orson Welles, acting in the radio play of H. G. Wells's "War of the Worlds," made many people hysterical because they believed there was an actual invasion from outer space. Those who said it could not happen again were proved wrong in 1977 when Swiss radio aired a program with mock news bulletins about neutron bombs being dropped in a war between East and West Germany. Mock casualty reports listed 480,000 people killed. Panicked listeners called the station and received an official apology over Swiss radio.

The broadcast media do not seem so credible when it comes to advertising. Asking a question similar to the one about news credibility, research has shown newspapers as a more credible advertising source than other media. A survey asked, "Some advertising seems honest and believable, while other advertising seems hard to believe. In this area, considering radio, newspapers, television, and magazines—which one is likely to carry the most believable advertising?" Newspapers were listed first by the different categories of respondents.[5]

Further research on media and source credibility can be invaluable to both the business and consumer communities. Our access to the media, its availability, our interests in specific programming, our attitudes, and the video or audio techniques used to communicate broadcast programming are all part of our perceptions.

categorizing the broadcast audience

We already have talked about some of the approaches used to examine the audience—individual differences, categories, and social relations. All of these approaches will eventually concentrate on two broad categories of audience classification: demographics and psychographics.

Demographics
Demographics refer to such things as the age, sex, education, income, and race of an audience. Partly because the data are easily obtainable, from everything from courthouse records to census figures, demographic characteristics of an audience are the most commonly used types of broadcast audience classification. The rating services, for example, use age and sex as their two principal categories, such as women eighteen to twenty-four years of age and men thirty-four to forty-nine years of age.

Arbitron Television developed special questionnaires approved by its legal counsel which provide demographic breakdowns of ethnic audiences. The questionnaire asked families "how you describe your family." It was then validated through personal interviews with respondents who had indicated their race/nationality characteristics on the Arbitron diary.

For advertisers, the demographic audience becomes something to reach, something to identify and something to persuade. Using data from rating services and broadcasting stations, advertisers make time buys based on such things as when the highest concentration of women is watching and what the cost-per-thousand (CPM) is for those women reached. Others need data to reach children, males, minorities, or teenagers. The key is to match the product to the audience, and the more specialized the product becomes, the greater the need is to reach a specialized audience.

Demographics will most likely continue to be the main identifier of broadcast audiences for four reasons. First, the information is easily obtained from different sources. Second, the industry is geared to using demographic data and feels comfortable with them. Although psychographics is becoming more important, the average radio station manager is much more at ease with data on age, sex, and income than with psychological constructs, media involvement scales, or value profiles. Third, advertisers are comfortable with demographic data. An account executive selling a local druggist a radio commercial is much better talking about the station's high income audience than attempting to teach a course in psychology while explaining the station's rate card. Fourth, information other than demographics is still subject to conflicting methodologies, not so much in the minds of the researchers collecting the data as in the minds of the industry that uses them. Although one research firm which might claim to analyze an audience's value structure while another which analyzes its personality may be correlated, the broadcast manager may simply groan and say, "Just tell me if it's a man or a woman and how much money they make."

Psychographics

Psychographics refers to such things as attitudes, values, beliefs, or opinions. Psychographics methodologies range from dividing an audience into attitudes about brand preferences to discovering that audience's subconscious reactions when interacting with broadcast programming. Asking consumers if they prefer brand X or brand Y is one form of psychographic information. Interviewing prison inmates to determine how they react to television crime shows entails a far more complex psychographic profile.

One example of widely used psychographic information is applied to television programming in adjacent time blocks. It is based on the theory of *cluster programming*. Using the work of clinical psychologists, a group of researchers at Ohio State University classified different types of audiences,[6] such as people who like situation comedies or people who like Westerns. The idea of program preferences among these "types" was then carried one step further by a member of the same research team that worked on the original study, Dr. Joseph Plummer, research executive with Leo Burnett, U.S.A. Plummer subjected data on viewing habits to statistical measures which showed different programs tended to "cluster" together. For example, if you like "Baretta," you probably will like the other police-action shows, "Starsky and Hutch" or "Kojak." If you like situation comedies, then "Three's Company," "Mork and Mindy," and "Angie" could be your favorites. "Sha, Na, Na," "Happy Days," and "Laverne and Shirley" are three similar programs. Basing decisions on both program preferences and how different programs cluster together, networks now find it profitable to schedule similar types of programs in blocks.

functional uses of broadcasting

By applying psychographics, we can study the functional uses of broadcasting. The term functional uses asks "What function does broadcasting play in our lives?" Examples of these uses are, "I use broadcasting to learn what's happening in the world," "I use broadcasting to be entertained," or "I use broadcasting to escape from reality." The three functions just mentioned are information, entertainment, and escape.

Stephenson's Play Theory

Using a data gathering procedure called Q-sort, William Stephenson completed extensive research on how different types of audiences, expressed as typical individuals, feel about the media. From this research has come Stephenson's Play Theory which, applied to broadcasting, suggests that we use radio and television as a means of escaping into a world of "play" not accessible at other times.[7] Those researchers familiar with Q methodology have supported Stephenson's theory as well as his methodologies. Others have been severely critical, like Professor David Chaney who contends that "Stephenson . . .fails to move beyond an individualistic level of description. While the importance of audience commitment is understood, his concern with finding a methodological demonstration of his argument leads his audience to be conceived as only a conglomeration of individuals."[8]

Professor Deanna Robinson offers a more generous view of Stephenson's methodologies. For conducting research on the uses of television and film by upper-middle class professionals, she suggests that Stephenson's technique could be used to directly examine people's attitudes toward the media and be able to demonstrate "(1) that within any single, demographically defined audience group several attitude or 'taste' groups exist and (2) that similar taste groups exist within other classes."[9] Schramm also supported Stephenson and generalized that Stephenson, with a style of writing like Marshall McLuhan's, could have been the guru of modern media.[10]

Uses and Gratifications

Stephenson's Play Theory is part of a wider body of research and theory centering on what uses we make of media and what gratification we get from exposing ourselves to them. Research has examined these uses and gratifications in populations ranging from farmers in less developed countries to American homemakers. The research has not escaped vigorous debate, however, not only on the different types of uses and gratifications, but on the very methodologies which attempt to identify them.

Part of the debate is a conflict between the individual differences approach and the social categories approach to the study of media effects. Consider a television program. We could argue that a soap opera provides certain role models for homemakers or college students. We also could contend that reaction to soap operas cannot be classified in demographic terms, but rather in psychographic terms. Soap operas have certain *uses* for people with specific motivations or certain psychological characteristics. We could argue that even this approach is unsatisfactory since each individual is different, and many different individuals may have many different uses for the same soap opera. How we learn what uses these many different people or groups of people make of the media is still another dilemma. Do we test them individually in tightly controlled laboratories, psychologically wiring them to get at the depths of their thought processes? Or do we sample a large population of respondents in a survey? We shall examine the conflict of which research design to use later.

What has the research told us about uses and gratifications? By sampling a few of these studies, Robinson examined upper-middle class professionals. She discovered the presence of "information absorbers," people who passively absorb information from television without actively interpreting it. Another group she labeled "analytical artists," who used television to increase their understanding of themselves, other people, and the world.[11] Researcher Neil T. Weintraub suggested that radio makes teenagers feel more aware, that it makes their day pass more quickly and that it also lets them know what is happening.[12] Researcher

Lawrence Wenner examined the elderly and found that one use of television among this group was companionship.[13]

One of the earliest studies on the uses of the broadcast media was conducted by Herta Herzog who examined the use of radio soap operas for listeners. Conducting in-depth interviews, Herzog found three uses: compensation, wish fulfillment, and advice. In the compensation category were people who wanted their own behavior compensated by identifying with a soap opera character. Others listened because they were living vicariously, having achieved in the soap opera what was missing in their own lives. The third group sought advice on how to conduct their own lives.[14] More recently, Professor Joseph Foley examined a viewing audience and found eight functions of television viewing to be: withdrawal, play, conversation, togetherness, para-social interaction, educational, background (the set on but not being watched), and normative function (learning about social norms).[15]

The list of studies and their audiences continues. No matter what methodology the studies use or theory they devise, they all add continuing fuel to the debate over what use we gain from broadcasting and how it should be studied[16]

The Agenda-Setting Function

Of all the recent research on the functional uses of media, some of the best and most systematic is on the *agenda-setting function.* Agenda setting argues that the media not only inform us but inform us about what we should be informed. In other words, media set an agenda for our thought processes; they tell us what is important and what we should know and need.

Sophisticated analyses have now made it possible to isolate those media which are dominant in the agenda-setting function, no small task since many communities have more than fifty different media. Newspapers, radio, television, books, and magazines all are important. By keeping track of which media are important to specific populations and then concentrating on those media, a theoretical base for agenda-setting can be built.

The agenda-setting function becomes more pertinent when we consider that media suddenly have become some of the main determinants of how we perceive our world. The media, in effect, actually structure our world, and we, in turn, reinforce this structure. Bernard Cohen summarized the agenda-setting concept when he said that the mass media may not be successful much of the time in telling people what to think, but the media are stunningly successful in telling their audiences what to think about.[17]

This agenda-setting function can also be divided into the *interpersonal agenda,* the things we not only think about but talk about, and an *intrapersonal agenda,* the things we merely think about.

The first empirical test of the agenda-setting function was in 1968 when researchers Maxwell McCombs and Donald Shaw examined the presidential elections.[18] Since then, McCombs and others have continued research on agenda-setting.

Still other research winds its way into media decision making, specifically, the study of how and why gatekeepers select the news they do and feed this to the public. The gatekeeper agenda appears to originate in the wire services. Thus, although the local press sets the public's agenda, the wire services set the agenda for the gatekeeper. Although researchers continue to search for remnants of media theory, agenda-setting research appears to have one of the firmest holds on an identifiable and consistently proveable phenomenon.

socialization

Closely aligned to how we use broadcasting is its effect on our social development in acquiring culture and social norms. Although a significant amount of research centers on broadcasting's effects on the socialization of children, we all know that socialization continues throughout our lives, and broadcasting can affect this socialization at any time. Again, as with other approaches to studying effects, the content of broadcast messages can mean different things to different people. For example, the effect of a violent television program on a group of male adults can be in sharp contrast to the effect of that same program on a group of small children, whose world and ideas are just being formed and whose socialization process is much less developed than that of the adults. The adult might go to bed thinking how great John Wayne was as the hero. The child may have frightening nightmares about evil forces affecting his or her ability to survive in the world.

Here again, research has opened up a Pandora's Box of results, theory, and debate, and different methodologies are used. As a responsible consumer of broadcasting in society we should recognize them. Since socialization does not occur simply by watching a single program, we must gather data from a wide body of research across many disciplines with which to begin to theorize exactly how media in general and broadcasting in particular affect our socialization process. Moreover that data must be gathered over time. Few studies examine socialization over time. Most ask a given group of individuals what meaning television or radio has for them and then group the results under the heading of socialization of uses and gratification research. Although examining a great deal of research about these different audiences is valuable, studying the same individuals over a longer time period is much more desirable.

Stages in Studying Effects on Socialization

Socialization research has three stages. First, numerous studies have examined the "content" of broadcast messages. Such elements as the image of women in television commercials, hero figures in prime-time television, and acts of violence have told us much about what we see or hear on radio and television. The second stage of this research tells us if people exposed to the broadcast message actually perceive or recognize the messages which are conveyed. Were children who saw a given program able to recognize examples of good behavior and pro-social messages? The third stage of investigation must determine what effect the messages have once they are received.

Studying the Results

From socialization research, we have learned that children can indentify certain pro-social content themes. For example, CBS has actively supported various research projects on ths issue, which have been beneficial to public relations. The research has been conducted under responsible surveillance.[19] In examining the program "Fat Albert and the Cosby Kids," research in three cities—Cleveland, Philadelphia, and Memphis—revealed that close to nine out of ten children who had seen an episode of "Fat Albert" received one or more messages of social value. Some of the pro-social messages reported being received included, "Take care of younger children," "Father's job is important," "Support a friend in trouble," "Be honest," and "Be friendly; don't be rude, nasty, jealous, or mean."

Similar research by CBS showed older children were more likely to receive more abstract messages than younger children were. For example, in studying the program "Shazam" about a Superman figure, about half of the seven to eight year-olds received the message "obey your parents," whereas about three-fourths of the ten to eleven year-olds and the thirteen to fourteen year-olds received that message. Only 4 percent of the seven to eight year-olds received the message to "be independent," whereas 11 percent of the ten to eleven year-olds and 25 percent of the thirteen to fourteen year-olds received the message. In examining the program "Isis," about a superhuman female figure, the research discovered that girls were more likely than boys to comment on Isis's concern for others and her beauty, while boys mentioned her superhuman qualities as often as girls did.

After analyzing the effects of broadcasting socialization, we can conclude that parents have a major responsibility in not permitting television to become a surrogate parent.[20] Watching television with very young children, then discussing the results and referring to possible pro-social lessons is one positive use of the medium. This same process was common in pretelevision times when parents read storybooks to children, then discussed the content of the books. Children apparently learn from television, and such broadcasting practices as stereotyping the roles of certain classes of people can form a child's perception of reality.

The amount of television and when and how it becomes part of children's lives also can influence how children relate to their environment. In studying three towns in Australia which had three different types of television programs available, researchers found that the content viewed was directly related to the context in which it was viewed.[21] When television experience was restricted to mostly an informative-educational context, children perceived it to be far more than just entertainment. When high levels of television viewing tended initially to reduce participation in such outside activities as sports, participation returned to normal levels after the "novelty" wore off.

Content and context variables also are included in research on the political socialization of children. Political knowledge, news discussion, public affairs interest, and seeking information about news events were investigated by Professors Charles K. Atkin and Walter Gantz.[22] They found the amount of news viewing to be somewhat associated with a child's political awareness, with the highest correlations being among older children. The amount of exposure to television news has some relationship to the child's knowledge of politics, but it has more among middle-class youngsters than among working-class youngsters. Many children in the research reported being stimulated to seek further information after watching television news, and to some degree this desire for more information increased with the amount of news exposure.

Advertising also can influence the socialization process. For example, one study showed children three different eyeglass commercials with a woman giving a testimonial.[23] One commercial showed her dressed as a court judge, another as a computer programmer, and the third as a television technician. The children who saw that woman as a particular role model were more apt to choose that occupation as appropriate for women.

There is still much to be learned about the effects of broadcast messages on the socialization process. In dealing with children, part of our knowledge will be gained from examining what psychologists long have taught about learning theory and formative development.

the violence debate

Of all broadcasting's effects, none has attracted more attention than the portrayal of violence on television. It has been grounds for research and debate for academicians, government agencies, local schools, and international research organizations alike.

Violence Gains Attention

The current attention to televised violence has been attracted not only by its very presence, but also by research indicating a possible causal relationship between violence and behavior. Two published articles appearing in the *Journal of Abnormal and Social Psychology* in 1968 set the stage. The two articles examined exposure to filmed violence and subsequent aggressive behavior.[24] Other studies followed. Yet the issue had been raised before, back in 1952 by Senator Estes Kefauver's committee on juvenile delinquency. But in 1952, television was too new to draw pertinent conclusions, despite testimony by recognized authorities and psychiatrists.

The issue remained mostly academic until 1969 when a letter from Senator John O. Pastore called for a blue-ribbon committee of leading scholars to examine the relationship in detail. In a letter to the secretary of health, education, and welfare, Pastore said, "I am exceedingly troubled by the lack of any definitive information which would help to resolve the question of whether there is a causal connection between televised crime and violence and antisocial behavior by individuals, especially children . . . ." Pastore then called for the formation of what became known as the Surgeon General's Scientific Advisory Committee on Television and Social Behavior. The list of experts selected to serve on the committee was chosen by reviewing names of experts on the subject. The final selection process was assigned to the three commercial networks and the NAB. CBS, seeing its own research director as a possible appointee, withdrew from the selection committee to avoid conflict of interest. Completed in 1972, the committee's report immediately drew praise, criticism, and varying interpretations. What did stand out as the most succinct summary statement of the committee report was:

> There is a convergence of the fairly substantial experimental evidence for short-run causation of aggression among some children by viewing violence on the screen and the much less certain evidence from field studies that extensive violence viewing precedes some long-run manifestations of aggressive behavior. This convergence of the two types of evidence constitutes some preliminary evidence of a causal relationship.

449

In other words, the report found the possibility that television violence could adversely affect some people. The report aroused widespread attention to the violence issue and encouraged new research to discover the heart of the "causal effect." Today, the research has not yet been finished, with various foundations and government agencies continuing to fund studies.

The networks have also become involved in violence studies. Undoubtedly because of the pressure to revamp American television, the networks and, for that matter, their affiliate stations are under fire. In some cases, the affiliates are even worse offenders than the networks, preempting bland network fare to insert more violent programming.

Theories of the Effects of Televised Violence

The relationship between televised violence and aggressive behavior is best understood using the various learning theories. There are four main theories. The *catharsis* theory suggests that we build up frustrations in our daily lives which are released vicariously by watching violent behavior. The catharsis theory claims that therefore there are actual benefits from televised violence. This theory is the least supported of the four, although the results of some studies have provided limited support for the idea.[25] The *aggressive cues theory* suggests that exposure to violence on television will raise the level of excitement in the viewer, and that televised violent acts may possibly be repeated in a real-life setting.[26] Closely aligned to the aggressive cues theory is the *reinforcement theory,* suggesting that televised violence will reinforce behavior already existing in an individual.[27] Inherent in such a theory is the probability that the violent person, because of violent tendencies, perceives violent behavior as a real-life experience, whereas the nonviolent person may perceive the violent program as entertainment without becoming "psychologically" involved in the program. The *observational learning theory* suggests that we can learn violent behavior from watching violent programs.[28]

Clearly, all of the theories have merits, and none should be discounted. Research is examining new variations on these four principal approaches. The observational learning theory, for example, could apply more to very young children in their formative years of growth, when their environment has a significant effect on what they learn. In essence, if television becomes a surrogate parent, it could certainly teach behavior. Later in the child's life when behavior is more firmly determined, violence learned in the formative years could be "reinforced." For the hyperactive or easily excitable child, the aggressive cues theory might be used to explain easily aroused emotions from exposure to televised violence. The catharsis theory even could apply to the business executive who uses television to unwind and vicariously vent his or her frustrations through the actions of others. We immediately begin to see all sides of the violence debate.

Current research is centering primarily on children, partly because of funding for such research and partly because of a general feeling that children may very well be those most affected by television violence. In this arena, the violence debate is becoming public with considerable pressure from and visibility of citizens' groups. Along with suggesting the causal relationship of televised violence to aggression, the widely quoted research of George Gerbner, Director of the Annenberg School of Communication at the University of Pennsylvania, is used to add fuel to the arguments.

For more than a decade, Gerbner and his associates have compared violence on television among the major networks, then have plotted their data over time, providing a running record of the number of violent acts representative of each new television season. Two often discussed measures are Gerbner's Violence Index, measuring the actual acts of violence, and the Risk Ratio, describing the risk of encountering violence. The index is used mostly to count violent acts on television; the ratio is a bit more complex. It measures the aggressors and the victims, dividing the larger by the smaller with the final figure preceded by a plus sign if the number of aggressors exceeds that of the victims and a minus sign if the number of victims exceeds that of the aggressors. CBS uses a different violence-measuring device, prompting the continuing debate over which measure is more accurate and more representative of actual violence.[29]

Effects of Portrayal on Aggressive Behavior

The amount of research on televised violence is now tremendous, with more studies in progress. What the research is telling us about the relationship between the portrayal of violence and aggressive behavior is summarized by Professor George Comstock in the *Journal of Communication*. He states that the evidence suggests:

1. Cartoon as well as live portrayals of violence can lead to aggressive performance on the part of the viewer.
2. Repeated exposure to cartoon and live portrayals of violence does not eliminate the possibility that new exposure will increase the likelihood of aggressive performance.
3. Aggressive performance is not dependent on a typical frustration, although frustration facilitates aggressive performance.
4. Although the "effect" in some experiments may be aggressive but not antisocial play, implications in regard to the contribution of television violence to antisocial aggression remain.

5. In ordinary language, the factors in a portrayal which increase the likelihood of aggressive performance are: the suggestion that aggression is justified, socially acceptable, motivated by malice, or pays off; a realistic depiction; highly exciting material; the presentation of conditions similar to those experienced by the young viewer, including a perpetrator similar to the viewer and circumstances like those of his environment, such as a target, implements, or other cues resembling those of the real-life milieu.

6. Although there is no evidence that prior repeated exposure to violent portrayals totally immunizes the young viewer against any influence on aggressive performance, exposure to television portrayals may desensitize young persons to responding to violence in their environment.[30]

Policy Dilemmas

Where all this leads is difficult to predict, but even if the evidence becomes conclusive, the end result may be a constitutional crisis of sizeable proportions. Although government has traditionally kept an open ear to complaints about violence, sympathy and rhetoric has been about as much as Congress or the FCC has been willing to offer. To offer more would collide head on with the First Amendment to the Constitution and with the Communications Act of 1934. Even if the broad "public interest" standard were applied to try to curtail violence, court tests would be necessary to keep it from encroaching on the Constitution. Although the debate and research will continue, the biggest battle of all may be fought in the political arena in which no medium yet has been successfully curtailed, except superficially. Nor have the courts been sympathetic to the violence issue. When a Florida teenager claimed that telvision violence caused him to commit murder, the jury did not accept the idea. Then there is the case of Japan, receiving more and more attention because of the high level of violence on Japanese television and the low crime rate. Some possible explanations are that students are too busy with school work to watch much television, Japan's strict gun laws, and citizens becoming involved in crime protection.[31] Another possible explanation is Japan's society, with its emphasis on collective (family, school, company) responsibility versus the U.S.'s emphasis on individual responsibility. Although the attacks on violence will continue, possible alternative causes to the ills of our society will also remain of keen interest to policy makers.

Some changes may come as advertisers begin to place economic pressure on the networks. But support among local advertisers is minimal, and stations wishing to preempt network programming to air locally originated programs with more violence are finding little to stop them. At least in the immediate future, if televised violence is a serious threat, the threat will be held at bay best by educating responsible consumers of broadcast communication, including parents, and hoping for restraint and responsible decisions on the part of the broadcast industry.

summary

Chapter 19 has examined the broadcast audience and approaches to studying it to identify the effects of broadcast programming. Early theories of the audience as a mass of unrelated individuals responding like sitting ducks to media messages has been greatly altered. Contemporary theorists view the audience as interacting with the media, permitting media to be an important part of their lives. Three approaches to studying the audience are individual differences, categories, and social relations. Information received from the media flows through the population in a multistep process called the two-step flow. Opinion leaders are at the heart of the two-step flow theory and are used to describe those individuals who influence other people's reactions to media content. How we react to media content can be determined by three factors: selective exposure, selective perception, and selective retention. The importance we place on a broadcast message can be based on source credibility, media credibility, or a combination of both.

Demographics and psychographics are used to classify the broadcast audience. Demographics refer to such factors as age, sex, income, occupation, and race. Psychographics refer to attitudes, beliefs, values, opinions, and the psychological characteristics of the audience. With these classifications in mind, three approaches to how we use media are Stephenson's Play Theory, uses and gratification, and agenda-setting. Radio and television also are important to the socialization process of acquiring culture and social norms. Finally, one of the most visible issues of the effects of broadcast programming is that of televised violence.

spotlight on further learning

BLUMLER, JAY G., and ELIHU KATZ, eds., *The Uses of Mass Communications: Current Perspectives on Gratifications Research.* Beverly Hills, Calif.: Sage Publications, Inc. 1974.

BOWER, ROBERT T., *Television and the Public.* New York: Holt, Rinehart & Winston, 1973.

CASSATA, MARY B., and MOLEFI K. ASANTE, eds., *The Social Uses of Mass Communication.* Buffalo: State University of New York, Department of Communication, Communication Research Center, 1977.

CATER, DOUGLASS, AND STEPHEN STRICKLAND, *TV Violence and the Child.* New York: Russell Sage Foundation, 1975.

CBS Office of Social Research, *Communicating with Children through Television.* New York: CBS, 1977.

CHAFFEE, STEVEN H., ed., *Political Communication: Issues and Strategies for Research.* Beverly Hills, Calif: Sage Publications, Inc., 1975.

Cisin, Ira and others, *Television and Growing Up: The Impact of Televised Violence.* Washington, D.C.: Superintendent of Documents, 1972.

Comstock, George and others, *Television and Human Behavior.* Santa Monica, Calif.: The Rand Corporation, 1975. (Three volumes—annotated bibliography.)

Kraus, Sidney, and Dennis Davis, *The Effects of Mass Communication on Political Behavior.* University Park, Pa.: Pennsylvania State University Press, 1976.

Lesser, Gerald S., *Children and Television: Lessons from Sesame Street.* New York: Random House, Inc., 1974.

Milgram, Stanley, and R. Lance Shotland, *Television and Antisocial Behavior* New York: Academic Press, Inc., 1973.

Piepe, Anthony, Miles Emerson, and Judy Lannon, *Television and the Working Class.* Lexington, Mass.: Lexington Books/D.C. Heath & Company, 1975.

Rubin, Bernard, *Political Television.* Belmont, Calif.: Wadsworth Publishing Co., Inc., 1967.

Severiges Radio, *Uses and Gratifications Studies: Theory and Methods.* Stockholm: Severiges Radio, 1974.

Shaw, Donald, and Maxwell E. McCombs, eds., *The Emergence of American Political Issues.* St. Paul: West Publishing Co., 1977.

Winick, Mariann, and Charles Winick, *The Television Experience: What Children See.* Beverly Hills, Calif.: Sage Publications, Inc., 1979.

appendix
using the library to learn about broadcasting

Information about broadcasting can be found in many library sources. The following guide is an introduction to selected major sources. Whether you are writing a paper, preparing an oral report, or just interested in expanding your knowledge, this guide will be useful.

For broadcasting, as for most disciplines, the library can be used for particular purposes, with a definite search strategy in mind. You can begin to build your knowledge of broadcasting using the simpler library sources and then move on to the more complex ones.

a strategy

The first and most important step in research on a particular topic is to find general background information. You can use several methods, including your course lectures and general textbooks. Another method is to use the library. Reference materials, such as general encyclopedias, are valuable in this "first" step.

Second, using the information found in the background sources, you can widen the search to include journals and books. Often the background information will give you possible search terms which will be useful in finding relevant information in other sources. The various

455

sources also may include bibliographies and footnotes which may be helpful.

Third, you can use more extensive sources such as abstracts, annual bibliographies, and literary reviews. Usually these sources cite more books, journal articles, and other sources than usually are included in any one library, but they cannot be overlooked if you hope to exhaust the available sources of information.

Finally, collections such as United States government documents and newspapers may be helpful.

This guide follows this general organizational pattern.

using the card catalogue

The card catalogue will probably be your first source of information. It can be used for finding suitable references to background information.

Using Subject Heading Cards

Subject heading cards are important to using the card catalogue efficiently. A newcomer to a discipline will use more subject heading cards than someone who has extensively studied a discipline and is familiar with specific authors and titles.

Try to think of terms which might be used as subject headings. Background information should give you some ideas.

A guide to finding subject headings in the card catalogue is the *Library of Congress Subject Headings*. This guide will refer you to the subject heading. For example, if you are looking for information on laws affecting television, you might find the following catalogue card:

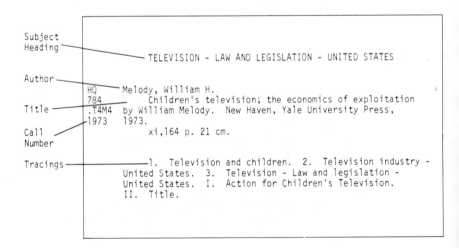

```
Subject
Heading
                    TELEVISION - LAW AND LEGISLATION - UNITED STATES

Author
        HQ      Melody, William H.
        784          Children's television; the economics of exploitation
Title   .T4M4   by William Melody.  New Haven, Yale University Press,
        1973    1973.
Call                 xi,164 p. 21 cm.
Number

Tracings               1.  Television and children.  2.  Television industry -
                    United States.  3.  Television - Law and legislation -
                    United States.  I.  Action for Children's Television.
                    II.  Title.
```

This particular catalogue card is called a *subject card,* but the book could also be found by using an author or a title card. In addition, tracings indicate that this book could be found under the other subject headings, such as "Television and children" and "Television industry— United States."

The following list is of common subject headings used in the card catalogues.

Community Antenna Television	Television
Radio	Television—Apparatus and
Radio—U.S.—Laws and	Supplies
Regulations	Television—Law and Legislation—
Radio Advertising	U.S.
Radio Announcing	Television—Production and
Radio as a Profession	Direction
Radio Audiences	Television and Children
Radio Authorship	Television Announcing
Radio Broadcasting	Television Audiences
Radio in Education	Television Authorship
Radio Journalism	Television in Education
Radio Plays	Television in Politics
Telecommunications	Television Plays
	Television Programs

You usually will use the subject cards more than the author and title cards. However, all three are in most card catalogues.

Using Call Numbers

The call number of the book is in the upper left hand corner of the catalogue card. It refers to the location of the book, and it is assigned so that books on similar subjects are shelved together.

Call numbers may be used to peruse the shelves and find relevant books on a topic. This is not as efficient as using the card catalogue but still may be useful. Usually a particular subject heading card is not assigned to a book unless a significant portion of the book is on that topic. You may want to use subject headings to find several areas of the library to examine. While doing so, check the index in the back of each book for terms relating to the topic of research.

using references

References provide definitions of terms, biographical information, bibliographies, and introductory essays. Information from references can aid you in searching for information in more extensive library sources.

Biographical Sources

Most libraries have numerous sources of biographical information. Three major ones are:

Contemporary Authors
A multivolume work containing biographical information on living authors from every nation, giving personal data, avocations, writings, and works in progress.

Current Biography 1940—to date
A multivolume set of biographies of people in the news, both in the United States and abroad, often with photographs and obituaries.

Who's Who in America
The standard source of current biographies of noted men and women in the United States.

Dictionaries and Encyclopedias

Because many disciplines include topics relating to radio and television broadcasting, specialized dictionaries and encyclopedias, such as the *Encyclopedia of Psychology*, the *International Encyclopedia of the Social Sciences*, and the *Encyclopedia of Educational Research* also can be helpful.

Other References

Other references more directly related to the study of broadcasting are:

Broadcasting Yearbook
This work attempts to be "the most comprehensive directory of the business of broadcasting." Included are: a short history of broadcasting, a section on the dominant areas of influence of television stations, a directory of AM and FM radio stations, a directory of advertising agencies, and an equipment and engineering section.

BBC Handbook
This is not only an annual report of the financial position of the British Broadcasting Corporation but also a review of its programs and other operations.

Television Factbook
Included in this book are general data on the industry, directories of communications lawyers and engineers, manufacturers of TV and radio receivers and station equipment, trade organizations and publications, CATV systems, pay-TV organizations, and satellite data.

The Standard Directory of Advertising Agencies
Issued three times a year, this source lists some 4,000 advertising agencies, 30,000 personnel by title, and 60,000 accounts.

using indexes

Indexes are used to find relevant articles in magazines, journals, books, and newspapers. Some good ones to start with are:

Reader's Guide to Periodical Literature
A subject and author index to over 100 periodicals of popular interest.

Business Periodicals Index 1958—to date
An index to over 100 journals on various aspects of business.

Education Index
An index to over 200 periodicals relating to education.

Humanities Index
Formerly part of the *Social Sciences and Humanities Index,* an index to over 260 journals in the humanities.

Social Sciences Index
Formerly part of the *Social Sciences and Humanities Index,* an index to over 260 journals in the social sciences.

Social Sciences and Humanities Index 1965—1972
(formerly *International Index* 1920—1965)
An index of over 200 journals which in 1973 was divided into two indexing services: *Humanities Index* and *Social Sciences Index.*

Public Affairs Information Services (P.A.I.S.)
A subject index to "current books, pamphlets, periodical articles, government documents, and other materials in the field of economics and public affairs."

The New York Times
The index of *The New York Times.* Check your library for available indexes of other newspapers.

Topicator
An article guide to broadcasting, advertising, communication, and marketing.

International Index to MultiMedia Information
An index of material found in audio-video services.

Television Sponsors Directory
Cross-reference guide of addresses of companies producing goods advertised on television.

Also consult the indexes in most scholarly journals and magazines. Each is designed for quick reference to that particular publication.

using abstracting services

An abstract contains the information to find a particular journal article, just as an indexing service does, but it also summarizes the article. Usually the abstract is short—50 to 150 words. An abstracting service often indexes and abstracts many different sources.

The following are important abstracts pertaining to the study of broadcasting.

Communicontents
Abstracts of publications dealing with all aspects of communication, including broadcasting.

Dissertation Abstracts International
A monthly compilation of abstracts of doctoral dissertations submitted to University Microfilms International by more than 375 cooperating institutions in the United States and Canada. Abstracts of dissertations on broadcasting can be found in Volume A under the section "Mass Communications." A key word index, based on the dissertation's title, is at the end of each volume.

Mass Media Booknotes
Issued monthly, this publication provides descriptions and brief reviews of new books on all subjects related to the mass media.

Psychological Abstracts
An abstracting service for the world's literature in psychology and related fields.

Sociological Abstracts 1953—to date
An abstracting service for articles on sociology from many journals.

using government documents

Many libraries contain U.S. government publications. To get a head start in using government documents, check the following:

Indexes

Monthly Catalog of United States Government Publications
This is the standard guide to most U.S. government publications. Documents are arranged by issuing agency. See also the:

Cumulative Subject Index to the Monthly Catalog of the United States Government Publications

Selected Documents

The federal government publishes a variety of statistical reports, hearings, studies, and other documents relating to broadcasting. The following are major examples of various types of government documents.

United States Statutes at Large
As each law is enacted, it is published. Later, the laws passed during each session of Congress are included in the *Statutes at Large*. They are arranged chronologically by date of passage.

United States Code
This is the official compilation of U.S. laws in force. Laws from the *Statutes at Large* are consolidated and codified in this work. A completely revised edition is issued every six years with annual supplements between revisions.

Media Law Reporter
This contains the texts of major decisions affecting mass media.

Federal Register
Statutory laws prescribe general intent; Congress delegates to the executive and the various departments the detailed task of administration. Administrative rules and regulations have legal force. The *Federal Register* provides the medium through which new rules and regulations are disseminated.

Code of Federal Regulations
The regulations and rules in the *Federal Register* are codified annually and published in this source. It lists, among other things, all the permanent and general rules and regulations established by the Federal Communications Commission.

Federal Communications Commission Reports 1934–
This contains the administrative decisions of the Federal Communications Com-

mission, including decisions regarding "the Fairness Doctrine," television station license renewal, cable television, and many more.

Federal Communications Commission Annual Report 1935–
This is a review of the major events for the year in the area of regulatory concern, including administrative matters, engineering, legal and legislative activities.

Congressional Hearings, Monthly Catalog
The *Monthly Catalog* is the index to use when researching congressional hearings.

journals and trade publications

Numerous journals and trade publications relate directly or indirectly to broadcasting. Usually, the best method of finding information in the various journals is by using indexing and abstracting sources already described. Some journals periodically publish their own indexes.

Ulrich's International Periodicals Directory. This volume is arranged by subject and includes an alphabetical index by title of periodicals. Some of the more important periodicals relating to broadcasting are:

Advertising Age
Published weekly and contains news of the advertising business.

AV Communications Review
Contains articles on educational media, educational television, and other aspects of the teacher-learning process related to technology and communications.

Broadcast Daily
Published for delegates at major broadcasting conventions.

Broadcast Engineering
A technology publication of the broadcast/communications industry.

Broadcast Management/Engineering
Contains articles of interest to broadcast management and engineering personnel.

Broadcasting
Weekly business news of the television and radio profession. Reports of interest to advertisers as well as to programmers, journalists, engineers, and others. Includes reports on the Federal Communication Commission's decisions, hearings, and procedures.

Broadcasting and the Law (Perry's)
A biweekly newsletter which reports and interprets current court and FCC rulings affecting broadcasting.

Cable News
A weekly magazine covering cable television matters.

Cablecasting
Covers engineering and technical aspects of cable television.

Cablevision
Directed toward those in the cable television industry responsible for managing, constructing and operating CATV systems.

CATV
A weekly trade publication on cable television.

Communication
A journal devoted to "conceptual/theoretical/philosophical approaches to the role of communications in human affairs."

Mass Media Booknotes

© 1979 by Christopher H. Ster[...]
Temple University

ISSN 0045-3188   10:5 ('[...]

JANUARY 1979

Book of the Month
Comstock, et al. Television and Human Behavior ....

Broadcasting
Clift-Greer, Broadcast Programming: The Current Perspecti[...]
Local community video and cable production, four publica[...]
Price, The Best Thing on TV: Commercials................
Public and educational broadcasting, three publications[...]
Studies of Business Regulation (series from Duke Unive[...]

Media Regulation/Responsibility
Casebier, Social Responsibilities of the Mass Media.[...]
Cole-Oettinger, Reluctant Regulator (revised ed.)..[...]
Dennis, et al. Justice Hugo Black and the First Ame[...]
Murray, The Media Law Dictionary...................
Rubin, Questioning Media Doctrine and the Media....[...]
Simmons, The Fairness Doctrine and the Media....[...]
Will, Telecommunications Structure and M[...]

Feedback

NTIA's Henry Geller:
At the pinnacle of policymaking

Broadcasting⚡Feb 19

Our 48th Year 1979

The newsweekly of broadcasting and allied arts

NEWSPAPER

This is ROOTS week .

Alex Haley's

ROOTS
THE NEXT GENER[...]
A David L. Wolper Pr[...]

THE 14-HOUR CONTINUATION[...]
EVENT IN TELEVISIO[...]

Warner Bros. Television Distribution

WINTER 1979
VOL. 23: 1

Journal of Broadcasting

FEATURED
TOPIC:

Television and Children

Selective Exposure to Televised Violence
....Charles Atkin, Bradley Greenberg, Felipe Korzenny and Steven McDermott

Viewing Rules as Mediating Factors of Children's Responses to Commercials
....Leonard N. Reid

Television's Impact on Preferences for Non-White Playmates: Canadian Sesame
Street Inserts
...........................Marvin E. Goldberg and Gerald J. Gorn

Communication Education (formerly *Speech Teacher*)
Designed to cover all aspects of teaching and learning speech communication on all levels of education. Includes articles on mass communication, public address, and instructional materials.

Communication Monographs (formerly *Speech Monographs*)
Publishes articles on general communication studies. Some articles on broadcasting and mass communication.

Communication News
Summarizes recent developments in all areas of broadcasting and telecommunication.

Communication Research
Concerned with the study of communication processes on all levels.

Communicontents (see Abstracting Services)

CPB Report
Newsletter of the Corporation for Public Broadcasting.

db
Trade publication concentrating on high fidelity and stereo recording.

Educational & Instructional Television
Contains articles and notes on recent developments and uses of television in both industry and education.

Educational Broadcasting
Contains both scholarly and general articles on ETV.

Educational Technology
Includes articles on various aspects of educational technology.

Educational Television
Covers closed circuit television in education, business, and industry.

Feedback
Articles on broadcast education.

Journal of Advertising Research
Devoted to research on advertising strategies and effects.

Journal of Broadcasting
Devoted to all aspects of broadcasting.

Journal of College Radio
Contains general articles of interest to management and staff of college radio stations.

Journal of Marketing Research
Research on effectiveness and development of marketing strategies.

Journal of Communication
On the study of communication theory, practice, and policy. Articles on such topics as television violence, censorship in broadcasting, and radio programming.

Journalism Quarterly
Covers all areas of journalism and mass communication.

Mass Media Booknotes (see Abstracting Services)

Mass Comm Review
Devoted to the study of mass communication, including broadcasting.

NAB Highlights
Affiliate newsletter of the National Association of Broadcasters.

Public Telecommunications Review
A journal of articles on public television.

The Quill
Publication of the Society of Professional Journalists, Sigma Delta Chi. Contains articles of interest to both print and broadcast journalists.

RTNDA Communicator
Newsletter of the Radio-Television News Directors Association.

Satellite Communications
For users, systems designers, common carriers, and manufacturers in the international satellite communications industry.

Television/Radio Age
Similar in some ways to *Broadcasting* but with longer, more substantial articles.

TV Communications
Devoted to cable television, it deals with topics on management systems design, finance, engineering, pay-cable, and others.

TV Guide
Local program listings and articles about radio and television.

With a good strategy and understanding of the library's organization and content, research about broadcasting will be an exciting and enjoyable experience.

glossary of terms

A.A.A.A. American Association of Advertising Agencies.

ABC (1) African Broadcasting Company; (2) American Broadcasting Company; (3) Australian Broadcasting Company.

Access channels cable television channels for general public use.

Accountable programming term used in educational television to describe a program meeting a specified set of instructional objectives.

ACT Action for Children's Television.

ADI rating term used to describe area of dominant influence.

AEJ Association for Education in Journalism.

Affiliate a broadcasting station bound by contract to associate with a particular broadcasting network or wire service.

All-channel receiver receiver capable of receiving AM and FM radio signals.

Alternator developed by Ernst Alexanderson at the General Electric Laboratories. Used to modulate early voice broadcasting.

AM amplitude modulation.

AM stereo dual-channel broadcasting on AM frequencies. A common method is to use one channel as amplitude modulation and the other channel as frequency modulation.

Anik   Canadian satellite system.

Annual billings   broadcast station's bill to advertisers for commercials carried over a one-year period.

AP Radio   radio network of the Associated Press.

ARB   Arbitron rating survey.

ARD   Federal Coordinator of West German Radio Broadcasting.

Armature   revolving iron core of the alternator.

ARRL   American Radio Relay League.

ASCAP   American Society of Composers, Authors, and Publishers.

ATS   Application Technology Satellite.

Audio actuality   the recording of the "actual" sounds in the news for incorporation into radio newscasts.

Audion   three-element vacuum tube invented by Lee de Forest.

AWRT   American Women in Radio and Television.

Banks   or "buses," groups of control switches on a master control console used to program various portions of an audio or video production.

Barter   trade-out advertising agreement.

BBC   British Broadcasting Corporation.

BEA   Broadcast Education Association.

BMI   Broadcast Music Incorporated.

BRC   Broadcast Rating Council.

Buses   see banks.

Cash-flow statement   amount of money necessary to keep the station running smoothly without going into debt.

CATV   community antenna television, or cable TV.

CBC   Canadian Broadcasting Corporation.

CCTV   closed-circuit television.

Chain broadcasting   early term used for network broadcasting.

Chart of accounts   ledger used to record station finances.

Coaxial cable   cable consisting of an inner wire core surrounded by a layer of plastic, metal-webbed insulation, and a third layer of plastic.

Coherer   small glass tube used in Marconi's experiments to create and break an electrical connection.

Columbia   term used in early broadcasting which included an identifying label for such companies as the Columbia Broadcasting System (CBS) and the Columbia Phonograph Broadcasting System, Incorporated.

Communication Model   a pictorial representation of the communicative process.

Comparative balance sheet   compares assets and liabilities between two points in time.

Comparative income statement   compares income between two points in time.

COMSAT   Communications Satellite Corporation. Formed by the Communication Satellite Act of 1962.

COMSTAR   satellite system launched by NASA and leased by AT&T.

Conduction   the use of ground or water, both electrical conductors, to replace a second wire in a telegraph hookup.

Construction permit   permission granted by FCC to begin construction of a broadcast facility.

Co-op advertising   a split in the cost of advertising usually between a retail outlet and the manufacturer.

Co-ops   (1) broadcast news networks, also called "infomal networks," created by a group of radio or TV news personnel, (2) trade-out advertising agreements between advertisers and the individual advertising outlet.

CPB   Corporation for Public Broadcasting.

CPM   cost per thousand.

CPS   cycles per second.

CRTC   Canadian Radio-Television Commission.

CTNC   Catholic Television Network of Chicago.

CTS   Communication Technology Satellite.

CTV   abbreviation for Canada's CTV Television Network, Ltd.

Cume   (or cumulative audience), the number of different persons or households watching or listening to a given station or program during a certain time period.

Cume persons   the number of different persons listening to a station at least once during a given time period.

CWA   Communication Workers of America.

Daytimers   radio stations required by the FCC to sign off at sunset.

Demographics   age, sex, education level, income, and ethnic background of an audience.

Diary method   method of data collection in a rating survey utilizing a diary.

Directional antennas   a group of strategically placed broadcast antennas transmitting a signal in a specific direction to form an irregular rather than a circular contour.

Directional stations   radio stations primarily in the AM band with directional antennas to keep their signals from interfering with those of other stations.

Director the person responsible for the entire production of a program.

Direct-wave propagation radio wave pattern of signals in direct line-of-sight transmission.

Dissolve smoothly changing from an image produced by one television camera to an image produced by a second television camera, film, slide, or videotape.

DMA Designated Market Area.

Double billing fraudulent practice of double charging advertisers.

Drop cable cable from the subtrunk of a cable system to the home terminal.

Earth station satellite earth-based receiving station.

Electromagnetic spectrum the range of levels of electromagnetic energy, called frequency.

Electromagnetic waves energy traveling through space at the speed of light. Used to transmit radio and television signals.

ENG Electronic News Gathering.

ERI Electronic Response Indicator.

ETV educational television.

FACTS Federation of Commercial Television Stations (Australia).

FBC Federal Broadcasting Corporation. (Rhodesia).

Fairness Doctrine FCC rule requiring equal air time for controversial issues.

Fiber optics the use of thin strands of glass to carry as many as 1,000 or more cable channels. Also used for data communication.

Filament element in a three-element vacuum tube. The other two are plate and grid. Early tubes used just a plate and filament.

Floor manager the person communicating the commands of the director to the performers.

FM frequency modulation.

FR3 French Regional Broadcasting Service.

Frequency (1) broadcast rating term indicating how often a viewer has tuned to a given station, (2) position on the electromagnetic spectrum.

Gatekeeper the person directly involved in relaying or transferring information from one individual to another through a mass medium.

GBC Ghana Broadcasting Corporation.

Geostationary or "synchronous," an orbiting satellite traveling at a speed proportional to that of the earth's rotation, thus appearing to remain stationary over one point on the earth.

gHz gigahertz. One billion hertz or cycles per second. (See kHz and MHz.)

Grid element in a three-element vacuum tube. The other two are filament and plate.

Ground wave waves adhering to the earth's surface.

Ham slang for amateur radio operators.

Headend the human and hardware combination responsible for originating, controlling and processing signals over the cable system.

Hertz(Hz) last name of Heinrich Rudolph Hertz commonly used as abbreviation for "cycles per second" in referring to electromagnetic frequencies.

Hetrodyne Circuit improved detector of radio waves invented by Reginald A. Fessenden.

Home terminal (1) receiving set for cable TV transmissions, either one-way or two-way, (2) device connecting the drop cable of a cable system to the receiving set.

Homophily the extent to which such things as beliefs, experiences, background, culture etc. are shared by two different communicators.

HUT households using television.

Hypoing promotion efforts used to increase the size of an audience during a rating period.

IBA Independent Broadcasting Authority. (British Television Network).

IBEW International Brotherhood of Electrical Workers.

IBM International Business Machines Corporation.

ICA International Communication Association.

ILR Independent Local Radio. (British Radio Network).

Image orthicon one of the first pickup tubes used in early television broadcasting.

INA Institute Nationale de l'Audiovisuel (France).

Induction process by which a current in one antenna produces a current in a nearby antenna.

Informal networks broadcast news networks created by a professional group of radio or TV news personnel. These networks also are called co-ops.

INTELSAT International Telecommunications Satellite Organization.

Ionosphere upper level of the atmosphere reflecting radio waves back to earth.

ITU International Telecommunication Union.

ITV (1) instructional television, programming specifically designed for direct or supplemental teaching, (2) the Independent Television Network.

JCET   Joint Committee on Educational Television.

kHz   kilohertz. One thousand hertz or cycles per second. Measurement of a position on the electromagnetic spectrum.

Long lines   term used by AT&T to describe long-distance telephone communication links.

Lowest Unit Charge   minimum charge on a station rate card.

Mass audience   audience reached by the mass media.

Master control console   heart of a television control room operation through which both the audio and video images are fed, joined together, and improved, perhaps by special effects, for the "on-air" image.

MBS   Mutual Broadcasting System.

Media   plural of medium.

Media credibility   the effect of various media on how mass communication messages are perceived.

Medium   channel of communication such as radio or television. Singular of media.

Message intensity   the value or importance of an event or its potential impact in relation to other events or potential news stories.

Meter method   a broadcast ratings measurement in which a monitoring device installed on TV sets is connected to a central computer, which then records channel selection at different times of the day.

MGM   Metro Goldwyn Mayer.

MHz   megahertz. One million hertz or cycles per second. (See kHz and gHz.)

Microwave   a very short wave of higher frequency than that of standard broadcast transmission. Usually measured in billions of cycles per second or gigahertz.

Mix   to join and separate the pictures of various television cameras for a composite "on-air" image.

MPATI   Midwest Program for Airborne Television Instruction.

MSA   rating term used to describe metro survey area.

NAB   National Association of Broadcasters.

NAEB   National Association of Educational Broadcasters.

NARB   National Association of Radio Broadcasters.

NARBA   North American Regional Broadcasting Agreement.

NASA   National Aeronautics and Space Administration.

NBC   National Broadcasting Company.

NCTA   National Cable Television Association.

NET   National Educational Television.

NHK Nippon Hōsō Kyōkai (Japanese broadcasting system).

NIS now defunct radio network formerly operated by NBC.

NOS Netherlands Broadcasting Foundation.

NPR National Public Radio.

Opinion leader person interpreting messages originally disseminated by the mass media.

ORTF Office de Radiodiffusion-Television Française.

Oscillation valve term used by inventor J. Ambrose Fleming to describe an early tube which constituted the main component in a wireless receiver.

OT Office of Telecommunication.

Pay cable a system in which cable subscribers pay a fee in addition to the standard monthly rental fee in order to receive special programming.

PBS Public Broadcasting Service.

Perigee the closest point to the earth of a satellite's orbit.

Physical noise breakdown in communication caused by some physical quality or object interfering with the communicative process.

Plate one of three elements in a three-element vacuum tube. The other two are filament and grid. Early tubes used just a plate and filament.

Plumbicon superseding both the image orthicon and vidicon tube, the plumbicon can capture color images with the sensitivity of the human eye. It is also a trade marked name of the Amperex Corporation.

Prime time time of largest audience when station charges highest price for advertising. 7 to 11 P.M. for TV, 7 to 9 A.M. and 4 to 6 P.M. for radio. (Radio prime time can vary depending on market and lifestyle trends.)

Production companies commonly called production houses, these businesses produce broadcasting programs for adoption either by networks or individual stations through syndication.

Program 1,2,3 Swedish Radio Networks.

Program managers persons responsible for selecting programs for airing, scheduling their air time, and overseeing the production and direction of locally produced programs.

Projection an estimate of the characteristics of a total universe based on a sample of that universe.

Psychographics study of the psychological characteristics of the mass audience.

PTA National Congress of Parents and Teachers.

Public broadcasting the operation of the various noncommercial radio and television stations in the United States.

Public service advertising (PSA)   designed to support a nonprofit cause or organization. Most of the time or space for this advertising is provided free as a service to the public by the print or broadcast media.

Quad   abbreviation for quadraphonic.

Quadraphonic   four-channel sound.

RAB   Radio Advertising Bureau.

Random sampling   selection process in which each unit of the larger portion has an equal chance of being selected.

Rating   percentage of people in a given population who are tuned to a radio or television station during a given time period.

RCA   Radio Corporation of America.

Relay satellite   A device capable of bouncing messages back to earth. Echo I was the first.

Repeater satellite   Satellite that could both receive and retransmit signals back to earth. The United States's Courier I-B was the first of the series.

ROS   run of schedule.

RTNDA   Radio Television News Directors Association.

SABC   South African Broadcasting Corporation.

Sales networks   a group of broadcasting stations linked together by a financial agreement to benfit all member stations by offering advertisers a joint rate.

Sampling   the process of examining a small portion of something to estimate the characteristics of the larger portion.

Satcom   domestic satellite system operated by RCA American Communications, Incorporated.

SBC   Swedish Broadcasting Corporation.

SCA   Speech Communication Association; also Subsidiary Communication Authority.

Selective exposure   exposing oneself to communication believed to coincide with preconceived ideas.

Selective perception   perceiving only those things which agree with preconceived ideas.

Selective retention   remembering only those things which agree with preconceived ideas.

Semantic noise   breakdown in communication caused by misunderstanding the meaning of words.

Share   (1) (or rating), a percentage of the total number of households or people tuned to a station or program during a certain time period; or

(2) the percentage of listeners to a particular station in comparison to listeners to all other stations or programs during a given time period.

Silicone detector crystal used in early radio receiving sets to detect radio waves.

SIN Spanish International Network.

Sky wave propagation radio-wave transmission pattern in which the signals travel up, bounce off the ionosphere, and rebound from the earth in a continuing process.

SPC Station Program Cooperative. Program acquisition method used in public television.

SPF Société Française de Production et de Creation Audiovisuelles.

SPJ, SDX Society of Professional Journalists, Sigma Delta Chi.

Subtrunk secondary cables branching out from the main trunk in a cable TV system to carry the signal to outlying areas.

Super hetrodyne circuit improvement on Fessenden's hetrodyne circuit. Developed by Edwin H. Armstrong.

Supering positioning a picture from one television camera on top of another picture from a second camera. This special effect is controlled by the master control console.

Sweep rating survey period.

Switcher or "technical director," person responsible for operating the master control console.

Synchronous or "geostationary," a satellite traveling at a speed in proportion to that of the earth's rotation, thus appearing to remain stationary over one point of the earth.

Syndicator company supplying syndicated programming to networks or local stations.

Talent raid CBS's "raid" on other network talent in 1948. Term sometimes refers to similar actions by ABC in 1976.

TDF Télédiffusion de Française.

Telestar early satellite used for the first transatlantic television broadcast.

Television household a broadcast rating term used for any home merely having a television set, as distinguished from a household actually using television.

TF1 Télévision Français 1.

The Freeze term used to describe the FCC's decision to stop allocating television frequencies between 1948 and 1952.

Toll broadcasting early term for commercial broadcasting first started on WEAF.

Trade-out an agreement in which a product or service is traded for advertising on a station.

Transistor wafer-thin crystal in three layers used extensively in electronic equipment. Performs many of the functions of the three-element vacuum tube.

Translators television transmitting antennas, usually located on high natural terrain.

TSA rating term used for total survey area.

Two-step flow process by which information disseminated by mass media is (1) received by a direct audience and then (2) relayed to other persons.

Two-way cable cable system capable of both sending and receiving data.

UHF Ultra High Frequency.

Universe the whole from which a sample is being chosen. In broadcast ratings, this can be the sample area, metro area, or rating area.

UPI Audio audio network of United Press International.

Valve term used for an early two-element vacuum tube.

VHF Very High Frequency.

Vidicon sensitive television tube which followed the image orthicon.

VOA Voice of America.

VTR video tape recorder.

WARC World Administrative Radio Conference.

Wave length distance between two waves.

Westar Western Union Satellite System.

WICI Women In Communication, Inc.

Wireless term used for early radio.

Wireless telephone term used for an early invention by Nathan B. Stubblefield.

YLE YLEISRADO. Finnish broadcasting state-controlled monopoly.

notes

chapter 1

[1] Thomas W. Hoffer, "Nathan B. Stubblefield and his Wireless Telephone," in *American Broadcasting: A Source Book on the History of Radio and Television,* ed. Lawrence W. Lichty and Malachi C. Topping (New York: Hastings House, Publishers, 1975), pp. 32–38.

[2] *The American Heritage Dictionary of the English Language* (New York: American Heritage Publishing Co., Inc. and Houghton Mifflin Company, 1973), p. 167.

[3] Robert A. Dutch, ed., *Roget's Thesaurus* (New York: St. Martin's Press, Inc., 1965), p. 711.

[4] Minow made the remarks in a speech to the 1961 meeting of the National Association of Broadcasters.

[5] *American Heritage Dictionary,* p. 269.

[6] Stewart L. Tubbs and Sylvia Moss, *Human Communication: An Interpersonal Perspective* (New York: Random House, Inc., 1974), p. 6.

[7] Kenneth E. Anderson, *Introduction to Communication Theory and Practice* (Menlo Park, Calif.: Cummings Publishing Company, 1972), p. 8.

[8] Wilbur Schramm, "The Nature of Communication between Humans," in *The Process and Effects of Mass Communication,* rev. ed., ed. Wilbur Schramm and Donald F. Roberts (Urbana: University of Illinois Press, 1971), p. 8. Schramm posited the "sharing" emphasis of communication in the first edition of *The Process and Effects of Mass Communication,* published in 1954.

[9] Karlyn Kohrs Campbell, *Critiques of Contemporary Rhetoric* (Belmont, Calif.: Wadsworth Publishing Co., Inc., 1972), p. 33. This is Campbell's interpretation of Kenneth Burke, *A Grammar of Motives and a Rhetoric of Motives* (New York: Meridian Books, 1962).

[10] James C. McCroskey and Lawrence Wheeless, *Introduction to Human Communication* (Boston: Allyn & Bacon, Inc., 1976), (sender) added. The term's conceptual application is credited to Paul F. Lazarsfeld and Robert K. Merton, "Friendship as a Social Process; A Substantive and Methodological Analysis," in *Freedom and Control in Modern Society,* ed. Monroe Berger and others. (New York: Octagon, 1964), p. 23.

[11] Kurt Lewin, "Channels of Group Life; Social Planning and Action Research," *Human Relations,* (1947), 143–53.

[12] Kenneth Starck and John Soloski, "Effect of Reporter Predispositions in Covering a Controversial Story," *Journalism Quarterly,* 55 (Spring 1977), 120–25; Bradley S. Greenberg and Percy H. Tannenbaum, "Communicator Performance under Cognitive Stress," *Journalism Quarterly,* 39 (Spring 1962), 169–78; Jean S. Kerrick, Thomas E. Anderson, and Luita B. Swales, "Balance and Writers' Attitude in News Stories and Editorials," *Journalism Quarterly,* 41 (Spring 1964) 207–15; Ithiel de Sola Pool and Irwin Shulman, "Newsmen's Fantasies, Audiences, and Newswriting," and Walter Gieber, "News Is What Newspapermen Make It" in *People, Society, and Mass Communications,* ed. Lewis Anthony and David Manning White (Glencoe, N. Y.: The Free Press, 1964), pp. 141–59 and 173–82.

[13] Although many studies have examined the various relationships, the concept was first reported and applied to current mass communication in Paul F. Lazarsfeld, Bernard Berelson, and H. Gaudet, *The People's Choice* (New York: Columbia University Press, 1948). Opinion leaders also can act in strictly interpersonal communication. However, it is in reference to mass media that the term is applied here.

[14] Wilbur Schramm and Janet Alexander, "Broadcasting" in *Handbook of Communication,* ed. Ithiel de Sola Pool, Frederick W. Frey, Wilbur Schramm, Nathan Maccoby, and Edwin B. Parker (Chicago: Rand McNally College Publishing Company, 1973), p. 586.

chapter 2

[1] Readers interested in the pre-Hertz era should consult such works as Hugh G. J. Aitken, *Syntony and Spark—The Origins of Radio* (New York: John Wiley & Sons, Inc., 1976); George G. Blake, *History of Radio Telegraphy and Telephony* (London: Chapman & Hall Ltd., 1928, reprinted by Arno Press, Inc., 1974); Silvanus P. Thompson, *Michael Faraday: His Life and Work* (Macmillian Publishing Co., Inc., 1898); John J. Fahie, *The History of Wireless Telegraphy* (New York: Dodd, Mead & Company, 1901, reprinted by Arno Press, Inc., 1971); W. Rupert MacLaurin, *Invention and Innovation in the Radio Industry* (New York: Macmillan Publishing Co., Inc., 1949, reprinted by Arno Press, Inc., 1971); Gleason L. Archer, *History of Radio to 1926* (New York: The American Historical Society, Inc., 1938); J. A. Fleming, *The Principles of Electric Wave Telegraphy* (London: Longmans, Green, and Co., 1908); Richard T. Glazebrook, *James Clerk Maxwell and Modern Physics* (New York: Macmillan Publishing Co., Inc., 1896); E. A. Marland, *Early Electrical Communication* (London: Abelard Schuman, 1964).

[2] Accounts of Marconi's life are found in Orrin E. Dunlap, *Marconi: The Man and His Wireless* (New York: Macmillan Publishing Co., Inc., 1937, reprinted by Arno Press, Inc., 1971); W. P. Jolly, *Marconi* (Briarcliff Manor, N.Y.: Stein & Day Publishers, 1972); Degna Marconi, *My Father Marconi* (New York: McGraw-Hill Book Company, 1962); Niels H. de V. Heathcote, *Nobel Prize Winners in Physics: 1901–1950* (New York: Henry Schuman, 1953); R. N. Vyvyan, *Marconi and Wireless* (East Ardsley, England: E P Publishing Limited, 1974).

[3] Marconi, p. 27.

[4] *Ibid.,* p. 28.

[5] *Ibid.,* pp. 38–39.

[6] *Ibid.,* p. 36.

[7] *Ibid.,* pp. 100–104.

[8] "Wireless Signals across the Ocean," *New York Times,* December 15, 1901, pp. 1–2; "Wireless Telegraphy across the Atlantic," *Times* (London), December 16, 1901, p. 5.

[9] P. T. McGrath, "Marconi and His Transatlantic Signal," *Century Magazine,* 63 (March, 1902), 769; George Iles, "Marconi's Triumph," *The World's Work* (February 1902), 1784; Ray Stannard Baker, *McClure's Magazine,* 18 (February 1902), 291.

[10] McGrath, p. 781.

[11] "Signor Marconi's Experiments," *Times* (London), December 19, 1901, p. 5.

[12] Although there is much written on the early development of the Marconi companies, particularly useful to this text were W. J. Baker, *A History of the Marconi Company* (New York: St. Martins Press, Inc., 1971); Gleason L. Archer, *History of Radio to 1926* (New York: The American Historical Society, Inc., 1938); L. S. Howeth, *History of Communications—Electronics in the United States Navy* (Washington, D. C.: Bureau of Ships and Office of Naval History, 1963); Hiram L. Jome, *Economics of the Radio Industry* (London: A. W. Shaw Company, 1925); Thorn Mayes, "History of the American Marconi Company," *The Old Timer's Bulletin*, 13 (June 1972), 11–18, as cited in Lawrence W. Lichty and Malachi C. Topping, ed., *A Source Book on the History of Radio and Television* (New York: Hastings House, Publishers, 1975); Jolly, *Marconi* (cited above).

[13] The first proposed title of this company was Marconi's Patent Telegraphs Ltd., to which Guglielmo Marconi himself had objected. Source: Baker, p. 35.

[14] *Investors World*, October 7, 1898, p. 484.

[15] Howeth, p. 36.

[16] Archer, pp. 81–82.

[17] Lee de Forest, *Father of Radio* (Chicago: Wilcox & Follett Co., 1950). Fleming's work is discussed in J. A. Fleming, *An Elementary Manual of Radio Telegraphy and Radio Telephony* (London: Longmans, Green, and Co., 1908), pp. 204–11; Blake, pp. 238–40.

[18] Harlow, *Old Wires and New Waves*, pp. 462–63, as cited in Archer, p. 92.

[19] "October Meeting of the American Institute of Electrical Engineers," *Electrical World*, 43 (November 3, 1906), 836–37; also published as Lee de Forest, "The Audion, A New Receiver for Wireless Telegraphy," *The Scientific American Supplement*, November 30, 1907, pp. 348–56.

[20] J. A. Fleming, "Wireless Telegraph Receiver," 43 *Electrical World* (December 8, 1906), 1117.

[21] Lee de Forest, "Wireless Telegraph Receiver," 43 *Electrical World* (December 22, 1906), 1206. See also footnote 33.

[22] Jome, p. 208; *Marconi Wireless Telegraphy Company of America* v. *de Forest Radio Telephone and Telegraph Company*, 236 Fd. 942, affirmed by the Circuit Court of Appeals in 243 Fd. 560.

[23] de Forest, pp. 325–26.

[24] de Forest, p. 457.

[25] Charles Susskind, "de Forest, Lee," in *Dictionary of Scientific Biography Volume 3*, ed. Charles Coulston Gillispie (New York: Charles Scribner's Sons, 1975), pp. 6–7.

[26] Elliot N. Sivowitch, "A Technological Survey of Broadcasting's Prehistory," *Journal of Broadcasting* 5 (Winter 1970–71), 1–20. For the most part, research into the "induction–conduction" developments in radio's development has been overlooked by historians, partly because such approaches became scientifically obsolete in later years. Research on especially the political, economic, and social implications, still is needed. It would provide some important "micro" insights into how new scientific knowledge is applied to the invention process. Sources for such research studies are still available and more easily accessible than those of inventors in the early nineteenth century.

[27] Sivowitch, pp. 20–22; Thomas W. Hoffer, "Nathan B. Stubblefield and His Wireless Telephone," *Journal of Broadcasting*, 15 (Summer 1971), 317–29; Harvey Geller, "The Man History Overheard," *Circular-Warner/Reprise*, 7 (December 8, 1975), 1–4.

[28] Geller, p. 2.

[29] *Washington Post*, August 10, 1940: as cited in Hoffer.

[30] Hoffer, p. 322.

[31] Details of Fessenden's experiments can be found in Sivowitch, Howeth, Archer, Blake, and Barnow, as well as in other works on early wireless.

[32] Archer, pp. 102–103.

[33] de Forest, p. 268.

[34] *Ibid.*, p. 260.

[35] Information on ham radio is from Clinton B. DeSota, *Two Hundred Meters and Down: The Story of Amateur Radio* (West Hartford, Conn.: The American Radio Relay League, Inc., 1936).

chapter 3

[1] R. Franklin Smith, "Oldest Station in the Nation," in *American Broadcasting: A Sourcebook on the History of Radio and Television*, ed. Lawrence W. Lichty and Malachi C. Topping (New York: Hastings House, Publishers, 1975), pp. 114–16. (Originally published in *Journal of Broadcasting*, Winter 1959–60.

[2] Gordon R. Greb, "The Golden Anniversary of Broadcasting," in *American Broadcasting: A Sourcebook on the History of Radio and Television*, ed. Lawrence W. Lichty and Malachi C. Topping (New York: Hastings House, Publishers, 1975), pp. 95–96. (Originally published in *Journal of Broadcasting*, Winter 1958–59, pp. 3–13.

[3] *Ibid.*, p. 98.

[4] *Ibid.*, p. 102.

[5] Two accounts of WHA's early history are found in Werner J. Severin, "WHA-Madison 'Oldest Station in the Nation' and the Wisconsin State Broadcasting Service." Paper presented at the meeting of the Association for Education in Journalism, Madison, Wisconsin, 1977; *The First 50 Years of University of Wisconsin Broadcasting: WHA 1919–1969, and a Look ahead to the Next 50 Years* (Madison: University of Wisconsin, 1970.)

[6] The accounts of WWJ's early history are from R. J. McLauchlin, "What the Detroit *News* Has Done in Broadcasting," in Lichty and Topping, *American Broadcasting*, pp. 110–13. Originally published in *Radio Broadcast*, June 1922, pp. 136–41, *WWJ* (Detroit: WWJ, 1936); and brochures "WWJ Broadcasting Firsts" and "WWJ Radio One" published by WWJ in 1970, commemorating the station's sixtieth anniversary.

[7] The account of KDKA's history is from *The History of KDKA Radio and Broadcasting* (Pittsburgh: KDKA). Also found in *American Broadcasting*, pp. 13–110.

[8] *The History of KDKA*, p. 10.

[9] Gleason L. Archer, *History of Radio to 1926.* (New York: The American Historical Society, Inc., 1938), p. 164.

[10] *Ibid.*, p. 157.

[11] *Ibid.*, pp. 162–63.

[12] *Ibid.*, pp. 112–13.

[13] There are many accounts of the formation of RCA. These include, along with Archer's *History of Radio to 1926*, his *Big Business and Radio* (New York: The American Historical Society, 1939, reprinted by the Arno Press, Inc., 1971) pp. 3–22, Eric Barnouw, *A Tower in Babel* (New York: Oxford University Press, Inc., 1966), pp. 52–61.

[14] Barnouw, pp. 44–45.

[15] *Ibid.*, p. 49.

[16] W. R. MacLaurin, *Invention and Innovation in the Radio Industry* (New York: The Macmillan Publishing Co., Inc., 1949, reprinted by Arno Press, Inc., 1971), p. 123.

[17] *Ibid.*, p. 106.

[18] *Report on Chain Broadcasting* (Washington, D.C.: Federal Communications Commission, 1941), p. 10.

[19] Barnouw, p. 181.

[20] Archer, *History of Radio*, p. 276.

[21] *American Radio Journal*, 1 (June 15, 1922), 4; cited in William Peck Banning, *Commercial Broadcasting Pioneer: The WEAF Experiment* (Cambridge, Mass.: Harvard University Press, 1946), p. 94.

[22] Banning, p. 93.

[23] *Ibid.*, pp. 231–36.

[24] Archer, *Big Business and Radio*, pp. 133–65.

[25] *Ibid.*, p. 169.

[26] *Ibid.*, p. 173.

[27] *Report on Chain Broadcasting*, p. 17 (material in parentheses added).

[28] *Ibid.*, p. 92.

[29] Manuel Rosenberg, *The Advertiser*, 14 (August, 1943), 1–2, 24.

[30] FCC release (71159), October 12, 1943.

[31] *Ibid.*

[32] "Blue Sales Record an Outstanding One," Press release, Blue Network, 1942.

[33] Press release, American Broadcasting Company, March 29, 1945.

[34] Press release, American Broadcasting Company, June 15, 1945.

[35] *FCC Report on Chain Broadcasting,* p. 23.

[36] "The Way We've Been ... and Are," *Columbine,* 2 (April-May 1974), 1. Columbine is a corporate publication of CBS.

[37] *FCC Report on Chain Broadcasting,* pp. 26–28.

[38] Eric Barnouw, *The Golden Web: The History of Broadcasting in the United States* (New York: Oxford University Press, Inc., 1968), p. 40.

[39] MacLaurin, p. 186.

[40] Archer, *Big Business and Radio,* p. 424.

[41] Barnouw, *The Golden Web,* p. 242.

[42] *Cox Looks at FM Radio: Past, Present and Future* (Atlanta: Cox Broadcasting Corporation), pp. 81–82.

[43] *Ibid.,* p. 82.

[44] Farnsworth's contributions to television have been overlooked by many broadcast historians, perhaps because of the familiarity of RCA and the encouragement by Sarnoff to push ahead "publicly" for television's development.

[45] Romaine Galey Hon, ed., *Headlines Idaho Remembers,* (Boise: Friends of the Bishops' House, Inc., 1977), p. 39. Reported in the *Idaho Statesman,* July 13, 1953.

[46] *Columbine,* 2 (April/May 1974), 8.

[47] Sources of the history of television recording include Albert Abramson, "A Short History of Television Recording," *Journal of SMPTE,* 64 (February 1955), 72–76; Albert Abramson, "A Short History of Television Recording: Part II" *Journal of SMPTE,* 82 (March 1973), 188–98; Joseph Roizen, "Video-Tape Recorders: A Never-Ending Revolution," *Broadcast Engineering,* April 1976, pp. 26–30; Joseph Roizen, "The Video-tape Recorder Revolution," *Broadcasting Engineering,* May 1976, pp. 50, 52–53.

[48] Roizen, May 1976.

[49] Ron Whittaker, "Super 8 in Broadcasting, CATV and CCTV—Current Technology and Applications" (unpublished paper, University of Florida, Gainesville, 1975).

chapter 4

[1] *American Heritage Dictionary* (New York: American Heritage Publishing Co., Inc. and Houghton Mifflin Company, 1973), p. 45. There are many other applications of the term other than to physics and electronics. Mathematicians apply it as "the maximum ordinate value of a periodic curve," and astronomers view it as "the angular distance along the horizon from true east or west to the intersection of the vertical circle of a celestial body with the horizon." Source: *Ibid.*

[2] *Ibid.,* p. 696.

[3] See "FM Broadcast Channel Frequency Spacing," FCC/OCE RS 75–80.

[4] Lou Dorren, "Editorial," *FM 4-Channel Forum,* June, July 1976, p. 1.

[5] "California Quad 'Network' Broadcast A Resounding Success," *FM 4-Channel Forum,* 1(October/November 1976), 4.

[6] *Ibid.*

[7] The example of the car salesman is credited to "Radio Production—Four Times Better in Quad," *FM 4-Channel Forum, 1 (October/November 1976), 1.*

[8] "Do We Want Discrete Four-Channel Stereo for FM?" *Broadcast Management/Engineering,* 12 (February 1976), 40–46.

[9] *Ibid.,* p. 3. The emphasis on AM stereo was voiced at the 1976 meeting of the National Association of Radio Broadcasters.

[10] "The Road ahead Looks Smooth for AM Stereo," *Broadcast Management/Engineering,* 12 (February 1976), 48–50. If a station already has FM stereo equipment, simulcasting stereo AM is possible, although the FCC has strict nonduplication-of-programming rules.

[11] The video portion of the signal is sent over AM; the audio signal via FM. "The effective radiated power of the aural transmitter shall not be less than 10 percent nor more than 20 percent of the peak radiated power of the visual transmitter"; FCC Rules 73.682 (a) (15). The television broadcast band in the FCC Rules (73.881) are defined as "the frequencies in the band extending from 54-890 megahertz which are assignable to television stations. These frequencies are 54 to 72 megahertz (channels 2 and 4), 76 to 88 megahertz (channels 5 and 6), 174 to 216 megahertz (channels 7 through 13), and 470 to 890 megahertz (channels 14 through 83)." Because channel 6 is part of the FM broadcast band, it can be heard on the lower end of most FM radios. Approximately 4 MHz of the frequency range allocated to television stations is used for video transmission; FCC Rules, 73.699, figure 5.

[12] Not to be confused with the three primary "pigment" colors of red, yellow, and blue.

[13] FCC Rules, section 73.11(a). There can be interference from either groundwave or skywave propagation. Section 73.184 of the FCC Rules in presenting charts to measure the coverage of ground waves, defines ground wave field intensity as "that part of the vertical component of the electric field received on the ground which has not been reflected from the ionosphere nor the troposphere."

[14] FCC Rules, Section 73.11(b): "The signal is subject to intermittent variations in intensity."

[15] FCC Rules 73.11(c).

[16] FCC Rules 73.21(a).

[17] The specific definition in FCC Rules 73.21(b) (1) is "a regional channel is one on which several stations may operate with powers not in excess of 5 kilowatts. The primary service area of a station operating on any such channel may be limited to a given field intensity contour as a consequence of interference."

[18] The frequency 89.1 in New York City is reserved for the United Nations station. In Alaska, frequencies 88–100 MHz are allocated exclusively to government radio services. Frequencies 11.1–107.9 MHz are allocated to Alaskan, noncommercial broadcast use. FCC Rules, 73.501(a).

[19] Information on ITU activities are found in annual ITU reports and the *Yearbook of the United Nations*.

[20] FCC Rules, 73.183(b), 73.183(c).

chapter 5

[1] "Video Entertainment Offers Chicago Its First Pay TV Channel via Microwave," *Communications News,* 13 (September 1975), 14.

[2] D. Dean VanUitert, "Microwave Expands Campus Borders," *Educational and Industrial Television,* 6 (November 1974), 58–59, 60–63.

[3] Indiana Higher Education Telecommunications System (IHETS).

[4] Richard Witkin, "Live Images Transmitted across Ocean First Time," *New York Times,* July 11, 1962, p. 16. Also, Richard Witkin, "Europeans Beam First Television to Screens in U.S.," *New York Times,* July 12, 1962, pp. 1, 12. The specific agreement, "Cooperative Agreement between the National Aeronautics and Space Administration and the American Telephone and Telegraph Company for the Development and Experimental Testing of Active Communications Satellites" provided back-up launching systems, and all data resulting from the experiments were to be made available to NASA.

[5] Anthony Lewis, "Sarnoff Suggests Industry Merger," *New York Times,* August 8, 1962, pp. 1, 14.

[6] Eventually, Western Union operated its own satellite system.

[7] Jack Gould, "TV: Telstar and World Broadcasting," *New York Times,* July 11, 1962, p. 71.

[8] Leonard H. Marks summarized these events in an article in the *Journal of Broadcasting* entitled "Communication Satellites: New Horizons for Broadcasters," 9 (Spring 1965), 97–101. The article also summarized issues and asked probing questions.

[9] Source of the account of the launch and Hughes Aircraft Company's part in it. "Mr. Watson I Want You," *Vectors,* 15 (Summer, Fall, 1973), 7–9.

[10] "Interactive Satellite ATS-6 Brings People Together," *Broadcast Management Engineering,* 10 (November 1974), 30–44.

[11] International development is cited in "A Television Station Goes into Orbit," *Business Week,* February 16, 1976, p. 36H.

[12] Marks, p. 100.

[13] Dallas W. Smythe, "Space-Satellite Broadcasting: Threat or Promise?" *Journal of Broadcasting,* 4 (Summer 1960), 193–94.

[14] "Federal Regulatory System Seen as Inadequate to Requirement for Orderly Telecommunications Change," Cambridge, Mass.: Arthur D. Little Inc., August 27, 1976.

[15] For example Michael Kinsley, "Is AT&T Hamstringing Comsat?" *New York Times,* June 13, 1976, sec. F, p.11.

[16] AT&T, *1975 Annual Report to Shareholders,* p.5.

chapter 6

[1] Susan Q. Kelly, Public Affairs Coordinator, National Cable Television Association. Letter to the author, December 8, 1976. Source of Oregon and Pennsylvania beginnings.

[2] Figures on the size of the industry are compiled from material supplied by the National Cable Television Association, as well as from *Cable Sourcebook, 1979* (Washington, D.C.: Broadcasting Publications, Inc., 1978).

[3] Examples of cable use in secondary education are found in *Cable Television and Education: A Report from the Field* (Washington, D.C.: National Cable Television Association, 1973).

[4] *Ibid.,* p. 8. Bond issues.

[5] *Communication Properties, Inc.,* in *1973 Annual Report,* p. 11.

[6] Rolland C. Johnson and Donald Agostino, "The Columbus Video Access Center: A Research Evaluation of Audience and Public Attitudes," (Bloomington: Institute for Communication Research. Indiana University, 1974).

[7] Rudy Bretz, "Public-Access Cable TV: Audiences," *Journal of Communication,* 25 (Summer 1975), 29.

[8] *Ibid.,* p. 30.

[9] Pamela Doty, "Public-Access Cable TV: Who Cares?" *Journal of Communication,* 25 (Summer 1975), 33–41. Sometimes, however, there is controversy. See Alan Wurtzel, "Public-Access Cable TV: Programming," *Journal of Communication,* 25 (Summer 1975), 20.

[10] *Ibid.*

[11] *Ibid.*

[12] Clifford M. Kirtland, Jr., "Room for All," Speech delivered to the annual meeting of the Institute of Broadcasting Financial Management, Boston, September 14, 1976.

[13] Anne W. Branscomb, "The Cable Fable: Will it Come True?" *Journal of Communication,* 25 (Winter 1975), 52.

[14] Federal Regulatory System Seen as Inadequate to Requirement for Orderly Telecommunication Change" (Cambridge, Mass.: Arthur D. Little, Inc., 1976).

[15] The categories included are based upon Walter S. Baer, *Cable Television: A Handbook for Decision Making* (Santa Monica, Calif.: The Rand Corporation, 1973), p. 46.

[16] *Ibid.,* p. 57.

[17] A very conservative estimate based on TelePrompTer's average as listed in their *1973 Report,* p. 8.

[18] Source of illustration and data: *Cox Looks at the Future* (Atlanta: Cox Broadcasting Corporation, 1975).

[19] *Ibid.*

[20] Rolland C. Johnson and Robert T. Blau, "Single Versus Multiple-System Cable Television," *Journal of Broadcasting,* 18 (Summer 1974), 326.

[21] *Ibid.,* p. 324.

²² R.E. Park, *Prospects for Cable in the 100 Largest Television Markets* (Santa Monica, Calif.: The Rand Corporation, 1971).

²³ References to the Arthur D. Little, Inc. study in the section *Future of Cable,* are from Arthur D. Little, Inc., 1976 (press release).

²⁴ *Cox Looks at the Future.*

chapter 7

¹ The author is grateful for the material furnished by the three commercial networks, numerous syndication companies, and the wire services. Executives of these organizations and station management who willingly participated in interviews with the author also are acknowledged.

² From the UPI Advisory Board by-laws provided the author by UPI, "In the Matter of: Inquiry into Subscription Agreements between Radio Broadcast Stations and Musical Format Service Companies," Docket No. 19743, FCC, November 7, 1975.

³ *Ibid.*

chapter 8

¹ The account of the experimental broadcast is from C. C. Clark, "Television in Education," *School and Society,* 48 (October 1, 1938), 431–32.

² "Metropolitan Art Is to Be Televised," *New York Times,* May 26, 1941, p. 21.

³ The account of the series is from "NBC's Educational Television Series," *School and Society,* 63 (February 16, 1947), 110.

⁴ William M. Dennis, "Transition to Visual Education," *NEA Journal,* 35 (October, 1946), 424.

⁵ Amo DeBernardis and James W. Brown, "A Study of Teacher Skills and Knowledge Necessary for the Use of Audio-Visual Aids," *Elementary School Journal,* 46 (June 1946), 550–56.

⁶ "A Research Fellowship in Television Education," *School and Society,* 69 (April 16, 1949), 278.

⁷ As reported in "The U.S. Commissioner of Education on Television," *School and Society,* 72 (December 23, 1950), 427.

⁸ "Colleges and Universities Prepare Television Programs," *School and Society,* 72 (September 2, 1950), 155–56.

⁹ *Ibid.,* October 3, 1951, p. 44.

¹⁰ A general overview and summary of these issues appeared in the *Journal* of the National Education Association, by Vivian Powell, then president of the NEA Department of Classroom Teachers. See Vivian Powell, "Here's How Teachers Look at ITV," *NEA Journal,* (November 1957), 506.

¹¹ For a history of the MPATI program, see Norman Felsenthal, "MPATI: A History 1959–1971," *Educational Broadcasting Review,* 5 (December 1971), 36–44.

¹² As discussed in Richard J. Stonesifer, "The Separation Needed between ETV and ITV," *AV Communication Review,* 14 (Winter 1966), 489–97.

¹³ *Ibid.,* p. 490, citing Doris Willens, "ETV: An Uncertain Trumpet," *Television Magazine,* 21 (February 1964).

¹⁴ Jerome S. Bruner, *Toward A Theory of Instruction* (Cambridge, Mass.: Harvard University Press, 1966), pp.10–11. As cited in Edgar Dale, *Audiovisual Methods in Teaching* (Hinsdale, Ill.: Dryden Press, 1969), p. 108.

¹⁵ *Ibid.,* p. 110.

[16] See also Walter Wager, "Media Selection in the Affective Domain: A Further Interpretation of Dale's Cone of Experience for Cognitive and Affective Learning," *Educational Technology,* July 1975, pp. 9–13.

[17] P. Kenneth Komoski. Statement to the U.S. House of Representatives Committee on Education and Labor in "Hearings: To Establish a National Institute of Education," 92nd Congress, First Session (Washington, D.C.: U.S. Government Printing Office, 1971). Cited in "Evaluating Instructional TV," *Educational Broadcasting,* 7 (May/June 1974), 12. Although Komoski made the statement in 1971, there still is criticism of ITV's accountability. See George Hall, "Is It Time to Turn ITV Off?" *Public Telecommunication Review,* 4 (May/June 1976), 15–20.

[18] Harold B. McCarty, "Educational Radio's Role," *NAEB Journal,* 18 (October 1958), 3–6, 26–29.

[19] "Carnegie Commission II," *Public Telecommunication Review,* 5 (May/June 1977), 13–14.

chapter 9

[1] Based on 600 companies using television in 1976, having grown twofold in three years, and there being somewhat more than 700 commercial television stations in 1976. Source of corporate estimate: Judith M. Brush, "Private Television Communications," *Matrix,* 62 (Winter 1976–77), 14–15, 30.

[2] *Ibid.,* p. 15.

[3] *Ibid.,* p. 15.

[4] William L. Cathcart, "Television and Industry: How a New Trend Relates to Students," *Feedback,* 18 (May 1976), 11–14.

[5] "Image Building Begins at Home," *Chemical Week,* (November 19, 1975), 5.

[6] *Ibid.*

[7] The account of the First National City Bank's programming is from Eugene Marlow, "Programming for a Company Television News Show," *Educational and Industrial Television,* 6 (April 1974), 30, 33–37.

[8] Greg Stark and Rod Rightmier, "Around the Clock Video for Employee Communication," *Educational & Industrial Television,* 7 (February 1975), 18, 20–21.

[9] Source: "What's on JDTV?" *JD Journal,* 5 (Summer 1976), 7–9. Also: "Deere and Company: All out for Quality," *IVC Field Report.* International Video Corporation, 1975.

[10] "Tube Power: Reaching New Frontiers with Reliance Television," *Intercom,* (February 1976), 4–5.

[11] "Television Turns on at Owens-Corning," *Dialogue,* (October 1974), 4–5.

[12] *Honeywell Education: Multimedia Instructional Systems* (Wellesley Hills, Mass.: Honeywell, 1976). Corporate Brochure.

[13] "Holiday Inn's Vidnet System Helps in Training Employees," *Communications News,* 14 (October 1976), 36.

[14] Warren R. Wille, "The Dana Approach—A Management Information System," *Educational & Industrial Television,* 6 (January 1974), 10–12.

[15] "Extending Our Vision with the Video Communication System," *Viewpoints,* (November/December 1976), 16–18.

chapter 10

[1] The author is deeply indebted to stations, networks, governments, and individuals who have generously contributed information for this chapter not only in books and literature but in private correspondence.

[2] The CFCF-AM case is chronicled in both the 1975–76 and 1976–77 CRTC annual reports.

[3] Phil Gibson, "Canadians Fence off TV Violence," *Christian Science Monitor,* February 10, 1976.

[4] *BBC Handbook* (London: BBC, 1977), p. 207.

[5] The author acknowledges the generous help of both MTV and YLE in preparing the paragraphs on Finnish broadcasting. One of the most concise summaries of French broadcasting is found in Milton Hollstein, "French Broadcasting after the Split," *Public Telecommunication Review,* 6 (January/February 1978), 15–19.

[6] David E. Powell, "Television in the U.S.S.R.," *Public Opinion Quarterly,* 39 (Fall 1975), 287–300.

[7] Australian Broadcasting Commission *Annual Report,* 1975–1976.

[8] "Demonstration of the Experiment of Hi-OVIS Prototype," Publication of the Visual Information System Development Association, November 1976.

[9] *NHK 1976–77* (Tokyo: NHK, 1977), p. 2.

[10] *40 Years of Broadcasting in Ghana* (ACCRA: GBC, 1975).

[11] *Ibid.,* p. 8.

[12] *Broadcasting in the Seventies* (Salisbury, Rhodesia, n.d.). Correspondence from the RBC is also especially acknowledged.

[13] Peter B. Orlik, "Co-opting the Messenger: The Afrikaner Take-Over of the South African Broadcasting Corporation." Paper presented at the 1977 meeting of the Association for Education in Journalism, Madison, Wisconsin.

[14] Emile G. McAnany, *Radio's Role in Development: Five Strategies of Use* (Washington, D.C.: Academy for Educational Development, 1973), pp. 5–21. McAnany's review of international literature is comprehensive, and the student doing serious research on instructional radio will find it valuable.

[15] Wilbur Schramm, *Instructional Television in the Educational Reform of El Salvador* (Washington, D.C.: Academy for Educational Development, 1973).

[16] *Ibid.,* pp. 4–5. Gains were for seventh- and eighth-grade classes in 1970 and seventh grade in 1971 but not for eighth and ninth in 1971. Albeit with cautious optimism, p. 45.

[17] *Ibid.,* pp. 6–7.

[18] *Educational Technology and the Developing Countries* (Washington, D.C.: Agency for International Development, 1972), p. 81. As cited in Lynne Masland and Grant Masland, "The Samoan ETV Project," *Educational Broadcasting,* 8 (March/April 1975), 13–16.

[19] Masland and Masland, p. 15.

[20] *Ibid.,* p. 16.

[21] The Ivory Coast Republic, *Education by Television—Volume 3, Report of the Missions for the Evaluation of Educational Television in Niger, El Salvador, and American Samoa,* Ministry of National Education, approximate date of publication: 1969, p. 26. As cited in Masland and Masland. Another source on the early use of ETV in American Samoa is Wilbur Schramm, "Educational Television in American Samoa," in *New Educational Media In Action: Case Studies for Planners—Volume 1,* ed. Wilbur Schramm and others (Paris: UNESCO, 1IEP, 1967), p. 16.

[22] Reference to ETV in Malaysia is from Charles B. Klasek, "ITV in a Developing Nation," *Educational Broadcasting* (November/December 1975), 33–5, 46.

chapter 11

[1] The Wireless Ship Act of 1910. Public Law 262, 61st Congress, June 24, 1910.

[2] The Radio Act of 1912, Public Law 264, 62nd Congress, August 13, 1912, sec. 1.

[3] See Edward F. Sarno, Jr., "The National Radio Conferences," *Journal of Broadcasting,* 13 (Spring 1969), 189–202.

[4] *Ibid.* For a summary of the Department of Commerce's action during this period see Marvin R. Bensman, "Regulation of Broadcasting by the Department of Commerce, 1921–1927," in *American Broadcasting: A Source Book on the History of Radio and Television,* ed. Lawrence W. Lichty and Malachi C. Topping (New York: Hastings House, Publishers, 1975), pp. 544–55.

[5] *Hoover* v. *Intercity Radio Co., Inc.* 286 F. 1003 (D. C. Cir), February 25, 1923.

[6] *United States* v. *Zenith Radio Corporation et al.* 12F. 2d 614 (N. D. Ill.), April 16, 1926.

[7] Eric Barnouw, *A Tower in Babel: A History of Broadcasting in the United States* (New York: Oxford University Press, Inc., 1966), p. 175.

[8] Attorney General's Opinion, 35 Ops. Att'y Gen. 126, July 8, 1926, As cited in Kahn.

[9] H. Doc. 483, 69th Congress, 2nd Session. As cited in Kahn.

[10] The Radio Act of 1927, Public Law 632, 69th Congress, February 23, 1927, Sec. 3.

[11] S. Doc. 144, 73d Congress, 2d Session, February 26, 1934. President Franklin D. Roosevelt's message to Congress suggesting the formation of the Federal Communications Commission.

[12] *Ibid.,* Sec. 326.

chapter 12

[1] *FCC Annual Report, 1974,* pp. 2–3.

[2] Based on "The FCC and Broadcasting," FCC Broadcast Bureau publication # 8310-100.

[3] However, "staged" news events are not considered to be in the public interest.

[4] "The FCC and Broadcasting."

[5] A report of the Commission's first open meeting is chronicled in "Like a Day with the Sunshine at the FCC," *Broadcasting,* 46 (March 28, 1977), 29. Procedural policy was announced in "FCC in the Sunshine," NAB *Highlights,* 3 (March 7, 1977), 2.

[6] Nicholas Johnson and John Jay Dystel, *A Day in the Life: The Federal Communications Commission,* 82 *Yale Law Journal,* (1973), 1575–1634.

[7] *Ibid.*

[8] Nicholas Johnson and John Dystel are critical of the rule permitting a maximum of seven AM, FM, or TV stations to be owned by the same company. What was intended as a "per se maximum" has been converted into a "presumptively permissible number." Source: *Ibid.*

[9] On December 13, 1972. Source: *Ibid.*

[10] See Lawrence W. Lichty, "Members of the Federal Radio Commission and the Federal Communications Commission 1927–1961," *Journal of Broadcasting,* 6 (Winter 1961–62), 23–24; Lawrence W. Lichty, "The Impact of FRC and FCC Commissioners' Background on the Regulation of Broadcasting," *Journal of Broadcasting,* 6 (Spring 1962), 97–110.

[11] Wenmouth Williams Jr., "Impact of Commissioner Background on FCC Decisions: 1962–1975," *Journal of Broadcasting,* 20 (Spring 1976), 239–60.

[12] As discussed in FCC publications, *FCC Annual Reports, Broadcasting Yearbook,* and "How the FCC Is Organized into Offices and Bureaus," *Communication News,* 14 (January 1977), 46–48; "FCC Makes over Broadcast Bureau," *Broadcasting,* 45 (April 5, 1976), 53.

[13] See "FCC Lab Tests Radios for Rule Compliance," *Communication News,* 14 (January 1977), 50.

[14] *FCC Annual Report, 1974.*

[15] The Policy and Rules Division was formed in 1976 as a consolidation of the Rules and Standards Division and the Research and Education Division.

[16] The importance of cable as a "developing" medium is evident in the presence of two divisions directed toward future growth issues, the Research Division and the Policy Review and Development Division.

[17] *FCC Annual Report, 1974,* p. 78.

¹⁸ Donald M. Gilmor and Jerome A. Barron, *Mass Communication Law* (St. Paul: West Publishing Company, 1974), p. 889, citing Richard Sneed, 15 P. &. F. Radio Reg. 158 (1967).

¹⁹ *Ibid.*, p. 78, citing Mile High Stations, Inc., 28 FCC 795, 20 P. & F. Radio Reg. 345 (1960).

²⁰ As reported in *Broadcasting*, 46 (June 20, 1977), 68.

²¹ *FCC Annual Report*, 1974, pp. 37–38.

²² See Charles Clift, III., Fredric A. Weiss, and John D. Abel, "Ten Years at Forfeitures by the Federal Communications Commission," *Journal of Broadcasting*, 15 (Fall 1971), 379–85.

²³ *Ibid.* Categories are as defined in the Communications Act. The period covered was 1961 through June 1971.

²⁴ Authority granted by the same statute permitting forfeitures.

²⁵ *FCC Annual Report*, 1974, p. 37.

²⁶ Maurice E. Shelby, Jr., "Short-Term License Renewals: 1960–1972," *Journal of Broadcasting*, 18 (Summer 1974), 277–88.

²⁷ *Ibid.*, p. 282.

²⁸ Johnson and Dystel.

²⁹ *Ibid.*

³⁰ Erwin G. Krasnow and Lawrence D. Longley, *The Politics of Broadcast Regulation* (New York: St. Martin's Press, Inc. 1973), p. 25.

³¹ *Ibid.*, p. 25.

³² Marc C. Franklin, *The First Amendment and the Fourth Estate* (Mineola, N.Y.: The Foundation Press, Inc., 1977), pp. 465–66.

³³ *Ibid.*, p. 466. It is interesting to note that although not adhering to "regional" assignments of frequencies on a domestic scale, international agreements on frequency management are regional. Franklin points out the peculiar nature of the electromagnetic spectrum as a resource. When the overall territory (the world) is big enough, regional allocations are practical. Moreover, the political realities of trying to localize spectrum management on a world scale makes the task almost impossible.

³⁴ *A Study of the Federal Communications Commission's Equal Employment Opportunity Regulation—An Agency in Search of a Standard* (Washington, D.C.: Citizens Communications Center, 1976).

³⁵ *Window Dressing on the Set: Women and Minorities in Television* (Washington, D.C.: United States Commission on Civil Rights, 1977).

³⁶ Joseph A. Grundfest, *Citizen Participation in FCC Decision Making* (Santa Monica, Calif: The Rand Corporation, 1976).

³⁷ Erwin G. Krasnow and Lawrence D. Longley, *The Politics of Broadcast Regulation* (New York: St. Martin's Press, Inc., 1973), p. 24.

³⁸ Robert R. Smith and Paul T. Prince, "WHDH: The Unconscionable Delay," *Journal of Broadcasting*, 18 (Winter 1973–74), 85–86.

³⁹ Sterling Quinlan, *The Hundred Million Dollar Lunch* (Chicago: J. Philip O'Hara, Inc., 1974), p. 4.

⁴⁰ "FCC Berated for Policy on Stockholdings of the Employees," *Broadcasting*, 46 (May 30, 1977), 28, 30.

⁴¹ Johnson and Dystel.

chapter 13

¹ Harold L. Nelson and Dwight L. Teeter, Jr., *Law of Mass Communications* (Mineola, N.Y.: The Foundation Press, Inc., 1973), p. 1.

² Lee Loevinger, "The Role of Law in Broadcasting," *Journal of Broadcasting*, 8 (Spring 1964), 115–17.

³ *Ibid.*

[4] Blasi, *The Newsman's Privilege: An Empirical Study,* 70 *Michigan Law Review,* 233 (December, 1971).

[5] Loevinger, pp. 115–17.

[6] Section 73.120. Two publications have updated rules and regulations and have provided guidelines to broadcasters in interpreting Section 315. These are "Uses of Broadcast and Cablecast Facilities by Candidates for Public Office," Fed. Reg. 5796; and "Licensee Responsibility Under Amendments to the Communications Act of 1971," FCC *Public Notice,* June 5, 1974, 47 FCC 516 (1974).

[7] Gilmor and Barron, p. 230. *Farmers Educational and Cooperative Union of America, North Dakota Division* v. *WDAY,* 89 N. W. 2d 102, 109 (N. D. 1958).

[8] *Farmers Educational and Cooperative Union of America* v. *WDAY Inc.,* 360 U. S. 525, 79 S. Ct. 1302, 3 L. Ed. 2d 1407 (1959).

[9] "Reinterpretation of Equal Time Passes First Court Challenge," *Broadcasting,* 45 (April 19, 1976), 26–27.

[10] 74 Sta. 554 (1960). Gilmor and Barron, p. 797.

[11] "Use of Broadcast and Cablecast Facilities by Candidates for Public Office," FCC *Public Notice,* March 16, 1972; 37 *Fed. Reg.* 5804, March 21, 1972.

[12] *In the Matter of Editorializing by Broadcast Licensees,* 13 FCC 1246, June 1, 1949.

[13] Great Lakes Broadcasting Co., 3 F. R. C. Ann. Rep. 32 (1929), modified on other grounds, 37, F. 2d 993 (D. C. Cir.) certiorari dismissed, 281 U. S. 706 (1930). As cited in Franklin, p. 601.

[14] *In the Matter of the Mayflower Broadcasting Corporation and The Yankee Network, Inc. (WAAB),* 8 FCC 333, 338, January 16, 1941.

[15] *Ibid.*

[16] *Ibid.*

[17] In reference to United Broadcasting Co. (WHKC) 10 FCC 515 June 26, 1945.

[18] In reference to Petition of Robert Harold Scott for Revocation of Licenses of Radio Stations KQW, KPO and KFRC, 11 FCC 372, July 19, 1946.

[19] *In the Matter of Editorializing by Broadcast Licensees,* 13 FCC 1246, June 1, 1949.

[20] "Applicability of the Fairness Doctrine in the Handling of Controversial Issues of Public Importance", 29 *Fed. Reg.* 10416, July 25, 1964.

[21] *Red Lion Broadcasting Co.* v. *Federal Communications Commission.* 127 U. S. App. D. C. 129, 381 F. 2d 908 (1967). *Red Lion Broadcasting Co., Inc.* v. *Federal Communications Commission. United States* v. *Radio-Television News Directors Association.* 395 U.S. 367, 89 S. Ct. 1794, 23 L. Ed. 2d 371 (1969). The cases are well documented in numerous legal texts. The reader is referred to the latest edition of Gilmor and Barron for a detailed discussion as well as for pertinent questions on the decision. (Further citations of the case in this book are listed as "Red Lion.")

[22] *Radio-Television News Directors Association* v. *United States,* 400 F. 2d 1002 (7th Cir. 1968).

[23] *Red Lion.*

[24] "Fairness Case Goes against Eight California Radio Stations," *Broadcasting,* 45 (May 24, 1976), 40, 42. A discussion of the legal aspects of editorial advertising is found in Milan D. Meeske, "Editorial Advertising and the First Amendment," *Journal of Broadcasting,* 17 (Fall 1973), 417–26.

[25] For the early development of legal precedent in the area of regulating obscene, indecent, and profane programming, see James Walter Wesolowski, "Obscene, Indecent, or Profane Broadcast Language as Construed by the Federal Courts," *Journal of Broadcasting,* 13 (Spring 1969), 203–19.

[26] Title 18, United States Code (Codified June 25, 1948, Ch. 645, 62 Stat. 769).

[27] Sonderling Broadcasting Corporation, WGLD-FM, 27 Radio Reg. 2d 285 (FCC, 1973). The appeals case affirming the FCC ruling is *Illinois Citizens Committee for Broadcasting* v. *Federal Communications Commission,* 515 F. 2d 397 (D. C. Cir. 1975). The *Ginzburg* ruling cited in the text is *Ginzburg* v. *United States,* 383 U. S. 463, 86 S. Ct. 942, 16 L. Ed. 2d 31 (1966). See also Charles Feldman and Stanley Tickton, "Obscene/Indecent Programming: Regulation of Ambiguity," *Journal of Broadcasting,* 20 (Spring 1976), 273–82.

[28] *Pacifica Foundation,* 56 FCC 2d 94 (1975).

[29] *Ibid.*

[30] *Third Report and Order* in Docket No. 19622, FCC 75-542, May 13, 1975. The different times result from different network feed times to affiliates in the various time zones. For two perspectives on the sports antiblackout issue, see Ira Horowitz, "Sports Telecasts: Rights and Regulations," *Journal of Communication,* 27 (Summer 1977), 160–68; John J. Siegfried and C. Elton Hinshaw, "Professional Football and the Anti-Blackout Law," *Journal of Communication,* 27 (Summer 1977), 169–74.

[31] *National Association of Independent Television Producers and Distributors et al v. FCC,* CA No. 75-4021, April 21, 1975.

[32] *Rideau v. Louisiana,* 373 U. S. 723, 10 L. Ed. 2d 663, 83 S. Ct. 1417 (1963).

[33] *Estes v. State of Texas* 381 U. S. 532, 85 S. Ct. 1628, 14 L. Ed. 2d 543 (1965).

chapter 14

[1] The text does not quote, except where indicated, verbatim from FCC regulations. Logging requirements vary among types of stations. Some provide more detailed information than others (although staying within FCC guidelines), and many stations use automatic logging systems. The reader wishing to check the exact language of the law should consult sections 73.111 (AM), 73.281 (FM), and 73.669 (TV) for exact wording.

[2] FCC Field Operations Bureau. *Broadcast Operators Handbook* (Washington, D.C.: Federal Communications Commission, 1976), p. 79. Along with FCC rules, a discussion of station identification announcements can be found in Frederick W. Ford and Lee G. Lovett, "Station Identification Announcements," *Broadcast Management/Engineering,* 13 (December 1976), 24, 27.

[3] FCC rules section 73.112 (AM), 73.282 (FM), 73.670 (TV). For further discussion of program log requirements, consult Frederick W. Ford and Lee G. Lovett, "Program Log Requirements," *Broadcast Management/Engineering,* 13 (April 1976), 22, 24, 26. Note also the National Association of Broadcasters program log recommendations.

[4] FCC Rules 73.112 (AM), 73.282 (FM), 73.670 (TV).

[5] *Perry's Broadcasting and the Law,* August 1, 1976, p. 2.

[6] FCC Rules 73.111 (general), 73.115 (retention), 73.116 (availability), 73.113 (AM), 73.283 (FM), 73.671 (TV).

[7] Assuming there is no interruption.

[8] *Broadcast Operators Handbook,* p. 84.

[9] Thomas F. Baldwin and Stuart H. Surlin, "A Study of Broadcast Station License Application Exhibits on Ascertainment of Community Needs," *Journal of Broadcasting,* 14 (Spring 1970), 157–70. A further perspective of the ascertainment process is found in Stuart H. Surlin and Less Bradley, "Ascertainment through Community Leaders," *Journal of Broadcasting,* 18 (Winter 1973–74), 97–107; Joseph M. Foley, "Ascertaining Ascertainment: Impact of the FCC Primer on TV Renewal Applications," *Journal of Broadcasting,* 16 (Fall 1972), 387–406; Stuart H. Surlin, "Ascertainment of Community Needs by Black-Oriented Radio Stations," *Journal of Broadcasting,* 16 (Fall 1972), 421–29; Kenneth W. Hirsch and John C. Hwang, "Community Problems Measurement and Policy Setting" (Sacramento, Calif: Department of Communication Studies, California State University).

[10] "In the Matter of Primer on Ascertainment of Community Problems by Broadcast Applicants," *Federal Register,* 36 (March 3, 1971). Three different terms tend to be interchanged when discussing ascertainment: "problems," "issues," and "needs." All refer to what is wrong with a community and how the broadcaster can help to correct it.

[11] "Ascertainment of Community Problems by Broadcast Renewal Applicants Primer," *Federal Register,* 41 (January 7, 1976), adopted in 1975; "Ascertainment of Community Problems by Noncommercial Educational Broadcast Applicants, Permittees, and Licensees," *Federal Register,* 41 (March 25, 1976).

[12] Ascertainment Primer, 1382.

[13] The object, from management's perspective, is not only to learn what the needs of the community are but also to show that more than adequate measures have been taken to assure that these groups are represented in the survey and to avoid any question about procedure should a license challenge develop. Also see Frederick W. Ford and Lee G. Lovett, "New Community Ascertainment Guidelines for Broadcast Renewals," *Broadcast Management/Engineering*, 12 (March 1976), 26, 28, 32, 34.

[14] Herschel Shosteck, "Dangers of Mail Surveys in Ascertainment Proceedings," *Journal of Broadcasting*, 16 (Fall 1972), 431–39. Also David J. LeRoy and Donald F. Ungurait, "Ascertainment Surveys: Problem Perception and Voluntary Station Contact," *Journal of Broadcasting*, 19 (Winter 1975), 23–30.

[15] Along with the FCC Ascertainment Primer, other sources are available to assist broadcasters in conducting the various surveys used in ascertainment. Two of these are Bradley S. Greenberg, Thomas F. Baldwin, Byron Reeves, Lee Thornton, and Jack Wakshlag, *An Ascertainment Handbook for Public Broadcasting Facilities* (Washington, D.C.: Corporation for Public Broadcasting, 1975), prepared under a grant to the Departments of Communication and Telecommunication at Michigan State University. Material from the Greenberg *et al.* report is included in a report of the same title, copyrighted by CPB in 1976. Also, NAB stations received an NAB Legal *Memorandum* on "Ascertainment of Community Needs," in June 1976.

[16] Each radio frequency represents a "channel" on the AM or FM band.

[17] The concern here is mostly with the percentage ratio between entertainment and sufficient local (5 percent), informational (5 percent), and nonentertainment (10 percent) programming. Promised percentages which are less than the minimum are examined closely.

[18] Permitted to be exceeded by 10% at election time.

[19] For a perspective of how this affects public broadcasting see Robert K. Avery, "Public Broadcasting and the Duopoly Rule," *Public Telecommunications Review*, January/February 1977, pp. 29–37.

[20] *National Citizens Committee for Broadcasting* v. *Federal Communications Commission et al.* No. 75–1064 (D.C. Cir.).

chapter 15

[1] Lee Loevinger, "The Future of Television," *Television Quarterly*, 4 (Fall 1965), 41–52.

[2] A discussion of compromise in the 1972 rules and an example of the issues confronting a local change of service are found, respectively, in Harvey Jassem, "The Selling of the Cable TV Compromise," *Journal of Broadcasting*, 17 (Fall 1973), 427–36; and Norman Felsenthal, "Cherry-Picking, Cable, and the FCC," *Journal of Broadcasting*, 19 (Winter 1975), 43–53. Citation of the *Southwestern* case is: *United States* v. *Southwestern Cable Co.*, 392 U.S. 157 (1968). The 1972 rules were issued in: *Cable Television Report and Order*, 36 FCC 2d 143 (1972).

[3] Discussion of specific cable rules can be found in *Regulatory Developments in Cable Television*. (Washington, D.C.: Federal Communications Commission, May, 1977)*;* definitions cited on pp. 7–8. See also the most recent editions of the *Cable Sourcebook*.

[4] Vernone Sparkes, "Local Regulatory Agencies for Cable Television," *Journal of Broadcasting*, 19 (Spring 1975), 228–29.

[5] *FCC Information Bulletin* "Cable Television", March, 1979.

[6] *Regulatory Developments in Cable Television*, p. 6.

[7] *Ibid.*, pp. 8–10.

[8] *Ibid.*, p. 10.

[9] *Ibid.*

[10] *Ibid.*, p. 11. FCC Opinion is stated in: *Report and Order in Docket 20028*, 48 FCC 2d 699, 39 *Fed. Reg.* 33528 (1974), and *Memorandum Opinion and Order in Docket 20028*, 54 FCC 2d 1182, 40 Fed. Reg. 39509 (1975).

[11] *Ibid.* See *Report and Order in Docket 19859,* 57 FCC 2d 68 (1975), 41 *Fed. Reg.* 1063 (1976).

[12] *Ibid.* See *Report and Order in Docket 19859,* 57 FCC 2d 68 (1975), 41 *Fed. Reg.* 1063 (1976).

[13] *Ibid.,* pp. 14–15. *Notice of Inquiry in Docket 20988,* 41 *Fed. Reg.* 50055 (1976).

[14] Frederick W. Ford and Lee G. Lovett, "State Regulation of Cable Television, Part I: Current Statutes," *Broadcast Management/Engineering,* 10 (June 1974), 18, 21, 50; Frederick W. Ford and Lee G. Lovett, "State Regulation of Cable TV, Part II: States with No CATV Statutes; Short-Term and Long-Term Trends," *Broadcast Management/Engineering,* 30 (June 1974) 20, 21, 22.

[15] Cited in Benno Signitzer, *Regulation of Direct Broadcasting from Satellites: The U.N. Involvement* (New York: Praeger Publishers, Inc., 1976), p. 21, footnoting a quotation by Eilene Galloway in "Broadcast Satellites." Paper presented at Seventh Colloquium on the Law of Outer Space, International Institute of Space Law, Amsterdam, Netherlands, September 30–October 5, 1974.

[16] Benno, p. 22.

[17] Paul L. Laskin and Abram Chayes, "A Brief History of the Issues," in *Control of the Direct Broadcast Satellites: Values in Conflict* (Palo Alto, Calif.: Aspen Institute, 1974), pp. 3–14.

[18] *Ibid.,* p. 7.

[19] *Ibid.,* p. 8.

chapter 16

[1] Dick Stein, "Co-Op Q & A, Part I," *Radioactive,* 3 (May 1977), 16.

[2] Brenda Fox, "AM-FM Advertising Packages: Are They Legal?" *Radioactive,* 3 (April 1977), 6–7; NAB *Counsel* (September, 1977—L-711).

[3] John Coughlan, *Accounting Manual for Radio Stations* (Washington, D.C.: National Association of Broadcasters, 1975), p. 21.

[4] Author's interview with Richard A. Shaheen.

[5] These fees vary among brokers.

[6] For example, see Barry J. Dickstein, "True Station Value is Key Ingredient in Broadcast Financing," *Broadcast Management/Engineering,* (September 1976), pp. 41–42; Harold Poole, "What's Your Station's Worth," *Radioactive,* (July 1977), 16–17.

[7] For example, see Ginger Carnahan, "Logos: Your Station's Visible Voice," *Radioactive,* (January 1977), 8–9.

chapter 17

[1] From RKO Radio's "Breakthrough Course in Radio Selling," as cited in *Radioactive,* 2 (July 1976), 1.

[2] At a 95% level of confidence. Detailed discussions of sampling can be found in various statistics books. One detailed source is William L. Hays, *Statistics for the Social Sciences,* 2nd ed. (New York: Holt, Rinehart & Winston, 1973). A good general description of the sampling process is Maxwell McCombs, "Sampling Opinions and Behaviors," in *Handbook of Reporting Methods,* ed. Maxwell McCombs, Donald Lewis Shaw, and David Grey (Boston: Houghton Mifflin Company, 1976), pp. 123–38.

[3] Peter D. Fox, "Television Ratings and Cultural Programs," *Industrial Management Reivew,* 5 (Fall 1963), 37–43.

[4] John Colombotos, "Personal versus Telephone Interviews: Effect of Responses," *Public Health Reports,* 84 (September 1969), 773–82; D. A. Dillman and others, "Reducing Refusal Rates for Telephone Interviews," *Public Opinion Quarterly,* 40 (Spring 1976), 66–67; T. F. Rogers, "Interviews by Telephone and in Person: Quality of Responses and Field Performance," *Public Opinion Quarterly,* 40 (Spring 1976), 51–65. The actual questions asked in surveys also can affect the results: Bradley S. Greenberg, Brenda Dervin, and Joseph Dominick, "Do People Watch 'Television' as 'Programs'?" *Journal of Broadcasting,* 12 (Fall 1968), 367–76.

[5] *Television Ratings Revisited* (New York: Television Information Office, 1971).

[6] Many sources are useful in compiling information on interpreting the ratings. Although the formulas listed in Arbitron Publications are used as a basis for defining **some** of the terms, in no way should they be construed as directly applicable to every rating company. Like the terms themselves, the formulas are common methodology among different rating companies, although there is considerable diversity between services. The reader will want to keep this in mind. Helpful sources for this section include *A Broadcast Research Primer* (Washington, D.C.: National Association of Broadcasters, 1971); *Standard Definitions of Broadcast Research Terms* (Washington, D.C.: National Association of Broadcasters, 1973); *Probability Sampling* (Princeton, N.J.: Opinion Research Corporation, 1973); *How Arbitron Measures Radio* (Beltsville, Md.: American Research Bureau, 1974); *Description of Methodology: The Arbitron Television Market Reports* (Beltsville, Md.: American Research Bureau, 1975); *Understanding and Using Radio Audience Estimates* (Beltsville, Md.: American Research Bureau, 1976); *Arbitron Radio: Dayton, Ohio* (Beltsville, Md.: American Research Bureau, October/November 1976); William L. Hays, *Statistics for the Social Sciences,* 2nd ed. (New York: Holt Rinehart & Winston, 1973); *Nielsen Station Index: Methodology* (Northbrook, Ill.; A. C. Nielsen Co. 1973–74); *Nielsen Station Index: Demonstration Report and User's Manual* (Northbrook, Ill.; A. C. Nielsen Co., 1974–75); *Improvements in Your Arbitron Television Report* (Beltsville, Md.: American Research Bureau, 1975).

[7] *Understanding and Using Radio Audience Estimates,* p. 4.

[8] Adapted from *Ibid.,* p. 6.

[9] "Spanish-Language Net Still Battling Ratings," *Advertising Age,* February 26, 1973.

[10] "Language Problem," *Broadcasting,* 45 (April 26, 1976), 5.

[11] *Ibid.*

[12] Spanish International Network, press release, 1974.

[13] Letter from A. Howard to A. C. Nielsen Company, February 10, 1977. Given to the author by A. Howard.

[14] Neil Hickey, "The Case of the Missing Viewers," *TV Guide,* 24 (May 8, 1976), 5.

[15] Ad agencies also become concerned as reflected in "Television Viewing—an update on Hypocrisy," *Media Message* (Ogilvy & Mather, February 1, 1977).

[16] Associated Press, released to subscribers during the week of May 1, 1976.

[17] Based on listenership surveys for ABC Radio by Arbitron.

[18] Examples of joint efforts were seen in the Committee on Nationwide Television Audience Measurement (CONTAM), a joint effort of ABC, CBS, NBC, and the National Association of Broadcasters. Three well publicized studies by Martin Mayer are *How Good are Television Ratings?* 1966; *How Good are Television Ratings?* (continued . . .), 1969, and *Television Ratings Revisited,* 1969–70. In 1974, Arbitron commissioned a study which examined the reliability of the ratings: R. D. Altizer and R. R. Ridgeway Jr., *Arbitron Replication: A Study of the Reliability of Broadcast Ratings* (Beltsville, Md.: Arbitron/American Research Bureau, April 1974).

[19] Robert Edward Balon, "The Future of Broadcast Audience Measurement." Paper presented to the Southern States Communication Convention, April 1977.

chapter 18

[1] Thomas E. Coffin, "Progress to Date in Radio and Television Research." Paper presented to the 1960 conference of the American Association for Public Opinion Research.

2 Thomas E. Coffin, "What's Been Happening to the Media Research Budget?: A Case History." Proceedings of the 1974 conference of the Advertising Research Foundation.

3 Letter to the author from Lyn Garson of the PBS Research Office, October 31, 1977.

4 *A Broadcast Research Primer* (Washington, D.C.: National Association of Broadcasters, 1974).

5 Orville C. Walker, Jr. and William Rudelius, "Ascertaining Programming Needs of 'Voiceless' Community Groups," *Journal of Broadcasting*, 20 (Winter 1976), 89–99.

6 One method would be factor analysis. R. J. Rummel, *Applied Factor Analysis* (Evanston, Ill.: Northwestern University Press, 1970).

7 Charles E. Osgood, George J. Suci, and Percy H. Tannenbaum, *The Measurement of Meaning* (Urbana: University of Illinois Press, 1957). A related study of "image" on broadcast media is Joseph C. Philport and Robert E. Balon, "Candidate Image in a Broadcast Debate," *Journal of Broadcasting*, 19 (Spring 1975), 181–93. Two studies examining newscaster credibility are James C. McCroskey and Thomas A. Jenson, "Image of Mass Media News Sources," *Journal of Broadcasting*, 19 (Spring 1975), 169–80; D. Markham, "The Dimensions of Source Credibility of Television Newscasters," *Journal of Communication*, 18 (March 1968), 57–64.

8 M. Wayne DeLozier, *The Marketing Communications Process* (New York: McGraw-Hill Book Company, 1976), pp. 31–32.

9 D. Charles Whitney, "The Poll is Suspect," *The Quill*, 65 (July–August 1976), 23; G. Cleveland Wilhoit and Maxwell McCombs, "Reporting Surveys and Polls," in *Handbook of Reporting Methods*, ed. Maxwell McCombs, Donald Lewis Shaw, and David Grey (Boston: Houghton Mifflin Company, 1976), pp. 81–95; Philip Meyer, *Precision Journalism* (Bloomington: Indiana University Press, 1973).

10 The criteria listed here are adapted from D. Charles Whitney, p. 25.

11 Samuel L. Becker, "Reaction Profiles: Studies in Methodology," *Journal of Broadcasting*, 4 (Summer 1960), 253–68.

12 "Tell if Speaker Is Boring," *Science News Letter*, May 24, 1952, p. 325; Elwood A. Kretsinger, "Gross Bodily Movement as an Index of Audience Interest," *Speech Monographs*, 19 (1952), 244–48.

13 "Sweaty Palms over TV Commercials," *Broadcasting*, 44 (November 17, 1975), 48.

14 Claude Hall, "Doomsday Machine Will Evaluate," *Billboard*, (December 24, 1974), 1, 25–28.

15 Ralph Schoenstein, "Watching Howard Cosell for the Love of Science," *TV Guide*, 24 (February 21, 1976), 18–21.

16 Typical of such studies and approaches are: James T. Lull, "Ethnomethods of Television Viewers," paper presented at the annual meeting of the International Communication Association, Chicago, 1978; James A. Anderson, Timothy P. Meyer, and Thomas Donohue, "Ethnography and the Sphere of Effects," paper presented to the annual meeting of the International Communication Association, Chicago, 1978; Paul J. Traudt, "Families and Television: Implications from an Ethnomethodological View," paper presented at the annual meeting of the Southern Speech Communication Association, Biloxi, Mississippi, 1979.

chapter 19

1 Melvin De Fleur and Sandra Ball-Rokeach, *Theories of Mass Communication*, 3rd ed. (New York: David McKay Company, Inc., 1975), p. 205.

2 A discussion of how the categories approach evolved from the bullet theory and how it fits into current communication theory is found in Wilbur Schramm and Donald Roberts, *The Process and Effects of Mass Communication* (Urbana: University of Illinois Press, 1971), pp. 4–53.

3 Paul Lazarsfeld, Bernard Berelson, and H. Gaudet, *The Peoples' Choice* (New York: Columbia University Press, 1948).

[4] Neil Vidmar and Milton Rokeach, "Archie Bunker's Bigotry: A Study in Selective Perception and Exposure," *Journal of Communication,* 24 (Winter 1974), 43–44.

[5] Lee B. Becker, Raymond A. Martino, and Wayne M. Towers, "Media Advertising Credibility," *Journalism Quarterly,* 53 (Summer 1976), 216–22.

[6] Robert Monaghan, Joseph T. Plummer, David L. Rarick, and Dwight Williams, "Predicting Viewer Preference for New TV Program Concepts," *Journal of Broadcasting,* 18 (Spring 1974), 131–42.

[7] William Stephenson, *The Play Theory of Mass Communication* (Chicago: University of Chicago Press, 1967).

[8] David Chaney, *Processes of Mass Communication.* (London: The Macmillan Press Ltd., 1972), pp. 20–21.

[9] Deanna Campbell Robinson, "Television/Film Attitudes of Upper-Middle Class Professionals," *Journal of Broadcasting,* 19 (Spring 1975), 196.

[10] Wilbur Schramm, *Men, Messages, and Media: A Look at Human Communication* (New York: Harper & Row, Publishers, Inc., 1973).

[11] Robinson, p. 199.

[12] Neil T. Weintraub, "Some Meanings Radio Has for Teenagers," *Journal of Broadcasting,* 2 (Spring 1971), 147–52.

[13] Lawrence Wenner, "Functional Analysis of TV Viewing for Older Adults," *Journal of Broadcasting,* 20 (Winter 1976), 77–88.

[14] Herta Herzog, "What Do We Really Know about Daytime Serial Listeners," in *Radio Research, 1942–1943,* ed. Paul F. Lazarsfeld and Frank Stanton (New York: Duell, Sloan and Pearce, 1944).

[15] Joseph Foley, "A Functional Analysis of Television Viewing," (unpublished PhD. dissertation, University of Iowa, 1968), as cited in Wenner, p. 79.

[16] Characteristic of the debate is a series of articles appearing in the Winter and Fall 1975 issues of the *Journal of Broadcasting:* James A. Anderson and Timothy P. Meyer, "Functionalism and the Mass Media," *Journal of Broadcasting,* 19 (Winter 1975), 11–22; Calvin Pryluck, "Functions of Functional Analysis: Comments on Anderson-Meyer," *Journal of Broadcasting,* 19 (Fall 1975), 413–20; James A. Anderson and Timothy P. Meyer, "A Response to Pryluck," *Journal of Broadcasting,* 19 (Fall 1975), 421–23; Calvin Pryluck, "Rejoinder to Anderson-Meyer," *Journal of Broadcasting,* 19 (Fall 1975), 424–25.

[17] Bernard Cohen, *The Press, the Public, and Foreign Policy* (Princeton, N.J.: Princeton University Press, 1963), p. 13.

[18] Maxwell E. McCombs and Donald Shaw, "The Agenda Setting Function of Mass Media," *Public Opinion Quarterly,* 36 (1972), 176–87.

[19] CBS Office of Social Research, *Communicating with Children through Television* (New York: CBS, 1977).

[20] For example: Charles R. Corder-Bolz, "Television Content and Children's Social Attitudes," *Progress Report to the Office of Child Development* (Washington, D.C.: Department of Health, Education, and Welfare, 1976).

[21] John P. Murray and Susan Kippax, "Children's Social Behavior in Three Towns with Differing Television Experience," *Journal of Communication,* 28 (Winter 1978), 19–29.

[22] Charles K. Atkin and Walter Gantz, "The Role of Television News in the Political Socialization of Children." Paper presented at the 1975 meeting of the International Communication Association.

[23] Charles Atkin and Mark Miller, "The Effects of Television Advertising on Children: Experimental Evidence." Paper presented at the 1975 meeting of the International Communication Association. A review of recent research on television and advertising can be found in the *Journal of Communication,* Winter 1977.

[24] A. Bandura, D. Ross, and A. Ross, "Imitation of Film Mediated Aggressive Models," *Journal of Abnormal and Social Psychology,* 66 (1963), 3–11; L. Berkowitz and E. Rawlings, "Effects of Film Violence on Inhibitions against Subsequent Aggression," *Journal of Abnormal and Social Psychology,* 66 (1963), 405–12.

[25] Seymour Feshbach, "The Stimulating vs. Cathartic Effects of a Vicarious Aggressive Experience," *Journal of Abnormal and Social Psychology,* 63 (1961), 381–85.

[26] Leonard Berkowitz, *Aggression: A Social Psychological Analysis* (New York: McGraw-Hill Book Company, 1962).

[27] Joseph Klapper, *The Effects of Mass Communication* (New York: The Free Press, 1960).

[28] Albert Bandura and Richard Walters, *Social Learning and Personality Development* (New York: Holt Rinehart & Winston, 1963).

[29] The *Journal of Broadcasting*, Summer 1977, features a discussion of the Gerbner methodology, the CBS criticism, and the Gerbner response.

[30] George Comstock, "Types of Portrayal and Aggressive Behavior," *Journal of Communication*, 27 (Summer 1977), 189–98.

[31] Charles N. Barnard, "An Oriental Mystery," *TV Guide*, 26 (January 28, 1978), 2–4, 6, 8.

index

1941
Mayflower decision discourages
editorializing.

1954
Edward R. Murrow confronts
Senator Joseph McCarthy's "Red
Scare" tactics on *See It Now.*

1943
Edward G. Nobel buys NBC Blue,
becomes ABC in 1945.

1953
KUHT signs on as first educational
television station under new FCC
allocations for ETV.

1956
Ampex engineers demonstrate
videotape recording.

1935
Dr. Frank Stanton leaves position
at The Ohio State University to
join CBS's research efforts.

1953
ABC merges with United Paramount
Theatres, Inc.

1961-1968
Midwest Program
Airborne Televisic
Instruction (MPAT

1948-1953
FCC freezes television allocations.

1938
First over–the–air ETV
programming.

1956
United Press International launches
audio service, first for a wire service.

1950
First successful recording of color
television.

1941
FCC issues *Report on Chain
Broadcasting.*

1955
Association for Professional
Broadcasting Education is founded.
Later becomes Broadcast Education
Association (B.E.A.).

1948
CBS conducts the first of various
"talent raids" the networks
participate in during the late 1940s.

1939
Television is introduced at the
World's Fair.

1960
First televised Presidential deba
between John F. Kennedy and
Richard M. Nixon.

1947
Transistor is invented at Bell
Laboratories.

1957
RCA demonstrates color videotape
recording.

1948
Cable systems begin in Oregon and
Pennsylvania.

1945
Blue network changes name to
American Broadcasting Company.

1955
Radio programming begins to
specialize. Rock and roll formats
develop.

1938
Orson Welles makes famous "War
of the Worlds" broadcast.

1949
Fairness Doctrine is issued.

1961
Spanish International
Network (SIN) begin
operation.